PROBABILITY THEORY, AN ANALYTIC VIEW

This book is intended for graduate students who have a good undergraduate introduction to probability theory, a reasonably sophisticated introduction to modern analysis, and who now want to learn what these two topics have to say about each other. By modern standards, the topics treated here are classical and the techniques used far-ranging. No attempt has been made to present the subject as a monolithic structure resting on a few basic principles.

The first part of the book deals with independent random variables, Central Limit phenomena, the general theory of weak convergence and several of its applications, as well as elements of both the Gaussian and Markovian theory of measures on function space. The introduction of conditional expectation values is postponed until the second part of the book, where it is applied to the study of martingales. This section also explores the connection between martingales and various aspects of classical analysis, and the connections between Wiener's measure and classical potential theory.

Although the book is primarily intended for students and practitioners of probability theory and analysis, it will also be a valuable reference for those in fields as diverse as physics, engineering, and economics.

PROBABILITY THEORY, AN ANALYTIC VIEW

DANIEL W. STROOCK
Massachusetts Institute of Technology

CAMBRIDGE
UNIVERSITY PRESS

Published by the Press Syndicate of the University of Cambridge
The Pitt Building, Trumpington Street, Cambridge CB2 1RP
40 West 20th Street, New York, NY 10011-4211, USA
10 Stamford Road, Oakleigh, Victoria 3166, Australia

First published 1993

Printed in the United States of America

Library of Congress Cataloging-in-Publication Data

Stroock, Daniel W.

Probability theory : an analytic view / Daniel W. Stroock.

p. cm.

ISBN 0-521-43123-9 (hc)

1. Probabilities. I. Title.

QA273.S763 1993

519.2 – dc20 92–39118

CIP

A catalog record for this book is available from the British Library.

ISBN 0-521-43123-9 hardback

This book is dedicated to my teachers:

M. Kac, H.P. McKean, Jr., and S.R.S. Varadhan

Contents

Preface

When writing a graduate level mathematics book during the last decade of the twentieth century, one probably ought not be inquire too closely into one's motivation. In fact, if one's own pleasure from the exercise is not sufficient to justify the effort, then one should seriously consider dropping the project. Thus, to those who (either before or shortly after opening it) ask *for whom was this book written*, my pale answer is *me*; and, for this reason, I thought that I should preface this preface with an explanation of who I am and what were the peculiar educational circumstances which eventually gave rise to this somewhat peculiar book.

My own introduction to probability theory began with a private lecture from H.P. McKean, Jr. At the time, I was a (more accurately, *the*) graduate student of mathematics at what was then called The Rockefeller Institute for Biological Sciences. My official mentor there was M. Kac, whom I had cajoled into becoming my advisor after a year during which I had failed to insert even one micro-electrode into the optic nerves of innumerable limuli. However, as I soon came to realize, Kac had accepted his rôle on the condition that it would not become a burden. In particular, he had no intention of wasting much of his own time on a reject from the neurophysiology department. On the other hand, he was most generous with the time of his younger associates, and that is how I wound up in McKean's office. Never one to bore his listeners with a lot of dull preliminaries, McKean launched right into a wonderfully lucid explanation of P. Lévy's interpretation of the infinitely divisible laws. I have to admit that my appreciation of the lucidity of his lecture arrived nearly a decade after its delivery, and I can only hope that my reader will reserve judgment of my own presentation for an equal length of time.

In spite of my perplexed state at the end of McKean's lecture, I was sufficiently intrigued to delve into the readings which he suggested at its conclusion. Knowing that the only formal mathematics courses which I would be taking during my graduate studies would be given at N.Y.U. and guessing that those courses would be oriented toward partial differential equations, McKean directed me to material which would help me understand the connections between partial differential equations and probability theory. In particular, he suggested that I start with the, then recently translated, two articles by E.B. Dynkin which had appeared originally in the famous 1956 volume of *Teoriya Veroyatnostei i ee Primeneniya*. Dynkin's articles turned out to be a godsend. They were beautifully crafted to tell the reader enough so that he could understand the ideas and not so much that he would become bored by them. In addition, they gave me an introduction to a host of ideas and techniques (e.g., stopping times and the strong Markov property), all of which Kac himself consigned to the category of overelaborated measure theory. In fact, it would be reasonable to say that my thesis was simply the application of techniques which I picked up from Dynkin to a problem which I picked up by reading some notes by Kac. Of course, along the way I profited immeasurably from continued contact with McKean, a large number of courses at N.Y.U. (particularly one's taught by M. Donsker, F. John, and L. Nirenberg), and my increasingly animated conversations with S.R.S. Varadhan.

As I trust the preceding description makes clear, my graduate education was anything but deprived; I had ready access to some of the very best analysts of the day. On the other hand, I never had a *proper* introduction to my field, probability theory. The first time that I ever summed independent random variables was when I was summing them in front of a class at N.Y.U. Thus, although I now admire the magnificent body of mathematics created by A.N. Kolmogorov, P. Lévy, and the other twentieth-century heroes of the field, I am not a *dyed-in-the-wool* probabilist (i.e., what Donsker would have called a true *coin-tosser*). In particular, I have never been able to develop sufficient sensitivity to the distinction between a *proof* and a *probabilistic proof*. To me, a proof is clearly *probabilistic* only if its punch-line comes down to an argument like $P(A) \leq P(B)$ because $A \subseteq B$; and there are breathtaking examples of such arguments. However, in order to base an entire book on these examples would require a level of genius which I do not possess. In fact, I myself enjoy probability theory best when it is inextricably interwoven with other branches of mathematics and not when it is presented as an entity unto itself. For this reason, the reader should not be surprised to discover that he finds some of the material presented in this book *does not belong here*; but I hope that he will make an effort to figure out why I disagree with him. Be that as it may, I will devote the rest of this preface to a summary of what lies ahead.

Summary

I: Chapter I contains a sampling of the standard, point-wise convergence theorems dealing with partial sums of independent random variables. These include the Weak and Strong Laws of Large Numbers as well as Hartman–Wintner's Law of the Iterated Logarithm. In preparation for the law of the iterated logarithm, Cramér's theory of large deviations from the law of large numbers is developed in §1.4. Everything here is very standard, although I feel that the passage from the bounded to the general case of the law of the iterated logarithm has been considerably smoothed by the ideas which I learned in conversation with M. Ledoux.

II: The whole of Chapter II is devoted to the classical Central Limit Theorem. After an initial (and slightly flawed) derivation of the basic result via moment considerations, first Lindeberg's general version and then the Berry–Esseen estimate are derived in Section 2.1. Although Lindeberg's result has become a *sine qua non* in the writing of probability texts, the Berry–Esseen estimate has not. Indeed, until recently, the Berry–Esseen estimate required a good many somewhat tedious calculations with characteristic functions (i.e., Fourier transforms), and most recent authors seem to have decided that the rewards did not justify the effort. I was inclined to agree with them until P. Diaconis brought to my attention E. Bolthausen's adaptation of C. Stein's techniques (the so-called *Stein's method*) to give a proof which is not only brief but also, to me, aesthetically pleasing. In any case, no use of Fourier methods is made in Section 2.1. On the other hand, Fourier techniques are introduced in Section 2.2, where it is shown that even elementary Fourier analytic tools lead to important extensions of the basic Central Limit Theorem to more than one dimension. Finally, in Section 2.3, the Central Limit Theorem is applied to the study of Hermite multipliers and (following Wm. Beckner) is used to derive both E. Nelson's hypercontraction estimate for the Mehler kernel as well as Beckner's own estimate for the Fourier transform. I am afraid that, with this flagrant example of *the sort of thing that does not belong here*, I may be trying the patience of my purist colleagues. However, I hope that their indignation will be somewhat assuaged by the fact that rest of the book is essentially independent of the material in Section 2.3.

III: In preparation for the analysis of probability measures on function spaces, the general theory of weak convergence on Polish spaces is presented in Section 3.1. Although the main reason for my covering this material is to construct Wiener's measure and derive Donsker's Invariance Principle, I have postponed these applications until Section 3.3 in order to first develop a context in which to fit them. For this reason, I have devoted Section 3.2 to a proof of the Lévy–Khinchine formula as well as Lévy's explanation of this formula in terms of the path structure of the associated independent increment processes. That is, in Section 3.2, I construct all the Gaussian-free independent increment processes so that, when it is constructed in Section 3.3, Wiener's measure completes Lévy's program.

IV: Because Wiener's measure is quite possibly the single most important object in all of modern probability theory, I have decided that even an introductory text should contain at least one chapter devoted to an exposition of its marvelous and many-faceted nature. Hence, in Section 4.1, I have touched on a few of its elementary properties, namely: scaling invariance and basic continuity properties of Wiener paths. Following the Segel school, Wiener's measure is presented in Section 4.2 from the Gaussian point of view. This discussion includes a brief excursion into the concept of abstract Wiener spaces, introduces *pinned* Wiener paths, and concludes with one of Wiener's own constructions of his measure via random trigonometric series. The final section of Chapter IV, Section 4.3, takes up the Markov aspects of Wiener's measure. Specifically, after proving the strong Markov property, I have devoted the rest of this section to an exposition of the Feynman–Kac formula, and, following Kac, have used their formula to derive Lévy's *arcsine law.*

V: Because they are not needed earlier, conditional expectations do not appear until Chapter V. The advantage gained by this postponement is that, by the time it is introduced, I have an ample supply of examples to which conditioning can be applied; the disadvantage is that, with considerable justice, many probabilists feel that one is not doing *probability theory* until one is conditioning. In any case, Kolmogorov's definition is given in Section 5.1 and is shown to extend naturally both to σ-finite measure spaces as well as random variables with values in a Banach space. Section 5.2 develops Doob's basic theory of real-valued, discrete parameter martingales: Doob's inequality and his martingale convergence theorem; and, in Section 5.3, I have first pointed out that, without any particular difficulty, a good deal of martingale theory extends to σ-finite measure spaces and some of it applies to Banach space-valued random variables. I have tried to justify the consideration of martingales in the σ-finite context by showing that both the Hardy–Littlewood Inequality and the Calderón–Zygmund decomposition lemma can be obtained as rather easy applications of Doob's inequality.

VI: This chapter is a hodgepodge of results tied together (at least in my mind) by the ideas in Section 5.3. Section 6.1 is devoted to Birkhoff's famous Individual Ergodic Theorem, which is presented in the multidimensional context. The connection between this material and that in §5.3 comes in my derivation of the Maximal Ergodic Lemma, which I have given as an application of the Hardy–Littlewood Inequality (more precisely, Hardy's Inequality). For those who previously objected to the contents of Section 2.3, Section 6.2 is going to be a real affront. Indeed, even I find it difficult to justify the inclusion of Section 6.2 in an introduction to probability theory. On the other hand, once one knows both subjects, the similarity between the cancellations underlying both the theories of martingales and singular integral operators is too compelling to ignore. In fact, there is considerable evidence that, on the analytic side, J. Marcinkiewicz, A. Zygmund, E.M. Stein, and C. Fefferman were all guided at times by this similarity; and, on the probabilistic side, certainly both D. Burkholder and S. Gundy owe a great deal to the analytic antecedents of what they were doing. For this reason, in Section 6.2 I have provided the reader with a brief survey of a few of those analytic antecedents. In particular, I (borrowing heavily from E. Stein's masterful treatise†) have developed the elements of Calderón and Zygmund's theory of singular integrals, based on the use of their decomposition lemma; and, because my goal was, in part, to prepare the reader for Burkholder's inequality, I end Section 6.2 with an application of their theory to the analysis of the Littlewood–Paley square function. Armed with the background developed in Section 6.2, I give two derivations of Burkholder's inequality in Section 6.3. The first of these is, more or less, a reproduction of Burkholder's original proof and very much resembles the sort of reasoning used in Section 6.2. The second proof is entirely different and, in some ways, more elementary. Once again, it is taken, with nearly no alteration, from Burkholder, only this time it is based on the line of reasoning which he has been perfecting over the past decade.

VII: I return once more to the probabilistic fold in Chapter VII and begin by transferring to the continuous parameter setting only those parts of martingale theory which lend themselves best to the transition. In particular, except that in Exercise 7.1.32 I have outlined its proof in the simplest case, I have assiduously avoided Meyer's continuous parameter version of Doob's decomposition for submartingales. Instead, I have emphasized from the outset the important rôle which martingales can play in the characterization and analysis of diffusions. Thus, in Section 7.2, martingales are used first to prove some of the basic facts about the long-time behavior of Wiener paths and then to see how similar considerations apply to give recurrence and transience criteria for more general Markov processes. Section 7.3 begins with an introduction to the class of diffusions which are obtained by subjecting Wiener paths to a drift and ends with an application of the results in Section 7.2 to an analysis of the effect which the drift has on transience and recurrence properties. This program is continued in Section 7.4, where the ergodic theory developed in Section 6.1 is adapted to the Markov setting and then applied to the diffusions discussed in Section 7.2. Finally, in Section 7.5, I have specialized to drifts which are gradient fields, in which case the analysis becomes much easier and the results much more complete.

VIII: The final chapter is a piece of self-indulgence, pure and simple. Namely, I turn there to the miraculous connections between Wiener paths and classical potential theory. At one time, these connections were touted as the archetypical example of the way in which probability theory could enter mainstream analysis; and, in spite of occasional excesses by its practitioners, the subject has had a beneficial effect on potential theory as well

†See E. Stein's *Singular Integrals*, publ. by Princeton Univ. Press.

as probability theory. Furthermore, it remains the single most startling example I know of a perfect fit between a branch of classical analysis and a topic in probability theory: here the connection is simultaneously complete and useful. Thus, I have been able to rationalize that there may be some purpose, in addition to self-gratification, served by its inclusion in an introductory text. I begin with a little parabolic theory and, in Section 8.1, construct the heat kernel with Dirichlet boundary conditions for an arbitrary region. The complete description of the sense in which this kernel solves the problem (in particular, satisfies the boundary data) forces me to introduce and discuss here the notion of a regular point, which, in turn, leads naturally to the Dirichlet problem taken up in Section 8.2. In Section 8.3, I introduce the Green's function for regions in $\mathbb{R}^N$. Perhaps the most interesting aspect of this analysis is the rôle played here by dimension. In particular, when $N = 2$ and the region is unbounded, more care is required than one might expect. Finally, in Section 8.4, I present Riesz's Decomposition Theorem for excessive (i.e., nonnegative, superharmonic) functions, give K.L. Chung's beautiful interpretation of the capacitory distribution, and derive Wiener's test for regularity.

Suggestions

As anyone who attempts to base a course on it will soon discover, this book presents some real challenges and should not be adopted whole. On the other hand, both I and my friend Persi Diaconis have found that, if one is willing to do some judicious editing, one can carve either a one semester or a full year course out of it. For instance, assuming that the students have already had a thorough introduction to measure theory, a first semester of graduate level probability theory can be extracted out of §§1.1–1.5, §§2.1–2.2, and §§3.1–3.3. The major objection to this selection is that it does not include anything about conditioning and is therefore completely inappropriate for students who will not be taking a second semester of the subject. To avoid this objection, one can replace any or all of §1.4, §1.5, the Berry–Esseen Theorem in §2.1, and §3.2 with §§5.1–5.2. The choices of topics for a second semester course will, of course, be more or less dictated by those which were covered in the first semester. If one is attempting to get to diffusion theory as quickly as possible, then one should start with Chapter IV, cover §§5.1–5.2, skip to §§7.1–7.2, and do as much of Chapter VII as time permits. Unfortunately, the ergodic theory in §§7.3–7.4 relies on §6.1, which, in turn uses results from §5.3. However, the route which I have adopted to the ergodic theorem is, by no means, the most efficient one and §6.1 can be replaced by any one of the many standard treatments of basic ergodic theory.

Whatever route one takes through this book, it will be a great favor to your students to suggest that they consult other texts. Indeed, out of respect for the adage that the third book one reads on a subject is always the most lucid, one should suggest at least two other books. Among the many excellent choices available, I mention: Wm. Feller's *An Introduction to Probability Theory and Its Applications, Vol. II* and M. Loéve's classic *Probability Theory.* In addition, for background, precision (including accuracy of attribution), and supplementary material, R. Dudley's recent *Real Analysis and Probability* is superb. Finally, because it would be a shame to waste LAMS-TEX's diagram-making facility on a subject with no commutative diagrams, I have provided a (not entirely accurate) table of dependence on the following page. In this table, an arrow indicates that the sections at its point depend on material in the sections at its tail.

Acknowledgments

Besides the students in my classes who have suffered with various earlier metamorphoses of this book, there are several friends whom I now thank for the trouble they took in reading portions of the text and making suggestions on what they read. These include Mark Pinsky, whose question prompted me to write Exercise 2.1.18, and Bob Fefferman, who issued the Calderón–Zygmund–Stein school's stamp of approval on my treatment of the material covered in Section 6.2. However, of the many comments, encouraging and otherwise, which I have received about this manuscript, only those made by my friend Persi Diaconis have resulted in my making significant alterations. To paraphrase the sentiment expressed by former President Bush, Persi lobbied for a *kinder, gentler* book; and for this, my readers owe him a considerable debt of gratitude. After all, Bush's success was nothing to rave about either.

Table of Dependence

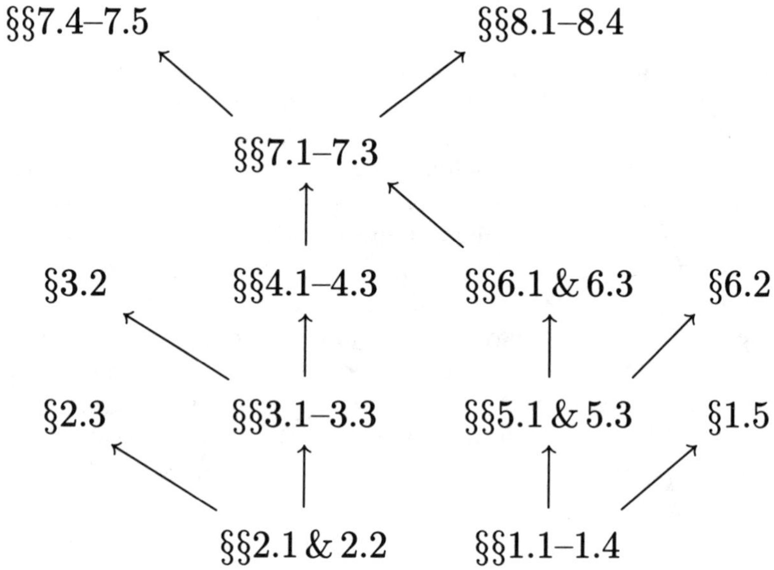

Chapter I:
Sums of Independent Random Variables

§1.1 Independence.

In one way or another, most probabilistic analysis entails the study of large families of random variables. The key to such analysis is an understanding of the relations among the family members; and of all the possible ways in which members of a family can be related, by far the simplest is when the relationship does not exist at all! For this reason, we will begin by looking at families of *independent* random variables.

Let $(\Omega, \mathcal{F}, P)$ be a **probability space** (i.e., Ω is a nonempty set, $\mathcal{F}$ is a σ-algebra over Ω, and P is a measure on the measurable space $(\Omega, \mathcal{F})$ having total mass 1); and, for each i from the (nonempty) index set $\mathcal{I}$, let $\mathcal{F}_i$ be a subσ-algebra of $\mathcal{F}$. We say that the σ-algebras $\mathcal{F}_i$, $i \in \mathcal{I}$, are **mutually P-independent** or, less precisely, **P-independent** if, for every finite subset $\{i_1, \dots, i_n\}$ of distinct elements of $\mathcal{I}$ and every choice of $A_{i_1} \in \mathcal{F}_{i_1}, \dots,$ and $A_{i_n} \in \mathcal{F}_{i_n}$,

$$P(A_{i_1} \cap \dots \cap A_{i_n}) = P(A_{i_1}) \cdots P(A_{i_n}). \tag{1.1.1}$$

In particular, if $\{A_i : i \in \mathcal{I}\}$ is a family of sets from $\mathcal{F}$, we say that A_i, $i \in \mathcal{I}$, are **P-independent** if the associated σ-algebras $\mathcal{F}_i = \{\emptyset, A_i, A_i\complement, \Omega\}$, $i \in \mathcal{I}$, are. To gain an the appreciation for the intuition on which this definition is based, it is important to notice that independence of the pair A_1 and A_2 in the present sense is equivalent to

$$P(A_1 \cap A_2) = P(A_1)P(A_2),$$

the classical definition which one encounters in elementary treatments. Thus, the notion of independence just introduced is no more than a simple generalization of the classical notion of *independent pairs of sets* encountered in nonmeasure theoretic presentations; and therefore, the intuition which underlies the elementary notion applies equally well to the definition given here. (See Exercise 1.1.10 below for more information about the connection between the present definition and the classical one.)

As will be become increasing evident as we proceed, infinite families of independent objects possess surprising and beautiful properties. In particular, mutually independent σ-algebras tend to *fill up space* in a sense which is made precise by the following beautiful thought experiment designed by A.N. Kolmogorov. Let $\mathcal{I}$ be any index set, take $\mathcal{F}_\emptyset = \{\emptyset, \Omega\}$, and for each nonempty subset $\Lambda \subseteq \mathfrak{I}$, let

$$\mathcal{F}_\Lambda = \bigvee_{i \in \Lambda} \mathcal{F}_i$$

be the σ-algebra generated by $\bigcup_{i\in\Lambda} \mathcal{F}_i$ (i.e., the smallest σ-algebra containing all of the $\mathcal{F}_i$'s). Next, define the **tail σ-algebra** $\mathcal{T}$ to be the intersection over finite $\Lambda \subseteq \mathcal{I}$ of the σ-algebras $\mathcal{F}_{\Lambda\complement}$. When $\mathcal{I}$ itself is finite, $\mathcal{T} = \{\emptyset, \Omega\}$ and is therefore P-**trivial** in the sense that $P(A) \in \{0,1\}$ for every $A \in \mathcal{T}$. The interesting remark made by Kolmogorov is that even when $\mathcal{I}$ is infinite, $\mathcal{T}$ is P-trivial whenever the original $\mathcal{F}_i$'s are P-independent. To see this, first note that, by assumption, $\mathcal{F}_{F_1}$ is P-independent of $\mathcal{F}_{F_2}$ whenever F_1 and F_2 are finite, disjoint subsets of $\mathcal{I}$. Since for any (finite or not) $\Lambda \subseteq \mathcal{I}$, $\mathcal{F}_\Lambda$ is generated by the algebra

$$\bigcup \{\mathcal{F}_F : F \text{ is a finite subset of } \Lambda\},$$

it follows (cf. Exercise 1.1.12) first that $\mathcal{F}_\Lambda$ is P-independent of $\mathcal{F}_{\Lambda\complement}$ for every $\Lambda \subseteq \mathcal{I}$ and then that $\mathcal{T}$ is P-independent of $\mathcal{F}_\mathcal{I}$. But $\mathcal{T} \subseteq \mathcal{F}_\mathcal{I}$, which means that $\mathcal{T}$ is *independent of itself*; that is, $P(A \cap B) = P(A)P(B)$ for all A, $B \in \mathcal{T}$. Hence, for every $A \in \mathcal{T}$, $P(A) = P(A)^2$, or, equivalently, $P(A) \in \{0,1\}$; and so we have now proved the following famous result.

1.1.2 Theorem (Kolmogorov's 0–1 Law). *Let $\{\mathcal{F}_i : i \in \mathcal{I}\}$ be a family of P-independent sub-σ-algebras of $(\Omega, \mathcal{F}, P)$, and define the tail σ-algebra $\mathcal{T}$ as above. Then, for every $A \in \mathcal{T}$, $P(A)$ is either 0 or 1.*

To get a feeling for the kind of conclusions which can be drawn from Kolmogorov's 0–1 Law (cf. Exercises 1.1.18 and 1.1.19 below as well), let $\{A_n\}_1^\infty$ be a sequence of subsets of Ω, and recall the notation

$$\begin{aligned}\overline{\lim_{n\to\infty}}\, A_n &\equiv \bigcap_{m=1}^{\infty} \bigcup_{n\ge m} A_n \\ &= \{\omega : \omega \in A_n \text{ for infinitely many } n \in \mathbb{Z}^+\}.\end{aligned} \tag{1.1.3}$$

Obviously, $\overline{\lim}_{n\to\infty} A_n$ is measurable with respect to the tail field determined by the sequence of σ-algebras $\{\emptyset, A_n, A_n\complement, \Omega\}$, $n \in \mathbb{Z}^+$; and therefore, if the A_n's are P-independent elements of $\mathcal{F}$, then

$$P\left(\overline{\lim_{n\to\infty}}\, A_n\right) \in \{0,1\}.$$

In words, this conclusion can be summarized as the statement that: *for any sequence of P-independent events A_n, $n \in \mathbb{Z}^+$, either P-almost every $\omega \in \Omega$ is in infinitely many A_n's or P-almost every $\omega \in \Omega$ is in at most finitely many A_n's.* A more quantitative statement of this same fact is contained in the second part the following useful result.

1.1.4 Borel–Cantelli Lemma. *Let $\{A_n : n \in \mathbb{Z}^+\} \subseteq \mathcal{F}$ be given. Then*

$$\text{(1.1.5)} \qquad \sum_{n=1}^{\infty} P(A_n) < \infty \implies P\left(\varlimsup_{n\to\infty} A_n\right) = 0.$$

Conversely, if the A_n's are P-independent sets, then

$$\text{(1.1.6)} \qquad \sum_{n=1}^{\infty} P(A_n) = \infty \implies P\left(\varlimsup_{n\to\infty} A_n\right) = 1.$$

(See part (**iii**) *of Exercise 5.2.34 and Lemma 8.4.46 for variations on this theme.)*

PROOF: The first assertion is an easy application of countable additivity. Namely, by countable additivity,

$$P\left(\varlimsup_{n\to\infty} A_n\right) = \lim_{m\to\infty} P\left(\bigcup_{n\geq m} A_n\right) \leq \lim_{m\to\infty} \sum_{n\geq m} P(A_n) = 0$$

if $\sum_{n=1}^{\infty} P(A_n) < \infty$.

To prove (1.1.6), note that, by countable additivity, $P\left(\varlimsup_{n\to\infty} A_n\right) = 1$ if and only if

$$\lim_{m\to\infty} P\left(\bigcap_{n\geq m} A_n\complement\right) = P\left(\bigcup_{m=1}^{\infty}\bigcap_{n\geq m} A_n\complement\right) = P\left(\left(\varlimsup_{n\to\infty} A_n\right)\complement\right) = 0.$$

But, again by countable additivity, for given $m \geq 1$ we have that:

$$P\left(\bigcap_{n=m}^{\infty} A_n\complement\right) = \lim_{N\to\infty} \prod_{n=m}^{N} (1 - P(A_n)) \leq \lim_{N\to\infty} \exp\left[-\sum_{n=m}^{N} P(A_n)\right] = 0$$

if $\sum_{n=1}^{\infty} P(A_n) = \infty$. (In the preceding, we have used the trivial inequality $1 - t \leq e^{-t}$, $t \in [0,\infty)$.) □

Another, and perhaps more dramatic, statement of the conclusion drawn in the second part of the preceding is the following. Let $\mathbf{N}(\omega) \in \mathbb{Z}^+ \cup \{\infty\}$ be the number of $n \in \mathbb{Z}^+$ such that $\omega \in A_n$. If the A_n's are independent, then Tonelli's Theorem implies that (1.1.6) is equivalent to†

$$P(\mathbf{N} < \infty) > 0 \implies \mathbb{E}^P[\mathbf{N}] < \infty.$$

† Throughout this book, we use $\mathbb{E}^P[X, A]$ to denote the expected value under P of X over the set A. That is, $\mathbb{E}^P[X, A] = \int_A X\,dP$. Finally, when $A = \Omega$ we will write $\mathbb{E}^P[X]$.

Having described what it means for the σ-algebras to be P-independent, we can now transfer the notion to random variables on $(\Omega, \mathcal{F}, P)$. Namely, for each $i \in \mathcal{I}$, let X_i be a **random variable** (i.e., a measurable function on $(\Omega, \mathcal{F})$) with values in the measurable space $(E_i, \mathcal{B}_i)$. We will say that the random variables X_i, $i \in \mathcal{I}$, are **(mutually) P-independent** if the σ-algebras

$$\sigma(X_i) = X_i^{-1}(\mathcal{B}_i) \equiv \{X_i^{-1}(B_i) : B_i \in \mathcal{B}_i\}, \; i \in \mathcal{I},$$

are P-independent. Using

$$B(E;\mathbb{R}) = B\big((E,\mathcal{B});\mathbb{R}\big)$$

to denote the space of bounded measurable $\mathbb{R}$-valued functions on the measurable space $(E, \mathcal{B})$, notice that P-independence of $\{X_i : i \in \mathcal{I}\}$ is equivalent to the statement that

$$\text{(1.1.7)} \qquad \mathbb{E}^P\big[f_{i_1} \circ X_{i_1} \cdots f_{i_n} \circ X_{i_n}\big] = \mathbb{E}^P\big[f_{i_1} \circ X_{i_1}\big] \cdots \mathbb{E}^P\big[f_{i_n} \circ X_{i_n}\big]$$

for all finite subsets $\{i_1, \dots, i_n\}$ of distinct elements of $\mathcal{I}$ and all choices of $f_{i_1} \in B(E_{i_1};\mathbb{R}), \dots,$ and $f_{i_n} \in B(E_{i_n};\mathbb{R})$. Finally, if we use $\mathbf{1}_A$ given by

$$\mathbf{1}_A(\omega) \equiv \begin{cases} 1 & \text{if} \quad \omega \in A \\ 0 & \text{if} \quad \omega \notin A \end{cases}$$

to denote the **indicator function** of the set $A \subseteq \Omega$, notice that the family of sets $\{A_i : i \in \mathcal{I}\} \subseteq \mathcal{F}$ is P-independent if and only if the random variables $\mathbf{1}_{A_i}$, $i \in \mathcal{I}$, are P-independent.

Thus far we have discussed only the abstract notion of independence and have yet to show that the concept is not vacuous. In the modern literature, the standard way to construct lots of independent quantities is to take products of probability spaces. Namely, if $(E_i, \mathcal{B}_i, \mu_i)$ is a probability space for each $i \in \mathcal{I}$, one sets $\Omega = \prod_{i \in \mathcal{I}} E_i$, defines $\pi_i : \Omega \longrightarrow E_i$ to be the natural projection map for each $i \in \mathcal{I}$, takes $\mathcal{F}_i = \pi_i^{-1}(\mathcal{B}_i)$, $i \in \mathcal{I}$, and $\mathcal{F} = \bigvee_{i \in \mathcal{I}} \mathcal{F}_i$, and shows that there is a unique probability measure P on $(\Omega, \mathcal{F})$ with the properties that

$$P(\pi_i^{-1}\Gamma_i) = \mu_i(\Gamma_i) \quad \text{for all} \quad i \in \mathcal{I} \text{ and } \Gamma_i \in \mathcal{B}_i$$

and the σ-algebras $\mathcal{F}_i$, $i \in \mathcal{I}$, are P-independent. Although this procedure is extremely powerful, it is rather mechanical. For this reason, we have chosen to defer the details of the product construction to Exercise 1.1.14 below and to, instead, spend the rest of this section developing a more hands-on approach to constructing independent sequences of real-valued random variables. Indeed, although the product method is more ubiquitous and has become the construction of choice, the one which we are about to present has the advantage that it shows independent random variables can arise "naturally" and even in a familiar context.

Until further notice, we take $(\Omega, \mathcal{F}) = \big([0,1), \mathcal{B}_{[0,1)}\big)$ (when E is a metric space, we use $\mathcal{B}_E$ to denote the Borel field over E) and P to be the restriction $\lambda_{[0,1)}$ of Lebesgue's measure $\lambda_{\mathbb{R}}$ to $[0,1)$. We next define the **Rademacher functions** R_n, $n \in \mathbb{Z}^+$, on Ω as follows. Define the **integer part** $[t]$ of $t \in \mathbb{R}$ to be the largest integer dominated by t and consider the function $R : \mathbb{R} \longrightarrow \{-1, 1\}$ given by

$$R(t) = \begin{cases} -1 & \text{if } \; t - [t] \in \big[0, \frac{1}{2}\big) \\ 1 & \text{if } \; t - [t] \in \big[\frac{1}{2}, 1\big) \end{cases}.$$

The function R_n is then defined on $[0,1)$ by

$$R_n(\omega) = R\big(2^{n-1}\omega\big), \qquad n \in \mathbb{Z}^+ \text{ and } \omega \in [0,1).$$

We will now show that the Rademacher functions are P-independent. To this end, first note that every real-valued function f on $\{-1,1\}$ is of the form $\alpha + \beta x$, $x \in \{-1,1\}$, for some pair of real numbers α and β. Thus, all that we have to show is that

$$\mathbb{E}^P\big[(\alpha_1 + \beta_1 R_1) \cdots (\alpha_n + \beta_n R_n)\big] = \alpha_1 \cdots \alpha_n$$

for any $n \in \mathbb{Z}^+$ and $(\alpha_1, \beta_1), \dots, (\alpha_n, \beta_n) \in \mathbb{R}^2$. Since this is obvious when $n = 1$, we will assume that it holds for n and will deduce that it must also hold for $n+1$; and clearly this comes down to checking that

$$\mathbb{E}^P\big[F(R_1, \dots, R_n)\, R_{n+1}\big] = 0$$

for any $F : \{-1,1\}^n \longrightarrow \mathbb{R}$. But $(R_1, \dots, R_n)$ is constant on each interval

$$I_{m,n} \equiv \left[\frac{m}{2^n}, \frac{m+1}{2^n}\right), \quad 0 \le m < 2^n$$

whereas R_{n+1} integrates to 0 on each $I_{m,n}$. Hence, by writing the integral over Ω as the sum of integrals over the $I_{m,n}$'s, we get the desired result.

At this point we have produced a countably infinite sequence of independent **Bernoulli random variables** (i.e., two-valued random variables whose range is usually either $\{-1,1\}$ or $\{0,1\}$) with mean-value 0. In order to get more general random variables, we combine our Bernoulli random variables together in a clever way.

Recall that a random variable U is said to be **uniformly distributed** on the finite interval $[a,b)$ if $P\big(U \in [a,b)\big) = 1$ and

$$P(U \le t) = \frac{t-a}{b-a} \quad \text{for } t \in [a,b).$$

1.1.8 Lemma. *Let $\{Y_\ell : \ell \in \mathbb{Z}^+\}$ be a sequence of P-independent $\{0,1\}$-valued Bernoulli random variables with mean-value $\frac{1}{2}$ on some probability space $(\Omega, \mathcal{F}, P)$, and set*

$$U = \sum_{\ell=1}^{\infty} \frac{X_\ell}{2^\ell}.$$

Then U is uniformly distributed on $[0,1)$.

PROOF: Because the assertion only involves properties of distributions, it will be proved in general as soon as we prove it for a particular realization of independent, mean-value $\frac{1}{2}$, $\{0,1\}$-valued Bernoulli random variables. In particular, by the preceding discussion, we need only consider the random variables

$$\epsilon_n(\omega) \equiv \frac{1+R_n(\omega)}{2}, \qquad n \in \mathbb{Z}^+ \text{ and } \omega \in [0,1)$$

on $\big([0,1), \mathcal{B}_{[0,1)}, \lambda_{[0,1)}\big)$. But, as is easily checked, for each $\omega \in [0,1)$, $\{\epsilon_n(\omega)\}_1^\infty$ is the sequence of coefficients in the **binary expansion** of ω: that is, $\{\epsilon_n(\omega)\}_1^\infty$ is the unique $\{0,1\}$-valued sequence with the property that

$$0 \le \omega - \sum_{m=1}^{n} \frac{\epsilon_m(\omega)}{2^m} < \frac{1}{2^n} \quad \text{for each } n \in \mathbb{Z}^+.$$

Hence,

$$\omega = \sum_{m=1}^{\infty} \frac{\epsilon_m(\omega)}{2^m}, \qquad \omega \in [0,1),$$

and so the desired conclusion is trivial in this case. □

Now let $(k,\ell) \in \mathbb{Z}^+ \times \mathbb{Z}^+ \longmapsto n(k,\ell) \in \mathbb{Z}^+$ be any one-to-one mapping of $\mathbb{Z}^+ \times \mathbb{Z}^+$ onto $\mathbb{Z}^+$, and set

$$Y_{k,\ell} = \frac{1+R_{n(k,\ell)}}{2}, \qquad (k,\ell) \in \left(\mathbb{Z}^+\right)^2.$$

Clearly, each $Y_{k,\ell}$ is a $\{0,1\}$-valued Bernoulli random variable with mean-value $\frac{1}{2}$, and the family $\left\{Y_{k,\ell} : (k,\ell) \in \left(\mathbb{Z}^+\right)^2\right\}$ is P-independent. Hence, by Lemma 1.1.8, each of the random variables

$$U_k \equiv \sum_{\ell=1}^{\infty} \frac{Y_{k,\ell}}{2^\ell}, \qquad k \in \mathbb{Z}^+,$$

is uniformly distributed on $[0,1)$. In addition, the U_k's are obviously mutually independent. Hence, we have now produced a sequence of mutually independent random variables, each of which is uniformly distributed on $[0,1)$. To complete our program, we use the time-honored transformation which takes a uniform random variable into an arbitrary one. Namely, given a **distribution function** F on $\mathbb{R}$ (i.e., F is a right-continuous, nondecreasing function which tends to 0 at $-\infty$ and 1 at $+\infty$), define F^{-1} on $[0,1]$ to be the left-continuous inverse of F. That is,

$$F^{-1}(t) = \inf\{s \in \mathbb{R} : F(s) \geq t\}, \qquad t \in [0,1].$$

(Throughout, the infimum over the empty set is taken to be $+\infty$.) It is then an easy matter to check that when U is uniformly distributed on $[0,1)$ the random variable $X = F^{-1} \circ U$ has distribution function F:

$$P(X \leq t) = F(t), \qquad t \in \mathbb{R}.$$

Hence, after combining this with what we already know, we have now completed the proof of the following theorem.

1.1.9 Theorem. *Let* $\Omega = [0,1)$, $\mathcal{F} = \mathcal{B}_{[0,1)}$, *and* $P = \lambda_{[0,1)}$. *Then for any sequence* $\{F_k : k \in \mathbb{Z}^+\}$ *of distribution functions on* $\mathbb{R}$ *there exists a sequence* $\{X_k : k \in \mathbb{Z}^+\}$ *of* P*-independent random variables on* $(\Omega, \mathcal{F}, P)$ *with the property that* $P\big(X_k \leq t\big) = F_k(t)$, $t \in \mathbb{R}$, *for each* $k \in \mathbb{Z}^+$.

Exercises

1.1.10 Exercise: As we pointed out, $P\big(A_1 \cap A_2\big) = P\big(A_1\big)P\big(A_2\big)$ if and only if the σ-algebra generated by A_1 is P-independent of the one generated by A_2. Construct an example to show that the analogous statement is false when dealing with three, instead of two, sets. That is, just because $P\big(A_1 \cap A_2 \cap A_3\big) = P\big(A_1\big)P\big(A_2\big)P\big(A_3\big)$, it is not necessarily true that the three σ-algebras generated by A_1, A_2, and A_3 are P-independent.

1.1.11 Exercise: In this exercise we point out two elementary, but important, properties of independent random variables. Throughout, $(\Omega, \mathcal{F}, P)$ is a given probability space.

(i) Let X_1 and X_2 be a pair of P-independent random variables with values in the measurable spaces $(E_1, \mathcal{B}_1)$ and $(E_2, \mathcal{B}_2)$, respectively. Given a $\mathcal{B}_1 \times \mathcal{B}_2$-measurable function $F : E_1 \times E_2 \longrightarrow \mathbb{R}$ which is either nonnegative or bounded, use Tonelli's or Fubini's Theorem to show that

$$x_2 \in E_2 \longmapsto f(x_2) \equiv \mathbb{E}^P\Big[F\big(X_1, x_2\big)\Big] \in \mathbb{R}$$

is $\mathcal{B}_2$-measurable and that

$$\mathbb{E}^P\Big[F(X_1, X_2)\Big] = \mathbb{E}^P\Big[f(X_2)\Big].$$

(ii) Suppose that $X_1, \dots, X_n$ are P-independent, real-valued random variables. If each of the X_m's is P-integrable, show that $X_1 \cdots X_n$ is also P-integrable and that

$$\mathbb{E}^P\big[X_1 \cdots X_n\big] = \mathbb{E}^P\big[X_1\big] \cdots \mathbb{E}^P\big[X_n\big].$$

1.1.12 Exercise: Given a nonempty set Ω, recall† that a collection $\mathcal{C}$ of subsets of Ω is called a **π-system** if $\mathcal{C}$ is closed under finite intersections. At the same time, recall that a collection $\mathcal{L}$ is called a **λ-system** if $\Omega \in \mathcal{L}$, $A \cup B \in \mathcal{L}$ whenever A and B are disjoint members of $\mathcal{L}$, $B \setminus A \in \mathcal{L}$ whenever A and B are members of $\mathcal{L}$ with $A \subseteq B$, and $\bigcup_1^\infty A_n \in \mathcal{L}$ whenever $\{A_n\}_1^\infty$ is a nondecreasing sequence of members of $\mathcal{L}$. Finally, recall (cf. Lemma 3.1.3 in *ibid.*) that if $\mathcal{C}$ is a π-system, then the σ-algebra $\sigma(\mathcal{C})$ is the smallest $\mathcal{L}$-system $\mathcal{L} \supseteq \mathcal{C}$.

Now, let $(\Omega, \mathcal{F}, P)$ be a probability space, and, for each element i of the index set $\mathcal{I}$, let $\mathcal{C}_i \subseteq \mathcal{F}$ be a π- system. Show that the σ-algebras $\mathcal{F}_i$ generated by the $\mathcal{C}_i$'s are P-independent if and only if (1.1.1) holds for all choices of $A_{i_1} \in \mathcal{C}_{i_1}, \dots,$ and $A_{i_n} \in \mathcal{C}_{i_n}$.

1.1.13 Exercise: In this exercise we discuss two criteria for determining when random variables on the probability space $(\Omega, \mathcal{F}, P)$ are independent.

(i) Let $X_1, \dots,$ and X_n be bounded, real-valued random variables. Using Weierstrass's approximation theorem, show that the X_m's are P-independent if and only if

$$\mathbb{E}^P\big[X_1^{m_1} \cdots X_n^{m_n}\big] = \mathbb{E}^P\big[X_1^{m_1}\big] \cdots \mathbb{E}^P\big[X_n^{m_n}\big]$$

for all $m_1, \dots, m_n \in \mathbb{N}$.

(ii) Let $\mathbf{X} : \Omega \longrightarrow \mathbb{R}^m$ and $\mathbf{Y} : \Omega \longrightarrow \mathbb{R}^n$ be random variables. Show that $\mathbf{X}$ and $\mathbf{Y}$ are P-independent if and only if

$$\begin{aligned}\mathbb{E}^P\Big[\exp\Big[\sqrt{-1}\,\Big(\big(\boldsymbol{\alpha}, \mathbf{X}\big)_{\mathbb{R}^m} + \big(\boldsymbol{\beta}, \mathbf{Y}\big)_{\mathbb{R}^n}\Big)\Big]\Big] \\ = \mathbb{E}^P\Big[\exp\Big[\sqrt{-1}\,\big(\boldsymbol{\alpha}, \mathbf{X}\big)_{\mathbb{R}^m}\Big]\Big]\,\mathbb{E}^P\Big[\exp\Big[\sqrt{-1}\,\big(\boldsymbol{\beta}, \mathbf{Y}\big)_{\mathbb{R}^n}\Big]\Big]\end{aligned}$$

for all $\boldsymbol{\alpha} \in \mathbb{R}^m$ and $\boldsymbol{\beta} \in \mathbb{R}^n$.

---

† See, for example, §3.1 in the author's *A Concise Introduction to the Theory of Integration*, World Scientific Series in Pure Math., Vol. **12** (1990).

Hint: The *only if* assertion is obvious. To prove the *if* assertion, first check that $\mathbf{X}$ and $\mathbf{Y}$ are independent if

$$\mathbb{E}^P\big[f(\mathbf{X})\,g(\mathbf{Y})\big] = \mathbb{E}^P\big[f(\mathbf{X})\big]\,\mathbb{E}^P\big[g(\mathbf{Y})\big]$$

for all $f \in C_c^\infty\big(\mathbb{R}^m;\mathbb{C}\big)$ and $g \in C_c^\infty\big(\mathbb{R}^n;\mathbb{C}\big)$. Second, given such f and g, apply elementary Fourier analysis to write

$$f(\mathbf{x}) = \int_{\mathbb{R}^m} e^{\sqrt{-1}\,(\boldsymbol{\alpha},\mathbf{x})_{\mathbb{R}^m}}\,\varphi(\boldsymbol{\alpha})\,d\boldsymbol{\alpha} \quad \text{and} \quad g(\mathbf{y}) = \int_{\mathbb{R}^n} e^{\sqrt{-1}\,(\boldsymbol{\beta},\mathbf{y})_{\mathbb{R}^n}}\,\psi(\boldsymbol{\beta})\,d\boldsymbol{\beta},$$

where φ and ψ are smooth functions with **rapidly decreasing** (i.e., tending to 0 as $|\mathbf{x}| \to \infty$ faster than any power of $(1+|\mathbf{x}|)^{-1}$) derivatives of all orders. Finally, apply Fubini's Theorem.

1.1.14 Exercise: Given a pair of measurable spaces $(E_1,\mathcal{B}_1)$ and $(E_2,\mathcal{B}_2)$, recall that their product is the measurable space $\big(E_1\times E_2, \mathcal{B}_1\times\mathcal{B}_2\big)$, where $\mathcal{B}_1\times\mathcal{B}_2$ is the σ-algebra over the Cartesian product space $E_1\times E_2$ generated by the sets $\Gamma_1\times\Gamma_2$, $\Gamma_i\in\mathcal{B}_i$. Further, recall that, for any probability measures μ_i on $(E_i,\mathcal{B}_i)$ there is a unique probability measure $\mu_1\times\mu_2$ on $\big(E_1\times E_2,\mathcal{B}_1\times\mathcal{B}_2\big)$ such that

$$(\mu_1\times\mu_2)\big(\Gamma_1\times\Gamma_2\big) = \mu_1(\Gamma_1)_2(\Gamma_2) \quad \text{for } \Gamma_i\in\mathcal{B}_i.$$

More generally, for any $n\ge 2$ and measurable spaces $\big\{(E_i,\mathcal{B}_i)\big\}_1^n$, one takes $\prod_1^n \mathcal{B}_i$ to be the σ-algebra over $\prod_1^n E_i$ generated by the sets $\prod_1^n \Gamma_i$, $\Gamma_i\in\mathcal{B}_i$. In particular, since $\prod_1^{n+1} E_i$ and $\prod_1^{n+1}\mathcal{B}_i$ can be identified with $(\prod_1^n E_i)\times E_{n+1}$ and $(\prod_1^n \mathcal{B}_i)\times\mathcal{B}_{n+1}$, respectively, one can use induction to show that, for every choice of probability measures μ_i on $(E_i,\mathcal{B}_i)$ there is a unique probability measure $\prod_1^n \mu_i$ on $(\prod_1^n E_i, \prod_1^n \mathcal{B}_i)$ such that

$$\left(\prod_1^n \mu_i\right)\left(\prod_1^n \Gamma_i\right) = \prod_1^n \mu_i(\Gamma_i), \quad \Gamma_i\in\mathcal{B}_i.$$

The purpose of this exercise is to generalize the preceding construction to infinite collections. Thus, let $\mathfrak{I}$ be an infinite index set, and, for each $i\in\mathfrak{I}$, let $(E_i,\mathcal{B}_i)$ be a measurable space. Given $\emptyset\ne\Lambda\subseteq\mathfrak{I}$, we will use $\mathbf{E}_\Lambda$ to denote the Cartesian product space $\prod_{i\in\mathfrak{I}} E_i$ and π_Λ to denote the natural projection map taking $\mathbf{E}_\mathfrak{I}$ onto $\mathbf{E}_\Lambda$. Further, we use $\mathcal{B}_\mathfrak{I} = \prod_{i\in\mathfrak{I}}\mathcal{B}_i$ to stand for the σ-algebra over $\mathbf{E}_\mathfrak{I}$ generated by the collection $\mathcal{C}$ of subsets

$$\pi_F^{-1}\left(\prod_{i\in F}\Gamma_i\right), \quad \Gamma_i\in\mathcal{B}_i,$$

as F varies over nonempty, finite subsets of $\mathfrak{I}$ (abbreviated by: $\emptyset \neq F \subset\subset \mathfrak{I}$). In the following steps, we will outline a proof that, for every choice of probability measures μ_i on $(E_i, \mathcal{B}_i)$, there is a unique probability measure $\prod_{i\in\mathfrak{I}} \mu_i$ on $(\mathbf{E}_\mathfrak{I}, \mathcal{B}_\mathfrak{I})$ with the property that

$$\left(\prod_{i\in\mathfrak{I}} \mu_i\right)\left(\pi_F^{-1}\left(\prod_{i\in F} \Gamma_i\right)\right) = \prod_{i\in F} \mu_i(\Gamma_i), \quad \Gamma_i \in \mathcal{B}_i, \tag{1.1.15}$$

for every $\emptyset \neq F \subset\subset \mathfrak{I}$. Not surprisingly, the probability space

$$\left(\prod_{i\in\mathfrak{I}} E_i, \prod_{i\in\mathfrak{I}} \mathcal{B}_i, \prod_{i\in\mathfrak{I}} \mu_i\right)$$

is called the **product** over $\mathfrak{I}$ of the spaces $(E_i, \mathcal{B}_i, \mu_i)$; and when all the factors are the same space $(E, \mathcal{B}, \mu)$, it is customary to denote it by $(E^\mathfrak{I}, \mathcal{B}^\mathfrak{I}, \mu^\mathfrak{I})$, and if, in addition, $\mathfrak{I} = \{1, \ldots, N\}$, one uses $(E^N, \mathcal{B}^N, \mu^N)$.

(i) After noting that two probability measures which agree on a π-system agree on the σ-algebra generated by that π-system, show that there is at most one probability measure on $(\mathbf{E}_\mathfrak{I}, \mathcal{B}_\mathfrak{I})$ which satisfies the condition in (1.1.15). Hence, the problem is purely one of existence.

(ii) Let $\mathcal{A}$ be the *algebra* over $\mathbf{E}_\mathfrak{I}$ generated by $\mathcal{C}$, and show that there is a *finitely* additive $\mu : \mathcal{A} \longrightarrow [0,1]$ with the property that

$$\mu\left(\pi_F^{-1}(\Gamma_F)\right) = \left(\prod_{i\in F} \mu_i\right)(\Gamma_F), \quad \Gamma_F \in \mathcal{B}_F,$$

for all $\emptyset \neq F \subset\subset \mathfrak{I}$. Hence, all that we have to do is check that μ admits a σ-additive extension to $\mathcal{B}_\mathfrak{I}$, and, by Carathéodory's Extension Theorem, this comes down to checking that $\mu(A_n) \searrow 0$ whenever $\{A_n\}_1^\infty \subseteq \mathcal{A}$ and $A_n \searrow \emptyset$. Thus, let $\{A_n\}_1^\infty$ be a nonincreasing sequence from $\mathcal{A}$, and assume that $\mu(A_n) \geq \epsilon$ for some $\epsilon > 0$ and all $n \in \mathbb{Z}^+$. We must show that $\bigcap_1^\infty A_n \neq \emptyset$.

(iii) Referring to the last part of **(ii)**, show that there is no loss in generality if we assume that $A_n = \pi_{F_n}^{-1}(\Gamma_{F_n})$, where, for each $n \in \mathbb{Z}^+$, $\emptyset \neq F_n \subset\subset \mathfrak{I}$ and $\Gamma_{F_n} \in \mathcal{B}_{F_n}$. In addition, show that we may assume that $F_1 = \{i_1\}$ and that $F_n = F_{n-1} \cup \{i_n\}$, $n \geq 2$, where $\{i_n\}_1^\infty$ is a sequence of distinct elements of $\mathfrak{I}$. Now, make these assumptions and show that it suffices for us to find $a_\ell \in E_{i_\ell}$, $\ell \in \mathbb{Z}^+$, with the property, for each $m \in \mathbb{Z}^+$, $(a_1, \ldots, a_m) \in \Gamma_{F_m}$.

(iv) Continuing **(iii)**, for each $m, n \in \mathbb{Z}^+$, define $g_{m,n} : \mathbf{E}_{F_m} \longrightarrow [0,1]$ so that

$$g_{m,n}(\mathbf{x}_{F_m}) = \mathbf{1}_{\Gamma_{F_n}}(x_{i_1}, \ldots, x_{i_n}) \quad \text{if } n \leq m$$

and

$$g_{m,n}(\mathbf{x}_{F_m}) = \int_{\mathbf{E}_{F_n \setminus F_m}} \mathbf{1}_{\Gamma_{F_n}}(\mathbf{x}_{F_m}, \mathbf{y}_{F_n \setminus F_m}) \left(\prod_{\ell=m+1}^{n} \mu_{i_\ell} \right) (d\mathbf{y}_{F_n \setminus F_m})$$

if $n > m$. After noting that, for each m and n, $g_{m,n+1} \le g_{m,n}$ and

$$g_{m,n}(\mathbf{x}_{F_m}) = \int_{E_{i_{m+1}}} g_{m+1,n}(\mathbf{x}_{F_m}, y_{i_{m+1}})\, \mu_{i_{m+1}}(dy_{i_{m+1}}),$$

set $g_m = \lim_{n\to\infty} g_{m,n}$ and conclude that

$$g_m(\mathbf{x}_{F_m}) = \int_{E_{i_{m+1}}} g_{m+1}(\mathbf{x}_{F_m}, y_{i_{m+1}})\, \mu_{i_{m+1}}(dy_{i_{m+1}}).$$

In addition, note that

$$\begin{aligned}\int_{E_{i_1}} g_1(x_{i_1})\, \mu_{i_1}(dx_{i_1}) &= \lim_{n\to\infty} \int_{E_{i_1}} g_{1,n}(x_{i_1})\, \mu_{i_1}(dx_{i_1}) \\ &= \lim_{n\to\infty} \mu(A_n) \ge \epsilon,\end{aligned}$$

and proceed by induction to produce $a_\ell \in E_{i_\ell}$, $\ell \in \mathbb{Z}^+$, so that

$$g_m\big((a_1, \dots, a_m)\big) \ge \epsilon \quad \text{for all } m \in \mathbb{Z}^+.$$

Finally, check that $\{a_m\}_1^\infty$ is a sequence of the sort for which we were looking at the end of part **(iii)**.

1.1.16 Exercise: Recall that if Φ is a measurable map from one measurable space $(E, \mathcal{B})$ into a second one $(E', \mathcal{B}')$, then the **distribution** of Φ under a measure μ on $(E, \mathcal{B})$ is the **pushforward** measure $\Phi^*\mu$ (also denoted by $\mu \circ \Phi^{-1}$) defined on $(E', \mathcal{B}')$ by

$$\Phi^*\mu(\Gamma) = \mu(\Phi^{-1}(\Gamma)) \quad \text{for} \quad \Gamma \in \mathcal{B}'.$$

Given an nonempty index set $\mathfrak{I}$ and, for each $i \in \mathfrak{I}$, a measurable space $(E_i, \mathcal{B}_i)$ and an E_i-valued random variable X_i on the probability space $(\Omega, \mathcal{F}, P)$, define $\mathbf{X} : \Omega \longrightarrow \prod_{i\in\mathfrak{I}} E_i$ so that $\mathbf{X}(\omega)_i = X_i(\omega)$ for each $i \in \mathfrak{I}$ and $\omega \in \Omega$. Show that $\{X_i : i \in \mathfrak{I}\}$ is a family of P-independent random variables if and only if $\mathbf{X}^*P = \prod_{i\in\mathfrak{I}} X_i^*P$. In particular, given probability measures μ_i on $(E_i, \mathcal{B}_i)$, set

$$\Omega = \prod_{i\in\mathfrak{I}} E_i, \quad \mathcal{F} = \prod_{i\in\mathfrak{I}} \mathcal{B}_i, \quad P = \prod_{i\in\mathfrak{I}} \mu_i,$$

let $X_i : \Omega \longrightarrow E_i$ be the natural projection map from Ω onto E_i, and show that $\{X_i : i \in \mathfrak{I}\}$ is a family of mutually P-independent random variables such that, for each $i \in \mathfrak{I}$, X_i has distribution μ_i.

1.1.17 Exercise: Although it does not entail infinite product spaces, an interesting example of the way in which the preceding type of construction can be effectively applied is given in the following elementary version of a *coupling* argument.

(i) Let $(\Omega, \mathcal{B}, P)$ be a probability space and X and Y a pair of square P-integrable random variables with the property that

$$\big(X(\omega) - X(\omega')\big)\big(Y(\omega) - Y(\omega')\big) \ge 0 \quad \text{for all } (\omega, \omega') \in \Omega^2.$$

Show that

$$\mathbb{E}^P\big[X\,Y\big] \ge \mathbb{E}^P[X]\,\mathbb{E}^P[Y].$$

Hint: Define X_i and Y_i on Ω^2 for $i \in \{1,2\}$ so that $X_i(\boldsymbol{\omega}) = X(\omega_i)$ and $Y_i(\boldsymbol{\omega}) = Y(\omega_i)$ when $\boldsymbol{\omega} = (\omega_1, \omega_2)$, and integrate the inequality

$$0 \le \big(X(\omega_1) - X(\omega_2)\big)\big(Y(\omega_1) - Y(\omega_2)\big) = \big(X_1(\boldsymbol{\omega}) - X_2(\boldsymbol{\omega})\big)\big(Y_1(\boldsymbol{\omega}) - Y_2(\boldsymbol{\omega})\big)$$

with respect to P^2.

(ii) Suppose that $n \in \mathbb{Z}^+$ and that f and g are $\mathbb{R}$-valued, Borel measurable functions on $\mathbb{R}^n$ which are nondecreasing with respect to each coordinate (separately). Show that if $\mathbf{X} = \big(X_1, \dots, X_n\big)$ is an $\mathbb{R}^n$-valued random variable on a probability space $(\Omega, \mathcal{B}, P)$ whose coordinates are mutually P-independent, then

$$\mathbb{E}^P\big[f(\mathbf{X})\,g(\mathbf{X})\big] \ge \mathbb{E}^P\big[f(\mathbf{X})\big]\,\mathbb{E}^P\big[g(\mathbf{X})\big]$$

so long as $f(\mathbf{X})$ and $g(\mathbf{X})$ are both square P-integrable.

Hint: First check that the case when $n = 1$ reduces to an application of **(i)**. Next, describe the general case in terms of a multiple integral, apply Fubini's Theorem, and make repeated use of the case when $n = 1$.

1.1.18 Exercise: A σ-algebra is said to be **countably generated** if it contains a countable collection of sets which generate it. In this exercise, we will show that just because a σ-algebra is itself countably generated does not mean that all its sub-σ-algebras are.

Let $(\Omega, \mathcal{F}, P)$ be a measurable space and $\{\mathcal{F}_n : n \in \mathbb{Z}^+\}$ be a sequence of P-independent sub-σ-algebras of $\mathcal{F}$. Further, assume that, for each $n \in \mathbb{Z}^+$, there is an $A_n \in \mathcal{F}_n$ which satisfies $\alpha \le P\big(A_n\big) \le 1 - \alpha$ for some fixed $\alpha \in (0, \frac{1}{2})$. Show that the tail σ-algebra $\mathcal{T}$ determined by $\{\mathcal{F}_n : n \in \mathbb{Z}^+\}$ cannot be countably generated.

Hint: First, reduce to the case when each $\mathcal{F}_n$ is generated by the set A_n. After making this reduction, show that C is an **atom** in $\mathcal{T}$ (i.e., $B = C$ whenever $B \in \mathcal{T} \setminus \{\emptyset\}$ is contained in C) only if one can write

$$C = \varliminf_{n\to\infty} C_n \equiv \bigcup_{m=1}^{\infty} \bigcap_{n\geq m} C_n$$

where, for each $n \in \mathbb{Z}^+$, either C_n equals A_n or $A_n\complement$. Conclude that every atom in $\mathcal{T}$ must have P-measure 0. Now suppose that $\mathcal{T}$ is generated by $\{B_\ell : \ell \in \mathbb{N}\}$. By Kolmogorov's 0–1 Law (cf. Theorem 1.1.2), $P(B_\ell) \in \{0,1\}$ for every $\ell \in \mathbb{N}$. Take

$$\hat{B}_\ell = \begin{cases} B_\ell & \text{if } P(B_\ell) = 1 \\ B_\ell\complement & \text{if } P(B_\ell) = 0 \end{cases} \quad \text{and set} \quad C = \bigcap_{\ell\in\mathbb{N}} \hat{B}_\ell.$$

Note that, on the one hand, $P(C) = 1$, while, on the other hand, C is an atom in $\mathcal{T}$.

1.1.19 Exercise: Here is an application of Kolmogorov's 0–1 Law to Lebesgue's measure on $[0,1)$.

(i) Referring to the discussion preceding Lemma 1.1.8, define the transformations $T_n : [0,1) \longrightarrow [0,1)$ for $n \in \mathbb{Z}^+$ so that

$$T_n(\omega) = \omega - \frac{1 + R_n(\omega)}{2^n}, \quad \omega \in [0,1),$$

and notice (cf. the proof of Lemma 1.1.8) that $T_n(\omega)$ simply *flips* the nth coefficient in the binary expansion ω. Next, let $\Gamma \in \mathcal{B}_{[0,1)}$, and show that Γ is measurable with respect of the σ-algebra $\sigma(R_n : n > m)$ generated by $\{R_n : n > m\}$ if and only if $T_n(\Gamma) = \Gamma$ for each $1 \leq n \leq m$. In particular, conclude that $\lambda_{[0,1)}(\Gamma) \in \{0,1\}$ if $T_n\Gamma = \Gamma$ for every $n \in \mathbb{Z}^+$.

(ii) Let $\mathfrak{F}$ denote the set of all finite subsets of $\mathbb{Z}^+$, and for each $F \in \mathfrak{F}$, define $T^F : [0,1) \longrightarrow [0,1)$ so that $T^\emptyset$ is the identity mapping and

$$T^{F\cup\{m\}} = T^F \circ T_m \quad \text{for each } F \in \mathfrak{F} \text{ and } m \in \mathbb{Z}^+ \setminus F.$$

As an application of (**i**), show that for every $\Gamma \in \mathcal{B}_{[0,1)}$ with $\lambda_{[0,1)}(\Gamma) > 0$,

$$\lambda_{[0,1)}\left(\bigcup_{F\in\mathfrak{F}} T^F(\Gamma)\right) = 1.$$

In particular, this means that if Γ has positive measure, then almost every $\omega \in [0,1)$ can be moved to Γ by *flipping* a finite number of the coefficients in the binary expansion of ω.

§1.2: The Weak Law of Large Numbers.

Starting with this section and for the rest of this chapter, we will be studying what happens when one averages P-independent, real-valued random variables. The remarkable fact, which will be confirmed repeatedly, is that the limiting behavior of such averages depends hardly at all on the variables involved. Intuitively, one can explain this phenomenon by pretending that the random variables are building blocks which, in the averaging process, first get homothetically shrunk and then reassembled according to a regular pattern. Hence, by the time that one passes to the limit, the peculiarities of the original blocks get lost.

Throughout our discussion, $(\Omega, \mathcal{F}, P)$ will be a probability space on which we have a sequence $\{X_n\}_1^\infty$ of real-valued random variables. Given $n \in \mathbb{Z}^+$, we will use S_n to denote the partial sum $X_1 + \cdots + X_n$ and $\overline{S}_n$ to denote the average

$$\frac{S_n}{n} = \frac{1}{n}\sum_{\ell=1}^{n} X_\ell.$$

Our first result is a very general one; in fact, it even applies to random variables which are not necessarily independent and do not necessarily have mean 0.

1.2.1 Lemma. *Assume that*

$$\text{(1.2.2)} \qquad \mathbb{E}^P\big[X_n^2\big] < \infty \text{ for } n \in \mathbb{Z}^+ \quad \text{and} \quad \mathbb{E}^P\big[X_k X_\ell\big] = 0 \text{ if } k \neq \ell.$$

Then, for each $\epsilon > 0$,

$$\text{(1.2.3)} \qquad \epsilon^2 P\Big(\big|\overline{S}_n\big| \geq \epsilon\Big) \leq \mathbb{E}^P\big[\overline{S}_n^2\big] = \frac{1}{n^2}\sum_{\ell=1}^{n} \mathbb{E}^P\big[X_\ell^2\big] \quad \text{for } n \in \mathbb{Z}^+.$$

In particular, if

$$\text{(1.2.4)} \qquad M \equiv \sup_{n\in\mathbb{Z}^+} \mathbb{E}^P\big[X_n^2\big] < \infty,$$

then

$$\text{(1.2.5)} \qquad \epsilon^2 P\Big(\big|\overline{S}_n\big| \geq \epsilon\Big) \leq \mathbb{E}^P\big[\overline{S}_n^2\big] \leq \frac{M}{n}, \qquad n \in \mathbb{Z}^+ \text{ and } \epsilon > 0;$$

and so $\overline{S}_n \longrightarrow 0$ *in* $L^2(P)$ *and also in* P*-probability.*

PROOF: To prove the equality in (1.2.3), note that, by (1.2.2),

$$\mathbb{E}^P\big[S_n^2\big] = \sum_{\ell=1}^{n} \mathbb{E}^P\big[X_\ell^2\big].$$

The rest is just an application of **Chebyshev's inequality**, the estimate which results after integrating the inequality

$$\epsilon^2 \mathbf{1}_{[\epsilon,\infty)}(|Y|) \le Y^2 \mathbf{1}_{[\epsilon,\infty)}(|Y|) \le Y^2$$

for any random variable Y. □

Obviously, Lemma 1.2.1 has less to do with the property of independence than it does with Bessels's inequality for general orthogonal functions. On the other hand, independent random variables provide a ready source of orthogonal functions. Indeed, recall that for any P-integrable random variable X, its **variance** $\operatorname{var}(X)$ satisfies

$$\text{(1.2.6)} \qquad \operatorname{var}(X) \equiv \mathbb{E}^P\left[\left(X - \mathbb{E}^P[X]\right)^2\right] = \mathbb{E}^P\left[X^2\right] - \left(\mathbb{E}^P[X]\right)^2 \le \mathbb{E}^P\left[X^2\right].$$

In particular, if the random variables X_n, $n \in \mathbb{Z}^+$, are P-independent and satisfy the first part of (1.2.2), then the random variables

$$\hat{X}_n \equiv X_n - \mathbb{E}^P\left[X_n\right] \qquad n \in \mathbb{Z}^+,$$

are still square P-integrable, now have mean-value 0, and therefore satisfy the whole of (1.2.2). Hence, the following statement is an immediate consequence of Lemma 1.2.1.

1.2.7 Theorem. *Let $\{X_n : n \in \mathbb{Z}^+\}$ be a sequence of P-independent, square P-integrable random variables with mean-value m and variance dominated by σ^2. Then, for every $n \in \mathbb{Z}^+$ and $\epsilon > 0$:*

$$\text{(1.2.8)} \qquad \epsilon^2\, P\left(\left|\overline{S}_n - m\right| \ge \epsilon\right) \le \frac{1}{n^2}\mathbb{E}^P\left[\left(\overline{S}_n - m\right)^2\right] \le \frac{\sigma^2}{n}.$$

In particular, $\overline{S}_n \longrightarrow m$ in $L^2(P)$ and therefore in P-probability.

As yet we have only made minimal use of independence: all that we have done is subtract off the mean of independent random variables and thereby made them orthogonal. In order to bring the full force of independence into play, one has to exploit the fact that one can compose independent random variables with any (measurable) functions without destroying their independence; in particular, *truncating* independent random variables does not destroy independence. To see how such a property can be brought to bear, we will now consider the problem of extending the last part of Theorem 1.2.7 to X_n's which are less than square P-integrable. In order to understand the statement, recall that a family $\{X_i : i \in \mathcal{I}\}$ of random variables is said to be **uniformly P-integrable** if

$$\text{(1.2.9)} \qquad \lim_{R \nearrow \infty} \sup_{i \in \mathcal{I}} \mathbb{E}^P\left[|X_i|,\, |X_i| \ge R\right] = 0.$$

As the proof of the following theorem illustrates, the importance of this condition is that it allows one to simultaneously approximate the random variables X_i, $i \in \mathcal{I}$ by bounded random variables.

1.2.10 Theorem (The Weak Law of Large Numbers). *Let $\{X_n : n \in \mathbb{Z}^+\}$ be a uniformly P-integrable sequence of P-independent random variables. Then*

$$\frac{1}{n}\sum_1^n \left(X_m - \mathbb{E}^P[X_m]\right) \longrightarrow 0 \text{ in } L^1(P)$$

and, therefore, also in P-probability. In particular, if $\{X_n : n \in \mathbb{Z}^+\}$ is a sequence of P-independent, P-integrable random variables which are identically distributed, then $\overline{S}_n \longrightarrow \mathbb{E}^P\big[X_1\big]$ in $L^1(P)$ and P-probability. (Cf. Exercise 1.2.15 below.)

PROOF: Without loss in generality, we will assume that $\mathbb{E}^P[X_n] = 0$ for every $n \in \mathbb{Z}^+$.

For each $R \in (0,\infty)$, define $f_R(t) = t\,\mathbf{1}_{[-R,R]}(t)$, $t \in \mathbb{R}$,

$$m_n^{(R)} = \mathbb{E}^P\big[f_R \circ X_n\big], \quad X_n^{(R)} = f_R \circ X_n - m_n^{(R)}, \quad \text{and} \quad Y_n^{(R)} = X_n - X_n^{(R)},$$

and set

$$\overline{S}_n^{(R)} = \frac{1}{n}\sum_{\ell=1}^n X_\ell^{(R)} \quad \text{and} \quad \overline{T}_n^{(R)} = \frac{1}{n}\sum_{\ell=1}^n Y_\ell^{(R)}.$$

Clearly,

$$\begin{aligned}
\mathbb{E}^P\Big[\big|\overline{S}_n\big|\Big] &\le \mathbb{E}^P\Big[\big|\overline{S}_n^{(R)}\big|\Big] + \mathbb{E}^P\Big[\big|\overline{T}_n^{(R)}\big|\Big] \\
&\le \mathbb{E}^P\Big[\big|\overline{S}_n^{(R)}\big|^2\Big]^{\frac{1}{2}} + 2\max_{1\le \ell\le n} \mathbb{E}^P\Big[\big|X_\ell\big|,\, |X_\ell| \ge R\Big] \\
&\le \frac{R}{\sqrt{n}} + 2\max_{\ell\in\mathbb{Z}^+} \mathbb{E}^P\Big[\big|X_\ell\big|,\, |X_\ell| \ge R\Big];
\end{aligned}$$

and therefore, for each $R > 0$,

$$\varlimsup_{n\to\infty} \mathbb{E}^P\Big[\big|\overline{S}_n\big|\Big] \le 2\sup_{\ell\in\mathbb{Z}} \mathbb{E}^P\Big[\big|X_\ell\big|,\, |X_\ell| \ge R\Big].$$

Hence, because the X_ℓ's are uniformly P-integrable, we get the desired convergence in $L^1(P)$ by letting $R \nearrow \infty$. □

The name of Theorem 1.2.10 comes from a somewhat invidious comparison with the result in Theorem 1.4.11. The reason why the appellation *weak* is not entirely fair is that, although The Weak Law is indeed less *refined* than the result in Theorem 1.4.11, it is every bit as useful as the one in Theorem 1.4.11 and maybe even more important when it comes to applications. Indeed, what The Weak Law does is provide us with a ubiquitous technique for constructing an **approximate identity** (i.e., a sequence of measures which approximate a point mass) and measuring how fast the approximation is taking place. To illustrate how clever selection of the random variables entering The Weak Law can lead to interesting applications, we will spend the rest of this section discussing S. Bernstein's approach to Weierstrass's approximation theorem.

For a given $p \in [0,1]$, let $\{X_n : n \in \mathbb{Z}^+\}$ be a sequence of P-independent $\{0,1\}$-valued Bernoulli random variables with mean-value p. Then

$$P(S_n = \ell) = \binom{n}{\ell} p^\ell (1-p)^{n-\ell} \quad \text{for} \quad 0 \le \ell \le n.$$

Hence, for any $f \in C([0,1];\mathbb{R})$, the nth **Bernstein polynomial**

$$B_n(p;f) \equiv \sum_{\ell=0}^{n} \binom{n}{\ell} f\left(\frac{\ell}{n}\right) p^\ell (1-p)^{n-\ell} \tag{1.2.11}$$

of f at p is equal to

$$\mathbb{E}^P\left[f \circ \overline{S}_n\right].$$

In particular,

$$\begin{aligned} \left|f(p) - B_n(p;f)\right| &= \left|\mathbb{E}^P\left[f(p) - f \circ \overline{S}_n\right]\right| \le \mathbb{E}^P\left[\left|f(p) - f \circ \overline{S}_n\right|\right] \\ &\le 2\|f\|_{\mathrm{u}} P\left(\left|\overline{S}_n - p\right| \ge \epsilon\right) + \rho(\epsilon; f), \end{aligned}$$

where $\|f\|_{\mathrm{u}}$ is the **uniform norm** of f (i.e., the supremum of $|f|$ over the domain of f) and

$$\rho(\epsilon; f) \equiv \sup\left\{|f(t) - f(s)| : 0 \le s < t \le 1 \text{ with } t - s \le \epsilon\right\}$$

is the modulus of continuity of f. Noting that $\mathrm{var}(X_n) = p(1-p) \le \frac{1}{4}$ and applying (1.2.8), we conclude that, for every $\epsilon > 0$,

$$\left\|f(p) - B_n(p;f)\right\|_{\mathrm{u}} \le \frac{\|f\|_{\mathrm{u}}}{2n\epsilon^2} + \rho(\epsilon; f)$$

In other words, for all $n \in \mathbb{Z}^+$,

$$\left\|f - B_n(\,\cdot\,;f)\right\|_{\mathrm{u}} \le \beta(n;f) \equiv \inf\left\{\frac{\|f\|_{\mathrm{u}}}{2n\epsilon^2} + \rho(\epsilon;f) : \epsilon > 0\right\}. \tag{1.2.12}$$

Obviously, (1.2.12) not only shows that, as $n \to \infty$, $B_n(\,\cdot\,;f) \longrightarrow f$ uniformly on $[0,1]$, but it even provides a rate of convergence in terms of the modulus of continuity of f. Thus, we have done more than simply prove Weierstrass's theorem; we have produced a rather explicit and tractable sequence of approximating polynomials, the sequence $\{B_n(\,\cdot\,;f) : n \in \mathbb{Z}^+\}$. Although this sequence is, by no means, the most efficient one,† as we are about to see, the Bernstein polynomials have a lot to recommend them. In particular, they have the feature that

† See G.G. Lorentz's *Bernstein Polynomials*, Chelsea Publ. Co., New York (1986) for a lot more information.

they provide nonnegative polynomial approximants to nonnegative functions. In fact, the following discussion reveals much deeper nonnegativity preservation properties of the Bernstein approximation scheme.

In order to bring out the virtues of the Bernstein polynomials, it is important to replace (1.2.11) with an expression in which the coefficients of $B_n(\,\cdot\,; f)$ (as polynomials) are clearly displayed. To this end, introduce the **difference operator** Δ_h for $h > 0$ given by

$$\big[\Delta_h f\big](t) = \frac{f(t+h) - f(t)}{h}.$$

A straightforward inductive argument (using Pascal's identity for the binomials coefficients) shows that

$$(-h)^m \big[\Delta_h^{(m)} f\big](t) = \sum_{\ell=0}^{m} (-1)^\ell \binom{m}{\ell} f(t + \ell h) \quad \text{for} \quad m \in \mathbb{Z}^+,$$

where $\Delta_h^{(m)}$ denotes the mth iterate of the operator Δ_h. Taking $h = \frac{1}{n}$, we now see that

$$\begin{aligned}
B_n(p; f) &= \sum_{\ell=0}^{n} \sum_{k=0}^{n-\ell} \binom{n}{\ell} \binom{n-\ell}{k} (-1)^k f(\ell h) p^{\ell+k} \\
&= \sum_{r=0}^{n} p^r \sum_{\ell=0}^{r} \binom{n}{\ell} \binom{n-\ell}{r-\ell} (-1)^{r-\ell} f(\ell h) \\
&= \sum_{r=0}^{n} (-p)^r \binom{n}{r} \sum_{\ell=0}^{r} \binom{r}{\ell} (-1)^\ell f(\ell h) \\
&= \sum_{r=0}^{n} \binom{n}{r} (ph)^r \big[\Delta_h^{(r)} f\big](0),
\end{aligned}$$

where $\big[\Delta_h^0 f\big] \equiv f$. Hence, we have proved that

$$B_n(p; f) = \sum_{\ell=0}^{n} n^{-\ell} \binom{n}{\ell} \Big[\Delta_{\frac{1}{n}}^{(\ell)} f\Big](0) p^\ell \quad \text{for} \quad p \in [0,1]. \tag{1.2.13}$$

The marked resemblance between the expression on the right-hand side of (1.2.13) and a Taylor polynomial is more than coincidental. To demonstrate how one can exploit the relationship between Bernstein and Taylor polynomials, say that a function $\varphi \in C^\infty\big((a,b); \mathbb{R}\big)$ is **absolutely monotone** if its mth derivative $D^m\varphi$ is nonnegative for every $m \in \mathbb{N}$. Also, say that $\varphi \in C^\infty\big([0,1]; [0,1]\big)$ is a

probability generating function if there exists a $\{u_n : n \in \mathbb{N}\} \subseteq [0,1]$ such that

$$\sum_{n=0}^{\infty} u_n = 1 \quad \text{and} \quad \varphi(t) = \sum_{n=0}^{\infty} u_n t^n \quad \text{for} \quad t \in [0,1].$$

Obviously, every probability generating function is absolutely monotone. The somewhat surprising (remember that most infinitely differentiable functions do not admit power series expansions) fact which we are about to prove is that, apart from a multiplicative constant, the converse is also true. In fact, we do not need to know, *a priori*, that the function is smooth so long as it satisfies a discrete version of absolute monotonicity.

1.2.14 Theorem. *Let $\varphi \in C([0,1];\mathbb{R})$ with $\varphi(1) = 1$ be given. Then the following are equivalent:*
(i) *φ is a probability generating function,*
(ii) *the restriction of φ to $(0,1)$ is absolutely monotone;*
(iii) *$\big[\Delta^{(m)}_{\frac{1}{n}}\varphi\big](0) \geq 0$ for every $n \in \mathbb{N}$ and $0 \leq m \leq n$.*

PROOF: The implication **(i)** $\Longrightarrow$ **(ii)** is trivial. To see that **(ii)** implies **(iii)**, first observe that if ψ is absolutely monotone on (a,b) and $h \in (0, b-a)$, then $[\Delta_h\psi]$ is absolutely monotone on $(a, b-h)$. Indeed, because $[D \circ \Delta_h \psi] = [\Delta_h \circ D\psi]$ on $(a, b-h)$, we see that

$$h\big[D^m \circ \Delta_h \psi\big](t) = \int_t^{t+h} D^{m+1}\psi(s)\,ds \geq 0, \quad t \in (a, b-h),$$

for any $m \in \mathbb{N}$. Returning to the function φ, we now know that $\big[\Delta^{(m)}_h\varphi\big]$ is absolutely monotone on $(0, 1-mh)$ for all $m \in \mathbb{N}$ and $h > 0$ with $mh < 1$. In particular,

$$\big[\Delta^{(m)}_h\varphi\big](0) = \lim_{t \searrow 0}\big[\Delta^{(m)}_h\varphi\big](t) \geq 0 \quad \text{if} \quad mh < 1,$$

and so $\big[\Delta^{(m)}_h\varphi\big](0) \geq 0$ when $h = \frac{1}{n}$ and $0 \leq m < n$. Moreover, since

$$\big[\Delta^{(n)}_{\frac{1}{n}}\varphi\big](0) = \lim_{h \nearrow \frac{1}{n}}\big[\Delta^{(n)}_h\varphi\big](0),$$

we also know that $\big[\Delta^n_h\varphi\big](0) \geq 0$ when $h = \frac{1}{n}$, and this completes the proof that **(ii)** implies **(iii)**.

Finally, assume that **(iii)** holds and set $\varphi_n = B_n(\,\cdot\,;\varphi)$. Then, by (1.2.13) and the equality $\varphi_n(1) = \varphi(1) = 1$, we see that each φ_n is a probability generating function. Thus, in order to complete the proof that **(iii)** implies **(i)**, all that we have to do is check that a uniform limit of probability generating functions is itself a probability generating function. To this end, write

$$\varphi_n(t) = \sum_{\ell=0}^{\infty} u_{n,\ell}t^\ell, \quad t \in [0,1] \text{ for each } n \in \mathbb{Z}^+.$$

Because the $u_{n,\ell}$'s are all elements of $[0,1]$, we can use a diagonalization procedure to choose $\{n_k : k \in \mathbb{Z}^+\}$ so that

$$\lim_{k\to\infty} u_{n_k,\ell} = u_\ell \in [0,1]$$

exists for each $\ell \in \mathbb{N}$. But, by Lebesgue's Dominated Convergence Theorem, this means that

$$\varphi(t) = \lim_{k\to\infty} \varphi_{n_k}(t) = \sum_{\ell=0}^{\infty} u_\ell t^\ell \quad \text{for every} \quad t \in [0,1).$$

Finally, by the Monotone Convergence Theorem, the preceding extends immediately to $t = 1$, and so φ is a probability generating function. (Notice that the argument just given does not even use the assumed uniform convergence and shows that the pointwise limit of probability generating functions is again a probability generating function.) □

The preceding is only one of many examples in which The Weak Law leads to useful ways of forming an *approximate identity.* Our treatment of Bernstein's ideas is based the exposition by Wm. Feller,† who provides several other similar applications of The Weak Law, including the ones in the following exercises.

Exercises

1.2.15 Exercise: Although, for historical reasons, The Weak Law is usually thought of as a theorem about convergence in P-probability, the forms in which we have presented it are clearly results about convergence in either P-mean or even square P-mean. Thus, it is interesting to discover that one can replace the uniform integrability assumption made in Theorem 1.2.10 with a *weak uniform integrability* assumption if one is willing to settle for convergence in P-probability. Namely, let $X_1, \dots, X_n, \dots$ be mutually P-independent random variables, assume that

$$F(R) \equiv \sup_{n\in\mathbb{Z}^+} RP\Big(|X_n| \geq R\Big) \longrightarrow 0 \quad \text{as } R \nearrow \infty, \tag{1.2.16}$$

and set

$$m_n = \frac{1}{n}\sum_{\ell=1}^{n} \mathbb{E}^P\Big[X_\ell,\ |X_\ell| \leq n\Big], \quad n \in \mathbb{Z}^+.$$

† Wm. Feller, *An Introduction to Probability Theory and Its Applications, Vol. II*, J. Wiley Series in Probability and Math. Stat. (1968).

Show that, for each $\epsilon > 0$,

$$P\left(\left|\overline{S}_n - m_n\right| \geq \epsilon\right) \leq \frac{1}{(n\epsilon)^2}\sum_{\ell=1}^{n} \mathbb{E}^P\left[X_\ell^2,\, |X_\ell| \leq n\right] + P\left(\max_{1\leq \ell\leq n} |X_\ell| > n\right)$$
$$\leq \frac{2}{n\epsilon^2}\int_0^n F(t)\,dt + F(n);$$

and conclude that $\left|\overline{S}_n - m_n\right| \longrightarrow 0$ in P-probability. (See part **(ii)** of Exercises 1.4.29 and 1.5.14 for a partial converse to this statement.)

Hint: Use the formula

$$\operatorname{var}(Y) \leq E^P\left[Y^2\right] = 2\int_{[0,\infty)} t\,P\big(|Y| > t\big)\,dt.$$

1.2.17 Exercise: Show that, for each $T \in [0,\infty)$ and $t \in (0,\infty)$,

$$\lim_{n\to\infty} e^{-nt}\sum_{k\leq nT} \frac{(nt)^k}{k!} = \begin{cases} 1 & \text{if } \; T > t \\ 0 & \text{if } \; T < t. \end{cases} \tag{1.2.18}$$

Hint: Let $X_1, \dots, X_n, \dots$ be P-independent **Poisson random variables on** $\mathbb{N}$ with mean-value t. That is, the X_n's are P-independent and

$$P\big(X_n = k\big) = e^{-t}\frac{t^k}{k!} \quad \text{for} \quad k \in \mathbb{N}.$$

Show that S_n is a Poisson random variable on $\mathbb{N}$ with mean-value nt, and conclude that, for each $T \in [0,\infty)$ and $t \in (0,\infty)$,

$$e^{-nt}\sum_{k\leq nT} \frac{(nt)^k}{k!} = P\big(\overline{S}_n \leq T\big).$$

1.2.19 Exercise: Given a right-continuous function $F : [0,\infty) \longrightarrow \mathbb{R}$ of bounded variation with $F(0) = 0$, define its **Laplace transform** $\varphi(\lambda)$, $\lambda \in [0,\infty)$, by the Riemann–Stieltjes integral

$$\varphi(\lambda) = \int_{[0,\infty)} e^{-\lambda t}\,dF(t).$$

Using Exercise 1.2.17, show that

$$\sum_{k\leq nT} \frac{(-n)^k}{k!}\left[D^k\varphi\right](n) \longrightarrow F(T) \quad \text{as} \quad n \to \infty$$

for each $T \in [0,\infty)$ at which F is continuous. Conclude, in particular, that F can be recovered from its Laplace transform.

§1.3: Cramér's Theory of Large Deviations.

From Theorem 1.2.7, we know that if $\{X_n : n \in \mathbb{Z}^+\}$ is a sequence of P-independent, square P-integrable random variables with mean-value 0, and if the averages $\overline{S}_n$, $n \in \mathbb{Z}^+$, are defined accordingly, then, for every $\epsilon > 0$,

$$P\left(\left|\overline{S}_n\right| \geq \epsilon\right) \leq \frac{\max_{1\leq m\leq n} \operatorname{var}(X_m)}{n\epsilon^2}, \qquad n \in \mathbb{Z}^+.$$

Thus, so long as

$$\frac{\operatorname{var}(X_n)}{n} \longrightarrow 0 \text{ as } n \to \infty,$$

the $\overline{S}_n$'s are becoming more and more concentrated near 0, and the rate at which this concentration is occurring can be estimated in terms of the variances $\operatorname{var}(X_n)$. In this section, we will show that, by placing more stringent integrability requirements on the X_n's, one can gain more information about the rate at which the $\overline{S}_n$'s are concentrating.

In all of this analysis, the trick is to see how independence can be combined with 0 mean-value to produce unexpected cancellations; and as a preliminary warm-up exercise, we begin with the following.

1.3.1 Theorem. *Let $\{X_n : n \in \mathbb{Z}^+\}$ be a sequence of P-independent, P-integrable random variables with mean-value 0, and assume that*

$$M_4 \equiv \sup_{n\in\mathbb{Z}^+} \mathbb{E}^P\left[X_n^4\right] < \infty.$$

Then, for each $\epsilon > 0$,

$$\epsilon^4 P\left(\left|\overline{S}_n\right| \geq \epsilon\right) \leq \mathbb{E}^P\left[\overline{S}_n^{\,4}\right] \leq \frac{3M_4}{n^2}, \quad n \in \mathbb{Z}^+; \tag{1.3.2}$$

In particular, $\overline{S}_n \longrightarrow 0$ P-almost surely.

Proof: Obviously, in order to prove (1.3.2), it suffices to check the second inequality. To this end, note that there is nothing to do when $n = 1$. Moreover, for any $n \in \mathbb{Z}^+$,

$$S_{n+1}^4 = S_n^4 + 4S_n^3 X_{n+1} + 6S_n^2 X_{n+1}^2 + 4S_n X_{n+1}^3 + X_{n+1}^4,$$

which, because S_n is independent of X_{n+1} and $\mathbb{E}^P[X_{n+1}] = \mathbb{E}^P[S_n] = 0$, means that

$$\begin{aligned}
\mathbb{E}^P\left[S_{n+1}^4\right] &= \mathbb{E}^P\left[S_n^4\right] + 6\mathbb{E}^P\left[S_n^2\right]\mathbb{E}^P\left[X_{n+1}^2\right] + \mathbb{E}^P\left[X_{n+1}^4\right] \\
&= \mathbb{E}^P\left[S_n^4\right] + 6\sum_{m=1}^{n} \mathbb{E}^P\left[X_m^2\right]\mathbb{E}^P\left[X_{n+1}^2\right] + \mathbb{E}^P\left[X_{n+1}^4\right] \\
&\le \mathbb{E}^P\left[S_n^4\right] + (6n+1)M_4,
\end{aligned}$$

where, in passing to the final line, we have used Schwarz's inequality. Hence, assuming the result for n, we find that

$$\mathbb{E}^P\left[S_{n+1}^4\right] \le 3M_4\left(n^2 + 2n + \tfrac{1}{3}\right) \le 3(n+1)^2 M_4.$$

Given (1.3.2), the proof of the last part becomes an easy application of the Borel–Cantelli Lemma. Indeed, for any $\epsilon > 0$, we know from (1.3.2) that

$$\sum_{n=1}^{\infty} P\left(\left|\overline{S}_n\right| \ge \epsilon\right) < \infty,$$

and therefore, by (1.1.5), that

$$P\left(\varlimsup_{n\to\infty} \left|\overline{S}_n\right| \ge \epsilon\right) = 0. \quad \square$$

1.3.3 Remark. The final assertion in Theorem 1.3.1 is a primitive version of The Strong Law of Large Numbers and represents the first time that we have actually used the simultaneous existence of infinitely many mutually independent random variables (previously, and for the rest of this section, it will be enough to know that there are, at any given moment, an arbitrary but finite number). Although The Strong Law will be taken up again, and considerably refined, in Section 1.4, the principle on which its proof here was based is an important one: namely, *control more moments and you will get better estimates; get better estimates and you will reach more interesting conclusions.*

With the preceding adage in mind, we will devote the rest of this section to examining what one can say when one has all moments at one's disposal. In fact, from now on, we will be assuming that $X_1, \dots, X_n, \dots$ are independent random variables with common distribution μ having the property that the **moment generating function**

(1.3.4) $$M_\mu(\xi) \equiv \int_{\mathbb{R}} e^{\xi x}\,\mu(dx) < \infty \quad \text{for all } \xi \in \mathbb{R}.$$

Obviously, (1.3.4) is more than sufficient to guarantee that the X_n's have moments of all orders. In fact, as an application of Lebesgue's Dominated Convergence Theorem, one sees that $\xi \in \mathbb{R} \longmapsto M(\xi) \in (0, \infty)$ is infinitely differentiable and that

$$\mathbb{E}^P\left[X_1^n\right] = \int_{\mathbb{R}} x^n \, \mu(dx) = \frac{d^n M}{d\xi^n}(0) \quad \text{for all } n \in \mathbb{N}.$$

In the discussion which follows, we will use m and σ^2 to denote, respectively, the mean-value and variance of the X_n's.

In order to develop some intuition for the considerations which follow, we first consider an example, which, for many purposes, is *the canonical example* in probability theory. Namely, let $\gamma : \mathbb{R} \longrightarrow (0, \infty)$ be the **Gauss kernel**

$$\text{(1.3.5)} \qquad \gamma(y) \equiv \frac{1}{\sqrt{2\pi}} \exp\left[-\frac{|y|^2}{2}\right], \quad y \in \mathbb{R};$$

and recall that a random variable X is **standard normal** if

$$P\big(X \in \Gamma\big) = \int_\Gamma \gamma(y)\, dy, \quad \Gamma \in \mathcal{B}_{\mathbb{R}}.$$

In spite of their somewhat insultingly bland appellation, standard normal random variables are the building blocks for the most honored family in all of probability theory. Indeed, given $m \in \mathbb{R}$ and $\sigma \in [0, \infty)$, the random variable Y is said to be **normal** (or **Gaussian**) **with mean-value** m **and variance** σ^2 (often this is abbreviated by saying that X is an $\mathfrak{N}(m, \sigma^2)$**-random variable**) if and only if the distribution of Y is the same as that of the random variable $\sigma X + m$, where X is standard normal. That is, Y is an $\mathfrak{N}(m, \sigma^2)$ random variable if, when $\sigma = 0$, $P(Y = m) = 1$ and, when $\sigma > 0$,

$$\text{(1.3.6)} \qquad P\big(Y \in \Gamma\big) = \int_\Gamma \frac{1}{\sigma}\gamma\left(\frac{y-m}{\sigma}\right) dy \quad \text{for } \Gamma \in \mathcal{B}_{\mathbb{R}}.$$

There are two obvious reasons for the honored position held by Gaussian random variables. In the first place, they certainly have finite moment generating functions. In fact, since

$$\int_{\mathbb{R}} e^{\xi\, y}\, \gamma(y)\, dy = \exp\left(\frac{\xi^2}{2}\right), \quad \xi \in \mathbb{R},$$

it is clear that

$$\text{(1.3.7)} \qquad M_{\gamma_{m,\sigma^2}}(\xi) = \exp\left[\xi m + \frac{\sigma^2 \xi^2}{2}\right].$$

Secondly, *they add nicely.* To be precise, it is a familiar fact from elementary probability theory that *if X is an $\mathfrak{N}(m,\sigma^2)$ random variable and $\hat{X}$ is an $\mathfrak{N}(\hat{m},\hat{\sigma}^2)$ random variable which is independent of X, then $X+\hat{X}$ is an $\mathfrak{N}(m+\hat{m},\sigma^2+\hat{\sigma}^2)$ random variable.* In particular, if $X_1,\dots,X_n$ are mutually independent standard normal random variables, then $\overline{S}_n$ is an $\mathfrak{N}\left(0,\frac{1}{n}\right)$ random variable. That is,

$$P(\overline{S}_n \in \Gamma) = \sqrt{\frac{n}{2\pi}}\int_\Gamma \exp\left[-\frac{n|y|^2}{2}\right]\,dy.$$

Thus (cf. Exercise 1.3.17 below), for any Γ we see that

$$\lim_{n\to\infty}\frac{1}{n}\log\Big[P\big(\overline{S}_n\in\Gamma\big)\Big] = -\text{ess inf}\left\{\frac{|y|^2}{2} : y\in\Gamma\right\}. \tag{1.3.8}$$

where the *ess* in (1.3.8) stands for *essential* and means that what follows is taken *modulo a set of measure* 0. (Hence, apart from a minus sign, the right-hand side of (1.3.8) is the greatest number dominated by $\frac{|y|^2}{2}$ for Lebesgue-almost every $y\in\Gamma$.) In fact, because (after comparing the derivatives of the quantities involved) one knows that

$$\big(x^{-1}-x^{-3}\big)\gamma(x) \le \int_x^\infty \gamma(y)\,dy \le x^{-1}\gamma(x) \quad \text{for all } x\in(0,\infty),$$

we have the rather precise estimate

$$\sqrt{\frac{2}{n\pi\epsilon^2}}\left(1-\frac{1}{n\epsilon^2}\right)\exp\left(-\frac{n\epsilon^2}{2}\right) \le P\big(|\overline{S}_n|\ge\epsilon\big) \le \sqrt{\frac{2}{n\pi\epsilon^2}}\exp\left(-\frac{n\epsilon^2}{2}\right),$$

More generally, if $X_1,\dots,X_n$ are mutually independent $\mathfrak{N}(m,\sigma^2)$ random variables and $\sigma>0$, then the above get replaced by

$$\lim_{n\to\infty}\frac{1}{n}\log\Big[P\big(\overline{S}_n-m\in\Gamma\big)\Big] = -\text{ess inf}\left\{\frac{|y|^2}{2\sigma^2} : y\in\Gamma\right\}$$

and

$$\sqrt{\frac{2\sigma^2}{n\pi\epsilon^2}}\left(1-\frac{\sigma^2}{n\epsilon^2}\right)\exp\left(-\frac{n\epsilon^2}{2\sigma^2}\right) \le P\big(|\overline{S}_n-m|\ge\epsilon\big) \le \sqrt{\frac{2\sigma^2}{n\pi\epsilon^2}}\exp\left(-\frac{n\epsilon^2}{2\sigma^2}\right),$$

respectively.

Of course, in general, one cannot hope to get such explicit expressions for the distribution of $\overline{S}_n$. Nonetheless, on the basis of the preceding, one can start to see what is going on. Namely, when the distribution μ falls off rapidly outside of compacts, averaging n independent random variables with distribution μ has the effect of *building a well in which the mean-value m lies at the bottom.* More precisely, if one believes that the Gaussian random variables *are normal* in the sense that they are typical, then one should conjecture that, even when the random variables are not normal, the behavior of $P\big(|\overline{S}_n - m| \geq \epsilon\big)$ for large n's should resemble that of Gaussians with the same variance; and it is in the verification of this conjecture that the moment generating function M_μ plays a central rôle. Indeed, although an expression in terms of μ for the distribution of S_n is seldom readily available, the moment generating function for S_n is easily expressed in terms of M_μ. Namely, as a trivial application of independence, we have:

$$\mathbb{E}^P\left[e^{\xi S_n}\right] = M_\mu(\xi)^n, \quad \xi \in \mathbb{R}.$$

Hence, by Markov's inequality applied to $e^{\xi S_n}$, we see that, for any $a \in \mathbb{R}$,

$$P\big(\overline{S}_n \geq a\big) \leq e^{-n\xi a}\, M_\mu(\xi)^n = \exp\big[-n\big(\xi a - \Lambda_\mu(\xi)\big)\big], \quad \xi \in [0,\infty),$$

where

$$\Lambda_\mu(\xi) \equiv \log\big(M_\mu(\xi)\big) \tag{1.3.9}$$

is the **logarithmic moment generating function of** μ. The preceding relation is one of those lovely situations in which a single quantity is dominated by a whole family of quantities, with the result that one should optimize by minimizing over the dominating quantities. Thus, we now have

$$P\big(\overline{S}_n \geq a\big) \leq \exp\left[-n \sup_{\xi\in[0,\infty)} \big(\xi a - \Lambda_\mu(\xi)\big)\right]. \tag{1.3.10}$$

Notice that (1.3.10) is really very good. For instance, when the X_n's are $\mathfrak{N}(m,\sigma^2)$-random variables and $\sigma > 0$, then (cf. (1.3.7)) the preceding leads quickly to the estimate

$$P\big(\overline{S}_n - m \geq \epsilon\big) \leq \exp\left(-\frac{n\epsilon^2}{2\sigma^2}\right),$$

which is essentially the upper bound at which we arrived before.

Taking a hint from the preceding, we now introduce the **Legendre transform**

$$I_\mu(x) \equiv \sup\big\{\xi x - \Lambda_\mu(\xi) : \xi \in \mathbb{R}\big\}, \qquad x \in \mathbb{R}, \tag{1.3.11}$$

of Λ_μ and, before proceeding further, make some elementary observations about the structure of the functions Λ_μ and I_μ.

1.3.12 Lemma. *The function Λ_μ is infinitely differentiable. In addition, for each $\xi \in \mathbb{R}$, the probability measure ν_ξ on $\mathbb{R}$ given by*

$$\nu_\xi(\Gamma) = \frac{1}{M_\mu(\xi)} \int_\Gamma e^{\xi x}\,\mu(dx) \quad \text{for } \Gamma \in \mathcal{B}_{\mathbb{R}},$$

has ν_ξ has moments of all orders,

$$\int_{\mathbb{R}} x\,\nu_\xi(dx) = \Lambda_\mu'(\xi) \quad \text{and} \quad \int_{\mathbb{R}} x^2\,\nu_\xi(dx) - \left(\int_{\mathbb{R}} x\,\nu_\xi(dx)\right)^2 = \Lambda_\mu''(\xi).$$

Next, the function I_μ is a $[0,\infty]$-valued, lower semicontinuous, convex function which vanishes at m . Moreover,

$$I_\mu(x) = \sup\{\xi x - \Lambda_\mu(\xi) : \xi \geq 0\} \quad \text{for} \quad x \in [m,\infty)$$

and

$$I_\mu(x) = \sup\{\xi x - \Lambda_\mu(\xi) : \xi \leq 0\} \quad \text{for} \quad x \in (-\infty,m].$$

Finally, if

$$\alpha = \inf\{x \in \mathbb{R} : \mu\big((-\infty,x]\big) > 0\}$$

and

$$\beta = \sup\{x \in \mathbb{R} : \mu\big([x,\infty)\big) > 0\},$$

then I_μ is smooth on (α,β) and identically $+\infty$ off of $[\alpha,\beta]$. In fact, either $\mu(\{m\}) = 1$ and $\alpha = m = \beta$; or $m \in (\alpha,\beta)$ and ${\Lambda_\mu}'$ is a smooth, strictly increasing mapping from $\mathbb{R}$ onto (α,β),

$$I_\mu(x) = \Xi_\mu(x)\,x - \Lambda_\mu\big(\Xi_\mu(x)\big),\ x \in (\alpha,\beta), \quad \text{where} \quad \Xi_\mu = \left({\Lambda_\mu}'\right)^{-1}$$

is the inverse of ${\Lambda_\mu}'$, $\mu(\{\alpha\}) = e^{-I_\mu(\alpha)}$ if $\alpha > -\infty$, and $\mu(\{\beta\}) = e^{-I_\mu(\beta)}$ if $\beta < \infty$.

PROOF: For notational convenience, we will drop the subscript "μ" during the proof. Further, we remark that the smoothness of Λ follows immediately from the positivity and smoothness of M, and the identification of $\Lambda'(\xi)$ and $\Lambda''(\xi)$ with the mean and variance of ν_ξ is elementary calculus combined with the remark following (1.3.4). Thus, we will concentrate on the properties of the function I.

As the pointwise supremum of functions which are linear, I is certainly lower semicontinuous and convex. Also, because $\Lambda(0) = 0$, it is obvious that $I \geq 0$. Next, by Jensen's inequality,

$$\Lambda(\xi) \geq \xi \int_{\mathbb{R}} x\,\mu(dx) = \xi\, m,$$

and, therefore, $\xi x - \Lambda(\xi) \leq 0$ if $x \leq m$ and $\xi \geq 0$ or if $x \geq m$ and $\xi \leq 0$. Hence, because I is nonnegative, this proves the one-sided extremal characterization of $I_\mu(x)$.

Turning to the final part, note first that there is nothing more to do in the case when $\mu(\{m\}) = 1$. Thus, we will assume that $\mu(\{m\}) < 1$, in which case it is clear that $m \in (\alpha, \beta)$ and that none of the measures ν_ξ is degenerate. In particular, because $\Lambda''(\xi)$ is the variance of the ν_ξ, we know that $\Lambda'' > 0$ everywhere. Hence, Λ' is strictly increasing and therefore admits a smooth inverse Ξ on its image. Furthermore, because $\Lambda'(\xi)$ is the mean of ν_ξ, it is clear that the image of Λ' is contained in (α, β). At the same time, given an $x \in [m, \beta)$, choose $y \in (x, \beta)$ and note that, for $\xi \geq 0$,

$$\Lambda(\xi) \geq \xi\, y - \kappa \quad \text{where} \quad \kappa = -\log\bigl[\mu\bigl([y,\infty)\bigr)\bigr].$$

After combining this with the fact (already established) that $\xi x - \Lambda(\xi) \leq 0$ for $\xi \leq 0$, we conclude that $\xi \in \mathbb{R} \longmapsto \xi x - \Lambda(\xi)$ achieves its absolute maximum somewhere in the interval $\left[0, \frac{\kappa}{y-x}\right]$ and therefore that $\Lambda'(\xi) = x$ for some ξ in that interval. Since an analogous argument applies when $x \in (\alpha, m]$, we now know that (α, β) is precisely the image of Λ'. Finally, because (by convexity) $I(x) = \xi x - \Lambda(\xi)$ if and only if $\Lambda'(\xi) = x$, we have also proved that I is given on (α, β) by the asserted expression.

To complete the proof, suppose that $\beta < \infty$. Then

$$e^{\xi\beta}\mu(\{\beta\}) \leq M(\xi), \quad \xi \in \mathbb{R}.$$

Thus, on the one hand, we have that $\mu(\{\beta\}) \leq e^{-I(\beta)}$. On the other hand, because

$$e^{-I(\beta)} \leq \int_{\mathbb{R}} e^{\xi(x-\beta)}\,\mu(dx) \quad \text{for} \quad \xi \in [0,\infty)$$

and

$$\int_{\mathbb{R}} e^{\xi(x-\beta)}\,\mu(dx) \searrow \mu(\{\beta\}) \quad \text{as} \quad \xi \nearrow \infty,$$

we also see that $\mu(\{\beta\}) \geq e^{-I(\beta)}$. On the other hand, if $x \in (\beta, \infty)$, then $I(x) = \infty$ follows immediately from $\Lambda(\xi) \leq \xi\beta$, $\xi \in [0, \infty)$.

Since the same reasoning applies when $\alpha > -\infty$, we are done. □

1.3.13 Theorem (Cramér's Theorem). *Let $\{X_n\}_1^\infty$ be a sequence of P-independent random variables with common distribution μ, assume that the associated moment generating function M_μ satisfies (1.3.4), set $m = \int_{\mathbb{R}} x\,\mu(dx)$, and define I_μ accordingly, as in (1.3.11). Then,*

$$P\big(\overline{S}_n \geq a\big) \leq e^{-nI_\mu(a)} \quad \text{for all } a \in [m,\infty),$$
$$P\big(\overline{S}_n \leq a\big) \leq e^{-nI_\mu(a)} \quad \text{for all } a \in (-\infty,m].$$

Moreover, for $a \in (\alpha,\beta)$ (cf. Lemma 1.3.12), $\epsilon > 0$, and $n \in \mathbb{Z}^+$,

$$P\Big(\big|\overline{S}_n - a\big| < \epsilon\Big) \geq \left(1 - \frac{\Lambda_\mu{}''\big(\Xi_\mu(a)\big)}{n\epsilon^2}\right) \exp\Big[-n\Big(I_\mu(a) + \epsilon|\Xi_\mu(a)|\Big)\Big],$$

where Λ_μ is the function given in (1.3.9) and $\Xi_\mu \equiv \big(\Lambda_\mu{}'\big)^{-1}$.

PROOF: To prove the first part, suppose that $a \in [m,\infty)$ and apply the second part of Lemma 1.3.12 to see that the exponent in (1.3.10) equals $I_\mu(a)$, and, after replacing $\{X_n\}_1^\infty$ by $\{-X_n\}_1^\infty$, also get the desired estimate when $a \leq m$.

To prove the lower bound, let $a \in [m,\beta)$ be given and set $\xi = \Xi_\mu(a) \in [0,\infty)$. Next, recall the probability measure ν_ξ described in Lemma 1.3.12, and remember that ν_ξ has mean $a = \Lambda'_\mu(\xi)$ and variance $\Lambda''_\mu(\xi)$. Further, if $\{Y_n : n \in \mathbb{Z}^+\}$ is a sequence of independent, identically distributed random variables with common distribution ν_ξ, then it is an easy matter to check that, for any $n \in \mathbb{Z}^+$ and every $\mathcal{B}_{\mathbb{R}^n}$-measurable $F : \mathbb{R}^n \longrightarrow [0,\infty)$,

$$\mathbb{E}^P\Big[F(Y_1,\dots,Y_n)\Big] = \frac{1}{M(\xi)_\mu^n}\mathbb{E}^P\Big[e^{\xi S_n}\,F(X_1,\dots,X_n)\Big].$$

In particular, if

$$T_n = \sum_{\ell=1}^n Y_\ell \quad \text{and} \quad \overline{T}_n = \frac{T_n}{n},$$

then, because $I_\mu(a) = \xi a - \Lambda_\mu(\xi)$,

$$\begin{aligned} P\Big(\big|\overline{S}_n - a\big| < \epsilon\Big) &= M(\xi)^n \mathbb{E}^P\Big[e^{-\xi T_n},\ \big|\overline{T}_n - a\big| < \epsilon\Big] \\ &\geq e^{-n\xi(a+\epsilon)}\,M(\xi)^n\,P\Big(\big|\overline{T}_n - a\big| < \epsilon\Big) \\ &= \exp\Big[-n\Big(I_\mu(a) + \xi\epsilon\Big)\Big]P\Big(\big|\overline{T}_n - a\big| < \epsilon\Big). \end{aligned}$$

But, because the mean-value and variance of the Y_n's are, respectively, a and $\Lambda''_\mu(\xi)$, (1.2.8) leads to

$$P\Big(\big|\overline{T}_n - a\big| \geq \epsilon\Big) \leq \frac{\Lambda''_\mu(\xi)}{n\epsilon^2}.$$

The case when $a \in (\alpha,m]$ is handled in the same way. □

Results like the ones obtained in Theorem 1.3.13 are examples of a class of results known as **large deviations estimates.** Although large deviation estimates are available in a variety of circumstances,† in general one has to settle for the cruder sort of information contained in the following.

1.3.14 Corollary. *For any $\Gamma \in \mathcal{B}_{\mathbb{R}}$,*

$$-\inf_{x\in\Gamma^\circ} I_\mu(x) \le \varliminf_{n\to\infty} \frac{1}{n}\log\Big[P\big(\overline{S}_n \in \Gamma\big)\Big]$$
$$\le \varlimsup_{n\to\infty} \frac{1}{n}\log\Big[P\big(\overline{S}_n \in \Gamma\big)\Big] \le -\inf_{x\in\overline{\Gamma}} I_\mu(x).$$

(We use Γ° and $\overline{\Gamma}$ to denote the interior and closure of a set Γ. Also, recall that we take the infimum over the empty set to be $+\infty$.)

PROOF: To prove the upper bound, let Γ be a closed set and define $\Gamma_+ = \Gamma \cap [m,\infty)$ and $\Gamma_- = \Gamma \cap (-\infty, m]$. Clearly,

$$P\big(\overline{S}_n \in \Gamma\big) \le 2P\big(\overline{S}_n \in \Gamma_+\big) \vee P\big(\overline{S}_n \in \Gamma_-\big).$$

Moreover, if $\Gamma_+ \ne \emptyset$ and $a_+ = \min\{x : x \in \Gamma_+\}$, then, by Lemma 1.3.12 and Theorem 1.3.13,

$$I_\mu(a_+) = \inf\big\{I_\mu(x) : x \in \Gamma_+\big\} \quad \text{and} \quad P\big(\overline{S}_n \in \Gamma_+\big) \le e^{-nI_\mu(a_+)}.$$

Similarly, if $\Gamma_- \ne \emptyset$ and $a_- = \max\{x : x \in \Gamma_-\}$, then

$$I_\mu(a_-) = \inf\big\{I_\mu(x) : x \in \Gamma_-\big\} \quad \text{and} \quad P\big(\overline{S}_n \in \Gamma_-\big) \le e^{-nI_\mu(a_-)}.$$

Hence, either $\Gamma = \emptyset$, and there is nothing to do anyhow, or

$$P\big(\overline{S}_n \in \Gamma\big) \le 2\exp\big[-n\inf\big\{I_\mu(x) : x \in \Gamma\big\}\big], \quad n \in \mathbb{Z}^+,$$

which certainly implies the asserted upper bound.

To prove the lower bound, assume that Γ is a nonempty open set. What we have to show is that

$$\varliminf_{n\to\infty} \frac{1}{n}\log\Big[P\big(\overline{S}_n \in \Gamma\big)\Big] \ge -I_\mu(a)$$

for every $a \in \Gamma$. If $a \in \Gamma \cap (\alpha,\beta)$, choose $\delta > 0$ so that $(a-\delta, a+\delta) \subseteq \Gamma$ and use the second part of Theorem 1.3.13 to see that

$$\varliminf_{n\to\infty} \frac{1}{n}\log\Big[P\big(\overline{S}_n \in \Gamma\big)\Big] \ge -I_\mu(a) - \epsilon\big|\Xi_\mu(a)\big|$$

† In fact, see, for example, J.-D. Deuschel and D. Stroock, *Large Deviations*, Academic Press Pure Math Series, **137** (1989); some people have written entire books on the subject.

for every $\epsilon \in (0, \delta)$. If $a \notin [\alpha, \beta]$, then $I_\mu(a) = \infty$, and so there is nothing to do. Finally, if $a \in \{\alpha, \beta\}$, then $\mu(\{a\}) = e^{-I_\mu(a)}$ and therefore

$$P(\overline{S}_n \in \Gamma) \geq P(\overline{S}_n = a) \geq e^{-nI_\mu(a)}. \quad \Box$$

1.3.15 Remark. The upper bound in Theorem 1.3.13 is often called **Chernoff's Inequality**. The idea underlying this estimate is rather mundane by comparison to the subtle one used in the proof of the lower bound. Indeed, it may not be immediately obvious what that idea was! Thus, consider once again the second part of the proof of Theorem 1.3.13. What we had to do is estimate the probability that $\overline{S}_n$ lies in a neighborhood of a. When a is the mean-value m, such an estimate is provided by The Weak Law. On the other hand, when $a \neq m$, The Weak Law for the X_n's has very little to contribute. Thus, what we did is replace the original X_n's by random variables Y_n, $n \in \mathbb{Z}^+$, whose mean-value is a. Furthermore, the transformation from the X_n's to the Y_n's was sufficiently simple that it was easy to estimate X_n-probabilities in terms of Y_n-probabilities. Finally, The Weak Law applied to the Y_n's gave strong information about the rate of approach of $\frac{1}{n}\sum_{\ell=1}^n Y_\ell$ to a.

We close this section by verifying the conjecture (cf. the discussion preceding Lemma 1.3.12) that the Gaussian case is *normal*. In particular, we want to check that the *well around* m in which the distribution of $\overline{S}_n$ becomes concentrated looks Gaussian, and, in view of Theorem 1.3.13, this comes down to the following.

1.3.16 Theorem. *Let everything be as in Lemma 1.3.12 and assume that the variance $\sigma^2 > 0$. There exists a $\delta > 0$ and a $K \in (0, \infty)$ such that $[m-\delta, m+\delta] \subseteq (\alpha, \beta)$ (cf. Lemma 1.3.12) and*

$$\left|\Xi_\mu(x)\right| \leq K|x-m| \quad \text{and} \quad \left|I_\mu(x) - \frac{(x-m)^2}{2\sigma^2}\right| \leq K|x-m|^3$$

for all $x \in [m-\delta, m+\delta]$.

PROOF: Without loss in generality (cf. Exercise 1.3.19 below), we will assume that $m = 0$ and $\sigma^2 = 1$. Since, in this case, $\Lambda_\mu(0) = \Lambda_\mu'(0) = 0$ and $\Lambda_\mu''(0) = 1$, it follows that $\Xi_\mu(0) = 0$ and $\Xi_\mu'(0) = 1$. Hence, we can find an $M \in (0, \infty)$ and a $\alpha < -\delta < \delta < \beta$ for which

$$\left|\Xi_\mu(x) - x\right| \leq M|x|^2 \quad \text{and} \quad \left|\Lambda_\mu(\xi) - \frac{\xi^2}{2}\right| \leq M|\xi|^3$$

whenever $|x| \leq \delta$ and $|\xi| \leq (M+1)\delta$, respectively. In particular, this leads immediately to $\left|\Xi_\mu(x)\right| \leq (M+1)|x|$ for $|x| \leq \delta \wedge 1$; and the estimate for I_μ comes easily from the preceding combined with equation $I_\mu(x) = \Xi(x)x - \Lambda_\mu\bigl(\Xi_\mu(x)\bigr)$. $\Box$

Exercises

1.3.17 Exercise: Let $(E, \mathcal{F}, \mu)$ be a measurable space and f a nonnegative, $\mathcal{F}$-measurable function. If either $\mu(E) < \infty$ or f is μ-integrable, show that

$$\|f\|_{L^p(\mu)} \longrightarrow \|f\|_{L^\infty(\mu)} \quad \text{as} \quad p \to \infty.$$

Hint: Handle the case $\mu(E) < \infty$ first, and handle the case when $f \in L^1(\mu)$ by considering the measure $\nu(dx) = f(x)\,\mu(dx)$.

1.3.18 Exercise: It is interesting to see how the proof of estimates like the one in (1.3.2) simplify if one uses moment generating functions. Thus, let $X_1, \dots, X_n$ be independent, identically distributed random variables with mean-value 0 and common distribution μ. Assuming that (1.3.4) holds, show that (1.3.2) is equivalent to the inequality

$$\frac{d^4}{d\xi^4}\left(M_\mu(\xi)^n\right)\Big|_{\xi=0} \le 3n^2 \frac{d^4 M_\mu}{d\xi^4}(0),$$

and use elementary calculus followed by Schwarz's inequality to check this. Finally, argue that, by truncation (cf. the beginning of the proof of Theorem 1.2.10), one can remove the hypothesis in (1.3.4).

1.3.19 Exercise: Referring to the notation used in this section, assume that μ is a nondegenerate (i.e., it is not concentrated at a single point) probability measure on $\mathbb{R}$ for which (1.3.4) holds. Next, let ν denote the distribution of

$$x \in \mathbb{R} \longmapsto \frac{x-m}{\sigma} \in \mathbb{R} \quad \text{under} \quad \mu,$$

and define Λ_ν, I_ν, and Ξ_ν accordingly. Show that

$$\begin{aligned} \Lambda_\mu(\xi) &= \xi m + \Lambda_\nu(\sigma\xi), && \xi \in \mathbb{R}, \\ I_\mu(x) &= I_\nu\left(\frac{x-m}{\sigma}\right), && x \in \mathbb{R}, \\ \text{Image}\big(\Lambda'_\mu\big) &= m + \sigma\,\text{Image}\big(\Lambda'_\nu\big), \end{aligned}$$

and

$$\Xi_\mu(x) = \frac{1}{\sigma}\Xi_\nu\left(\frac{x-m}{\sigma}\right), \quad x \in \text{Image}\big(\Lambda'_\mu\big).$$

1.3.20 Exercise: Continue with the same notation.

(i) Show that $I_\nu \le I_\mu$ if $M_\mu \le M_\nu$.

(ii) Show that

$$I_\mu(x) = \frac{(x-m)^2}{2\sigma^2}, \qquad x \in \mathbb{R},$$

when μ is the $\mathfrak{N}(m,\sigma^2)$ distribution and show that

$$I_\mu(x) = \frac{x-a}{b-a}\log\frac{x-a}{(1-p)(b-a)} + \frac{b-x}{b-a}\log\frac{b-x}{p(b-a)}, \qquad x \in (a,b),$$

when $a < b$, $p \in (0,1)$, and $\mu(\{a\}) = 1 - \mu(\{b\}) = p$.

(iii) When μ is the centered Bernoulli distribution given by $\mu(\{\pm 1\}) = \frac{1}{2}$, show that $M_\mu(\xi) \le \exp\left[-\frac{\xi^2}{2}\right]$, $\xi \in \mathbb{R}$, and conclude that $I_\mu(x) \ge \frac{x^2}{2}$, $x \in \mathbb{R}$. More generally, given $n \in \mathbb{Z}^+$, $\{\sigma_k\}_1^n \subseteq \mathbb{R}$, and independent random variables $X_1, \dots, X_n$ with this μ as their common distribution, let ν denote the distribution of $S \equiv \sum_1^n \sigma_k X_k$ and show that $I_\nu(x) \ge \frac{x^2}{2\Sigma^2}$, where $\Sigma^2 \equiv \sum_1^n \sigma_k^2$. In particular, conclude that

$$P\Big(|S| \ge a\Big) \le 2\exp\left[-\frac{a^2}{2\Sigma^2}\right], \qquad a \in [0,\infty).$$

1.3.21 Exercise: Although it is not exactly the direction in which we have been going, it seems appropriate to include here a derivation of **Stirling's formula**. Namely, recall **Euler's Gamma function**

$$\Gamma(t) \equiv \int_{[0,\infty)} x^{t-1} e^{-x}\, dx, \qquad t \in (-1,\infty). \tag{1.3.22}$$

What we want to prove is that

$$\Gamma(t+1) \sim \sqrt{2\pi t}\left(\frac{t}{e}\right)^t \quad \text{as} \quad t \nearrow \infty, \tag{1.3.23}$$

where the *tilde* "$\sim$" means that the two sides are **asymptotic** to one another in the sense that their ratio tends to 1. (See Exercise 2.1.44 for another approach.)

The first step is to make the problem look like one to which Exercise 1.3.17 is applicable. Thus, make the substitution $x = ty$ to see that

$$\frac{\Gamma(t+1)}{t^{t+1}} = \int_{[0,\infty)} y^t\, e^{-ty}\, dy.$$

In particular, use Exercise 1.3.17 to conclude from this that

$$\left(\frac{\Gamma(t+1)}{t^{t+1}}\right)^{\frac{1}{t}} \longrightarrow e^{-1} \quad \text{as} \quad t \nearrow \infty.$$

This is, of course, far less than we need to know. However, it does show that all the *action* is going to take place near $y = 1$ and that the principal factor in the asymptotics of $\frac{\Gamma(t+1)}{t^{t+1}}$ is e^{-t}. In order to highlight these observations, make the substitution $y = z + 1$ and obtain

$$\frac{\Gamma(t+1)}{t^{t+1}e^{-t}} = \int_{(-1,\infty)} (1+z)^t\, e^{-tz}\, dz.$$

Next, show that for $\delta \in (0,1)$ and $t \in (1,\infty)$:

$$\int_{-1}^{-\delta} (1+z)^t\, e^{-tz}\, dz \le \exp\left[-\frac{t\delta^2}{2}\right],$$

$$\int_{\delta}^{\infty} (1+z)^t\, e^{-tz}\, dz \le 2\Big[(1+\delta)e^{-\delta}\Big]^{t-1} \le 2e^{\delta}\exp\left[-t\left(\frac{\delta^2}{2} - \frac{\delta^3}{3(1-\delta)}\right)\right],$$

and

$$\begin{aligned}\int_{|z|\le\delta} (1+z)^t\, e^{-tz}\, dz &= \int_{|z|\le\delta} e^{-t\frac{z^2}{2}}\, e^{tR(z)}\, dz \\ &= \int_{|z|\le\delta} e^{-t\frac{z^2}{2}}\, dz + t\int_{|z|\le\delta} e^{-t\frac{z^2}{2}} R(z)\, dz + \int_{|z|\le\delta} e^{-t\frac{z^2}{2}} e_2\big(tR(z)\big)\, dz,\end{aligned}$$

where

$$R(z) = \log(1+z) - z - \frac{z^2}{2} = -\sum_{n=3}^{\infty}\frac{(-z)^n}{n} \quad\text{and}\quad e_2(\xi) = e^{\xi} - 1 - \xi.$$

After noting that

$$\int_{|z|\le\delta} e^{-t\frac{z^2}{2}} R(z)\, dz = -\int_{|z|\le\delta} e^{-t\frac{z^2}{2}} z^4 \sum_{m=0}^{\infty}\frac{z^{2m}}{2m+4}\, dz$$

$\big|e_2(\xi)\big| \le \frac{\xi^2}{2}e^{|\xi|}$, and $\int_{|x|\ge r} e^{-\frac{x^2}{2}}\, dx \le \frac{2}{r}e^{-\frac{r^2}{2}}$, conclude that, for $\delta \in \left(0, \frac{3}{7}\right)$,

$$\begin{aligned}&\left|\frac{\Gamma(t+1)}{\sqrt{2\pi t}\left(\frac{t}{e}\right)^t} - 1\right| \\ &\quad\le \sqrt{\frac{t}{2\pi}}(1+2e^{\delta})e^{-\frac{\delta^2 t}{4}} + \frac{1}{\sqrt{2\pi}}\int_{|x|\ge\delta\sqrt{t}} e^{-\frac{x^2}{2}}\, dx \\ &\quad\quad + \frac{(1-\delta^2)^{-1}}{4\sqrt{2\pi}\,t}\int_{\mathbb{R}} x^4 e^{-\frac{x^2}{2}}\, dx + \frac{(1-\delta)^{-2}}{18\sqrt{2\pi}\,t}\int_{\mathbb{R}} x^6 e^{-\frac{x^2}{2}}\, dx \\ &\quad\le \frac{1}{t}\left[\frac{1}{\sqrt{2\pi}}\left((1+e^{\delta})t^{\frac{3}{2}} + \frac{2}{\delta\sqrt{t}}\right)e^{-\frac{\delta^2 t}{2}} + \frac{19}{12}(1-\delta)^{-2}\right].\end{aligned}$$

Finally, take $\delta(t) = (4t)^{-\frac{2}{5}}$, and arrive at the estimate

$$\left|\frac{\Gamma(t+1)}{\sqrt{2\pi t}\left(\frac{t}{e}\right)^t} - 1\right| \le \frac{C}{t}, \quad t \in (2,\infty),$$

for some $C \in (0,\infty)$.

Hint: Note that $x \in (0,\infty) \longmapsto (1+x)e^{-x}$ is decreasing and $x \in (-1,0) \longmapsto (1+x)e^{-x}$ is increasing.

1.3.24 Exercise: Here is a rather different sort of application of large deviation estimates. Namely, inspired by T.H. Carne,† we will show that for each $n \in \mathbb{Z}^+$ and $1 \le m < n$ there exists an $(m-1)$st order polynomial $p_{m,n}$ with the property that

$$|x^n - p_{m,n}(x)| \le 2\exp\left[-\frac{m^2}{2n}\right] \quad \text{for } x \in [-1,1]. \tag{1.3.25}$$

(i) Given a $\mathbb{C}$-valued f on $\mathbb{Z}$, define $\Delta f : \mathbb{Z} \longrightarrow \mathbb{C}$ by

$$\Delta f(n) = \frac{f(n+1)+f(n-1)}{2}, \quad n \in \mathbb{Z}.$$

Show that, for each $z \in \mathbb{C}$, there is a unique sequence $\{Q(m,z) : m \in \mathbb{Z}\} \subseteq \mathbb{C}$ satisfying

$$Q(0,z) = 1 \quad \text{and} \quad \big[\Delta Q(\,\cdot\,,z)\big](m) = zQ(m,z) \text{ for all } m \in \mathbb{Z}.$$

In fact, show that, for each $m \in \mathbb{Z}^+$: $Q(-m,z) = Q(m,z)$, $Q(m,\,\cdot\,)$ is a polynomial of degree m, and

$$Q(m,\cos\theta) = \cos(m\theta), \quad \theta \in \mathbb{C}.$$

In particular, this means that $|Q(n,x)| \le 1$ for all $x \in [-1,1]$. (It also means that $Q(n,\,\cdot\,)$ is the nth **Chebychev polynomial**.)

(ii) Using induction on $n \in \mathbb{Z}^+$, show that

$$\big[\Delta^n Q(\,\cdot\,,z)\big](m) = z^n Q(m,z), \quad m \in \mathbb{Z} \text{ and } z \in \mathbb{C},$$

and conclude that

$$z^n = \mathbb{E}\Big[Q(S_n,z)\Big], \quad n \in \mathbb{Z}^+ \quad \text{and} \quad z \in \mathbb{C},$$

where S_n is the sum of n mutually independent, standard, $\{-1,1\}$-valued Bernoulli random variables. In particular, if

$$p_{m,n}(z) \equiv \mathbb{E}\Big[Q(S_n,z),\, |S_n| < m\Big] = 2^{-n} \sum_{\ell=-m+1}^{m-1} \binom{n}{\ell} Q(\ell,z),$$

† T.H. Carne, "A transformation formula for Markov chains," *Bull. Sc. Math.*, **109**: 399–405 (1985). As Carne points out, what he is doing is discrete analog of Hadamard's representation, via the Weierstrass transform, of solutions to heat equations in terms of solutions to the wave equations.

conclude that (cf. Exercise 1.3.20)

$$|x^n - p_{m,n}(x)| \le P\big(|S_n| \ge m\big) \le 2\exp\left[-\frac{m^2}{2n}\right] \quad \text{for all } 1 \le m \le n.$$

(iii) Suppose that A is a self-adjoint contraction on the Hilbert space H (i.e., $(f, Ag)_H = \overline{(g, Af)_H}$ and $\|Af\|_H \le \|f\|_H$ for all $f,\, g \in H$). Next, assume that $\big(f, A^\ell g\big)_H = 0$ for some $f,\, g \in H$ and each $0 \le \ell < m$. Show that

$$\big|\big(f, A^n g\big)_H\big| \le 2\|f\|_H\|g\|_H \exp\left[-\frac{m^2}{2n}\right] \quad \text{for } n \ge m.$$

(See Exercise 2.2.36 for an application.)

Hint: Note that $\big(f, p_{m,n}(A)g\big)_H = 0$, and use the Spectral Theorem to see that, for any polynomial p,

$$\|p(A)f\|_H \le \sup_{x\in[-1,1]} |p(x)|\, \|f\|_H, \quad f \in H.$$

§1.4: The Strong Law of Large Numbers.

In this section we will discuss a few almost sure convergence properties of partial sums of independent random variables. Thus, once again, $\{X_n\}_1^\infty$ will be a sequence of independent random variables on a probability space $\big(\Omega, \mathcal{F}, P\big)$, and S_n and $\overline{S}_n$ will be, respectively, the sum and average of $X_1, \dots, X_n$. Throughout this section, the reader should notice how much more immediately important a rôle independence (as opposed to orthogonality) plays than it did in Section 1.2.

To get started, we point out that, for both $\{S_n\}_1^\infty$ and $\{\overline{S}_n\}_1^\infty$, the set on which convergence occurs has P-measure either 0 or 1. In fact, we have the following simple application of Kolmogorov's 0–1 Law (Theorem 1.1.2).

1.4.1 Lemma. *For any sequence $\{a_n : n \in \mathbb{Z}^+\} \subseteq \mathbb{R}$ and any sequence $\{b_n : n \in \mathbb{Z}^+\} \subseteq (0,\infty)$ which converges to an element of $(0,\infty]$, the set on which*

$$\lim_{n\to\infty} \frac{S_n - a_n}{b_n} \quad \text{exists in} \quad \mathbb{R}$$

has P-measure either 0 or 1. In fact, if $b_n \longrightarrow \infty$ as $n \to \infty$, then both

$$\overline{\lim_{n\to\infty}}\, \frac{S_n - a_n}{b_n} \quad \text{and} \quad \underline{\lim_{n\to\infty}}\, \frac{S_n - a_n}{b_n}$$

are P-almost surely constant.

PROOF: Simply observe that all of the events and functions involved are tail-measurable. □

Our basic result about the almost sure convergence properties of both $\{S_n\}_1^\infty$ and $\{\overline{S}_n\}_1^\infty$ is the following beautiful statement, which was proved originally by Kolmogorov.

1.4.2 Theorem. *If the X_n's are independent, square P-integrable random variables and if*

$$\sum_{n=1}^{\infty} \operatorname{var}(X_n) < \infty, \tag{1.4.3}$$

then

$$\sum_{n=1}^{\infty} \left(X_n - \mathbb{E}^P[X_n]\right) \quad \textit{converges } P\textit{-almost surely.}$$

Note that, since

$$\sup_{n \geq N} P\left(\left| \sum_{\ell=N}^{n} \left(X_\ell - \mathbb{E}^P[X_\ell]\right) \right| \geq \epsilon \right) \leq \frac{1}{\epsilon^2} \sum_{\ell=N}^{\infty} \operatorname{var}(X_\ell), \tag{1.4.4}$$

(1.4.3) certainly implies that

$$\sum_{n=1}^{\infty} \left(X_n - \mathbb{E}^P[X_n]\right)$$

converges in P-measure. Thus, all that we are trying to do here is replace a convergence in measure statement with an almost sure one. Obviously, this replacement would be trivial if the "$\sup_{n \geq N}$" in (1.4.4) appeared on the other side of P. The remarkable fact which we are about to prove is that, in the present situation, the "$\sup_{n \geq N}$" can be brought inside!

1.4.5 Theorem (Kolmogorov's Inequality). *If the X_n's are independent and square P-integrable, then*

$$P\left(\sup_{n \geq 1} \left| \sum_{\ell=1}^{n} \left(X_\ell - \mathbb{E}^P[X_\ell]\right) \right| \geq \epsilon \right) \leq \frac{1}{\epsilon^2} \sum_{n=1}^{\infty} \operatorname{var}(X_n) \tag{1.4.6}$$

for each $\epsilon > 0$.

PROOF: Without loss in generality, we will assume that each X_n has mean-value 0.

Given $1 \leq n < N$, note that

$$S_N^2 - S_n^2 = (S_N - S_n)^2 + 2(S_N - S_n)S_n \geq 2(S_N - S_n)S_n;$$

and therefore, since $S_N - S_n$ has mean-value 0 and is independent of the σ-algebra $\sigma(X_1, \dots, X_n)$,

$$\text{(1.4.7)} \qquad \mathbb{E}^P\big[S_N^2,\, A_n\big] \ge \mathbb{E}^P\big[S_n^2,\, A_n\big] \quad \text{for any } A_n \in \sigma\big(X_1, \dots, X_n\big).$$

In particular, if $A_1 = \{|S_1| \ge \epsilon\}$ and

$$A_{n+1} = \left\{\big|S_{n+1}\big| \ge \epsilon \text{ and } \max_{1 \le \ell \le n} \big|S_\ell\big| < \epsilon\right\}, \qquad n \in \mathbb{Z}^+,$$

then, the A_n's are mutually disjoint,

$$B_N \equiv \left\{\max_{1 \le n \le N} \big|S_n\big| \ge \epsilon\right\} = \bigcup_{n=1}^N A_n,$$

and so (1.4.7) implies that

$$\begin{aligned}\mathbb{E}^P\big[S_N^2,\, B_N\big] &= \sum_{n=1}^N \mathbb{E}^P\big[S_N^2,\, A_n\big] \ge \sum_{n=1}^N \mathbb{E}^P\big[S_n^2,\, A_n\big] \\ &\ge \epsilon^2 \sum_{n=1}^N P\big(A_n\big) = \epsilon^2 P\big(B_N\big).\end{aligned}$$

In particular,

$$\begin{aligned}\epsilon^2 P\left(\sup_{n \ge 1} \big|S_n\big| \ge \epsilon\right) &= \lim_{N \to \infty} \epsilon^2 P\big(B_N\big) \\ &\le \lim_{N \to \infty} \mathbb{E}^P\big[S_N^2\big] \le \sum_{n=1}^\infty \mathbb{E}^P\big[X_n^2\big]. \quad \square\end{aligned}$$

PROOF OF THEOREM 1.4.2: Again we assume that the X_n's have mean-value 0. By (1.4.6) applied to $\{X_{N+n} : n \in \mathbb{Z}^+\}$, we see that (1.4.3) implies

$$P\left(\sup_{n > N} \big|S_n - S_N\big| \ge \epsilon\right) \le \frac{1}{\epsilon^2} \sum_{n=N+1}^\infty \mathbb{E}^P\big[X_n^2\big] \longrightarrow 0 \quad \text{as} \quad N \to \infty$$

for every $\epsilon > 0$; and this is equivalent to the P-almost sure Cauchy convergence of $\{S_n\}_1^\infty$. $\square$

In order to convert the conclusion in Theorem 1.4.2 into a statement about $\{\overline{S}_n\}_1^\infty$, we will need the following elementary *summability* fact about sequences of real numbers.

1.4.8 Lemma (Kronecker). *Let $\{b_n : n \in \mathbb{Z}^+\}$ be a nondecreasing sequence of positive numbers which tends to ∞, and set $\beta_n = b_n - b_{n-1}$, where $b_0 \equiv 0$. If $\{s_n\}_1^\infty \subseteq \mathbb{R}$ is a sequence which converges to $s \in \mathbb{R}$, then*

$$\frac{1}{b_n}\sum_{\ell=1}^{n}\beta_\ell s_\ell \longrightarrow s.$$

In particular, if $\{x_n\}_1^\infty \subseteq \mathbb{R}$, then

$$\sum_{n=1}^{\infty}\frac{x_n}{b_n} \text{ converges in } \mathbb{R} \implies \frac{1}{b_n}\sum_{\ell=1}^{n} x_\ell \longrightarrow 0 \text{ as } n \to \infty.$$

PROOF: To prove the first part, assume that $s = 0$, and for given $\epsilon > 0$ choose $N \in \mathbb{Z}^+$ so that $|s_\ell| < \epsilon$ for $\ell \geq N$. Then, with $M = \sup_{n\geq 1}|s_n|$,

$$\left|\frac{1}{b_n}\sum_{\ell=1}^{n}\beta_\ell s_\ell\right| \leq \frac{Mb_N}{b_n} + \epsilon \longrightarrow \epsilon$$

as $n \to \infty$.

Turning to the second part, set $y_\ell = \frac{x_\ell}{b_\ell}$, $s_0 = 0$, and $s_n = \sum_{\ell=1}^{n} y_\ell$. After summation by parts,

$$\frac{1}{b_n}\sum_{\ell=1}^{n} x_\ell = s_n - \frac{1}{b_n}\sum_{\ell=1}^{n}\beta_\ell s_{\ell-1};$$

and so, since $s_n \longrightarrow s \in \mathbb{R}$ as $n \to \infty$, the first part gives the desired conclusion. □

After combining Theorem 1.4.2 with Lemma 1.4.8, we arrive at the following interesting statement.

1.4.9 Corollary. *Assume that $\{b_n\}_1^\infty \subseteq (0,\infty)$ increase to infinity as $n \to \infty$, and suppose that $\{X_n\}_1^\infty$ is a sequence of independent, square P-integrable random variables. If*

$$\sum_{n=1}^{\infty}\frac{\operatorname{var}(X_n)}{b_n^2} < \infty, \tag{1.4.10}$$

then

$$\frac{1}{b_n}\sum_{\ell=1}^{n}\Big(X_\ell - \mathbb{E}^P[X_\ell]\Big) \longrightarrow 0 \quad P\text{-almost surely.}$$

As an immediate consequence of the preceding, we see that $\overline{S}_n \longrightarrow m$ P-almost surely if the X_n's are identically distributed and square P-integrable. In fact, without very much additional effort, we can also prove the following much more significant refinement of the last part of Theorem 1.3.1.

1.4.11 Theorem (Kolmogorov's Strong Law). *Let $\{X_n : n \in \mathbb{Z}^+\}$ be a sequence of P-independent, identically distributed random variables. If X_1 is P-integrable and has mean-value m, then, as $n \to \infty$, $\overline{S}_n \longrightarrow m$ P-almost surely and in $L^1(P)$. Conversely, if $\overline{S}_n$ converges (in $\mathbb{R}$) on a set of positive P-measure, then X_1 is P-integrable.*

PROOF: Assume that X_1 is P-integrable and that $\mathbb{E}^P[X_1] = 0$. Next, set $Y_n = X_n \mathbf{1}_{[0,n]}(|X_n|)$, and note that

$$\sum_{n=1}^\infty P(Y_n \neq X_n) = \sum_{n=1}^\infty P(|X_n| > n) \leq \sum_{n=1}^\infty \int_{n-1}^n P(|X_1| > t)\, dt = \mathbb{E}^P[|X_1|] < \infty.$$

Thus, by the first part of the Borel–Cantelli Lemma,

$$P\Big((\exists n \in \mathbb{Z}^+)(\forall N \geq n) Y_N = X_N\Big) = 1.$$

In particular, if $\overline{T}_n = \frac{1}{n}\sum_{\ell=1}^n Y_\ell$ for $n \in \mathbb{Z}^+$, then, for P-almost every $\omega \in \Omega$, $\overline{T}_n(\omega) \longrightarrow 0$ if and only if $\overline{S}_n(\omega) \longrightarrow 0$. Finally, to see that $\overline{T}_n \longrightarrow 0$ P-almost surely, first observe that, because $\mathbb{E}^P[X_1] = 0$, by the first part of Lemma 1.4.8,

$$\lim_{n\to\infty} \frac{1}{n}\sum_{\ell=1}^n \mathbb{E}^P[Y_\ell] = \lim_{n\to\infty} \mathbb{E}^P\Big[X_1,\, |X_1| \leq n\Big] = 0,$$

and therefore, by Corollary 1.4.9, it suffices for us to check that

$$\sum_{n=1}^\infty \frac{\mathbb{E}^P[Y_n^2]}{n^2} < \infty.$$

To this end, set

$$C = \sup_{\ell \in \mathbb{Z}^+} \ell \sum_{n=\ell}^\infty \frac{1}{n^2},$$

and note that

$$\begin{aligned}\sum_{n=1}^\infty \frac{\mathbb{E}^P[Y_n^2]}{n^2} &= \sum_{n=1}^\infty \frac{1}{n^2}\sum_{\ell=1}^n \mathbb{E}^P\Big[X_1^2,\, \ell-1 < |X_1| \leq \ell\Big] \\ &= \sum_{\ell=1}^\infty \mathbb{E}^P\Big[X_1^2,\, \ell-1 < |X_1| \leq \ell\Big]\sum_{n=\ell}^\infty \frac{1}{n^2} \\ &\leq C\sum_{\ell=1}^\infty \frac{1}{\ell}\mathbb{E}^P\Big[X_1^2,\, \ell-1 < |X_1| \leq \ell\Big] \leq C\,\mathbb{E}^P[|X_1|] < \infty.\end{aligned}$$

Thus, the P-almost sure convergence is now established, and the $L^1(P)$-convergence result was proved already in Theorem 1.2.10.

Turning to the converse assertion, first note that (by Lemma 1.4.1) if $\overline{S}_n$ converges in $\mathbb{R}$ on a set of positive P-measure, then it converges P-almost surely to some $m \in \mathbb{R}$. In particular,

$$\overline{\lim_{n\to\infty}} \frac{|X_n|}{n} = \overline{\lim_{n\to\infty}} \left|\overline{S}_n - \overline{S}_{n-1}\right| = 0 \quad P\text{-almost surely;}$$

and so, if $A_n \equiv \{|X_n| > n\}$, then $P\big(\overline{\lim}_{n\to\infty} A_n\big) = 0$. But the A_n's are mutually independent; and therefore, by the second part of the Borel–Cantelli Lemma, we now know that $\sum_{n=1}^{\infty} P(A_n) < \infty$. Hence,

$$\mathbb{E}^P\left[|X_1|\right] = \int_0^{\infty} P\big(|X_1| > t\big)\, dt \le 1 + \sum_{n=1}^{\infty} P\big(|X_n| > n\big) < \infty. \quad \square$$

Although Theorem 1.4.11 is the centerpiece of this section, we still want to give another approach to the study of the almost sure convergence properties of $\{S_n\}_1^{\infty}$. In fact, following P. Lévy, we are going to show that $\{S_n\}_1^{\infty}$ *converges P-almost surely if it converges in P*-measure. Hence, for example, Theorem 1.4.2 can be proved as a direct consequence of (1.4.4), without appeal to Kolmogorov's Inequality.

The key to Lévy's analysis lies in a version of the *reflection principle*, whose statement requires the introduction of a new concept. Namely, given an $\mathbb{R}$-valued random variable Y, we say that $\alpha \in \mathbb{R}$ is a **median** of Y and write $\alpha \in \text{med}(Y)$, if

$$P\big(Y \le \alpha\big) \wedge P\big(Y \ge \alpha\big) \ge \frac{1}{2}. \tag{1.4.12}$$

Notice that (as distinguished from a *mean-value*) every Y admits a median; for example, it is easy to check that

$$\alpha \equiv \inf\left\{t \in \mathbb{R} : P\big(Y \le t\big) \ge \frac{1}{2}\right\}$$

is a median of Y. In addition, it is clear that

$$\text{med}\,(\beta + Y) = \beta + \text{med}\,(Y) \quad \text{for all } \beta \in \mathbb{R}. \tag{1.4.13}$$

On the other hand, the notion of median is flawed by the fact that, in general, a random variable will admit an entire nondegenerate interval of medians. In addition, it is neither easy to compute the medians of a sum in terms of the medians of the summands nor to relate the *medians* of an integrable random variable to its *mean-value*. Nonetheless, at least if $Y \in L^p(P)$ for some $p \in [1,\infty)$, the following estimate provides some information. Namely, since, for $\alpha \in \operatorname{med}(Y)$ and $\beta \in \mathbb{R}$,

$$\frac{|\alpha-\beta|^p}{2} \le |\alpha-\beta|^p P(Y \ge \alpha) \wedge P(Y \le \alpha) \le \mathbb{E}^P\big[|Y-\beta|^p\big],$$

we see that, for any $p \in [1,\infty)$ and $Y \in L^p(P)$,

$$|\alpha-\beta| \le \Big(2\mathbb{E}^P\big[|Y-\beta|^p\big]\Big)^{\frac{1}{p}} \text{ for all } \beta \in \mathbb{R} \quad \text{and} \quad \alpha \in \operatorname{med}(Y).$$

In particular, if $Y \in L^2(P)$ and m is the mean-value of Y, then

$$|\alpha - m| \le \sqrt{2\operatorname{var}(Y)}. \tag{1.4.14}$$

1.4.15 Theorem (Lévy). *Let $\{X_n : n \in \mathbb{Z}^+\}$ be a sequence of P-independent random variables, and, for $k \le \ell$, choose $\alpha_{\ell,k} \in \operatorname{med}(S_\ell - S_k)$. Then, for any $N \in \mathbb{Z}^+$ and $\epsilon > 0$,*

$$P\left(\max_{1\le n\le N}(S_n + \alpha_{N,n}) \ge \epsilon\right) \le 2P(S_N \ge \epsilon); \tag{1.4.16}$$

and therefore

$$P\left(\max_{1\le n\le N}|S_n + \alpha_{N,n}| \ge \epsilon\right) \le 2P(|S_N| \ge \epsilon). \tag{1.4.17}$$

PROOF: Clearly (1.4.17) follows by applying (1.4.16) to both the sequences $\{X_n\}_1^\infty$ and $\{-X_n\}_1^\infty$ and then adding the two results.

To prove (1.4.16), set $A_1 = \{S_1 + \alpha_{N,1} \ge \epsilon\}$ and

$$A_{n+1} = \left\{\max_{1\le \ell\le n}(S_\ell + \alpha_{N,\ell}) < \epsilon \text{ and } S_{n+1} + \alpha_{N,n+1} \ge \epsilon\right\}$$

for $1 \le n < N$. Obviously, the A_n's are mutually disjoint and

$$\bigcup_{n=1}^N A_n = \left\{\max_{1\le n\le N}(S_n + \alpha_{N,n}) \ge \epsilon\right\}.$$

In addition,

$$\{S_N \ge \epsilon\} \supseteq A_n \cap \{S_N - S_n \ge \alpha_{N,n}\} \quad \text{for each } 1 \le n \le N.$$

Hence,

$$\begin{aligned} P(S_N \ge \epsilon) &\ge \sum_{n=1}^N P\Big(A_n \cap \{S_N - S_n \ge \alpha_{N,n}\}\Big) \\ &\ge \frac{1}{2}\sum_{n=1}^N P(A_n) = \frac{1}{2}P\left(\max_{1\le n\le N}(S_n + \alpha_{N,n}) \ge \epsilon\right), \end{aligned}$$

where, in the passage to the last line, we have used the independence of the sets A_n and $\{S_N - S_n \ge \alpha_{N,n}\}$. □

1.4.18 Corollary (Lévy). *Let $\{X_n : n \in \mathbb{Z}^+\}$ be a sequence of independent random variables, and assume that $\{S_n : n \in \mathbb{Z}^+\}$ converges in P-measure to an $\mathbb{R}$-valued random variable S. Then $S_n \longrightarrow S$ P-almost surely.*

PROOF: What we must show is that, for each $\epsilon > 0$, there is an $M \in \mathbb{Z}^+$ such that

$$\sup_{N\geq 1} P\Big(\max_{1\leq \ell\leq N} \big|S_{\ell+M} - S_M\big| \geq \epsilon\Big) < \epsilon.$$

To this end, first choose $\alpha_{\ell,k}$ as in the statement of Theorem 1.4.15 and note that, because $\{S_n\}_1^\infty$ converges in P-measure, there is an $M_1 \in \mathbb{Z}^+$ such that $|\alpha_{\ell,k}| \leq \frac{\epsilon}{2}$ for all $M_1 \leq k \leq \ell$. In addition, there is an $M_2 \in \mathbb{Z}^+$ for which

$$\sup_{N\geq 1} P\Big(\big|S_{N+M} - S_M\big| \geq \frac{\epsilon}{2}\Big) < \frac{\epsilon}{2} \quad \text{for each} \quad M \geq M_2.$$

Now set $M = M_1 \vee M_2$. Then, for $N \geq 1$,

$$\begin{aligned}
&P\left(\max_{1\leq \ell\leq N} \big|S_{\ell+M} - S_M\big| \geq \epsilon\right) \\
&\qquad \leq P\left(\max_{1\leq \ell\leq N} \big|S_{\ell+M} - S_M + \alpha_{N+M,\ell+M}\big| \geq \frac{\epsilon}{2}\right) \\
&\qquad \leq 2P\Big(\big|S_{N+M} - S_M\big| \geq \frac{\epsilon}{2}\Big) < \epsilon,
\end{aligned}$$

where, in passing to the last line, we have applied Theorem 1.4.15 to the sequence $\{X_{n+M} : n \in \mathbb{Z}^+\}$. □

1.4.19 Remark. The most beautiful and startling feature of Lévy's line of reasoning is that it requires *no* integrability assumptions. Of course, in many applications of Corollary 1.4.18, integrability considerations enter into the proof that $\{S_n\}_1^\infty$ converges in P-measure. Finally, a word of caution may be in order. Namely, the result in Corollary 1.4.18 applies to the quantities S_n themselves; it does *not* apply to associated quantities like $\overline{S}_n$! Indeed, suppose that $\{X_n\}_1^\infty$ is a sequence of independent random variables with common distribution satisfying

$$P(X_n \leq -t) = P(X_n \geq t) = \Big(\big(1+t^2\big)\log\big(e^4+t^2\big)\Big)^{-\frac{1}{2}} \quad \text{for all } t \geq 0.$$

On the one hand, by Exercise 1.2.15, we know that the associated averages $\overline{S}_n$ tend to 0 in probability. On the other hand, by the second part of Theorem 1.4.11, we know that the sequence $\{\overline{S}_n\}_1^\infty$ diverges almost surely.

Exercises

1.4.20 Exercise: Let X and Y be nonnegative random variables, and suppose that

$$P(X \geq t) \leq \frac{1}{t}\mathbb{E}^P\Big[Y,\, X \geq t\Big], \quad t \in (0,\infty). \tag{1.4.21}$$

Show that

$$\left(\mathbb{E}^P\big[X^p\big]\right)^{\frac{1}{p}} \leq \frac{p}{p-1}\left(\mathbb{E}^P\big[Y^p\big]\right)^{\frac{1}{p}}, \quad p \in (1,\infty). \tag{1.4.22}$$

Hint: Recall that, for any measure space $(E, \mathcal{F}, \mu)$, any nonnegative, measurable f on $(E, \mathcal{F})$, and any $\alpha \in (0, \infty)$,

$$\int_E f(x)^\alpha \,\mu(dx) = \alpha \int_{(0,\infty)} t^{\alpha-1}\,\mu(f > t)\,dt. \tag{1.4.23}$$

Use this together with (1.4.21) to justify the relation

$$\mathbb{E}^P\big[X^p\big] \leq p \int_{(0,\infty)} t^{p-2}\,\mathbb{E}^P\Big[Y,\, X \geq t\Big] = \frac{p}{p-1}\mathbb{E}^P\Big[X^{p-1}Y\Big];$$

and arrive at (1.4.22) after an application of Hölder's inequality.

1.4.24 Exercise: Let $\{X_n\}_1^\infty$ be a sequence of mutually independent, square P-integrable random variables with mean value 0, and assume that $\sum_1^\infty E\big[X_n^2\big] < \infty$. Let S denote the random variable (guaranteed by Theorem 1.4.2) to which $\{S_n\}_1^\infty$ converges P-almost surely, and, using elementary orthogonality considerations, check that $S_n \longrightarrow S$ in $L^2(P)$ as well. Next, after examining the proof of Kolmogorov's inequality (cf. (1.4.6)), show that

$$P\left(\sup_{n\in\mathbb{Z}^+} |S_n| \geq \epsilon\right) \leq \frac{1}{\epsilon^2}\mathbb{E}^P\left[S^2,\, \sup_{n\in\mathbb{Z}^+}|S_n| \geq \epsilon\right], \quad \epsilon > 0. \tag{1.4.25}$$

Finally, by applying (1.4.22), show that

$$\mathbb{E}^P\left[\sup_{n\in\mathbb{Z}^+} |S_n|^{2p}\right] \leq \left(\tfrac{p}{p-1}\right)^p \mathbb{E}^P\Big[|S|^{2p}\Big], \quad p \in (1,\infty); \tag{1.4.26}$$

and conclude from this that, for each $p \in (2,\infty)$, $\{S_n\}_1^\infty$ converges to S in $L^p(P)$ if and only if $S \in L^p(P)$.

1.4.27 Exercise: The following variant of (1.4.17) is sometimes useful and has the advantage that it avoids the introduction of medians. Namely show that for any $t \in (0,\infty)$ and $n \in \mathbb{Z}^+$:

$$P\Big(\max_{1\leq m\leq n} |S_n| \geq 2t\Big) \leq \frac{P\big(|S_m| > t\big)}{1 - \max\limits_{1\leq m\leq n} P\big(|S_n - S_m| > t\big)}. \tag{1.4.28}$$

Note that (1.4.28) can be used in place of (1.4.17) when proving results like the one in Corollary 1.4.18.

1.4.29 Exercise: A random variable X is said to by **symmetric** if $-X$ has the same distribution as X itself. Obviously, the most natural choice of median for a symmetric random variable is 0; and thus, because sums of independent, symmetric random variables are again symmetric, (1.4.16) and (1.4.17) are particularly interesting when the X_n's are symmetric, since the $\alpha_{\ell,k}$'s can then be taken to be 0. In this connection, we present the following interesting variation on the theme of Theorem 1.4.15.

(i) Let $X_1, \dots, X_n, \dots$ be independent, symmetric random variables, set $M_n(\omega) = \max_{1\le\ell\le n} |X_\ell(\omega)|$, let $\tau_n(\omega)$ be the smallest $1 \le \ell \le n$ with the property that $\big|X_\ell(\omega)\big| = M_n(\omega)$, and define

$$Y_n(\omega) = X_{\tau_n(\omega)}(\omega) \quad \text{and} \quad \hat{S}_n = S_n - Y_n.$$

Show that

$$\omega \in \Omega \longmapsto \big(\hat{S}_n(\omega), Y_n(\omega)\big) \in \mathbb{R}^2 \quad \text{and} \quad \omega \in \Omega \longmapsto \big(-\hat{S}_n(\omega), Y_n(\omega)\big) \in \mathbb{R}^2$$

have the same distribution, and conclude first that

$$\begin{aligned} P(Y_n \ge t) &\le P\Big(Y_n \ge t \;\&\; \hat{S}_n \ge 0\Big) + P\Big(Y_n \ge t \;\&\; \hat{S}_n \le 0\Big) \\ &= 2P\Big(Y_n \ge t \;\&\; \hat{S}_n \ge 0\Big) \le 2P(S_n \ge t), \end{aligned}$$

for all $t \in \mathbb{R}$; and then that

$$P\left(\max_{1\le\ell\le n} |X_\ell| \ge t\right) \le 2P\Big(\big|S_n\big| \ge t\Big), \quad t \in [0,\infty). \tag{1.4.30}$$

(ii) Continuing in the same setting, add the assumption that the X_n's are identically distributed, and use (1.4.30) to show that

$$\begin{aligned} &\lim_{n\to\infty} P\Big(\big|\overline{S}_n\big| \le C\Big) = 1 \quad \text{for some } C \in (0,\infty) \\ &\qquad \Longrightarrow \lim_{n\to\infty} nP\Big(\big|X_1\big| \ge n\Big) = 0. \end{aligned}$$

Hence, (cf. Exercise 1.2.15) *if* $\{X_n\}_1^\infty$ *is a sequence of independent, identically distributed symmetric random variables, then* $\overline{S}_n \longrightarrow 0$ *in* P*-probability if and only if* $\lim_{n\to\infty} nP\big(|X_1| \ge n\big) = 0$.

1.4.31 Exercise: Let $X_1, \dots, X_n, \dots$ be a sequence of mutually independent, identically distributed, P-integrable random variables with mean-value m. As we already know, when $m > 0$, the partial sums S_n tend, P-almost surely, to $+\infty$ at an asymptotic linear rate m; and, of course, when $m < 0$ the situation is similar at $-\infty$. Moreover, when $m = 0$, we know that, if $|S_n|$ tends to ∞ at all, then, P-almost surely, it does so at a strictly sublinear rate. In this exercise, we will sharpen this statement by proving that

$$m = 0 \implies \varliminf_{n\to\infty} |S_n| < \infty \quad P\text{-almost surely.}$$

The beautiful argument given below is due to Y. Guivarc'h, but it's full power cannot be appreciated in the present context (cf. Exercise 6.1.34). Indeed, as we will see in the next section, at least when they are square integrable, much more precise information can be given about sums of independent random variables.

In order to prove the assertion here, assume that $\lim_{n\to\infty} |S_n| = \infty$ with positive P-probability, use Kolmogorov's 0–1 Law to see that $|S_n| \longrightarrow \infty$ P-almost surely, and proceed as follows.

(**i**) Show that there must exist an $\epsilon > 0$ with the property that

$$P\Big(\forall \ell > k \ |S_\ell - S_k| \ge \epsilon\Big) \ge \epsilon$$

for some $k \in \mathbb{Z}^+$ and therefore that

$$P(A) \ge \epsilon \quad \text{where } A \equiv \Big\{\omega : \forall \ell \in \mathbb{Z}^+ \ |S_\ell(\omega)| \ge \epsilon\Big\}.$$

(**ii**) For each $\omega \in \Omega$ and $n \in \mathbb{Z}^+$, set

$$\Gamma_n(\omega) = \Big\{t \in \mathbb{R} : \exists 1 \le \ell \le n \ |t - S_\ell(\omega)| < \tfrac{\epsilon}{2}\Big\}$$

and

$$\Gamma_n'(\omega) = \Big\{t \in \mathbb{R} : \exists 1 \le \ell \le n \ |t - S_\ell'(\omega)| < \tfrac{\epsilon}{2}\Big\},$$

where $S_n' \equiv \sum_{\ell=1}^n X_{\ell+1}$. Next, let $R_n(\omega)$ and $R_n'(\omega)$ denote the Lebesgue measure of $\Gamma_n(\omega)$ and $\Gamma_n'(\omega)$, respectively; and, using the translation invariance of Lebesgue's measure, show that

$$R_{n+1}(\omega) - R_n'(\omega) \ge \epsilon \mathbf{1}_{A'}(\omega),$$
$$\text{where } A' \equiv \Big\{\omega : \forall \ell \ge 2 \ |S_\ell(\omega) - S_1(\omega)| \ge \epsilon\Big\}.$$

On the other hand, show that

$$\mathbb{E}^P\big[R_n'\big] = \mathbb{E}^P\big[R_n\big] \quad \text{and} \quad P(A') = P(A);$$

and conclude first that

$$\epsilon P(A) \leq \mathbb{E}^P\big[R_{n+1} - R_n\big], \quad n \in \mathbb{Z}^+,$$

and then that

$$\epsilon P(A) \leq \varliminf_{n\to\infty} \frac{1}{n}\mathbb{E}^P\big[R_n\big].$$

(iii) In view of parts **(i)** and **(ii)**, we will be done once we show that

$$m = 0 \implies \lim_{n\to\infty} \frac{1}{n}\mathbb{E}^P\big[R_n\big] = 0.$$

But clearly, $0 \leq R_n(\omega) \leq n\epsilon$. Thus, it is enough for us to show that, when $m = 0$, $\frac{R_n}{n} \longrightarrow 0$ P-almost surely; and, to this end, first check that

$$\frac{S_n(\omega)}{n} \longrightarrow 0 \implies \frac{R_n(\omega)}{n} \longrightarrow 0,$$

and, finally, apply The Strong Law of Large Numbers.

§1.5: Law of the Iterated Logarithm.

Let $X_1, \ldots, X_n, \ldots$ be a sequence of independent, identically distributed random variables with mean-value 0 and variance 1. In this section, we will investigate exactly how large $\{S_n : n \in \mathbb{Z}^+\}$ can become as $n \to \infty$. To get a feeling for what we should be expecting, first note that, by Corollary 1.4.9, for any nondecreasing $\{b_n\}_1^\infty \subseteq (0,\infty)$,

$$\frac{S_n}{b_n} \longrightarrow 0 \quad P\text{-almost surely if} \quad \sum_{n=1}^\infty \frac{1}{b_n^2} < \infty.$$

Thus, for example, S_n grows more slowly than $n^{\frac{1}{2}}\log n$. On the other hand, if the X_n's are $\mathfrak{N}(0,1)$-random variables, then so are the random variables $\frac{S_n}{\sqrt{n}}$; and therefore, for every $R \in (0,\infty)$,

$$\begin{aligned} P\left(\varlimsup_{n\to\infty} \frac{S_n}{\sqrt{n}} \geq R\right) &= \lim_{N\to\infty} P\left(\bigcup_{n\geq N}\left\{\frac{S_n}{\sqrt{n}} \geq R\right\}\right) \\ &\geq \lim_{N\to\infty} P\left(\frac{S_N}{\sqrt{N}} \geq R\right) > 0. \end{aligned}$$

Hence, at least for normal random variables, we can use Lemma 1.4.1 to see that

$$\overline{\lim_{n\to\infty}} \frac{S_n}{\sqrt{n}} = \infty \quad P\text{-almost surely;}$$

and so S_n grows faster than $n^{\frac{1}{2}}$.

If, as we did in Section 1.3, we proceed on the assumption that Gaussian random variables are typical, we should expect the growth rate of the S_n's to be something between $n^{\frac{1}{2}}$ and $n^{\frac{1}{2}} \log n$. What, in fact, turns out to be the precise growth rate is

$$\Lambda_n \equiv \sqrt{2n \log_{(2)}(n \vee 3)} \tag{1.5.1}$$

where $\log_{(2)} x \equiv \log\bigl(\log x\bigr)$ (*not* the logarithm with base 2) for $x \in [e, \infty)$. That is, one has the **Law of the Iterated Logarithm**:

$$\overline{\lim_{n\to\infty}} \frac{S_n}{\Lambda_n} = 1 \quad P\text{-almost surely.} \tag{1.5.2}$$

This remarkable fact was discovered first for Bernoulli random variables by Khinchine, was extended by Kolmogorov to random variables possessing $2 + \epsilon$ moments, and eventually achieved its final form in the work of Hartman and Wintner. The approach which we will adopt here is based on ideas (taught to the author by M. Ledoux) introduced originally to handle generalizations of (1.5.2) to random variables with values in a Banach space.† This approach consists of two steps. The first establishes a preliminary version of (1.5.2) which, although it is far cruder than (1.5.2) itself, will allow us to justify a reduction of the general case to the case of bounded random variables. In the second step, we deal with bounded random variables and more or less follow Khinchine's strategy for deriving (1.5.2) once one has estimates like the ones provided by Theorem 1.3.13.

In what follows, we will use $[\beta] \equiv \max\bigl\{n \in \mathbb{Z} : n \le \beta\bigr\}$ to denote the integer part of $\beta \in \mathbb{R}$ and will define

$$\Lambda_\beta = \Lambda_{[\beta]} \quad \text{and} \quad \tilde{S}_\beta = \frac{S_{[\beta]}}{\Lambda_\beta} \quad \text{for} \quad \beta \in [3, \infty).$$

1.5.3 Lemma. *Let $\bigl\{X_n\bigr\}_1^\infty$ be a sequence of independent, identically distributed random variables with mean-value* 0 *and variance* 1. *Then, for any $a \in (0, \infty)$ and $\beta \in (1, \infty)$,*

$$\overline{\lim_{n\to\infty}} \bigl|\tilde{S}_n\bigr| \le a \quad \text{(a.s., } P\text{)} \quad \textit{if} \quad \sum_{m=1}^{\infty} P\Bigl(\bigl|\tilde{S}_{\beta^m}\bigr| \ge a\,\beta^{-\frac{1}{2}}\Bigr) < \infty.$$

† See M. Ledoux and M. Talagrand, *Probability in Banach Spaces*, Springer–Verlag Ergebnisse Series 3.Folge·Band 23 (1991).

PROOF: Let $\beta \in (1, \infty)$ be given and, for each $m \in \mathbb{N}$ and $1 \le n \le \beta^m$, let $\alpha_{m,n}$ be a median (cf. (1.4.12)) of $S_{[\beta^m]} - S_n$. Noting that, by (1.4.14), $|\alpha_{m,n}| \le \sqrt{2\beta^m}$, we see that

$$\varlimsup_{n\to\infty} |\tilde{S}_n| = \varlimsup_{m\to\infty} \max_{\beta^{m-1}\le n\le\beta^m} |\tilde{S}_n|$$
$$\le \beta^{\frac{1}{2}} \varlimsup_{m\to\infty} \max_{\beta^{m-1}\le n\le\beta^m} \frac{|S_n|}{\Lambda_{\beta^m}}$$
$$\le \beta^{\frac{1}{2}} \varlimsup_{m\to\infty} \max_{n\le\beta^m} \frac{|S_n + \alpha_{m,n}|}{\Lambda_{\beta^m}};$$

and therefore,

$$P\left(\varlimsup_{n\to\infty} |\tilde{S}_n| \ge a\right) \le P\left(\varlimsup_{m\to\infty} \max_{n\le\beta^m} \frac{|S_n + \alpha_{m,n}|}{\Lambda_{\beta^m}} \ge a\,\beta^{-\frac{1}{2}}\right).$$

But, by Theorem 1.4.15,

$$P\left(\max_{n\le\beta^m} \frac{|S_n + \alpha_{m,n}|}{\Lambda_{\beta^m}} \ge a\beta^{-\frac{1}{2}}\right) \le 2P\left(|\tilde{S}_{\beta^m}| \ge a\beta^{-\frac{1}{2}}\right),$$

and so the desired result follows from the Borel–Cantelli Lemma. □

1.5.4 Lemma. *For any sequence $\{X_n\}_1^\infty$ of independent, identically distributed random variables with mean-value 0 and variance σ^2,*

$$\varlimsup_{n\to\infty} |\tilde{S}_n| \le 8\sigma \quad \text{(a.s., } P). \tag{1.5.5}$$

PROOF: Without loss in generality, we assume throughout that $\sigma = 1$; and, for the moment, we will also assume that the X_n's are symmetric (cf. Exercise 1.4.29). By Lemma 1.5.3, we will know that (1.5.5) holds with 8 replaced by 4 once we show that

$$\sum_{m=0}^{\infty} P\left(|\tilde{S}_{2^m}| \ge 2^{\frac{3}{2}}\right) < \infty. \tag{1.5.6}$$

In order to take maximal advantage of symmetry, let $(\Omega, \mathcal{F}, P)$ be the probability space on which the X_n's are defined, use $\{R_n\}_1^\infty$ to denote the sequence of Rademacher functions on $[0,1)$ introduced in Section 1.1, and set $Q = \lambda_{[0,1)} \times P$ on $\big([0,1) \times \Omega, \mathcal{B}_{[0,1)} \times \mathcal{F}\big)$. It is then an easy matter to check that symmetry of the X_n's is equivalent to the statement that

$$\omega \in \Omega \longrightarrow \big(X_1(\omega), \dots, X_n(\omega), \dots\big) \in \mathbb{R}^{\mathbb{Z}^+}$$

has the same distribution under P as

$$(t,\omega) \in [0,1) \times \Omega \longmapsto \big(R_1(t)X_1(\omega), \dots, R_n(t)X_n(\omega), \dots\big) \in \mathbb{R}^{\mathbb{Z}^+}$$

does under Q. Next, using the last part of **(iii)** in Exercise 1.3.20 with $\sigma_k = X_k(\omega)$, note that:

$$\begin{aligned}\lambda_{[0,1)}\left(\left\{t \in [0,1) : \left|\sum_{n=1}^{2^m} R_n(t)X_n(\omega)\right| \geq a\right\}\right)&\\ \leq 2\exp\left[-\frac{a^2}{2\sum_{n=1}^{2^m} X_n(\omega)^2}\right]&, \quad a \in [0,\infty) \text{ and } \omega \in \Omega.\end{aligned}$$

Hence, if

$$A_m \equiv \left\{\omega \in \Omega : \frac{1}{2^m}\sum_{n=1}^{2^m} X_m(\omega)^2 \geq 2\right\}$$

and

$$F_m(\omega) \equiv \lambda_{[0,1)}\left(\left\{t \in [0,1) : \left|\sum_{n=1}^{2^m} R_n(t)X_n(\omega)\right| \geq 2^{\frac{3}{2}}\Lambda_{2^m}\right\}\right),$$

then, by Tonelli's Theorem,

$$\begin{aligned}P\Big(\Big\{\omega \in \Omega : \big|S_{2^m}(\omega)\big| \geq 2^{\frac{3}{2}}\Lambda_{2^m}\Big\}\Big) &= \int_\Omega F_m(\omega)\, P(d\omega)\\ \leq 2\int_\Omega \exp\left[-\frac{8\Lambda_{2^m}^2}{2\sum_{n=1}^{2^m} X_n(\omega)^2}\right] P(d\omega) &\leq 2\exp\Big[-4\log_{(2)} 2^m\Big] + 2P\big(A_m\big).\end{aligned}$$

Thus, (1.5.6) comes down to proving that $\sum_{m=0}^{\infty} P\big(A_m\big) < \infty$, and in order to check this we argue in much the same way as we did when we proved the converse statement in Kolmogorov's Strong Law. Namely, set

$$T_m = \sum_{n=1}^{2^m} X_n^2, \quad B_m = \left\{\frac{T_{m+1} - T_m}{2^m} \geq 2\right\}, \quad \text{and} \quad \overline{T}_m = \frac{T_m}{2^m}$$

for $m \in \mathbb{N}$. Clearly, $P\big(A_m\big) = P\big(B_m\big)$. Moreover, the sets B_m, $m \in \mathbb{N}$, are mutually independent; and therefore, by the Borel–Cantelli Lemma, we need only check that

$$P\Big(\varlimsup_{m\to\infty} B_m\Big) = P\left(\varlimsup_{m\to\infty} \frac{T_{m+1} - T_m}{2^m} \geq 2\right) = 0.$$

But, by The Strong Law, we know that $\overline{T}_m \longrightarrow 1$ (a.s., P), and therefore it is clear that

$$\frac{T_{m+1} - T_m}{2^m} \longrightarrow 1 \quad (\text{a.s.}, P).$$

We have now proved (1.5.5) with 4 replacing 8 for symmetric random variables. To eliminate the symmetry assumption, again let $(\Omega, \mathcal{F}, P)$ be the probability space on which the X_n's are defined, let $(\Omega', \mathcal{F}', P')$ be a second copy of the same space, and consider the random variables

$$(\omega, \omega') \in \Omega \times \Omega' \longmapsto Y_n(\omega, \omega') \equiv \frac{X_n(\omega) - X_n(\omega')}{\sqrt{2}}$$

under the measure $Q \equiv P \times P'$. Since the Y_n's are obviously symmetric, the result which we have already proved says that

$$\varlimsup_{n\to\infty} \frac{|S_n(\omega) - S_n(\omega')|}{\Lambda_n} \le 2^{\frac{5}{2}} \le 8 \quad \text{for } Q\text{-almost every } (\omega, \omega') \in \Omega \times \Omega'.$$

Now suppose that $\varlimsup_{n\to\infty} \frac{|S_n|}{\Lambda_n} > 8$ on a set of positive P-measure. Then, by Kolmogorov's 0–1 Law, there would exist an $\epsilon > 0$ such that

$$\varlimsup_{n\to\infty} \frac{|S_n(\omega)|}{\Lambda_n} \ge 8 + \epsilon \quad \text{for } P\text{-almost every } \omega \in \Omega;$$

and so, by Fubini's Theorem,† we would have that, for Q-almost every $(\omega, \omega') \in \Omega \times \Omega'$, there is a $\{n_m(\omega) : m \in \mathbb{Z}^+\} \subseteq \mathbb{Z}^+$ such that $n_m(\omega) \nearrow \infty$ and

$$\begin{aligned}
&\varliminf_{m\to\infty} \frac{|S_{n_m(\omega)}(\omega')|}{\Lambda_{n_m(\omega)}} \\
&\qquad \ge \lim_{m\to\infty} \frac{|S_{n_m(\omega)}(\omega)|}{\Lambda_{n_m(\omega)}} - \varlimsup_{m\to\infty} \frac{|S_{n_m(\omega)}(\omega) - S_{n_m(\omega)}(\omega')|}{\Lambda_{n_m(\omega)}} \ge \epsilon.
\end{aligned}$$

But, again by Fubini's Theorem, this would mean that there exists a $\{n_m : m \in \mathbb{Z}^+\} \subseteq \mathbb{Z}^+$ such that $n_m \nearrow \infty$ and $\varliminf_{m\to\infty} \frac{|S_{n_m}(\omega')|}{\Lambda_{n_m}} \ge \epsilon$ for P'-almost every $\omega' \in \Omega'$; and obviously this contradicts

$$\mathbb{E}^{P'}\left[\left(\frac{S_n}{\Lambda_n}\right)^2\right] = \frac{1}{2\log_{(2)} n} \longrightarrow 0. \quad \square$$

We have now got the *crude statement* alluded to above. In order to get the more precise statement contained in (1.5.2), we will need the following application of the results in Section 3.

---

† This is Fubini at his best and subtlest. Namely, we are using Fubini to switch between *horizontal* and *vertical* sets of measure 0.

1.5.7 Lemma. *Let $\{X_n\}_1^\infty$ be a sequence of independent random variables with mean-value 0, variance 1, and common distribution μ. Further, assume that (1.3.4) holds. Then, for each $R \in (0,\infty)$ and $\gamma \in (0,1)$, there is an $N(R,\gamma) \in \mathbb{Z}^+$ such that*

$$P\Big(|\tilde{S}_n| \geq R\Big) \leq 2\exp\Big[-\gamma R^2 \log_{(2)} n\Big] \quad \text{for } n \geq N(R,\gamma). \tag{1.5.8}$$

In addition, for each $\epsilon \in (0,1]$ and $\gamma \in (1,\infty)$, there is an $N(\epsilon,\gamma) \in \mathbb{Z}^+$ such that, for all $n \geq N(\epsilon,\gamma)$ and $|a| \leq \frac{1}{\epsilon}$,

$$P\Big(|\tilde{S}_n - a| < \epsilon\Big) \geq \frac{1}{2}\exp\Big[-|a|\Big(\gamma|a| + 2K\epsilon\Big)\log_{(2)} n\Big], \tag{1.5.9}$$

where the constant $K \in (0,\infty)$ is the one in Theorem 1.3.16.

PROOF: Set

$$\lambda_n = \frac{\Lambda_n}{n} = \left(\frac{2\log_{(2)}(n \vee 3)}{n}\right)^{\frac{1}{2}}.$$

To prove (1.5.8), simply apply the first part of Theorem 1.3.13 to see that, for all $n \in \mathbb{Z}^+$,

$$\begin{aligned} P\Big(|\tilde{S}_n| \geq R\Big) &= P\Big(|\overline{S}_n| \geq R\lambda_n\Big) \\ &\leq 2\exp\Big[-n\, I_\mu(R\lambda_n) \wedge I_\mu(-R\lambda_n)\Big]. \end{aligned}$$

Since $\lambda_n \longrightarrow 0$ as $n \to \infty$ and, by Theorem 1.3.16, $\frac{I_\mu(x)}{x^2} \longrightarrow \frac{1}{2}$ as $x \to 0$, (1.5.8) follows.

To prove (1.5.9), first note that

$$P\Big(|\tilde{S}_n - a| < \epsilon\Big) = P\Big(|\overline{S}_n - a_n| < \epsilon_n\Big),$$

where $a_n = a\lambda_n$ and $\epsilon_n = \epsilon\lambda_n$. Next, take K and δ to be as in the statement of Theorem 1.3.16, and choose $N_1 \geq 3$ so that $|\lambda_n| < \epsilon\,\delta$ for $n \geq N_1$. Then, by the second part of Theorem 1.3.13 combined with Theorem 1.3.16

$$\begin{aligned} &P\Big(|\overline{S}_n - a_n| < \epsilon_n\Big) \\ &\geq \left(1 - \frac{\Lambda_\mu''(\Xi_\mu(a_n))}{n\epsilon_n^2}\right)\exp\Big[-n\Big(I_\mu(a_n) + \epsilon_n\,|\Xi_\mu(a_n)|\Big)\Big] \\ &\geq \tfrac{1}{2}\exp\left[-a^2\log_{(2)} n\left(1 + K\sqrt{\frac{8\log_{(2)} n}{n}}|a|\right) - 2K|a|\epsilon\log_{(2)} n\right] \end{aligned}$$

so long as $n \geq N_2$, where $N_2 \geq N_1$ has been chosen so that

$$\sup_{x\in[-\delta,\delta]} \frac{\left|\Lambda''_\mu(\Xi_\mu(x))\right|}{\log_{(2)} n} < \epsilon^2 \quad \text{for} \quad n \geq N_2.$$

Hence, all that we have to do is choose $N(\epsilon,\gamma) \geq N_2$ so that

$$1 + \frac{K}{\epsilon}\sqrt{\frac{8\log_{(2)} n}{n}} \leq \gamma \quad \text{for } n \geq N(\epsilon,\gamma). \quad \square$$

1.5.10 Theorem (Law of Iterated Logarithm). *The equation* (1.5.2) *holds for any sequence* $\{X_n\}_1^\infty$ *of independent, identically distributed random variables with mean-value* 0 *and variance* 1. *In fact, P-almost surely, the set of limit points of* $\left\{\frac{S_n}{\Lambda_n}\right\}_1^\infty$ *coincides with the entire interval* $[-1,1]$. *Equivalently, for any* $f \in C(\mathbb{R};\mathbb{R})$,

$$\overline{\lim_{n\to\infty}} f\left(\frac{S_n}{\Lambda_n}\right) = \sup_{t\in[-1,1]} f(t) \quad \text{(a.s., } P). \tag{1.5.11}$$

(Cf. Exercise 1.5.15 below for a converse statement.)

PROOF: We begin with the observation that, because of (1.5.5), we may restrict our attention to the case when the X_n's are bounded random variables. Indeed, for any X_n's and any $\epsilon > 0$, an easy truncation procedure allows us to find an $f \in C_b(\mathbb{R};\mathbb{R})$ such that $Y_n \equiv f \circ X_n$ again has mean-value 0 and variance 1 while $Z_n \equiv X_n - Y_n$ has variance less than ϵ^2. Hence, if the result is known when the random variables are bounded, then, by (1.5.5) applied to the Z_n's:

$$\overline{\lim_{n\to\infty}} \left|\tilde{S}_n(\omega)\right| \leq 1 + \overline{\lim_{n\to\infty}} \left|\frac{\sum_{m=1}^n Z_m(\omega)}{\Lambda_n}\right| \leq 1 + 8\epsilon,$$

and, for $a \in [-1,1]$,

$$\underline{\lim_{n\to\infty}} \left|\tilde{S}_n(\omega) - a\right| \leq \overline{\lim_{n\to\infty}} \left|\frac{\sum_{m=1}^n Z_m(\omega)}{\Lambda_n}\right| \leq 8\epsilon$$

for P-almost every $\omega \in \Omega$.

In view of the preceding, from now on we may and will assume that the X_n's are bounded. To prove that $\overline{\lim}_{n\to\infty} \tilde{S}_n \leq 1$ (a.s., P), let $\beta \in (1,\infty)$ be given and use (1.5.8) to see that

$$P\left(\left|\tilde{S}_{\beta^m}\right| \geq \beta^{\frac{1}{2}}\right) \leq 2\exp\left[-\beta^{\frac{1}{2}} \log_{(2)}\left[\beta^m\right]\right]$$

for all sufficiently large $m \in \mathbb{Z}^+$. Hence, by Lemma 1.5.3 with $a = \beta$, we see that $\overline{\lim}_{n\to\infty} |\tilde{S}_n| \le \beta$ (a.s., P) for every $\beta \in (1,\infty)$. To complete the proof, we must still show that, for every $a \in (-1,1)$ and $\epsilon > 0$,

$$P\Big(\varliminf_{n\to\infty} |\tilde{S}_n - a| < \epsilon\Big) = 1.$$

Because we want to get this conclusion as an application of the second part of the Borel–Cantelli Lemma, it is important that we be dealing with independent events; and for this purpose, we use the result just proved to see that, for every integer $k \ge 2$,

$$\begin{aligned}\varliminf_{n\to\infty} |\tilde{S}_n - a| &\le \varlimsup_{k\to\infty} \varliminf_{m\to\infty} |\tilde{S}_{k^m} - a| \\ &= \varlimsup_{k\to\infty} \varliminf_{m\to\infty} \left|\frac{S_{k^m} - S_{k^{m-1}}}{\Lambda_{k^m}} - a\right| \quad P\text{-almost surely.}\end{aligned}$$

Thus, because the events

$$A_{k,m} \equiv \left\{\left|\frac{S_{k^m} - S_{k^{m-1}}}{\Lambda_{k^m}} - a\right| < \epsilon\right\}, \quad m \in \mathbb{Z}^+,$$

are independent for each $k \ge 2$, all that we need to do is check that

$$\sum_{m=1}^{\infty} P(A_{k,m}) = \infty \quad \text{for each} \quad k \ge 2.$$

But

$$P(A_{k,m}) = P\left(\left|\tilde{S}_{k^m - k^{m-1}} - \frac{\Lambda_{k^m} a}{\Lambda_{k^m - k^{m-1}}}\right| < \frac{\Lambda_{k^m}\epsilon}{\Lambda_{k^m - k^{m-1}}}\right),$$

and, because

$$\frac{\Lambda_{k^m}}{\Lambda_{k^m - k^{m-1}}} \longrightarrow 1 \quad \text{as} \quad m \to \infty,$$

everything reduces to showing that

$$\sum_{m=1}^{\infty} P\Big(\big|\tilde{S}_{k^m - k^{m-1}} - a\big| < \epsilon\Big) = \infty \tag{1.5.12}$$

for each $k \ge 2$, $a \in (-1,1)$, and $\epsilon > 0$. Finally, referring to (1.5.9), choose $\epsilon_0 > 0$ so small that $\rho \equiv |a|\big(|a| + 2K\epsilon_0\big) < 1$ and take $\gamma = \rho^{-\frac{1}{2}}$. Then, by (1.5.9), for each $0 < \epsilon \le \epsilon_0$, we know that

$$P\Big(|\tilde{S}_n - a| < \epsilon\Big) \ge \frac{1}{2}\exp\Big[-\rho^{\frac{1}{2}} \log_{(2)} n\Big]$$

for all sufficiently large n's; and so (1.5.12) is proved. □

1.5.13 Remark. The reader should notice that the Law of the Iterated Logarithm provides a *naturally occurring* sequence of functions which converge in measure but not almost everywhere. Indeed, it is obvious that $\tilde{S}_n \longrightarrow 0$ in $L^2(P)$, but the Law of the Iterated Logarithm says that $\{\tilde{S}_n\}_1^\infty$ is wildly divergent when looked at in terms of P-almost sure convergence.

Exercises

1.5.14 Exercise: Let X and X' be a pair of independent random variables which have the same distribution, let α be a median of X, and set $Y = X - X'$.

(i) Show that

$$P\Big(|X-\alpha| \geq t\Big) \leq 2P\Big(|Y| \geq t\Big) \quad \text{for all} \quad t \in [0,\infty),$$

and conclude that, for any $p \in [1,\infty)$,

$$\tfrac{1}{2}\mathbb{E}^P\big[Y^p\big]^{\frac{1}{p}} \leq \mathbb{E}^P\big[X^p\big]^{\frac{1}{p}} \leq \big(2\mathbb{E}^P\big[Y^p\big]\big)^{\frac{1}{p}} + |\alpha|.$$

In particular, $X \in L^p(P)$ if and only if Y is.

(ii) As an initial application of **(i)**, we give our final refinement of The Weak Law of Large Numbers. Namely, let $\{X_n\}_1^\infty$ be a sequence of independent, identically distributed random variables. By combining Exercise 1.2.15, part **(ii)** in Exercise 1.4.29, and part **(i)** above, show that†

$$\begin{aligned} \lim_{n\to\infty} P\Big(|\overline{S}_n| \leq C\Big) = 1 \quad &\text{for some } C \in (0,\infty) \\ &\Longrightarrow \lim_{n\to\infty} nP(|X_1| \geq n) = 0 \\ \Longrightarrow \overline{S}_n - \mathbb{E}^P\big[X_1,\, |X_1| \leq n\big] &\longrightarrow 0 \text{ in } P\text{-probability.} \end{aligned}$$

1.5.15 Exercise: Let $X_1, \dots, X_n, \dots$ be a sequence of independent, identically distributed random variables for which

$$P\left(\varlimsup_{n\to\infty} \frac{|S_n|}{\Lambda_n} < \infty\right) > 0. \tag{1.5.16}$$

† These ideas are taken from the book by Wm. Feller's cited at the end of §1.2. They become even more elegant when combined with a theorem due to E.J.G. Pitman (cf. *ibid.*).

In this exercise we will show† that X_1 is square P-integrable, $\mathbb{E}^P\big[X_1\big] = 0$, and

$$\text{(1.5.17)} \qquad \overline{\lim_{n\to\infty}} \frac{S_n}{\Lambda_n} = -\underline{\lim_{n\to\infty}} \frac{S_n}{\Lambda_n} = \mathbb{E}^P\big[X_1^2\big]^{\frac{1}{2}} \quad \text{(a.s., } P).$$

(i) Using Lemma 1.4.1, show that there is a $\sigma \in [0,\infty)$ such that

$$\text{(1.5.18)} \qquad \overline{\lim_{n\to\infty}} \frac{|S_n|}{\Lambda_n} = \sigma \quad \text{(a.s., } P).$$

Next, assuming that X_1 is square P-integrable, use The Strong Law of Large Numbers together with Theorem 1.5.10 to show that $\mathbb{E}^P\big[X_1\big] = 0$ and

$$\sigma = \mathbb{E}^P\big[X_1^2\big]^{\frac{1}{2}} = \overline{\lim_{n\to\infty}} \frac{S_n}{\Lambda_n} = -\underline{\lim_{n\to\infty}} \frac{S_n}{\Lambda_n} \quad \text{(a.s., } P).$$

In other words, everything comes down to proving that (1.5.16) implies that X_1 is square P-integrable.

(ii) Assume that the X_n's are symmetric. For $t \in (0,\infty)$, set

$$\check{X}_n^t = X_n\,\mathbf{1}_{[0,t]}\big(|X_n|\big) - X_n\,\mathbf{1}_{(t,\infty)}\big(|X_n|\big),$$

and show that

$$\big(\check{X}_1^t,\dots,\check{X}_n^t,\dots\big) \quad \text{and} \quad \big(X_1,\dots,X_n,\dots\big)$$

have the same distribution. Conclude first that, for all $t \in [0,1)$,

$$\overline{\lim_{n\to\infty}} \frac{\big|\sum_{m=1}^n X_n\,\mathbf{1}_{[0,t]}\big(|X_n|\big)\big|}{\Lambda_n} \le \sigma \quad \text{(a.s., } P),$$

where σ is the number in (1.5.18), and second that

$$\mathbb{E}^P\big[X_1^2\big] = \lim_{t\nearrow\infty} \mathbb{E}^P\Big[X_1^2,\,|X_1| \le t\Big] \le \sigma^2.$$

Hint: Use the equation

$$X_n\,\mathbf{1}_{[0,t]}\big(|X_n|\big) = \frac{X_n + \check{X}_n^t}{2},$$

and apply part **(i)**.

† We follow Wm. Feller "An extension of the law of the iterated logarithm," *J. Math. Mech.* **18**, although V. Strassen was the first to prove the result.

(iii) For general $\{X_n\}_1^\infty$, produce an independent copy $\{X'_n\}_1^\infty$ (as in the proof of Lemma 1.5.4), and set $Y_n = X_n - X'_n$. After checking that

$$\varlimsup_{n\to\infty} \frac{|\sum_{m=1}^n Y_m|}{\Lambda_n} \le 2\sigma \quad \text{(a.s., } P\text{)},$$

conclude first that $\mathbb{E}^P\left[Y_1^2\right] \le 4\sigma^2$ and then (cf. part **(i)** of Exercise 1.5.14) that $\mathbb{E}^P\left[X_1^2\right] < \infty$. Finally, apply **(i)** to arrive at $\mathbb{E}^P\left[X_1\right] = 0$ and (1.5.17).

1.5.19 Exercise: Let $\{\tilde{s}_n\}_1^\infty$ be a sequence of real numbers which possess the properties that

$$\varlimsup_{n\to\infty} \tilde{s}_n = 1, \quad \varliminf_{n\to\infty} \tilde{s}_n = -1, \quad \text{and} \quad \lim_{n\to\infty} \left|\tilde{s}_{n+1} - \tilde{s}_n\right| = 0.$$

Show that the set of subsequential limit points of $\{\tilde{s}_n\}_1^\infty$ coincides with $[-1, 1]$. Apply this observation to show that in order to get the final statement in Theorem 1.5.8, we need only have proved (1.5.10) for the function $f(x) = x$, $x \in \mathbb{R}$.

Hint: In proving the last part, use the square integrability of X_1 to see that

$$\sum_{n=1}^\infty P\left(\frac{X_n^2}{n} \ge 1\right) < \infty,$$

and apply the Borel–Cantelli Lemma to conclude that $\tilde{S}_n - \tilde{S}_{n-1} \longrightarrow 0$ (a.s., P).

Chapter II:
The Central Limit Theorem

§2.1: The Theorems of Lindeberg and Berry–Esseen.

Up to this point, the only reason for our believing that Gaussian random variables deserve the nickname *normal* is that on two occasions (cf. Sections 1.3 and 1.5) it turned out that their behavior led to accurate predictions about sums of more general, independent, square integrable random variables. In this section, we will provide a mathematically rigorous explanation why *Gaussians are so normal.*

Given a sequence $\{X_n : n \in \mathbb{Z}^+\}$ of mutually independent, identically distributed random variables with mean-value 0 and variance 1, set $S_n = \sum_1^n X_k$. In Chapter I we discussed the limit behavior, as $n \to \infty$, of S_n divided by various weights. In particular, we saw (cf. Theorem 1.5.10) that the precise rate at which S_n grows is given by the numbers Λ_n in (1.5.1). Hence, from the point of view of either almost sure convergence or even convergence in probability, there is no hope of getting $\hat{S}_n \equiv \frac{S_n}{\sqrt{n}}$ to converge. On the other hand, the random variables $\hat{S}_n$ do possess some basic stability: their mean-value is always 0 and their variance is always 1. In other words, for any quadratic function φ, $\mathbb{E}^P\big[\varphi(\hat{S}_n)\big]$ is independent of both $n \in \mathbb{Z}^+$ and the particular random variables out of which $\hat{S}_n$ is built. At first sight, this stability may not appear to be very significant. Nonetheless, the remarkable fact which emerges from the considerations below is that, together with independence, this stability for quadratic functions leads to the existence, for a much larger class of functions φ, of the limit $\lim_{n\to\infty} \mathbb{E}^P\big[\varphi(\hat{S}_n)\big]$, where (and, perhaps, this is the most surprising part) *the limit does not depend on the particular choice of the original random variables* X_n.

Before getting down to serious business, it may be helpful to start with a rather hands-on approach; namely, we suppose that X_1 has moments of all orders and attempt to compute $\lim_{n\to\infty} \mathbb{E}^P\big[\hat{S}_n^m\big]$ for integers $m \geq 3$. To this end, first note that

$$\begin{aligned}
\mathbb{E}^P\big[S_n^{m+1}\big] &= n\mathbb{E}^P\big[X_n(X_n+S_{n-1})^m\big] \\
&= n\sum_{j=0}^{m}\binom{m}{j}\mathbb{E}^P\big[X_n^{j+1}\big]\mathbb{E}^P\big[S_{n-1}^{m-j}\big] \\
&= nm\mathbb{E}^P\big[S_{n-1}^{m-1}\big] + n\sum_{j=2}^{m}\binom{m}{j}\mathbb{E}^P\big[X_n^{j+1}\big]\mathbb{E}^P\big[S_{n-1}^{m-j}\big].
\end{aligned}$$

Thus, after dividing through by $n^{\frac{m+1}{2}}$, one can use induction to prove that $L_m \equiv \lim_{n\to\infty}\mathbb{E}^P\big[\hat{S}_n^m\big]$ not only exists for all $m \in \mathbb{N}$ but also satisfies $L_{m+1} = mL_{m-1}$ for $m \in \mathbb{Z}^+$. Therefore, since $L_0 = 1$ and $L_1 = 0$, we have now proved that

$$\lim_{n\to\infty}\mathbb{E}^P\big[\hat{S}_n^{2m-1}\big] = 0 \quad \text{and} \quad \lim_{n\to\infty}\mathbb{E}^P\big[\hat{S}_n^{2m}\big] = \frac{(2m)!}{2^m m!}, \tag{2.1.1}$$

for all $m \in \mathbb{Z}^+$. In other words, at least when X_1 has moments of all orders and φ is a polynomial, $\lim_{n\to\infty}\mathbb{E}^P\big[\varphi(\hat{S}_n)\big]$ exists and is independent of the particular choice of random variables.

In order to carry our analysis to the next stage, it will be essential to find another expression for what we expect the limit $\lim_{n\to\infty}\mathbb{E}^P\big[\varphi(\hat{S}_n)\big]$ to be; and, because we expect the limit to be independent of the particular choice of X_n's, it is only reasonable to carry out this computation with X_n's for which the distribution of $\hat{S}_n$ is as simple as possible. In particular, it is reasonable to guess that, if such a random variable exists, one should take X_1 to have the property that

$$\mathbb{E}^P\big[X_1^m\big] = \lim_{n\to\infty}\mathbb{E}^P\big[\hat{S}_n^m\big], \quad m \in \mathbb{N}.$$

That is, in view of (2.1.1), we are guessing (somewhat formally) that

$$\mathbb{E}^P\Big[e^{\alpha X_1}\Big] = \sum_{\ell=0}^{\infty}\frac{\alpha^{2\ell}}{2^\ell \ell!} = \exp\left[\frac{\alpha^2}{2}\right], \quad \alpha \in \mathbb{C}.$$

But this is exactly the expression which one gets when X_1 is an $\mathfrak{N}(0,1)$-random variable, and so we are now led to take the X_n's to be mutually independent $\mathfrak{N}(0,1)$-random variables. Moreover, even though our route to this choice was somewhat suspect, once we make it there can be no doubt that it is precisely the right one! Indeed, if the X_n's are mutually independent $\mathfrak{N}(0,1)$-random variables, then S_n is an $\mathfrak{N}(0,n)$-random variable and so $\hat{S}_n$ is again an $\mathfrak{N}(0,1)$-random variable. In other words, we are guessing that the correct formulation of the result for which we are looking is that, for a large class of independent, identically distributed random variables and a large class of functions φ,

$$\lim_{n\to\infty}\mathbb{E}^P\big[\varphi(\hat{S}_n)\big] = \int_{\mathbb{R}}\varphi(y)\,\gamma(dy), \tag{2.1.2}$$

where we have allowed ourselves the nearly unforgivable abuse of notation involved in letting γ denote the *probability measure* on $\mathbb{R}$ whose *density*, the Gauss kernel, we denoted by γ in (1.3.5).

Because of the pivotal position that it occupies in probability theory, we will devote the rest of this section to two quite different derivations of (2.1.2). In fact, in order to indicate how amazingly robust a phenomenon underlies the computation, we will, from now on, be dealing with the following more general setting. Namely, $(\Omega, \mathcal{F}, P)$ is a probability space on which $\{X_m\}_1^n$ will be mutually independent square P-integrable random variables with mean-value 0 and $\sigma_m \equiv \left(\mathbb{E}^P\left[X_m^2\right]\right)^{\frac{1}{2}} > 0$. Also, we will use the notation

$$S_n \equiv \sum_1^n X_m, \quad \Sigma_n \equiv \left(\sum_1^n \sigma_m^2\right)^{\frac{1}{2}}, \quad \text{and } \hat{S}_n \equiv \frac{S_n}{\Sigma_n}. \tag{2.1.3}$$

Notice that when the X_k's are identically distributed and have variance 1, the $\hat{S}_n$ in (2.1.3) is consistent with the notation used above. Finally, we set

$$r_n = \max_{1 \le m \le n} \frac{\sigma_m}{\Sigma_n} \quad \text{and} \quad g_n(\epsilon) = \frac{1}{\Sigma_n^2} \sum_{m=1}^n \mathbb{E}^P\left[X_m^2, \, |X_m| \ge \epsilon \Sigma_n\right] \tag{2.1.4}$$

for $\epsilon > 0$. Clearly, in the identically distributed case, $r_n = n^{-\frac{1}{2}}$ and

$$g_n(\epsilon) = \sigma_1^{-2} \mathbb{E}^P\left[X_1^2, \, |X_1| \ge n^{\frac{1}{2}} \sigma_1 \epsilon\right] \longrightarrow 0 \quad \text{as } n \to \infty \text{ for each } \epsilon > 0.$$

2.1.5 Theorem (Lindeberg). *Refer to the preceding and let φ be an element of $C^3(\mathbb{R};\mathbb{R})$ with bounded second and third order derivatives. Then, for each $\epsilon > 0$,*

$$\left| \mathbb{E}^P\left[\varphi(\hat{S}_n)\right] - \int_{\mathbb{R}} \varphi \, d\gamma \right| \le \left(\frac{\epsilon}{6} + \frac{r_n}{2}\right) \|\varphi'''\|_{\mathrm{u}} + g_n(\epsilon) \|\varphi''\|_{\mathrm{u}}. \tag{2.1.6}$$

In particular, because

$$r_n^2 \le \epsilon^2 + g_n(\epsilon), \quad \epsilon > 0, \tag{2.1.7}$$

(2.1.2) holds if $g_n(\epsilon) \longrightarrow 0$ as $n \to \infty$ for each $\epsilon > 0$.[†]

[†] The condition that $g_n(\epsilon) \longrightarrow 0$ for each $\epsilon > 0$ is often called *Lindeberg's condition*, because it was Lindeberg who introduced it and proved that it is a sufficient condition for (2.1.2). Later, Feller proved that (2.1.2) plus $r_n \to 0$ implies that Lindeberg's condition hold. Together, these two results are known as the **Lindeberg–Feller Theorem**. For a proof of Feller's part, see Sec. 20 of M. Loéve's *Probability Theory*, publ. originally in the University Series in Higher Math. by van Norstrand, Inc. (1963), and now by Springer–Verlag.

PROOF: Choose $\mathfrak{N}(0,1)$-random variables $Y_1, \dots, Y_n$ which are both mutually independent and independent of the X_m's. (After changing the probability spaces, if necessary, this can be done as an application of Theorem 1.1.9 or Exercise 1.1.14.) Next, set

$$\hat{Y}_k = \frac{\sigma_k Y_k}{\Sigma_n} \quad \text{and} \quad \hat{T}_n = \sum_1^n \hat{Y}_k,$$

and observe that $\hat{T}_n$ is again an $\mathfrak{N}(0,1)$-random variable and therefore that

$$\Delta \equiv \left|\mathbb{E}^P\left[\varphi(\hat{S}_n)\right] - \int_{\mathbb{R}} \varphi \, d\gamma\right| = \left|\mathbb{E}^P\left[\varphi(\hat{S}_n)\right] - \mathbb{E}^P\left[\varphi(\hat{T}_n)\right]\right|. \tag{2.1.8}$$

Further, set $\hat{X}_k = \frac{X_k}{\Sigma_n}$, and define

$$U_m = \sum_{k=1}^{m-1} \hat{X}_k + \sum_{k=m+1}^{n} \hat{Y}_k \quad \text{for } 1 \le m \le n,$$

where the first sum is taken to be 0 if $m = 1$ and the second sum is 0 if $m = n$. It is then clear, from (2.1.8), that

$$\Delta \le \sum_1^n \Delta_m \quad \text{where } \Delta_m \equiv \left|\mathbb{E}^P\left[\varphi(U_m + \hat{X}_m)\right] - \mathbb{E}^P\left[\varphi(U_m + \hat{Y}_m)\right]\right|.$$

Moreover, if

$$R_m(\xi) \equiv \varphi(U_m + \xi) - \varphi(U_m) - \xi\varphi'(U_m) - \tfrac{\xi^2}{2}\varphi''(U_m), \quad \xi \in \mathbb{R},$$

then (because both $\hat{X}_m$ and $\hat{Y}_m$ are independent of U_m and have the same first two moments)

$$\begin{aligned}\Delta_m &= \left|\mathbb{E}^P\left[R_m(\hat{X}_m)\right] - \mathbb{E}^P\left[R_m(\hat{Y}_m)\right]\right| \\ &\le \left|\mathbb{E}^P\left[R_m(\hat{X}_m)\right]\right| + \left|\mathbb{E}^P\left[R_m(\hat{Y}_m)\right]\right|.\end{aligned}$$

In order to complete the derivation of (2.1.6), note that, by Taylor's Theorem,

$$|R_m(\xi)| \le \left(\|\varphi'''\|_{\mathrm{u}} \tfrac{|\xi|^3}{6}\right) \wedge \left(\|\varphi''\|_{\mathrm{u}} |\xi|^2\right);$$

and therefore, for each $\epsilon > 0$,

$$\begin{aligned}\sum_1^n \mathbb{E}^P\Big[\big|R_m(\hat{X}_m)\big|\Big] &\le \frac{\|\varphi'''\|_{\mathrm{u}}}{6}\sum_1^n \mathbb{E}^P\Big[|\hat{X}_m|^3,\ |X_m| \le \epsilon\Sigma_n\Big] \\ &\quad + \|\varphi''\|_{\mathrm{u}}\sum_1^n \mathbb{E}^P\Big[\hat{X}_m^2,\ |X_m| \ge \epsilon\Sigma_n\Big] \\ &\le \frac{\epsilon\|\varphi'''\|_{\mathrm{u}}}{6}\sum_1^n \frac{\sigma_m^2}{\Sigma_n^2} + \|\varphi''\|_{\mathrm{u}} g_n(\epsilon) = \frac{\epsilon\|\varphi'''\|_{\mathrm{u}}}{6} + \|\varphi''\|_{\mathrm{u}} g_n(\epsilon);\end{aligned}$$

while

$$\sum_1^n \mathbb{E}^P\Big[\big|R_m(\hat{Y}_n)\big|\Big] \le \frac{\|\varphi'''\|_{\mathrm{u}}}{6}\,\mathbb{E}^P\Big[|Y_1|^3\Big]\sum_1^n \frac{\sigma_m^3}{\Sigma_n^3} \le \frac{3^{\frac{3}{4}} r_n \|\varphi'''\|_{\mathrm{u}}}{6}.$$

Hence, (2.1.6) is now proved.

Given (2.1.6), all that remains is to prove (2.1.7). However, for any $1 \le m \le n$ and $\epsilon > 0$,

$$\sigma_m^2 = \mathbb{E}^P\Big[X_m^2,\ |X_m| < \epsilon\Sigma_n\Big] + \mathbb{E}^P\Big[X_m^2,\ |X_m| \ge \epsilon\Sigma_n\Big] \le \Sigma_n^2\big(\epsilon^2 + g_n(\epsilon)\big). \quad \square$$

If one is not concerned about rates of convergence, then the differentiability requirement can be dropped from the last part of Theorem 2.1.5. In fact, with essentially no further effort, we have the following version of the famous **Central Limit Theorem**.

2.1.9 Corollary (Central Limit Theorem). *With the setting the same as it was in Theorem 2.1.5, assume that $g_n(\epsilon) \longrightarrow 0$ as $n \to \infty$ for each $\epsilon > 0$. Then (2.1.2) holds for every $\varphi \in C(\mathbb{R};\mathbb{C})$ satisfying*

$$\sup_{y\in\mathbb{R}} \frac{|\varphi(y)|}{1+|y|^2} < \infty. \tag{2.1.10}$$

In particular, for every pair $-\infty \le a < b \le \infty$,

$$\lim_{n\to\infty} P\Big(a \le \hat{S}_n \le b\Big) = \frac{1}{\sqrt{2\pi}}\int_a^b \exp\left[-\frac{y^2}{2}\right]dy. \tag{2.1.11}$$

(See Exercise 2.1.38 below for more information in the identically distributed case.)

PROOF: To prove the first part, choose a function $\rho \in C_c^\infty\big((-1,1);[0,\infty)\big)$ with $\int_{\mathbb{R}} \rho(y)\,dy = 1$. Then, for each $k \in \mathbb{Z}^+$, the function φ_k given by

$$\varphi_k(x) = k\int_{-k}^{k} \rho\big(k(x-y)\big)\,\varphi(y)\,dy, \quad x \in \mathbb{R},$$

is a smooth, compactly supported function. Furthermore, as $k \to \infty$, $\varphi_k \longrightarrow \varphi$ uniformly on compacts; and, because of (2.1.10), there is a $K \in (0,\infty)$ such that

$$\sup_{k\in\mathbb{Z}^+} \big|(\varphi-\varphi_k)(y)\big| \le K\big(1+y^2\big) \quad \text{for all} \quad y \in \mathbb{R}.$$

Thus, by Theorem 2.1.5 applied to φ_k,

$$\begin{aligned}\varlimsup_{n\to\infty}&\left|\mathbb{E}^P\Big[\varphi\big(\hat{S}_n\big)\Big] - \int_{\mathbb{R}}\varphi(y)\,\gamma(dy)\right| \\ &\le \varlimsup_{k\to\infty}\,\varlimsup_{n\to\infty}\,\mathbb{E}^P\Big[\big|(\varphi-\varphi_k)\big|\big(\hat{S}_n\big)\Big] \\ &\qquad + \varlimsup_{k\to\infty}\int_{\mathbb{R}}\big|(\varphi-\varphi_k)\big|(y)\,\gamma(dy) \\ &\le K\,\varlimsup_{n\to\infty}\,\mathbb{E}^P\Big[1+\hat{S}_n^2,\,\big|\hat{S}_n\big|\ge R\Big] + K\int_{|y|\ge R}(1+y^2)\,\gamma(dy)\end{aligned}$$

for every $R \in (0,\infty)$. Obviously, the second term in the final line tends to 0 as $R \nearrow \infty$. To handle the first term, choose $\eta \in C_b^\infty\big(\mathbb{R};[0,1]\big)$ so that $\eta = 0$ on $\big[-\frac{1}{2},\frac{1}{2}\big]$ and $\eta = 1$ off $(-1,1)$; and define

$$\psi_R(y) = \big(1+y^2\big)\eta\left(\tfrac{y}{R}\right) \quad \text{for} \quad y \in \mathbb{R}.$$

Then, by Theorem 2.1.5 applied to ψ_R,

$$\begin{aligned}\varlimsup_{n\to\infty}\,\mathbb{E}^P\Big[1+\hat{S}_n^2,\,\big|\hat{S}_n\big|\ge R\Big] &\le \varlimsup_{n\to\infty}\,\mathbb{E}^P\Big[\psi_R\big(\hat{S}_n\big)\Big] \\ &= \int_{\mathbb{R}}\psi_R(y)\,\gamma(dy) \longrightarrow 0\end{aligned}$$

as $R \nearrow \infty$. Hence, we have now proved that (2.1.2) holds for every $\varphi \in C(\mathbb{R};\mathbb{R})$ which satisfies (2.1.10).

Turning to the second assertion, let $a < b$ be given. To prove (2.1.11), choose $\{\varphi_k\}_1^\infty \subseteq C_b(\mathbb{R};\mathbb{R})$ and $\{\psi_k\}_1^\infty \subseteq C_b(\mathbb{R};\mathbb{R})$ so that $0 \le \varphi_k \nearrow \mathbf{1}_{(a,b)}$ and $1 \ge \psi_k \searrow \mathbf{1}_{[a,b]}$ as $k \to \infty$. Then,

$$\begin{aligned}\varliminf_{n\to\infty} P\Big(a < \hat{S}_n < b\Big) &\ge \lim_{n\to\infty}\mathbb{E}^P\Big[\varphi_k\big(\hat{S}_n\big)\Big] \\ &= \int_{\mathbb{R}}\varphi_k(y)\,\gamma(dy) \longrightarrow \gamma\big((a,b)\big)\end{aligned}$$

as $k \to \infty$; and, similarly,

$$\varlimsup_{n\to\infty} P\left(a \le \hat{S}_n \le b\right) \le \lim_{n\to\infty} \mathbb{E}^P\left[\psi_k(\hat{S}_n)\right]$$
$$= \int_{\mathbb{R}} \psi_k(y)\,\gamma(dy) \longrightarrow \gamma([a,b]).$$

Hence, since $\gamma((a,b)) = \gamma([a,b])$, we are done. $\square$

As we will see in the next section, the principles underlying the passage from Theorem 2.1.5 to Corollary 2.1.9 are very general. In fact, as we will see in Chapter 3, some of these principles can be formulated in such a way that they extend to a very abstract setting. However, before we start delving into such extensions, we will devote the rest of this section to a closer examination of the situation at hand. In particular, we are going to see how to improve the quantitative statements contained in Lindeberg's Theorem.

From (2.1.6), we get a rate of convergence in terms of the second and third derivatives of φ. In fact, if we assume that

$$\tau_k \equiv \left(\mathbb{E}^P\left[|X_k|^3\right]\right)^{\frac{1}{3}} < \infty, \quad 1 \le k \le n, \tag{2.1.12}$$

then (cf. the proof of Theorem 2.1.5) by using the estimates

$$\left|R_m(\xi)\right| \le \frac{\|\varphi'''\|_{\mathrm{u}}|\xi|^3}{6} \quad \text{and} \quad \sigma_k \le \tau_k,$$

one sees that (2.1.6) can be replaced by

$$\left|\mathbb{E}^P\left[\varphi(\hat{S}_n)\right] - \int_{\mathbb{R}} \varphi\, d\gamma\right| \le \frac{2\|\varphi'''\|_{\mathrm{u}}}{3} \frac{\sum_1^n \tau_k^3}{\Sigma_n^3} \tag{2.1.13}$$

when the X_k's have third moments.

Although both (2.1.6) and (2.1.13) are interesting, neither one of them can be used to give very much information about the rate at which the distribution functions

$$x \in \mathbb{R} \longmapsto F_n(x) \equiv P\left(\hat{S}_n \le x\right) \in [0,1] \tag{2.1.14}$$

are tending to the **error function**

$$G(x) \equiv \gamma\left((-\infty, x]\right) = \frac{1}{\sqrt{2\pi}} \int_{-\infty}^{x} e^{-\frac{t^2}{2}}\, dt. \tag{2.1.15}$$

To see how (2.1.6) and (2.1.13) must be modified in order to gain such information, first observe that

$$\begin{aligned} &\int_{\mathbb{R}} \varphi'(x)\left(G(x) - F_n(x)\right) dx \\ &\quad = \mathbb{E}^P\left[\varphi(\hat{S}_n)\right] - \int_{\mathbb{R}} \varphi(y)\,\gamma(dy), \quad \varphi \in C_{\mathrm{b}}^1(\mathbb{R};\mathbb{R}). \end{aligned} \tag{2.1.16}$$

(To see (2.1.16), reduce to the case in which $\varphi \in C_c^1(\mathbb{R};\mathbb{R})$ and $\varphi(0) = 0$; and for this case apply integration by parts over the intervals $(-\infty, 0]$ and $[0, \infty)$ separately.) Hence, in order to get information about the distance between F_n and G, we will have to learn how to replace the right-hand sides of (2.1.6) and (2.1.13) with expressions which depend only on the first derivative of φ. For example, if the dependence is on $\|\varphi'\|_{\mathrm{u}}$, then we get information about the $L^1(\mathbb{R})$ distance between F_n and G; whereas if the dependence is on $\|\varphi'\|_{L^1(\mathbb{R})}$, then the information will be about the uniform distance between F_n and G.

The basic idea which we will use to get our estimates in terms of φ' was introduced by C. Stein and is an example of a procedure known as **Stein's method.**† In the case at hand, Stein's method rests on the simple observation that if $\psi \in C(\mathbb{R};\mathbb{R})$ has no more than linear growth at infinity, then the only obstruction to finding a boundedly differentiable solution f to the equation $f'(x) - xf(x) = \psi(x)$ is that $\int \psi\, d\gamma = 0$. More precisely, we will use the following.

2.1.17 Lemma. *Let $\varphi \in C^1(\mathbb{R};\mathbb{R})$, assume that $\|\varphi'\|_{\mathrm{u}} < \infty$, set*

$$\tilde{\varphi} = \varphi - \int_{\mathbb{R}} \varphi\, d\gamma,$$

and define

$$x \in \mathbb{R} \longmapsto f(x) \equiv e^{\frac{x^2}{2}} \int_{-\infty}^{x} \tilde{\varphi}(t) e^{-\frac{t^2}{2}}\, dt. \tag{2.1.18}$$

Then $f \in C_{\mathrm{b}}^2(\mathbb{R};\mathbb{R})$,

$$\|f\|_{\mathrm{u}} \le 2\|\varphi'\|_{\mathrm{u}}, \quad \|f'\|_{\mathrm{u}} \le 3\sqrt{\tfrac{\pi}{2}}\|\varphi'\|_{\mathrm{u}}, \quad \|f''\|_{\mathrm{u}} \le 4\|\varphi'\|_{\mathrm{u}}, \tag{2.1.19}$$

and

$$f'(x) - xf(x) = \tilde{\varphi}(x), \quad x \in \mathbb{R}. \tag{2.1.20}$$

PROOF: The facts that $f \in C^1(\mathbb{R};\mathbb{R})$ and that (2.1.20) holds are elementary applications of the Fundamental Theorem of Calculus. Moreover, knowing that $f \in C^1(\mathbb{R};\mathbb{R})$ and using (2.1.20), we see that $f \in C^2(\mathbb{R};\mathbb{R})$ and, in fact, that

$$f''(x) - xf'(x) = f(x) + \varphi'(x), \quad x \in \mathbb{R}. \tag{2.1.21}$$

To prove the estimates in (2.1.19), first note that, because $\tilde{\varphi}$ and therefore f are unchanged when φ is replaced by $\varphi - \varphi(0)$, we may and will assume that $\varphi(0) = 0$ and therefore that $|\varphi(t)| \le \|\varphi'\|_{\mathrm{u}}|t|$. In particular, this means that

$$\left|\int_{\mathbb{R}} \varphi\, d\gamma\right| \le \|\varphi'\|_{\mathrm{u}} \int_{\mathbb{R}} |t|\, \gamma(dt) = \|\varphi'\|_{\mathrm{u}} \sqrt{\tfrac{2}{\pi}}.$$

† Stein provided an introduction, by way of examples, to his own method in *Approximate Computation of Expectations*, IMS Lec. Notes & Monograph Series **7** (1986).

Next, observe that, because

$$\int_{\mathbb{R}} \tilde{\varphi}(t)e^{-\frac{t^2}{2}}\,dt = 0,$$

an alternative expression for f is

$$f(x) = -e^{\frac{x^2}{2}} \int_x^\infty \tilde{\varphi}(t)e^{-\frac{t^2}{2}}\,dt, \quad x \in \mathbb{R}.$$

Thus, by using the original expression for $f(x)$ when $x \in (-\infty, 0)$ and the alternative one when $x \in [0, \infty)$, we see first that

$$\text{(2.1.22)} \qquad |f(x)| \le e^{\frac{x^2}{2}} \int_{|x|}^\infty \left|\tilde{\varphi}(-t\,\mathrm{sgn}(x))\right| e^{-\frac{t^2}{2}}\,dt, \quad x \in \mathbb{R},$$

and then that

$$|f(x)| \le \|\varphi'\|_{\mathrm{u}} e^{\frac{x^2}{2}} \int_{|x|}^\infty \left(t + \sqrt{\tfrac{2}{\pi}}\right) e^{-\frac{t^2}{2}}\,dt.$$

But, since

$$\frac{d}{dx}\left(e^{\frac{x^2}{2}} \int_x^\infty e^{-\frac{t^2}{2}}\,dt\right) \le e^{\frac{x^2}{2}} \int_x^\infty t\,e^{-\frac{t^2}{2}}\,dt - 1 = 0 \quad \text{for } x \in [0, \infty),$$

we have that

$$\text{(2.1.23)} \qquad e^{\frac{x^2}{2}} \int_{|x|}^\infty te^{-\frac{t^2}{2}}\,dt = 1 \text{ and } e^{\frac{x^2}{2}} \int_{|x|}^\infty e^{-\frac{t^2}{2}}\,dt \le \sqrt{\tfrac{\pi}{2}}, \quad x \in \mathbb{R};$$

which means that we have now proved the first estimate in (2.1.19). To prove the other two estimates there, derive from (2.1.21) that

$$\frac{d}{dx}\left(e^{-\frac{x^2}{2}} f'(x)\right) = e^{-\frac{x^2}{2}}\left(f(x) + \varphi'(x)\right)$$

and therefore that

$$\begin{aligned} f'(x) &= e^{\frac{x^2}{2}} \int_{-\infty}^x \left(f(t) + \varphi'(t)\right)e^{-\frac{t^2}{2}}\,dt \\ &= -e^{\frac{x^2}{2}} \int_x^\infty \left(f(t) + \varphi'(t)\right)e^{-\frac{t^2}{2}}\,dt, \quad x \in \mathbb{R}. \end{aligned}$$

Thus, reasoning as we did above and using the first estimate in (2.1.19), the estimates in (2.1.23), and the equation in (2.1.21), we arrive at the second and third estimates in (2.1.19). □

2.1.24 Theorem. *Continuing in the setting of Theorem 2.1.5, one has that for all $\epsilon > 0$ (cf. (2.1.4), (2.1.14), and (2.1.15))*

$$\|F_n - G\|_{L^1(\mathbb{R})} \le 4(r_n + \epsilon) + 3\sqrt{2\pi}\, g_n(2\epsilon). \tag{2.1.25}$$

Moreover, if (cf. (2.1.12)) $\tau_m < \infty$ for each $1 \le m \le n$, then

$$\|F_n - G\|_{L^1(\mathbb{R})} \le \left(4r_n + \frac{2\sum_{m=1}^n \tau_m^3}{\Sigma_n^3}\right) \wedge \left(\frac{6\sum_{m=1}^n \tau_m^3}{\Sigma_n^3}\right). \tag{2.1.26}$$

In particular, if $\sigma_m^2 = 1$ and $\tau_m \le \tau < \infty$ for each $1 \le m \le n$, then

$$\|F_n - G\|_{L^1(\mathbb{R})} \le \frac{4 + 2\tau^3}{\sqrt{n}} \le \frac{6\tau^3}{\sqrt{n}}.$$

PROOF: Let $\varphi \in C^1(\mathbb{R};\mathbb{R})$ with bounded first derivative be given, and define f accordingly, as in (2.1.18). Everything turns on the equality in (2.1.20). Indeed, because of that equality, we know that the right-hand side of (2.1.16) is equal to

$$\mathbb{E}^P\big[f'(\hat{S}_n)\big] - \mathbb{E}^P\big[\hat{S}_n f(\hat{S}_n)\big] = \sum_{m=1}^n \Big(\hat{\sigma}_m^2 \mathbb{E}^P\big[f'(\hat{S}_n)\big] - \mathbb{E}^P\big[\hat{X}_m f(\hat{S}_n)\big]\Big),$$

where we have set $\hat{\sigma}_m = \frac{\sigma_m}{\Sigma_n}$ and $\hat{X}_m = \frac{X_m}{\Sigma_n}$. Next, define

$$\hat{T}_{n,m}(t) = \hat{S}_n + (t-1)\hat{X}_m \quad \text{for } t \in [0,1],$$

note that $\hat{T}_{n,m}(0)$ is independent of $\hat{X}_m$, and conclude that

$$\begin{aligned}\mathbb{E}^P\Big[\hat{X}_m f(\hat{S}_n)\Big] &= \int_0^1 \mathbb{E}^P\Big[\hat{X}_m^2 f'\big(\hat{T}_{n,m}(t)\big)\Big]\,dt \\ &= \hat{\sigma}_m^2 \mathbb{E}^P\Big[f'\big(\hat{T}_{n,m}(0)\big)\Big] + \int_0^1 \mathbb{E}^P\Big[\hat{X}_m^2\big(f'(\hat{T}_{n,m}(t)) - f'\big(\hat{T}_{n,m}(0)\big)\Big]\,dt\end{aligned}$$

for each $1 \le m \le n$. Hence, we now see that

$$\mathbb{E}^P\big[\varphi(\hat{S}_n)\big] - \int_{\mathbb{R}} \varphi\,d\gamma = \sum_{m=1}^n \hat{\sigma}_m^2 A_m - \sum_{m=1}^n \int_0^1 B_m(t)\,dt \tag{2.1.27}$$

where

$$A_m \equiv \mathbb{E}^P\Big[f'(\hat{S}_n) - f'\big(\hat{T}_{n,m}(0)\big)\Big]$$

and

$$B_m(t) \equiv \mathbb{E}^P\Big[\hat{X}_m^2\Big(f'(\hat{T}_{n,m}(t)) - f'\big(\hat{T}_{n,m}(0)\big)\Big)\Big].$$

Obviously, for each $1 \le m \le n$,

$$|A_m| \le \hat{\sigma}_m \|f''\|_{\mathrm{u}} \le \left(r_n \wedge \frac{\tau_m}{\Sigma_n}\right) \|f''\|_{\mathrm{u}} \tag{2.1.28}$$

while, for each $t \in [0,1]$ and $\epsilon > 0$,

$$\big|B_m(t)\big| \le 2\epsilon t \hat{\sigma}_m^2 \|f''\|_{\mathrm{u}} + 2\frac{\|f'\|_{\mathrm{u}}}{\Sigma_n^2} \mathbb{E}^P\Big[X_m^2,\, |X_m| \ge 2\epsilon\Sigma_n\Big].$$

Thus, after summing over $1 \le m \le n$, integrating with respect to $t \in [0,1]$, and using (2.1.16), we arrive at

$$\left|\int_{\mathbb{R}} \varphi'(x)\big(F_n(x) - G(x)\big)\, dx\right| \le (r_n + \epsilon)\|f''\|_{\mathrm{u}} + 2g_n(2\epsilon)\|f'\|_{\mathrm{u}},$$

which, in conjunction with the estimates in (2.1.19), leads immediately to the one in (2.1.25). In order to get (2.1.26), simply note that

$$\big|B_m(t)\big| \le t\int_0^1 \mathbb{E}^P\Big[|\hat{X}_m|^3 \big|f''\big(\hat{T}_{n,m}(st)\big)\big|\Big]\, ds \le t\|f''\|_{\mathrm{u}} \frac{\tau_m^3}{\Sigma_n^3},$$

and again use (2.1.27), (2.1.19), and (2.1.28). □

The preceding argument already displays the power of Stein's method in that it has allowed us to replace Lindeberg's estimate in terms of $\|\varphi'''\|_{\mathrm{u}}$ by an estimate in terms of $\|\varphi'\|_{\mathrm{u}}$. However, it does not tell us how to replace $\|\varphi'\|_{\mathrm{u}}$ by $\|\varphi'\|_{L^1(\mathbb{R})}$; and, in fact, this replacement will require us to use a clever inductive procedure which was introduced into this context by E. Bolthausen.† However, before we can use Bolthausen's argument, we will need the following simple lemma.

2.1.29 Lemma. *Let $\varphi \in C^1\big(\mathbb{R};[0,1]\big)$ be nonincreasing, and define f accordingly as in* (2.1.18). *Then $\|f\|_{\mathrm{u}} \le 1$ and $\|f'\|_{\mathrm{u}} \le 2$.*

PROOF: Since $\|\tilde{\varphi}\|_{\mathrm{u}} \le 1$ and therefore (cf. (2.1.22) and (2.1.23)) $|xf(x)|$ is dominated by 1, the estimate on $\|f'\|_{\mathrm{u}}$ follows immediately from (2.1.20). Moreover, again because $|xf(x)| \le 1$, all that remains is to check that $|f(x)| \le 1$ when $x \in [-1,1]$. To this end, set

$$\psi(x) = \frac{e^{-\frac{x^2}{2}} f(x)}{\sqrt{2\pi}} = \int_{(-\infty,x]} \tilde{\varphi}(t)\,\gamma(dt) \quad \text{and} \quad a = \inf\big\{t : \tilde{\varphi}(t) = 0\big\}.$$

† The Berry–Esseen Theorem appears as a warm-up exercise in Bolthausen's "An estimate of the remainder term in a combinatorial central limit theorem," *Z. Wahr. Gebiete* **66**.

It is then clear that ψ is nonincreasing on $(-\infty, a]$ and nondecreasing on $[a, \infty)$. Hence, since ψ tends to 0 at both ends of $\mathbb{R}$,

$$\begin{aligned}\|\psi\|_{\mathrm{u}} = \psi(a) &= \int_{(-\infty,a]} \varphi(t)\,\gamma(dt) - G(a)\int_{\mathbb{R}} \varphi\,d\gamma \\ &\le \left(G(a)\int_{\mathbb{R}} \varphi\,d\gamma\right)^{\frac{1}{2}} - G(a)\int_{\mathbb{R}} \varphi\,d\gamma \le \tfrac{1}{4},\end{aligned}$$

where the first inequality comes from Schwarz's inequality and the fact that φ takes its values in $[0,1]$, and the second inequality is an elementary quadratic relation. Finally, since we now have

$$|f(x)| \le \sqrt{2\pi e}\,|\psi(x)| \le \sqrt{\tfrac{\pi e}{8}} \le 1 \quad \text{for } x \in [-1,1],$$

the proof is complete. □

2.1.30 Theorem (Berry–Esseen). *Let everything be as in Theorem 2.1.5, and assume that (cf. (2.1.12)) $\tau_m < \infty$ for each $1 \le m \le n$. Then (cf. (2.1.14) and (2.1.15))*

$$\|F_n - G\|_{\mathrm{u}} \le 10\,\frac{\sum_1^n \tau_m^3}{\Sigma_n^3}. \tag{2.1.31}$$

In particular, if $\sigma_m = 1$ for all $1 \le m \le n$, then (2.1.31) can be replaced by

$$\|F_n - G\|_{\mathrm{u}} \le 10\,\frac{\sum_1^n \tau_m^3}{n^{\frac{3}{2}}} \le 10\,\frac{\max_{1\le m\le n} \tau_m^3}{\sqrt{n}}. \tag{2.1.32}$$

PROOF: For each $n \in \mathbb{Z}^+$ let β_n denote the smallest number β with the property that

$$\|F_n - G\|_{\mathrm{u}} \le \beta\frac{\sum_1^n \tau_m^3}{\Sigma_n^3}$$

for all choices of random variables satisfying the hypotheses under which (2.1.31) is to be proved. Our goal is to give an inductive proof that $\beta_n \le 10$ for all $n \in \mathbb{Z}^+$; and, because $\Sigma_1 \le \tau_1$ and therefore $\beta_1 \le 1$, we need only be concerned with $n \ge 2$.

Given $n \ge 2$ and $X_1, \dots, X_n$, define $\hat{X}_m$, $\hat{\sigma}_m$, and $\hat{T}_{n,m}(t)$ for $1 \le m \le n$ and $t \in [0,1]$ as in the proof of Theorem 2.1.24. Next, for each $1 \le m \le n$, set

$$\Sigma_{n,m} = \sqrt{\Sigma_n^2 - \sigma_m^2}, \quad \hat{\tau}_m = \frac{\tau_m}{\Sigma_n}, \quad \rho_n = \sum_1^n \hat{\tau}_m^3, \quad \text{and } \rho_{n,m} = \sum_{\substack{1\le \ell\le n\\ \ell\ne m}} \left(\frac{\tau_\ell}{\Sigma_{n,m}}\right)^3.$$

Finally, set

$$S_{n,m} = \sum_{\substack{1\le \ell\le n\\ \ell\neq m}} X_\ell \quad \text{and} \quad \hat{S}_{n,m} = \frac{S_{n,m}}{\Sigma_{n,m}},$$

and let $x \in \mathbb{R} \longmapsto F_{n,m}(x) \equiv P(\hat{S}_{n,m} \le x) \in [0,1]$ denote the distribution of $\hat{S}_{n,m}$. Notice that, by definition, $\|F_{n,m} - G\|_{\mathrm{u}} \le \beta_{n-1}\rho_{n,m}$ for each $1 \le m \le n$. Furthermore, because (cf. (2.1.4))

$$\frac{\Sigma_{n,m}^2}{\Sigma_n^2} = 1 - \hat{\sigma}_m^2 \ge 1 - r_n^2 \quad \text{and} \quad \rho_{n,m} \le \left(\frac{\Sigma_n}{\Sigma_{n,m}}\right)^3 \rho_n,$$

we see first that

$$\rho_{n,m} \le \frac{\rho_n}{(1-r_n^2)^{\frac{3}{2}}}, \qquad 1 \le m \le n,$$

and therefore that

$$\max_{1\le m\le n} \|F_{n,m} - G\|_{\mathrm{u}} \le \frac{\rho_n\beta_{n-1}}{(1-r_n^2)^{\frac{3}{2}}}. \tag{2.1.33}$$

Now let $\varphi \in C_{\mathrm{b}}^2(\mathbb{R};\mathbb{R})$ with $\|\varphi''\|_{L^1(\mathbb{R})} < \infty$ be given, define f accordingly as in (2.1.18), and let

$$\{A_m : 1 \le m \le n\} \quad \text{and} \quad \{B_m(t) : 1 \le m \le n \;\&\; t \in [0,1]\}$$

be the associated quantities appearing in (2.1.27). By (2.1.20), we have that

$$\begin{aligned}
|A_m| &\le \left|\mathbb{E}^P\left[\hat{X}_m f(\hat{S}_n)\right]\right| + \left|\mathbb{E}^P\left[\hat{T}_{n,m}(0)\Big(f(\hat{S}_n) - f\big(\hat{T}_{n,m}(0)\big)\Big)\right]\right| \\
&\qquad + \left|\mathbb{E}^P\left[\varphi(\hat{S}_n) - \varphi\big(\hat{T}_{n,m}(0)\big)\right]\right| \\
&\le \mathbb{E}^P\left[|\hat{X}_m|\right]\|f\|_{\mathrm{u}} + \mathbb{E}^P\left[|\hat{X}_m\,\hat{T}_{n,m}(0)|\right]\|f'\|_{\mathrm{u}} \\
&\qquad + \int_0^1 \left|\mathbb{E}^P\left[\hat{X}_m\varphi'\big(\hat{T}_{n,m}(\xi)\big)\right]\right| d\xi \\
&\le \hat{\sigma}_m\left(\|f\|_{\mathrm{u}} + \frac{\Sigma_{n,m}}{\Sigma_n}\|f'\|_{\mathrm{u}}\right) + \max_{\xi\in[0,1]}\left|\mathbb{E}^P\left[\hat{X}_m\varphi'\big(\hat{T}_{n,m}(\xi)\big)\right]\right| \\
&\le \hat{\sigma}_m\Big(\|f\|_{\mathrm{u}} + \|f'\|_{\mathrm{u}}\Big) + \max_{\xi\in[0,1]}\left|\mathbb{E}^P\left[\hat{X}_m\varphi'\big(\hat{T}_{n,m}(\xi)\big)\right]\right|.
\end{aligned}$$

Similarly (from (2.1.20)), one sees that $|B_m(t)|$ is dominated by:

$$
\begin{aligned}
t\Big|\mathbb{E}^P\Big[\hat{X}_m^3 f\big(\hat{T}_{n,m}(t)\big)\Big]\Big| &+ \Big|\mathbb{E}^P\Big[\hat{X}_m^2\hat{T}_{n,m}(0)\Big(f\big(\hat{T}_{n,m}(t)\big)-f\big(\hat{T}_{n,m}(0)\big)\Big)\Big]\Big| \\
&+ \Big|\mathbb{E}^P\Big[\hat{X}_m^2\Big(\varphi\big(\hat{T}_{n,m}(t)\big)-\varphi\big(\hat{T}_{n,m}(0)\big)\Big)\Big]\Big| \\
\le t\mathbb{E}^P\Big[\big|\hat{X}_m\big|^3\Big]\|f\|_{\mathrm{u}} &+ t\mathbb{E}^P\Big[\big|\hat{X}_m\big|^3\Big]\mathbb{E}^P\Big[\big|\hat{T}_{n,m}(0)\big|\Big]\|f'\|_{\mathrm{u}} \\
&+ t\int_0^1\Big|\mathbb{E}^P\Big[\hat{X}_m^3\varphi'\big(\hat{T}_{n,m}(t\xi)\big)\Big]\Big|\,d\xi \\
\le t\hat{\tau}_m^3\Big(\|f\|_{\mathrm{u}}+\|f'\|_{\mathrm{u}}\Big) &+ t\max_{\xi\in[0,1]}\Big|\mathbb{E}^P\Big[\hat{X}_m^3\varphi'\big(\hat{T}_{n,m}(\xi)\big)\Big]\Big|
\end{aligned}
$$

In order to handle the second term in the last line of each of these calculations, we introduce the function

$$
(\xi,\omega,y)\in[0,1]\times\Omega\times\mathbb{R}\longmapsto\psi(\xi,\omega,y)\equiv\varphi'\left(\xi\hat{X}_m(\omega)+\frac{\Sigma_{n,m}}{\Sigma_n}\,y\right).
$$

Next, using the independence of $\hat{X}_m$ and $\hat{T}_{n,m}(0)$, note that

$$
\begin{aligned}
&\left|\mathbb{E}^P\Big[\hat{X}_m^k\varphi'\big(\hat{T}_{n,m}(\xi)\big)\Big]-\int_\Omega\hat{X}_m(\omega)^k\left(\int_{\mathbb{R}}\psi(\xi,\omega,y)\,\gamma(dy)\right)P(d\omega)\right| \\
&\le\int_\Omega\big|\hat{X}_m(\omega)\big|^k\left|\int_{\mathbb{R}}\psi(\xi,\omega,y)\,dF_{n,m}(y)-\int_{\mathbb{R}}\psi(\xi,\omega,y)\,dG(y)\right|P(d\omega) \\
&=\int_\Omega\big|\hat{X}_m(\omega)\big|^k\left|\int_{\mathbb{R}}\psi'(t,\omega,y)\big(G(y)-F_{n,m}(y)\big)\,dy\right|P(d\omega) \\
&\le\frac{\beta_{n-1}\rho_n}{(1-r_n^2)^{\frac{3}{2}}}\mathbb{E}^P\Big[\big|\hat{X}_m\big|^k\Big]\|\varphi''\|_{L^1(\mathbb{R})}\le\frac{\hat{\tau}_m^k\beta_{n-1}\|\varphi''\|_{L^1(\mathbb{R})}\rho_n}{(1-r_n^2)^{\frac{3}{2}}},\qquad k\in\{1,3\},
\end{aligned}
$$

where we have used $\psi'(t,\omega,y)$ to denote the first derivative of $y\in\mathbb{R}\longmapsto\psi(\xi,\omega,y)$, applied (2.1.16) and (2.1.33), and noted that, for all $(\xi,\omega)\in[0,1]\times\Omega$, $\|\psi'(\xi,\omega,\cdot)\|_{L^1(\mathbb{R})}=\|\varphi''\|_{L^1(\mathbb{R})}$. At the same time, because

$$
\|\psi(\xi,\omega,\,\cdot\,)\|_{L^1(\mathbb{R})}=\frac{\Sigma_n}{\Sigma_{n,m}}\|\varphi'\|_{L^1(\mathbb{R})}\quad\text{for all }(\xi,\omega)\in[0,1]\times\Omega,
$$

we have that, for each $\xi\in[0,1]$,

$$
\left|\int_\Omega\hat{X}_m(\omega)^k\left(\int_{\mathbb{R}}\psi(\xi,\omega,y)\,\gamma(dy)\right)P(d\omega)\right|\le\frac{\|\varphi'\|_{L^1(\mathbb{R})}\,\hat{\tau}_m^k}{\big(2\pi(1-r_n^2)\big)^{\frac{1}{2}}}.
$$

Hence, by combining these estimates, we arrive at

$$|A_m| \le \hat{\tau}_m \left(\|f\|_{\mathrm{u}} + \|f'\|_{\mathrm{u}} + \frac{\|\varphi'\|_{L^1(\mathbb{R})}}{\left(2\pi(1-r_n^2)\right)^{\frac{1}{2}}} + \frac{\beta_{n-1}\rho_n}{(1-r_n^2)^{\frac{3}{2}}} \|\varphi''\|_{L^1(\mathbb{R})} \right)$$

and

$$|B_m(t)| \le t\hat{\tau}_m^3 \left(\|f\|_{\mathrm{u}} + \|f'\|_{\mathrm{u}} + \frac{\|\varphi'\|_{L^1(\mathbb{R})}}{\left(2\pi(1-r_n^2)\right)^{\frac{1}{2}}} + \frac{\beta_{n-1}\rho_n}{(1-r_n^2)^{\frac{3}{2}}} \|\varphi''\|_{L^1(\mathbb{R})} \right)$$

for all $1 \le m \le n$ and $t \in [0,1]$. Thus, after putting these together with (2.1.16) and (2.1.27), we conclude that

$$\begin{aligned} &\left| \int_{\mathbb{R}} \varphi'(y)\big(G(y) - F_n(y)\big)\,dy \right| \\ (2.1.34) \qquad &\le \frac{3}{2}\Bigg(\|f\|_{\mathrm{u}} + \|f'\|_{\mathrm{u}} \\ &\qquad + \frac{\|\varphi'\|_{L^1(\mathbb{R})}}{\left(2\pi(1-r_n^2)\right)^{\frac{1}{2}}} + \frac{\beta_{n-1}\|\varphi''\|_{L^1(\mathbb{R})}\rho_n}{(1-r_n^2)^{\frac{3}{2}}} \Bigg)\rho_n. \end{aligned}$$

We next apply (2.1.34) to a special class of φ's. Namely, set

$$h(x) = \begin{cases} 1 & \text{if } \ x < 0 \\ 1 - x & \text{if } \ x \in [0,1] \\ 0 & \text{if } \ x > 1, \end{cases}$$

and define

$$h_\epsilon(x) = \epsilon^{-1}\int_{\mathbb{R}} \eta(\epsilon^{-1}y)h(x-y)\,dy \quad \text{for } \epsilon > 0 \text{ and } x \in \mathbb{R},$$

where $\eta \in C_c^\infty\big(\mathbb{R};[0,\infty)\big)$ satisfies $\int_{\mathbb{R}} \eta(y)\,dy = 1$. Finally, let $a \in \mathbb{R}$ be given, and set

$$\varphi_{\epsilon,L}(x) = h_\epsilon\left(\tfrac{x-a}{L\rho_n}\right), \quad x \in \mathbb{R} \text{ and } \epsilon,\ L > 0.$$

It is then an easy matter to check that each $\varphi_{\epsilon,L}$ satisfies the hypotheses of Lemma 2.1.29 and that $\|\varphi'_{\epsilon,L}\|_{L^1(\mathbb{R})} = 1$ while $\|\varphi''_{\epsilon,L}\|_{L^1(\mathbb{R})} \le \frac{2}{L\rho_n}$. Hence, by plugging the estimates from Lemma 2.1.29 into (2.1.34) and then letting $\epsilon \searrow 0$, we find that, for each $L > 0$,

$$\begin{aligned} &\sup_{a\in\mathbb{R}} \left| \frac{1}{L\rho_n}\int_a^{a+L\rho_n} \big(G(y) - F_n(y)\big)\,dy \right| \\ (2.1.35) \qquad &\le \frac{3}{2}\left(3 + \frac{1}{\left(2\pi(1-r_n^2)\right)^{\frac{1}{2}}} + \frac{2\beta_{n-1}}{(1-r_n^2)^{\frac{3}{2}}L} \right)\rho_n. \end{aligned}$$

But

$$\frac{1}{L\rho_n}\int_{a-L\rho_n}^{a} F_n(y)\,dy \le F_n(a) \le \frac{1}{L\rho_n}\int_a^{a+L\rho_n} F_n(y)\,dy,$$

while

$$0 \le \frac{1}{L\rho_n}\int_a^{a+L\rho_n} G(y)\,dy - G(a) = \frac{1}{L\rho_n}\int_a^{a+L\rho_n} (a+L\rho_n - y)\,\gamma(dy) \le \frac{L\rho_n}{\sqrt{8\pi}},$$

and, similarly,

$$0 \le G(a) - \int_{a-L\rho_n}^{a} G(y)\,dy \le \frac{L\rho_n}{\sqrt{8\pi}}.$$

Thus, from (2.1.35), we first obtain, for each $L \in (0,\infty)$,

$$\|F_n - G\|_{\mathrm{u}} \le \left(\frac{9}{2} + \frac{3}{\left(8\pi(1-r_n^2)\right)^{\frac12}} + \frac{3\beta_{n-1}}{(1-r_n)^{\frac32}L} + \frac{L}{(8\pi)^{\frac12}}\right)\rho_n,$$

and then, after minimizing with respect to $L \in (0,\infty)$,

$$\begin{aligned}\|F_n - G\|_{\mathrm{u}} \le \Bigg(\frac{9}{2} + \left(\frac{9}{8\pi}\right)^{\frac12}&\left(1-r_n^2\right)^{-\frac12}\\ &+ \left(\frac{18}{\pi}\right)^{\frac14}\beta_{n-1}^{\frac12}\left(1-r_n^2\right)^{-\frac34}\Bigg)\rho_n.\end{aligned} \tag{2.1.36}$$

In order to complete the proof starting from (2.1.36), we have to consider the two cases determined by whether $r_n \ge \frac{1}{10}$ or $r_n < \frac{1}{10}$. Because $\rho_n \ge r_n$, in the first case it is trivial to check that $\|F_n - G\|_{\mathrm{u}} \le 10\rho_n$. On the other hand, if we assume that $\beta_{n-1} \le 10$ and that $r_n < \frac{1}{10}$, then (2.1.36) says that

$$\|F_n - G\|_{\mathrm{u}} \le \left(4.5 + .6\left(1+\frac{1}{63}\right)^{\frac12} + 4.8\left(1+\frac{1}{63}\right)^{\frac34}\right)\rho_n \le 10\rho_n.$$

Hence, in either case,

$$\beta_{n-1} \le 10 \implies \beta_n \le 10\rho_n. \quad \square$$

2.1.37 Remark. It is clear from the preceding derivation (in particular, the final step) that the constant 10 appearing in (2.1.31) and (2.1.32) can be replaced by the unique $\beta \ge 4.5$ which satisfies the equation

$$\beta = 4.5 + \frac{3\beta}{\left(8\pi(\beta^2-1)\right)^{\frac12}} + \left(\frac{18\beta^8}{\pi(\beta^2-1)^3}\right)^{\frac14}.$$

Numerical experimentation indicates that 10 is quite a good approximation to the actual solution to this equation. However, it should be recognized that, with sufficient diligence and entirely different techniques, one can show that the 10 in (2.1.31) can, in fact, be replaced by a number which is less than 1. Thus, we do not claim that Stein's method gives the best result, only that it gives whatever it gives with relatively little pain.

In this connection, it is important to remark that, at least qualitatively, one cannot do better than Berry-Esseen. Indeed, consider independent, standard Bernoulli random variables, and define F_n accordingly. Next, set $t_n = -(2n+1)^{-\frac{1}{2}}$ and observe that

$$F_{2n+1}(t_n) - G(t_n) = \frac{1}{\sqrt{2\pi}} \int_{t_n}^{0} e^{-\frac{x^2}{2}}\, dx$$

and therefore that

$$\varlimsup_{n\to\infty} n^{\frac{1}{2}} \|F_n - G\|_{\mathrm{u}} \geq \frac{1}{\sqrt{2\pi}}.$$

In particular, since $\tau_m = 1$ for these Bernoulli random variables, we conclude that the constant in the Berry–Esseen estimate cannot be smaller than $(2\pi)^{-\frac{1}{2}}$.

Exercises

2.1.38 Exercise: Let $\{X_n\}_1^\infty$ be a sequence of independent, identically distributed random variables, define $\{\hat{S}_n : n \in \mathbb{Z}^+\}$ accordingly, and assume that

$$\varlimsup_{n\to\infty} \mathbb{E}^P\left[\hat{S}_n^2 \wedge R^2\right] \leq 1 \quad \text{for every } R \in [0,\infty).$$

In particular, this will certainly be the case whenever (2.1.2) holds for every $\varphi \in C_c(\mathbb{R};\mathbb{R})$. The purpose of this exercise is to show that the X_n's are square P-integrable, have mean-value 0, and variance no more than 1; and the method which we will use is based on the same line of reasoning as was given in Exercise 1.5.15.

(i) Assuming that $X_1 \in L^2(P)$, show that $\mathbb{E}^P[X_1] = 0$ and $\mathbb{E}^P[X_1^2] \leq 1$. In particular, use this together with the result in part **(i)** of Exercise 1.5.14 to see that it suffices to handle the case when the X_n's are symmetric.

(ii) In this, and the succeeding parts of this exercise, we will be assuming that the X_n's are symmetric. Following the same route as we took in **(ii)** of Exercise

1.5.15, we set

$$\check{X}_n^t = X_n \mathbf{1}_{[0,t]}\big(|X_n|\big) - X_n \mathbf{1}_{(t,\infty)}\big(|X_n|\big), \quad n \in \mathbb{Z}^+,$$

and recall that

$$\Big(\check{X}_1^t, \dots, \check{X}_n^t, \dots\Big) \quad \text{and} \quad \Big(X_1, \dots, X_n, \dots\Big)$$

have the same distribution for each $t \in (0,\infty)$. Use this together with our basic assumption to see that

$$\lim_{R\to\infty} \sup_{\substack{n\in\mathbb{Z}^+ \\ t\in(0,\infty)}} P\Big(A_n(t,R)\Big) = 0,$$

where

$$A_n(t,R) \equiv \left\{ \left|\sum_1^n X_k\right| \vee \left|\sum_1^n \check{X}_k^t\right| \geq n^{\frac{1}{2}} R \right\}.$$

(iii) Continuing in the setting of part **(ii)**, set

$$\widehat{S_n^t} = \frac{1}{n^{\frac{1}{2}}} \sum_1^n X_k \mathbf{1}_{[0,t]}\big(|X_k|\big).$$

After noting that the $X_n \mathbf{1}_{[0,t]}\big(|X_n|\big)$'s are symmetric, check that

$$\mathbb{E}^P\Big[\big|\widehat{S_n^t}\big|^2\Big] \leq t^2,$$

and use this and induction on $n \in \mathbb{Z}^+$ to see (cf. the proof of Theorem 1.3.1) that

$$\mathbb{E}^P\Big[\big|\widehat{S_n^t}\big|^4\Big] \leq 3t^4.$$

In particular, conclude that, for each $t \in (0,\infty)$, there is an $R(t) \in (0,\infty)$ such that

$$\mathbb{E}^P\Big[\big|\widehat{S_n^t}\big|^2, A_n\big(t, R(t)\big)\Big] \leq 3^{\frac{1}{2}} t^2 P\Big(A_n\big(t, R(t)\big)\Big)^{\frac{1}{2}} \leq 1$$

for all $n \in \mathbb{Z}^+$.

(iv) Given $t \in (0,\infty)$, choose $R(t) \in (0,\infty)$ as in the preceding. Taking into account the identity

$$\widehat{S_n^t} = \frac{\sum_1^n X_k + \sum_1^n \check{X}_k^t}{2n^{\frac{1}{2}}},$$

show that

$$\mathbb{E}^P\left[X_1^2,\, |X_1| \le t\right] = \mathbb{E}^P\left[\left|\widehat{S_n^t}\right|^2\right] \le \mathbb{E}^P\left[\left|\widehat{S_n^t}\right|^2,\, A_n\big(t, R(t)\big)\complement\right] + 1$$
$$\le \mathbb{E}^P\left[\hat{S}_n^2 \wedge R(t)^2\right] + 1$$

for all $n \in \mathbb{Z}^+$ and $t \in (0,\infty)$. In particular, use this and our basic hypothesis to conclude first that

$$\mathbb{E}^P\left[X_1^2,\, |X_1| \le t\right] \le 2$$

for all $t \in (0,\infty)$ and then that X_1 is square P-integrable.

(v) After combining the preceding with the Central Limit Theorem, we see that, in the case of independent, identically distributed random variables, X_1 is square P-integrable with $\mathbb{E}^P[X_1] = 0$ and $\mathbb{E}^P\left[X_1^2\right] \le \sigma^2$ if and only if

$$\varlimsup_{n\to\infty} \mathbb{E}^P\left[\hat{S}_n^2 \wedge R^2\right] \le \sigma^2 \quad \text{for all } R \in (0,\infty).$$

2.1.39 Exercise: An interesting way in which to interpret The Central Limit Theorem is as the solution to a certain *fixed point problem.* Namely, let $\mathcal{P}$ denote the set of probability measures μ on $(\mathbb{R}; \mathcal{B}_{\mathbb{R}})$ with the properties that

$$\int_{\mathbb{R}} x^2\, \mu(dx) = 1 \quad \text{and} \quad \int_{\mathbb{R}} x\, \mu(dx) = 0.$$

Next, define $T\mu$ for $\mu \in \mathcal{P}$ to be the probability measure on $(\mathbb{R}, \mathcal{B}_{\mathbb{R}})$ given by

$$T\mu(\Gamma) = \iint_{\mathbb{R}^2} \mathbf{1}_\Gamma\left(\frac{x+y}{\sqrt{2}}\right) \mu(dx)\mu(dy) \quad \text{for} \quad \Gamma \in \mathcal{B}_{\mathbb{R}}.$$

After checking that T maps $\mathcal{P}$ into itself, use The Central Limit Theorem to show that, for every $\mu \in \mathcal{P}$,

$$\lim_{n\to\infty} \int_{\mathbb{R}} \varphi\, dT^n\mu = \int_{\mathbb{R}} \varphi\, d\gamma, \quad \varphi \in C_b(\mathbb{R}; \mathbb{C}).$$

Conclude, in particular, that γ is the one and only element μ of $\mathcal{P}$ with the property that $T\mu = \mu$ and that this fixed point is *attracting.*

2.1.40 Exercise: An important result which is closely related to The Central Limit Theorem is the following observation, which occupies a central position in the development of classical statistical mechanics.

For each $n \in \mathbb{Z}^+$, let λ_n denote the normalized surface measure on the $n-1$ dimensional sphere

$$\mathbf{S}^{n-1}(\sqrt{n}) = \{\mathbf{x} \in \mathbb{R}^n : |\mathbf{x}| = n^{\frac{1}{2}}\},$$

and denote by $\lambda_n^{(1)}$ the distribution of the coordinate x_1 under λ_n. Check that, when $n \geq 2$, $\mu_n(dt) = f_n(t)\,dt$, where

$$f_n(t) = \frac{\omega_{n-2}}{n^{\frac{1}{2}}\omega_{n-1}}\left(1 - \frac{t^2}{n}\right)^{\frac{n-3}{2}} \mathbf{1}_{(-1,1)}\big(n^{-\frac{1}{2}}t\big),$$

and ω_{k-1} denotes the surface area of the $(k-1)$-dimensional unit sphere in $\mathbb{R}^k$. Using polar coordinates to compute the right-hand side of

$$(2\pi)^{\frac{k}{2}} = \int_{\mathbb{R}^k} e^{-\frac{|\mathbf{x}|^2}{2}}\,dx,$$

first check that

$$\omega_{k-1} = \frac{2\pi^{\frac{k}{2}}}{\Gamma\big(\frac{k}{2}\big)},$$

where $\Gamma(t)$ is Euler's Γ-function (cf. (1.3.22)), and then apply Stirling's formula (cf. Exercise 1.3.23) to see that

$$\frac{\omega_{n-2}}{n^{\frac{1}{2}}\omega_{n-1}} \longrightarrow \frac{1}{\sqrt{2\pi}} \quad \text{as} \quad n \to \infty.$$

Now, using γ to denote the density for the standard Gauss distribution (i.e., the Gauss kernel in (1.3.5)), apply these computations to show that

$$\sup_{n\geq 3}\sup_{t\in\mathbb{R}} \frac{f_n(t)}{\gamma(t)} < \infty \quad \text{and that} \quad \frac{f_n(t)}{\gamma(t)} \longrightarrow 1 \text{ uniformly on compacts.}$$

In particular, conclude that, for any $\varphi \in L^1(\gamma;\mathbb{R})$:

$$\int_{\mathbb{R}} \varphi\,d\lambda_n^{(1)} \longrightarrow \int_{\mathbb{R}} \varphi\,d\gamma, \tag{2.1.41}$$

where, this time, γ is the Gauss distribution.†

† Although E. Borel seems to have thought he was the first to discover this result in "Sur les principes de la cinétique des gaz," *Ann. l'École Norm. sup.*, 3e **t. 23**, and probably was the first one to see its significance for statistical mechanics, it appears already in the 1866 article "Über die Entwicklungen einer Funktion von beliebig vielen Variabeln nach Laplaceshen Funktionen höherer Ordnung," *J. Reine u. Angewandte Math.* by F. Mehler. Actually, the preceding is only a small part of what Mehler discovered.

A less computational approach to the same question is the following. Let $X_1, \dots, X_n, \dots$ be a sequence of independent $\mathfrak{N}(0,1)$ random variables, and set $R_n = \sqrt{X_1^2 + \cdots + X_n^2}$. First note that $P(R_n = 0) = 0$ and then that the distribution of

$$\boldsymbol{\theta}_n \equiv \frac{n^{\frac{1}{2}}(X_1, \dots, X_n)}{R_n}$$

is λ_n. Next, use the Strong Law of Large Numbers to see that $\frac{R_n^2}{n} \longrightarrow 1$ (a.s., P) and conclude that, for any $N \in \mathbb{Z}^+$,

$$\lim_{n\to\infty} \mathbb{E}^P\Big[\varphi\big(\boldsymbol{\theta}_n^{(N)}\big)\Big] = \mathbb{E}^P\Big[\varphi(X_1, \dots, X_N)\Big], \quad \varphi \in C_c(\mathbb{R}^N;\mathbb{R}),$$

where, for $n \geq N$, $\boldsymbol{\theta}_n^{(N)} \in \mathbb{R}^N$ denotes the projection of $\boldsymbol{\theta}_n \in \mathbb{R}^n$ onto its first N coordinates. Conclude that if $\lambda_n^{(N)}$ on $(\mathbb{R}^N, \mathcal{B}_{\mathbb{R}^N})$ denotes the distribution of $\mathbf{x} \longmapsto \mathbf{x}^{(N)} \equiv (x_1, \dots, x_N) \in \mathbb{R}^N$ under λ_n, then

$$\lim_{n\to\infty} \int_{\mathbb{R}^N} \varphi \, d\lambda_n^{(N)} = \int_{\mathbb{R}^N} \varphi \, d\gamma^N \quad \text{for all} \quad \varphi \in C_b(\mathbb{R}^N;\mathbb{C}).$$

In particular, by considering the case when $N = 2$, show that, for any $\varphi \in C_b(\mathbb{R};\mathbb{R})$,

$$\lim_{n\to\infty} \int_{\mathbf{S}^{n-1}(\sqrt{n})} \left(\frac{1}{n}\sum_{k=1}^n \varphi(x_k) - \int_{\mathbb{R}} \varphi \, d\gamma\right)^2 \lambda_n(d\mathbf{x}) = 0. \tag{2.1.42}$$

Notice that the noncomputational argument has the advantage that it immediately generalizes the earlier result to cover $\lambda_n^{(N)}$ for all $N \in \mathbb{Z}^+$, not just $N = 1$ (cf. Exercise 2.2.25). On the other hand, the conclusion is weaker in the sense that convergence of the densities has been replaced by convergence of integrals with bounded continuous integrands and that no estimate on the rate of convergence is provided. More work is required to restore the stronger statements.

When couched in terms of statistical mechanics, this result can be interpreted as a derivation of the Maxwell distribution of velocities for a gas of free particles of mass 2 and having average energy 1.

2.1.43 Exercise: Because the derivation of Theorem 2.1.24 is so elegant and simple, one cannot help wondering whether (2.1.26) cannot be used as the starting point for a proof of the Berry–Esseen estimate in Theorem 2.1.30. Unfortunately, the following naïve idea falls considerably short of the mark.

Let $X_1, \dots, X_n$ satisfy the hypotheses of Theorem 2.1.30. Starting from (2.1.26) and proceeding as we did in the passage from (2.1.35) to (2.1.36), show that for every $L > 0$

$$\|F_n - G\|_{\mathrm{u}} \le \frac{6\sum_1^n \tau_m^3}{L\Sigma_n^3} + \frac{L}{\sqrt{8\pi}},$$

and conclude that

$$\|F_n - G\|_{\mathrm{u}} \le \left(\frac{72}{\pi}\right)^{\frac{1}{4}} \left(\frac{\sum_1^n \tau_m^3}{\Sigma_n^3}\right)^{\frac{1}{2}}.$$

Obviously, this is unacceptably poor when $\Sigma_n^{-3}\sum_1^n \tau_m^3$ is small.

2.1.44 Exercise: The most frequently encountered applications of Stirling's formula (cf. (1.3.23)) are to cases when $t \in \mathbb{Z}^+$. That is, one is usually interested in the formula

$$n! \sim \sqrt{2\pi n}\left(\frac{n}{e}\right)^n. \tag{2.1.45}$$

Here is a derivation of (2.1.45) as an application of the Central Limit Theorem. Namely, take $\{X_n\}_1^\infty$ to be a sequence of independent, random variables with $P(X_n > x) = \exp\big((x+1)^+\big)$, $x \in \mathbb{R}$ for all $n \in \mathbb{Z}^+$. For $n \ge 2$, note that

$$\begin{aligned} P\left(\hat{S}_n \in \left[0, \tfrac{1}{4}\right]\right) &= \frac{1}{n!}\int_n^{4^{-1}\sqrt{n}+n} x^n e^{-x}\,dx \\ &= \frac{n^{n+\frac{1}{2}}e^{-n}}{n!}\int_0^{\frac{1}{4}} \left(1 + n^{-\frac{1}{2}}y\right)^n e^{-\sqrt{n}\,y}\,dy. \end{aligned}$$

By the Central Limit Theorem,

$$P\left(\hat{S}_n \in \left[0, \tfrac{1}{4}\right]\right) \longrightarrow \frac{1}{\sqrt{2\pi}}\int_0^{\frac{1}{4}} e^{-\frac{x^2}{2}}\,dx.$$

At the same time, an elementary computation shows that

$$\int_0^{\frac{1}{4}} \left(1 + n^{-\frac{1}{2}}y\right)^n e^{-\sqrt{n}\,y}\,dy \longrightarrow \int_0^{\frac{1}{4}} e^{-\frac{x^2}{2}}\,dx,$$

and clearly (2.1.45) follows from these. In fact, if one applies the Berry–Esseen estimate, one finds that

$$\frac{\sqrt{2\pi n}\left(\frac{n}{e}\right)^n}{n!} = 1 + \mathcal{O}\left(n^{-\frac{1}{2}}\right).$$

However, this is last observation is not very interesting since we saw in Exercise 1.3.21 that the true correction term is of order t^{-1}.†

§2.2: Some Extensions of The Central Limit Theorem.

There are various directions in which The Central Limit Theorem can be extended, and in this section we will investigate two of these directions. In the first place we will show that there is natural multidimensional analogue of The Central Limit Theorem, and, secondly, we will show that as one increases the integrability assumptions on the original X_n's one may reduce the growth restrictions on the φ's for which (2.1.2) can be shown to hold.

In carrying out both the extensions mentioned above, we will be using the same general principle as the one on which the proof of Corollary 2.1.9 rests. Namely, because we are dealing here with probability measures (and not arbitrary generalized functions), weak convergence tested against compactly supported, smooth functions is sufficient to imply weak convergence with respect to test functions which are only continuous and reasonably bounded. To be more precise, we give the following general formulation of this principle.

2.2.1 Lemma. *Let $\{\mu_n : n \in \mathbb{Z}^+\}$ be a sequence of probability measures on $(\mathbb{R}^N, \mathcal{B}_{\mathbb{R}^N})$. Further, assume that there is a probability measure μ on $(\mathbb{R}^N, \mathcal{B}_{\mathbb{R}^N})$ to which the μ_n's converge in the sense that*

$$\lim_{n\to\infty} \int_{\mathbb{R}^N} \varphi\, d\mu_n \longrightarrow \int_{\mathbb{R}^N} \varphi\, d\mu \quad \text{for each} \quad \varphi \in C_c^\infty(\mathbb{R}^N; \mathbb{C}).$$

Then, for any $\varphi \in C(\mathbb{R}^N; [0,\infty))$,

$$\int_{\mathbb{R}^N} \varphi(y)\, \mu(dy) \le \varliminf_{n\to\infty} \int_{\mathbb{R}^N} \varphi(y)\, \mu_n(dy). \tag{2.2.2}$$

Moreover, if $\varphi \in C(\mathbb{R}^N; \mathbb{C})$ satisfies

$$\lim_{R\to\infty} \sup_{n\in\mathbb{Z}^+} \int_{|\varphi(y)|\ge R} |\varphi(y)|\, \mu_n(dy) = 0, \tag{2.2.3}$$

† For more information, see, for example, Wm. Feller's discussion of Stirling's formula in his *Introduction to Probability Theory and Its Applications, Vol. I*, J. Wiley Series in Probability and Math. Stat. (1968)

then φ is μ-integrable and

$$\lim_{n\to\infty}\int_{\mathbb{R}^N}\varphi\,d\mu_n=\int_{\mathbb{R}^N}\varphi\,d\mu. \tag{2.2.4}$$

In particular, (2.2.4) *holds for any $\varphi\in C(\mathbb{R}^N;\mathbb{C})$ which satisfies*

$$\sup_{n\in\mathbb{Z}^+}\int_{\mathbb{R}^N}|\varphi|^{1+\alpha}\,d\mu_n<\infty\quad\textit{for some } \alpha\in(0,\infty). \tag{2.2.5}$$

PROOF: We begin by proving (2.2.4) for every $\varphi\in C_b(\mathbb{R}^N;\mathbb{C})$. To this end, choose $\rho\in C_c^\infty\big(B_{\mathbb{R}^N}(0,1);[0,\infty)\big)$ so that $\int_{\mathbb{R}^N}\rho(\mathbf{y})\,d\mathbf{y}=1$, and set

$$\varphi_k(\mathbf{x})=k\int\limits_{|\mathbf{y}|\le k}\rho\big(k(\mathbf{x}-\mathbf{y})\big)\varphi(\mathbf{y})\,d\mathbf{y}\quad\text{for}\quad k\in\mathbb{Z}^+.$$

Clearly, each φ_k is an element of $C_c^\infty(\mathbb{R}^N;\mathbb{C})$ and $\|\varphi_k\|_u\le\|\varphi\|_u$. In addition, $\varphi_k\longrightarrow\varphi$ uniformly on compacts as $k\to\infty$. Thus,

$$\begin{aligned}\overline{\lim_{n\to\infty}}\left|\int_{\mathbb{R}^N}\varphi\,d\mu_n-\int_{\mathbb{R}^N}\varphi\,d\mu\right| &\le \overline{\lim_{k\to\infty}}\;\overline{\lim_{n\to\infty}}\int_{\mathbb{R}^N}|\varphi-\varphi_k|\,d\mu_n\\ &\le 2\|\varphi\|_u\,\overline{\lim_{n\to\infty}}\,\mu_n\big(B_{\mathbb{R}^N}(0,R)\complement\big)\end{aligned}$$

for every $R\in(0,\infty)$. But, for any $R\in(0,\infty)$, we can choose a ψ_R from $C_c^\infty\big(B_{\mathbb{R}^N}(0,R);[0,1]\big)$ so that $\psi_R=1$ on $B_{\mathbb{R}^N}\big(0,\frac{R}{2}\big)$; and therefore,

$$\begin{aligned}\overline{\lim_{n\to\infty}}\,\mu_n\big(B_{\mathbb{R}^N}(0,R)\complement\big) &\le 1-\underline{\lim_{n\to\infty}}\int_{\mathbb{R}^N}\psi_R\,d\mu_n\\ &\le \mu\big(B_{\mathbb{R}^N}\big(0,\tfrac{R}{2}\big)\complement\big)\longrightarrow 0\quad\text{as}\quad R\nearrow\infty.\end{aligned}$$

Hence, we have now proved that (2.2.4) holds for every $\varphi\in C_b(\mathbb{R}^N;\mathbb{C})$.

Given the preceding, the proof of (2.2.2) is easy. Namely, if φ is a nonnegative, continuous function on $\mathbb{R}^N$, set $\varphi_R=\varphi\wedge R$, $R\in(0,\infty)$. Then, by the Monotone Convergence Theorem,

$$\int_{\mathbb{R}^N}\varphi\,d\mu=\lim_{R\nearrow\infty}\int_{\mathbb{R}^N}\varphi_R\,d\mu=\lim_{R\nearrow\infty}\lim_{n\to\infty}\int_{\mathbb{R}^N}\varphi_R\,d\mu_n\le\underline{\lim_{n\to\infty}}\int_{\mathbb{R}^N}\varphi\,d\mu_n.$$

Finally, to prove (2.2.4) for $\varphi\in C(\mathbb{R}^N;\mathbb{C})$ satisfying (2.2.3), it suffices to handle the case when φ is nonnegative. But then, by (2.2.2), we know that φ is μ-integrable. Hence, for any $\epsilon>0$, we can choose an $R\in(0,\infty)$ so that

$$\sup_{n\in\mathbb{Z}^+}\int\big|\varphi(\mathbf{y})-\varphi_R(\mathbf{y})\big|\,\mu_n(d\mathbf{y})\vee\int\big|\varphi(\mathbf{y})-\varphi_R(\mathbf{y})\big|\,\mu(d\mathbf{y})<\epsilon;$$

and so (2.2.4) for φ follows easily from (2.2.4) for the φ_R's. □

As Lemma 2.2.1 makes explicit, to test whether (2.2.4) holds for all $\varphi \in C_b(\mathbb{R}^N;\mathbb{C})$ requires only that we test it for all $\varphi \in C_c^\infty(\mathbb{R}^N;\mathbb{C})$; and, in conjunction with elementary Fourier analysis, this means that we need only test it for φ's which are imaginary exponential functions. To be precise, for a given probability measure μ on $(\mathbb{R}^N, \mathcal{B}_{\mathbb{R}^N})$, the **characteristic function** $\hat{\mu}$ of μ is its Fourier transform given by

$$\hat{\mu}(\boldsymbol{\xi}) = \int_{\mathbb{R}^N} \exp\left[\sqrt{-1}\,(\boldsymbol{\xi},\mathbf{x})_{\mathbb{R}^N}\right]\mu(d\mathbf{x}) \quad \text{for} \quad \boldsymbol{\xi}\in\mathbb{R}^N. \tag{2.2.6}$$

Also, if $\varphi \in L^1(\mathbb{R}^N;\mathbb{C})$, we will use

$$\hat{\varphi}(\boldsymbol{\xi}) = \int_{\mathbb{R}^N} \exp\left[\sqrt{-1}\,(\boldsymbol{\xi},\mathbf{x})_{\mathbb{R}^N}\right]\varphi(\mathbf{x})\,d\mathbf{x} \quad \text{for} \quad \boldsymbol{\xi}\in\mathbb{R}^N \tag{2.2.7}$$

to denote its Fourier transform. Obviously, $\hat{\mu}$ is a continuous function which is bounded by 1; and only slightly less obvious is the fact that, for $\varphi \in C_c^\infty(\mathbb{R}^N;\mathbb{C})$, $\hat{\varphi} \in C^\infty(\mathbb{R}^N;\mathbb{C})$ and that $\hat{\varphi}$ as well as all its derivatives are **rapidly decreasing** (i.e., they tend to 0 at infinity faster than $(1+|\mathbf{x}|^2)^{-1}$ to any power).

2.2.8 Lemma. *Let μ be a probability measure on $(\mathbb{R}^N, \mathcal{B}_{\mathbb{R}^N})$. Then, for every $\varphi \in C_b(\mathbb{R}^N;\mathbb{C}) \cap L^1(\mathbb{R}^N;\mathbb{C})$ with $\hat{\varphi} \in L^1(\mathbb{R}^N;\mathbb{C})$:*

$$\int_{\mathbb{R}^N} \varphi\,d\mu = \frac{1}{(2\pi)^N}\int_{\mathbb{R}^N} \hat{\varphi}(\boldsymbol{\xi})\,\hat{\mu}(\boldsymbol{\xi})\,d\boldsymbol{\xi}. \tag{2.2.9}$$

Moreover, given a sequence $\{\mu_n : n \in \mathbb{Z}^+\}$ of Borel probability measures on $\mathbb{R}^N$, (2.2.4) holds for every $\varphi \in C(\mathbb{R}^N,\mathbb{C})$ satisfying (2.2.3) if and only if

$$\hat{\mu}_n(\boldsymbol{\xi}) \longrightarrow \hat{\mu}(\boldsymbol{\xi}) \quad \text{for every} \quad \boldsymbol{\xi}\in\mathbb{R}^N.$$

(Cf. Exercise 3.1.19 for more information on this subject.)

PROOF: Choose $\rho \in C_c^\infty(\mathbb{R}^N;[0,\infty))$ to be an even function which satisfies $\int_{\mathbb{R}^N}\rho\,d\mathbf{x} = 1$, and set $\rho_\epsilon(\mathbf{x}) = \epsilon^{-N}\rho(\epsilon^{-1}\mathbf{x})$ for $\epsilon \in (0,\infty)$. Next, define ψ_ϵ for $\epsilon \in (0,\infty)$ by

$$\psi_\epsilon(\mathbf{x}) = \int_{\mathbb{R}^N}\rho_\epsilon(\mathbf{x}-\mathbf{y})\,\mu(d\mathbf{y}) \quad \text{for} \quad \mathbf{x}\in\mathbb{R}^N.$$

It is then an easy matter to check that $\psi_\epsilon \in C_b(\mathbb{R}^N;\mathbb{R})$ and $\|\psi_\epsilon\|_{L^1(\mathbb{R}^N)} = 1$ for every $\epsilon \in (0,\infty)$. In addition, one sees (by Fubini's Theorem) that $\hat{\psi}_\epsilon(\boldsymbol{\xi}) = \hat{\rho}(\epsilon\boldsymbol{\xi})\hat{\mu}(\boldsymbol{\xi})$. Thus, for any $\varphi \in C_b(\mathbb{R}^N;\mathbb{C})\cap L^1(\mathbb{R}^N;\mathbb{C})$, Fubini's Theorem followed by the classical Parseval identity (cf. Exercise 2.3.36 below) yields

$$\int_{\mathbb{R}^N}\varphi_\epsilon\,d\mu = \int_{\mathbb{R}^N}\varphi(\mathbf{x})\,\psi_\epsilon(\mathbf{x})\,d\mathbf{x} = \frac{1}{(2\pi)^N}\int_{\mathbb{R}^N}\hat{\rho}(\epsilon\boldsymbol{\xi})\,\hat{\varphi}(\boldsymbol{\xi})\,\hat{\mu}(\boldsymbol{\xi})\,d\boldsymbol{\xi},$$

where $\varphi_\epsilon \equiv \rho_\epsilon \bigstar \varphi$ is the convolution of ρ_ϵ with φ. Since, as $\epsilon \searrow 0$, $\varphi_\epsilon \longrightarrow \varphi$ while $\hat{\rho}(\epsilon\, \boldsymbol{\xi}) \longrightarrow 1$ boundedly and pointwise, (2.2.9) now follows from Lebesgue's Dominated Convergence Theorem.

Turning to the second part of the theorem, note that the *only if* assertion is trivial and that, by Lemma 2.2.1, we need only check (2.2.4) when $\varphi \in C_c^\infty(\mathbb{R}^N;\mathbb{C})$. But, for such a φ, $\hat{\varphi}$ is smooth and rapidly decreasing, and therefore the result follows immediately from the first part of the present theorem together with Lebesgue's Dominated Convergence Theorem. □

The observation made in Lemma 2.2.8 enables one to capitalize on properties of the μ_n's which come from properties of Euclidian space. For example, here is a simple proof of The Central Limit Theorem in the case when the random variables are identically distributed. Namely, if μ_n is the distribution of $\hat{S}_n$, then

$$\hat{\mu}_n(\xi) = \left(\hat{\mu}\left(\tfrac{\xi}{\sqrt{n}}\right)\right)^n = \left(1 - \frac{\xi^2}{2n} + o\left(\tfrac{1}{n}\right)\right)^n \longrightarrow e^{-\frac{\xi^2}{2}} = \hat{\gamma}(\xi)$$

for every $\xi \in \mathbb{R}$. Actually, as we are about to see, a slight variation on the preceding will allow us to lift the results which we already have for real-valued random variables to random variables with values in $\mathbb{R}^N$. However, before we can state this result, we must introduce the analogs of the mean-value and variance for vector-valued random variables. Thus, given a P-integrable, $\mathbb{R}^N$-valued random variable $\mathbf{X}$ on the probability space $(\Omega, \mathcal{F}, P)$, the **mean-value** $\mathbb{E}^P[\mathbf{X}]$ of $\mathbf{X}$ is that $\mathbf{m} \in \mathbb{R}^N$ which is determined by the property that

$$(\boldsymbol{\xi}, \mathbf{m})_{\mathbb{R}^N} = \mathbb{E}^P\left[(\boldsymbol{\xi}, \mathbf{X})_{\mathbb{R}^N}\right] \quad \text{for all} \quad \boldsymbol{\xi} \in \mathbb{R}^N.$$

Similarly, if $\mathbf{X}$ is square P-integrable, then the **covariance** $\mathbf{cov}(\mathbf{X})$ of $\mathbf{X}$ is the $N \times N$-matrix $\mathbf{C}$ determined by

$$(\boldsymbol{\xi}, \mathbf{C}\,\boldsymbol{\eta})_{\mathbb{R}^N} = \mathbb{E}^P\left[\left(\boldsymbol{\xi}, \mathbf{X} - \mathbb{E}^P[\mathbf{X}]\right)_{\mathbb{R}^N}\left(\boldsymbol{\eta}, \mathbf{X} - \mathbb{E}^P[\mathbf{X}]\right)_{\mathbb{R}^N}\right] \quad \text{for } \boldsymbol{\xi}, \boldsymbol{\eta} \in \mathbb{R}^N.$$

Notice that $\mathbf{cov}(\mathbf{X})$ is both symmetric and nonnegative: in fact, for each $\boldsymbol{\xi} \in \mathbb{R}^N$, $(\boldsymbol{\xi}, \mathbf{cov}(\mathbf{X})\,\boldsymbol{\xi})_{\mathbb{R}^N}$ is nothing but the variance of $(\boldsymbol{\xi}, \mathbf{X})_{\mathbb{R}^N}$. Finally, given $\mathbf{m} \in \mathbb{R}^N$ and a symmetric, nonnegative $\mathbf{C} \in \mathbb{R}^N \otimes \mathbb{R}^N$, we use $\gamma_{\mathbf{m},\mathbf{C}}$ to denote the probability measure on $\mathbb{R}^N$ which is determined by the property that

$$\text{(2.2.10)} \qquad \int_{\mathbb{R}^N} \varphi\, d\gamma_{\mathbf{m},\mathbf{C}} = \int_{\mathbb{R}^N} \varphi\left(\mathbf{m} + \mathbf{C}^{\frac{1}{2}}\mathbf{y}\right) \gamma^N(d\mathbf{y}), \quad \varphi \in C_b(\mathbb{R}^N;\mathbb{R}),$$

where (cf. Exercise 1.1.14) γ^N is the N-fold product of the measure γ with itself. Clearly, an $\mathbb{R}^N$-valued random variable $\mathbf{Y}$ has distribution $\gamma_{\mathbf{m},\mathbf{C}}$ if and only if, for each $\boldsymbol{\xi} \in \mathbb{R}^N$, $(\boldsymbol{\xi}, \mathbf{Y})_{\mathbb{R}^N}$ is a normal random variable with mean-value

$(\boldsymbol{\xi}, \mathbf{m})_{\mathbb{R}^N}$ and variance $(\boldsymbol{\xi}, \mathbf{C}\boldsymbol{\xi})_{\mathbb{R}^N}$. For this reason, $\gamma_{\mathbf{m},\mathbf{C}}$ is called the **normal or Gaussian distribution** with mean-value $\mathbf{m}$ and covariance $\mathbf{C}$. For the same reason, a random variable with $\gamma_{\mathbf{m},\mathbf{C}}$ as its distribution is called a **normal or Gaussian random variable** with mean-value $\mathbf{m}$ and covariance $\mathbf{C}$, or, more briefly, an $\mathfrak{N}(\mathbf{m}, \mathbf{C})$-random variable.

In the following statements, we will be assuming that $\{\mathbf{X}_n : n \in \mathbb{Z}^+\}$ is a sequence of mutually independent, square P-integrable, $\mathbb{R}^N$-valued random variables on the probability space $(\Omega, \mathcal{F}, P)$. Further, we will assume that, for each $n \in \mathbb{Z}^+$, $\mathbf{X}_n$ has mean-value $\mathbf{0}$ and strictly positive covariance $\mathbf{cov}(\mathbf{X}_n)$. Finally, for $n \in \mathbb{Z}^+$, we set

$$\mathbf{S}_n = \sum_{m=1}^{n} \mathbf{X}_m, \quad \mathbf{C}_n \equiv \mathbf{cov}(\mathbf{S}_n) = \sum_{m=1}^{n} \mathbf{cov}(\mathbf{X}_m),$$

$$\Sigma_n = \big(\det(\mathbf{C}_n)\big)^{\frac{1}{2N}} \text{ and } \hat{\mathbf{S}}_n = \frac{\hat{S}_n}{\Sigma_n}.$$

Notice that when $N = 1$, the above use of the notation Σ_n and $\hat{\mathbf{S}}_n$ is consistent with that of Section 2.1.

With these preparations, we are ready to prove the following multidimensional generalization of Corollary 2.1.9.

2.2.11 Theorem. *Referring to the preceding, assume that the limit*

$$\mathbf{A} \equiv \lim_{n\to\infty} \frac{\mathbf{C}_n}{\Sigma_n^2} \tag{2.2.12}$$

exists and that

$$\lim_{n\to\infty} \frac{1}{\Sigma_n^2} \sum_{m=1}^{n} \mathbb{E}^P\Big[|\mathbf{X}_m|^2,\ |\mathbf{X}_m| \ge \epsilon\Sigma_n\Big] = 0 \quad \textit{for each } \epsilon > 0. \tag{2.2.13}$$

Then, for every $\varphi \in C(\mathbb{R}^N; \mathbb{R})$ which satisfies

$$\sup_{\mathbf{y}\in\mathbb{R}^N} \frac{|\varphi(\mathbf{y})|}{1 + |\mathbf{y}|^2} < \infty \tag{2.2.14}$$

one has that

$$\lim_{n\to\infty} \mathbb{E}^P\Big[\varphi(\hat{\mathbf{S}}_n)\Big] = \int_{\mathbb{R}^N} \varphi \, d\gamma_{\mathbf{0},\mathbf{A}}. \tag{2.2.15}$$

In particular, when the $\mathbf{X}_n$ are uniformly square P-integrable random variables with mean-value $\mathbf{0}$ and common covariance $\mathbf{C}$, then

$$\lim_{n\to\infty} \mathbb{E}^P\left[\varphi\left(\frac{\mathbf{S}_n}{\sqrt{n}}\right)\right] = \int_{\mathbb{R}^N} \varphi \, d\gamma_{\mathbf{0},\mathbf{C}}$$

for each $\varphi \in C(\mathbb{R}^N; \mathbb{R})$ satisfying (2.2.14).

PROOF: We begin by reducing to the case in which $\mathbf{A}$ is the identity $\mathbf{I}$. To this end, observe that $\mathbf{A}$ is symmetric, nonnegative, and has determinant 1. Thus, we can introduce the random variables $\mathbf{Y}_n = \mathbf{A}^{-\frac{1}{2}}\mathbf{X}_n$, $n \in \mathbb{Z}^+$; and, clearly, when everything is defined with these $\mathbf{Y}_n$'s in place of the original $\mathbf{X}_n$'s, the only change in the hypotheses is that $\mathbf{A}$ gets replaced by $\mathbf{I}$. Hence, without loss in generality, we will assume from now on that $\mathbf{A} = \mathbf{I}$. Equivalently, we are now assuming that

$$\lim_{n\to\infty} \frac{(\boldsymbol{\xi}, \mathbf{C}_n\,\boldsymbol{\xi})_{\mathbb{R}^N}}{\Sigma_n^2} = |\boldsymbol{\xi}|^2, \quad \boldsymbol{\xi} \in \mathbb{R}^N. \tag{2.2.16}$$

Let μ_n denote the distribution of $\hat{\mathbf{S}}_n$. Our first step will be to show that

$$\lim_{n\to\infty} \hat{\mu}_n(\boldsymbol{\xi}) = \widehat{\gamma^N}(\boldsymbol{\xi}), \quad \boldsymbol{\xi} \in \mathbb{R}^N. \tag{2.2.17}$$

Since there is nothing to do when $\boldsymbol{\xi} = \mathbf{0}$, let $\boldsymbol{\xi} \in \mathbb{R}^N \setminus \{\mathbf{0}\}$ be given, and define

$$\mathbf{e} = \tfrac{\boldsymbol{\xi}}{|\boldsymbol{\xi}|} \quad \text{and} \quad \boldsymbol{\xi}_n = \frac{\Sigma_n}{\sqrt{(\mathbf{e}, \mathbf{C}_n\mathbf{e})_{\mathbb{R}^N}}}\,\boldsymbol{\xi}, \quad n \in \mathbb{Z}^+.$$

Then, by Corollary 2.1.9 applied to the real-valued random variables

$$\{(\mathbf{e}, \mathbf{X}_n)_{\mathbb{R}^N},\ n \in \mathbb{Z}^+\},$$

we know that

$$\begin{aligned}\hat{\mu}_n(\boldsymbol{\xi}_n) &= \mathbb{E}^P\left[\exp\left(\frac{\sqrt{-1}\,|\boldsymbol{\xi}|}{\sqrt{(\mathbf{e}, \mathbf{C}_n\mathbf{e})_{\mathbb{R}^N}}}\sum_1^n (\mathbf{e}, \mathbf{X}_m)_{\mathbb{R}^N}\right)\right] \\ &\longrightarrow e^{-\frac{|\boldsymbol{\xi}|^2}{2}} = \widehat{\gamma^N}(\boldsymbol{\xi}) \quad \text{as } n \to \infty.\end{aligned}$$

At the same time, if $\boldsymbol{\theta}_n \equiv \boldsymbol{\xi}_n - \boldsymbol{\xi}$, then, by (2.2.16), $|\boldsymbol{\theta}_n| \longrightarrow 0$ and therefore

$$\begin{aligned}&\left|\hat{\mu}_n(\boldsymbol{\xi}_n) - \hat{\mu}_n(\boldsymbol{\xi})\right|^2 \\ &\quad = \left|\mathbb{E}^P\left[\exp\left[\sqrt{-1}(\boldsymbol{\xi}_n, \hat{\mathbf{S}}_n)_{\mathbb{R}^N}\right] - \exp\left[\sqrt{-1}(\boldsymbol{\xi}, \hat{\mathbf{S}}_n)_{\mathbb{R}^N}\right]\right]\right|^2 \\ &\quad \le \mathbb{E}^P\left[\left|1 - \exp\left[\sqrt{-1}\,(\boldsymbol{\theta}_n, \hat{S}_n)_{\mathbb{R}^N}\right]\right|^2\right] \\ &\quad = 2\mathbb{E}^P\left[1 - \cos(\boldsymbol{\theta}_n, \hat{S}_n)_{\mathbb{R}^N}\right] \le \frac{(\boldsymbol{\theta}_n, \mathbf{C}_n\boldsymbol{\theta}_n)_{\mathbb{R}^N}}{\Sigma_n^2} \longrightarrow 0\end{aligned}$$

as $n \to \infty$. Hence, we have now proved (2.2.17); and therefore, by Lemma 2.2.8, (2.2.15) holds for all $\varphi \in C(\mathbb{R}^N; \mathbb{R})$ which satisfy (2.2.3).

In view of the preceding paragraph, all that remains is to check that (2.2.3) holds when φ satisfies (2.2.14). Thus, let $\varphi \in C(\mathbb{R}^N;\mathbb{R})$ satisfying $|\varphi(\mathbf{y})| \le 1+|\mathbf{y}|^2$, $\mathbf{y}\in\mathbb{R}^N$, be given. For $\epsilon \in (0,\infty)$, choose $R_\epsilon \in (0,\infty)$ so that

$$\int_{|t|\ge R_\epsilon} (1+t^2)\,\gamma(dt) < \frac{\epsilon}{2N^2}.$$

Next, choose $\eta \in C^\infty\big(\mathbb{R};[0,1]\big)$ so that η equals 0 on $\left[-\frac{1}{2},\frac{1}{2}\right]$ and 1 off of $(-1,1)$, let $\big(\mathbf{e}_1,\dots,\mathbf{e}_N\big)$ be an orthonormal basis in $\mathbb{R}^N$, and set

$$\hat{S}_{n,j} = \frac{1}{\sqrt{\big(\mathbf{e}_j,\mathbf{C}_n\mathbf{e}_j\big)_{\mathbb{R}^N}}}\sum_{m=1}^{n}\big(\mathbf{e}_j,\mathbf{X}_m\big)_{\mathbb{R}^N} \quad \text{for } 1\le j\le N \text{ and } n\in\mathbb{Z}^+.$$

Then, by (2.2.16) and Corollary 2.1.9, we have:

$$\begin{aligned}
&\varlimsup_{n\to\infty}\mathbb{E}^P\Big[1+\big|\hat{\mathbf{S}}_n\big|^2,\ \big|\hat{\mathbf{S}}_n\big|>R\Big]\\
&\quad\le \varlimsup_{n\to\infty}\mathbb{E}^P\left[1+N\max_{1\le j\le N}\big|\hat{S}_{n,j}\big|^2,\ \max_{1\le j\le N}\big|\hat{S}_{n,j}\big|\ge\frac{R}{\sqrt{N}}\right]\\
&\quad\le N\sum_{j=1}^{N}\varlimsup_{n\to\infty}\mathbb{E}^P\left[1+\big|\hat{S}_{n,j}\big|^2,\ \big|\hat{S}_{n,j}\big|\ge\frac{R}{\sqrt{N}}\right]\\
&\quad\le N\sum_{j=1}^{N}\varlimsup_{n\to\infty}\mathbb{E}^P\left[\Big(1+\big|\hat{S}_{n,j}\big|^2\Big)\,\eta\left(\frac{\sqrt{N}\hat{S}_{n,j}}{R}\right)\right]\\
&\quad= N^2\int_{\mathbb{R}^N}(1+t^2)\eta\left(\frac{\sqrt{N}t}{R}\right)\gamma(dt)\le N^2\int_{|t|\ge(4N)^{-\frac{1}{2}}R}(1+t^2)\,\gamma(dt)<\frac{\epsilon}{2}
\end{aligned}$$

for all $R\ge(4N)^{\frac{1}{2}}R_\epsilon$. Since this means that

$$\begin{aligned}
&\varlimsup_{n\to\infty}\mathbb{E}^P\Big[\big|\varphi(\hat{\mathbf{S}}_n)\big|,\ \big|\varphi(\hat{\mathbf{S}}_n)\big|\ge R\Big]\\
&\qquad\le \varlimsup_{n\to\infty}\mathbb{E}^P\Big[1+\big|\hat{\mathbf{S}}_n\big|^2,\ \big|\hat{\mathbf{S}}_n\big|\ge(R-1)^{\frac{1}{2}}\Big]<\frac{\epsilon}{2}
\end{aligned}$$

whenever $R\ge R_\epsilon\equiv 4NR_\epsilon^2+1$, we now know that there is an $n_\epsilon\in\mathbb{Z}^+$ such that

$$\sup_{n\ge n_\epsilon}\mathbb{E}^P\Big[\big|\varphi(\hat{\mathbf{S}}_n)\big|,\ \big|\varphi(\hat{\mathbf{S}}_n)\big|\ge R\Big]<\epsilon$$

for all $R \geq R_\epsilon$. Finally, choose $R \geq R_\epsilon$ so that

$$\max_{1\leq n\leq n_\epsilon} \mathbb{E}^P\left[|\varphi(\hat{\mathbf{S}}_n)|,\ |\varphi(\hat{\mathbf{S}}_n)| \geq R\right] < \epsilon. \quad \square$$

Having obtained a reasonable multidimensional version of The Central Limit Theorem, we will devote the remainder of this section to a proof that when the original random variables are as integrable as a Gaussian random variable, then (2.2.15) will hold as soon as $\log(1+|\varphi|)$ has subquadratic growth at ∞. (See Exercise 2.2.33 below for related results.)

2.2.18 Lemma. *Suppose that* $\mathbf{Y}$ *is a* P*-integrable,* $\mathbb{R}^N$*-valued random variable with mean-value* $\mathbf{0}$*. If*

$$\mathbb{E}^P\left[\exp\left[\alpha|\mathbf{Y}|^2\right]\right] \leq M$$

for some $\alpha \in (0,\infty)$ *and* $M \in (0,\infty)$*, then there is a* $\beta = \beta(\alpha, M) \in (0,\infty)$ *with the property that*

$$\mathbb{E}^P\left[\exp\left((\boldsymbol{\xi},\mathbf{Y})_{\mathbb{R}^N}\right)\right] \leq \exp\left[\frac{\beta|\boldsymbol{\xi}|^2}{2}\right], \quad \boldsymbol{\xi} \in \mathbb{R}^N. \tag{2.2.19}$$

PROOF: First note that, because $\mathbb{E}^P[\mathbf{Y}] = \mathbf{0}$,

$$\mathbb{E}^P\left[\exp\left[(\boldsymbol{\xi},\mathbf{Y})_{\mathbb{R}^N}\right]\right] \leq 1 + \frac{|\boldsymbol{\xi}|^2}{2}\mathbb{E}^P\left[|\mathbf{Y}|^2\right] + \frac{|\boldsymbol{\xi}|^3}{6}\mathbb{E}^P\left[|\mathbf{Y}|^3 e^{|\boldsymbol{\xi}||\mathbf{Y}|}\right].$$

Hence, since

$$\mathbb{E}^P\left[|\mathbf{Y}|^2\right] \leq \tfrac{1}{\alpha}\mathbb{E}^P\left[e^{\alpha|\mathbf{Y}|^2}\right] \leq \frac{M}{\alpha}$$

and

$$\left(\mathbb{E}^P\left[|\mathbf{Y}|^3 e^{|\boldsymbol{\xi}||\mathbf{Y}|}\right]\right)^2 \leq \mathbb{E}^P\left[|\mathbf{Y}|^6\right]\mathbb{E}^P\left[e^{2|\boldsymbol{\xi}||\mathbf{Y}|}\right] \leq \frac{6M^2 e^{\frac{|\boldsymbol{\xi}|^2}{\alpha}}}{\alpha^3},$$

we see that there is a $\delta = \delta(\alpha, M) > 0$ such that

$$\mathbb{E}^P\left[\exp\left[(\boldsymbol{\xi},\mathbf{Y})_{\mathbb{R}^N}\right]\right] \leq 1 + \frac{M|\boldsymbol{\xi}|^2}{\alpha} \leq e^{\frac{M|\boldsymbol{\xi}|^2}{\alpha}}$$

so long as $|\boldsymbol{\xi}| \leq \delta$. On the other hand, for any $\boldsymbol{\xi} \in \mathbb{R}^N$,

$$\begin{aligned}
&\mathbb{E}^P\left[\exp\left[(\boldsymbol{\xi},\mathbf{Y})_{\mathbb{R}^N}\right]\right] \\
&\quad = \mathbb{E}^P\left[\exp\left[(\boldsymbol{\xi},\mathbf{Y})_{\mathbb{R}^N}\right],\ |\boldsymbol{\xi}| < \alpha|\mathbf{Y}|\right] \\
&\qquad\quad + \mathbb{E}^P\left[\exp\left[(\boldsymbol{\xi},\mathbf{Y})_{\mathbb{R}^N},\ |\boldsymbol{\xi}| \geq \alpha|\mathbf{Y}|\right]\right] \\
&\quad \leq \mathbb{E}^P\left[e^{\alpha|\mathbf{Y}|^2}\right] + e^{\frac{|\boldsymbol{\xi}|^2}{\alpha}} \leq M + e^{\frac{|\boldsymbol{\xi}|^2}{\alpha}}.
\end{aligned}$$

Hence, if we choose $\beta = \beta(\alpha, M, \delta) \in (0,\infty)$ so that

$$M + e^{\frac{|\boldsymbol{\xi}|^2}{\alpha}} \leq \exp\left[\frac{\beta|\boldsymbol{\xi}|^2}{2}\right] \quad \text{for } |\boldsymbol{\xi}| \geq \delta,$$

then it is clear that (2.2.19) holds for this choice of β. $\square$

2.2.20 Theorem. *Let $\{\mathbf{X}_n : n \in \mathbb{Z}^+\}$ be a sequence of mutually independent, P-integrable, $\mathbb{R}^N$-valued random variables with mean-value $\mathbf{0}$ and nondegenerate covariance; and define $\mathbf{S}_n$, $\mathbf{C}_n$, Σ_n, and $\hat{\mathbf{S}}_n$ accordingly, as in Theorem 2.2.11. Next, set*

$$\sigma_m = \left(\mathbb{E}^P\left[|\mathbf{X}_m|^2\right]\right)^{\frac{1}{2}}, \quad m \in \mathbb{Z}^+,$$

and assume that

$$\mathbb{E}^P\left[\exp\left(\frac{\alpha}{\sigma_m^2}|\mathbf{X}_m|^2\right)\right] \le M, \quad m \in \mathbb{Z}^+, \tag{2.2.21}$$

for some $\alpha \in (0,\infty)$ and $M \in (0,\infty)$. Then there exists an $\rho = \epsilon(\alpha, M) \in (0,\infty)$ with the property that

$$\sup_{n\in\mathbb{Z}^+} P\left(|\mathbf{S}_n| \ge t\left(\mathrm{Trace}(\mathbf{C}_n)\right)^{\frac{1}{2}}\right) \le 2N\exp\left[-\frac{\epsilon t^2}{2N}\right] \tag{2.2.22}$$

for all $t \in (0,\infty)$. Hence, if, in addition, (2.2.12) holds for some $\mathbf{A}$ and

$$r_n \equiv \max_{1\le m\le n} \frac{\sigma_m}{\Sigma_n} \longrightarrow 0 \quad \text{as } n \to \infty, \tag{2.2.23}$$

then (2.2.15) holds for all $\varphi \in C(\mathbb{R}^N;\mathbb{R})$ which satisfy the condition

$$\lim_{|\mathbf{y}|\to\infty} \frac{\log\left(1+|\varphi(\mathbf{y})|\right)}{|\mathbf{y}|^2} = 0. \tag{2.2.24}$$

In particular, if the $\mathbf{X}_n$'s all have covariance $\mathbf{C}$, then (2.2.21) implies that

$$\lim_{n\to\infty} \mathbb{E}^P\left[\varphi\left(\frac{\mathbf{S}_n}{\sqrt{n}}\right)\right] = \int_{\mathbb{R}^N} \varphi(\mathbf{y})\,\gamma_{\mathbf{0},\mathbf{C}}(d\mathbf{y})$$

for each $\varphi \in C(\mathbb{R}^N;\mathbb{R})$ satisfying (2.2.24).

PROOF: Set

$$T_n \equiv \left(\mathrm{Trace}(\mathbf{C}_n)\right)^{\frac{1}{2}} = \left(\sum_1^n \sigma_m^2\right)^{\frac{1}{2}}.$$

Then, by (2.2.19),

$$\begin{aligned}
&\mathbb{E}^P\left[\exp\left[(\boldsymbol{\xi}, \mathbf{S}_n)_{\mathbb{R}^N}\right]\right] \\
&\quad = \prod_1^n \mathbb{E}^P\left[\exp\left[\left(\sigma_m\boldsymbol{\xi}, \frac{\mathbf{X}_m}{\sigma_m}\right)_{\mathbb{R}^N}\right]\right] \\
&\quad \le \exp\left[\frac{\beta|\boldsymbol{\xi}|^2}{2}\sum_1^n \sigma_m^2\right] = \exp\left[\frac{\beta T_n^2|\boldsymbol{\xi}|^2}{2}\right] \quad \text{for all } \boldsymbol{\xi} \in \mathbb{R}^N.
\end{aligned}$$

Hence, for any $\mathbf{e} \in \mathbf{S}^{N-1}$ and $t \in (0,\infty)$,

$$P\Big((\mathbf{e},\mathbf{S}_n)_{\mathbb{R}^N} \geq t\,T_n\Big) \leq e^{-\lambda t}\mathbb{E}^P\left[\exp\left(\frac{\lambda}{T_n}(\mathbf{e},\mathbf{S}_n)_{\mathbb{R}^N}\right)\right] \leq \exp\left[-\lambda t + \frac{\beta\lambda^2}{2}\right]$$

for every $\lambda \in (0,\infty)$; and so, after minimizing with respect to λ, we arrive at

$$P\Big((\mathbf{e},\mathbf{S}_n)_{\mathbb{R}^N} \geq t\,T_n\Big) \leq \exp\left[-\frac{t^2}{2\beta}\right], \quad t \in (0,\infty) \text{ and } \mathbf{e} \in \mathbf{S}^{N-1}.$$

Finally, choose an orthonormal basis $(\mathbf{e}_1,\dots,\mathbf{e}_N)$ in $\mathbb{R}^N$, and note that, by the preceding,

$$\begin{aligned} P\Big(|\mathbf{S}_n| \geq t\,T_n\Big) &\leq \sum_{j=1}^N P\Big(|(\mathbf{e}_j\mathbf{S}_n)_{\mathbb{R}^N}| \geq N^{-\frac{1}{2}}t\,T_n\Big) \\ &\leq 2N\exp\left[-\frac{t^2}{2N\beta}\right]. \end{aligned}$$

In other words, we can take $\rho = \beta^{-1}$ in (2.2.22).

We now add the assumptions made in the second part of the theorem and begin by checking that (2.2.13) holds. But, by (2.2.21) and (2.2.23),

$$\begin{aligned} \frac{1}{\Sigma_n^2}\sum_1^n \mathbb{E}^P\Big[|\mathbf{X}_m|^2,\ |\mathbf{X}_m| \geq \epsilon\Sigma_n\Big] &\leq \frac{1}{\epsilon^2\Sigma_n^4}\sum_1^n \sigma_m^4\mathbb{E}^P\left[\left|\frac{\mathbf{X}_m}{\sigma_m}\right|^4\right] \\ \leq \frac{1}{\epsilon^2\Sigma_n^4}\sum_1^n \frac{2M\sigma_m^4}{\alpha^2} &\leq \frac{2Mr_n^2}{\epsilon^2\alpha^2} \longrightarrow 0 \quad \text{as } n\to\infty. \end{aligned}$$

Finally, because (2.2.12) implies the existence of a $\kappa \in (0,\infty)$ for which $T_n \leq \kappa\Sigma_n$, $n \in \mathbb{Z}^+$, (2.2.22) leads to

$$\sup_{n\in\mathbb{Z}^+} P\Big(|\hat{\mathbf{S}}_n| \geq t\Big) \leq 2N\exp\left[-\frac{\rho t^2}{2N\kappa^2}\right], \quad t \in (0,\infty);$$

and therefore (2.2.3) certainly holds for any φ which satisfies (2.2.24). □

Exercises

2.2.25 Exercise: Return to the setting of Exercise 2.1.40. After noting that, so long as $\mathbf{e} \in \mathbf{S}^{n-1}$, the distribution of

$$\mathbf{x} \in \mathbf{S}^{n-1}(\sqrt{n}) \longmapsto (\mathbf{e},\mathbf{x})_{\mathbb{R}^n} \in \mathbb{R}$$

is independent of **e**, use Lemma 2.2.8 to prove that the assertion in (2.1.42) follows as a consequence of the one in (2.1.41).

2.2.26 Exercise: Let $Y_1, \dots, Y_n$ be mutually independent, square P-integrable random variables with mean-value 0 and variance 1. Given real numbers $\{a_m\}_1^n$, set

$$S = \sum_1^n a_m Y_m \quad \text{and} \quad A = \left(\sum_1^n a_m^2\right)^{\frac{1}{2}}.$$

Clearly, $\mathbb{E}^P\left[S^2\right] = A^2$. The purpose of this exercise is to examine what can be said about the relation between $\mathbb{E}^P\left[|S|^p\right]$ and A^p for $p \neq 2$. (See Section 6.3 for a significant generalization of these considerations.)

(i) If the Y_m's are normal, note that S is then an $\mathfrak{N}(0, A^2)$-random variable and conclude that

$$\mathbb{E}^P\left[|S|^p\right] = \mu_p A^p, \quad p \in [0, \infty),$$

where

$$\mu_p \equiv \int_{\mathbb{R}} |y|^p \, \gamma(dy) = \sqrt{\tfrac{2^p}{\pi}}\, \Gamma\left(\tfrac{p+1}{2}\right).$$

($\Gamma(t)$ is the quantity defined in (1.3.22).)

(ii) Next, suppose that the Y_m's are **sub-Gaussian** in the sense that

$$\max_{1 \le m \le n} \mathbb{E}^P\left[e^{\xi Y_m}\right] \le \exp\left[\frac{\beta \xi^2}{2}\right], \quad \xi \in \mathbb{R}, \tag{2.2.27}$$

for some $\beta \in (0, \infty)$. (In the case when the Y_m's are $\mathfrak{N}(0, \sigma^2)$-random variables, (2.2.27) is an equality with $\beta = \sigma^2$, and other examples are provided by Lemma 2.2.18.) Show that

$$\mathbb{E}^P\left[e^{\xi S}\right] \le e^{\beta A^2 \xi^2}, \quad \xi \in \mathbb{R},$$

and conclude that

$$P\big(|S| \ge t\big) \le 2 \exp\left[-\frac{t^2}{2\beta A^2}\right], \quad t \in [0, \infty). \tag{2.2.28}$$

(iii) Continuing in the setting of **(ii)**, show that, for each $p \in (0, \infty)$,

$$K_{4-p\wedge 2}^{-1}\big(\beta^{\frac{1}{2}} A\big)^p \le \mathbb{E}^P\left[|S|^p\right] \le K_p\big(\beta^{\frac{1}{2}} A\big)^p \quad \text{with } K_p \equiv p 2^p \Gamma\left(\tfrac{p}{2}\right), \tag{2.2.29}$$

In the case when the Y_m's are symmetric Bernoulli (i.e., $P(Y_m = \pm 1) = \frac{1}{2}$), Exercise 1.3.20 says that $\beta = 1$; and in this special, but important, case, (2.2.29) is called **Khinchine's inequality**.

Hint: First, reduce to the case when $\beta = 1$. Next, starting from (2.2.28) with $\beta = 1$, use (1.4.23) to show that the right-hand side of (2.2.29) holds with the required choice of K_p. To prove the left-hand side of (2.2.29) when $p \in [2,\infty)$, simply apply Jensen's inequality. Finally, to get the left-hand side when $p \in (0,2)$, note that

$$A^2 = \mathbb{E}^P\left[S^2\right] \le \left(\mathbb{E}^P\left[|S|^p\right]\right)^{\frac{1}{2}} \left(\mathbb{E}^P\left[|S|^{4-p}\right]\right)^{\frac{1}{2}}.$$

2.2.30 Exercise: Let $X_1, \dots, X_n$ be mutually independent random variables, and set $S_n = \sum_1^n X_m$.

(i) Assuming that the X_m's are symmetric (cf. Exercise 1.4.29), show that, for each $p \in (0,\infty)$:

$$K_{4-p\wedge 2}^{-1}\mathbb{E}^P\left[\left(\sum_1^n X_m^2\right)^{\frac{p}{2}}\right] \le \mathbb{E}^P\left[|S_n|^p\right] \le K_p \mathbb{E}^P\left[\left(\sum_1^n X_m^2\right)^{\frac{p}{2}}\right], \tag{2.3.31}$$

where K_p is the same as it is in (2.2.29).

Hint: Refer to the beginning of the proof of Lemma 1.5.4 and let $R_1, \dots, R_n$ be the Rademacher functions on $[0,1)$, set $Q = \lambda_{[0,1)} \times P$ on $\left([0,1)\times\Omega, \mathcal{B}_{[0,1)} \times \mathcal{F}\right)$, and observe that

$$\omega \in \Omega \longmapsto S_n(\omega) \equiv \sum_1^n X_m(\omega)$$

has the same distribution under P as

$$(t,\omega) \in [0,1)\times\Omega \longmapsto T_n(t,\omega) \equiv \sum_1^n R_m(t) X_m(\omega)$$

does under Q. Next, apply Khinchine's inequality to see that, for each $\omega \in \Omega$,

$$K_{4-p\wedge 2}^{-1}\left(\sum_1^n X_m(\omega)^2\right)^{\frac{p}{2}} \le \int_{[0,1)} |T_n(t,\omega)|^p\,dt \le K_p \left(\sum_1^n X_m(\omega)^2\right)^{\frac{p}{2}},$$

and complete the proof by taking the P-integral of this with respect to ω.

(ii) Now drop the symmetry assumption made in part **(i)**, but add the assumption that the X_m's are P-integrable and still have mean-value 0. Show that, for each $p \in [2,\infty)$:

$$\mathbb{E}^P\left[|S_n|^p\right] \le C_p \mathbb{E}^P\left[\left(\sum_{m=1}^n X_m^2\right)^{\frac{p}{2}}\right] \quad \text{where } C_p \equiv \left(4K_p^{\frac{1}{p}} + 2^{\frac{1}{2}}\right)^p. \tag{2.2.32}$$

Hint: After augmenting the underlying probability space, if necessary, construct new random variables $X_1',\dots,X_n'$ so that they are independent of the original X_m's but $(X_1',\dots,X_n')$ has the same distribution as $(X_1,\dots,X_n)$. Next, set $S_n'=\sum_1^n X_m'$, and use part (**i**) of Exercise 1.5.14 to see that

$$\mathbb{E}^P\Big[|S_n|^p\Big]^{\frac{1}{p}} \le \Big(2\mathbb{E}^P\Big[|S_n - S_n'|^p\Big]\Big)^{\frac{1}{p}} + |\alpha_n|,$$

where α_n is a median of S_n. As an application of part (**i**) above plus Minkowski's inequality, show that

$$\begin{aligned}\mathbb{E}^P\Big[|S_n - S_n'|^p\Big]^{\frac{1}{p}} &\le \left(K_p\mathbb{E}^P\left[\left(\sum_{m=1}^n (X_m - X_m')^2\right)^{\frac{p}{2}}\right]\right)^{\frac{1}{p}}\\ &\le 2\left(K_p\mathbb{E}^P\left[\left(\sum_{m=1}^n X_m^2\right)^{\frac{p}{2}}\right]\right)^{\frac{1}{p}}.\end{aligned}$$

Finally, note that, by (1.4.14),

$$|\alpha_n|^2 \le 2\sum_{m=1}^n \mathbb{E}^P\big[X_m^2\big] \le 2\mathbb{E}^P\left[\left(\sum_{m=1}^n X_m^2\right)^{\frac{p}{2}}\right]^{\frac{2}{p}}.$$

2.2.33 Exercise: Suppose that $\{\mathbf{X}_n : n \in \mathbb{Z}^+\}$ is a sequence of mutually independent, $\mathbb{R}^N$-valued, P-integrable random variables with mean-value $\mathbf{0}$, and set $\mathbf{S}_n = \sum_{m=1}^m \mathbf{X}_m$. Show that, for each $p\in[2,\infty)$,

$$\mathbb{E}^P\Big[|\mathbf{S}_n|^p\Big] \le C_p N^{\frac{p-2}{2}}\mathbb{E}^P\left[\left(\sum_{m=1}^n |\mathbf{X}_m|^2\right)^{\frac{p}{2}}\right], \tag{2.2.34}$$

where C_p is the same as it was in (2.2.32). Next, assume that $\sigma_m \equiv \mathbb{E}^P\big[|\mathbf{X}_m|^2\big]^{\frac{1}{2}} \in (0,\infty)$ for each $1\le m\le n$, and, starting from (2.2.34), show that

$$\mathbb{E}^P\Big[|\mathbf{S}_n|^p\Big] \le C_p N^{\frac{p-2}{2}}\left(\sum_{m=1}^n \sigma_m^2\right)^{\frac{p}{2}} \max_{1\le m\le n}\mathbb{E}^P\left[\left(\frac{|\mathbf{X}_m|}{\sigma_m}\right)^p\right]. \tag{2.2.35}$$

Finally, add the assumptions that both (2.2.12) and (2.2.23) hold, and suppose that

$$\sup_{m\in\mathbb{Z}^+}\mathbb{E}^P\left[\left(\frac{|\mathbf{X}_m|}{\sigma_m}\right)^p\right] < \infty$$

for some $p \in (2,\infty)$. Show that (2.2.15) holds for every $\varphi \in C(\mathbb{R}^N;\mathbb{R})$ which satisfies

$$\lim_{|\mathbf{y}|\to\infty} \frac{|\varphi(\mathbf{y})|}{1+|\mathbf{y}|^p} = 0.$$

2.2.36 Exercise: Given $h \in L^2(\mathbb{R}^N;\mathbb{C})$, recall that $h^{\star\,(n+2)}$ is a bounded continuous function for each $n \in \mathbb{N}$. Next, assume that $h(-\mathbf{x}) = \overline{h(\mathbf{x})}$ for almost every $\mathbf{x} \in \mathbb{R}^N$ and that $h \equiv 0$ off of $B_{\mathbb{R}^N}(\mathbf{0},1)$. As an application of part **(iii)** in Exercise 1.3.24, show that

$$\left|h^{\star\,(n+2)}(\mathbf{x})\right| \le 2\|h\|^2_{L^2(\mathbb{R}^N)}\|h\|^n_{L^1(\mathbb{R}^N)} \exp\left[-\frac{|\mathbf{x}|^2}{2n}\right].$$

Hint: Note that $h \in L^1(\mathbb{R}^N;\mathbb{C})$, assume that $M \equiv \|h\|_{L^1(\mathbb{R}^N)} > 0$, and define $Af = M^{-1}h\bigstar f$ for $f \in L^2(\mathbb{R}^N;\mathbb{C})$. Show that A is a self-adjoint contraction on $L^2(\mathbb{R}^N;\mathbb{C})$, check that

$$h^{\star\,(n+2)}(\mathbf{x}) = M^n\left(\tau_{\mathbf{x}}h, A^n h\right)_{L^2(\mathbb{R}^N;\mathbb{C})},$$

where $\tau_{\mathbf{x}}h \equiv h(\,\cdot+\mathbf{x})$, and note that

$$\left(\tau_{\mathbf{x}}h, A^\ell h\right)_{L^2(\mathbb{R}^N;\mathbb{C})} = 0 \quad \text{if } \ell \le |\mathbf{x}|.$$

§2.3: An Application to Hermite Multipliers.

This section does not really belong here and should probably be skipped by those readers who want to restrict their attention to purely probabilistic matters. On the other hand, for those who want to see how probability theory interacts with other aspects of mathematical analysis, the present section may prove to be something of a revelation.

The topic of this section will be a class of linear operators called Hermite multipliers, and what will be discussed are certain boundedness properties of these operators. The setting is as follows. Once again, γ will be the standard Gauss measure whose density is given in (1.3.5). Next, for $n \in \mathbb{N}$, define

$$H_n(x) = (-1)^n e^{\frac{x^2}{2}} \frac{d^n}{dx^n}\left(e^{-\frac{x^2}{2}}\right), \quad x \in \mathbb{R}. \tag{2.3.1}$$

Clearly, H_n is an nth order, real, monic (i.e., 1 is the coefficient of the highest order term) polynomial. Moreover, if we define the **raising operator** A_+ on $C^1(\mathbb{R};\mathbb{C})$ by

$$[A_+\varphi](x) = -e^{\frac{x^2}{2}}\frac{d}{dx}\left(e^{-\frac{x^2}{2}}\varphi(x)\right) = -\frac{d\varphi}{dx}(x) + x\varphi(x), \quad x \in \mathbb{R},$$

then

$$H_{n+1} = A_+ H_n \quad \text{for all } n \in \mathbb{N}. \tag{2.3.2}$$

At the same time, if φ and ψ are continuously differentiable functions whose first derivatives are **tempered** (i.e., have at most polynomial growth at infinity), then

$$\left(\varphi, A_+\psi\right)_{L^2(\gamma)} = \left(A_-\varphi, \psi\right)_{L^2(\gamma)}, \tag{2.3.3}$$

where A_- is the **lowering operator** given by $A_-\varphi = \frac{d\varphi}{dx}$. After combining (2.3.2) with (2.3.3), we see that, for all $0 \le m \le n$,

$$\left(H_m, H_n\right)_{L^2(\gamma)} = \left(H_m, A_+^n H_0\right)_{L^2(\gamma)} = \left(A_-^n H_m, H_0\right)_{L^2(\gamma)} = m!\,\delta_{m,n},$$

where, at the last step, we have used the fact that H_m is a monic mth order polynomial. Hence, the (normalized) **Hermite polynomials**

$$\overline{H}_n(x) = \frac{H_n(x)}{\sqrt{n!}} = \frac{(-1)^n}{\sqrt{n!}} e^{\frac{x^2}{2}} \frac{d^n}{dx^n}\left(e^{-\frac{x^2}{2}}\right), \quad x \in \mathbb{R}$$

form an orthonormal set in $L^2(\gamma;\mathbb{C})$. (Indeed, they are one choice of the orthogonal polynomials relative to the Gauss weight.)

2.3.4 Lemma. *For each $\lambda \in \mathbb{C}$, set*

$$H(x;\lambda) = \exp\left[\lambda x - \frac{\lambda^2}{2}\right], \quad x \in \mathbb{R}.$$

Then

$$H(x;\lambda) = \sum_{n=0}^{\infty} \frac{\lambda^n}{n!} H_n(x), \quad x \in \mathbb{R}, \tag{2.3.5}$$

where the convergence is both uniform on compact subsets of $\mathbb{R}\times\mathbb{C}$ and, for λ's in compact subsets of $\mathbb{C}$, uniform in $L^2(\gamma;\mathbb{C})$. In particular, $\left\{\overline{H}_n : n \in \mathbb{N}\right\}$ is an orthonormal basis in $L^2(\gamma;\mathbb{C})$.

PROOF: By (2.3.1) and the Taylor's expansion for the function $e^{-\frac{x^2}{2}}$, it is clear that (2.3.5) holds for each (x,λ) and that the convergence is uniform on compact subsets of $\mathbb{R}\times\mathbb{C}$. Furthermore, because the H_n's are orthogonal, the asserted uniform convergence in $L^2(\gamma;\mathbb{C})$ comes down to checking that

$$\lim_{m\to\infty}\sup_{|\lambda|\le R}\sum_{n=m}^{\infty}\left|\frac{\lambda^n}{n!}\right|^2\|H_n\|^2_{L^2(\gamma)}=0$$

for every $R\in(0,\infty)$, and obviously this follows from our earlier calculation that $\|H_n\|^2_{L^2(\gamma)}=n!$.

To prove the assertion that $\{\overline{H}_n:n\in\mathbb{N}\}$ forms an orthonormal basis, it suffices to check that any $\varphi\in L^2(\gamma;\mathbb{C})$ which is orthogonal to all of the H_n's must be 0. But, because of the $L^2(\gamma)$-convergence in (2.3.5), we would have that

$$\int_{\mathbb{R}}\varphi(x)\,e^{\lambda x}\,\gamma(dx)=0,\quad \lambda\in\mathbb{C},$$

for such a φ. Hence, if we set

$$\psi(x)=\frac{e^{-\frac{x^2}{2}}\varphi(x)}{\sqrt{2\pi}},\qquad x\in\mathbb{R},$$

then $\|\psi\|_{L^1(\mathbb{R})}=\|\varphi\|_{L^1(\gamma)}\le\|\varphi\|_{L^2(\gamma)}<\infty$, and (cf. (2.2.7)) $\hat{\psi}\equiv 0$; which, by the $L^1(\mathbb{R})$-Fourier inversion formula

$$\frac{1}{2\pi}\int_{\mathbb{R}}e^{-\alpha|\xi|}\,e^{-\sqrt{-1}\,x\xi}\hat{\psi}(\xi)\,d\xi\overset{\alpha\searrow 0}{\longrightarrow}\psi\quad\text{in } L^1(\mathbb{R}),$$

means that ψ and therefore φ vanish Lebesgue-almost everywhere. □

Now that we know $\{\overline{H}_n:n\in\mathbb{N}\}$ is an orthonormal basis, we can uniquely determine a normal operator $\mathcal{H}_\theta$ for each $\theta\in\mathbb{C}$ by specifying that

$$\mathcal{H}_\theta H_n=\theta^n\,H_n\quad\text{for each } n\in\mathbb{N}.$$

The operator $\mathcal{H}_\theta$ is called the **Hermite multiplier** with parameter θ, and clearly

$$\mathrm{Dom}(\mathcal{H}_\theta)=\left\{\varphi\in L^2(\gamma;\mathbb{C}):\sum_{n=1}^{\infty}|\theta|^{2n}\left|\left(\varphi,\overline{H}_n\right)_{L^2(\gamma)}\right|^2<\infty\right\}$$

$$\mathcal{H}_\theta\varphi=\sum_{n=0}^{\infty}\theta^n\left(\varphi,\overline{H}_N\right)_{L^2(\gamma)}\overline{H}_n,\quad\varphi\in\mathrm{Dom}(\mathcal{H}_\theta).$$

In particular, $\mathcal{H}_\theta$ is a contraction if and only if θ is an element of the closed unit disk $\mathbf{D}$ in $\mathbb{C}$, and it is unitary precisely when $\theta \in \mathbf{S}^1 \equiv \partial\mathbf{D}$. Also, the adjoint of $\mathcal{H}_\theta$ is $\mathcal{H}_{\overline{\theta}}$, and so it is self-adjoint if and only if $\theta \in \mathbb{R}$.

As we are about to see, there are special choices of θ for which the corresponding Hermite multiplier has interesting alternative interpretations and unexpected additional properties. For example, consider the **Mehler kernel**†

$$M(x,y;\theta) = \frac{1}{\sqrt{1-\theta^2}} \exp\left[-\frac{(\theta x)^2 - 2\theta xy + (\theta y)^2}{2(1-\theta^2)}\right]$$

for $\theta \in (0,1)$ and $x,\, y \in \mathbb{R}$. By a straightforward Gaussian computation (i.e., "complete the square" in the exponential) one can easily check that

$$\int_{\mathbb{R}} H(y;\lambda)\, M(x,y;\theta)\, \gamma(dy) = H(x;\theta\lambda)$$

for all $\theta \in (0,1)$ and $(x,\lambda) \in \mathbb{R}\times\mathbb{C}$. In conjunction with (2.3.5), this means that

$$\mathcal{H}_\theta\varphi = \int_{\mathbb{R}} M(\,\cdot\,,y;\theta)\,\varphi(y)\,\gamma(dy), \quad \theta \in (0,1) \text{ and } \varphi \in L^2(\gamma;\mathbb{C}), \tag{2.3.6}$$

and from here it is not very difficult to prove the following properties of $\mathcal{H}_\theta$ for $\theta \in (0,1)$.

2.3.7 Lemma. *For each $\varphi \in L^2(\gamma;\mathbb{C})$, $(\theta,x) \in (0,1)\times\mathbb{R} \longmapsto \mathcal{H}_\theta\varphi(x) \in \mathbb{C}$ may be chosen to be a continuous function which is nonnegative if $\varphi \geq 0$ Lebesgue-almost everywhere. In addition, for each $\theta \in (0,1)$ and every $p \in [1,\infty]$,*

$$\left\|\mathcal{H}_\theta\varphi\right\|_{L^p(\gamma)} \leq \|\varphi\|_{L^p(\gamma)}. \tag{2.3.8}$$

PROOF: The first assertions are immediate consequences of the representation in (2.3.6). To prove the second assertion, observe that, as a special case of (2.3.6),

$$\int_{\mathbb{R}} M(x,y;\theta)\,\gamma(dy) = 1 \quad \text{for all} \quad \theta \in (0,1) \text{ and } x \in \mathbb{R}.$$

Hence, by (2.3.6) and Jensen's inequality, for any $p \in [1,\infty)$,

$$\left|\left[\mathcal{H}_\theta\varphi\right](x)\right|^p \leq \int_{\mathbb{R}} M(x,y;\theta)\,|\varphi(y)|^p\,\gamma(dy).$$

† This kernel appears in the 1866 article by Mehler referred to in the footnote following (2.1.41). It arises there as the generating function for spherical harmonics on the sphere $\mathbf{S}^\infty\left(\sqrt{\infty}\right)$.

At the same time, by symmetry, $\int_{\mathbb{R}} M(x,y;\theta)\,\gamma(dx) = 1$ for all $(\theta,y) \in (0,1)\times\mathbb{R}$, and therefore

$$\int_{\mathbb{R}} \big|[\mathcal{H}_\theta\varphi](x)\big|^p\,\gamma(dx) \le \iint\limits_{\mathbb{R}\times\mathbb{R}} M(x,y;\theta)\,|\varphi(y)|^p\,\gamma(dx)\gamma(dy) = \int_{\mathbb{R}} |\varphi|^p\,d\gamma.$$

Hence, (2.3.8) is now proved for $p \in [1,\infty)$. The case when $p=\infty$ is even easier and is left to the reader. □

The consequences drawn in Lemma 2.3.7 from the Mehler representation in (2.3.6) are interesting but not very deep (cf. Exercise 2.3.35 below). A deeper fact is the relationship between Hermite multipliers and the Fourier transform. For the purposes of this analysis, it is best to define the **Fourier operator** $\mathcal{F}$ by

$$[\mathcal{F}f](\xi) = \int_{\mathbb{R}} e^{\sqrt{-1}\,2\pi\xi x}\,f(x)\,dx, \quad \xi\in\mathbb{R}, \tag{2.3.9}$$

for $f \in L^1(\mathbb{R};\mathbb{C})$. The advantage of this choice is, that without the introduction of any further factors of $\sqrt{2\pi}$, the Parseval identity (cf. Exercise 2.3.36) becomes the statement that $\mathcal{F}$ determines a unitary operator on $L^2(\mathbb{R};\mathbb{C})$. In order to relate $\mathcal{F}$ to Hermite multipliers, observe that, after analytically continuing the result of another simple Gaussian computation,

$$\frac{1}{\sqrt{2\pi p}}\int_{\mathbb{R}} \exp\left[(\lambda+\sqrt{-1}\,\eta)y - \frac{y^2}{2p}\right]dy = \exp\left[\frac{p}{2}(\lambda+\sqrt{-1}\,\eta)^2\right]$$

for all $p\in(1,\infty)$ and all complex numbers λ and η. Hence, after making the change of variables $y=\sqrt{2\pi p}\,x$ and $\eta = \sqrt{\frac{2\pi}{p}}\,\xi$, we see from (2.3.5) that

$$\begin{aligned}
&\sum_{n=0}^{\infty}\frac{\lambda^n}{n!}\int_{\mathbb{R}} e^{\sqrt{-1}\,2\pi\xi x}H_n\big(\sqrt{2\pi p}\,x\big)e^{-\pi x^2}\,dx\\
&\qquad = e^{-\pi\xi^2}\exp\left[\frac{(p-1)\lambda^2}{2} + \sqrt{-1}\,\lambda\sqrt{2\pi p}\,\xi\right]\\
&\qquad = e^{-\pi\xi^2}\sum_{n=0}^{\infty}\frac{\lambda^n}{n!}\theta_p^n H_n\big(\sqrt{2\pi p'}\,\xi\big),
\end{aligned}$$

where $p' = \frac{p}{p-1}$ is the **Hölder conjugate** of p and $\theta_p \equiv \sqrt{-1}\,(p-1)^{\frac{1}{2}}$. Thus, we have now proved that, for each $p\in(1,\infty)$ and $n\in\mathbb{N}$,

$$\int_{\mathbb{R}} e^{\sqrt{-1}\,2\pi\xi x}H_n\big(\sqrt{2\pi p}\,x\big)e^{-\pi x^2}\,dx = \theta_p^n\,H_n\big(\sqrt{2\pi p'}\,x\big)\,e^{-\pi\xi^2}. \tag{2.3.10}$$

In particular, when $p = 2$, (2.3.10) says that

$$\mathcal{F}h_n = \left(\sqrt{-1}\right)^n h_n, \quad n \in \mathbb{N}, \tag{2.3.11}$$

where h_n is the nth (un-normalized) **Hermite function** given by

$$h_n(x) = H_n\left(2\pi^{\frac{1}{2}}x\right)e^{-\pi x^2}, \quad n \in \mathbb{N} \text{ and } x \in \mathbb{R}. \tag{2.3.12}$$

More generally, (2.3.10) leads to the following relationship between $\mathcal{F}$ and Hermite multipliers. Namely, for each $p \in (1, \infty)$, define $\mathcal{U}_p$ on $L^p(\gamma; \mathbb{C})$ by

$$\left[\mathcal{U}_p\varphi\right](x) = p^{\frac{1}{2p}}\varphi\left((2\pi p)^{\frac{1}{2}}x\right)e^{-\pi x^2}, \quad x \in \mathbb{R}.$$

It is then an easy matter to check that $\mathcal{U}_p$ is an isometric surjection from $L^p(\gamma; \mathbb{C})$ onto $L^p(\mathbb{R}; \mathbb{C})$. In addition, (2.3.10) can now be interpreted as the statement that, for every $p \in (1, \infty)$ and every polynomial φ,

$$\mathcal{U}_{p'}^{-1} \circ \mathcal{F} \circ \mathcal{U}_p\varphi = A_p\mathcal{H}_{\theta_p}\varphi \quad \text{where} \quad A_p \equiv \left(\frac{p^{\frac{1}{p}}}{(p')^{\frac{1}{p'}}}\right)^{\frac{1}{2}}. \tag{2.3.13}$$

(See Exercise 2.3.34 below for more information about A_p.)

With this brief introduction to Hermite multipliers complete, we will now address the problem to which The Central Limit Theorem has something to contribute here. Namely, we want to get a handle on the set of $(\theta, p, q) \in \mathbf{D} \times (0, \infty) \times (0, \infty)$ with the property that $\mathcal{H}_\theta$ determines a contraction from $L^p(\gamma; \mathbb{C})$ into $L^q(\gamma; \mathbb{C})$. In view of the preceding discussion, when $\theta \in (0, 1)$, a solution to this problem has implications for the Mehler transform; and, when $q = p'$, the solution tells us about the Fourier operator. The rôle that The Central Limit Theorem plays in this analysis is hidden in the following beautiful criterion, which was first discovered by Wm. Beckner.†

2.3.14 Theorem (Beckner). *Let $\theta \in \mathbf{D}$ and $1 \le p \le q < \infty$ be given. Then*

$$\left\|\mathcal{H}_\theta\varphi\right\|_{L^q(\gamma)} \le \|\varphi\|_{L^p(\gamma)} \quad \text{for all} \quad \varphi \in L^2(\gamma; \mathbb{C}) \tag{2.3.15}$$

if and only if

$$\left(\frac{|1-\theta\zeta|^q + |1+\theta\zeta|^q}{2}\right)^{\frac{1}{q}} \le \left(\frac{|1-\theta\zeta|^p + |1+\theta\zeta|^q}{2}\right)^{\frac{1}{p}} \tag{2.3.16}$$

for every $\zeta \in \mathbb{C}$.

† See Beckner's "Inequalities in Fourier analysis," *Ann. Math.* **102**.

That (2.3.15) implies (2.3.16) is trivial: simply take

$$\varphi(x) = \begin{cases} 1-\zeta & \text{if } x \in (-\infty, 0) \\ 1+\zeta & \text{if } x \in [0,\infty). \end{cases}$$

On the other hand, the opposite implication is remarkable! Indeed, it takes a problem in infinite dimensional analysis and reduces it to a calculus question about functions on the complex plane. Even though, as we will see later, this reduction leads to highly nontrivial problems in calculus, Theorem 2.3.14 has to be considered a major step toward understanding the contraction properties of Hermite multipliers.

The first step in the proof of Theorem 2.3.14 is to interpret (2.3.16) in operator theoretic language. For this purpose, let β denote the standard Bernoulli probability measure on $(\mathbb{R}, \mathcal{B}_{\mathbb{R}})$. That is, $\beta(\{\pm 1\}) = \frac{1}{2}$. Next, use $\chi_\emptyset$ to denote the function on $\mathbb{R}$ which is constantly equal to 1 and $\chi_{\{1\}}$ to stand for the identity function on $\mathbb{R}$ (i.e., $\chi_{\{1\}}(x) = x$, $x \in \mathbb{R}$). It is then clear that $\chi_\emptyset$ and $\chi_{\{1\}}$ constitute an orthonormal basis in $L^2(\beta;\mathbb{C})$; in fact, they are the orthogonal polynomials there. Hence, for each $\theta \in \mathbb{C}$, we can determine the **Bernoulli multiplier** $\mathcal{K}_\theta$ to be the unique normal operator on $L^2(\beta;\mathbb{C})$ prescribed by

$$\mathcal{K}_\theta \chi_F = \begin{cases} \chi_\emptyset & \text{if } F = \emptyset \\ \theta\chi_{\{1\}} & \text{if } F = \{1\}. \end{cases}$$

Furthermore, (2.3.16) is equivalent to the statement that

$$\text{(2.3.17)} \qquad \left\|\mathcal{K}_\theta \varphi\right\|_{L^q(\beta)} \le \|\varphi\|_{L^p(\beta)} \quad \text{for all} \quad \varphi \in L^2(\beta;\mathbb{C}).$$

Indeed, it is obvious that (2.3.16) is equivalent to (2.3.17) restricted to φ's of the form $x \in \mathbb{R} \longmapsto 1 + \zeta x$ as ζ runs over $\mathbb{C}$; and from this, together with the observation that every element of $L^2(\beta;\mathbb{C})$ can be represented in the form $a\chi_\emptyset + b\chi_{\{1\}}$ as (a,b) runs over $\mathbb{C}^2$, one quickly concludes that (2.3.16) implies (2.3.17) for general $\varphi \in L^2(\beta;\mathbb{C})$.

We next want to show that (2.3.17) can be parlayed into a seemingly more general statement. To this end, we define the n-fold tensor product operator $\mathcal{K}_\theta^{\otimes n}$ on $L^2(\beta^n;\mathbb{C})$ as follows. For $F \subseteq \{1,\dots,n\}$ set $\chi_F \equiv 1$ if $F = \emptyset$ and define

$$\chi_F(\mathbf{x}) = \prod_{j\in F} \chi_{\{1\}}(x_j) \quad \text{for} \quad \mathbf{x} = (x_1,\dots,x_n) \in \mathbb{R}^n$$

if $F \ne \emptyset$. Note that $\{\chi_F : F \subseteq \{1,\dots,n\}\}$ is an orthonormal basis for $L^2(\beta^n;\mathbb{C})$, and define $\mathcal{K}_\theta^{\otimes n}$ to be the unique normal operator on $L^2(\beta^n;\mathbb{C})$ for which

$$\text{(2.3.18)} \qquad \mathcal{K}_\theta^{\otimes n}\chi_F = \theta^{|F|}\chi_F, \quad F \subseteq \{1,\dots,n\},$$

where $|F|$ is used to denote the number of elements in the set F. Alternatively, one can describe $\mathcal{K}_\theta^{\otimes n}$ inductively on $n \in \mathbb{Z}^+$ by saying that $\mathcal{K}_\theta^{\otimes 1} = \mathcal{K}_\theta$ and that, for $\Phi \in C(\mathbb{R}^{n+1};\mathbb{C})$ and $(\mathbf{x}, y) \in \mathbb{R}^n \times \mathbb{R}$,

$$\left[\mathcal{K}_\theta^{\otimes(n+1)}\Phi\right](\mathbf{x}, y) = \left[\mathcal{K}_\theta\Psi(\mathbf{x}, \cdot\,)\right](y) \quad \text{where} \quad \Psi(\mathbf{x}, y) = \left[\mathcal{K}_\theta^{\otimes n}\Phi(\,\cdot\,, y)\right](\mathbf{x}).$$

It is this alternative description which makes it easiest to see the extension of (2.3.17) alluded to above. Namely, what we will now show is that, for every $n \in \mathbb{Z}^+$,

$$\text{(2.3.19)} \qquad \text{(2.3.17)} \implies \left\|\mathcal{K}_\theta^{\otimes n}\Phi\right\|_{L^q(\beta^n)} \le \|\Phi\|_{L^p(\beta^n)}, \quad \Phi \in L^2(\beta^n;\mathbb{C}).$$

Obviously, there is nothing to do when $n = 1$. Next, assume (2.3.19) for n, let $\Phi \in C(\mathbb{R}^{n+1};\mathbb{C})$ be given, and define Ψ as in the second description of $\mathcal{K}_\theta^{\otimes(n+1)}\Phi$. Then, by (2.3.17) applied to $\Psi(\mathbf{x}, \cdot\,)$ for each $\mathbf{x} \in \mathbb{R}^n$ and by the induction hypothesis applied to $\Phi(\,\cdot\,, y)$ for each $y \in \mathbb{R}$, we have that

$$\begin{aligned}
\left\|\mathcal{K}_\theta^{\otimes(n+1)}\Phi\right\|_{L^q(\beta^{n+1})}^q &= \int_{\mathbb{R}^n}\left(\int_{\mathbb{R}}\left|\left[\mathcal{K}_\theta\Psi(\mathbf{x}, \cdot\,)\right](y)\right|^q \beta(dy)\right)\beta^n(d\mathbf{x}) \\
&\le \int_{\mathbb{R}^n}\left(\int_{\mathbb{R}}|\Psi(\mathbf{x}, y)|^p\,\beta(dy)\right)^{\frac{q}{p}}\beta^n(d\mathbf{x}) \\
&= \left\|\int_{\mathbb{R}^n}|\Psi(\,\cdot\,, y)|^p\,\beta(dy)\right\|_{L^{\frac{q}{p}}(\beta^n)}^{\frac{q}{p}} \\
&\le \left(\int_{\mathbb{R}^n}\left\|\,|\Psi(\,\cdot\,, y)|^p\,\right\|_{L^{\frac{q}{p}}(\beta^n)}\beta(dy)\right)^{\frac{q}{p}} \\
&= \left(\int_{\mathbb{R}}\|\Psi(\,\cdot\,, y)\|_{L^q(\beta^n)}^p\,\beta(dy)\right)^{\frac{q}{p}} \\
&\le \left(\int_{\mathbb{R}}\|\Phi(\,\cdot\,, y)\|_{L^p(\beta^n)}^p\,\beta(dy)\right)^{\frac{q}{p}} = \|\Phi\|_{L^p(\beta^{n+1})}^q,
\end{aligned}$$

where, in the passage to the fourth line, we have used the continuous form of Minkowski's equality (it is at this point that the only essential use of the hypothesis $p \le q$ is made).

We are now ready to take the main step in the proof of Theorem 2.3.14.

2.3.20 Lemma. *Define* $\mathcal{A}_n : L^2(\beta;\mathbb{C}) \longrightarrow L^2(\beta^n;\mathbb{C})$ *by*

$$\left[\mathcal{A}_n\varphi\right](\mathbf{x}) = \varphi\left(\frac{\sum_{\ell=1}^n x_\ell}{\sqrt{n}}\right) \quad \textit{for} \quad \mathbf{x} \in \mathbb{R}^n.$$

Then, for every pair of tempered φ and ψ from $C(\mathbb{R};\mathbb{C})$,

$$\|\varphi\|_{L^p(\gamma)} = \lim_{n\to\infty} \|\mathcal{A}_n\varphi\|_{L^p(\beta^n)} \quad \text{for every} \quad p \in [1,\infty) \tag{2.3.21}$$

and

$$\left(\mathcal{H}_\theta\varphi,\psi\right)_{L^2(\gamma)} = \lim_{n\to\infty}\left(\mathcal{K}_\theta^{\otimes n}\circ\mathcal{A}_n\varphi,\mathcal{A}_n\psi\right)_{L^2(\beta^n)} \tag{2.3.22}$$

for every $\theta \in (0,1)$. Moreover, if, in addition, either φ or ψ is a polynomial, then (2.3.22) continues to hold for all $\theta \in \mathbb{C}$.

PROOF: Let φ and ψ be tempered elements of $C(\mathbb{R};\mathbb{C})$, and define

$$f_n(\theta) = \left(\mathcal{K}_\theta^{\otimes n}\circ\mathcal{A}_n\varphi,\mathcal{A}_n\psi\right)_{L^2(\beta^n)} \quad \text{and} \quad f(\theta) = \left(\mathcal{H}_\theta\varphi,\psi\right)_{L^2(\gamma)}$$

for $n \in \mathbb{Z}^+$ and $\theta \in \mathbb{C}$. We begin by showing that

$$\lim_{n\to\infty} f_n(\theta) = f(\theta), \quad \theta \in (0,1). \tag{2.3.23}$$

Notice that (2.3.23) is (2.3.22) for $\theta \in (0,1)$ and therefore that (2.2.21) follows immediately from (2.3.23) by repacing φ and ψ, respectively, with $\mathbf{1}$ and $|\varphi|^p$.

In order to prove (2.3.23), we will need to introduce other expressions for $f(\theta)$ and the $f_n(\theta)$'s. To this end, set

$$\mathbf{C}_\theta = \begin{bmatrix} 1 & \theta \\ \theta & 1 \end{bmatrix},$$

and, using (2.3.6), observe (cf. (2.2.10)) that

$$f(\theta) = \int_{\mathbb{R}^2} \varphi(x)\,\overline{\psi(\mathbf{y})}\,\gamma_{\mathbf{0},\mathbf{C}_\theta}(dx \times dy).$$

Next, let $\boldsymbol{\beta}_\theta$ be the probability measure on $\mathbb{R}^2$ determined by

$$\boldsymbol{\beta}_\theta(\{\pm1,\pm1\}) = \tfrac{1+\theta}{4} \quad \text{and} \quad \boldsymbol{\beta}_\theta(\{\pm1,\mp1\}) = \tfrac{1-\theta}{4};$$

and note that, because

$$\left(\mathcal{K}_\theta\varphi,\psi\right)_{L^2(\beta)} = \int_{\mathbb{R}^2} \varphi(x)\,\overline{\psi(y)}\,\boldsymbol{\beta}_\theta(dx\times dy),$$

one has that

$$\left(\mathcal{K}_\theta^{\otimes n}\Phi,\Psi\right)_{L^2(\beta)} = \int_{\mathbb{R}^2}\cdots\int_{\mathbb{R}^2} \Phi(\mathbf{x})\,\overline{\Psi(\mathbf{y})}\,\boldsymbol{\beta}_\theta(dx_1\times dy_1)\cdots\boldsymbol{\beta}_\theta(dx_n\times dy_n)$$

for all $\Phi, \Psi \in C(\mathbb{R}^n;\mathbb{C})$. Hence, if (cf. Exercise 1.1.14) $\Omega = (\mathbb{R}^2)^{\mathbb{Z}^+}$, $\mathcal{F} = \mathcal{B}_\Omega$, and $P_\theta = (\beta_\theta)^{\mathbb{Z}^+}$, then

$$f_n(\theta) = \mathbb{E}^{P_\theta}\left[F\left(\frac{\sum_1^n \mathbf{Z}_n}{\sqrt{n}}\right)\right],$$

where $F(\mathbf{z}) \equiv \varphi(x)\overline{\psi(y)}$ for $\mathbf{z} = (x,y) \in \mathbb{R}^2$ and $\mathbf{Z}_n(\omega) = \mathbf{z}_n$, $n \in \mathbb{Z}^+$, for $\omega = (\mathbf{z}_1, \dots, \mathbf{z}_n, \dots) \in \Omega$. Note that, under P_θ, the $\mathbf{Z}_n$'s are mutually independent, identically distributed $\mathbb{R}^2$-valued random variables with mean-value $\mathbf{0}$ and covariance $\mathbf{C}_\theta$. In addition, $\|\mathbf{Z}_n\|^2_{L^\infty(P_\theta;\mathbb{R}^2)} = 2$, and therefore (2.2.21) is certainly satisfied. Hence, (2.3.23) now follows as an application of the last part of Theorem 2.2.20.

To complete the proof, suppose that φ is a polynomial of degree k. It is then an easy matter to check that

$$\left(\mathcal{A}_n\varphi, \chi_F\right)_{L^2(\beta^n)} = 0 \quad \text{if } |F| > k,$$

and therefore (cf. (2.3.18)) $\theta \in \mathbb{C} \longmapsto f(\theta) \in \mathbb{C}$ is also a polynomial of degree no more than k. Moreover, because

$$\left|f_n(\theta)\right| = \left|\sum_F \theta^{|F|}\left(\mathcal{A}_n\varphi, \chi_F\right)_{L^2(\beta^n)}\left(\chi_F, \mathcal{A}_n\psi\right)_{L^2(\beta^n)}\right|,$$

this also means that

$$\left|f_n(\theta)\right| \le \left(|\theta| \vee 1\right)^k \left\|\mathcal{A}_n\varphi\right\|_{L^2(\beta^n)} \left\|\mathcal{A}_n\psi\right\|_{L^2(\beta^n)}, \qquad n \in \mathbb{Z}^+ \text{ and } \theta \in \mathbb{C}.$$

Hence, because of (2.3.21) with $p = 2$, $\{f_n : n \in \mathbb{Z}^+\}$ is a family of entire functions on $\mathbb{C}$ which are uniformly bounded on compact subsets. At the same time, because $(\varphi, H_m)_{L^2(\gamma)} = 0$ for $m > k$, f_n is also a polynomial of degree $k \wedge n$; and therefore (2.3.23) already implies that the convergence extends to the whole of $\mathbb{C}$ and is uniform on compacts. Finally, in the case when ψ, instead of φ, is a polynomial, simply note that

$$\left(\mathcal{K}_\theta^{\otimes n} \circ \mathcal{A}_n\varphi, \mathcal{A}_n\psi\right)_{L^2(\beta^n)} = \overline{\left(\mathcal{K}_{\bar\theta}^{\otimes n} \circ \mathcal{A}_n\psi, \mathcal{A}_n\varphi\right)_{L^2(\beta^n)}},$$

and apply the preceding. □

Proof of Theorem 2.3.14: Assume that (2.3.16) holds for a given pair $1 < p \le q < \infty$ and $\theta \in \mathbb{C}$. We then know that (2.3.19) holds for every $n \in \mathbb{Z}^+$.

Hence, by Lemma 2.3.20, if φ and ψ are tempered elements of $C(\mathbb{R};\mathbb{C})$ and at least one of them is a polynomial, then

$$\begin{aligned}\left|\left(\mathcal{H}_\theta\varphi,\psi\right)_{L^2(\gamma)}\right| &= \lim_{n\to\infty}\left|\left(\mathcal{K}_\theta^{\otimes n}\circ\mathcal{A}_n\varphi,\mathcal{A}_n\psi\right)_{L^2(\beta^n)}\right| \\ &\le \lim_{n\to\infty}\left\|\mathcal{A}_n\varphi\right\|_{L^p(\beta^n)}\left\|\mathcal{A}_n\psi\right\|_{L^{q'}(\beta^n)} = \|\varphi\|_{L^p(\gamma)}\|\psi\|_{L^{q'}(\gamma)}.\end{aligned}$$

In other words, we now know that for all tempered φ and ψ from $C(\mathbb{R};\mathbb{C})$

$$\left|\left(\mathcal{H}_\theta\varphi,\psi\right)_{L^2(\gamma)}\right| \le \|\varphi\|_{L^p(\gamma)}\|\psi\|_{L^{q'}(\gamma)} \tag{2.3.24}$$

so long as one or the other is a polynomial.

We next complete the proof in the case when $p \in (1,2]$. To this end, note that, for any fixed polynomial φ, (2.3.24) for every tempered $\psi \in C(\mathbb{R};\mathbb{C})$ guarantees that the inequality in (2.3.15) holds for that φ. At the same time, because $p \in (1,2]$ and the polynomials are dense in $L^2(\gamma;\mathbb{C})$, (2.3.15) follows immediately from its own restriction to polynomials.

Finally, assume that $p \in [2,\infty)$ and therefore that $q' \in (1,2]$. Then, again because the polynomials are dense in $L^2(\gamma;\mathbb{C})$, (2.3.24) for a fixed tempered $\varphi \in C(\mathbb{R};\mathbb{C})$ and all polynomials ψ implies (2.3.15) first for all tempered continuous ψ's and thence for all $\psi \in L^2(\gamma;\mathbb{C})$. □

We will now apply Theorem 2.3.14 to two important examples. The first example involves the case when $\theta \in (0,1)$ and shows that the contraction property proved in Lemma 2.3.7 can be improved to say that, for each $p \in (1,\infty)$ and $\theta \in (0,1)$, there is a $q = q(p,\theta) \in (p,\infty)$ such that $\mathcal{H}_\theta$ is a contraction on $L^p(\gamma;\mathbb{C})$ into $L^q(\gamma;\mathbb{C})$. Such an operator is said to be **hypercontractive**, and the fact that $\mathcal{H}_\theta$ is hypercontractive was first proved by E. Nelson in connection with his renowned construction of a nontrivial, two-dimensional quantum field.† The proof which we will give is entirely different from Nelson's and is much closer to the ideas introduced by L. Gross‡ as they were developed by Beckner.

2.3.25 Theorem (Nelson). *Let $\theta \in (0,1)$ and $p \in (1,\infty)$ be given, and set*

$$q(p,\theta) = 1 + \frac{p-1}{\theta^2}.$$

Then

$$\left\|\mathcal{H}_\theta\varphi\right\|_{L^q(\gamma)} \le \|\varphi\|_{L^p(\gamma)}, \quad \varphi \in L^2(\gamma;\mathbb{C}), \tag{2.3.26}$$

† Nelson's own proof appeared in his "The free Markov field," *J. Fnal. Anal.* **12**.

‡ See Gross's "Logarithmic Sobolev inequalities," *Amer. J. Math.* **97**. Also, have a look at Exercise 2.3.37 below.

for every $1 \le q \le q(p,\theta)$. Moreover, if $q > q(p,\theta)$, then

$$\sup\left\{\left\|\mathcal{H}_\theta\varphi\right\|_{L^q(\gamma)} : \varphi \in L^2(\gamma;\mathbb{C})\right\} = \infty. \tag{2.3.27}$$

PROOF: We will leave the proof of (2.3.27) as an exercise. (Try taking φ's of the form $e^{\lambda x^2}$.) Also, because γ is a probability measure and therefore the left-hand side of (2.3.26) is nondecreasing as a function of q, we will restrict our attention to the proof of (2.3.26) for $q = q(p,\theta)$. Hence, by Theorem 2.3.14, what we have to do is prove (2.3.16) for every $1 < p < q < \infty$ and $\theta \in (0,1)$ which are related by

$$\theta = \left(\frac{p-1}{q-1}\right)^{\frac{1}{2}}. \tag{2.3.28}$$

We begin with the case when $1 < p < q \le 2$; and we first consider $\zeta \in [0,1)$. Introducing the generalized binomial coefficients

$$\binom{r}{\ell} \equiv \frac{r(r-1)\cdots(r-\ell+1)}{\ell!} \quad \text{for} \quad r \in \mathbb{R} \text{ and } \ell \in \mathbb{N},$$

we can write

$$\frac{|1-\theta\zeta|^q + |1+\theta\zeta|^q}{2} = 1 + \sum_{k=1}^{\infty}\binom{q}{2k}(\theta\zeta)^{2k}$$

and

$$\frac{|1-\zeta|^p + |1+\zeta|^p}{2} = 1 + \sum_{k=1}^{\infty}\binom{p}{2k}\zeta^{2k}.$$

Noting that, because $q \le 2$, $\binom{q}{2k} \ge 0$ for every $k \in \mathbb{Z}^+$, and using the fact that, because $\frac{p}{q} \in (0,1)$, $(1+x)^{\frac{p}{q}} \le 1 + \frac{p}{q}x$ for all $x \ge 0$, we see that

$$\left(\frac{|1-\theta\zeta|^q + |1+\theta\zeta|^q}{2}\right)^{\frac{p}{q}} \le 1 + \frac{p}{q}\sum_{k=1}^{\infty}\binom{q}{2k}(\theta\zeta)^{2k}.$$

Hence, we will have completed the case under consideration once we check that

$$\frac{p}{q}\sum_{k=1}^{\infty}\binom{q}{2k}(\theta\zeta)^{2k} \le \sum_{k=1}^{\infty}\binom{p}{2k}\zeta^{2k};$$

and clearly this will follow if we show that

$$\frac{p}{q}\binom{q}{2k}\theta^{2k} \le \binom{p}{2k} \quad \text{for each} \quad k \in \mathbb{Z}^+.$$

But the choice of θ in (2.3.28) makes the preceding an equality when $k = 1$; and, when $k \ge 2$,

$$\frac{\frac{p}{q}\binom{q}{2k}\theta^{2k}}{\binom{p}{2k}} \le \prod_{j=2}^{2k-1} \frac{j-q}{j-p} \le 1,$$

since $1 < p < q \le 2$.

At this point, we have proved (2.3.16) for $1 < p < q \le 2$ and θ given by (2.3.28) when $\zeta \in (0,1)$. Continuing with this choice of p, q, and θ, note that (2.3.16) extends immediately to $\zeta \in [-1,1]$ by continuity and symmetry. Finally, for general $\zeta \in \mathbb{C}$, set

$$a = \frac{|1-\zeta| + |1+\zeta|}{2}, \quad b = \frac{|1-\zeta| - |1+\zeta|}{2}, \quad \text{and } c = \frac{b}{a}.$$

Then,

$$|1 \pm \theta\zeta| = \left|\tfrac{1\pm\theta}{2}(1\pm\zeta) + \tfrac{1-\theta}{2}(1\mp\zeta)\right| \le a \pm \theta b,$$

and therefore, by the preceding applied to $c \in [-1,1]$, we have that

$$\begin{aligned}
\left(\frac{|1-\theta\zeta|^q + |1+\theta\zeta|^q}{2}\right)^{\frac{1}{q}} &\le a\left(\frac{|1-\theta c|^q + |1+\theta c|^q}{2}\right)^{\frac{1}{q}} \\
\le a\left(\frac{|1-c|^p + |1+c|^p}{2}\right)^{\frac{1}{p}} &= \left(\frac{|a-b|^p + |a+b|^p}{2}\right)^{\frac{1}{p}} \\
&= \left(\frac{|1-\zeta|^p + |1+\zeta|^p}{2}\right)^{\frac{1}{p}}.
\end{aligned}$$

Hence, we have now completed the case when $1 < p < q \le 2$ and θ is given by (2.3.28).

To handle the other cases, we use the equivalence of (2.3.16) and (2.3.17). Thus, what we already know is that (2.3.17) holds for $1 < p < q \le 2$ and the θ in (2.3.28). Next, suppose that $2 \le p < q < \infty$. Then, since $1 < q' < p' \le 2$ and

$$\frac{p-1}{q-1} = \frac{q'-1}{p'-1},$$

an application to q' and p' of the result which we already have yields

$$\begin{aligned}
\|\mathcal{K}_\theta\varphi\|_{L^q(\beta)} &= \sup\left\{\left(\mathcal{K}_\theta\varphi, \psi\right)_{L^2(\beta)} : \psi \in L^2(\beta;\mathbb{C}) \text{ with } \|\psi\|_{L^{q'}(\beta)} = 1\right\} \\
&= \sup\left\{\left(\varphi, \mathcal{K}_\theta\psi\right)_{L^2(\beta)} : \psi \in L^2(\beta;\mathbb{C}) \text{ with } \|\psi\|_{L^{q'}(\beta)} = 1\right\} \\
&\le \|\varphi\|_{L^p(\beta)},
\end{aligned}$$

where the θ is the one given in (2.3.28). Thus, the only case which remains is the one when $1 < p \le 2 \le q < \infty$. But, in this case, set $\xi = (p-1)^{\frac{1}{2}}$, $\eta = (q-1)^{-\frac{1}{2}}$, and observe that, because the associated θ in (2.3.28) is the product of ξ with η, $\mathcal{K}_\theta = \mathcal{K}_\eta \circ \mathcal{K}_\xi$ and therefore

$$\left\|\mathcal{K}_\theta\varphi\right\|_{L^q(\beta)} \le \left\|\mathcal{K}_\xi\varphi\right\|_{L^2(\beta)} \le \|\varphi\|_{L^p(\beta)}. \quad \square$$

As our second, and final, application of Theorem 2.3.14, we present the theorem of Beckner for which Theorem 2.3.14 was concocted in the first place.

2.3.29 Theorem (Beckner). *For each $p \in [1,2]$,*

$$\|\mathcal{F}f\|_{L^{p'}(\mathbb{R})} \le A_p\|f\|_{L^p(\mathbb{R})}, \quad f \in L^1(\mathbb{R};\mathbb{C}) \cap L^2(\mathbb{R};\mathbb{C}), \tag{2.3.30}$$

where $\mathcal{F}$ is the Fourier operator in (2.3.9) and A_p is the constant in (2.3.13). Moreover, if f is the Gauss kernel $e^{-\pi x^2}$, then (2.3.30) is an equality.

Proof: Because of (2.3.11), the second part is a straightforward computation which we leave to the reader. Also, we will only consider (2.3.30) when $p \in (1,2)$, the other cases being well-known (cf. Exercise 2.3.36).

Because of (2.3.13), the proof of (2.3.30) comes down to showing that

$$\left\|\mathcal{H}_{\theta_p}\varphi\right\|_{L^{p'}(\gamma)} \le \|\varphi\|_{L^p(\gamma)}, \quad \varphi \in L^2(\gamma;\mathbb{C}), \tag{2.3.31}$$

where $\theta_p = \sqrt{-1}\,(p-1)^{\frac{1}{2}}$; and, by Theorem 2.3.14, (2.3.31) will follow as soon as we prove (2.3.16) for θ_p. For this purpose, write

$$\zeta = \xi + \sqrt{-1}\,(p-1)^{-\frac{1}{2}}\eta \quad \text{where} \quad \xi,\ \eta \in \mathbb{R}.$$

Then, proving (2.3.16) for θ_p becomes the problem of checking that

$$\begin{aligned} &\left(\frac{\left[(1-\eta)^2 + (p-1)\xi^2\right]^{\frac{p'}{2}} + \left[(1+\eta)^2 + (p-1)\xi^2\right]^{\frac{p'}{2}}}{2}\right)^{\frac{1}{p'}} \\ &\qquad \le \left(\frac{\left[(1-\xi)^2 + (p-1)\eta^2\right]^{\frac{p}{2}} + \left[(1+\xi)^2 + (p-1)\eta^2\right]^{\frac{p}{2}}}{2}\right)^{\frac{1}{p}} \end{aligned} \tag{2.3.32}$$

for all ξ, $\eta \in \mathbb{R}$.

To prove (2.3.32), consider, for each $\alpha \in (0,\infty)$, the function $g_\alpha : [0,\infty)^2 \longrightarrow [0,\infty)$ defined by $g_\alpha(x,y) = \left[x^{\frac{1}{\alpha}} + y^{\frac{1}{\alpha}}\right]^\alpha$. It is an easy matter to check that g_α is concave or convex depending on whether $\alpha \in [1,\infty)$ or $\alpha \in (0,1)$. In particular, since $\frac{p'}{2} \in (1,\infty)$, when we set $\alpha = \frac{p'}{2}$, $x_\pm = |1 \pm \eta|^{p'}$, and $y = (p-1)^{\frac{p'}{2}}|\xi|^{p'}$, we get

$$\begin{aligned}
&\frac{\left[(1-\eta)^2 + (p-1)\xi^2\right]^{\frac{p'}{2}} + \left[(1+\eta)^2 + (p-1)\xi^2\right]^{\frac{p'}{2}}}{2} \\
&\quad = \frac{g_\alpha(x_-,y) + g_\alpha(x_+,y)}{2} \le g_\alpha\left(\frac{x_- + x_+}{2}, y\right) \\
&\quad = \left[\left(\frac{|1-\eta|^{p'} + |1+\eta|^{p'}}{2}\right)^{\frac{p}{2}} + (p-1)\xi^2\right]^{\frac{p'}{2}};
\end{aligned}$$

and similarly, because $\frac{p}{2} \in (0,1)$,

$$\begin{aligned}
&\frac{\left[(1-\xi)^2 + (p-1)\eta^2\right]^{\frac{p}{2}} + \left[(1+\xi)^2 + (p-1)\eta^2\right]^{\frac{p}{2}}}{2} \\
&\quad \ge \left[\left(\frac{|1-\xi|^p + |1+\xi|^p}{2}\right)^{\frac{2}{p}} + (p-1)\eta^2\right]^{\frac{p}{2}}.
\end{aligned}$$

Thus, (2.3.32) will be proved if we show that

$$\begin{aligned}
&\left(\frac{|1-\eta|^{p'} + |1+\eta|^{p'}}{2}\right)^{\frac{2}{p'}} + (p-1)\xi^2 \\
&\qquad \le \left(\frac{|1-\xi|^p + |1+\xi|^p}{2}\right)^{\frac{2}{p}} + (p-1)\eta^2.
\end{aligned} \tag{2.3.33}$$

But because (cf. Theorems 2.3.14 and 2.3.25) we know that (2.3.16) holds with p replaced by 2, $q = p'$, and $\theta = (p'-1)^{\frac{1}{2}}$, the left-hand side of (2.3.33) is dominated by

$$(p-1)\xi^2 + \frac{\left(1 - \frac{\eta}{(p'-1)^{\frac{1}{2}}}\right)^2 + \left(1 + \frac{\eta}{(p'-1)^{\frac{1}{2}}}\right)^2}{2} = 1 + (p-1)(\xi^2 + \eta^2).$$

At the same time, again by (2.3.16), only this time with p, 2, and $\theta = (p-1)^{-\frac{1}{2}}$, we see that the right-hand side of (2.3.33) dominates

$$(p-1)\eta^2 + \frac{\left(1-(p-1)^{\frac{1}{2}}\xi\right)^2 + \left(1+(p-1)^{\frac{1}{2}}\xi\right)^2}{2} = 1 + (p-1)(\xi^2 + \eta^2). \quad \square$$

Exercises

2.3.34 Exercise: Because the Fourier operator $\mathcal{F}$ (cf. (2.3.9)) is a contraction from $L^1(\mathbb{R})$ to $L^\infty(\mathbb{R})$ as well as from $L^2(\mathbb{R})$ into $L^2(\mathbb{R})$, the Riesz–Thorin Interpolation Theorem guarantees that it is a contraction from $L^p(\mathbb{R})$ into $L^{p'}(\mathbb{R})$ for each $p \in (0,1)$. Hence, we know, from Theorem 2.3.29, that the number A_p in (2.3.13) must be less than or equal to 1. However, the preceding is a rather convoluted line of reasoning to what must be a far more elementary fact. Indeed, show that

$$t \in \left(\tfrac{1}{2}, 1\right) \longmapsto \log A_{\frac{1}{t}} \in \mathbb{R}$$

is a strictly convex function which tends to 0 at both end points and is therefore strictly negative. In other words, what Beckner's result proves is that the Fourier operator is one for which interpolation fails to give the best result.

2.3.35 Exercise: The inequality in (2.3.8) is an example of a general principle. Namely, if $(E, \mathcal{B})$ is any measurable space, then a map $(x, \Gamma) \in E \times \mathcal{B} \longmapsto \Pi(x, \Gamma) \in [0,1]$ is called a **transition probability** whenever $x \in E \longmapsto \Pi(x, \Gamma)$ is $\mathcal{B}$-measurable for each $\Gamma \in \mathcal{B}$ and $\Gamma \in \mathcal{B} \longmapsto \Pi(x, \Gamma)$ is a probability measure on $(E, \mathcal{B})$ for each $x \in E$. Given a transition probability $\Pi(x, \cdot\,)$, we define the linear operator Π on $B(E;\mathbb{C})$ by

$$\left[\Pi\varphi\right](x) = \int_E \varphi(y)\, \Pi(x, dy), \quad x \in E, \quad \text{for} \quad \varphi \in B(E;\mathbb{C}).$$

Check that Π takes $B(E;\mathbb{C})$ into itself and that $\|\Pi\varphi\|_{\mathrm{u}} \le \|\varphi\|_{\mathrm{u}}$. Next, given a σ-finite measure μ on $(E, \mathcal{B})$, we say that μ is **Π-invariant** if

$$\mu(\Gamma) = \int_E \Pi(x, \Gamma)\, \mu(dx) \quad \text{for all} \quad \Gamma \in \mathcal{B}.$$

Using Jensen's inequality, first show that, for each $p \in [1, \infty)$,

$$\left|\left[\Pi\varphi\right](x)\right|^p \le \left[\Pi|\varphi|^p\right](x), \quad x \in E,$$

and then that, for any Π-invariant μ,

$$\|\Pi\varphi\|_{L^p(\mu)} \le \|\varphi\|_{L^p(\mu)}, \quad \varphi \in B(E;\mathbb{C}).$$

Finally, show that μ is Π-invariant if it is **Π-reversing** in the sense that

$$\int_{\Gamma_1} \Pi(x, \Gamma_2)\, \mu(dx) = \int_{\Gamma_2} \Pi(y, \Gamma_1)\, \mu(dy) \quad \text{for all} \quad \Gamma_1, \Gamma_2 \in \mathcal{B}.$$

2.3.36 Exercise: Recall the Hermite functions h_n, $n \in \mathbb{N}$, in (2.3.12) and define the **normalized Hermite functions** $\overline{h}_n$, $n \in \mathbb{N}$ by

$$\overline{h}_n = \frac{2^{\frac{1}{4}}}{(n!)^{\frac{1}{2}}} h_n, \quad n \in \mathbb{N}.$$

By noting that (cf. the discussion following (2.3.12)) $\overline{h}_n = \mathcal{U}_2 \overline{H}_n$, show that $\{\overline{h}_n : n \in \mathbb{N}\}$ constitutes an orthonormal basis in $L^2(\mathbb{R};\mathbb{C})$; and from this together with (2.3.11), arrive at **Parseval's Identity**

$$\|\mathcal{F}f\|_{L^2(\mathbb{R})} = \|f\|_{L^2(\mathbb{R})}, \quad f \in L^1(\mathbb{R};\mathbb{C}) \cap L^2(\mathbb{R};\mathbb{C}),$$

and conclude that $\mathcal{F}$ determines a unique unitary operator $\overline{\mathcal{F}}$ on $L^2(\mathbb{R};\mathbb{C})$ such that $\overline{\mathcal{F}}f = \mathcal{F}f$ for $f \in L^1(\mathbb{R};\mathbb{C}) \cap L^2(\mathbb{R};\mathbb{C})$. Finally, use this to verify the $L^2(\mathbb{R})$-Fourier inversion formula $\overline{\mathcal{F}}^{-1} = \tilde{\mathcal{F}}$, where $[\tilde{\mathcal{F}}f](x) \equiv [\mathcal{F}f](-x)$, $x \in \mathbb{R}$, for $f \in L^1(\mathbb{R};\mathbb{C}) \cap L^2(\mathbb{R};\mathbb{C})$.

2.3.37 Exercise: L. Gross had a somewhat different approach to the proof of (2.3.26). As in the proof which we have given, he reduced everything to the checking (2.3.17). However, he did this in a different way. Namely, given $b \in (0,1)$ he set $f(x) = 1 + bx$ and introduced the functions

$$f_t(x) \equiv [\mathcal{K}_{e^{-t}} f](x) = \tfrac{1+e^{-t}}{2} f(x) + \tfrac{1-e^{-t}}{2} f(-x), \quad (t,x) \in [0,\infty) \times \mathbb{R},$$

and $q(t) = 1 + (p-1)e^{2t}$, $t \in [0,\infty)$, and proved that

$$\frac{d}{dt}\|f_t\|_{L^{q(t)}(\beta)} \leq 0. \tag{2.3.38}$$

Following the steps below, see if you can reproduce Gross's calculation.

(**i**) Set

$$F(t) = \|f_t\|_{L^{q(t)}(\beta)},$$

and, by somewhat tedious but completely elementary differential calculus, show that

$$\begin{aligned}\tfrac{dF}{dt}(t) = \frac{F(t)^{1-q(t)}}{q(t)^2}\Big[&-\dot{q}(t)\int_{\mathbb{R}} f^{q(t)} \log\left(\tfrac{f_t}{F(t)}\right)^{q(t)} d\beta \\ &+ \tfrac{q(t)^2}{2}\int_{\mathbb{R}} f_t(x)^{q(t)-1}\big(f_t(-x) - f_t(x)\big)\,\beta(dx)\Big].\end{aligned}$$

Next, check that

$$\begin{aligned}&\int_{\mathbb{R}} f_t(x)^{q(t)-1}\big(f_t(-x) - f_t(x)\big)\,\beta(dx) \\ &\quad = -\tfrac{1}{2}\int_{\mathbb{R}}\big(f_t(x)^{q(t)-1} - f_t(-x)^{q(t)-1}\big)\big(f_t(x) - f_t(-x)\big)\,\beta(dx),\end{aligned}$$

and, after verifying that

$$\left(\xi^{q-1}-\eta^{q-1}\right)(\xi-\eta) \geq \frac{4(q-1)\left(\xi^{\frac{q}{2}}-\eta^{\frac{q}{2}}\right)^2}{q^2}, \quad \xi, \eta \in (0,\infty) \text{ and } q \in (1,\infty),$$

conclude that

$$\begin{aligned}\frac{dF}{dt}(t) \leq \frac{F(t)^{1-q(t)}}{q(t)^2}\Bigg[&-\dot{q}(t)\int_{\mathbb{R}} f^{q(t)} \log\left(\frac{f_t}{F(t)}\right)^{q(t)} d\beta \\ &+ \left(q(t)-1\right)\int_{\mathbb{R}}\left(f_t(x)^{\frac{q(t)}{2}} - f_t^{\frac{q(t)}{2}}(-x)\right)^2 \beta(dx)\Bigg].\end{aligned} \tag{2.3.39}$$

(ii) Prove the **logarithmic Sobolev inequality**

$$\int_{\mathbb{R}} \varphi^2 \log\left(\frac{\varphi}{\|\varphi\|_{L^2(\beta)}}\right)^2 d\beta \leq 2\int_{\mathbb{R}}\left(\varphi(x)-\varphi(-x)\right)^2 \beta(dx) \tag{2.3.40}$$

for strictly positive φ's on $\mathbb{R}$.

Hint: Reduce to the case when $\varphi(x) = 1 + bx$ for some $b \in (0,1)$, and, in this case, check that (2.3.40) is the elementary calculus inequality

$$(1+b)^2\log(1+b) + (1-b)^2\log(1-b) - (1+b^2)\log(1+b^2) \leq 2b^2, \quad b \in (0,1).$$

(iii) By plugging (2.3.40) into (2.3.39), arrive at (2.3.38), and conclude that (2.3.17) holds for $\theta \in (0,1)$ and $q = 1 + \frac{p-1}{\theta^2}$.

2.3.41 Exercise: In this exercise we will use the Mehler kernel to solve two famous evolution equations.

(i) Starting from (2.3.2), show that $A_- H_0 = 0$ and $A_- H_n = nH_{n-1}$, $n \in \mathbb{Z}^+$; and conclude that

$$\frac{d^2 H_n}{dx^2} - x\frac{dH_n}{dx} = -nH_n \quad \text{for each} \quad n \in \mathbb{N}.$$

In particular, if φ is a polynomial on $\mathbb{R}$ and $u_\varphi(t,\,\cdot\,) = \mathcal{H}_{e^{-t}}\varphi$, then u_φ is a smooth solution to the initial value Cauchy problem

$$\frac{\partial u}{\partial t} = \frac{\partial^2 u}{\partial x^2} - x\frac{\partial u}{\partial x}, \quad t \in (0,\infty), \quad \text{and} \quad u(0,\,\cdot\,) = \varphi. \tag{2.3.42}$$

Next, using (2.3.6), reexpress u_φ as

$$u_\varphi(t,x) = \int_{\mathbb{R}} M\left(x,y;e^{-t}\right)\varphi(y)\,\gamma(dy), \quad (t,x) \in (0,\infty)\times\mathbb{R}, \tag{2.3.43}$$

and show directly from (2.3.43) that, for any $\varphi \in C_b(\mathbb{R};\mathbb{R})$, the function u_φ extends to $[0,\infty)\times\mathbb{R}$ as an element of $C^{1,2}\left((0,\infty)\times\mathbb{R};\mathbb{R}\right) \cap C_b(\mathbb{R};\mathbb{R})$ which solves (2.3.42).

(ii) Define $U : L^2(\gamma;\mathbb{C}) \longrightarrow L^2(\mathbb{R};\mathbb{C})$ so that

$$[U\varphi](x) = \frac{\exp\left[-\frac{x^2}{4}\right]\varphi(x)}{(2\pi)^{\frac{1}{4}}}, \qquad x \in \mathbb{R},$$

and observe that U is unitary from $L^2(\gamma;\mathbb{R})$ onto $L^2(\mathbb{R};\mathbb{C})$. Next, define $\tilde{h}_n = U^{-1}\overline{H}_n$, $n \in \mathbb{N}$. Clearly $\{\tilde{h}_n;\, n \in \mathbb{N}\}$ is an orthonormal basis for $L^2(\mathbb{R};\mathbb{C})$. In addition, if $\tilde{A}_\pm = U \circ A_\pm \circ U^{-1}$, check first that

$$\tilde{A}_+\tilde{h}_n = (n+1)^{\frac{1}{2}}\tilde{h}_{n+1} \quad \text{and} \quad \tilde{A}_-\tilde{h}_n = (n-1)^{\frac{1}{2}}\tilde{h}_{n-1} (\equiv 0 \text{ if } n = 0)$$

and second that

$$\frac{d^2\tilde{h}_n}{dx^2} - \frac{x^2}{4}\tilde{h}_n = -\left(n + \frac{1}{2}\right)\tilde{h}_n, \qquad n \in \mathbb{N}.$$

Hence, if $f = U\varphi$, where φ is a polynomial, then

$$(t,x) \in [0,\infty) \times \mathbb{R} \longmapsto \tilde{u}_{\tilde{\varphi}}(t,x) = \sum_{n=1}^{\infty} e^{-(n+\frac{1}{2})t}\left(f,\tilde{h}_n\right)_{L^2(\mathbb{R})}\tilde{h}_n(x)$$

is a smooth solution to the Cauchy initial value problem

(2.3.44) $$\frac{\partial \tilde{u}}{\partial t} = \frac{\partial^2 \tilde{u}}{\partial x^2} - \frac{x^2}{4}u, \quad t \in (0,\infty), \quad \text{with } \tilde{u}(0,\,\cdot\,) = f.$$

Finally, set

$$h(t,x,y) = \frac{e^{-\frac{t}{2}}}{\sqrt{2\pi}}\, e^{-\frac{x^2}{4}}\, M\left(x,y;e^{-t}\right) e^{-\frac{y^2}{4}}$$

for $(t,x,y) \in (0,\infty) \times \mathbb{R}^2$, and show by direct computation that, for each $f \in C_b(\mathbb{R};\mathbb{R})$, the function $\tilde{u}_f$ given by

$$\tilde{u}_f(t,x) = \int_{\mathbb{R}} f(y)\, h(t,x,y)\, dy, \quad (t,x) \in (0,\infty) \times \mathbb{R},$$

extends as an element of $C^{1,2}\left((0,\infty) \times \mathbb{R};\mathbb{R}\right) \cap C_b\left([0,\infty) \times \mathbb{R};\mathbb{R}\right)$ which solves (2.3.44).

2.3.45 Exercise: By the same reasoning as we used to prove Theorem 2.3.29, show that, for any pair $1 < p \le 2 \le q < \infty$ and any complex number $\theta = \xi + \sqrt{-1}\,\eta$, (2.3.16), and therefore (2.3.15), holds if and only if† both $(q-1)\eta^2 + \xi^2 \le 1$ and

$$(q-2)(\xi\eta)^2 \le \left[1 - \xi^2 - (q-1)\eta^2\right]\left[(p-1) - (q-1)\alpha^2 - \beta^2\right].$$

† Recently, in his article "Gaussian kernels have only Gaussian maximizers," *Invent. Math.* **12** (1990), E. Lieb has essentially killed this line of research. His argument, which is entirely different from the one discussed here, handles not only the Hermite multipliers but essentially every operator whose kernel can be represented as the exponential of a second order polynomial.

Chapter III:

Convergence of Measures, Infinite Divisibility, and Processes with Independent Increments

§3.1: Convergence of Probability Measures.

In this section we are going to develop the convergence theory of probability measures in what will, at first sight, appear to be an absurd degree of generality. In particular, for reasons which will not become clear until Section 3.3, we want a theory which does not rely on our underlying space being locally compact. We hope that the reader will bear with us and, when he gets to Section 3.3, will even forgive us for making this rather lengthy excursion into the realm of *abstract nonsense.*

When discussing the convergence of probability measures on a measurable space $(E, \mathcal{B})$, one always has at least two senses in which the convergence may take place, and (depending on additional structure the space may possess) one may have more. To be more precise, let $B(E;\mathbb{R}) \equiv B\big((E,\mathcal{B});\mathbb{R}\big)$ be the space of bounded, $\mathbb{R}$-valued, $\mathcal{B}$-measurable functions on E, use $\mathbf{M}_1(E) \equiv \mathbf{M}_1(E,\mathcal{B})$ to denote the space of all probability measures on $(E,\mathcal{B})$, and define the **duality relation**

$$\langle \varphi, \mu \rangle = \int_E \varphi \, d\mu \quad \text{for } \varphi \in B(E;\mathbb{R}) \text{ and } \mu \in \mathbf{M}_1(E).$$

Next, again using $\|\varphi\|_{\mathrm{u}} \equiv \sup_{x\in E} |\varphi(x)|$ to denote the uniform norm of $\varphi \in B(E;\mathbb{R})$, set $B_1(E) = \{\varphi \in B(E;\mathbb{R}) : \|\varphi\|_{\mathrm{u}} \le 1\}$. Finally, given $\mu \in \mathbf{M}_1(E)$, consider the neighborhood basis at μ given by the sets

$$U(\mu,\delta) = \left\{ \nu \in \mathbf{M}_1(E) : \sup_{\varphi \in B_1(E)} \Big| \langle \varphi, \nu \rangle - \langle \varphi, \mu \rangle \Big| < \delta \right\}$$

as δ runs over $(0,\infty)$. For obvious reasons, the topology defined by these neighborhoods U is called the **uniform topology** on $\mathbf{M}_1(E)$. In order to develop some feeling for the uniform topology, we will begin by examining a few of its elementary properties.

3.1.1 Lemma. *Define*

$$\|\nu - \mu\|_{\mathrm{var}} = \sup\Big\{ \big|\langle \varphi, \mu \rangle - \langle \varphi, \nu \rangle\big| : \varphi \in B_1(E) \Big\}$$

for μ and ν from $\mathbf{M}_1(E)$. Then $(\mu,\nu) \in \mathbf{M}_1(E)^2 \longmapsto \|\mu-\nu\|_{\text{var}}$ is a metric on $\mathbf{M}_1(E)$ which is compatible with the uniform topology. Moreover, if μ, $\nu \in \mathbf{M}_1(E)$ are two elements of $\mathbf{M}_1(E)$ and λ is any element of $\mathbf{M}_1(E)$ with respect to which both μ and ν are absolutely continuous (e.g., $\frac{\mu+\nu}{2}$), then

$$\|\mu-\nu\|_{\text{var}} = \|g-f\|_{L^1(\lambda)} \quad \text{where} \quad f = \frac{d\mu}{d\lambda} \text{ and } g = \frac{\partial\nu}{\partial\lambda}. \tag{3.1.2}$$

In particular, $\|\mu-\nu\|_{\text{var}} \le 2$, and equality holds precisely when $\nu \perp \mu$ (i.e., they are singular). Finally, the metric $(\mu,\nu) \in \mathbf{M}_1(E)^2 \longmapsto \|\mu-\nu\|_{\text{var}}$ is complete.

PROOF: The first assertion needing comment is the one in (3.1.2). But, for every $\varphi \in B_1(E)$,

$$\left|\langle\varphi,\nu\rangle - \langle\varphi,\mu\rangle\right| = \left|\int_E \varphi(g-f)\,d\lambda\right| \le \|g-f\|_{L^1(\lambda)},$$

and equality holds when $\varphi = \operatorname{sgn}\circ(g-f)$. To prove the assertion which follows (3.1.2), note that

$$\|g-f\|_{L^1(\lambda)} \le \|f+g\|_{L^1(\lambda)} = 2$$

and that the inequality is strict if and only if $fg > 0$ on a set of strictly positive λ-measure or, equivalently, $\mu \not\perp \nu$. Thus, all that remains is to check the completeness assertion. To this end, let $\{\mu_n\}_1^\infty \subseteq \mathbf{M}_1(E)$ satisfying

$$\lim_{m\to\infty} \sup_{n\ge m} \|\mu_n - \mu_m\|_{\text{var}} = 0$$

be given, and set

$$\lambda = \sum_{n=1}^{\infty} \frac{\mu_n}{2^n}.$$

Clearly, λ is an element of $\mathbf{M}_1(E)$ with respect to which each μ_n is absolutely continuous. Moreover, if $f_n = \frac{d\mu_n}{d\lambda}$, then, by (3.1.2), $\{f_n\}_1^\infty$ is a Cauchy convergent sequence in $L^1(\lambda)$. Hence, since $L^1(\lambda)$ is complete, there is an $f \in L^1(\lambda)$ to which the f_n's converge in $L^1(\lambda)$. Moreover, we may choose f to be nonnegative, and certainly it has λ-integral 1. Thus, the measure μ given by $d\mu = f\,d\lambda$ is an element of $\mathbf{M}_1(E)$, and, by (3.1.2), $\|\mu_n - \mu\|_{\text{var}} \longrightarrow 0$. □

As a consequence of Lemma 3.1.1, we see that the uniform topology on $\mathbf{M}_1(E)$ admits a complete metric and that convergence in this topology is intimately related to L^1-convergence in the L^1-space of an appropriate element of $\mathbf{M}_1(E)$. In fact, $\mathbf{M}_1(E)$ looks in the uniform topology like a *galaxy* which is broken into many *constellations*, each constellation consisting of measures which are all absolutely continuous with respect to some fixed measure. In particular, there will usually be too many *constellations* for $\mathbf{M}_1(E)$ in the uniform topology to be separable. For example, if E is uncountable and $\{x\} \in \mathcal{B}$ for every $x \in E$, then the **point masses** δ_x, $x \in E$, (i.e., $\delta_x(\Gamma) = \mathbf{1}_\Gamma(x)$) form an uncountable subset of $\mathbf{M}_1(E)$ and $\|\delta_y - \delta_x\|_{\text{var}} = 2$ for $y \neq x$. Hence, in this case, $\mathbf{M}_1(E)$ cannot be covered by a countable collection of open $\|\cdot\|_{\text{var}}$-balls of radius 1.

As we said at the beginning of this section, the uniform topology is not the only one available. Indeed, for many purposes and, in particular, for probability theory, it is often *too rigid* a topology to be useful. For this reason, it is often convenient to consider a more lenient topology on $\mathbf{M}_1(E)$. The first one which comes to mind is the one which results from eliminating the *uniformity* in the uniform topology. That is, given a $\mu \in \mathbf{M}_1(E)$, define

$$(3.1.3) \qquad S(\mu, \delta; \varphi_1, \dots, \varphi_n) \equiv \left\{\nu \in \mathbf{M}_1(E) : \max_{1 \le k \le n} \left|\langle \varphi_k, \nu\rangle - \langle \varphi_k, \mu\rangle\right| < \delta\right\}$$

for $n \in \mathbb{Z}^+$, $\varphi_1, \dots, \varphi_n \in B(E;\mathbb{R})$, and $\delta > 0$. Clearly these sets S determine a Hausdorff topology on $\mathbf{M}_1(E)$ in which the net $\{\mu_\alpha : \alpha \in A\}$ converges to μ if and only if

$$\lim_\alpha \langle \varphi, \mu_\alpha\rangle = \langle \varphi, \mu\rangle \quad \text{for every} \quad \varphi \in B(E;\mathbb{R}).$$

For historical reasons, in spite of the fact that it is obviously *weaker* than the uniform topology, this topology on $\mathbf{M}_1(E)$ is usually called the **strong topology**; although, in some of the statistics literature, it is also known as the **$\boldsymbol{\tau}$-topology**.

A good understanding of the relationship between the strong and uniform topologies is most easily gained through functional analytic considerations and will not be particularly important for what follows. Nonetheless, it will be useful to recognize that, except in very special circumstances, it is *strictly weaker* than the uniform topology. For example, take $E = [0,1]$ with its Borel field and consider the probability measures $\mu_n(dt) = \big(1 + \sin(n\pi t)\big)\,dt$ for $n \in \mathbb{Z}^+$. Noting that, since $|\sin(n\pi t) - \sin(m\pi t)| \le 2$ and therefore

$$\begin{aligned}
\tfrac{1}{2}\|\mu_n - \mu_m\|_{\text{var}} &= \int_0^1 \frac{|\sin(n\pi t) - \sin(m\pi t)|}{2}\,dt \\
&\ge \tfrac{1}{4}\int_0^1 \big(\sin(n\pi t) - \sin(m\pi t)\big)^2\,dt = \frac{1}{4}
\end{aligned}$$

for $m \neq n$, one sees that $\{\mu_n\}_1^\infty$ not only fails to converge in the uniform topology, it does not even have any limit points as $n \to \infty$. On the other hand, because $\big\{2^{\frac{1}{2}}\sin(n\pi t)\big\}_1^\infty$ is orthonormal in $L^2\big(\lambda_{[0,1]}\big)$, Bessel's inequality says that

$$2\sum_{n=1}^\infty \left(\int_{[0,1]} \varphi(t)\,\sin(n\pi t)\,dt\right)^2 \le \|\varphi\|^2_{L^2(\lambda_{[0,1]})} \le \|\varphi\|_{\text{u}} < \infty$$

and therefore $\langle \varphi, \mu_n\rangle \longrightarrow \langle \varphi, \lambda_{[0,1]}\rangle$ for every $\varphi \in B\big([0,1];\mathbb{R}\big)$. In other words, *$\{\mu_n\}_1^\infty$ converges to $\lambda_{[0,1]}$ in the strong topology, but it converges to nothing at all in the uniform topology.*

Although the strong topology is weaker than the uniform and can be effectively used in various applications, it is still not weak enough for most probabilistic applications. Indeed, even when E possesses a good topological structure and $\mathcal{B} = \mathcal{B}_E$ is the Borel field over E, the strong topology on $\mathbf{M}_1(E)$ shows no respect for the topology on E. For example, suppose that E is a metric space and, for each $x \in E$, consider the point mass δ_x on $\mathcal{B}_E$. Then, no matter how *close* x gets to y in the metric topology on E, δ_x is not getting *close* to δ_y in the strong topology on $\mathbf{M}_1(E)$. More generally (cf. Exercise 3.1.16 below), measures cannot be close in the strong topology unless their *sets of small measure* are essentially the same. Thus, for example, the convergence which is occurring in The Central Limit Theorem (cf. Corollary 2.1.9) cannot, in general, be taking place in the strong topology; and since The Central Limit Theorem is an archetypical example of the sort of convergence result at which probabilists look, it is only sensible for us to take a hint from the result which we got there. That is, let E be a metric space, set $\mathcal{B} = \mathcal{B}_E$, and consider the neighborhood basis at $\mu \in \mathbf{M}_1(E)$ given by the sets in (3.1.3) when the φ_k's are *restricted to be elements of* $C_b(E;\mathbb{R})$. The topology which results is *much weaker* than the strong topology, and is therefore justifiably called the **weak topology** on $\mathbf{M}_1(E)$. (The reader who is familiar with the language of functional analysis will, with considerable justice, complain about this terminology. Indeed, if one thinks of $C_b(E;\mathbb{R})$ as a Banach space and of $\mathbf{M}_1(E)$ as a subspace of its dual space $C_b(E;\mathbb{R})^*$, then the topology which we are calling the weak topology is what a functional analyst would call the weak* topology. However, because it is the most commonly accepted choice of probabilists, we will continue to use the term *weak* instead of the more correct term *weak**.) In particular, the weak topology respects the topology on E: δ_y tends to δ_x in the weak topology on $\mathbf{M}_1(E)$ if and only if $y \longrightarrow x$ in E. As another indication that the weak topology is well adapted to the sort of analysis encountered in probability theory, let $\{\mu_n\} \cup \{\mu\} \subseteq \mathbf{M}_1(\mathbb{R}^N)$, and (cf. (2.2.6)) use Lemma 2.2.8 to see that

$$\mu_n \Longrightarrow \mu \quad \text{if and only if} \quad \hat{\mu}_n(\boldsymbol{\xi}) \longrightarrow \hat{\mu}(\boldsymbol{\xi}), \quad \boldsymbol{\xi} \in \mathbb{R}^N,$$

where we have introduced the notation $\mu_n \Longrightarrow \mu$ to indicate **weak convergence** (i.e., convergence in the weak topology). (See Exercise 3.1.19 for more information on this topic.)

Besides being well adapted to probabilistic analysis, the weak topology turns out to have many intrinsic virtues which are not shared by either the uniform or strong topologies. In particular, as we will see shortly, when E is a separable metric space, the weak topology on $\mathbf{M}_1(E)$ is not only a metric topology, which (cf. Exercise 3.1.16) the strong topology seldom is, but it is even separable, which the uniform topology seldom is. In order to check these properties, we will first have to review some elementary facts about separable metric spaces.

Given a metric ρ for a topological space E, we will use $U_b^\rho(E;\mathbb{R})$ to denote the space of bounded, ρ-uniformly continuous $\mathbb{R}$-valued functions on E and will endow $U_b^\rho(E;\mathbb{R})$ with the topology determined by the uniform metric. Thus, $U_b^\rho(E;\mathbb{R})$ becomes in this way a closed subspace of $C_b(E;\mathbb{R})$.

3.1.4 Lemma. *Let E be a separable metric space. Then E is homeomorphic to a subset of $[0,1]^{\mathbb{Z}^+}$. In particular:*

(i) *If E is compact, then the space $C(E;\mathbb{R})$ is separable with respect to the uniform metric.*

(ii) *Even when E is not compact, it nonetheless admits a metric $\hat\rho$ with respect to which it becomes a totally bounded metric space.*

(iii) *If $\hat\rho$ is a totally bounded metric on E, then $U_b^{\hat\rho}(E;\mathbb{R})$ is separable.*

PROOF: Let ρ be any metric on E and choose $\{p_n\}_1^\infty$ to be a countable, dense subset of E. Next, define $\mathbf{h} : E \longrightarrow [0,1]^{\mathbb{Z}^+}$ to be the mapping whose nth coordinate is given by

$$h_n(x) = \frac{\rho(x,p_n)}{1+\rho(x,p_n)}, \quad x \in E.$$

It is then an easy matter to check that $\mathbf{h}$ is a homeomorphism.

To prove (**i**), we first check it for compact subsets K of $E = [0,1]^{\mathbb{Z}^+}$. To this end, denote by $\mathcal{P}$ the space of functions $\mathbf{p} : [0,1]^{\mathbb{Z}^+} \longrightarrow \mathbb{R}$ of the form $\mathbf{p}(\boldsymbol{\xi}) = \prod_{m=1}^n p_m(t_m)$ as n runs over $\mathbb{Z}^+$ and each of the factors p_m is a polynomial on $[0,1]$ with rational coefficients. Clearly $\mathcal{P}$ is a countable subalgebra of $C\big([0,1]^{\mathbb{Z}^+};\mathbb{R}\big)$. Furthermore, given distinct points $\boldsymbol{\xi}$ and $\boldsymbol{\eta}$ from $[0,1]^{\mathbb{Z}^+}$, it is an easy (in fact, a one dimensional) matter to see that there is a $p \in \mathcal{P}$ for which $p(\boldsymbol{\xi}) \neq p(\boldsymbol{\eta})$. Hence, by the famous Stone–Weierstrass Approximation Theorem, we know that $\big\{p \restriction K : p \in \mathcal{P}\big\}$ is dense in $C(K;\mathbb{R})$ for every $K \subset\subset [0,1]^{\mathbb{Z}^+}$. Finally, for an arbitrary compact metric space E, define $\mathbf{h} : E \longrightarrow [0,1]^{\mathbb{Z}^+}$ as above, note that $K \equiv \mathbf{h}(E)$ is compact, and conclude that the map $\varphi \in C(K;\mathbb{R}) \longmapsto \varphi \circ \mathbf{h} \in C(E;\mathbb{R})$ is a homeomorphism between the uniform topologies on these spaces. Since we already know that $C(K;\mathbb{R})$ is separable, this completes (**i**).

The proof of (**ii**) is easy. Namely, define

$$D(\boldsymbol{\xi},\boldsymbol{\eta}) = \sum_{n=1}^\infty \frac{|\xi_n - \eta_n|}{2^n} \quad \text{for } \boldsymbol{\xi},\, \boldsymbol{\eta} \in [0,1]^{\mathbb{Z}^+}.$$

Clearly, D is a metric for $[0,1]^{\mathbb{Z}^+}$, and therefore

$$(x,y) \in E^2 \longmapsto \hat\rho(x,y) \equiv D\big(\mathbf{h}(x),\mathbf{h}(y)\big)$$

is a metric for E. At the same time, since $[0,1]^{\mathbb{Z}^+}$ is compact and therefore the restriction of D to any subset is totally bounded, it is clear that $\hat\rho$ is totally bounded on E.

To prove **(iii)**, let $\hat{E}$ denote the completion of E with respect to the totally bounded metric $\hat{\rho}$. Then, because E is dense in $\hat{E}$, $\hat{E}$ is both complete and totally bounded; and, therefore, it is compact. In addition, $\hat{\varphi} \in C(\hat{E};\mathbb{R}) \longmapsto \hat{\varphi} \upharpoonright E \in U_b^{\hat{\rho}}(E;\mathbb{R})$ is a surjective homeomorphism; and so **(iii)** now follows from **(i)**. □

One of the main reasons why Lemma 3.1.4 will be important to us is that it will enable us to show that, for separable metric spaces E, the weak topology on $\mathbf{M}_1(E)$ is also a separable metric topology. However, thus far we do not even know that the neighborhood bases are countably generated, and so, for a moment longer, we must continue to consider nets when discussing convergence.

3.1.5 Theorem. *Let E be any metric space and $\{\mu_\alpha : \alpha \in A\}$ a net in $\mathbf{M}_1(E)$. Given any $\mu \in \mathbf{M}_1(E)$, the following statements are equivalent:*

(i) $\mu_\alpha \Longrightarrow \mu$.

(ii) *If ρ is any metric for E, then $\langle \varphi, \mu_\alpha\rangle \longrightarrow \langle \varphi, \mu\rangle$ for every $\varphi \in U_b^{\rho}(E;\mathbb{R})$.*

(iii) *For every closed set $F \subseteq E$, $\overline{\lim_\alpha}\, \mu_\alpha(F) \le \mu(F)$.*

(iv) *For every open set $G \subseteq E$, $\underline{\lim_\alpha}\, \mu_\alpha(G) \ge \mu(G)$.*

(v) *For every upper semicontinuous function $f : E \longrightarrow \mathbb{R}$ which is bounded above, $\overline{\lim_\alpha} \langle f, \mu_\alpha\rangle \le \langle f, \mu\rangle$.*

(vi) *For every lower semicontinuous function $f : E \longrightarrow \mathbb{R}$ which is bounded below, $\underline{\lim_\alpha} \langle f, \mu_\alpha\rangle \ge \langle f, \mu\rangle$.*

(vii) *For every $f \in B(E;\mathbb{R})$ which is continuous at μ-almost every $x \in E$, $\langle f, \mu_\alpha\rangle \longrightarrow \langle f, \mu\rangle$.*

Next, assume that E is separable, and let $\hat{\rho}$ be a totally bounded metric for E. Then there exists a countable subset $\{\varphi_n\}_1^\infty \subseteq U_b^{\hat{\rho}}(E;[0,1])$ whose span is dense in $U_b^{\hat{\rho}}(E;\mathbb{R})$, and therefore the mapping $\mathbf{H} : \mathbf{M}_1(E) \longrightarrow [0,1]^{\mathbb{Z}^+}$ given by $\mathbf{H}(\mu) = (\langle\varphi_1,\mu\rangle, \dots, \langle\varphi_n,\mu\rangle, \dots)$ is a homeomorphism from the weak topology on $\mathbf{M}_1(E)$ into $[0,1]^{\mathbb{Z}^+}$. In particular, when E is separable, $\mathbf{M}_1(E)$ with the weak topology is itself a separable metric space and, in fact, one can take

$$(\mu,\nu) \in \mathbf{M}_1(E)^2 \longmapsto R(\mu,\nu) \equiv \sum_{n=1}^{\infty} \frac{|\langle\varphi_n,\mu\rangle - \langle\varphi_n,\nu\rangle|}{2^n}$$

to be a metric for $\mathbf{M}_1(E)$.

PROOF: The implications

$$\textbf{(vii)} \implies \textbf{(i)} \implies \textbf{(ii)}, \qquad \textbf{(iii)} \iff \textbf{(iv)}, \quad \text{and } \textbf{(v)} \iff \textbf{(vi)}$$

are all trivial. Thus, the first part will be complete once we check that **(ii)** $\implies$ **(iii)**, **(iv)** $\implies$ **(vi)**, and that **(v)** together with **(vi)** imply **(vii)**. To see the first of these, let F be a closed subset of E and set

$$\psi_n(x) = 1 - \left(\frac{\rho(x,F)}{1+\rho(x,F)}\right)^{\frac{1}{n}} \quad \text{for} \quad n \in \mathbb{Z}^+ \text{ and } x \in E.$$

It is then clear that $\psi_n \in U_b^\rho(E;\mathbb{R})$ for each $n \in \mathbb{Z}^+$ and that $1 \geq \psi_n(x) \searrow \mathbf{1}_F(x)$ as $n \to \infty$ for each $x \in E$. Thus, countable additivity followed by **(ii)** imply that

$$\mu(F) = \lim_{n\to\infty} \langle \psi_n, \mu\rangle = \lim_{n\to\infty} \lim_{\alpha} \langle \psi_n, \mu_\alpha\rangle \geq \overline{\lim_{\alpha}}\, \mu_\alpha(F).$$

In proving that **(iv)** $\implies$ **(vi)**, we may and will assume that f is a nonnegative, lower semicontinuous function. For $n \in \mathbb{N}$, define

$$f_n = \sum_{\ell=1}^{\infty} \frac{\ell \wedge 4^n}{2^n} \mathbf{1}_{I_{\ell,n}} \circ f = \frac{1}{2^n} \sum_{\ell=1}^{4^n} \mathbf{1}_{J_{\ell,n}} \circ f,$$

where

$$I_{\ell,n} = \left(\frac{\ell}{2^n}, \frac{\ell+1}{2^n}\right] \quad \text{and} \quad J_{\ell,n} = \left(\frac{\ell}{2^n}, \infty\right).$$

It is then clear that $0 \leq f_n \nearrow f$ and therefore that $\langle f_n, \mu\rangle \longrightarrow \langle f, \mu\rangle$ as $n \to \infty$. At the same time, by lower semicontinuity and **(iv)** applied to the open sets $\{f \in J_{\ell,n}\}$,

$$\langle f_n, \mu\rangle \leq \underline{\lim_{\alpha}} \langle f_n, \mu_\alpha\rangle \leq \underline{\lim_{\alpha}} \langle f, \mu_\alpha\rangle$$

for each $n \in \mathbb{Z}^+$; and so, after letting $n \to \infty$, we have now shown that **(iv)** $\implies$ **(vi)**.

Turning to the proof that **(v)** & **(vi)** $\implies$ **(vii)**, suppose that $f \in B(E;\mathbb{R})$ is continuous at μ-almost every $x \in E$, and define

$$\underline{f}(x) = \underline{\lim_{y\to x}} f(y) \quad \text{and} \quad \overline{f}(x) = \overline{\lim_{y\to x}} f(y) \quad \text{for } x \in E.$$

It is then an easy matter to check that $\underline{f} \leq f \leq \overline{f}$ everywhere and that equality holds μ-almost surely. Furthermore, $\underline{f}$ is lower semicontinuous, $\overline{f}$ is upper semicontinuous, and both are bounded. Hence, by **(v)** and **(vi)**,

$$\begin{aligned}\overline{\lim_{\alpha}} \langle f, \mu_\alpha\rangle &\leq \overline{\lim_{\alpha}} \langle \overline{f}, \mu_\alpha\rangle \leq \langle \overline{f}, \mu\rangle = \langle f, \mu\rangle \\ &= \langle \underline{f}, \mu\rangle \leq \underline{\lim_{\alpha}} \langle \underline{f}, \mu_\alpha\rangle \leq \underline{\lim_{\alpha}} \langle f, \mu_\alpha\rangle;\end{aligned}$$

and so we have now completed the proof that conditions (**i**) through (**vii**) are equivalent.

Now assume that E is separable, and let $\hat{\rho}$ be a totally bounded metric for E. By (**iii**) of Lemma 3.1.4, $U_b^{\hat{\rho}}(E;\mathbb{R})$ is separable. Hence, we can find a countable set $\{\varphi_n\}_1^\infty \subseteq U_b^{\hat{\rho}}(E;[0,1])$ which spans a dense subset of $U_b^{\hat{\rho}}(E;\mathbb{R})$. In particular, by the equivalence of (**i**) and (**ii**) above, we see that $\langle\varphi_n,\mu_\alpha\rangle \longrightarrow \langle\varphi_n,\mu\rangle$ for all $n\in\mathbb{Z}^+$ if and only if $\mu_\alpha \Longrightarrow \mu$; which is to say that the corresponding map $\mathbf{H}: E \longrightarrow [0,1]^{\mathbb{Z}^+}$ is a homeomorphism. Since $[0,1]^{\mathbb{Z}^+}$ is a compact metric space and D (cf. the proof of (**ii**) in Lemma 3.1.4) is a metric for it, we also see that the R in (3.1.5) is a totally bounded metric for $\mathbf{M}_1(E)$. Finally, since, by (**ii**) in Lemma 3.1.4, it is always possible to find a totally bounded metric for E, the last assertion needs no further comment. □

The reader would do well to pay close attention to what (**iii**) and (**iv**) say about the nature of weak convergence. Namely, even though $\mu_\alpha \Longrightarrow \mu$, it is possible that some or all of the mass which the μ_α's assign to the interior of a set may gravitate to the boundary in the limit. This phenomenon is most easily understood by taking $E=\mathbb{R}$, μ_α to be the unit point mass δ_α at $\alpha\in[0,1)$, checking that $\mu_\alpha \Longrightarrow \delta_1$, and noting that $0=\delta_1\big((0,1)\big) < 1 = \mu_\alpha\big((0,1)\big)$ for each $\alpha\in[0,1)$.

3.1.6 Remark. Those who find nets distasteful will be pleased to learn that, from now on, we will be restricting our attention to separable metric spaces E and therefore need only discuss sequential convergence when working with the weak topology on $\mathbf{M}_1(E)$. Furthermore, unless the contrary is explicitly stated, *we will always be thinking of the weak topology when working with* $\mathbf{M}_1(E)$.

Given a separable metric space E, we next want to find conditions which guarantee that a subset of $\mathbf{M}_1(E)$ is compact; and at this point it will be convenient to have introduced the notation $K \subset\subset E$ to indicate that K is a compact subset of E. Also, we will say that E is a **Polish space** if E is a separable metric space which admits a complete metric.

The key here is the following variation on the renowned Riesz Representation Theorem combined with an important observation made by Ulam.†

3.1.7 Lemma. *Let E be a separable metric space, ρ a metric for E, and Λ a nonnegative linear functional on $U_b^\rho(E;\mathbb{R})$ (i.e., Λ is a linear map which assigns nonnegative numbers to nonnegative $\varphi\in U_b^\rho(E;\mathbb{R})$) with $\Lambda(\mathbf{1})=1$. Then in order for there to be a (necessarily unique) $\mu\in\mathbf{M}_1(E)$ satisfying $\Lambda(\varphi)=\langle\varphi,\mu\rangle$*

† It is no accident that Ulam was the first to make this observation. Indeed, the term *Polish space* was coined by Bourbaki in recognition of the contribution made to this subject by the Polish school in general and C. Kuratowski in particular (cf. Kuratowski's *Topologie, Vol. I*, Warszawa–Lwow, (1933)). Ulam had been Kuratowski's student.

for all $\varphi \in U_b^\rho(E;\mathbb{R})$, it is sufficient that, for every $\epsilon > 0$, there exist a $K \subset\subset E$ such that

$$|\Lambda(\varphi)| \le \sup_{x\in K} |\varphi(x)| + \epsilon \sup_{x\notin K} |\varphi(x)|, \quad \varphi \in U_b^\rho(E;\mathbb{R}). \tag{3.1.8}$$

Conversely, if E is a Polish space and $\mu \in \mathbf{M}_1(E)$, then for every $\epsilon > 0$ there is a $K \subset\subset E$ such that $\mu(K) \ge 1-\epsilon$. In particular, if $\mu \in \mathbf{M}_1(E)$ and $\Lambda(\varphi) = \langle \varphi, \mu\rangle$ for $\varphi \in C_b(E;\mathbb{R})$, then, for each $\epsilon > 0$ (3.1.8) holds for some $K \subset\subset E$.

PROOF: We begin with the trivial observation that, because Λ is nonnegative and $\Lambda(\mathbf{1}) = 1$, $|\Lambda(\varphi)| \le \|\varphi\|_u$. Next, according to the Daniell theory of integration, the first statement will be proved as soon as we show that $\Lambda(\varphi_n) \searrow 0$ whenever $\{\varphi_n\}_1^\infty$ is a nonincreasing sequence of functions from $U_b^\rho(E;[0,\infty))$ which tend pointwise to 0 as $n \to \infty$. To this end, let $\epsilon > 0$ be given and choose $K \subset\subset E$ so that (3.1.8) holds. One then has that

$$\varlimsup_{n\to\infty} |\Lambda(\varphi_n)| \le \lim_{n\to\infty} \sup_{x\in K} |\varphi_n(x)| + \epsilon\|\varphi_1\|_u = \epsilon\|\varphi_1\|_u,$$

since, by Dini's Lemma, $\varphi_n \searrow 0$ uniformly on compact subsets of E.

Turning to the second part, assume that E is Polish and use $B_E(x,r)$ to denote the open ball of radius $r > 0$ around $x \in E$, computed with respect to the complete metric ρ for E. Next, let $\{p_k\}_1^\infty$ be a countable dense subset of E, and set $B_{k,n} = B_E\left(p_k, \frac{1}{n}\right)$ for $k,\ n \in \mathbb{Z}^+$. Given $\mu \in \mathbf{M}_1(E)$ and $\epsilon > 0$, we can choose, for each $n \in \mathbb{Z}^+$, an $\ell_n \in \mathbb{Z}^+$ so that

$$\mu\left(\bigcup_{k=1}^{\ell_n} B_{k,n}\right) \ge 1 - \frac{\epsilon}{2^n}.$$

Hence, if

$$C_n \equiv \bigcup_{k=1}^{\ell_n} \overline{B}_{k,n} \quad \text{and} \quad K = \bigcap_{n=1}^{\infty} C_n,$$

then $\mu(K) \ge 1-\epsilon$. At the same time, it is obvious that, on the one hand, K is closed (and therefore ρ-complete) and that, on the other hand, $K \subseteq \bigcup_{k=1}^{\ell_n} B_E\left(p_k, \frac{2}{n}\right)$ for every $n \in \mathbb{Z}^+$. Hence, K is both complete and totally bounded with respect to ρ and is, therefore, compact. □

As Lemma 3.1.7 makes clear, probability measures on a Polish space like to be *nearly concentrated on a compact set.* Following Prohorov and Varadarajan,†

† See Yu. V. Prohorov's article "Convergence of random processes and limit theorems in probability theory," *Theory of Prob. & Appl.*, which appeared in 1956. Independently, V.S. Varadarajan developed essentially the same theory in "Weak convergence of measures on a separable metric spaces," *Sankhyă*, which was published in 1958. Although Prohorov got into print first, subsequent expositions, including this one, rely heavily on Varadarajan.

what we are about to show is that, for a Polish space E, relatively compact subsets of $\mathbf{M}_1(E)$ are those whose elements are *nearly concentrated on the same compact set of* E. More precisely, given a separable metric space E, we will say that $M \subseteq \mathbf{M}_1(E)$ is **tight** if, for every $\epsilon > 0$, there exists a $K \subset\subset E$ such that $\mu(K) \geq 1 - \epsilon$ for all $\mu \in M$.

3.1.9 Theorem. *Let E be a separable metric space and $M \subseteq \mathbf{M}_1(E)$. Then $\overline{M}$ is compact if M is tight. Conversely, when E is Polish, M is tight if $\overline{M}$ is compact.*†

PROOF: Since it is clear, from (**iii**) in Theorem 3.1.5, that $\overline{M}$ is tight if and only if M is, we will assume throughout that M is closed in $\mathbf{M}_1(E)$.

To prove the first statement, take $\hat{\rho}$ to be a totally bounded metric on E, choose $\{\varphi_n\}_1^\infty \subseteq U_{\mathrm{b}}^{\hat{\rho}}(E;[0,1])$ accordingly, as in the last part of Theorem 3.1.5, and let $\varphi_0 = \mathbf{1}$. Given a sequence $\{\mu_\ell\}_1^\infty \subseteq \mathbf{M}_1(E)$, we can use a standard diagonalization procedure to extract a subsequence $\{\mu_{\ell_k}\}_{k=1}^\infty$ such that

$$\Lambda(\varphi_n) \equiv \lim_{k\to\infty} \langle \varphi_n, \mu_{\ell_k} \rangle$$

exists for each $n \in \mathbb{N}$. Since

$$\Lambda(\varphi) \equiv \lim_{k\to\infty} \langle \varphi, \mu_{\ell_k} \rangle$$

continues to exist for every φ in the uniform closure of the span of $\{\varphi_n\}_1^\infty$, we now see that Λ determines a nonnegative linear functional on $U_{\mathrm{b}}^{\hat{\rho}}(E;\mathbb{R})$ and that $\Lambda(\mathbf{1}) = 1$. Moreover, because M is tight, we can use (**iii**) in Theorem 3.1.5 to find, for any $\epsilon > 0$, a $K \subset\subset E$ such that $\mu(K) \geq 1 - \epsilon$, $\mu \in M$; and therefore (3.1.8) holds with this choice of K. Hence, by Lemma 3.1.7, we know that there is a $\mu \in \mathbf{M}_1(E)$ for which $\Lambda(\varphi) = \langle \varphi, \mu \rangle$, $\varphi \in U_{\mathrm{b}}^{\hat{\rho}}(E;\mathbb{R})$. Because this means that $\langle \varphi, \mu_{\ell_k} \rangle \longrightarrow \langle \varphi, \mu \rangle$ for every $\varphi \in U_{\mathrm{b}}^{\hat{\rho}}(E;\mathbb{R})$, the equivalence of (**i**) and (**ii**) in Theorem 3.1.5 allows us to conclude that $\mu_{\ell_k} \Longrightarrow \mu$.

Finally, suppose that E is Polish and that M is compact in $\mathbf{M}_1(E)$. To see that M must be tight, we repeat the argument used to prove the second part of Lemma 3.1.7. Thus, choose $B_{k,n}$, k, $n \in \mathbb{Z}^+$ as in the proof there, and set

$$f_{\ell,n}(\mu) = \mu\left(\bigcup_{k=1}^{\ell} B_{k,n}\right) \quad \text{for } \ell,\ n \in \mathbb{Z}^+.$$

† For the reader who wishes to investigate just how far these results can be pushed before they start of break down, a good place to start is Appendix III in P. Billingsley's *Convergence of Probability Measures*, publ. by J. Wiley, (196?). In particular, although it is reasonably clear that completeness is more or less essential for the necessity, the havoc which results from dropping separability may come as a surprise.

By **(iv)** in Theorem 3.1.5, $\mu \in \mathbf{M}_1(E) \longmapsto f_{\ell,n}(\mu) \in [0,1]$ is lower semicontinuous. Moreover, for each $n \in \mathbb{Z}^+$, $f_{\ell,n} \nearrow 1$ as $\ell \nearrow \infty$. Thus, by Dini's Lemma, we can choose, for each $n \in \mathbb{Z}^+$, one $\ell_n \in \mathbb{Z}^+$ so that $f_{\ell_n,n}(\mu) \geq 1 - \frac{\epsilon}{2^n}$ for all $\mu \in M$; and at this point the rest of the argument is precisely the same as the one given at the end of the proof of Lemma 3.1.7. □

We have now seen that $\mathbf{M}_1(E)$ inherits properties from E. To be more specific, if E is a metric space, then $\mathbf{M}_1(E)$ is separable or compact if E itself is. What we want to show next is that completeness also gets transferred. That is, we will show that $\mathbf{M}_1(E)$ is Polish if E is. In order to see this, we will need a lemma which is of considerable importance in its own right.

3.1.10 Lemma. *Let E be a Polish space and Φ a bounded subset of $C_b(E;\mathbb{R})$ which is equicontinuous at each $x \in E$. (That is, for each $x \in E$, $\sup_{\varphi\in\Phi} |\varphi(y) - \varphi(x)| = 0$ as $y \to x$.) If $\{\mu_n\}_1^\infty \cup \{\mu\} \subseteq \mathbf{M}_1(E)$ and $\mu_n \Longrightarrow \mu$, then*

$$\lim_{n\to\infty} \sup_{\varphi\in\Phi} \Big| \langle \varphi, \mu_n \rangle - \langle \varphi, \mu \rangle \Big| = 0.$$

PROOF: Let $\epsilon > 0$ be given and use the second part of Theorem 3.1.8 to choose $K \subset\subset E$ so that

$$\sup_{\varphi\in\Phi} \|\varphi\|_u \sup_{n\in\mathbb{Z}^+} \mu_n(K\complement) < \frac{\epsilon}{4}.$$

By **(iii)** of Theorem 3.1.5, $\mu(K\complement)$ satisfies the same estimate. Next, choose a metric ρ for E and a countable dense set $\{p_k\}_1^\infty$ in K. Using equicontinuity together with compactness, find $\ell \in \mathbb{Z}^+$ and $\delta_1, \dots, \delta_\ell > 0$ so that $K \subseteq \{x : \rho(x, p_k) < \delta_k \text{ for some } 1 \leq k \leq \ell\}$ and

$$\sup_{\varphi\in\Phi} \big|\varphi(x) - \varphi(p_k)\big| < \frac{\epsilon}{4} \quad \text{for } 1 \leq k \leq \ell \text{ and } x \in K \text{ with } \rho(x, p_k) < 2\delta_k.$$

Because

$$r \in (0,\infty) \longmapsto \mu\Big(\big\{y \in K : \rho(y,x) \leq r\big\}\Big) \in [0,1]$$

is nondecreasing for each $x \in K$, we can find an $r_k \in \big(\delta_k, 2\delta_k\big)$ so that, when $B_k \equiv \big\{x \in K : \rho(x, p_k) < r_k\big\}$, $\mu\big(\partial B_k\big) = 0$ for each $1 \leq k \leq \ell$. Finally, set $A_1 = B_1$ and $A_{k+1} = B_{k+1} \setminus \bigcup_{j=1}^k B_j$ for $1 \leq k < \ell$. Then, $K \subseteq \bigcup_{k=1}^\ell A_k$, the A_k's are disjoint, and, for each $1 \leq k \leq \ell$,

$$\sup_{\varphi\in\Phi} \sup_{x\in A_k} \big|\varphi(x) - \varphi(p_k)\big| < \frac{\epsilon}{4} \quad \text{and} \quad \mu\big(\partial A_k\big) = 0.$$

Hence, by **(vii)** in Theorem 3.1.5 applied to the $\mathbf{1}_{A_k}$'s:

$$\varlimsup_{n\to\infty} \sup_{\varphi\in\Phi} \Big|\langle\varphi,\mu_n\rangle - \langle\varphi,\mu\rangle\Big|$$

$$< \epsilon + \varlimsup_{n\to\infty} \sum_{k=1}^{\ell} \sup_{\varphi\in\Phi} \big|\varphi(p_k)\big| \big|\mu_n(A_k) - \mu(A_k)\big| = \epsilon. \quad \square$$

3.1.11 Theorem. *Let E be a Polish space and ρ a complete metric for E. Given $(\mu,\nu) \in \mathbf{M}_1(E)^2$, define*

$$\begin{aligned} L(\mu,\nu) = \inf\Big\{\delta : \mu(F) \le \nu\big(F^{(\delta)}\big) + \delta \\ \text{and } \nu(F) \le \mu\big(F^{(\delta)}\big) + \delta \text{ for all closed } F \subseteq E\Big\}, \end{aligned} \tag{3.1.12}$$

where we use $F^{(\delta)}$ to denote the set of $x \in E$ which lie a ρ-distance less than δ from F. Then L is a complete metric for $\mathbf{M}_1(E)$, and therefore $\mathbf{M}_1(E)$ is Polish.

PROOF: It is clear that L is symmetric and that it satisfies the triangle inequality. Thus, we will know that it is a metric for $\mathbf{M}_1(E)$ as soon as we show that $L(\mu_n,\mu) \longrightarrow 0$ if and only if $\mu_n \Longrightarrow \mu$. To this end, first suppose that $L(\mu_n,\mu) \longrightarrow 0$. Then, for every closed F, $\mu\big(F^{(\delta)}\big) + \delta \ge \overline{\lim}_{n\to\infty} \mu_n(F)$ for all $\delta > 0$; and therefore, by countable additivity, $\mu(F) \ge \overline{\lim}_{n\to\infty} \mu_n(F)$ for every closed F. Hence, by the equivalence of **(i)** and **(iii)** in Theorem 3.1.5, $\mu_n \Longrightarrow \mu$. Next, suppose that $\mu_n \Longrightarrow \mu$ and let $\delta > 0$ be given. Given a closed F in E, define

$$\psi_F(x) = \frac{\rho\big(x, F^{(\delta)}\complement\big)}{\rho\big(x, F^{(\delta)}\complement\big) + \rho(x,F)} \quad \text{for} \quad x \in E.$$

It is then an easy matter to check that both

$$\mathbf{1}_F \le \psi_F \le \mathbf{1}_{F^{(\delta)}} \quad \text{and} \quad \big|\psi_F(x) - \psi_F(y)\big| \le \frac{\rho(x,y)}{\delta}.$$

In particular, by Lemma 3.1.10, we can choose $m \in \mathbb{Z}^+$ so that

$$\sup_{n\ge m} \sup\Big\{\big|\langle\psi_F,\mu_n\rangle - \langle\psi_F,\mu\rangle\big| : F \text{ closed in } E\Big\} < \delta;$$

from which it is an easy matter to see that, for all $n \ge m$,

$$\mu(F) \le \mu_n\big(F^{(\delta)}\big) + \delta \quad \text{and} \quad \mu_n(F) \le \mu\big(F^{(\delta)}\big) + \delta.$$

In other words, $\sup_{n\ge m} L(\mu_n,\mu) \le \delta$; and, since $\delta > 0$ was arbitrary, we have shown that $L(\mu_n,\mu) \longrightarrow 0$.

In order to finish the proof, we must show that if $\{\mu_n\}_1^\infty \subseteq \mathbf{M}_1(E)$ is L-Cauchy convergent, then it is tight. Thus, let $\epsilon > 0$ be given and choose, for each $\ell \in \mathbb{Z}^+$, an $m_\ell \in \mathbb{Z}^+$ and a $K_\ell \subset\subset E$ so that

$$\sup_{n\ge m_\ell} L(\mu_n,\mu_{m_\ell}) \le \frac{\epsilon}{2^{\ell+1}} \quad \text{and} \quad \max_{1\le n\le m_\ell} \mu_n\big(K_\ell\complement\big) \le \frac{\epsilon}{2^{\ell+1}}.$$

Setting $\epsilon_\ell = \frac{\epsilon}{2^\ell}$, one then has that:

$$\sup_{n\in\mathbb{Z}^+} \mu_n\big(K_\ell^{(\epsilon_\ell)}\complement\big) \le \epsilon_\ell \quad \text{for each} \quad \ell \in \mathbb{Z}^+.$$

In particular, if

$$K \equiv \bigcap_{\ell=1}^{\infty} \overline{K_\ell^{(\epsilon_\ell)}},$$

then $\mu_n(K) \ge 1-\epsilon$ for all $n \in \mathbb{Z}^+$. In addition, because each K_ℓ is compact, it is easy to see that K is both ρ-complete and totally bounded and therefore also compact. □

When $E = \mathbb{R}$, P. Lévy was the first one to construct a complete metric on $\mathbf{M}_1(E)$, and it is for this reason that we will call the metric L in (3.1.12) the **Lévy metric** determined by ρ. Using an abstract argument, Varadarajan showed that $\mathbf{M}_1(E)$ must be Polish whenever E is, and the explicit construction which we have used was produced first by Prohorov.

Before closing this section, it seems appropriate to introduce and explain some of the more classical terminology connected with the applications of weak convergence to probability theory. For this purpose, let $(\Omega, \mathcal{F}, P)$ be a probability space and E a metric space. Given E-valued random variables $\{X_n\}_1^\infty \cup \{X\}$ on $(\Omega, \mathcal{F}, P)$, one says that the sequence X_n **tends to** X **in law** (or **distribution**) and writes $X_n \xrightarrow{\mathcal{L}} X$ if (cf. Exercise 1.1.16) $X_n^*P \Longrightarrow X^*P$. The idea here is that, when the measures under consideration are the distributions of random variables, one wants to think of weak convergence of the distributions as determining a kind of convergence of the corresponding random variables. Thus, one can add *convergence in law* to the list of possible ways in which random variables might converge. In order to elucidate the relationship between convergence in law, P-almost sure convergence, and convergence in P-measure, it will be useful to have the following lemma.

3.1.13 Lemma. *Let $(\Omega, \mathcal{F}, P)$ be a probability space and E a metric space. Given any E-valued random variables $\{X_n\}_1^\infty \cup \{X\}$ on $(\Omega, \mathcal{F}, P)$ and any pair of topologically equivalent metrics ρ and σ for E, $\rho\big(X_n, X\big) \longrightarrow 0$ in P-measure if and only if $\sigma\big(X_n, X\big) \longrightarrow 0$ in P-measure. In particular, convergence in P-measure does not depend on the choice of metric; and so we can write $X_n \longrightarrow X$ in P-measure without specifying a metric. Moreover, if $X_n \longrightarrow X$ in P-measure, then $X_n \xrightarrow{\mathcal{L}} X$.*

Next, given any metric ρ for E, set

$$\sigma(x,y) = \frac{\rho(x,y)}{1+\rho(x,y)} \quad \text{for} \quad x,\, y \in E,$$

and define

$$R(X,Y) = \mathbb{E}^P\big[\sigma(X,Y)\big]^{\frac{1}{2}}$$

for E-valued random variables X and Y. Then σ is again a metric for E, R is a pseudo-metric for P-convergence in the space of all E-valued random variables on $(\Omega, \mathcal{F}, P)$, and both σ and R are complete if ρ is complete.
Finally, assume that E is Polish, let ρ be a complete metric for E, define σ and R from ρ as in the preceding, and let L be the Lévy metric determined by σ. Then

$$L\big(X^*P, Y^*P\big) \le R(X,Y)$$

for all E-valued random variables X and Y on $(\Omega, \mathcal{F}, P)$.

PROOF: To prove the first assertion, suppose that

$$\rho(X_n, X) \longrightarrow 0 \text{ in } P\text{-measure but that } \sigma(X_n, X) \not\longrightarrow 0 \text{ in } P\text{-measure.}$$

After passing to a subsequence if necessary, we could then arrange that $\rho(X_n, X) \longrightarrow 0$ (a.s., P) but $P\big(\sigma(X_n, X) \ge \epsilon\big) \ge \epsilon$ for all $n \in \mathbb{Z}^+$ and some $\epsilon > 0$. But this is impossible, since then we would have that $\sigma(X_n, X) \longrightarrow 0$ P-almost surely but not in P-measure. Hence, we have now proved that convergence in P-measure does not depend on the choice of metric. To complete the first part, suppose that $\rho(X_n, X) \longrightarrow 0$ in P-measure. Then, for every $\varphi \in U_{\text{b}}^{\rho}(E;\mathbb{R})$ and $\delta > 0$,

$$\begin{aligned}\varlimsup_{n\to\infty} \left|\mathbb{E}^P\big[\varphi(X_n)\big] - \mathbb{E}^P\big[\varphi(X)\big]\right| &\le \varlimsup_{n\to\infty} \mathbb{E}^P\Big[\big|\varphi(X_n) - \varphi(X)\big|\Big] \\ &\le \epsilon(\delta) + \|\varphi\|_{\text{u}} \varlimsup_{n\to\infty} P\Big(\rho(X_n, X) \ge \delta\Big) = \epsilon(\delta),\end{aligned}$$

where

$$\epsilon(\delta) \equiv \sup\big\{|\varphi(y) - \varphi(x)| : \rho(x,y) \le \delta\big\} \longrightarrow 0 \quad \text{as} \quad \delta \searrow 0.$$

Hence, by **(ii)** in Theorem 3.1.5, $X_n^*P \Longrightarrow X^*P$.

Turning to the second part, note that the only assertion about σ needing comment is that it satisfies the triangle inequality. However, this is easily seen from the triangle inequality for ρ together with the trivial facts that $t \in [0,\infty) \longmapsto \frac{t}{1+t}$ is strictly increasing and that $\frac{a+b}{1+a+b} \le \frac{a}{1+a} + \frac{b}{1+b}$ for all a, $b \in [0,\infty)$. Once one knows that σ is a metric, it is immediate that $(X,Y) \longmapsto R(X,Y)^2$ is a pseudo-metric, and therefore the same conclusion for R itself follows from the inequality $\sqrt{a+b} \le \sqrt{a} + \sqrt{b}$ for all a, $b \in [0,\infty)$. Furthermore, the equivalence between $R(X_n, X) \longrightarrow 0$ and $X_n \longrightarrow X$ in P-measure is an easy application of elementary measure theory.

To prove the final assertion, suppose that $\delta \equiv R(X,Y) > 0$. Then, for any closed F in E,

$$\begin{aligned}X^*P(F) = P\big(X \in F\big) &\le P\big(\sigma(Y,F) < \delta\big) + P\big(\sigma(X,Y) \ge \delta\big) \\ &\le Y^*P\big(F^{(\delta)}\big) + \frac{R(X,Y)^2}{\delta} = Y^*P\big(F^{(\delta)}\big) + \delta,\end{aligned}$$

where $F^{(\delta)}$ is computed relative to the metric σ. Since the same relation holds when the rôles of X and Y are reversed, it follows that, when L is defined relative to σ, $L\big(X^*P, Y^*P\big) \le \delta$. $\square$

As a demonstration of the sort of use to which one can put these ideas, we present the following version of the **principle of accompanying laws**.

3.1.14 Theorem. *Let E be a Polish space and, for each $k \in \mathbb{Z}^+$, let $\{Y_{k,n}\}_{n=1}^{\infty}$ be a sequence of E-valued random variables on the probability space $(\Omega, \mathcal{F}, P)$. Further, assume that, for each $k \in \mathbb{Z}^+$, there is a $\mu_k \in \mathbf{M}_1(E)$ such that $Y_{k,n}^* P \Longrightarrow \mu_k$ as $n \to \infty$. Finally, let ρ be a complete metric for E and suppose that $\{X_n\}_1^{\infty}$ is a sequence of E-valued random variables on $(\Omega, \mathcal{F}, P)$ with the property that*

$$\lim_{k\to\infty} \varlimsup_{n\to\infty} P\Big(\rho\big(X_n, Y_{k,n}\big) \geq \epsilon\Big) = 0 \quad \text{for every } \epsilon > 0. \tag{3.1.15}$$

Then there is a $\mu \in \mathbf{M}_1(E)$ such that $\mu_k \Longrightarrow \mu$ as $k \to \infty$ and $X_n^* P \Longrightarrow \mu$ as $n \to \infty$.

PROOF: Clearly, (3.1.15) continues to hold after ρ is replaced by $\sigma = \frac{\rho}{1+\rho}$. Thus, with R and L as in Lemma 3.1.13, we see that, for any $k \in \mathbb{Z}^+$,

$$\begin{aligned}
\sup_{\ell\geq k} L\big(\mu_\ell, \mu_k\big) &= \sup_{\ell\geq k} \lim_{n\to\infty} L\big(Y_{\ell,n}^* P, Y_{k,n}^* P\big) \\
&\leq \sup_{\ell\geq k} \varlimsup_{n\to\infty} \Big(L\big(Y_{\ell,n}^* P, X_n^* P\big) + L\big(X_n^* P, Y_{k,n}^* P\big)\Big) \\
&\leq 2 \sup_{\ell\geq k} \varlimsup_{n\to\infty} R\big(Y_{\ell,n}, X_n\big) \longrightarrow 0 \quad \text{as} \quad k \to \infty.
\end{aligned}$$

Hence, because L is complete, we know that $\mu_k \Longrightarrow \mu$ for some $\mu \in \mathbf{M}_1(E)$. In addition,

$$\begin{aligned}
\varlimsup_{n\to\infty} L\big(\mu, X_n^* P\big) &= \lim_{k\to\infty} \varlimsup_{n\to\infty} L\big(Y_{k,n}^* P, X_n^* P\big) \leq \varlimsup_{k\to\infty} \varlimsup_{n\to\infty} R\big(Y_{k,n}, X_n\big) \\
&\leq \varlimsup_{k\to\infty} \varlimsup_{n\to\infty} \left(\epsilon^{\frac{1}{2}} + P\Big(\sigma\big(Y_{k,n}, X_n\big) \geq \epsilon\Big)^{\frac{1}{2}}\right) = \epsilon^{\frac{1}{2}}
\end{aligned}$$

for every $\epsilon > 0$. □

Exercises

3.1.16 Exercise: Let $(E, \mathcal{B})$ be a measurable space. In this exercise, we investigate the strong topology in a little more detail. In particular, in part (**iii**) below, we will show that when $\mu \in \mathbf{M}_1(E)$ is **nonatomic** (i.e., $\mu(\{x\}) = 0$ for every $x \in E$), then there is no countable neighborhood basis of μ in the strong topology. Obviously, this means that *the strong topology for* $\mathbf{M}_1(E)$ *admits no metric whenever* $\mathbf{M}_1(E)$ *contains a nonatomic element.*

(**i**) Show that if $\{\mu_n\}_1^\infty$ is a sequence in $\mathbf{M}_1(E)$ which tends in the strong topology to $\mu \in \mathbf{M}_1(E)$, then

$$\mu \ll \sum_{n=1}^{\infty} \frac{\mu_n}{2^n}.$$

(**ii**) Given $\mu \in \mathbf{M}_1(E)$, show that μ admits a countable neighborhood basis in the strong topology if and only if there exists a countable $\{\varphi_k\}_1^\infty \subseteq B(E;\mathbb{R})$ such that, for every net $\{\mu_\alpha : \alpha \in A\} \subseteq \mathbf{M}_1(E)$, $\mu_\alpha \longrightarrow \mu$ in the strong topology as soon as

$$\lim_\alpha \langle \varphi_k, \mu_\alpha \rangle = \langle \varphi_k, \mu \rangle \quad \text{for every} \quad k \in \mathbb{Z}^+.$$

(**iii**) Referring to Exercises 1.1.14 and 1.1.16, set $\Omega = E^{\mathbb{Z}^+}$ and $\mathcal{F} = \mathcal{B}_E^{\mathbb{Z}^+}$. Next, let $\mu \in \mathbf{M}_1(E)$ be given, and define $P = \mu^{\mathbb{Z}^+}$ on $(\Omega, \mathcal{F})$. Show that, for any $\varphi \in B(E;\mathbb{R})$, the random variables

$$\mathbf{x} \in \Omega \longmapsto X_n^\varphi(\mathbf{x}) \equiv \varphi(x_n) \qquad n \in \mathbb{Z}^+$$

are mutually independent and all have distribution $\varphi^*\mu$. In particular, use the Strong Law of Large Numbers to conclude that

$$\lim_{n\to\infty} \frac{1}{n} \sum_{m=1}^{n} X_m^\varphi(\mathbf{x}) = \langle \varphi, \mu \rangle$$

for each $\mathbf{x}$ outside of a P-null set.

Now assume that μ is nonatomic and suppose that μ admitted a countable neighborhood basis in the strong topology. Choose $\{\varphi_k\}_1^\infty \subseteq B(E;\mathbb{R})$ accordingly, as in (**ii**), and (using the preceding) conclude that there exists at least one $\mathbf{x} \in \Omega$ for which the measures μ_n given by

$$\mu_n \equiv \frac{1}{n} \sum_{m=1}^{n} \delta_{x_m}, \quad n \in \mathbb{Z}^+,$$

converge in the strong topology to μ. Finally, apply (**i**) to see that this is impossible.

3.1.17 Exercise: Throughout this exercise, E is a separable metric space.

(**i**) We already know that $\mathbf{M}_1(E)$ is separable; however our proof was nonconstructive. Show that if $\{p_k\}_1^\infty$ is a dense subset of E, then the set of all convex combinations $\sum_{k=1}^n \alpha_k \delta_{p_k}$, where $n \in \mathbb{Z}^+$ and $\{\alpha_k\}_1^n \subset [0,1] \cap \mathbb{Q}$ with $\sum_1^n \alpha_k = 1$, form a countable dense set in $\mathbf{M}_1(E)$.

(ii) We have seen that $\mathbf{M}_1(E)$ is compact if E is. To see that the converse is also true, show that $x \in E \longmapsto \delta_x \in \mathbf{M}_1(E)$ is a homeomorphism whose image is closed.

(iii) Although it is a little off our track, it is amusing to show that E being compact is equivalent to $C_b(E;\mathbb{R})$ being separable; and, in view of **(i)** in Lemma 3.1.4, this comes down to checking that E is compact if $C_b(E;\mathbb{R})$ is separable.

Hint: Let $\hat{\rho}$ be a totally bounded metric on E and use $\hat{E}$ to denote the $\hat{\rho}$-completion of E. Show that if $\{x_n\}_1^\infty \subseteq E$ has the properties that $x_n \longrightarrow \hat{x} \in \hat{E}$ and $\lim_{n\to\infty} \varphi(x_n)$ exists for every $\varphi \in C_b(E;\mathbb{R})$, then $\hat{x} \in E$. (Suppose not, set

$$\psi(x) = \frac{1}{\hat{\rho}(x,\hat{x})},$$

and consider functions of the form $f \circ \psi$ for $f \in C_b(\mathbb{R};\mathbb{R})$.) Finally, assuming that $C_b(E;\mathbb{R})$ is separable, and using a diagonalization procedure, show that every sequence $\{x_n\}_1^n \subseteq E$ admits a subsequence $\{x_{n_m}\}_{m=1}^\infty$ which converges to some $\hat{x} \in \hat{E}$ and $\lim_{m\to\infty} \varphi(x_{n_m})$ exists for every $\varphi \in C_b(E;\mathbb{R})$.

(iv) Let $\{M_n\}_1^\infty$ be a sequence of finite measures on $(E,\mathcal{B})$. Assuming that $\{M_n\}_1^\infty$ is **tight** in the sense that $\{M_n(E)\}_1^\infty$ is bounded and that, for each $\epsilon > 0$, there is a $K \subset\subset E$ such that $\sup_n M_n(K\complement) \le \epsilon$, show that there is a subsequence $\{M_{n_k}\}_{k=1}^\infty$ and a finite measure M such that

$$\int_E \varphi\, dM = \lim_{k\to\infty} \int_E \varphi\, dM_{n_k}, \quad \text{for all } \varphi \in C_b(E;\mathbb{R}).$$

Conversely, if there is a finite measure M such that $\int_E \varphi\, dM_n \longrightarrow \int_E \varphi\, dM$ for every $\varphi \in C_b(E;\mathbb{R})$, show that $\{M_n\}_1^\infty$ is tight.

3.1.18 Exercise: Let $\{E_\ell\}_1^\infty$ be a sequence of Polish spaces, set $\mathbf{E} = \prod_1^\infty E_\ell$, and give $\mathbf{E}$ the product topology.

(i) For each $\ell \in \mathbb{Z}^+$, let ρ_ℓ be a complete metric for E_ℓ, and define

$$\mathbf{R}(\mathbf{x},\mathbf{y}) = \sum_{\ell=1}^\infty \frac{1}{2^\ell} \frac{\rho_\ell(x_\ell,y_\ell)}{1+\rho_\ell(x_\ell,y_\ell)} \quad \text{for} \quad \mathbf{x},\, \mathbf{y} \in \mathbf{E}.$$

Show that $\mathbf{R}$ is a complete metric for $\mathbf{E}$, and conclude that $\mathbf{E}$ is a Polish space. In addition, check that $\mathcal{B}_{\mathbf{E}} = \prod_1^\infty \mathcal{B}_{E_\ell}$.

(ii) For $\ell \in \mathbb{Z}^+$, let π_ℓ be the natural projection map from $\mathbf{E}$ onto E_ℓ, and show that $\mathbf{K} \subset\subset \mathbf{E}$ if and only if

$$\mathbf{K} = \bigcap_{\ell\in\mathbb{Z}^+} \pi_\ell^{-1}(K_\ell) \quad \text{where} \quad K_\ell \subset\subset E_\ell \text{ for each } \ell \in \mathbb{Z}^+.$$

Also, show that the span of the functions

$$\prod_{k=1}^{\ell} \varphi_k \circ \pi_k \quad \text{where } \ell \in \mathbb{Z}^+ \text{ and } \varphi_k \in U_{\mathrm{b}}^{\rho_k}(E_k;\mathbb{R}),\ 1 \le k \le \ell,$$

is dense in $U_{\mathrm{b}}^{\mathbf{R}}(\mathbf{E};\mathbb{R})$. In particular, conclude from these that $\mathbf{A} \subseteq \mathbf{M}_1(\mathbf{E})$ is tight if and only if $\{\pi_\ell^*\mu : \mu \in \mathbf{A}\} \subseteq \mathbf{M}_1(E_\ell)$ is tight for every $\ell \in \mathbb{Z}^+$ and that $\mu_n \Longrightarrow \mu$ in $\mathbf{M}_1(\mathbf{E})$ if and only if

$$\left\langle \prod_{k=1}^{\ell} \varphi_k \circ \pi_k, \mu_n \right\rangle \longrightarrow \left\langle \prod_{k=1}^{\ell} \varphi_k \circ \pi_k, \mu \right\rangle$$

for every $\ell \in \mathbb{Z}^+$ and choice of $\varphi_k \in U_{\mathrm{b}}^{\rho_k}(E_k;\mathbb{R})$, $1 \le k \le \ell$.

(iii) For each $\ell \in \mathbb{Z}^+$, set $\mathbf{E}_\ell = \prod_{k=1}^{\ell} E_k$, and let $\boldsymbol{\pi}_\ell$ denote the natural projection map from $\mathbf{E}$ onto $\mathbf{E}_\ell$. Next, let $\mu_{[1,\ell]}$ be an element of $\mathbf{M}_1(\mathbf{E}_\ell)$, and assume that the $\mu_{[1,\ell]}$'s are **consistent** in the sense that, for every $\ell \in \mathbb{Z}^+$,

$$\mu_{[1,\ell+1]}\big(\Gamma \times E_{\ell+1}\big) = \mu_{[1,\ell]}(\Gamma) \quad \text{for all } \Gamma \in \mathcal{B}_{\mathbf{E}_\ell}.$$

Show that there is a unique $\mu \in \mathbf{M}_1(\mathbf{E})$ such that $\mu_{[1,\ell]} = \boldsymbol{\pi}_\ell{}^*\mu$ for every $\ell \in \mathbb{Z}^+$.

Hint: Choose and fix an $\mathbf{e} \in \mathbf{E}$, and define $\Phi_\ell : \mathbf{E}_\ell \longrightarrow \mathbf{E}$ so that

$$\Big(\Phi_\ell(x_1, \dots, x_\ell)\Big)_n = \begin{cases} x_n & \text{if } \ n \le \ell \\ e_n & \text{otherwise.} \end{cases}$$

Show that $\{\Phi_\ell^*\mu_{[1,\ell]} : \ell \in \mathbb{Z}^+\} \in \mathbf{M}_1(\mathbf{E})$ is tight and that any limit must be the desired μ.

The conclusion drawn in **(iii)** is the renowned **Kolmogorov Extension** (or **Consistency) Theorem**. Notice that, at least for Polish spaces, it represents a vast generalization of the result obtained in Exercise 1.1.14.

3.1.19 Exercise: For $\mu \in \mathbf{M}_1(\mathbb{R}^N)$, recall the definition of the characteristic function $\boldsymbol{\xi} \in \mathbb{R}^N \longmapsto \hat{\mu}(\boldsymbol{\xi}) \in \mathbb{C}$ given in (2.2.6). In this exercise we are going to develop the intimate relationship which exists between weak convergence in $\mathbf{M}_1(\mathbb{R}^N)$ and convergence of characteristic functions. In particular, by the end, we will have proved the famous **Continuity Theorem** of P. Lévy.

(i) Let $\{\mu_n\}_1^\infty \cup \{\mu\} \subseteq \mathbf{M}_1(\mathbb{R}^N)$. Show that $\hat{\mu}_n$ tends to $\hat{\mu}$ uniformly on compacts if $\mu_n \Longrightarrow \mu$.

(ii) Show that if $\{\mu_n\}_1^\infty \subseteq \mathbf{M}_1(\mathbb{R}^N)$ is tight and $f(\boldsymbol{\xi}) = \lim_{n\to\infty} \hat{\mu}_n(\boldsymbol{\xi})$ exists for each $\boldsymbol{\xi} \in \mathbb{R}^N$, then there is a $\mu \in \mathbf{M}_1(\mathbb{R}^N)$ such that $\mu_n \Longrightarrow \mu$ and $f = \hat{\mu}$.

(iii) Show that $A \subseteq \mathbf{M}_1(\mathbb{R}^N)$ is tight if and only if the family $\{\hat{\mu} : \mu \in A\} \subseteq C_b(\mathbb{R}^N;\mathbb{C})$ is equicontinuous at the origin.

Hint: In proving the sufficiency of the equicontinuity condition, first use Exercise 3.1.18 to show that one need only handle the case when $N = 1$. Working with μ's on $\mathbb{R}$, note that, for any $\epsilon > 0$, equicontinuity implies that there is a $\delta > 0$ such that

$$\int_{\mathbb{R}} \Big(1 - \cos(\xi x)\Big)\, \mu(dx) = 1 - \frac{\hat{\mu}(\xi) + \hat{\mu}(-\xi)}{2} \le \frac{\epsilon}{2} \quad \text{for all } |\xi| \le \delta \text{ and } \mu \in A.$$

After integrating the preceding with respect to ξ over $(0, \delta)$, conclude that

$$\sup_{\mu \in A} \int_{\mathbb{R}\setminus\{0\}} \left(1 - \frac{\sin(\delta x)}{\delta x}\right) \mu_n(dx) \le \frac{\epsilon}{2}.$$

In particular, arrive at

$$\sup_{\mu \in A} \mu\left(\left\{x \in \mathbb{R} : |x| > \tfrac{2}{\delta}\right\}\right) \le \epsilon.$$

(iv) Let $\{\mu_n\}_1^\infty \subseteq \mathbf{M}_1(\mathbb{R}^N)$ be a sequence with the property that the limit $f(\boldsymbol{\xi}) = \lim_{n\to\infty} \hat{\mu}_n(\boldsymbol{\xi})$ exists for each $\boldsymbol{\xi} \in \mathbb{R}^N$. If the convergence of $\{\hat{\mu}_n\}_1^\infty$ to f is uniform in a neighborhood of the origin, show that there is a $\mu \in \mathbf{M}_1(\mathbb{R}^N)$ with the properties that $\mu_n \Longrightarrow \mu$ and that $f = \hat{\mu}$.

3.1.20 Exercise: As we will see in the next section, Lévy's Continuity Theorem has a very important rôle to play in probability theory. However, to demonstrate that its importance is not restricted to probabilistic applications, we will devote the present exercise to a derivation of a purely analytic statement. Namely, given a function $f : \mathbb{R}^N \longrightarrow \mathbb{C}$, we will outline the proof of **Bochner's Theorem** which states that f is the characteristic function $\hat{\mu}$ of a $\mu \in \mathbf{M}_1(\mathbb{R}^N)$ if and only if $f \in C(\mathbb{R}^N;\mathbb{C})$, $f(\mathbf{0}) = 1$, and f is **nonnegative definite** in the sense that, for each $n \in \mathbb{Z}^+$ and $(\boldsymbol{\xi}_1, \dots, \boldsymbol{\xi}_n) \in \mathbb{C}^n$, the matrix $\Big(\big(f(\boldsymbol{\xi}_\ell - \boldsymbol{\xi}_k)\big)\Big)_{1 \le k,\ell \le n}$ is nonnegative definite.

(i) The "only if" assertion is the easy direction. (Ironically, it is also the direction used most in applications.) Check it by expressing

$$\sum_{k,\ell=1}^{n} f(\boldsymbol{\xi}_k - \boldsymbol{\xi}_\ell)\alpha_k\overline{\alpha_\ell}$$

in terms of an integral with respect to μ.

(ii) The first step in proving the "if" assertion is to use the nonnegative definiteness assumption to show that $f(-\boldsymbol{\xi}) = \overline{f(\boldsymbol{\xi})}$ and $|f(\boldsymbol{\xi})| \le f(\mathbf{0})$ for all $\boldsymbol{\xi} \in \mathbb{R}^N$. In particular, this proves that $\|f\|_{\mathrm{u}} \le 1$. Second, using an obvious Riemann approximation procedure, check that for any rapidly decreasing, continuous $\hat{\psi} : \mathbb{R}^N \longrightarrow \mathbb{C}$,

$$\iint_{\mathbb{R}^N \times \mathbb{R}^N} f(\boldsymbol{\xi} - \boldsymbol{\eta}) \hat{\psi}(\boldsymbol{\xi}) \overline{\hat{\psi}(\boldsymbol{\eta})}\, d\boldsymbol{\xi}\, d\boldsymbol{\eta} \ge 0.$$

In particular, when $f \in L^1(\mathbb{R}^N; \mathbb{C})$, set

$$m(\mathbf{x}) = (2\pi)^{-N} \int_{\mathbb{R}^N} e^{-\sqrt{-1}\,(\boldsymbol{\xi}, \mathbf{x})_{\mathbb{R}^N}} f(\boldsymbol{\xi})\, d\boldsymbol{\xi},$$

and use Parseval's Identity, Fubini's Theorem, and elementary manipulations to arrive at

$$(2\pi)^N \int_{\mathbb{R}^N} m(\mathbf{x})\, \psi(\mathbf{x})^2\, d\mathbf{x} = \iint_{\mathbb{R}^N \times \mathbb{R}^N} f(\boldsymbol{\xi} - \boldsymbol{\eta}) \hat{\psi}(\boldsymbol{\xi}) \overline{\hat{\psi}(\boldsymbol{\eta})}\, d\boldsymbol{\xi}\, d\boldsymbol{\eta} \ge 0$$

for all smooth $\psi : \mathbb{R}^N \longrightarrow \mathbb{R}$ with rapidly decreasing derivatives of all orders. Conclude that m is nonnegative, and use this to complete the proof in the case when $f \in L^1(\mathbb{R}^N; \mathbb{C})$.

(iii) It remains only to pass from the case when $f \in L^1(\mathbb{R}^N; \mathbb{C})$ to the general case; and it is at this point that Lévy's result comes into play. Namely, for each $t \in (0, \infty)$, set $f_t(\boldsymbol{\xi}) = e^{-t|\boldsymbol{\xi}|^2} f(\boldsymbol{\xi})$. Clearly, f_t is continuous and $f_t(\mathbf{0}) = 1$. In addition, after checking the identity (cf. (2.2.10))

$$e^{-t|\boldsymbol{\xi}|^2} = \int_{\mathbb{R}^N} e^{\sqrt{-1}\,(\boldsymbol{\xi}, \mathbf{x})_{\mathbb{R}^N}}\, \gamma_{\mathbf{0}, t\mathbf{I}}(d\mathbf{x}), \quad \boldsymbol{\xi} \in \mathbb{R}^N,$$

show that

$$\sum_{k,\ell=1}^{n} f_t(\boldsymbol{\xi}_k - \boldsymbol{\eta}_\ell) \alpha_k \overline{\alpha_\ell} = \int_{\mathbb{R}^N} \left(\sum_{k,\ell=1}^{n} f(\boldsymbol{\xi}_k - \boldsymbol{\xi}_\ell) \alpha_k(\mathbf{x}) \overline{\alpha_\ell(\mathbf{x})} \right) \gamma_{\mathbf{0}, t\mathbf{I}}(d\mathbf{x}) \ge 0,$$

where $\alpha_k(\mathbf{x}) \equiv \alpha_k e^{\sqrt{-1}\,(\boldsymbol{\xi}_k, \mathbf{x})_{\mathbb{R}^N}}$. Hence, f_t is also nonnegative definite; and so, because $f_t \in L^1(\mathbb{R}^N; \mathbb{C})$, part **(ii)** applies and shows that $f_t = \hat{\mu}_t$ for some $\mu_t \in \mathbf{M}_1(\mathbb{R}^N)$. Finally, apply Lévy's Continuity Theorem to show that there is a $\mu \in \mathbf{M}_1(\mathbb{R}^N)$ such that $\mu_t \Longrightarrow \mu$ as $t \searrow 0$, and conclude that $f = \hat{\mu}$.

3.1.21 Exercise: In this exercise we will use the theory of weak convergence to develop variations on The Strong Law of Large Numbers (cf. Theorem 1.4.11). Thus, let E be a Polish space, $(\Omega, \mathcal{F}, P)$ a probability space, and $\{X_n\}_1^\infty$ a sequence of mutually independent E-valued random variables on $(\Omega, \mathcal{F}, P)$ with common distribution $\mu \in \mathbf{M}_1(E)$. Next, define the *empirical distribution functional*

$$\omega \in \Omega \longrightarrow \mathbf{L}_n(\omega) \equiv \frac{1}{n} \sum_{m=1}^{n} \delta_{X_m(\omega)} \in \mathbf{M}_1(E); \tag{3.1.22}$$

and observe that, for any $\varphi \in B(E;\mathbb{R})$,

$$\Big\langle \varphi, \mathbf{L}_n(\omega) \Big\rangle = \frac{1}{n} \sum_{m=1}^{n} \varphi(X_m(\omega)), \quad n \in \mathbb{Z}^+ \text{ and } \omega \in \Omega.$$

As a consequence of The Strong Law in Theorem 1.4.11, show that

$$\mathbf{L}_n(\omega) \Longrightarrow \mu \quad \text{for} \quad P\text{-almost every } \omega \in \Omega. \tag{3.1.23}$$

Now suppose that E is a real, separable, Banach space with dual space E^*, and set $\overline{S}_n(\omega) = \frac{1}{n}\sum_1^n X_m(\omega)$ for $n \in \mathbb{Z}^+$ and $\omega \in \Omega$. What we want to do is prove that The Strong Law for $\mathbb{R}$-valued random variables extends, without change, to the present setting.

(i) As a preliminary step, we begin with the case when

$$\mu\Big(B_E(0,R)\complement\Big) = 0 \quad \text{for some} \quad R \in (0,\infty). \tag{3.1.24}$$

Choose $\eta \in C_b(\mathbb{R};\mathbb{R})$ so that $\eta(t) = t$ for $t \in [-R,R]$ and $\eta(t) = 0$ when $|t| \geq R+1$, and define $\psi_\lambda \in C_b(E;\mathbb{R})$ for $\lambda \in E^*$ by

$$\psi_\lambda(x) = \eta\Big(\langle x, \lambda\rangle\Big), \quad x \in E,$$

where $\langle x, \lambda \rangle$ is used here to denote the action of $\lambda \in E^*$ on $x \in E$. Taking (3.1.24) into account and applying (3.1.23) and Lemma 3.1.10, show that

$$\lim_{n\to\infty} \sup_{\|\lambda\|_{E^*}\leq 1} \left| \langle \psi_\lambda, \mathbf{L}_n(\omega)\rangle - \int_E \langle x, \lambda\rangle\, \mu(dx) \right| = 0$$

for P-almost every $\omega \in \Omega$; and conclude from this that

$$\lim_{n\to\infty} \left\|\overline{S}_n(\omega) - \mathbf{m}\right\|_E = 0 \quad \text{for } P\text{-almost every } \omega \in \Omega,$$

where $\mathbf{m} \in E$ is uniquely determined by the relation

$$\langle \mathbf{m}, \lambda \rangle = \int_E \langle x, \lambda \rangle \, \mu(dx), \quad \lambda \in E^*. \tag{3.1.25}$$

(ii) We next want to replace the boundedness assumption in (3.1.24) by the hypothesis

$$\int_E \|x\|_E \, \mu(dx) < \infty. \tag{3.1.26}$$

Assuming (3.1.26), define, for $R \in (0, \infty)$, $n \in \mathbb{Z}^+$, and $\omega \in \Omega$:

$$X_n^{(R)}(\omega) = \begin{cases} X_n(\omega) & \text{if } \|X_n(\omega)\|_E < R \\ 0 & \text{otherwise} \end{cases}$$

and $Y_n^{(R)}(\omega) = X_n(\omega) - X_n^{(R)}(\omega)$. Next, set $\overline{S}_n^{(R)} = \frac{1}{n} \sum_1^n X_m^{(R)}$, $n \in \mathbb{Z}^+$; and, from (i), note that $\{\overline{S}_n^{(R)}(\omega)\}_1^\infty$ converges in E for P-almost every $\omega \in \Omega$. In particular, if $\epsilon > 0$ is given and $R \in (0, \infty)$ is chosen so that

$$\int_{\{\|x\|_E \geq R\}} \|x\|_E \, \mu(dx) < \frac{\epsilon}{8},$$

use the preceding and Theorem 1.4.11 to verify the computation

$$\begin{aligned}
&\overline{\lim_{m\to\infty}} P\left(\sup_{n \geq m} \left\| \overline{S}_n - \overline{S}_m \right\|_E \geq \epsilon \right) \\
&\quad \leq \lim_{m\to\infty} P\left(\sup_{n \geq m} \left\| \overline{S}_n^{(R)} - \overline{S}_m^{(R)} \right\| \geq \frac{\epsilon}{2} \right) \\
&\qquad + 2 \overline{\lim_{m\to\infty}} P\left(\sup_{n \geq m} \left\| \frac{1}{n} \sum_1^n Y_k^{(R)} \right\|_E \geq \frac{\epsilon}{4} \right) \\
&\quad \leq 2 \overline{\lim_{m\to\infty}} P\left(\sup_{n \geq m} \frac{1}{n} \sum_1^n \left\| Y_k^{(R)} \right\|_E \geq \frac{\epsilon}{4} \right) = 0;
\end{aligned}$$

and from this, conclude that $\overline{S}_n \longrightarrow \mathbf{m}$ P-almost surely, with $\mathbf{m} \in E$ satisfying (3.1.25).

(iii) Finally, repeat the argument given in the proof of Theorem 1.4.11 to show that (3.1.26) must hold if $\{\overline{S}_n\}_1^\infty$ converges in E on a set of positive P-measure.

The beautiful argument which we have just given is due to Ranga Rao.† Notice that, as a dividend, we have proved that for any $\mu \in \mathbf{M}_1(E)$ satisfying (3.1.26) there is a **mean-value** $\mathbf{m} \in E$ satisfying (3.1.25). In keeping with the real-valued case, one uses $\int_E x\,\mu(dx)$ to denote this element of E. (See Lemma 5.1.20 for more information about integration of Banach space valued functions. Also, entirely different proofs of The Strong Law for Banach spaces are given in Exercise 5.3.38 and Exercise 6.1.34.)

§3.2: Infinitely Divisible Laws.

Certainly the first truly significant, and perhaps still the most beautiful, application of the theory of weak convergence to probability theory is the one which led eventually to the Lévy–Khinchine (cf. (3.2.6)) formula. In this section, we will first derive their formula and then examine what it says.

We begin by stating the problem to which the Lévy–Khinchine formula is the solution. Namely, we want a description of all *the random variables* Y *which can be expressed, for every* $n \in \mathbb{Z}^+$*, as the sum of* n *mutually independent, identically distributed random variables* $X_{1,n}, \dots, X_{n,n}$. Here, by the term *description*, we mean that we want to characterize the distributions of such Y's. Thus, an alternative statement of the same problem can be given without any reference to the random variables themselves. Indeed, introduce on $\mathbf{M}_1(\mathbb{R})$ the **convolution product** $(\mu,\nu) \in \mathbf{M}_1(\mathbb{R})^2 \longmapsto \mu\bigstar\nu \in \mathbf{M}_1(\mathbb{R})$ given by

$$\mu\bigstar\nu(\Gamma) = \iint_{\mathbb{R}\times\mathbb{R}} \mathbf{1}_\Gamma(x+y)\,\mu(d\mathbf{x})\,\nu(d\mathbf{y}), \quad \Gamma \in \mathcal{B}_{\mathbb{R}}.$$

Clearly, this product turns $\mathbf{M}_1(\mathbb{R})$ into a semigroup in which the identity element is the point-mass δ_0. Moreover, if X and Y are independent random variables with distributions μ and ν, respectively, then $X+Y$ has distribution $\mu\bigstar\nu$. Hence, the preceding problem is equivalent to the problem of characterizing the set $\mathcal{I}$ consisting of those $\mu \in \mathbf{M}_1(\mathbb{R})$ which are **infinitely divisible** in the sense that, for each $n \in \mathbb{Z}^+$, there exists a $\mu_{\frac{1}{n}} \in \mathbf{M}_1(\mathbb{R})$ with the property that

$$\mu = \mu_{\frac{1}{n}}{}^{\star n} \equiv \overbrace{\mu_{\frac{1}{n}}\bigstar\cdots\bigstar\mu_{\frac{1}{n}}}^{n\text{-times}}.$$

† Ranga Rao's 1963 article "The law of large numbers for $D[0,1]$-valued random variables," *Theory of Prob. & Appl.* **VIII, 1**, shows that this method applies even outside the separable context. His is not the first proof of the result given here. The Strong Law for separable Banach spaces was proved first in 1953 by E. Mourier in "Eléments aléatorires dans un espace de Banach," *Ann. Inst. Poincaré*.

Because the convolution product derives from the action of the translation group on $\mathbf{M}_1(\mathbb{R})$, it should come as no surprise that both the statement as well as the solution of this problem are greatly simplified when couched in Fourier language (i.e., in terms of characteristic functions). For example, because (cf. (2.2.6))

$$\widehat{\mu\star\nu}(\xi) = \hat{\mu}(\xi)\,\hat{\nu}(\xi), \quad \mu,\,\nu \in \mathbf{M}_1(\mathbb{R}) \text{ and } \xi \in \mathbb{R},$$

another formulation of the problem is that of describing those characteristic functions $\hat{\mu}$ which are **infinitely divisible** (i.e., for every $n \in \mathbb{Z}^+$, can be written as the nth-power of a characteristic function $\widehat{\mu_{\frac{1}{n}}}$); and it is the solution to this formulation which Lévy–Khinchine formula provides.

Rather than simply writing down their formula, it may be best to see how far one can get by guessing. Whether one looks at the problem directly or from the characteristic function point of view, it is clear that what one is seeking is a description of those probability measures which admit a *logarithm*. Thus, it is reasonable to start by *exponentiating* measures. More precisely, let $\mu \in \mathbf{M}_1(\mathbb{R})$ be given, and, for each $\alpha \in [0,\infty)$, consider the (compound) **Poisson measure** given by

$$\pi_{\alpha\mu} = e^{-\alpha}\sum_{n=0}^{\infty}\frac{\alpha^n}{n!}\mu^{\star n}, \quad \text{where } \mu^{\star 0} \equiv \delta_0. \tag{3.2.1}$$

(The factor $e^{-\alpha}$ in front is simply a renormalization.) As it turns out, this is a very good idea. In fact, what Lévy and Khinchine showed is that $\mathcal{I}$ is the closure of

$$\mathcal{P} \equiv \left\{\pi_{\alpha\mu} : \alpha \in (0,\infty) \text{ and } \mu \in \mathbf{M}_1(\mathbb{R})\right\};$$

and our first goal will be to describe the elements of this closure in terms of their characteristic functions and to check that they are infinitely divisible.

Obviously, we should begin by writing down the characteristic function of $\pi_{\alpha\mu}$. But clearly

$$\widehat{\pi_{\alpha\mu}}(\xi) = e^{-\alpha}\sum_{n=0}^{\infty}\frac{\left(\alpha\hat{\mu}(\xi)\right)^n}{n!},$$

and so

$$\widehat{\pi_{\alpha\mu}}(\xi) = \exp\left[\alpha\int_{\mathbb{R}^N}\left(e^{\sqrt{-1}\,\xi y} - 1\right)\mu(dy)\right].$$

In particular, this expression shows that

$$\pi_{\alpha\mu} = \left(\pi_{\frac{\alpha}{n}\mu}\right)^{\star n}, \quad n \in \mathbb{Z}^+,$$

and therefore that $\pi_{\alpha\mu} \in \mathcal{I}$. In addition, it indicates that it is foolish for us to continue separating the mass α from the distribution μ and that we should simply write π_M for the probability measure whose characteristic function is given by

$$\widehat{\pi_M}(\xi) = \exp\left[\int_{\mathbb{R}} \left(e^{\sqrt{-1}\,\xi y} - 1\right) M(dy)\right].$$

This change in notation has more than aesthetic benefits. Indeed, it points us in the right direction as we start constructing the closure of $\mathcal{P}$. More precisely, it shows that whatever mass M may have at $\{0\}$ does not contribute to π_M and therefore that we should think of M as a measure on $\mathbb{R} \setminus \{0\}$ and write

$$\widehat{\pi_M}(\xi) = \exp\left[\int_{\mathbb{R}\setminus\{0\}} \left(e^{\sqrt{-1}\,\xi y} - 1\right) M(dy)\right]$$

in place of the preceding. Secondly, it encourages us to abandon the assumption that M is finite. For example, the expression for $\widehat{\pi_M}$ makes perfectly good sense so long as M satisfies the condition

$$\text{(3.2.2)} \qquad \int_{\{0<|y|<1\}} |y|\, M(dy) + M\big((-1,1)\complement\big) < \infty.$$

In fact, by considering the measures

$$M_\epsilon(\Gamma) \equiv M\left(\Gamma \cap (-\epsilon,\epsilon)\complement\right), \quad \Gamma \in \mathcal{B}_{\mathbb{R}}, \quad \text{for} \quad \epsilon \in (0,1]$$

and applying Lévy's Continuity Theorem (cf. Exercise 3.1.19), we see that there is a $\pi_M \in \overline{\mathcal{P}}$ whose characteristic function is

$$\exp\left[\int_{\mathbb{R}\setminus\{0\}} \left(e^{\sqrt{-1}\xi y} - 1\right)\right] M(dy).$$

The point here is, of course, that because (3.2.2) guarantees the uniform M-integrability of the functions

$$y \in \mathbb{R} \setminus \{0\} \longmapsto e^{\sqrt{-1}\,\xi y} - 1 \in \mathbb{C}$$

as ξ varies over any compact, $\widehat{\pi_{M_\epsilon}}(\xi) \longrightarrow \widehat{\pi_M}(\xi)$ uniformly for ξ in each compact. Moreover, it is obvious that π_M is not only in $\overline{\mathcal{P}}$ but also in $\mathcal{I}$, since an nth root is obtained by replacing M with $\frac{M}{n}$.

The next step is a little more sophisticated and requires some preparations. Namely, we want to replace the condition in (3.2.2) by the condition that

$$\text{(3.2.3)} \qquad \int_{\{0<|y|<1\}} |y|^2\, M(dy) + M\big((-1,1)\complement\big) < \infty.$$

However, before we can do this, we have to first *squeeze* the class of measures M which satisfy (3.2.2) in order to get the translates $\pi_{a,M} \equiv \delta_a \bigstar \pi_M$. To this end, let $a \in \mathbb{R}$ be given, and set

$$M_\epsilon = \frac{\delta_{\epsilon a}}{\epsilon} + M \quad \text{for } \epsilon > 0.$$

Note that, although M_ϵ gets more singular as $\epsilon \searrow 0$, it does so in such a way that the quantity in (3.2.2) is staying bounded. Moreover, as $\epsilon \searrow 0$,

$$\widehat{\pi_{M_\epsilon}}(\xi) \longrightarrow \widehat{\pi_{a,M}}(\xi) = \exp\left[\sqrt{-1}\,\xi a + \int_{\mathbb{R}\setminus\{0\}} \left(e^{\sqrt{-1}\,\xi y} - 1\right) M(dy)\right]$$

uniformly fast as ξ varies over compacts. Hence, by another application of Lévy's Continuity Theorem (or even just Lemma 2.2.8) we see that $\pi_{a,M} \in \overline{\mathcal{P}}$; and clearly, for every $n \in \mathbb{Z}^+$, $\pi_{a,M} = \pi_{\frac{a}{n},\frac{M}{n}}{}^{\bigstar n}$, which means that $\pi_{a,M}$ is also an element of $\mathcal{I}$.

We now have the possibility of subtracting off the next term in the Taylor's expansion of $e^{\sqrt{-1}\,\xi y}$ in order to produce an integrand which vanishes to second order as $|y| \searrow 0$. Namely, having introduced translation, we now introduce the notation

$$\widehat{\tilde{\pi}_{a,M}}(\xi) \equiv \exp\left[\sqrt{-1}\,\xi a + \int_{\mathbb{R}\setminus\{0\}} e_2(\xi, y)\, M(dy)\right], \tag{3.2.4}$$

where

$$e_2(\xi, y) \equiv e^{\sqrt{-1}\,\xi y} - 1 - \frac{\sqrt{-1}\,\xi y}{1 + y^2}. \tag{3.2.5}$$

(The denominator in the third term is there to prevent possible integrability problems at ∞. As is explained in Exercise 3.2.40, its choice is quite arbitrary.) Clearly $\widehat{\tilde{\pi}_{a,M}} = \widehat{\pi_{\tilde{a},M}}$ when M satisfies (3.2.2) and

$$\tilde{a} \equiv a - \int_{\mathbb{R}\setminus\{0\}} \frac{y}{1 + y^2}\, M(dy).$$

Thus, at least when M satisfies (3.2.2), we already know that there is a unique $\tilde{\pi}_{a,M} \in \overline{\mathcal{P}} \cap \mathcal{I}$ with characteristic function $\widehat{\tilde{\pi}_{a,M}}$. Moreover, by exactly the same limit argument which before allowed us to replace finite M's by M's satisfying (3.2.2), we can now see that, for each $a \in \mathbb{R}$ and each Borel measure M on $\mathbb{R} \setminus \{0\}$ satisfying (3.2.3), there is a unique $\tilde{\pi}_{a,M} \in \overline{\mathcal{P}} \cap \mathcal{I}$ whose characteristic function is given by (3.2.4).

At this point one might think that the preceding scheme should continue indefinitely. That is, the next step is to *squeeze* again, only this time instead of squeezing in such a way that (3.2.2) remains bounded, we squeeze in such a way that (3.2.3) stays bounded. Thus, given $\sigma \in [0,\infty)$, $a \in \mathbb{R}$, and an M satisfying (3.2.3), set

$$M_\epsilon = \frac{\delta_{\epsilon\sigma} + \delta_{-\epsilon\sigma}}{2\epsilon^2} + M, \quad \epsilon > 0,$$

and conclude, that

$$\tilde{\pi}_{a,M_\epsilon} \Longrightarrow \pi_{a,\sigma^2,M},$$

where $\pi_{a,\sigma^2,M}$ is the measure whose characteristic function is given by

$$\text{(3.2.6)} \qquad \widehat{\pi_{a,\sigma^2,M}}(\xi) = \exp\left[\sqrt{-1}\,a\xi - \frac{\sigma^2\xi^2}{2} + \int_{\mathbb{R}\setminus\{0\}} e_2(\xi,y)\,M(dy)\right].$$

By construction, we know that $\pi_{a,\sigma^2,M} \in \overline{\mathcal{P}}$, and once again it is clear that one can get the nth-root of $\pi_{a,\sigma^2,M}$ by simply dividing a, σ^2, and M by n.

So far so good. However, the scheme breaks down as soon as one attempts take the next step. That is, if one tries to produce an integrand which vanishes to third order by repeating the same trick as we just used to go from first to second order, one discovers that there is no expression having the form $\sqrt{-1}\,\xi a - \frac{\sigma^2\xi^2}{2}$ at 0 whose introduction into the integrand will compensate for the term which we want to knock out. Indeed, although the desired term is quadratic in ξ, *it has the wrong sign*; and, for this reason, our procedure comes to a screeching halt. Actually, there is a far more compelling reason why our procedure breaks down: namely, as we will see below (cf. Theorem 3.2.20), (3.2.6) *is the* **Lévy–Khinchine formula** *for the characteristic function of the most general infinitely divisible probability measure on* $\mathbb{R}$*, and, as such, it is also the expression for the characteristic function of the most general element of* $\overline{\mathcal{P}}$. However, we still have work to do before we can justify these claims.

A measure M on $\mathbb{R}\setminus\{0\}$ which satisfies (3.2.3) is called a **Lévy measure**, and we will use $\mathfrak{M}$ to denote the set of all Lévy measures and $\mathfrak{L}$ to denote the set $\mathbb{R}\times[0,\infty)\times\mathfrak{M}$ of all **Lévy systems**. Thus, we can summarize our progress so far by saying that

$$\left\{\pi_{a,\sigma^2,M} : (a,\sigma,M) \in \mathfrak{L}\right\} \subseteq \overline{\mathcal{P}} \cap \mathcal{I}.$$

In order to refine this statement, we will need a little more preparation. In the first place, the preceding parameterization does not lend itself well to a discussion of limits. For one thing, it is inconvenient to be dealing with infinite measures. Secondly, although for each individual Lévy system the separation of σ from M is reasonably clear, these parameters have a disturbing propensity to get mixed up when limits are involved; for example, as we have just seen, what starts as mass in the Lévy measure M can, after a limit is taken, turn into σ. For these reasons, we will now reorganize the parameterization in such a way that σ and M are represented by a single, finite measure on $\mathbb{R}$. To be precise, given $\sigma \in [0,\infty)$ and $M \in \mathfrak{M}$, define the finite measure $\tilde{M}^\sigma$ on $(\mathbb{R}, \mathcal{B}_\mathbb{R})$ so that

$$(3.2.7) \qquad \int_\mathbb{R} \varphi(y)\, \tilde{M}^\sigma(dy) = \sigma^2\varphi(0) + \int_{\mathbb{R}\setminus\{0\}} \varphi(y)\rho(y)\, M(dy), \quad \varphi \in B(\mathbb{R};\mathbb{R}),$$

where ρ is the bounded, smooth function given by

$$(3.2.8) \qquad \rho(y) \equiv 6\left(1 - \frac{\sin y}{y}\right) \quad \text{for } y \in \mathbb{R}\setminus\{0\} \quad \text{and } \rho(0) \equiv 0.$$

Clearly, $(\sigma, M) \in [0,\infty) \times \mathfrak{M} \longmapsto \tilde{M}^\sigma$ is a one to one mapping onto the space of finite, Borel measures on $(\mathbb{R}, \mathcal{B}_\mathbb{R})$. Also, observe that

$$(3.2.9) \qquad \widehat{\pi_{a,\sigma^2,M}}(\xi) = \exp\left[\sqrt{-1}\, a\xi + \int_\mathbb{R} \tilde{e}_2(\xi, y)\, \tilde{M}^\sigma(d\xi)\right], \quad \xi \in \mathbb{R},$$

where, for each $\xi \in \mathbb{R}$, $\tilde{e}_2(\xi,\,\cdot\,)$ is the element of $C_b(\mathbb{R};\mathbb{C})$ determined by (cf. (3.2.5))

$$\tilde{e}_2(\xi, y) \equiv \frac{e_2(\xi,y)}{\rho(y)} \quad \text{for } y \in \mathbb{R}\setminus\{0\} \quad \text{and } \tilde{e}_2(\xi, 0) \equiv -\frac{\xi^2}{2}.$$

Thus, if we write

$$(a_n, \sigma_n, M_n) \xrightarrow{\mathfrak{L}} (a, \sigma, M)$$

when $\{(a_n, \sigma_n, M_n)\}_1^\infty \cup \{(a, \sigma, M)\} \subseteq \mathfrak{L}$ and

$$a = \lim_{n\to\infty} a_n \quad \text{and} \quad \int_\mathbb{R} \varphi\, d\tilde{M}^\sigma = \lim_{n\to\infty} \int_\mathbb{R} \varphi\, d\tilde{M}_n^{\sigma_n}, \quad \varphi \in C_b(\mathbb{R};\mathbb{R}),$$

then it is clear (cf. either Lemma 2.2.8 or Exercise 3.1.19) that

$$(3.2.10) \qquad (a_n, \sigma_n, M_n) \xrightarrow{\mathfrak{L}} (a, \sigma, M) \text{ implies that } \pi_{a_n,\sigma_n^2,M_n} \Longrightarrow \pi_{a,\sigma^2,M}.$$

In order to get an appropriate converse to (3.2.10), we will need the following.

3.2.11 Lemma. *For each $f \in C(\mathbb{R};\mathbb{C}\setminus\{0\})$ with $f(0) = 1$ there is a unique $\ell_f \in C(\mathbb{R};\mathbb{C})$ with the properties that $\ell_f(0) = 0$ and $f(\xi) = \exp[\ell_f(\xi)]$ for all $\xi \in \mathbb{R}$. Furthermore, if $\{f_n\}_1^\infty \subseteq C(\mathbb{R};\mathbb{C}\setminus\{0\})$, $f_n(0) = 1$ for each $n \in \mathbb{Z}^+$,*

$$\inf_{n\in\mathbb{Z}^+} \inf_{|\xi|\le R} |f_n(\xi)| > 0 \quad \text{for all } R \in (0,\infty),$$

and $f_n \longrightarrow f$ uniformly on compacts, then $f \in C(\mathbb{R};\mathbb{C}\setminus\{0\})$, $f(0) = 1$, and $\ell_{f_n} \longrightarrow \ell_f$ uniformly on compacts.

PROOF: Let f be given, set $\xi_0 = 0$, and define

$$\xi_{\pm n \pm 1} = \inf\left\{\pm\xi \geq \pm\xi_{\pm n} : \left|\frac{f(\xi)}{f(\xi_{\pm n})} - 1\right| \geq \frac{1}{2}\right\} \quad \text{for } n \in \mathbb{N}.$$

By continuity, it is clear that $\xi_{\pm n} \longrightarrow \pm\infty$ as $n \to \infty$; and so we can determine an $\ell_f \in C(\mathbb{R};\mathbb{C})$ inductively by setting $\ell_f(0) = 0$ and, for $n \in \mathbb{N}$,

$$\ell_f(\pm\xi) = \ell_f(\xi_{\pm n}) + \log\frac{f(\xi)}{f(\xi_{\pm n})} \quad \text{if } \pm\xi \in [\pm\xi_{\pm n}, \pm\xi_{\pm n \pm 1}],$$

where "log" denotes the principal branch of the logarithm function. Moreover, it is clear (again by induction) that this is the one and only choice of $\ell \in C(\mathbb{R};\mathbb{C})$ with $\ell(0) = 0$ for which $f(\xi) = e^{\ell(\xi)}$, $\xi \in \mathbb{R}$.

Now suppose that $\{f_n\}_1^\infty$ and f are as in the second part of the statement. Obviously $f \in C(\mathbb{R};\mathbb{C}\setminus\{0\})$. Next, set $\Delta_n = \ell_{f_n} - \ell_f$, $n \in \mathbb{Z}^+$, and, given $R \in (0,\infty)$, choose $N \in \mathbb{Z}^+$ so that

$$\left|\frac{f_n(\xi)}{f(\xi)} - 1\right| \leq \frac{1}{2} \quad \text{for } n \geq N \text{ and } |\xi| \leq R.$$

Then

$$\Delta_n(\xi) = \log\frac{f_n(\xi)}{f(\xi)} \mod \sqrt{-1}\,2\pi\mathbb{Z} \quad \text{for } n \geq N \text{ and } |\xi| \leq R,$$

and therefore, since $n \geq N$ implies $\Delta_n - \log\frac{f_n}{f}$ is continuous on $[-R,R]$ and vanishes at 0,

$$\Delta_n(\xi) = \log\frac{f_n(\xi)}{f(\xi)} \longrightarrow 0$$

uniformly on $[-R,R]$. □

Given (a,σ,M), set $f_{a,\sigma^2,M} = \widehat{\pi_{a,\sigma^2,M}}$, and observe that

$$\ell_{a,\sigma^2,M}(\xi) \equiv \ell_{f_{a,\sigma^2,M}}(\xi) = \sqrt{-1}\,a\xi + \int_{\mathbb{R}} \tilde{e}_2(\xi,y)\,\tilde{M}^\sigma(dy).$$

In order to see that $(a,\sigma,M) \in \mathfrak{L} \longmapsto \pi_{a,\sigma^2,M} \in \mathbf{M}_1(\mathbb{R})$ is one-to-one, we must see how $\tilde{M}^\sigma$, and therefore a, can be recovered from a knowledge of $f_{a,\sigma^2,M}$ or, equivalently, of $\ell_{a,\sigma^2,M}$. To this end, let $\varphi \in C_c^\infty(\mathbb{R};\mathbb{R})$ be given. Clearly,

$$\int_{\mathbb{R}} \varphi(y)\,\tilde{M}^\sigma(dy) = \sigma^2\varphi(0) + \int_{\mathbb{R}} \psi(y)\,M(dy) \quad \text{where } \psi \equiv \rho\varphi.$$

Since $\psi \in C_c^\infty(\mathbb{R};\mathbb{R})$ and therefore $\hat{\psi}$ is rapidly decreasing, the relation

$$\varphi(0) = -\frac{1}{4\pi}\int_{\mathbb{R}} \xi^2 \hat{\psi}(-\xi)\, dx$$

is an easy application of elementary Fourier analysis and the fact that $\psi''(0) = 2\varphi(0)$. At the same time, because $\psi(0) = 0 = \psi'(0)$, an application of (2.2.9) can be used to verify the calculation:

$$\begin{aligned}
\int_{\mathbb{R}} \psi(y)\, M(dy) &= \lim_{\epsilon \searrow 0} \int_{\{|y|\geq\epsilon\}} \psi(y)\, M(dy) \\
&= \lim_{\epsilon \searrow 0} \frac{1}{2\pi}\int_{\mathbb{R}} \hat{\psi}(-\xi)\left(\int_{\{|y|\geq\epsilon\}} e^{\sqrt{-1}\,\xi y}\, M(dy)\right) d\xi \\
&= \lim_{\epsilon \searrow 0} \frac{1}{2\pi}\int_{\mathbb{R}} \hat{\psi}(-\xi)\left(\sqrt{-1}\,\xi a + \int_{\{|y|\geq\epsilon\}} e_2(\xi, y)\, M(dy)\right) d\xi \\
&= \int_{\mathbb{R}} \hat{\psi}(-\xi)\ell_{a,0,M}(\xi)\, d\xi.
\end{aligned}$$

Hence, after combining these we arrive at the identity

$$\int_{\mathbb{R}} \varphi(y)\, \tilde{M}^\sigma(dy) = \frac{1}{2\pi}\int_{\mathbb{R}} \ell_{a,\sigma^2,M}(\xi)\, \widehat{\rho\varphi}(-\xi)\, d\xi, \quad \varphi \in C_c^\infty(\mathbb{R};\mathbb{R}). \tag{3.2.12}$$

In particular, this, together with Lemma 3.2.11, means that $\tilde{M}^\sigma$ and therefore also a can be recovered from $f_{a,\sigma^2,M}$. In other words, we now know that the map $(a,\sigma,M) \in \mathfrak{L} \longmapsto \pi_{a,\sigma^2,M} \in \overline{\mathcal{P}}$ is one-to-one. What we are going to show now is that it is also onto.

3.2.13 Lemma. *Given $\{(a_n, \sigma_n, M_n)\}_1^\infty \subseteq \mathfrak{L}$, there is a $\mu \in \mathbf{M}_1(\mathbb{R})$ such that $\pi_{a_n,\sigma_n^2,M_n} \Longrightarrow \mu$ if and only if there is an $(a,\sigma,M) \in \mathfrak{L}$ such that $(a_n,\sigma_n,M_n) \xrightarrow{\mathfrak{L}} (a,\sigma,M)$; in which case $\mu = \pi_{a,\sigma^2,M}$. In particular, the map*

$$(a,\sigma,M) \in \mathfrak{L} \longmapsto \pi_{a,\sigma^2,M} \in \overline{\mathcal{P}} \text{ is one-to-one and onto.} \tag{3.2.14}$$

PROOF: Because of (3.2.10), it is sufficient for us to discuss the *only if* assertion.

Set $f_n = f_{a_n,\sigma_n^2,M_n}$, $\ell_n = \ell_{a_n,\sigma_n^2,M_n}$, and $f = \hat{\mu}$. By part (**i**) of Exercise 3.1.19, we know that $f_n \longrightarrow f$ uniformly on compacts. At the same time,

$$\mathfrak{Re}\big(\ell_n(\xi)\big) = -\frac{\sigma_n^2\xi^2}{2} + \int_{\mathbb{R}\setminus\{0\}} \big(\cos(\xi y) - 1\big)\, M_n(dy);$$

and so (cf. (3.2.7) and (3.2.8)), for each $\delta > 0$,

$$\frac{1}{\delta}\int_0^\delta \mathfrak{Re}\bigl(\ell_n(\xi)\bigr)\,d\xi = -\frac{1}{6}\int_{\mathbb{R}} \beta_\delta(y)\,\tilde{M}_n(dy),$$

where $\tilde{M}_n \equiv \tilde{M}_n^{\sigma_n}$ and $\beta_\delta \in C_b\bigl(\mathbb{R};(0,\infty)\bigr)$ is defined by

$$\beta_\delta(y) = \frac{\rho(\delta y)}{\rho(y)} \quad \text{when } y \neq 0 \text{ and } \beta_\delta(0) = \delta^2.$$

In particular, since $f(0) = 1$ and $f_n \longrightarrow f$ uniformly on compacts, we can choose, for each $\epsilon > 0$, a $\delta(\epsilon) > 0$ so that

$$\epsilon \geq \int_{\mathbb{R}} \beta_\delta(y)\,\tilde{M}_n(dy) \quad \text{for all } n \in \mathbb{Z}^+ \text{ and } 0 < \delta \leq \delta(\epsilon). \tag{3.2.15}$$

As our first application of (3.2.15), we show that there is an $\ell \in C(\mathbb{R};\mathbb{C})$ to which $\{\ell_n\}_1^\infty$ converges uniformly on compacts. Indeed, note that, for each $\delta > 0$, $m(\delta) \equiv \inf_{y\in\mathbb{R}} \beta_\delta(y) > 0$, and therefore, by (3.2.15), that

$$\sup_{n\in\mathbb{Z}^+} \tilde{M}_n(\mathbb{R}) \leq \frac{1}{m\bigl(\delta(1)\bigr)} < \infty.$$

But, by (3.2.9),

$$|f_n(\xi)| = \exp\left[-\int_{\mathbb{R}} \frac{1-\cos(\xi y)}{\rho(y)}\,\tilde{M}_n(dy)\right] \geq \exp\bigl[-c\xi^2\tilde{M}_n(\mathbb{R})\bigr],$$

where

$$c \equiv \sup_{\substack{y\in\mathbb{R}\\ \xi\neq 0}} \frac{1-\cos\xi y}{\xi^2\rho(y)} < \infty;$$

and so we now know, first, that $\inf_n |f_n(\xi)|$ is uniformly positive on compact subsets and then, by Lemma 3.2.11, that $\ell_n \longrightarrow \ell$ uniformly on compacts, where $\ell \equiv \ell_f$.

Our next goal is to show that there is a finite measure $\tilde{M}$ with the property that

$$\int_{\mathbb{R}} \varphi\,d\tilde{M}_n \longrightarrow \int_{\mathbb{R}} \varphi\,d\tilde{M}, \quad \varphi \in C_b(\mathbb{R};\mathbb{R}). \tag{3.2.16}$$

To this end, we first observe that, for some $C \in (0,\infty)$,

$$\left|\int_{\mathbb{R}} \tilde{e}_2(\xi,y)\,\tilde{M}^{\sigma_n}(dy)\right| \leq C(1+\xi^2)\,\tilde{M}^{\sigma_n}(\mathbb{R})$$

and

$$|a_n| \le \left|\ell_n(1)\right| + \left|\int_{\mathbb{R}} \tilde{e}_2(1,y)\,\tilde{M}^{\sigma_n}(dy)\right|.$$

Hence, from what we already know, we can use (3.2.12) and Lebesgue's Dominated Convergence Theorem to see that

$$\lim_{n\to\infty}\int_{\mathbb{R}} \varphi\,\tilde{M}_n(dy) = \frac{1}{2\pi}\int_{\mathbb{R}} \ell(\xi)\widehat{\rho\varphi}(-\xi)\,d\xi \quad \text{for } \varphi \in C_{\mathrm{c}}^{\infty}(\mathbb{R};\mathbb{R}).$$

Thus, we will know that there is an $\tilde{M}$ for which (3.2.16) holds as soon as we check that every subsequence of $\{\tilde{M}_n\}_1^\infty$ admits a subsequence which converges in the sense required in (3.2.16). But, by **(iv)** of Exercise 3.1.17, this reduces to checking that $\{\tilde{M}_n(\mathbb{R})\}_1^\infty$ is bounded and that

$$\lim_{\delta\searrow 0}\sup_{n\in\mathbb{Z}^+} \tilde{M}_n\big([-\delta^{-1},\delta^{-1}]\big) = 0.$$

However, the first of these has already been noted, and the second is an easy application of (3.2.15).

To complete the proof, we must still show that $\{a_n\}_1^\infty$ converges to some a. But

$$\sqrt{-1}\,a_n = \ell_n(1) - \int_{\mathbb{R}} \tilde{e}_2(1,y)\,\tilde{M}_n(dy) \longrightarrow \ell(1) - \int_{\mathbb{R}} \tilde{e}_2(1,y)\,\tilde{M}(dy). \quad \Box$$

We are now in a position to prove that $\mathcal{I} = \overline{\mathcal{P}}$. Thus, let $\mu \in \mathcal{I}$ be given, and, for each $n \in \mathbb{Z}^+$, choose $\mu_{\frac{1}{n}} \in \mathbf{M}_1(\mathbb{R})$ so that $\mu = \mu_{\frac{1}{n}}{}^{\star n}$. The idea behind the proof is based on the intuition that, for large n's, $\mu_{\frac{1}{n}}$ must be nearly concentrated in a small neighborhood of 0; and therefore, $\lambda_n \equiv \mu_{\frac{1}{n}} - \delta_0$ ought to be *small enough* that the error in the approximation represented by

$$\mu_{\frac{1}{n}}{}^{\star n} = \left(\delta_0 + \frac{n\lambda_n}{n}\right)^{\star n} \approx \text{“}e^{n\lambda_n}\text{”} = e^{-n}\sum_{m=0}^{\infty}\frac{n^m\mu_{\frac{1}{n}}{}^{\star m}}{m!}$$

ought to become negligible as $n \to \infty$. With this in mind, we attempt to prove that

$$\pi_n \equiv e^{-n}\sum_{m=0}^{\infty}\frac{n^m\mu_{\frac{1}{n}}{}^{\star m}}{m!} \Longrightarrow \mu \quad \text{as } n \to \infty. \tag{3.2.17}$$

To this end, let $\varphi \in C_{\mathrm{b}}^1(\mathbb{R};\mathbb{R})$ be given, and define

$$\Delta_n(R) = e^{-n}\sum_{\substack{m\in\mathbb{N}\\ |n-m|\le\sqrt{n}\,R}} \frac{n^m}{m!}\left|\langle\varphi,\mu_{\frac{1}{n}}{}^{\star m}\rangle - \langle\varphi,\mu\rangle\right|$$

for $n \in \mathbb{Z}^+$ and $R \in (0,\infty)$. Clearly

$$\left|\langle\varphi,\pi_n\rangle - \langle\varphi,\mu\rangle\right| \le \Delta_n(R) + 2\|\varphi\|_{\mathrm{u}}\epsilon_n(R),$$

where

$$\epsilon_n(R) \equiv e^{-n} \sum_{\substack{m\in\mathbb{N}\\ |n-m|>\sqrt{n}\,R}} \frac{n^m}{m!}.$$

But

$$\epsilon_n(R) \le \frac{1}{nR^2} e^{-n} \sum_{m=0}^{\infty} (m-n)^2 \frac{n^m}{m!} = \frac{1}{R^2},$$

where the final equality involves an elementary computation. (In fact, when looked at correctly, this estimate for $\epsilon_n(R)$ is just an application of Chebyshev's inequality.) Hence, we have now proved that

$$\left|\langle\varphi,\pi_n\rangle - \langle\varphi,\mu\rangle\right| \le \Delta_n(R) + \frac{1}{R^2}, \quad n \in \mathbb{Z}^+ \text{ and } R \in (0,\infty),$$

and so we will be done once we check that, for each $R \in (0,\infty)$, $\Delta_n(R) \longrightarrow 0$ as $n \to \infty$; and clearly this will follow once we show that

$$\lim_{n\to\infty} \sup\left\{\left|\langle\varphi,\mu_{\frac{1}{n}}{}^{\star(m+\ell)}\rangle - \langle\varphi,\mu_{\frac{1}{n}}{}^{\star m}\rangle\right| : m\in\mathbb{N} \text{ and } 0\le\ell\le\sqrt{n}\,R\right\} = 0 \tag{3.2.18}$$

for every $R \in (0,\infty)$. But, for each $r \in (0,\infty)$,

$$\begin{aligned}
&\left|\langle\varphi,\mu_{\frac{1}{n}}{}^{\star(m+\ell)}\rangle - \langle\varphi,\mu_{\frac{1}{n}}{}^{\star m}\rangle\right| \\
&\quad = \left|\int_{\mathbb{R}}\left(\int_{\mathbb{R}}\big(\varphi(x+y)-\varphi(x)\big)\,\mu_{\frac{1}{n}}{}^{\star\ell}(dy)\right)\mu_{\frac{1}{n}}{}^{\star m}(dx)\right| \\
&\quad \le r\|\varphi'\|_{\mathrm{u}} + 2\|\varphi\|_{\mathrm{u}}\mu_{\frac{1}{n}}{}^{\star\ell}\big([-r,r]\complement\big),
\end{aligned}$$

and so (3.2.18) will follow from

$$\lim_{n\to\infty} \sup_{0\le\ell\le\sqrt{n}\,R} \mu_{\frac{1}{n}}{}^{\star\ell}\big([-r,r]\complement\big) = 0 \quad \text{for all } r\in(0,\infty). \tag{3.2.19}$$

Finally, to prove (3.2.19), choose $\delta > 0$ so that $|\hat{\mu}(\xi)-1| \le \frac{1}{2}$ for $\xi \in [-\delta,\delta]$, and note that, because both $\widehat{\mu_{\frac{1}{n}}}$ and $\hat{\mu}$ are continuous functions on $\mathbb{R}$ which equal 1 at $\xi = 0$ and satisfy $\widehat{\mu_{\frac{1}{n}}}^{\,n} = \hat{\mu}$ everywhere,

$$\widehat{\mu_{\frac{1}{n}}{}^{\star\ell}}(\xi) = g_{\frac{\ell}{n}}\big(\hat{\mu}(\xi)\big), \quad \text{for } |\xi| \le \delta,$$

where, for $\alpha \in (0,\infty)$, $g_\alpha(z) = e^{\alpha \log z}$ and log is the principal branch of the logarithm in the disk $\{z \in \mathbb{C} : |z-1| < 1\}$. In particular, since

$$\left|1 - g_\alpha(z)\right| \le 2\alpha|z-1| \quad \text{for } \alpha \in [0,1] \text{ and } |z-1| \le \tfrac{1}{2},$$

we have that

$$\begin{aligned}\int_{\mathbb{R}} \left(1 - \frac{\sin \delta y}{\delta y}\right) \mu_{\frac{1}{n}}{}^{\star \ell}(dy) &= \frac{1}{\delta}\int_0^\delta \left(1 - \mathfrak{Re}\left(\widehat{\mu_{\frac{1}{n}}{}^{\star \ell}}(\xi)\right)\right) d\xi \\ &\le \frac{1}{\delta}\int_0^\delta \left|1 - g_{\frac{\ell}{n}}\left(\hat{\mu}(\xi)\right)\right| d\xi \le \frac{\ell}{n}.\end{aligned}$$

Hence, if

$$m(r) \equiv \inf\left\{1 - \frac{\sin \delta y}{\delta y} : |y| \ge r\right\},$$

then $m(r) > 0$ for each $r > 0$ and

$$\mu_{\frac{1}{n}}{}^{\star \ell}\left([-r,r]\complement\right) \le \frac{2\ell}{m(r)n} \le \frac{2R}{m(r)\sqrt{n}} \quad \text{for } 0 \le \ell \le \sqrt{n}\,R;$$

which is more than enough to verify (3.2.19).

After combining the preceding with Lemma 3.2.13, we arrive at the following statement of Lévy and Khinchine's main result.

3.2.20 Theorem (Lévy–Khinchine). *Let $\mathcal{P}$ and $\mathcal{I}$ be, respectively, the sets of Poisson and infinitely divisible measures on $\mathbb{R}$. Then $\overline{\mathcal{P}} = \mathcal{I}$, and the map (cf. (3.2.6))*

$$(a,\sigma,M) \in \mathfrak{L} \longmapsto \pi_{a,\sigma^2,M} \in \mathcal{I} \text{ is one-to-one and onto.}$$

In fact, if $\mu \in \mathcal{I}$ and $\mu = \mu_{\frac{1}{n}}{}^{\star n}$ for each $n \in \mathbb{Z}^+$, then (3.2.17) holds. Hence, if $a_n \in \mathbb{R}$ and the finite measure $\tilde{M}_n$ on $\mathbb{R}$ are defined by (cf. (8.2.8))

$$a_n = n\int_{\mathbb{R}} \frac{y}{1+y^2}\,\mu_{\frac{1}{n}}(dy) \quad \text{and} \quad \tilde{M}_n(dy) = n\rho(y)\,\mu_{\frac{1}{n}}(dy),$$

then there exists an $a \in \mathbb{R}$ to which $\{a_n\}_1^\infty$ converges and a finite measure $\tilde{M}$ on $\mathbb{R}$ determined by

$$\int_{\mathbb{R}} \varphi(y)\,\tilde{M}(dy) = \lim_{n\to\infty}\int_{\mathbb{R}} \varphi(y)\,\tilde{M}_n(dy), \quad \varphi \in C_b(\mathbb{R};\mathbb{R}),$$

such that $\mu = \pi_{a,\sigma^2,M}$, where $\sigma^2 = \tilde{M}(\{0\})$ and

$$M(dy) = \frac{\mathbf{1}_{\mathbb{R}\setminus\{0\}}}{\rho(y)}\,\tilde{M}(dy).$$

Although, with Theorem 3.2.20, we have reached the goal toward which we have been heading, there is still a great deal more which ought[†] to be and has been[‡] said about this topic! However, rather than entering into a discussion of the many profound convergence results of which Theorem 3.2.20 is only the most rudimentary, we will devote the rest of this section as well as the next to finding a probabilistic interpretation for Lévy and Khinchine's description of an infinitely divisible distribution. The starting point of our analysis is the observation that, given their description, one knows that every infinitely divisible distribution μ is *continuously divisible*. That is, for every $t \in (0,1)$ (not just rational ones), there is a $\mu_{\frac{1}{t}} \in \mathbf{M}_1(\mathbb{R})$ with the property that "$\mu = \mu_{\frac{1}{t}}{}^{\star t}$" in the sense that $\hat{\mu} = \big(\widehat{\mu_{\frac{1}{t}}}\big)^t$. Indeed, if $\mu = \pi_{a,\sigma^2,M}$, then $\mu_{\frac{1}{t}} = \pi_{\frac{a}{t},\frac{\sigma^2}{t},\frac{M}{t}}$. Once this observation has been made, the obvious question is what it means probabilistically. Hence, given a Lévy system (a,σ,M), our goal will be to both construct and, to some extent, understand a *nice*, continuously parameterized family $\big\{Z_{a,\sigma^2,M}(t) : t \in [0,\infty)\big\}$ of random variables on some probability space $(\Omega, \mathcal{F}, P)$ with the properties that $Z_{a,\sigma^2,M}(0) = 0$ and

$$\text{(3.2.21)} \qquad \begin{aligned} &Z_{a,\sigma^2,M}(s+t) - Z_{a,\sigma^2,M}(s) \text{ has } P\text{-distribution } \pi_{ta,t\sigma^2,tM} \\ &\qquad\qquad \text{and is independent of } \sigma\big(Z_{a,\sigma^2,M}(\xi) : \xi \in [0,s]\big) \end{aligned}$$

for all $s \in [0,\infty)$ and $t \in (0,\infty)$. For historical reasons, a continuously parameterized family of random variables is called a **stochastic process** or, more briefly, a **process**. Furthermore, when a process $\{Z(t) : t \in [0,\infty)\}$ has the properties that, like $\big\{Z_{a,\sigma^2,M}(t) : t \in [0,\infty)\big\}$, each increment $Z(t) - Z(s)$ is independent of $\sigma\big(Z(\xi) : \xi \in [0,s]\big)$ and has a distribution depending only on $t-s$, the process is said to have **independent, identically distributed increments**. Finally, the ambiguous term *nice* will refer here to the quality of the **sample paths** $t \in [0,\infty) \longmapsto Z(t,\omega) \in \mathbb{R}$ for each sample point $\omega \in \Omega$. At the very least, we will want these paths to be right-continuous and have a left-limit at each $t \in (0,\infty)$. In particular, they should have no oscillatory discontinuities.

Before going any further, we should point out that, if one ignores the *niceness* criterion as well as the desire to understand what is happening, it is possible to make a relatively simple construction of $\big\{Z_{a,\sigma^2,M} : t \in [0,\infty)\big\}$. Namely, set $\Omega = \mathbb{R}^{[0,\infty)}$ and take $\mathcal{F} = \sigma\big(\omega(t) : t \in [0,\infty)\big)$. Next, given $S = \{s_n\}_1^\infty \subseteq (0,\infty)$, set $T_0 = 0$, $T_n = \sum_1^n s_m$ for $n \in \mathbb{Z}^+$, $\mathcal{F}_S = \sigma\big(\omega(T_n) : n \in \mathbb{N}\big)$, and define

[†] In particular, we have entirely avoided issues which arise from replacing the assumption that μ has an nth root for each n with the less rigid assumption that μ can be factored into factors which, in an appropriate sense, are arbitrarily small.

[‡] The classic book on this topic is B.V. Gnedenko & A.N. Kolmogorov's *Limit Distributions for Sums of Independent Random Variables*, publ. by Addison–Wesley.

$\Phi_S : \mathbb{R}^{\mathbb{Z}^+} \longrightarrow \Omega$ so that Φ_S is constant on each of the intervals $[T_n, T_{n+1})$ and

$$\big[\Phi_S(\mathbf{x})\big](T_n) = \sum_{m=1}^{n} x_m \quad \text{for } n \in \mathbb{N}.$$

Now, define the probability measure P_S on $(\Omega, \mathcal{F}_S)$ by (cf. Exercise 1.1.16)

$$P_S = \Phi_S{}^* \left(\delta_0 \times \prod_{n=1}^{\infty} \pi_{s_n a, s_n \sigma^2, s_n M} \right).$$

It is then an easy matter to check that $P_S(A) = P_{S'}(A)$ for $A \in \mathcal{F}_S \cap \mathcal{F}_{S'}$. Hence, since every element of $\mathcal{F}$ is an element of $\mathcal{F}_S$ for some sequence S, we can define $P^{a,\sigma^2,M} : \mathcal{F} \longrightarrow [0,1]$ so that $P^{a,\sigma^2,M} \restriction \mathcal{F}_S = P_S$. In fact, if $\{A_m\}_1^\infty \subseteq \mathcal{F}$, then we can choose S so that $\{A_m\}_1^\infty \subseteq \mathcal{F}_S$, and therefore $P^{a,\sigma^2,M}$ is a probability measure on $\mathcal{F}$. Finally, if $Z(t)$ is the random variable on $(\Omega, \mathcal{F})$ given by $Z(t,\omega) \equiv \omega(t)$, $\omega \in \Omega$, then process $\big\{Z(t) : t \in [0,\infty)\big\}$ on $\big(\Omega, \mathcal{F}, P^{a,\sigma^2,M}\big)$ satisfies both parts of (3.2.21).

Unfortunately, like many constructions based entirely on general principles, the preceding one reveals very little information about the nature of the object constructed. In particular, we have no idea how, or even whether, the parameters (a, σ, M) are reflected in the paths $t \longmapsto Z_{a,\sigma^2,M}(t) \in \mathbb{R}$. For this reason, we now adopt an entirely different approach. To begin with, we note that the task can be broken naturally into steps corresponding to the decomposition $\pi_{a,\sigma^2,M} = \pi_{0,\sigma^2,0} \bigstar \pi_{a,0,M}$. That is, if we can build versions of the processes $\big\{Z_{0,\sigma^2,0}(t) : t \in [0,\infty)\big\}$ and $\big\{Z_{a,0,M}(t) : t \in [0,\infty)\big\}$, then we get a version of $\big\{Z_{a,\sigma^2,M} : t \in [0,\infty)\big\}$ by making these independent of one another and taking $Z_{a,\sigma^2,M} = Z_{0,\sigma^2,0} + Z_{a,0,M}$. For the rest of this section we will be constructing a *nice* version of $\big\{Z_{a,0,M}(t) : t \in [0,\infty)\big\}$, and we postpone the construction of $\big\{Z_{0,\sigma^2,0}(t) : t \in [0,\infty)\big\}$ until the next section.

We start with the case when $M\big(\mathbb{R} \setminus \{0\}\big) \in (0,\infty)$ and therefore (cf. (3.2.1))

$$\text{(3.2.22)} \qquad \pi_M = \pi_{\alpha\mu} \quad \text{where } \alpha \equiv M\big(\mathbb{R} \setminus \{0\}\big) \text{ and } \mu = \tfrac{M}{\alpha}.$$

In fact, we begin with $\alpha = 1$ and $M = \delta_1$. For this purpose, we choose a sequence $\tau_1, \dots, \tau_n, \dots$ of mutually independent **unit exponential random variables** on some probability space $(\Omega, \mathcal{F}, P)$. That is,

$$P\big(\tau_1 > t_1, \dots, \tau_n > t_n\big) = \exp\left[-\sum_1^n t_m\right]$$
$$\text{for } n \in \mathbb{Z}^+ \text{ and } (t_1, \dots, t_n) \in (0,\infty)^n;$$

and, for convenience, we will assume that $\tau_n(\omega) > 0$ for all $n \in \mathbb{Z}^+$ and $\omega \in \Omega$. Next, we set $\mathbf{T}_0 = 0$, $\mathbf{T}_n = \sum_{m=1}^{n} \tau_m$ for $n \in \mathbb{Z}^+$, and define the **simple Poisson process** $\{\mathbf{N}(t) : t \in [0,\infty)\}$ by

$$\begin{aligned} \mathbf{N}(t,\omega) &= \sup\{n \in \mathbb{N} : \mathbf{T}_n(\omega) \le t\} \\ &= \sum_{n=1}^{\infty} \mathbf{1}_{[\mathbf{T}_n(\omega),\infty)}(t) \quad \text{for } (t,\omega) \in [0,\infty) \times \Omega. \end{aligned} \tag{3.2.23}$$

Thinking of τ_n as being *the time that it takes alarm clock n to ring after it has been set*, $\mathbf{T}_n$ becomes *the total time that has elapsed at the time when the nth alarm sounds* (assuming that the clocks are run in tandem: the nth being set at the instant when the $(n-1)$st one sounds), and $\mathbf{N}(t)$ is *the number of alarms that have sounded by time t*. In order to see what all this has to do with the Poisson measures, we apply an easy inductive argument to first check that

$$P(\mathbf{T}_n \le t) = \int_0^t \frac{s^{n-1}}{(n-1)!} e^{-s}\, ds \quad \text{for } n \in \mathbb{Z}^+ \text{ and } t \in [0,\infty),$$

and then that

$$\begin{aligned} P(\mathbf{N}(t) = n) &= P(\mathbf{T}_n \le t) - P(\mathbf{T}_{n+1} \le t) \\ &= e^{-t}\frac{t^n}{n!} \quad \text{for } n \in \mathbb{N} \text{ and } t \in [0,\infty). \end{aligned}$$

In other words, for each $t \in [0,\infty)$, the distribution of $\mathbf{N}(t)$ is precisely $\pi_{t\delta_1}$. In fact, we can say a great deal more.

3.2.24 Lemma. *For any $\ell \in \mathbb{N}$ and $0 \le t_0 < \cdots < t_{\ell+1}$,*

$$\begin{aligned} &P\Big(\mathbf{N}(t_{\ell+1}) - \mathbf{N}(t_\ell) = n \;\&\; \big(\mathbf{N}(t_0),\dots,\mathbf{N}(t_\ell)\big) = \mathbf{m}\Big) \\ &\qquad = e^{-(t_{\ell+1}-t_\ell)}\frac{(t_{\ell+1}-t_\ell)^n}{n!} P\Big(\big(\mathbf{N}(t_0),\dots,\mathbf{N}(t_\ell)\big) = \mathbf{m}\Big) \end{aligned}$$

for every $n \in \mathbb{N}$ and $\mathbf{m} = (m_0,\dots,m_\ell) \in \mathbb{N}^{\ell+1}$ with $m_0 \le m_1 \le \cdots \le m_\ell$. In other words, $\{\mathbf{N}(t) : t \in [0,\infty)\}$ is an independent increment process for which the increment $\mathbf{N}(s+t) - \mathbf{N}(s)$ has distribution $\pi_{t\delta_1}$.

PROOF: Set

$$A_\ell(\mathbf{m}) = \Big\{\big(\mathbf{N}(t_0),\dots,\mathbf{N}(t_\ell)\big) = \mathbf{m}\Big\},$$

and define

$$B_\ell(\mathbf{m}) = \begin{cases} \{\mathbf{T}_{m_0} \le t_0\} & \text{if } \ell = 0 \\ A_{\ell-1}\big((m_0,\dots,m_{\ell-1})\big) \cap \{\mathbf{T}_{m_\ell} \le t_\ell\} & \text{if } \ell \in \mathbb{Z}^+. \end{cases}$$

Clearly (cf. Exercise 1.1.11)

$$
\begin{aligned}
P\big(A_\ell(\mathbf{m})\big) &= P\Big(\big\{\tau_{m_\ell+1} > t_\ell - \mathbf{T}_{m_\ell}\big\} \cap B_\ell(\mathbf{m})\Big) \\
&= \mathbb{E}^P\Big[e^{-(t_\ell - \mathbf{T}_{m_\ell})},\, B_\ell(\mathbf{m})\Big].
\end{aligned} \tag{3.2.25}
$$

Similarly, for any $n \in \mathbb{Z}^+$ and $t > 0$:

$$
\begin{aligned}
&P\Big(\big\{\mathbf{N}(t_\ell + t) - \mathbf{N}(t_\ell) \ge n\big\} \cap A_\ell(\mathbf{m})\Big) \\
&= P\Big(\big\{\mathbf{T}_{m_\ell+n} - \mathbf{T}_{m_\ell} \le t_\ell + t - \mathbf{T}_{m_\ell}\big\} \cap \big\{\tau_{m_\ell+1} > t_\ell - \mathbf{T}_{m_\ell}\big\} \cap B_\ell(\mathbf{m})\Big) \\
&= \mathbb{E}\Big[f\big(\mathbf{T}_{m_\ell}\big),\, B_\ell(\mathbf{m})\Big],
\end{aligned}
$$

where

$$
\begin{aligned}
f(T) &\equiv P\Big(\big\{\mathbf{T}_n \le t_\ell + t - T\big\} \cap \big\{\tau_1 > t_\ell - T\big\}\Big) \\
&= \int_{t_\ell - T}^{t_\ell + t - T} e^{-s} P\big(\mathbf{T}_{n-1} \le t_\ell + t - T - s\big)\, ds \\
&= e^{-(t_\ell - T)} \int_0^t e^{-s} P\big(\mathbf{T}_{n-1} \le t - s\big)\, ds = e^{-(t_\ell - T)} P\big(\mathbf{T}_n \le t\big).
\end{aligned}
$$

Hence, after combining these with (3.2.25), we arrive at

$$
P\Big(\big\{\mathbf{N}(t_\ell + t) - \mathbf{N}(t_\ell) \ge n\big\} \cap A_\ell(\mathbf{m})\Big) = P\big(\mathbf{T}_n \le t\big) P\big(A_\ell(\mathbf{m})\big),
$$

from which the desired conclusion is an easy step. □

Obviously, $\mathbf{N}(\,\cdot\,,\omega)$ is always right-continuous and nondecreasing. In addition, by (3.2.23), $\mathbf{N}(t,\omega) < \infty$ (a.s., P) for each $t \in [0,\infty)$. Thus, without loss in generality, we will assume that $\mathbf{N}(t,\omega) < \infty$ for all $(t,\omega) \in [0,\infty) \times \Omega$.

In order to get the processes corresponding to general M's with $M(\mathbb{R}\setminus\{0\}) < \infty$, we write π_M as in (3.2.22) and, after expanding the original probability space $(\Omega, \mathcal{F}, P)$ if necessary, choose mutually independent random variables X_n, $n \in \mathbb{Z}^+$, which have common distribution μ and are independent of the τ_n's. Next, we define the **Poisson process** $Y_M(\,\cdot\,)$ by

$$
Y_M(t) = \sum_{n=1}^{\mathbf{N}(\alpha t)} X_n, \quad t \in [0,\infty), \quad \text{where } \sum_{n=1}^{0} X_n \equiv 0.
$$

Notice that when $M = \delta_1$, and therefore $X_n = 1$ almost surely, $Y_M(\,\cdot\,) = \mathbf{N}(\,\cdot\,)$ almost surely. In fact, what $Y_M(\,\cdot\,)$ is now modeling is a process which evolves by *jumping* a random amount X_n at the instant when the nth alarm clock sounds and is at rest between alarms. Furthermore, if $S_n \equiv \sum_{m=1}^n X_m$, then

$$\begin{aligned} P\big(Y_M(t) \in \Gamma\big) &= \sum_{n=0}^{\infty} P\big(S_n \in \Gamma \;\&\; \mathbf{N}(\alpha t) = n\big) \\ &= e^{-\alpha t} \sum_{n=0}^{\infty} \frac{(\alpha t)^n}{n!} \mu^{\star n}(\Gamma) = \pi_{tM}(\Gamma) \end{aligned}$$

and

$$\begin{aligned} &P\Big(Y_M(t_\ell) - Y_M(t_{\ell-1}) \in \Gamma \;\&\; \big(Y_M(t_0), \dots, Y_M(t_{\ell-1})\big) \in A\Big) \\ &= \sum_{n=0}^{\infty} \sum_{\mathbf{m} \in \mathbb{N}^\ell} P\Big(S_{m_{\ell-1}+n} - S_{m_{\ell-1}} \in \Gamma \;\&\; \big(S_{m_0}, \dots, S_{m_{\ell-1}}\big) \in A\Big) \\ &\qquad \times P\Big(\mathbf{N}(\alpha t_\ell) - \mathbf{N}(\alpha t_{\ell-1}) = n \;\&\; \big(\mathbf{N}(\alpha t_0), \dots, \mathbf{N}(\alpha t_{\ell-1})\big) = \mathbf{m}\Big) \\ &= \sum_{n=0}^{\infty} P(S_n \in \Gamma) P\big(\mathbf{N}(\alpha(t_\ell - t_{\ell-1})) = n\big) \\ &\qquad \times \sum_{\mathbf{m} \in \mathbb{N}^\ell} P\Big(\big(S_{m_0}, \dots, S_{m_{\ell-1}}\big) \in A\Big) P\Big(\big(\mathbf{N}(\alpha t_0), \dots, \mathbf{N}(\alpha t_{\ell-1})\big) = \mathbf{m}\Big) \\ &= P\Big(Y_M(t_\ell - t_{\ell-1}) \in \Gamma\Big)\, P\Big(\big(Y_M(t_0), \dots, Y_M(t_{\ell-1})\big) \in A\Big), \end{aligned}$$

for all $\ell \in \mathbb{Z}^+$, $0 \le t_0 < \cdots < t_\ell$, $\Gamma \in \mathcal{B}_{\mathbb{R}}$, and $A \in \mathcal{B}_{\mathbb{R}^\ell}$. Hence, $\{Y_M(t) : t \in [0, \infty)\}$ has independent increments and the increment $Y_M(s+t) - Y_M(s)$ has distribution π_{tM}. Finally, it is clear, from our construction, that the sample paths $t \in [0, \infty) \longmapsto Y_M(t) \in \mathbb{R}$ are right-continuous and piecewise constant, and, as such, have finite total variation on bounded intervals and a left limit at each $t \in (0, \infty)$. In fact, we have the following more quantitative statements about their behavior.

3.2.26 Lemma. *Let* $\operatorname{var}_{[0,T]}(\psi)$ *denote the total variation of the function* $\psi : [0, \infty) \longrightarrow \mathbb{R}$ *on the interval* $[0, T]$*. Then, for each* $T \in [0, \infty)$, $\omega \in \Omega \longmapsto \operatorname{var}_{[0,T]}\big(Y_M(\,\cdot\,, \omega)\big)$ *is* $\mathcal{F}$*-measurable and*

$$\mathbb{E}^P\Big[\operatorname{var}_{[0,T]}(Y_M)\Big] = T \int_{\mathbb{R}\setminus\{0\}} |y|\, M(dy). \tag{3.2.27}$$

Next, suppose that $\int_{\mathbb{R}\setminus\{0\}} |y|\, M(dy) < \infty$*, and set*

$$b = \int_{\mathbb{R}\setminus\{0\}} y\, M(dy) \quad \text{and} \quad \bar{Y}_M(t, \omega) = Y_M(t, \omega) - bt \quad \text{for } (t, \omega) \in [0, \infty) \times \Omega.$$

Then, for each $T \in (0,\infty)$,

$$\omega \in \Omega \longmapsto \bar{Y}^*_M(T) \equiv \sup_{t\in[0,T]} \left|\bar{Y}_M(t,\omega)\right|$$

is $\mathcal{F}$-measurable and

$$P\Big(\bar{Y}^*_M(T) \geq \epsilon\Big) \leq \frac{T}{\epsilon^2} \int_{\mathbb{R}\setminus\{0\}} y^2\, M(dy), \quad \epsilon > 0. \tag{3.2.28}$$

PROOF: To prove the first assertion, simply note that

$$\operatorname{var}_{[0,T]}\big(Y_M(\,\cdot\,,\omega)\big) = \sum_{n=0}^{\mathbf{N}(\alpha T,\omega)} |X_n(\omega)|,$$

and therefore both that $\operatorname{var}_{[0,T]}(Y_M)$ is $\mathcal{F}$-measurable and that

$$\begin{aligned}\mathbb{E}^P\Big[\operatorname{var}_{[0,t]}(Y_M)\Big] &= \mathbb{E}^P\Big[\mathbf{N}(\alpha T)\Big]\,\mathbb{E}^P\Big[|X_1|\Big] \\ &= \alpha T \int_{\mathbb{R}} |y|\,\mu(dy) = T\int_{\mathbb{R}} |y|\,M(dy).\end{aligned}$$

To prove the second part, set

$$L_n(\omega) = \max\Big\{\big|\bar{Y}_M\big(\tfrac{mT}{2^n},\omega\big)\big| : 0 \leq m \leq 2^n\Big\}$$

and

$$\Delta_{m,n}(\omega) = \bar{Y}_M\big(\tfrac{mT}{2^n},\omega\big) - \bar{Y}_M\Big(\tfrac{(m-1)T}{2^n},\omega\Big), \quad 1 \leq m \leq 2^n.$$

By right-continuity, we know that $L_n(\omega) \nearrow \bar{Y}^*_M(T,\omega)$ as $n \to \infty$, and this clearly proves measurability. Furthermore, for each $n \in \mathbb{N}$, the $\Delta_{m,n}$'s are mutually independent, identically distributed random variables with mean-value 0 and variance $2^{-n}T\int y^2\,M(dy)$. Hence, since

$$L_n(\omega) = \max\left\{\left|\sum_{\ell=1}^{m} \Delta_{\ell,n}(\omega)\right| : 1 \leq m \leq 2^n\right\},$$

Kolmogorov's inequality (cf. Theorem 1.4.5) says that

$$P\Big(L_n \geq \epsilon\Big) \leq \frac{T}{\epsilon^2}\int y^2\,M(dy), \quad n \in \mathbb{Z}^+;$$

and clearly this completes the proof of (3.2.28). □

In order to handle $M \in \mathfrak{M}$ for which $M(\mathbb{R}\setminus\{0\}) = \infty$, we will decompose M into manageable parts, apply the preceding to each part, and reassemble by super-position. For this purpose it will be useful to introduce the following partition of $\mathbb{R}\setminus\{0\}$. Namely, choose a decreasing sequence $\{r_k\}_0^\infty \subseteq (0,\infty)$ so that $r_k \searrow 0$, $\alpha_k \equiv M(A_k) > 0$, $k \in \mathbb{N}$, and

$$\begin{aligned} &\int_{A_k} |y|\, M(dy) \le \frac{1}{2^k},\ k \in \mathbb{Z}^+, \quad \text{when (3.2.2) holds} \\ &\int_{A_k} y^2\, M(dy) \le \frac{1}{2^k},\ k \in \mathbb{Z}^+, \quad \text{when (3.2.2) fails,} \end{aligned} \tag{3.2.29}$$

where $A_0 = \{y : |y| > r_0\}$ and $A_k = \{y : r_k < |y| \le r_{k-1}\}$ for $k \in \mathbb{Z}^+$. Then the preceding applies to each of the measures $M_k \in \mathfrak{M}$, $k \in \mathbb{N}$, given by $M_k(\Gamma) = M(\Gamma \cap A_k)$ for $\Gamma \in \mathcal{B}_{\mathbb{R}\setminus\{0\}}$; and so we can produce for each $k \in \mathbb{N}$ a process $\{Y_{M_k}(t) : t \in [0,\infty)\}$ built from the process $\{\mathbf{N}_k(t) : t \in [0,\infty)\}$ and the random variables $\{X_{n,k} : n \in \mathbb{Z}^+\}$. Furthermore, we can use Exercise 1.1.16 to arrange for all these quantities to live on the same probability space $(\Omega, \mathcal{F}, P)$ and that the quantities corresponding to distinct k's be mutually independent there.

3.2.30 Lemma. *Assume that $M \in \mathfrak{M}$ satisfies (3.2.2), and let π_M denote the Poisson measure whose characteristic function is*

$$\widehat{\pi_M}(\xi) = \exp\left[\int_{\mathbb{R}\setminus\{0\}} \left(e^{\sqrt{-1}\,\xi y} - 1\right) M(dy)\right].$$

Then there is an independent increment process $\{Y_M(t) : t \in [0,\infty)\}$ on a probability space $(\Omega, \mathcal{F}, P)$ with the properties that $Y_M(s+t) - Y_M(s)$ has distribution π_{tM} for each $s \in [0,\infty)$ and $t \in (0,\infty)$; and, for each $\omega \in \Omega$, $t \in [0,\infty) \longmapsto Y_M(t,\omega) \in \mathbb{R}$ is a right-continuous path with bounded variation on bounded intervals and $Y_M(0,\omega) = 0$. In fact, for each $(T,\omega) \in [0,\infty)\times\Omega$

$$Y_M(T,\omega) = \sum_{t \in (0,T]} \big(Y_M(t,\omega) - Y_M(t-,\omega)\big), \tag{3.2.31}$$

where $Y_M(t-,\omega) \equiv \lim_{s\nearrow t} Y_M(s,\omega)$ for $t \in (0,\infty)$.

PROOF: Referring to the preceding paragraph, set

$$S_\ell(t,\omega) = \sum_{k=0}^{\ell} Y_{M_k}(t,\omega) \quad \text{for } \ell \in \mathbb{N} \text{ and } (t,\omega) \in [0,\infty)\times\Omega.$$

By Lemma 3.2.26, we know that, for each $\ell \in \mathbb{N}$ and $\omega \in \Omega$, $S_\ell(\cdot,\omega)$ is a right-continuous, piecewise constant path. Moreover, by (3.2.27),

$$\mathbb{E}^P\left[\operatorname{var}_{[0,T]}(S_\ell - S_{\ell-1})\right] = T\int_{\mathbb{R}} |y|\, M_\ell(dy) = T\int_{A_\ell} |y|\, M(dy);$$

and therefore, by (3.2.29) and the Borel–Cantelli Lemma, there is a $\Lambda \in \mathcal{F}$ with $P(\Lambda) = 0$ such that

$$\sup_{\ell\in\mathbb{Z}^+} \ell^2 \operatorname{var}_{[0,T]}\big(S_\ell(\cdot,\omega) - S_{\ell-1}(\cdot,\omega)\big) < \infty \quad \text{for all } T \in (0,\infty) \text{ and } \omega \notin \Lambda.$$

Hence, for each $\omega \notin \Lambda$ there is a right-continuous $Y_M(\cdot,\omega)$ of bounded variation on bounded intervals such that

$$\lim_{\ell\to\infty} \operatorname{var}_{[0,T]}\big(Y_M(\cdot,\omega) - S_\ell(\cdot,\omega)\big) = 0$$

for all $T \in (0,\infty)$; and we complete the definition of Y_M by setting

$$Y_M(\cdot,\omega) \equiv 0 \quad \text{for } \omega \in \Lambda.$$

Notice that (3.2.31) is trivial for $\omega \in \Lambda$, and that, because the convergence is in variation, it follows from the analogous fact about the $S_\ell(\cdot,\omega)$'s when $\omega \notin \Lambda$. Similarly, the fact that $\{Y_M(t) : t \in [0,\infty)\}$ has independent increments comes from the independence of the increments of each $\{S_\ell(t) : t \in [0,\infty)\}$; and the distribution of $Y_M(s+t) - Y_M(s)$ is identified via the computation:

$$\begin{aligned}
&\mathbb{E}^P\left[\exp\Big(\sqrt{-1}\,\xi\big(Y_M(s+t) - Y_M(s)\big)\Big)\right] \\
&\quad = \lim_{\ell\to\infty} \mathbb{E}^P\left[\exp\Big(\sqrt{-1}\,\xi\big(S_\ell(s+t) - S_\ell(s)\big)\Big)\right] \\
&\quad = \lim_{\ell\to\infty} \prod_{k=0}^{\ell} \widehat{\pi_{tM_k}}(\xi) = \widehat{\pi_{tM}}(\xi). \quad \square
\end{aligned}$$

As the preceding lemma shows, so long as (3.2.2) holds, the sample paths of independent increment process determined by π_M can be taken to be right-continuous functions which are constant between simple jump discontinuities. The point is, of course, that although $M(\mathbb{R}\setminus\{0\}) = \infty$ means that there will be infinitely many jumps in each open interval, (3.2.2) says that the frequency and size of these jumps are balanced in such a way that the resulting path still has finite arc-length over finite time intervals. For this reason, it is reasonable to continue referring to such processes as a Poisson process. However, when (3.2.2) fails, the situation changes radically.

3.2.32 Lemma. *Assume that (3.2.2) fails but (3.2.3) holds. Then there is an independent increment process $\{\tilde{Y}_M(t) : t \in [0,\infty)\}$ on a probability space $(\Omega, \mathcal{F}, P)$ with the properties that $\tilde{Y}_M(s+t) - \tilde{Y}_M(s)$ has distribution $\pi_{0,0,tM}$ for each $s \in [0,\infty)$ and $t \in (0,\infty)$, and, for each $\omega \in \Omega$, $\tilde{Y}_M(\,\cdot\,,\omega)$ is a right-continuous path with $\tilde{Y}_M(0,\omega) = 0$ and a left limit at each $t \in (0,\infty)$. However, for every $T \in (0,\infty)$, $\operatorname{var}\big(\tilde{Y}_M(\,\cdot\,,\omega)\big) = \infty$ for P-almost every $\omega \in \Omega$.*

PROOF: Again referring to the paragraph preceding Lemma 3.2.30, set

$$b_k = \int_{A_k} y\, M(dy) \quad \text{and} \quad \bar{Y}_{M_k}(t) = Y_{M_k}(t) - b_k t \quad \text{for } k \in \mathbb{Z}^+,$$

and

$$\bar{S}_\ell(t) = Y_{M_0}(t) + \sum_{k=1}^{\ell} \bar{Y}_{M_k}(t), \quad \ell \in \mathbb{N}.$$

Then, by (3.2.28) and (3.2.29),

$$P\left(\sup_{t\in[0,T]} \big|\bar{S}_\ell(t) - \bar{S}_{\ell-1}(t)\big| \geq \epsilon \right) \leq \frac{T}{2^\ell \epsilon^2}, \quad \ell \in \mathbb{Z}^+,$$

and, therefore, another application of Borel–Cantelli proves that there exists a $\Lambda \in \mathcal{F}$ with $P(\Lambda) = 0$ and the property that

$$\sup_{\ell\in\mathbb{Z}^+} \ell^2 \sup_{t\in[0,T]} \big|\bar{S}_\ell(t,\omega) - \bar{S}_{\ell-1}(t,\omega)\big| < \infty, \quad T \in (0,\infty) \text{ and } \omega \notin \Lambda.$$

Thus, there is a measurable $\bar{Y}_M : [0,\infty) \times \Omega \longrightarrow \mathbb{R}$ such that $\bar{Y}_M(\,\cdot\,,\omega) \equiv 0$ when $\omega \in \Lambda$ and $\bar{S}_\ell(\,\cdot\,,\omega) \longrightarrow \bar{Y}_M(\,\cdot\,,\omega)$ uniformly on bounded time intervals when $\omega \notin \Lambda$. In particular, this means that $\bar{Y}_M(\,\cdot\,,\omega)$ is right-continuous and has a left limit at every $t \in (0,\infty)$. In addition, $\{\bar{Y}_M(t) : t \in [0,\infty)\}$ inherits the independent increment property from the $\bar{S}_\ell$'s; and

$$\begin{aligned}
&\mathbb{E}^P\Big[\exp\Big(\sqrt{-1}\,\xi\big(\bar{Y}_M(s+t) - \bar{Y}_M(s)\big)\Big)\Big] \\
&= \widehat{\pi_{tM_0}}(\xi) \prod_{k=1}^{\infty} e^{-\sqrt{-1}\,\xi t b_k} \widehat{\pi_{tM_k}}(\xi) \\
&= \exp\left[t \int_{\mathbb{R}\setminus\{0\}} \left(e^{\sqrt{-1}\,\xi y} - 1 - \sqrt{-1}\,\xi y \mathbf{1}_{[0,r_0]}(|y|) \right) M(dy) \right] \\
&= \widehat{\pi_{a,0,tM}}(\xi),
\end{aligned}$$

where

$$a = \int_{\mathbb{R}\setminus\{0\}} \left(\frac{y}{1+y^2} - y\mathbf{1}_{[0,r_0]}(|y|) \right) M(dy).$$

Thus, if we take $\tilde{Y}_M(t) = \bar{Y}_M(t) - at$, then $\{\tilde{Y}_M(t) : t \in [0,\infty)\}$ has independent increments and $\tilde{Y}_M(s+t) - \tilde{Y}_M(s)$ has distribution $\pi_{0,0,tM}$; and obviously the path regularity of $\tilde{Y}_M(\,\cdot\,,\omega)$ is the same as that of $\bar{Y}_M(\,\cdot\,,\omega)$.

In order to complete the proof, we must still show that $\tilde{Y}_M(\,\cdot\,,\omega)$ has unbounded variation on $[0,T]$ for almost every $\omega \in \Omega$. To this end, note that, from our construction, we know that

$$\begin{gathered}\left|\tilde{Y}_M(t,\omega) - \tilde{Y}_M(t-,\omega)\right| \in (r_k, r_{k-1}] \iff \\ \left|\tilde{Y}_M(t,\omega) - \tilde{Y}_M(t-,\omega)\right| = \left|Y_{M_k}(t,\omega) - Y_{M_k}(t-,\omega)\right| > 0\end{gathered}$$

for all $t \in (0,\infty)$ and $\omega \notin \Lambda$. Hence, for $\omega \notin \Lambda$,

$$\operatorname{var}_{[0,T]}\bigl(\tilde{Y}_M(\,\cdot\,,\omega)\bigr) \geq \sum_{k=1}^{\infty} \operatorname{var}_{[0,T]}\bigl(Y_{M_k}(\,\cdot\,,\omega)\bigr).$$

Thus, all that we need to do is check that, for all $T \in (0,\infty)$,

$$\sum_1^\infty V_k(T) = \infty \quad \text{(a.s., } P\text{)}, \quad \text{where } V_k(T,\omega) = \operatorname{var}_{[0,T]}\bigl(Y_{M_k}(\,\cdot\,,\omega)\bigr).$$

But, for each $T \in (0,\infty)$, the $V_k(T)$'s are mutually independent random variables. Moreover, since (cf. the paragraph preceding Lemma 3.2.26)

$$V_k(T) = \sum_{n=1}^{\mathbf{N}_k(\alpha_k T)} |X_{n,k}|,$$

one can easily check that

$$\mathbb{E}^P\bigl[V_k(T)\bigr] = T\,m_k \quad \text{where } m_k \equiv \int_{A_k} |y|\,M(dy)$$

and that

$$\mathbb{E}^P\Bigl[\bigl(V_k(T) - T\,m_k\bigr)^2\Bigr] = T\,\sigma_k^2 \quad \text{where } \sigma_k^2 \equiv \int_{A_k} y^2\,M(dy).$$

Hence, since, by (3.2.3),

$$\sum_{k=1}^{\infty} \sigma_k^2 = \int_{0<|y|\leq r_0} y^2\,M(dy) < \infty,$$

Theorem 1.4.2 tells us that the sum

$$\sum_{k=1}^{\infty} \bigl(V_k(T) - T\,m_k\bigr) \quad \text{converges (a.s., } P\text{)}.$$

On the other hand, because (3.2.2) fails and therefore $\sum_{k=1}^\infty m_k = \infty$, this implies the required divergence of $\sum_1^\infty V_k(T)$. $\square$

In order to formulate our findings as a single statement, we introduce the space $CL(\mathbb{R})$ of right-continuous paths $\psi : [0,\infty) \longrightarrow \mathbb{R}$ which possess a left limit at each $t \in (0,\infty)$ (the notation "CL" derives from the *anglfrish* expression **cádlág** which, in turn, derives from *continue á droite limites á gauche*) and set $\mathcal{F}_{CL(\mathbb{R})}$ equal to the σ-algebra $\sigma\big(\psi(t) : t \in [0,\infty)\big)$ generated by the maps $\psi \in CL(\mathbb{R}) \longrightarrow \psi(t) \in \mathbb{R}$, $t \in [0,\infty)$. Note that, because of the right-continuity, the space $BV(\mathbb{R})$ of $\psi \in CL(\mathbb{R})$ having bounded variation on each bounded interval is an element of $\mathcal{F}_{CL(\mathbb{R})}$. In addition, observe that even those $\psi \in CL(\mathbb{R})$ which have unbounded variation nonetheless have the property that, for any $\delta > 0$ and $T \in (0,\infty)$, there are at most a finite number of $t \in (0,T]$ with the property that $|\psi(t) - \psi(t-)| \geq \delta$.

3.2.33 Theorem. *For each $a \in \mathbb{R}$ and $M \in \mathfrak{M}$ there is a unique probability measure*

$$P^{a,0,M} \text{ on } \big(CL(\mathbb{R}), \mathcal{F}_{CL(\mathbb{R})}\big)$$

with the properties that $P^{a,0,M}\big(\psi(0) = 0\big) = 1$ and, for each $s \in [0,\infty)$ and $t \in (0,\infty)$, $\psi \in CL(\mathbb{R}) \longmapsto \psi(s+t) - \psi(s) \in \mathbb{R}$ is independent of $\sigma\big(\psi(\tau) : \tau \in [0,s]\big)$ and has distribution $\pi_{ta,0,tM}$. Moreover,

$$P^{a,0,M}\big(BV(\mathbb{R})\big) = \begin{cases} 1 & \text{if } (3.2.2) \text{ holds} \\ 0 & \text{otherwise.} \end{cases} \tag{3.2.34}$$

PROOF: The existence of $P^{a,0,M}$ as well as the equality in (3.2.34) are essentially trivial consequences of Lemma 3.2.30 and Lemma 3.2.32. Namely, in the case when (3.2.2) holds, define (cf. Lemma 3.2.30)

$$\tilde{Y}_{a,M}(t,\omega) = Y_M(t,\omega) + \left(a - \int_{\mathbb{R}\setminus\{0\}} \frac{y}{1+y^2}\, M(dy)\right) t, \quad (t,\omega) \in [0,\infty) \times \Omega,$$

and let $P^{a,0,M}$ be the distribution of

$$\omega \in \Omega \longmapsto \tilde{Y}_{a,M}(\,\cdot\,,\omega) \in BV(\mathbb{R})$$

under P. In the case when (3.2.2) fails, define (cf. Lemma 3.2.32)

$$\tilde{Y}_{a,M}(t,\omega) = \tilde{Y}_M(t,\omega) + at, \quad (t,\omega) \in [0,\infty) \times \Omega,$$

and again take $P^{a,0,M}$ to be the distribution of

$$\omega \in \Omega \longmapsto \tilde{Y}_{a,M}(\,\cdot\,,\omega) \in CL(\mathbb{R})$$

under P.

To prove the uniqueness assertion, note that two probability measures defined on $\big(CL(\mathbb{R}), \mathcal{F}_{CL(\mathbb{R})}\big)$ coincide if they are agree on the π-system of sets

$$A(t_1,\dots,t_n;\Gamma),\ n\in\mathbb{Z}^+,\quad 0=t_0<t_1<\cdots<t_n,\quad \text{and } \Gamma\in\mathcal{B}_{\mathbb{R}^{n+1}},$$

of the form

$$\big\{\psi:\ \big(\psi(0),\psi(t_1)-\psi(t_0),\dots,\psi(t_n)-\psi(t_{n-1})\big)\in\Gamma\big\}.$$

But, if P satisfies the prescribed conditions with respect to a and M, then

$$P\big(A(t_1,\dots,t_n;\Gamma)\big)=\left(\delta_0\times\prod_{m=1}^{n}\pi_{s_m a,0,s_m M}\right)(\Gamma)$$

where $s_m=t_m-t_{m-1}$, $1\le m\le n$; and therefore there is at most one such P. □

3.2.35 Remark. As the preceding sequence of results makes precise, the picture of the sample paths becomes murkier and murkier as M becomes more and more singular at 0. Thus, when $\alpha\equiv M\big(\mathbb{R}\setminus\{0\}\big)<\infty$, the sample paths, apart from the trivial linear drift

$$t\in[0,\infty)\longmapsto L_{a,M}(t)\equiv\left(a-\int_{\mathbb{R}\setminus\{0\}}\frac{y}{1+y^2}\,M(dy)\right)t, \tag{3.2.36}$$

are piecewise constant functions in which the rate of jumping is determined by α and the distribution of jump sizes is determined by $\frac{M}{\alpha}$. In the case when $\alpha=\infty$ but (3.2.2) holds, the only change is that the jumps come infinitely often, although the sample paths still admit a clean decomposition into the linear drift in (3.2.36) and jumps; that is:

$$\psi(t)=L_{a,M}(t)+\sum_{s\in(0,t]}\big(\psi(s)-\psi(s-)\big)\quad (\text{a.s.}, P^{a,0,M}).$$

Finally, when (3.2.2) fails, the jumps start coming so fast that the clean separation of jumps from drift breaks down, and the net effect of the infinitely many very small jumps gets *smeared out* and becomes indistinguishable from that of the drift. In fact, as it turns out, this elision of the small jumps with the drift is extremely important and, in the end, turns out to save the day by allowing us to *renormalize* with the now infinite drift in (3.2.36). However, as the following result makes clear, even when (3.2.2) fails, these complications occur only with the small jumps and never with large ones.

3.2.37 Corollary. *For $\Delta \in \mathcal{B}_{\mathbb{R}\setminus\{0\}}$ and $M \in \mathfrak{M}$, define $M_\Delta(\Gamma) = M(\Delta\cap\Gamma)$, $\Gamma \in \mathcal{B}_{\mathbb{R}\setminus\{0\}}$. Further, if Δ lies a positive distance from 0 (i.e., $\Delta \cap (-\delta,\delta) = \emptyset$ for some $\delta > 0$), define $a_\Delta \in \mathfrak{M}$ and $\psi \in CL(\mathbb{R}) \longmapsto \psi_\Delta \in CL(\mathbb{R})$ by*

$$a_\Delta = \int_\Delta \frac{y}{1+y^2}\, M(dy)$$

and

$$\psi_\Delta(t) = \sum_{s\in(0,t]} \big(\psi(s)-\psi(s-)\big)\mathbf{1}_\Delta\big(\psi(s)-\psi(s-)\big), \quad t \in [0,\infty).$$

Then, for any $(a, M) \in \mathbb{R}\times\mathfrak{M}$ and $\Delta \in \mathcal{B}_{\mathbb{R}\setminus\{0\}}$ lying a positive distance from 0, the distribution of

$$\psi \in CL(\mathbb{R}) \longmapsto \big(\psi_\Delta, \psi - \psi_\Delta\big) \in CL(\mathbb{R}) \times CL(\mathbb{R})$$

under $P^{a,0,M}$ is $P^{a_\Delta,0,M_\Delta} \times P^{a-a_\Delta,0,M-M_\Delta}$. Equivalently, the σ-algebras

$$\mathcal{F}(\Delta) \equiv \sigma\big(\psi_\Delta(t) : t \in [0,\infty)\big)$$

and

$$\mathcal{F}(\Delta\complement) \equiv \sigma\big(\psi(t) - \psi_\Delta(t) : t \in [0,\infty)\big)$$

are independent under $P^{a,0,M}$,

$$P^{a,0,M} \restriction \mathcal{F}(\Delta) = P^{a_\Delta,0,M_\Delta} \restriction \mathcal{F}(\Delta)$$

and

$$P^{a,0,M} \restriction \mathcal{F}(\Delta\complement) = P^{a-a_\Delta,0,M-M_\Delta} \restriction \mathcal{F}(\Delta\complement).$$

PROOF: The proof turns on two simple observations. The first observation is that, for any $\Delta \in \mathcal{B}_{\mathbb{R}\setminus\{0\}}$, $a \in \mathbb{R}$, and $M \in \mathfrak{M}$, $M\big(\Delta\complement\big) = 0$ implies that

$$P^{a,0,M}\Big(\big\{\psi : \psi(t) - \psi(t-) \notin \Delta \cup \{0\} \text{ for some } t \in [0,\infty)\big\}\Big) = 0. \tag{3.2.38}$$

To prove (3.2.38), one notes that it suffices to handle Δ's which lie a positive distance from 0 and that, given such a Δ, (3.2.38) follows immediately from our construction procedure. The second observation is that, for any pair $(a, M) \in \mathbb{R}\times\mathfrak{M}$, $(b, N) \in \mathbb{R}\times\mathfrak{M}$, and $A \in \mathcal{F}_{CL(\mathbb{R})}$:

$$\begin{aligned} &P^{a,0,M} \times P^{b,0,N}\big(\{(\varphi,\psi) \in CL(\mathbb{R})^2 : \varphi + \psi \in A\}\big) \\ &\quad = P^{a+b,0,M+N}(A). \end{aligned} \tag{3.2.39}$$

To see this, one need only note that, under $P^{a,0,M} \times P^{b,0,N}$, the increment

$$\big(\varphi(s+t), \psi(s+t)\big) - \big(\varphi(s), \psi(s)\big)$$

is independent of

$$\sigma\Big(\big(\varphi(\xi), \psi(\xi)\big) : \xi \in [0,s]\Big)$$

and that $(\varphi+\psi)(s+t) - (\varphi+\psi)(s)$ has distribution $\pi_{t(a+b),0,t(M+N)}$. Indeed, as soon as one checks these, (3.2.39) follows immediately from the uniqueness part of Theorem 3.2.33.

Given (3.2.38) and (3.2.39), one proceeds as follows. Let $A \in \mathcal{F}(\Delta)$ and $B \in \mathcal{F}(\Delta\complement)$ be given, and set $Q = P^{a_\Delta,0,M_\Delta} \times P^{a-a_\Delta,0,M-M_\Delta}$. Starting from (3.2.38), it is easy to check

$$(\varphi+\psi)_\Delta = \varphi_\Delta + \psi_\Delta = \varphi \quad \text{for } Q\text{-almost every } (\varphi,\psi) \in CL(\mathbb{R})^2.$$

Hence, by (3.2.39),

$$\begin{aligned} P^{a,0,M}(A \cap B) &= P^{a,0,M}\big(\{\psi : \psi_\Delta \in A \;\&\; \psi - \psi_\Delta \in B\}\big) \\ &= Q\big(\{(\varphi,\psi) : \varphi \in A \;\&\; \psi \in B\}\big) \\ &= P^{a_\Delta,0,M_\Delta}(A)\, P^{a-a_\Delta,0,M-M_\Delta}(B). \quad \square \end{aligned}$$

Exercises

3.2.40 Exercise: The difficulty (alluded to at the end of Remark 3.2.35) of distinguishing between the drift and small jumps when (3.2.2) fails has a purely analytic antecedent which is reflected by an inherent arbitrariness in the Lévy–Khinchine formula. Namely, there is nothing sacrosanct, or even particularly compelling, about the way in which we *corrected* $e^{\sqrt{-1}\,\xi y} - 1$ in order to accommodate M's for which (3.2.2) fails. Indeed, show that we could have equally well taken (cf. (3.2.5)) $e_2(\xi, y)$ to be any function of the form

$$e^{\sqrt{-1}\,\xi y} - 1 - \sqrt{-1}\,\xi\psi(y),$$

where $\psi : \mathbb{R} \longrightarrow \mathbb{R} \setminus \{0\}$ is any bounded measurable function with the property that $\sup_{y\neq 0}\left|\frac{\psi(y)-y}{y^2}\right| < \infty$. Further, show that each such ψ leads to a reparameterization of $\mathcal{I}$ in which a is replaced by

$$a + \int_{\mathbb{R}\setminus\{0\}} \left(\psi(y) - \frac{y}{1+y^2}\right) M(dy).$$

Finally, when M is symmetric (i.e. $M(-\Gamma) = M(\Gamma)$ for all $\Gamma \in \mathcal{B}_{\mathbb{R}\setminus\{0\}}$), check that

$$\widehat{\pi_{a,\sigma^2,M}}(\xi) = \exp\left[\sqrt{-1}\,a\xi - \int\limits_{\mathbb{R}\setminus\{0\}} \big(1-\cos(\xi y)\big)\,M(dy)\right];$$

thus partially alleviating some of the arbitrariness alluded to above.

3.2.41 Exercise: There are very few cases in which a good closed expression for the measure $\pi_{a,\sigma^2,M}$ is available. The two given here are probably the most famous. (See Exercise 4.3.46 as well.)

(i) Show that, for each $a \in \mathbb{R}$ and $\sigma \in (0,\infty)$,

$$\pi_{a,\sigma^2,0}(dx) = \frac{1}{\sqrt{2\pi\sigma^2}} \exp\left[-\frac{(x-a)^2}{2\sigma^2}\right] dx.$$

(ii) Suppose that

$$M(dy) = \frac{\mathbf{1}_{\mathbb{R}\setminus\{0\}}(y)}{\pi y^2}\,dy,$$

and show that

$$\pi_{a,0,tM}(dx) = \frac{t}{\pi\big(t^2 + (x-a)^2\big)}\,dx, \quad (a,t) \in \mathbb{R}\times(0,\infty).$$

When $a = 0$ and $t = 1$, this measure is the famous **Cauchy distribution**.

Hint: At this point, Cauchy's integral formula is probably the simplest route to these computations. See Exercise 4.3.49 for another look at this example.

3.2.42 Exercise: Let $I(\mathbb{R})$ denote the set of nondecreasing elements of $CL(\mathbb{R})$ (cf. the paragraph preceding Theorem 3.2.33). Referring to Theorem 3.2.33, show that $P^{a,0,M}\big(I(\mathbb{R})\big) = 1$ if (3.2.2) holds, $M\big((-\infty,0)\big) = 0$, and

$$a \geq \int_{\mathbb{R}\setminus\{0\}} \frac{y}{1+y^2}\,M(dy).$$

Conversely, if either of these conditions fails, show that $P^{a,0,M}\big(I(\mathbb{R})\big) = 0$.

3.2.43 Exercise: Let $(a,\sigma,M) \in \mathfrak{L}$ be given, and assume that

$$\int\limits_{\mathbb{R}\setminus\{0\}} y^2\,M(dy) < \infty.$$

(i) Show that $\pi_{a,\sigma^2,M}$ admits two finite moments, that

$$m_{a,\sigma^2,M} \equiv \int_{\mathbb{R}} x\,\pi_{a,\sigma^2,M}(dx) = a + \int\limits_{\mathbb{R}\setminus\{0\}} \frac{y^3}{1+y^2}\,M(dy),$$

and that

$$\int_{\mathbb{R}} \left(x - m_{a,\sigma^2,M}\right)^2 \pi_{a,\sigma^2,M}(dx) = \frac{\sigma^2}{2} + \int\limits_{\mathbb{R}\setminus\{0\}} y^2\,M(dy).$$

(ii) Assume that $\sigma = 0$ and set $\bar{\psi}(t) = \psi(t) - t\,m_{a,0,M}$ for $t \in [0,\infty)$ and $\psi \in CL(\mathbb{R})$. Proceeding as in the derivation of (3.2.28), show that

$$P^{a,0,M}\left(\sup_{t\in[0,T]} \left|\bar{\psi}(t)\right| \geq r\right) \leq \frac{T}{r^2} \int\limits_{\mathbb{R}\setminus\{0\}} y^2\,M(dy)$$

for all $r \in (0,\infty)$ and $T \in [0,\infty)$.

(iii) Referring to the simple Poisson process $\{\mathbf{N}(t) : t \in [0,\infty)\}$ in (3.2.23), use part **(ii)** to show that

$$\lim_{\alpha\to\infty} P\left(\sup_{t\in[0,T]} \left|\frac{\mathbf{N}(\alpha t)}{\alpha} - t\right| \geq r\right) = 0 \quad \text{for all } T \in [0,\infty) \text{ and } r \in (0,\infty).$$

(iv) Assume that

$$\int\limits_{\mathbb{R}\setminus\{0\}} y^2 e^{\lambda y}\,M(dy) < \infty \quad \text{for all } \lambda \in \mathbb{R},$$

and use analytic continuation to show that

$$\int_{\mathbb{R}} e^{\xi x}\,\pi_{a,\sigma^2,M}(dx) = \exp\left[a\xi + \frac{\sigma^2\xi^2}{2} + \int\limits_{\mathbb{R}\setminus\{0\}} \left(e^{\xi y} - 1 - \frac{\xi y}{1+y^2}\right) M(dy)\right]$$

for all $\xi \in \mathbb{C}$.

3.2.44 Exercise: Let μ be an infinitely divisible element of $\mathbf{M}_1(\mathbb{R})$.

(**i**) Show that

$$\mu\big((-\infty,0)\big)=0 \iff \hat{\mu}(\xi)=\exp\left[\sqrt{-1}\,\xi a+\int_{(0,\infty)}\left(e^{\sqrt{-1}\,\xi y}-1\right)M(dy)\right]$$

for some $a\in[0,\infty)$ and Lévy measure M satisfying (3.2.2). (**Hint**: Show that the associated independent increment process must be nondecreasing, and apply Theorem 3.2.33.)

(**ii**) Suppose that X is an $\mathfrak{N}(0,1)$ random variable, and let ν be the distribution of $|X|$. Using the part (**i**), show that ν *cannot* be infinitely divisible. (**Hint**: Note that $\mathfrak{Re}\big(\hat{\mu}(\xi)\big)=e^{-\frac{\xi^2}{2}}$.)

§3.3: Wiener's Measure and the Invariance Principle.

In the preceding section we constructed the independent increment processes corresponding to the Lévy systems $(a,0,M)$ and found that, apart from a linear drift, they all are, more or less easily, obtained by the superposition of Poisson jump processes on top of one another. At first glance, a look at the way in which we carried out the final step in the derivation of (3.2.6) might lead one to hope that the same procedure ought to work for the Lévy system $(0,\sigma^2,0)$. Indeed, take M to be the standard Bernoulli measure $\beta=\frac{\delta_{-1}+\delta_1}{2}$ and $\{Z(t):t\in[0,\infty)\}$ to be an independent increment process corresponding to $(0,0,M)$. As we have seen, we can represent the associated process as $Z(t)=\sum_{n=1}^{\mathbf{N}(t)}X_n$, where $\{\mathbf{N}(t):t\in[0,\infty)\}$ (cf. (3.2.23)) is the simple Poisson process and $\{X_n:n\in\mathbb{Z}^+\}$ is a sequence of mutually independent β-distributed random variables which are independent of $\sigma\big(\mathbf{N}(t):t\in[0,\infty)\big)$. Next, for $\epsilon>0$, set

$$Z_\epsilon(t)=\epsilon Z\big(\epsilon^{-2}t\big)=\epsilon\sum_{n=1}^{\mathbf{N}(\epsilon^{-2}t)}X_n,$$

and observe that $\big\{Z_\epsilon(t)\,:\,t\in[0,\infty)\big\}$ is the independent increment process corresponding to $(0,0,M_\epsilon)$, where

$$M_\epsilon=\frac{\delta_{-\epsilon}+\delta_\epsilon}{2\epsilon^2}.$$

In particular, the distribution of $Z_\epsilon(t)$ is $\pi_{0,0,tM_\epsilon}$ and therefore, as we saw in the last section via consideration of characteristic functions, the distribution of

$Z_\epsilon(t)$ tends (weakly) to the Gauss measure γ_t, where

$$\gamma_{\sigma^2}(dx) = \frac{1}{\sqrt{2\pi\sigma^2}} \exp\left[-\frac{x^2}{2\sigma^2}\right] dx, \quad \sigma \in (0,\infty).$$

On the other hand, it cannot be true that the random variables $Z_\epsilon(t)$ themselves are converging as $\epsilon \searrow 0$. Indeed, by part **(iii)** of Exercise 3.2.43, we know that

$$P\left(\left|Z_\epsilon(1) - \epsilon \sum_{1 \le n \le \epsilon^{-2}} X_n\right| \ge r\right) \longrightarrow 0 \quad \text{as } \epsilon \searrow 0$$

for each $r \in (0,\infty)$, and therefore the only way that $\{Z_\epsilon(1) : \epsilon > 0\}$ could be converging as random variables would be if the sequence $n^{-\frac{1}{2}} \sum_1^n X_m$ were to converge as $n \to \infty$. However, we already know (cf. the discussion in the second paragraph of Section 2.1), that this is not the case. Hence, unlike the construction (cf. the proofs of Lemma 3.2.30 and Lemma 3.2.32) of the processes corresponding to Lévy systems in which $\sigma = 0$, convergence in probability is out of the question here.

Although the discussion in the preceding paragraph may seem rather inconclusive, it does point us in the right direction. More specifically, it tells us that we ought to be looking for convergence in distribution and that the setting should be similar to the one in The Central Limit Theorem. In fact, what we are seeking is a *path-space version of The Central Limit Theorem*. But which path-space? In order to settle this question, notice that (cf. the preceding paragraph) the path $Z_\epsilon(\,\cdot\,)$ almost surely never has a jump of size larger than ϵ, and so it is reasonably certain that the distribution of these processes must tend to a measure which lives on *continuous paths*. As it turns out, this fact simplifies our lives a great deal. Indeed, a path-space version of The Central Limit Theorem entails weak convergence of measures on path-space, and so it will be important for us to be working on a path-space which admits a natural Polish structure. Thus, we are fortunate not to be forced to deal with the space $CL(\mathbb{R}^N)$ (cf. the paragraph preceding Theorem 3.2.33) which is rather difficult to interpret as a Polish space (for instance, it fails to be separable when given the topology of uniform convergence on compact intervals).

Before getting down to serious business, we begin with a few general remarks about spaces of continuous paths. Given a Polish space E, let $\mathfrak{P}(E)$ denote the space $C\big([0,\infty); E\big)$ of continuous paths $\psi : [0,\infty) \longmapsto E$, and endow $\mathfrak{P}(E)$ with the **topology of uniform convergence on compact intervals**. That is, given a metric ρ for E, we take the topology on $\mathfrak{P}(E)$ to be the one for which sets of the form

$$B(\psi; T, \epsilon) \equiv \left\{\varphi \in \mathfrak{P}(E) : \sup_{t\in[0,T]} \rho\big(\varphi(t), \psi(t)\big) < \epsilon\right\} \tag{3.3.1}$$

constitute a neighborhood basis of $\psi \in \mathfrak{P}(E)$ as T and ϵ vary over $(0,\infty)$. Notice that, because $\{\psi(t) : t \in [0,T]\} \subset\subset E$, this topology depends only on that of E and not on the particular choice of metric ρ. Next, define

$$D(\varphi,\psi) = \sum_{n=1}^{\infty} \frac{1}{2^n} \frac{\sup_{t\in[0,n]} \rho\big(\varphi(t),\psi(t)\big)}{1+\sup_{t\in[0,n]} \rho\big(\varphi(t),\psi(t)\big)},$$

and (cf. Lemma 3.1.13) note that D is a metric for $\mathfrak{P}(E)$ and that D is complete if ρ itself is. Thus, we will know that $\mathfrak{P}(E)$ is a Polish space once we check that it is separable. To this end, recall (cf. part **(ii)** of Exercise 3.1.17) that E can be embedded as a closed subset of $\mathbf{M}_1(E)$ and therefore that $\mathfrak{P}(E)$ can be embedded as a closed subset of $\mathfrak{P}\big(\mathbf{M}_1(E)\big)$. In particular, since subsets of second countable spaces are second countable in the induced topology, we need only show that $\mathfrak{P}\big(\mathbf{M}_1(E)\big)$ is separable. But clearly, if $\mathcal{C}$ is a countable dense subset of $\mathbf{M}_1(E)$ and, for $n \in \mathbb{Z}^+$, $\mathfrak{P}_n(\mathcal{C})$ denotes the space of $\psi \in \mathfrak{P}\big(\mathbf{M}_1(E)\big)$ with the properties that

$$\psi\left(\tfrac{m}{n}\right) \in \mathcal{C} \quad \text{and} \quad \psi(t) = \big(m+1-nt\big)\psi\left(\tfrac{m}{n}\right) + \big(nt-m\big)\psi\left(\tfrac{m+1}{n}\right)$$

for $t \in \left[\frac{m}{n}, \frac{m+1}{n}\right]$ and $m \in \mathbb{N}$, then $\bigcup_{n\in\mathbb{Z}^+} \mathfrak{P}_n(\mathcal{C})$ forms a countable dense subset of $\mathfrak{P}\big(\mathbf{M}_1(E)\big)$. In other words, we have already proved the first part of the following lemma.

3.3.2 Lemma. *For any Polish space E, the path-space $\mathfrak{P}(E)$ is again a Polish space. Moreover, if $\mathcal{C}$ is any dense subset of $[0,\infty)$, then $\mathcal{B}_{\mathfrak{P}(E)}$ coincides with the smallest σ-algebra over $\mathfrak{P}(E)$ with respect to which all of the maps $\psi \in \mathfrak{P}(E) \longmapsto \psi(t) \in E$, $t \in \mathcal{C}$, are measurable. In particular, if, for a finite subset F of $[0,\infty)$, $\pi_F : \mathfrak{P}(E) \longrightarrow E^F$ is defined by $\big(\pi_F(\psi)\big)_t = \psi(t)$, $t \in F$, then two elements μ and ν of $\mathbf{M}_1\big(\mathfrak{P}(E)\big)$ coincide if and only if ${\pi_F}^*\mu = {\pi_F}^*\nu$ for every finite subset F of some dense subset of $[0,\infty)$.*

PROOF: Clearly, the last assertion is an immediate consequence of the one preceding it. Furthermore, in proving that $\mathcal{B}_{\mathfrak{P}(E)}$ is generated by the maps $\psi \in \mathfrak{P}(E) \longmapsto \psi(t)$, $t \in \mathcal{C}$, we may and will assume that $\mathcal{C}$ is countable and must show that every set $B(\psi;T,\epsilon)$ in (3.3.1) can be expressed measurably in terms of these maps. But, by path-continuity,

$$B(\psi;T,\epsilon) = \bigcup_{n=1}^{\infty} \bigcap_{t\in\mathcal{C}(T)} \left\{\varphi : \rho\big(\varphi(t),\psi(t)\big) \le \frac{n\epsilon}{n+1}\right\},$$

where $\mathcal{C}(T) \equiv \mathcal{C} \cap [0,T]$. $\square$

Given a $\mu \in \mathbf{M}_1(\mathfrak{P}(E))$ and a finite $F \subseteq [0,\infty)$, the measure ${\pi_F}^*\mu \in \mathbf{M}_1(E^F)$ is called the **marginal distribution of μ corresponding to the times $t \in F$**, and the final statement in Lemma 3.3.2 says that *μ is uniquely determined by its finite dimensional time marginals.*

To conclude this brief introduction to the analysis of measures on $\mathfrak{P}(E)$, we give a criterion for determining when a subset of $\mathbf{M}_1(\mathfrak{P}(E))$ is tight. Actually, the criterion which we have in mind is no more than a trivial translation of the famous Ascoli–Arzela criterion for compact subsets of $\mathfrak{P}(E)$. Namely, their criterion says that a subset Ψ of $\mathfrak{P}(E)$ is relatively compact if and only if, for each $T \in (0,\infty)$, the set $\{\psi(t) : t \in [0,T] \text{ and } \psi \in \Psi\}$ is relatively compact in E and

$$\lim_{\delta \searrow 0} \sup_{\psi \in \Psi} \sup_{\substack{0 \le s < t \le T \\ t-s<\delta}} \rho\big(\psi(t), \psi(s)\big) = 0.$$

Hence, after combining this with Theorem 3.1.9, we have the following description of relatively compact subsets in $\mathbf{M}_1(\mathfrak{P}(E))$.

3.3.3 Lemma. *Let E be a Polish space and ρ a complete metric for E. A subset A of $\mathbf{M}_1(\mathfrak{P}(E))$ is tight, and therefore relatively compact, if and only if, for each $T \in (0,\infty)$ and $\epsilon > 0$, there is a $K_{T,\epsilon} \subset\subset E$ and a function $m_{T,\epsilon} : (0,\infty) \longrightarrow (0,\infty)$ satisfying $m_{T,\epsilon}(\delta) \longrightarrow 0$ as $\delta \searrow 0$ such that*

$$\inf_{\mu \in A} \mu\Big(\big\{\psi \in \mathfrak{P}(E) : \psi(t) \in K_{T,\epsilon} \text{ for all } t \in [0,T]\big\}\Big) \ge 1-\epsilon \tag{3.3.4}$$

and

$$\inf_{\mu \in A} \mu\Big(\big\{\psi \in \mathfrak{P}(E) : \rho\big(\psi(t),\psi(s)\big) \le m_{T,\epsilon}\big(|t-s|\big) \text{ for all } s,\, t \in [0,T]\big\}\Big) \ge 1-\epsilon. \tag{3.3.5}$$

In particular, if $E = \mathbb{R}^N$, then A is relatively compact if

$$\sup_{\mu \in A} \mu\Big(\big\{\psi \in \mathfrak{P}(\mathbb{R}^N) : |\psi(0)| \ge R\big\}\Big) \longrightarrow 0 \tag{3.3.6}$$

and, for some $\beta > 0$ and every $T \in (0,\infty)$,

$$\sup_{\mu \in A} \mu\left(\left\{\psi \in \mathfrak{P}(\mathbb{R}^N) : \sup_{0 \le s < t \le T} \frac{|\psi(t)-\psi(s)|}{(t-s)^\beta} \ge R\right\}\right) \longrightarrow 0 \tag{3.3.7}$$

as $R \to \infty$.

With these preliminaries in place, we are at last ready to confront the problem of providing a $\mathfrak{P}(\mathbb{R}^N)$-version of the Central Limit Theorem. Thus, let $\{\mathbf{X}_n\}_1^\infty$ be a sequence of independent $\mathbb{R}^N$-valued random variables on a probability space $(\Omega, \mathcal{F}, P)$, and assume that the $\mathbf{X}_n$'s have mean-value $\mathbf{0}$, covariance $\mathbf{I}$, and satisfy

$$\lim_{R \nearrow \infty} \sup_{n \in \mathbb{Z}^+} \mathbb{E}^P\Big[|\mathbf{X}_n|^2,\, |\mathbf{X}_n| \geq R\Big] = 0. \tag{3.3.8}$$

Next, for $n \in \mathbb{Z}^+$, define $\omega \in \Omega \longmapsto \mathbf{S}_n(\,\cdot\,,\omega) \in C\big([0,\infty);\mathbb{R}^N\big)$ so that $\mathbf{S}_n(0,\omega) = \mathbf{0}$ and, for each $m \in \mathbb{Z}^+$, $\mathbf{S}_n(\,\cdot\,,\omega)$ is linear on the interval $\big[\frac{m-1}{n}, \frac{m}{n}\big]$ with slope $n^{\frac{1}{2}}\,\mathbf{X}_m(\omega)$. That is,

$$\begin{gathered}\mathbf{S}_n(0,\omega) = \mathbf{0}, \quad \mathbf{S}_n\left(\tfrac{m}{n},\omega\right) = n^{-\frac{1}{2}} \sum_{k=1}^m \mathbf{X}_k, \ m \in \mathbb{Z}^+, \quad \text{and} \\ \mathbf{S}_n(t,\omega) = \big(m - nt\big)\mathbf{S}_n\left(\tfrac{m-1}{n},\omega\right) + \big(nt - m + 1\big)\mathbf{S}_n\left(\tfrac{m}{n},\omega\right)\end{gathered} \tag{3.3.9}$$

for $t \in \left(\frac{m-1}{n}, \frac{m}{n}\right)$. Finally, let

$$\mu_n \equiv \mathbf{S}_n{}^*P \in \mathbf{M}_1\big(\mathfrak{P}(\mathbb{R}^N)\big)$$

denote the distribution of

$$\omega \in \Omega \longmapsto \mathbf{S}_n(\,\cdot\,,\omega) \in \mathfrak{P}(\mathbb{R}^N)$$

under P. Our main goal in this section will be to prove that $\mu_n \Longrightarrow \mathcal{W}^{(N)}$, where $\mathcal{W}^{(N)} = \mathcal{W}^N$ and $\mathcal{W} \in \mathbf{M}_1\big(\mathfrak{P}(\mathbb{R})\big)$ (cf. Exercise 3.3.27) is the distribution of the independent increment process associated with the Lévy system $(0,1,0)$; and, as a first step in this direction, we now show that, if $\{\mu_n\}_1^\infty$ converges at all, it converges to the right thing.

3.3.10 Lemma. *Given $\ell \in \mathbb{Z}^+$ and $0 = t_0 < t_1 < \cdots < t_\ell$,*

$$\big(\mathbf{S}_n(t_1), \mathbf{S}_n(t_2) - \mathbf{S}_n(t_1), \dots, \mathbf{S}_n(t_\ell) - \mathbf{S}_n(t_{\ell-1})\big)^*P \Longrightarrow \gamma_{\tau_1}^N \times \gamma_{\tau_2}^N \times \cdots \times \gamma_{\tau_\ell}^N,$$

where $\tau_k = t_k - t_{k-1}$, $1 \leq k \leq \ell$.

PROOF: For $1 \leq k \leq \ell$ and $n > \frac{1}{\tau_k}$, define

$$\mathbf{\Delta}_n(k) = n^{-\frac{1}{2}} \sum_{j=[nt_{k-1}]+1}^{[nt_k]} \mathbf{X}_j,$$

where we use the notation $[t]$ to denote the largest $n \in \mathbb{Z}$ dominated by $t \in \mathbb{R}$. Noting that

$$\begin{aligned}
&\left|\mathbf{S}_n(t_k) - \mathbf{S}_n(t_{k-1}) - \boldsymbol{\Delta}_n(k)\right| \\
&\quad \leq \left|\mathbf{S}_n(t_k) - \mathbf{S}_n\left(\frac{[nt_k]}{n}\right)\right| + \left|\mathbf{S}_n(t_{k-1}) - \mathbf{S}_n\left(\frac{[nt_{k-1}]}{n}\right)\right| \\
&\quad \leq \frac{\left|\mathbf{X}_{[nt_k]+1}\right| + \left|\mathbf{X}_{[nt_{k-1}]+1}\right|}{n^{\frac{1}{2}}},
\end{aligned}$$

one sees that, for any $\epsilon > 0$,

$$\begin{aligned}
&P\left(\sum_{k=1}^{\ell}\left|\mathbf{S}_n(t_k) - \mathbf{S}_n(t_{k-1}) - \boldsymbol{\Delta}_n(k)\right|^2 \geq \epsilon^2\right) \\
&\quad \leq P\left(\sum_{k=0}^{\ell}\left|\mathbf{X}_{[nt_k]+1}\right|^2 \geq \frac{n\epsilon^2}{4}\right) \\
&\quad \leq \frac{4}{n\epsilon^2}\sum_{k=0}^{\ell}\mathbb{E}^P\left[\left|\mathbf{X}_{[nt_k]+1}\right|^2\right] = \frac{2(\ell+1)N}{n\epsilon^2} \longrightarrow 0
\end{aligned}$$

as $n \to \infty$. Hence, by the Principle of Accompanying Laws (cf. Theorem 3.1.19), we need only check that

$$\left(\boldsymbol{\Delta}_n(1), \dots, \boldsymbol{\Delta}_n(\ell)\right)^* P \Longrightarrow \gamma^N_{\tau_1} \times \cdots \times \gamma^N_{\tau_\ell}.$$

Moreover, since

$$\left(\boldsymbol{\Delta}_n(1), \dots, \boldsymbol{\Delta}_n(\ell)\right)^* P = \boldsymbol{\Delta}_n(1)^* P \times \cdots \times \boldsymbol{\Delta}_n(\ell)^* P$$

for all sufficiently large n's, it is clear (cf. part **(ii)** of Exercise 3.1.18) that this comes down to checking $\boldsymbol{\Delta}_n(k)^* P \Longrightarrow \gamma^N_{\tau_k}$ for each $1 \leq k \leq \ell$. Finally, given $1 \leq k \leq \ell$, set $M_n(k) = [nt_k] - [nt_{k-1}]$ and use Theorem 2.2.11 plus **(i)** of Exercise 3.1.14 to see that, as $n \to \infty$,

$$\mathbb{E}^P\left[\exp\left(\frac{\sqrt{-1}}{M_n(k)^{\frac{1}{2}}}\sum_{j=1}^{M_n(k)}\left(\boldsymbol{\xi}, \mathbf{X}_{[nt_k]+j}\right)_{\mathbb{R}^N}\right)\right] \longrightarrow \exp\left[-\frac{|\boldsymbol{\xi}|^2}{2}\right]$$

uniformly for $\boldsymbol{\xi}$ in compact subsets of $\mathbb{R}^N$. Hence, since $\frac{M_n(k)}{n} \longrightarrow \tau_k$, we now see that, for any fixed $\boldsymbol{\xi} \in \mathbb{R}^N$,

$$\mathbb{E}^P\left[\exp\left(\sqrt{-1}\left(\boldsymbol{\xi}, \boldsymbol{\Delta}_n(k)\right)_{\mathbb{R}^N}\right)\right] \longrightarrow \exp\left[-\frac{\tau_k|\boldsymbol{\xi}|^2}{2}\right] = \widehat{\gamma^N_{\tau_k}}(\boldsymbol{\xi});$$

and therefore, $\boldsymbol{\Delta}_n(k)^* P \Longrightarrow \gamma^N_{\tau_k}$ follows from Lemma 2.2.8. $\square$

As we said just before its statement, Lemma 3.3.10 (in conjunction with Lemma 3.3.2) shows that $\{\mu_n\}_1^\infty$ can have at most one limit and that this limit would possess the independent increment properties for which we are looking. Thus, all that is missing is a proof that $\{\mu_n\}_1^\infty$ is relatively compact; and for this proof we will use the criterion in (3.3.6) and (3.3.7). Notice that (3.3.6) presents no difficulty:

$$\mu_n\big(\boldsymbol{\psi} : \boldsymbol{\psi}(0) = \mathbf{0}\big) = 1 \quad \text{for all} \quad n \in \mathbb{Z}^+.$$

On the other hand, the verification of (3.3.7) has got to be hard. In fact, since

$$\dot{\mathbf{S}}_n(t) = n^{\frac{1}{2}}\mathbf{X}_k \quad \text{for} \quad t \in \left(\tfrac{k-1}{n}, \tfrac{k}{n}\right),$$

it is clear that control on the modulus of continuity of $\mathbf{S}_n(\,\cdot\,)$ must rely on cancellation properties coming from the mean-value $\mathbf{0}$ property of the $\mathbf{X}_n$'s. That is, we must learn how to take advantage of the fact that, although at any given instant $\mathbf{S}_n(\,\cdot\,)$ has a *speed* proportional to $n^{\frac{1}{2}}$, after each time interval of length $\frac{1}{n}$ its *velocity* changes direction in such a way that its *average velocity* over a time interval of length 1 will be quite moderate. With this in mind, suppose that we start by looking at the expected value of $\left|\mathbf{S}_n(t) - \mathbf{S}_n(s)\right|^2$, where $0 \le s < t < \infty$. If $k - 1 \le ns < nt \le k$ for some $k \in \mathbb{Z}^+$, then

$$\mathbb{E}^P\left[\left|\mathbf{S}_n(t) - \mathbf{S}_n(s)\right|^2\right] = nN(t-s)^2 \le N(t-s);$$

and, because we are dealing with such a short time interval, no cancellation is involved. If, on the other hand, $k - 1 \le ns \le k \le \ell \le nt \le \ell + 1$ for some $k, \ell \in \mathbb{Z}^+$, then

$$\begin{aligned}
&\mathbb{E}^P\left[\left|\mathbf{S}_n(t) - \mathbf{S}_n(s)\right|^2\right] \\
&\quad \le 3\mathbb{E}^P\left[\left|\mathbf{S}_n(t) - \mathbf{S}_n\left(\tfrac{\ell}{n}\right)\right|^2\right] + 3\mathbb{E}^P\left[\left|\mathbf{S}_n\left(\tfrac{\ell}{n}\right) - \mathbf{S}_n\left(\tfrac{k}{n}\right)\right|^2\right] \\
&\qquad + 3\mathbb{E}^P\left[\left|\mathbf{S}_n\left(\tfrac{k}{n}\right) - \mathbf{S}_n(s)\right|^2\right] \\
&\quad \le 3nN\left(t - \frac{\ell}{n}\right)^2 + \frac{3}{n}N\mathbb{E}^P\left[\left|\sum_{j=1}^{\ell-k}\mathbf{X}_{k+j}\right|^2\right] + 3nN\left(\frac{k}{n} - s\right)^2 \\
&\quad \le 3N(t-s) + 3N\tfrac{\ell-k}{n} + 3N(t-s) \le 9N(t-s);
\end{aligned}$$

and cancellation has been used to handle the time interval $\left[\frac{k}{n}, \frac{\ell}{n}\right]$. Thus, in any case, we have now shown that

$$\sup_{n\in\mathbb{Z}^+} \mathbb{E}^{\mu_n}\left[\left|\boldsymbol{\psi}(t) - \boldsymbol{\psi}(s)\right|^2\right] \le 9N|t-s| \quad \text{for all} \quad s, t \in [0,\infty). \tag{3.3.11}$$

At first sight, (3.3.11) might appear to be all that is needed in order check that $\{\mu_n\}_1^\infty$ satisfies (3.3.7) with $\beta = \frac{1}{2}$. In fact, (3.3.11) can be rewritten as

$$\sup_{n\in\mathbb{Z}^+}\sup_{s\neq t}\mathbb{E}^{\mu_n}\left[\left(\frac{|\psi(t)-\psi(s)|}{|t-s|^{\frac{1}{2}}}\right)^2\right]\leq 9N,$$

which, if one is sufficiently careless, looks quite close to

$$\sup_{n\in\mathbb{Z}^+}\mathbb{E}^{\mu_n}\left[\sup_{s\neq t}\left(\frac{|\psi(t)-\psi(s)|}{|t-s|^{\frac{1}{2}}}\right)^2\right]\leq 9N, \tag{3.3.12}$$

from which one would immediately derive

$$\sup_{n\in\mathbb{Z}^+}\mu_n\left(\left\{\psi:\sup_{0\leq s<t<\infty}\frac{|\psi(t)-\psi(s)|}{|t-s|^{\frac{1}{2}}}\geq R\right\}\right)\leq\frac{9N}{R^2}\longrightarrow 0$$

as $R\nearrow\infty$. Unfortunately, as soon as one stops being careless, it becomes clear that there is a world of difference between (3.3.11) and (3.3.12): as distinguished from the one in Kolmogorov's Inequality, the "sup" here simply cannot be brought inside this integral!

At this point it looks as if we have made little progress toward our goal. Nonetheless, as we are about to see, all is not lost, and, by being a little less naïve, we can rescue something from the preceding line of reasoning. Namely, although one cannot commute "sup" with integration, one can commute integration with integration. Thus, for example, from (3.3.11), one can certainly say that

$$\sup_{n\in\mathbb{Z}^+}\mathbb{E}^{\mu_n}\left[\iint\limits_{[0,T]^2}\left(\frac{|\psi(t)-\psi(s)|}{|t-s|^{\frac{1}{2}}}\right)^2 ds\,dt\right]\leq 9NT^2,\quad T\in(0,\infty);$$

and so we would be in good shape if we knew how to estimate some Hölder modulus of continuity for $\psi\restriction[0,T]$ in terms of the *integral Hölder norm*

$$\left(\iint\limits_{[0,T]^2}\left(\frac{|\psi(t)-\psi(s)|}{|t-s|^{\frac{1}{2}}}\right)^2 ds\,dt\right)^{\frac{1}{2}}.$$

The preceding quantity is what analysts call a Besov norm, and the general principles relating such integral quantities to pointwise estimates were developed originally by Besov's teacher Sobolev. As it turns out (cf. Exercise 3.3.27 below), the preceding Besov norm is not sufficiently strong to control ψ's modulus of continuity. Nonetheless, as a special case of the following clever lemma due to Garsia, Rademich, and Rumsey, we will see that a closely related Besov norm is sufficient.

3.3.13 Lemma. *Let p and Φ be strictly increasing, continuous functions on $[0,\infty)$ satisfying $p(0) = \Phi(0) = 0$ and $\lim_{\xi\to\infty}\Phi(\xi) = \infty$. If $T \in (0,\infty)$ and $\psi \in C([0,T];\mathbb{R}^N)$ satisfy*

$$B \equiv \iint_{[0,T]^2} \Phi\left(\frac{|\psi(t)-\psi(s)|}{p(|t-s|)}\right) ds\, dt < \infty,$$

then

$$|\psi(t)-\psi(s)| \le 8\int_0^{t-s} \Phi^{-1}\left(\frac{4B}{u^2}\right) p(du), \quad 0 \le s < t \le T,$$

where $p(du)$ is the measure on $[0,\infty)$ determined by $p([0,u]) = p(u)$ for all $u \in [0,\infty)$.

PROOF: We begin with the case when $T = t = 1$ and $s = 0$. Set

$$I(t) = \int_0^1 \Phi\left(\frac{|\psi(t)-\psi(s)|}{p(|t-s|)}\right) ds \quad \text{for} \quad t \in [0,1].$$

By assumption, $\int_0^1 I(t)\,dt = B$, and so there must exist a $t_0 \in (0,1)$ for which $I(t_0) \le B$. We next show that there exist positive numbers t_n, $n \in \mathbb{Z}^+$, such that, for $n \in \mathbb{N}$: $t_{n+1} \in (0, d_n)$,

$$(3.3.14) \qquad I(t_{n+1}) \le \frac{2B}{d_n} \quad \text{and} \quad \Phi\left(\frac{|\psi(t_{n-1})-\psi(t_n)|}{p(|t_{n-1}-t_n|)}\right) \le \frac{2I(t_n)}{d_n},$$

where d_n is determined by

$$(3.3.15) \qquad p(d_n) = \frac{p(t_n)}{2}.$$

Indeed, given t_n, define d_n by (3.3.15), and observe that the set of t_{n+1}'s in $(0,d_n)$ for which one or the other of the conditions in (3.3.14) fails must have measure strictly less than $\frac{d_n}{2}$. Thus, there must exist at least one $t_{n+1} \in (0,d_n)$ for which they both hold. Notice that $t_0 > d_0 > t_1 > d_1 > \cdots > t_n > d_n > \cdots$ and that, since $p(d_{n+1}) \le \frac{1}{2}p(d_n)$, $d_n \searrow 0$ and therefore $t_n \searrow 0$. In addition, because

$$\begin{aligned} p(t_n - t_{n+1}) &\le p(t_n) = 2p(d_n) \\ &= 4\left(p(d_n) - \tfrac{1}{2}p(d_n)\right) \le 4\left(p(d_n) - p(d_{n+1})\right), \end{aligned}$$

(3.3.14) leads to

$$\begin{aligned}\left|\psi(t_n) - \psi(t_{n+1})\right| &\le \Phi^{-1}\left(\frac{2I(t_n)}{d_n}\right) p(t_n - t_{n+1}) \\ &\le 4\Phi^{-1}\left(\frac{4B}{d_n\, d_{n-1}}\right)\left(p(d_n) - p(d_{n+1})\right) \\ &\le 4\int_{d_{n+1}}^{d_n} \Phi^{-1}\left(\frac{4B}{u^2}\right) p(du),\end{aligned}$$

where, when $n = 0$, we have used d_{-1} to denote 1. By summing these inequalities over $n \in \mathbb{N}$, we obtain the estimate

$$\left|\psi(t_0) - \psi(0)\right| \le 4\int_0^1 \Phi^{-1}\left(\frac{4B}{u^2}\right) p(du);$$

and, after repeating the preceding argument when $t \in [0,1] \longmapsto \psi(t)$ is replaced by $t \in [0,1] \longmapsto \psi(1-t)$, we also see that

$$\left|\psi(1) - \psi(t_0)\right| \le 4\int_0^1 \Phi^{-1}\left(\frac{4B}{u^2}\right) p(du).$$

Hence, we have now proved that

$$|\psi(1) - \psi(0)| \le 8\int_0^1 \Phi^{-1}\left(\frac{4B}{u^2}\right) p(du)$$

for all functions p, Φ, and ψ which satisfy our hypotheses.

To complete the proof, let general $0 \le s < t \le T < \infty$ be given, and set

$$\overline{\psi}(\tau) = \psi\big(s + (t-s)\tau\big) \quad \text{and} \quad \overline{p}(\tau) = p\big((t-s)\tau\big) \quad \text{for } \tau \in [0,1].$$

By what we already know,

$$|\psi(t) - \psi(s)| = |\overline{\psi}(1) - \overline{\psi}(0)| \le 8\int_0^1 \Phi^{-1}\left(\frac{4\overline{B}}{u^2}\right) \overline{p}(du),$$

where

$$\begin{aligned}\overline{B} &\equiv \iint_{[0,1]^2} \Phi\left(\frac{|\overline{\psi}(\tau) - \overline{\psi}(\sigma)|}{\overline{p}(\tau - \sigma)}\right) d\sigma\, d\tau \\ &= \frac{1}{(t-s)^2}\iint_{[s,t]^2} \Phi\left(\frac{|\psi(\tau) - \psi(\sigma)|}{p(\tau - \sigma)}\right) d\sigma\, d\tau \le \frac{B}{(t-s)^2},\end{aligned}$$

and therefore

$$\int_0^1 \Phi^{-1}\left(\frac{4\overline{B}}{u^2}\right) \overline{p}(du) \le \int_0^{t-s} \Phi^{-1}\left(\frac{4B}{u^2}\right) p(du). \quad \square$$

Obviously, the preceding lemma gives no information unless the function $u \in (0,T] \longmapsto \Phi^{-1}\left(\frac{4B}{u^2}\right)$ turns out to be p-integrable at 0. Thus, for example, when $\Phi(\xi) = \xi^r$ and $p(t) = t^\gamma$ with some positive r and γ, one needs to have $\gamma > \frac{2}{r}$ in order to get a meaningful estimate. In particular, one cannot take $\gamma = \frac{1}{2}$ and $r = 2$; which is the reason why we cannot use the Besov norm which arose in our earlier considerations. On the other hand, Lemma 3.3.13 does yield the following important criterion for checking when (3.3.7) holds.

3.3.16 Theorem (Kolmogorov's† Criterion). *Let $r \in [1,\infty)$ and $\alpha \in (0,\infty)$ be given, and suppose that $\mu \in \mathbf{M}_1\big(\mathfrak{P}(\mathbb{R}^N)\big)$ satisfies*

$$E^\mu\Big[\big|\psi(t) - \psi(s)\big|^r\Big] \le C|t-s|^{1+\alpha}, \quad 0 \le s \le t \le T, \tag{3.3.17}$$

for some $C < \infty$ and $T \in (0,\infty)$. Then, for each $\beta \in \left(0, \frac{\alpha}{r}\right)$, there exists a $K = K(r,\alpha,\beta,T) \in (0,\infty)$ such that

$$\mu\left(\left\{\psi : \sup_{0 \le s < t \le T} \frac{|\psi(t) - \psi(s)|}{(t-s)^\beta} \ge R\right\}\right) \le \frac{KC}{R^r}, \quad R \in (0,\infty). \tag{3.3.18}$$

In particular, if $A \subseteq \mathbf{M}_1\big(\mathfrak{P}(\mathbb{R}^N)\big)$ satisfies

$$\sup_{\mu \in A} E^\mu\Big[\big|\psi(t) - \psi(s)\big|^r\Big] \le C_T|t-s|^{1+\alpha}, \quad 0 \le s \le t \le T,$$

for each $T \in (0,\infty)$ and some $C_T < \infty$, then A satisfies (3.3.7).

PROOF: Set $\gamma = \frac{2}{r} + \beta$. By Fubini's Theorem, (3.3.17) implies

$$\int_{\mathfrak{P}(\mathbb{R}^N)} B(\psi)^r\, \mu(d\psi) \le C \iint_{[0,T]^2} |t-s|^{1+\alpha-\gamma r}\, ds\, dt < \infty,$$

where

$$B(\psi) \equiv \left(\iint_{[0,T]^2} \left(\frac{|\psi(t) - \psi(s)|}{|t-s|^\gamma}\right)^r ds\, dt\right)^{\frac{1}{r}}.$$

Since, by Lemma 3.3.19,

$$\big|\psi(t) - \psi(s)\big| \le \frac{32\gamma B(\psi)}{\beta}(t-s)^\beta, \quad 0 \le s < t \le T,$$

(3.3.18) is now an easy application of Markov's inequality. □

---

† There is no question that this attribution is correct. However, Kolmogorov's own derivation obscures the fact that the result is really an application of basic real-variable theory and has nearly no probabilistic component. Indeed, Garsia, Rademich, and Rumsey appear to have been the first ones to observe that, in this connection, probabilists were following the wrong Russian.

Before applying Kolmogorov's Criterion to the sequence $\{\mu_n\}_1^\infty$, we must first replace (3.3.11) with an estimate in which $(t-s)$ is raised to a power strictly larger than 1.

3.3.19 Lemma. *Assume that*

$$M_4 \equiv \sup_{n\in\mathbb{Z}^+} \mathbb{E}^P\left[|\mathbf{X}_n|^4\right] < \infty.$$

Then there exists a $C<\infty$ *such that*

$$\sup_{n\in\mathbb{Z}^+} \mathbb{E}^{\mu_n}\left[|\psi(t)-\psi(s)|^4\right] \le C(t-s)^2, \quad 0 \le s < t < \infty.$$

In particular, $\{\mu_n\}_1^\infty$ *is tight in* $\mathbf{M}_1(\mathfrak{P}(\mathbb{R}^N))$.

PROOF: We proceed in very much the same way as we did in the derivation of (3.3.11). Thus, when $k-1 \le ns < nt \le k$,

$$\mathbb{E}^P\left[|\mathbf{S}_n(t)-\mathbf{S}_n(s)|^4\right] = n^2(t-s)^4\mathbb{E}^P\left[|\mathbf{X}_k|^4\right] \le M_4(t-s)^2;$$

and, when $k-1 \le ns \le k \le \ell \le nt \le \ell+1$,

$$\begin{aligned}
&\mathbb{E}^P\left[|\mathbf{S}_n(t)-\mathbf{S}_n(s)|^4\right] \\
&\quad \le 27\mathbb{E}^P\left[\left|\mathbf{S}_n(t)-\mathbf{S}_n\left(\tfrac{\ell}{n}\right)\right|^4\right] + 27\mathbb{E}^P\left[\left|\mathbf{S}_n\left(\tfrac{\ell}{n}\right)-\mathbf{S}_n\left(\tfrac{k}{n}\right)\right|^4\right] \\
&\qquad + 27\mathbb{E}^P\left[\left|\mathbf{S}_n\left(\tfrac{k}{n}\right)-\mathbf{S}_n(s)\right|^4\right] \\
&\quad \le 27M_4n^2\left(t-\frac{\ell}{n}\right)^4 + \frac{27}{n^2}\mathbb{E}^P\left[\left|\sum_{j=1}^{\ell-k}\mathbf{X}_{k+j}\right|^4\right] + 27M_4n^2\left(\frac{k}{n}-s\right)^4 \\
&\quad \le 54M_4(t-s)^2 + \frac{81N^2M_4(\ell-k)^2}{n^2} \le 135N^2M_4(t-s)^2,
\end{aligned}$$

where, in the passage to the final line, we have chosen an orthonormal basis $\{\mathbf{e}_i\}_1^N$ for $\mathbb{R}^N$ and used the estimate

$$\begin{aligned}
\mathbb{E}^P\left[\left|\sum_{j=1}^{\ell-k}\mathbf{X}_{k+j}\right|^4\right] &= \mathbb{E}^P\left[\left(\sum_{i=1}^{N}\left(\sum_{j=1}^{\ell-k}(\mathbf{e}_i,\mathbf{X}_{k+j})_{\mathbb{R}^N}\right)^2\right)^2\right] \\
&\le N\sum_{i=1}^{N}\mathbb{E}^P\left[\left(\sum_{j=1}^{\ell-k}(\mathbf{e}_i,\mathbf{X}_{k+j})_{\mathbb{R}^N}\right)^4\right] \le 3N^2M_4(\ell-k)^2
\end{aligned}$$

coming from second inequality in (1.3.2). □

At last, we have all the ingredients needed to complete our program.

3.3.20 Theorem (Wiener & Donsker). *There is a unique probability measure $\mathcal{W}^{(N)}$ on $\mathbf{M}_1(\mathfrak{P}(\mathbb{R}^N))$ with the properties that $\psi(0) = \mathbf{0}$ for $\mathcal{W}^{(N)}$-almost every $\psi \in \mathfrak{P}(\mathbb{R}^N)$ and*

$$\begin{aligned} \mathcal{W}^{(N)}\Big(\Big\{\psi : \psi(t_1) - \psi(t_0) \in B_1, \dots, \psi(t_\ell) - \psi(t_{\ell-1}) \in B_\ell\Big\}\Big) \\ = \gamma^N_{t_1}(B_1) \times \gamma^N_{t_2-t_1}(B_2) \times \cdots \times \gamma^N_{t_\ell - t_{\ell-1}}(B_\ell), \end{aligned} \tag{3.3.21}$$

for all $\ell \in \mathbb{Z}^+$, $0 = t_0 < t_1 < \cdots < t_\ell$, and $B_1, \dots, B_\ell \in \mathcal{B}_{\mathbb{R}^N}$. Moreover, if $\{\mathbf{X}_n\}_1^\infty$ is a sequence of independent, uniformly square P-integrable random variables (i.e. (3.3.8) holds) with mean-value $\mathbf{0}$ and covariance $\mathbf{I}$ and if $\omega \in \Omega \longmapsto \mathbf{S}_n(\,\cdot\,,\omega) \in \mathfrak{P}(\mathbb{R}^N)$ is defined accordingly, as in (3.3.9), then $\mu_n \equiv \mathbf{S_n}^*P \Longrightarrow \mathcal{W}^{(N)}$ as $n \to \infty$.

PROOF: Only the existence, not the uniqueness, of $\mathcal{W}^{(N)}$ is in doubt. Moreover, if we assume that the $\mathbf{X}_n$'s satisfy the condition in Lemma 3.3.19, then we know (by that lemma) that $\{\mu_n\}_1^\infty$ is tight and (by Lemma 3.3.10) that every limit of $\{\mu_n\}_1^\infty$ must satisfy (3.3.21). Thus, we not only know that $\mathcal{W}^{(N)}$ exists, we already know that $\mu_n \Longrightarrow \mathcal{W}^{(N)}$ whenever the condition in Lemma 3.3.19 holds.

In order to complete the proof, we will apply the Principle of Accompanying Laws (cf. Theorem 3.1.14). Namely, because the $\mathbf{X}_n$'s are uniformly square P-integrable, we can use a truncation procedure to find functions $\{f_{n,\delta} : n \in \mathbb{Z}^+ \text{ and } \delta > 0\} \subseteq C_{\mathrm{b}}(\mathbb{R}^N, \mathbb{R}^N)$ with the properties that, for each $\delta > 0$, $\sup_{n\in\mathbb{Z}^+} \|f_{n,\delta}\|_{\mathrm{u}} < \infty$,

$$\sup_{n\in\mathbb{Z}^+} \mathbb{E}^P\Big[\big|\mathbf{X}_n - f_{n,\delta} \circ \mathbf{X}_n\big|^2\Big] < \delta,$$

and, for every $n \in \mathbb{Z}^+$, the random variable $\mathbf{X}_{n,\delta} \equiv f_{n,\delta} \circ \mathbf{X}_n$ has mean-value $\mathbf{0}$ and covariance $\mathbf{I}$. Next, for each $\delta > 0$, define the maps $\omega \in \Omega \longmapsto \mathbf{S}_{n,\delta}(\,\cdot\,,\omega) \in \mathfrak{P}(\mathbb{R}^N)$ relative to $\{\mathbf{X}_{n,\delta}\}_1^\infty$, and set $\mu_{n,\delta} = \mathbf{S}_{n,\delta}{}^*P$. Then, by the preceding, we know that $\mu_{n,\delta} \Longrightarrow \mathcal{W}^{(N)}$ for each $\delta > 0$. Hence, by Theorem 3.1.14, we will have proved that $\mu_n \Longrightarrow \mathcal{W}^{(N)}$ as soon as we show that

$$\varlimsup_{\delta\searrow 0} \sup_{n\in\mathbb{Z}^+} P\left(\sup_{0\le t\le T} |\mathbf{S}_n(t) - \mathbf{S}_{n,\delta}(t)| \ge \epsilon\right) = 0$$

for every $T \in \mathbb{Z}^+$ and $\epsilon > 0$. To this end, first observe that

$$\sup_{t\in[0,T]} \big|\mathbf{S}_n(t) - \mathbf{S}_{n,\delta}(t)\big| = \max_{1\le m\le nT} \frac{1}{n^{\frac{1}{2}}} \left|\sum_{k=1}^m \mathbf{Y}_{n,\delta}\right|,$$

where $\mathbf{Y}_{k,\delta} \equiv \mathbf{X}_k - \mathbf{X}_{k,\delta}$. Next, note that

$$P\left(\max_{1\le m\le nT} \frac{1}{n^{\frac{1}{2}}}\left|\sum_{k=1}^{m} \mathbf{Y}_{k,\delta}\right| \ge \epsilon\right)$$
$$\le N \max_{\mathbf{e}\in\mathbf{S}^{N-1}} P\left(\max_{1\le m\le nT}\left|\sum_{k=1}^{m}\left(\mathbf{e}, \mathbf{Y}_{k,\delta}\right)_{\mathbb{R}^N}\right| \ge \frac{n^{\frac{1}{2}}\epsilon}{N^{\frac{1}{2}}}\right).$$

Finally, by Kolmogorov's Inequality (cf. Theorem 1.4.5),

$$P\left(\max_{1\le m\le nT}\left|\sum_{k=1}^{m}\left(\mathbf{e}, \mathbf{Y}_{k,\delta}\right)_{\mathbb{R}^N}\right| \ge \frac{n^{\frac{1}{2}}\epsilon}{N^{\frac{1}{2}}}\right) \le \frac{NT\delta}{n\epsilon^2}$$

for every $\mathbf{e} \in \mathbf{S}^{N-1}$. □

The probability measure $\mathcal{W}^{(N)}$ discussed in Theorem 3.3.20 is the renowned **Wiener's measure** for $\mathbb{R}^N$-valued paths. The assertion in the second part of Theorem 3.3.20 is a version of the Donsker's **invariance principle**, the name given to what we previously called The Central Limit Theorem in path-space. In the chapters which follow, we will have a lot more to say about Wiener's measure and will see that it holds the same sort of distinguished position among measures on $\mathfrak{P}(\mathbb{R}^N)$ that Lebesgue's measure holds among measures on $\mathbb{R}^N$. However, for the moment, we will content ourselves with the observation that it provides us with the missing ingredient in our program to construct independent increment processes. In the following statement, as well as in everything which follows, we use $\mathcal{W}$ to denote $\mathcal{W}^{(1)}$.

3.3.22 Corollary. *Given a Lévy system (a, σ, M) with $\sigma > 0$, let $P^{a,0,M}$ be the measure on $CL(\mathbb{R})$ described in Theorem 3.2.33 and use $P^{a,\sigma^2,M}$ to denote the measure on $CL(\mathbb{R})$ which is the distribution of*

$$(\varphi, \psi) \in CL(\mathbb{R}) \times \mathfrak{P}(\mathbb{R}) \longmapsto \varphi + \sigma\psi \in CL(\mathbb{R})$$

under $P^{a,0,M} \times \mathcal{W}$. Then $P^{a,\sigma^2,M}$ is the one and only probability measure P on $CL(\mathbb{R})$ with the properties that $P\big(\psi(0) = 0\big) = 1$ and, for each $s \in [0, \infty)$ and $t \in (0, \infty)$, the increment $\psi(s + t) - \psi(s)$ under P is independent of $\sigma\big(\psi(\xi) : \xi \in [0, s]\big)$ and has distribution $\pi_{ta,t\sigma^2,tM}$.

Before we close this section, it seems only right that we complete the line of reasoning with which we started it. Actually, as the following statement shows, we can do even better than we had hoped when we started.

3.3.23 Theorem. *Let $\{\tau_n\}_1^\infty$ be a sequence of mutually independent unit exponential random variables and $\{\mathbf{X}_n\}_1^\infty$ a sequence of mutually independent*

$\mathbb{R}^N$-valued, uniformly square P-integrable random variables with mean-value $\mathbf{0}$ and covariance $\mathbf{I}$ on the probability space $(\Omega, \mathcal{F}, P)$. (Note that the $\mathbf{X}_n$'s are not assumed to be independent of the τ_n's.) Next, let $\{\mathbf{N}(t) : t \in [0,\infty)\}$ be the simple Poisson process in (3.2.23), and, for each $\epsilon \in (0,1]$, define

$$\mathbf{X}_\epsilon(t,\omega) = \sqrt{\epsilon} \sum_{m=1}^{\mathbf{N}_\epsilon(t,\omega)} \mathbf{X}_m, \quad (t,\omega) \in [0,\infty) \times \Omega,$$

where

$$\mathbf{N}_\epsilon(t,\omega) \equiv \mathbf{N}\left(\tfrac{t}{\epsilon}, \omega\right).$$

Then, for all $r \in (0,\infty)$ and $T \in [0,\infty)$,

$$\lim_{\epsilon \searrow 0} P\left(\sum_{t \in [0,T]} \left|\mathbf{X}_\epsilon(t) - \mathbf{S}_{n_\epsilon}(t)\right| \geq r \right) = 0 \tag{3.3.24}$$

where $n_\epsilon \equiv [\epsilon^{-1}]$.

PROOF: Note that

$$\mathbf{X}_\epsilon(t,\omega) - \mathbf{S}_{n_\epsilon}(t,\omega) = \left(\sqrt{\epsilon n_\epsilon} - 1\right) \mathbf{S}_{n_\epsilon}\left(\frac{\mathbf{N}_\epsilon(t,\omega)}{n_\epsilon}, \omega\right) + \left(\mathbf{S}_{n_\epsilon}\left(\frac{\mathbf{N}_\epsilon(t,\omega)}{n_\epsilon}, \omega\right) - \mathbf{S}_{n_\epsilon}(t,\omega)\right).$$

Hence, for every $\delta > 0$,

$$\begin{aligned} &P\left(\sup_{t\in[0,T]} \left|\mathbf{X}_\epsilon(t) - \mathbf{S}_{n_\epsilon}(t)\right| \geq r\right) \\ &\leq P\left(\sup_{t \in [0,T+\delta]} \left|\mathbf{S}_{n_\epsilon}(t)\right| \geq \frac{r}{2\epsilon}\right) + P\left(\sup_{t\in[0,T]} \left|\frac{\mathbf{N}_\epsilon(t)}{n_\epsilon} - t\right| \geq \delta\right) \\ &\quad + P\left(\sup_{s \in [0,T]} \sup_{|t-s|\leq\delta} \left|\mathbf{S}_{n_\epsilon}(t) - \mathbf{S}_{n_\epsilon}(s)\right| \geq \frac{r}{2}\right). \end{aligned}$$

But, by the converse statement in Theorem 3.1.9 combined with (3.3.4) and (3.3.5), we know that the first term tends to 0 as $\epsilon \searrow 0$ uniformly in $\delta \in [0,1]$ and that the third term tends to 0 as $\delta \searrow 0$ uniformly in $\epsilon \in (0,1]$. Finally, by part **(iii)** of Exercise 3.2.43, the second term tends to 0 as $\epsilon \searrow 0$ for each $\delta > 0$; and so (3.3.24) follows. □

Although Theorem 3.3.23 certainly shows that one can obtain $\mathcal{W}$ as a limit of independent increment processes corresponding to more and more jumps of smaller and smaller size, this fact is not the most intuitively appealing application of (3.3.24). To get a more intuitive statement, we introduce **Rayleigh's random flight model**. Namely, let $\{\tau_n\}_1^\infty$ and $\{\mathbf{N}(t) : t \in [0,\infty)\}$ be as in Theorem 3.3.23, and recall the random variables $\{\mathbf{T}_n\}_0^\infty$ on which the definition of $\mathbf{N}(t)$ was based in (3.2.23). Next, suppose that $\{\boldsymbol{\theta}_n\}_1^\infty$ is a sequence of mutually independent $\mathbb{R}^N$-valued random variables which satisfy the conditions that

$$M \equiv \sup_{n\in\mathbf{Z}^+} \mathbb{E}^P\left[|\tau_n\boldsymbol{\theta}_n|^4\right]^{\frac{1}{4}} < \infty,$$

$$\mathbb{E}^P\left[\tau_n\boldsymbol{\theta}_n\right] = \mathbf{0}, \quad \text{and} \quad \mathbb{E}^P\left[(\tau_n\boldsymbol{\theta}_n)\otimes(\tau_n\boldsymbol{\theta}_n)\right] = \mathbf{I}, \quad n \in \mathbb{Z}^+.$$

Finally, define $\omega \in \Omega \longmapsto \mathbf{R}(\,\cdot\,,\omega) \in \mathfrak{P}(\mathbb{R}^N)$ by

$$\mathbf{R}(t,\omega) = \big(t - \mathbf{T}_{\mathbf{N}(t,\omega)}(\omega)\big)\boldsymbol{\theta}_{\mathbf{N}(t,\omega)+1}(\omega) + \sum_{m=1}^{\mathbf{N}(t,\omega)} \tau_m(\omega)\boldsymbol{\theta}_m(\omega).$$

The process $\{\mathbf{R}(t) : t \in [0,\infty)\}$ models the path of a bird which leaves the origin at time 0 with the randomly chosen velocity $\boldsymbol{\theta}_1$ and travels in a straight line until *alarm* 1 sounds, at which instant it switches to velocity $\boldsymbol{\theta}_2$ and travels with this new velocity until *alarm* 2 sounds, etc. For example, a typical choice of the $\boldsymbol{\theta}_n$'s would be to make them independent of the *alarms* (i.e., the τ_n's) and to choose them to be uniformly distributed over the sphere $\mathbf{S}^{N-1}\left(\sqrt{N}\right)$.

3.3.25 Corollary. *Referring to the preceding, set*

$$\mathbf{R}_\epsilon(t,\omega) = \sqrt{\epsilon}\,\mathbf{R}\left(\tfrac{t}{\epsilon},\omega\right), \quad (t,\omega) \in [0,\infty)\times\Omega.$$

Then $\mathbf{R}_\epsilon^* P \Longrightarrow \mathcal{W}^{(N)}$ *as* $\epsilon \searrow 0$.

PROOF: Set $\mathbf{X}_n = \tau_n\boldsymbol{\theta}_n$, and, using the same notation as in Theorem 3.3.23, observe that

$$\left|\mathbf{R}_\epsilon(t) - \mathbf{X}_\epsilon(t)\right| \le \sqrt{\epsilon}\left|\mathbf{X}_{\mathbf{N}_\epsilon(t)+1}\right|.$$

Hence, by Theorem 3.3.20, (3.3.24), and the Principle of Accompanying Laws, all that we have to do is check that

$$\lim_{\epsilon\searrow 0} P\left(\sup_{t\in[0,T]} \left|\sqrt{\epsilon}\,\mathbf{X}_{\mathbf{N}_\epsilon(t)+1}\right| \ge r\right) = 0$$

for every $r \in (0,\infty)$ and $T \in [0,\infty)$. To this end, set $T_\epsilon = \frac{1+T}{\epsilon}$. Then, by part **(iii)** of Exercise 3.2.43, we have that

$$\lim_{\epsilon\searrow 0} P\left(\sup_{t\in[0,T]} \left|\sqrt{\epsilon}\,\mathbf{X}_{\mathbf{N}_\epsilon(t)+1}\right| \geq r\right) = \lim_{\epsilon\searrow 0} P\left(\max_{0\leq n\leq T_\epsilon} |\mathbf{X}_{n+1}| \geq \frac{r}{\sqrt{\epsilon}}\right)$$

$$\leq \lim_{\epsilon\searrow 0} \frac{\sqrt{\epsilon}}{r}\mathbb{E}^P\left[\left(\sum_{0\leq n\leq T_\epsilon} |\mathbf{X}_{n+1}|^4\right)^{\frac{1}{4}}\right] \leq \lim_{\epsilon\searrow 0} \frac{M\epsilon^{\frac{1}{4}}(2+T)^{\frac{1}{4}}}{r} = 0. \quad \square$$

Exercises

3.3.26 Exercise: Let E be a Polish space and $\{\mu_n\}_1^\infty \subseteq \mathbf{M}_1\big(\mathfrak{P}(E)\big)$, and, for each $T \in (0,\infty)$, let $\mu_n^T \in \mathbf{M}_1\big(C([0,T];E)\big)$ denote the distribution of

$$\psi \in \mathfrak{P}(E) \longmapsto \psi \restriction [0,T] \in C\big([0,T];E\big) \text{ under } \mu_n.$$

Starting from Lemma 3.3.3, show that there is a $\mu \in \mathbf{M}_1\big(\mathfrak{P}(E)\big)$ to which $\{\mu_n\}$ converges in $\mathbf{M}_1\big(\mathfrak{P}(E)\big)$ if and only if, for each $T \in (0,\infty)$, there is a $\mu^T \in \mathbf{M}_1\big(C([0,T];E)\big)$ with the property that

$$\mu_n^T \Longrightarrow \mu^T \quad \text{in } \mathbf{M}_1\big(C([0,T];E)\big);$$

in which case, μ^T is the distribution of

$$\psi \in (E) \longmapsto \psi \restriction [0,T] \in C([0,T];E) \text{ under } \mu.$$

In particular, weak convergence of measures on $\mathfrak{P}(E)$ is really a *local* property.

3.3.27 Exercise: For $r \in [1,\infty)$ and $\gamma \in (0,\infty)$, define the **Besov norm**

$$B_{r,\gamma}(\psi) = \left(\iint_{[0,1]^2} \left(\frac{|\psi(t)-\psi(s)|}{|t-s|^\gamma}\right)^r ds\,dt\right)^{\frac{1}{r}}, \quad \psi \in C\big([0,1];\mathbb{R}\big).$$

(i) Using Lemma 3.3.13, show that

$$\big|\psi(t)-\psi(s)\big| \leq 2^{3+\frac{2}{r}}\frac{\gamma}{\beta}B_{r,\gamma}(\psi)\,|t-s|^\beta, \quad \psi \in C\big([0,1];\mathbb{R}\big),$$

if $\beta \equiv \gamma - \frac{2}{r} > 0$. In particular, this means that

$$\bigl|\psi(1) - \psi(0)\bigr| \le 2^{3+\frac{2}{r}} \frac{\gamma}{\beta} B_{r,\gamma}(\psi), \quad \psi \in C\bigl([0,1];\mathbb{R}\bigr),$$

if $\gamma r > 2$. Show that if $r > 1$ and $\gamma r \le 2$, then there is no $C < \infty$ for which

$$\bigl|\psi(1) - \psi(0)\bigr| \le C\, B_{r,\gamma}(\psi), \quad \psi \in C\bigl([0,1];\mathbb{R}\bigr).$$

Hint: First observe that $B_{r,\gamma}(\psi)$ is a nondecreasing function of $\gamma \in (0,\infty)$ for each $r \in (0,\infty)$. Next, take $\gamma r = 2$, consider the paths

$$\psi_\epsilon(t) \equiv \log\bigl(-\log\bigl(\epsilon + \tfrac{t}{2}\bigr)\bigr), \quad t \in [0,1],$$

for $\epsilon \in \bigl(0, \frac{1}{4}\bigr]$, and show that

$$\sup_{\epsilon \in (0,\frac{1}{4}]} B_{r,\gamma}(\psi_\epsilon) < \infty.$$

(ii) Actually, there is another way of our seeing that $B_{2,\gamma}(\psi)$ cannot be used to control the continuity of ψ for any $\gamma \in (0,1)$. Namely, let $\mathbf{N}(\,\cdot\,)$ be the simple Poisson process appearing in (3.2.23), set $X(t) = \mathbf{N}(t) - t$, note that

$$\mathbb{E}^P\Bigl[\bigl|X(t) - X(s)\bigr|^2\Bigr] = t - s \quad \text{for } 0 \le s < t < \infty,$$

and conclude that, for every $\gamma \in (0,1)$, $B_{2,\gamma}\bigl(X(\,\cdot\,)\bigr) < \infty$ P-almost surely. On the other hand, we know that

$$P\Bigl(\exists t \in [0,1]\ X(t) - X(t-) = 1\Bigr) = P\bigl(\tau_1 \le 1\bigr) = 1 - e^{-1} < 1.$$

3.3.28 Exercise: When we first mentioned Wiener's measure, we somewhat casually alluded to the fact that $\mathcal{W}^{(N)} = \mathcal{W}^N$. To make this assertion more precise, let $\bigl(\mathbf{e}_1, \dots, \mathbf{e}_N\bigr)$ be an orthonormal basis for $\mathbb{R}^N$, define

$$\psi \in P(\mathbb{R}^N) \longmapsto \Phi(\psi) \equiv \Bigl(\bigl(\mathbf{e}_1, \psi\bigr)_{\mathbb{R}^N}, \dots, \bigl(\mathbf{e}_N, \psi\bigr)_{\mathbb{R}^N}\Bigr) \in \bigl(\mathfrak{P}(\mathbb{R})\bigr)^N,$$

and show that $\Phi^*\mathcal{W}^{(N)} = \mathcal{W}^N$. That is, the random variables $\psi \longmapsto \bigl(\mathbf{e}_i, \psi\bigr)_{\mathbb{R}^N}$, $1 \le i \le N$, are mutually independent, and each has distribution $\mathcal{W}$ under $\mathcal{W}^{(N)}$. In particular, this means that $\mathcal{W}^{(N)}$ is **rotation invariant** in the sense that if $\mathbf{R}$ is a rotation on $\mathbb{R}^N$ and $\mathbf{R} : \mathfrak{P}(\mathbb{R}^N) \longrightarrow \mathfrak{P}(\mathbb{R}^N)$ is given by $[\mathbf{R}\psi](t) = \mathbf{R}\bigl(\psi(t)\bigr)$, $t \in [0,\infty)$, then $\mathcal{W}^{(N)} = \mathbf{R}^*\mathcal{W}^{(N)}$.

3.3.29 Exercise: Show that for any $T \in (0,\infty)$,

$$\mathcal{W}^{(N)}\left(\sup_{t\in[0,T]} |\psi(t)| \geq R\right) \leq 2N \exp\left[-\frac{R^2}{2NT}\right], \quad R \in [0,\infty). \tag{3.3.30}$$

Hint: Using Exercise 3.3.28, reduce to the case when $N = 1$. Next, for fixed $T \in (0,\infty)$ and $n \in \mathbb{Z}^+$, write

$$\psi\left(\frac{\ell T}{n}\right) - \psi(0) = \sum_{k=1}^{n}\left[\psi\left(\frac{kT}{n}\right) - \psi\left(\frac{(k-1)T}{n}\right)\right], \quad \ell \in \mathbb{Z}^+,$$

and use Theorem 1.4.15 to see first that

$$\mathcal{W}\left(\max_{0\leq\ell\leq n} \left|\psi\left(\tfrac{\ell T}{n}\right)\right|\right) \leq 2\mathcal{W}\Big(|\psi(T)| \geq R\Big),$$

and then that

$$\mathcal{W}\left(\sup_{t\in[0,T]} |\psi(t)| \geq R\right) \leq 2\sqrt{\tfrac{2}{\pi}} \int_{T^{-\frac{1}{2}}R}^{\infty} e^{-\frac{x^2}{2}}\, dx.$$

Finally, apply the second estimate in (2.1.23). (See part (**iii**) of Exercise 7.1.23 for another derivation.)

3.3.31 Exercise: The purpose of this exercise is to prove **The Strong Law of Large Numbers** for Wiener paths. That is, show that

$$\lim_{t\to\infty} \frac{\boldsymbol{\psi}(t)}{t} = \mathbf{0} \quad \text{for } \mathcal{W}^{(N)}\text{-almost every } \boldsymbol{\psi} \in \mathfrak{P}(\mathbb{R}^N). \tag{3.3.32}$$

Hint: First note that, because of Exercise 3.3.28, it suffices to treat the case when $N = 1$. Second, observe that

$$\lim_{n\to\infty} \frac{\psi(n)}{n} = 0 \qquad (\text{a.s.}, \mathcal{W}) \tag{3.3.33}$$

as an application of The Strong Law for random variables. Finally, note that, for $n \in \mathbb{Z}^+$ and $t \in [n, n+1]$,

$$\left|\frac{\psi(t)}{t} - \frac{\psi(n)}{n}\right| \leq \frac{|\psi(t) - \psi(n)|}{t} + \frac{|\psi(n)|}{nt}$$

and, by (3.3.33), the second term tends to 0 (a.s., $\mathcal{W}$) as $n \to \infty$. At the same time, as an application of (3.3.30), one sees that, for each $r \in (0,\infty)$,

$$\sum_{n=1}^{\infty} \mathcal{W}\left(\left\{\psi : \sup_{t\in[n,n+1]} \frac{|\psi(t)-\psi(n)|}{n} \geq r\right\}\right) < \infty;$$

and therefore, by the Borel–Cantelli Lemma, it follows that

$$\overline{\lim_{n\to\infty}} \sup_{t\in[n,n+1]} \left|\frac{\psi(t)}{t} - \frac{\psi(n)}{n}\right| = 0 \quad \text{(a.s., } \mathcal{W}).$$

In particular, this means that (3.3.32) follows from (3.3.33).

3.3.34 Exercise: Here is another version of Rayleigh's random flight model. Again let $\{\tau_k\}_1^\infty$, $\{\mathbf{T}_m\}_0^\infty$, and $\{\mathbf{N}(t) : t \geq 0\}$ be as in (3.2.23); and set

$$R(t) = \int_0^t (-1)^{\mathbf{N}(s)}\, ds \quad \text{and} \quad R_\epsilon(t) = \sqrt{\epsilon}\, R\left(\tfrac{t}{\epsilon}\right).$$

Show that $R_\epsilon{}^*P \Longrightarrow \mathcal{W}$ as $\epsilon \searrow 0$.

Hint: Set $\beta_k = 0$ or 1 according to whether $k \in \mathbb{N}$ is even or odd, and note that

$$\sum_{k=1}^{n}(-1)^k \tau_k = \sum_{k=1}^{n} \beta_k\left(\tau_{k+1}-\tau_k\right) - \beta_n\tau_n = \sum_{1\leq k\leq \frac{n}{2}} \left(\tau_{2k}-\tau_{2k-1}\right) - \beta_n\tau_{n+1};$$

and now proceed as in the derivations of Theorem 3.3.23 and Corollary 3.3.25.

Chapter IV: A Celebration of Wiener's Measure

§4.1: Preliminary Results.

There are so many things that ought and have† to be said about Wiener's measure that anything short of an entire book on the subject must be woefully incomplete. Wiener's measure is such a rich subject because it enjoys the best of all worlds: its paths have independent increments, it is Gaussian, and it is Markovian. In this chapter, we will attempt to give a brief introduction to each of these aspects of Wiener's measure, and we concentrate in the present section on a few of its properties which follow more or less immediately from the fact that it is a process with independent, identically distributed Gaussian increments.

We asserted after Theorem 3.3.20 that $\mathcal{W}^{(N)}$ is *the Lebesgue measure for* $\mathfrak{P}(\mathbb{R}^N)$. Perhaps the best reason for our making this assertion is that $\mathcal{W}^{(N)}$ enjoys many invariance properties, one of the most important of which, namely, rotation invariance, we already discussed in Exercise 3.3.28. We begin this section with other invariance properties enjoyed by $\mathcal{W}^{(N)}$. To state the first of these, define, for each $\alpha \in (0,\infty)$, the **scaling map** $\mathbf{S}_\alpha : \mathfrak{P}(\mathbb{R}^N) \longrightarrow \mathfrak{P}(\mathbb{R}^N)$ by

$$[\mathbf{S}_\alpha \psi](t) = \alpha^{-\frac{1}{2}} \psi(\alpha t), \quad t \in [0,\infty). \tag{4.1.1}$$

Clearly, under $\mathcal{W}^{(N)}$, the process $\{\mathbf{S}_\alpha\psi(t) : t \in [0,\infty)\}$ again has independent, identically distributed increments. Furthermore, because the distribution of $\mathbf{x} \in \mathbb{R}^N \longmapsto \alpha^{-\frac{1}{2}}\mathbf{x} \in \mathbb{R}^N$ under $\gamma^N_{\alpha t}$ is γ^N_t, it follows that the distribution of $\psi \in \mathfrak{P}(\mathbb{R}^N) \longmapsto \mathbf{S}_\alpha\psi \in \mathfrak{P}(\mathbb{R}^N)$ under $\mathcal{W}^{(N)}$ is again $\mathcal{W}^{(N)}$. In other words,

$$\mathbf{S}_\alpha{}^*\mathcal{W}^{(N)} = \mathcal{W}^{(N)} \quad \text{for all } \alpha \in (0,\infty). \tag{4.1.2}$$

This property is called **Wiener scaling invariance.**

We next turn to an invariance property of $\mathcal{W}^{(N)}$ which is nothing more than a useful restatement the independent, identical increment property. For this purpose, let

$$\mathcal{B}_s = \sigma\Big(\psi(t) : t \in [0,s]\Big), \quad s \in [0,\infty), \tag{4.1.3}$$

† The classic account is K. Itô and H.P. McKean's *Diffusions and Their Sample Paths*, publ. by Springer–Verlag in 1965. A more recent account is given by D. Revuz and M. Yor in *Continuous Martingales and Brownian Motion*, which appeared in 1991 as **293** in the Grundlehren Series of the same publisher.

be the σ-algebra over $\mathfrak{P}(\mathbb{R}^N)$ generated by $\boldsymbol{\psi} \in \mathfrak{P}(\mathbb{R}^N) \longmapsto \boldsymbol{\psi}(t) \in \mathbb{R}^N$, $t \in [0, s]$, and define the **time increment map** $\boldsymbol{\delta}_s : \mathfrak{P}(\mathbb{R}^N) \longrightarrow \mathfrak{P}(\mathbb{R}^N)$ by

$$\left[\boldsymbol{\delta}_s \boldsymbol{\psi}\right](t) = \boldsymbol{\psi}(t+s) - \boldsymbol{\psi}(s), \quad t \in [0, \infty).$$

4.1.4 Lemma. *For every $s \in [0, \infty)$ and every bounded, $\mathcal{B}_s \times \mathcal{B}$-measurable $F : \mathfrak{P}(\mathbb{R}^N)^2 \longrightarrow \mathbb{R}$,*

$$\int_{\mathfrak{P}(\mathbb{R}^N)} F\big(\boldsymbol{\psi}, \boldsymbol{\delta}_s\boldsymbol{\psi}\big)\, \mathcal{W}^{(N)}(d\boldsymbol{\psi}) = \iint_{\mathfrak{P}(\mathbb{R}^N)^2} F(\boldsymbol{\varphi}, \boldsymbol{\psi})\, \mathcal{W}^{(N)}(d\boldsymbol{\varphi})\mathcal{W}^{(N)}(d\boldsymbol{\psi}). \tag{4.1.5}$$

In other words, $\mathcal{B}_s$ is independent of $\sigma\big(\boldsymbol{\delta}_s\psi(t) : t \in [0, \infty)\big)$ and $\boldsymbol{\delta}_s{}^*\mathcal{W}^{(N)} = \mathcal{W}^{(N)}$.

PROOF: Since, $\mathcal{W}^{(N)}\big(\boldsymbol{\psi}(0) = \mathbf{0}\big) = 1$, there is hardly anything to do when $s = 0$. To handle $s > 0$, let $0 = s_0 < s_1 < \cdots < s_k = s$ and $0 = t_0 < t_1 < \cdots < t_\ell < \infty$ be given, and set $\sigma_i = s_i - s_{i-1}$, $1 \leq i \leq k$, and $\tau_j = t_j - t_{j-1}$, $1 \leq j \leq \ell$. Next, suppose that g and h are bounded, measurable functions on $(\mathbb{R}^N)^k$ and $(\mathbb{R}^N)^\ell$, respectively, and set $F(\boldsymbol{\varphi}, \boldsymbol{\psi}) = G(\boldsymbol{\varphi})\, H(\boldsymbol{\psi})$, where

$$G(\boldsymbol{\psi}) = g\big(\boldsymbol{\psi}(s_1) - \boldsymbol{\psi}(s_0), \ldots, \boldsymbol{\psi}(s_k) - \boldsymbol{\psi}(s_{k-1})\big)$$

and

$$H(\boldsymbol{\psi}) = h\big(\boldsymbol{\psi}(t_1) - \boldsymbol{\psi}(t_0), \ldots, \boldsymbol{\psi}(t_\ell) - \boldsymbol{\psi}(t_{\ell-1})\big)$$

for $\boldsymbol{\psi} \in \mathfrak{P}(\mathbb{R}^N)$. Then

$$\begin{aligned}
&\int_{\mathfrak{P}(\mathbb{R}^N)} F\big(\boldsymbol{\psi}, \boldsymbol{\delta}_s\boldsymbol{\psi}\big)\, \mathcal{W}^{(N)}(d\boldsymbol{\psi}) \\
&\quad = \int_{(\mathbb{R}^N)^k} g\big(\mathbf{x}_1, \ldots, \mathbf{x}_k\big)\, \gamma^N_{\sigma_1}(d\mathbf{x}_1) \cdots \gamma^N_{\sigma_k}(d\mathbf{x}_k) \\
&\qquad\qquad \times \int_{(\mathbb{R}^N)^\ell} h\big(\mathbf{y}_1, \ldots, \mathbf{y}_\ell\big)\gamma^N_{\tau_1}(d\mathbf{y}_1) \cdots \times^N_{\tau_\ell}(d\mathbf{y}_\ell) \\
&\quad = \mathbb{E}^{\mathcal{W}^{(N)}}[G]\, \mathbb{E}^{\mathcal{W}^{(N)}}[H] = \iint_{\mathfrak{P}(\mathbb{R}^N)^2} F(\boldsymbol{\varphi}, \boldsymbol{\psi})\, \mathcal{W}^{(N)}(d\boldsymbol{\varphi})\mathcal{W}^{(N)}(d\boldsymbol{\psi}).
\end{aligned}$$

To complete the proof at this point, one need only note that the set of bounded $\mathcal{B}_s \times \mathcal{B}$-measurable F's for which (4.1.5) holds is closed under both linear operations and bounded pointwise convergence. Hence, since all bounded $\mathcal{B}_s \times \mathcal{B}$-measurable F's can be obtained with these operations starting from the F's for which we have just proved (4.1.5), we are done. □

The fact proved in Lemma 4.1.4 can be summarized by the statement that, under $\mathcal{W}^{(N)}$, *future increments are independent of the past and have the same distribution as the initial increment*; and, as we will see in Section 4.3 (cf. Theorem 4.3.3), the scope of this statement can be considerably expanded.

We next turn our attention to a couple of properties having to do with the regularity of Wiener paths. In view of the development given in Sections 3.2 and 3.3, it is clear that as the governing Lévy measure becomes more singular the paths become less regular, at least when regularity is measured in the sense of bounded variation. On the other hand, when the Lévy measures become more concentrated near the origin as they become more singular, then, as we saw in Section 3.3, the loss of bounded variation is, in the limit, compensated by the emergence of continuity; thus, it is reasonable to ask *just how continuous* are these paths.

4.1.6 Theorem. *Set* $\omega(\delta) = \sqrt{\delta \log \frac{1}{\delta}}$ *for* $\delta \in \left(0, \frac{1}{2}\right]$. *Then there exists a* $K \in (0,\infty)$ *such that, for each* $T \in [1,\infty)$ *and* $R \in (0,\infty)$,

$$\text{(4.1.7)} \qquad \begin{aligned} &\mathcal{W}^{(N)}\Big(\Big\{\psi : \big|\psi(t)-\psi(s)\big| \geq R\omega\big(|t-s|\big) \\ &\qquad\qquad \text{for some } s \in [0,T] \text{ and } s \leq t \leq s+\tfrac{1}{2}\Big\}\Big) \leq \frac{KT^2}{R}. \end{aligned}$$

Moreover (cf. (1.5.1)*),*

$$\text{(4.1.8)} \qquad \varlimsup_{t\to\infty} \frac{|\psi(t)|}{\sqrt{2t\log_{(2)} t}} = 1 \qquad (\text{a.s., } \mathcal{W}^{(N)}).$$

PROOF: The first assertion is a straightforward application of Lemma 3.3.15 in which we take

$$\Phi(\xi) = \exp\left[\frac{\xi^2}{4}\right] \quad \text{and} \quad p(u) = \sqrt{u},$$

and observe that

$$\int_{\mathfrak{P}(\mathbb{R}^N)} \Phi\left(\frac{|\psi(t)-\psi(s)|}{p(|t-s|)}\right) \mathcal{W}^{(N)}(d\psi) = 2^{\frac{N}{2}}, \quad 0 \leq s < t.$$

In order to prove the second assertion, we begin with the case when $N = 1$ and argue in much the same way as we suggested in the hint to Exercise 3.3.31. Thus, we first note that, by the Law of the Iterated Logarithm (cf. Theorem 1.5.10),

$$\text{(4.1.9)} \qquad \varlimsup_{n\to\infty} \frac{|\psi(n)|}{\Lambda(n)} = 1 \quad (\text{a.s., } \mathcal{W}^{(N)})$$

where $\Lambda(t) \equiv \sqrt{2t \log_{(2)}(t \vee 3)}$ for $t \in (0, \infty)$. Next, note that

$$\left| \frac{\psi(t)}{\Lambda(t)} - \frac{\psi(n)}{\Lambda(n)} \right| \leq \frac{|\psi(t) - \psi(n)|}{\Lambda(n)} + \frac{\Lambda(n+1) - \Lambda(n)}{\Lambda(n)} \frac{|\psi(n)|}{\Lambda(n)}$$

for $9 \leq n \leq t \leq n+1$ and, by (4.1.9), the second term tends to 0 (a.s., $\mathcal{W}$) as $n \to \infty$. At the same time, as an application of (3.3.30) and the last part of Lemma 4.1.4, one sees that

$$\sum_{n=9}^{\infty} \mathcal{W}\left(\left\{ \psi : \sup_{t \in [n,n+1]} \frac{|\psi(t) - \psi(n)|}{\Lambda(n)} \geq \frac{1}{n^{\frac{1}{4}}} \right\} \right) < \infty;$$

and therefore, by the Borel–Cantelli Lemma, it follows that

$$\varlimsup_{n \to \infty} \sup_{t \in [n,n+1]} \left| \frac{\psi(t)}{\Lambda(t)} - \frac{\psi(n)}{\Lambda(n)} \right| = 0 \quad (\text{a.s.}, \mathcal{W}).$$

In particular, this means that, when $N = 1$, (4.1.8) follows from (4.1.9). To handle $N \in \mathbb{Z}^+ \setminus \{1\}$, we combine the preceding with Exercise 3.3.28 to see that

$$\varlimsup_{t \to \infty} \frac{\left| \left(\psi(t), \mathbf{e} \right)_{\mathbb{R}^N} \right|}{\Lambda(t)} = 1 \quad (\text{a.s.}, \mathcal{W}^{(N)})$$

for each $\mathbf{e} \in \mathbf{S}^{N-1}$; and therefore that

$$\varlimsup_{t \to \infty} \frac{|\psi(t)|}{\Lambda(t)} \geq 1 \quad (\text{a.s.}, \mathcal{W}^{(N)}).$$

To prove the opposite inequality, let $\epsilon > 0$ be given, choose a finite set $\left\{ \mathbf{e}_m \right\}_1^{n(\epsilon)}$ from $\mathbf{S}^{N-1}$ so that

$$|\mathbf{v}| \leq (1 + \epsilon) \max_{1 \leq m \leq n(\epsilon)} \left| \left(\mathbf{v}, \mathbf{e}_m \right)_{\mathbb{R}^N} \right| \quad \text{for all} \quad \mathbf{v} \in \mathbb{R}^N,$$

and conclude that

$$\varlimsup_{t \to \infty} \frac{|\psi(t)|}{\Lambda(t)} \leq 1 + \epsilon \quad (\text{a.s.}, \mathcal{W}^{(N)}). \quad \square$$

Lévy showed† that the result in (4.1.7) can be sharpened to give

$$\lim_{\delta\searrow 0}\sup_{0<t-s<\delta}\frac{|\psi(t)-\psi(s)|}{\omega(t-s)}=\sqrt{2}\quad(\text{a.s.},\mathcal{W}^{(N)}).$$

In particular, Lévy's result proves, in a very definitive way, Wiener's famous conclusion that *almost no Wiener path is anywhere differentiable.* Rather than go into the details of Lévy's result (they are much the same as those required to prove the Law of the Iterated Logarithm for Gaussian random variables), we will content ourselves here with a statement which, although it is far cruder than Lévy's, is still more than enough to get Wiener's conclusion.‡

4.1.10 Theorem. *For each* $\epsilon>0$,

$$\mathcal{W}^{(N)}\left(\left\{\psi:\exists s\in[0,\infty)\ \overline{\lim_{t\searrow s}}\frac{|\psi(t)-\psi(s)|}{(t-s)^{\frac{1}{2}+\epsilon}}<\infty\right\}\right)=0.$$

In particular, $\mathcal{W}^{(N)}$*-almost no* ψ *is anywhere differentiable.*

PROOF: Clearly, it is sufficient to deal with the case when $N=1$. To this end, choose $L\in\mathbb{Z}^+$ so that $\epsilon L>1$, and note that

$$\left\{\psi:\exists s\in[0,\infty)\ \overline{\lim_{t\searrow s}}\frac{|\psi(t)-\psi(s)|}{(t-s)^{\frac{1}{2}+\epsilon}}<\infty\right\}\subseteq\bigcup_{M=1}^{\infty}\bigcap_{n=1}^{\infty}\bigcup_{m=1}^{nM}A(M,n,m),$$

where

$$A(M,n,m)=\left\{\psi:\max_{1\le k\le L}\left|\psi\left(\frac{m+k}{n}\right)-\psi\left(\frac{m+k-1}{n}\right)\right|\le M\,n^{-\frac{1}{2}-\epsilon}\right\}.$$

Hence, it suffices to show that

$$\lim_{n\to\infty}\mathcal{W}\left(\bigcup_{m=1}^{nM}A(M,n,m)\right)=0$$

for every $M\in\mathbb{Z}^+$. But, by independence of increments and Wiener scaling,

$$\mathcal{W}\left(\bigcup_{m=1}^{nM}A(M,n,m)\right)\le nM\Big(P\big(|\psi(1)|\le Mn^{-\epsilon}\big)\Big)^{L}\le\frac{M\,n^{1-\epsilon L}}{(2\pi)^{\frac{L}{2}}},$$

and so we are done. □

† A straightforward derivation is given in H.P. McKean's *Stochastic Integrals*, publ. in the Academic Press Probability & Math. Stat. Series (1969).

‡ This simple argument given below was discovered by A. Dvoretsky.

Exercises

4.1.11 Exercise: As we saw in Theorem 4.1.10, almost no Wiener path is anywhere differentiable. Of course, this means that almost no Wiener path has bounded variation on any open interval. Another, and perhaps more interesting, route to the same fact is the following. Assume that $N = 1$, define

$$V_n(t,\psi) = \sum_{m=1}^{[nt]} \left(\psi\left(\frac{m}{n}\right) - \psi\left(\frac{m-1}{n}\right)\right)^2$$

for $n \in \mathbb{Z}^+$ and $(t,\psi) \in [0,\infty) \times \mathfrak{P}(\mathbb{R})$, and show that

$$\text{(4.1.12)} \qquad \lim_{n\to\infty} \sup_{t\in[0,T]} \left|V_n(t,\psi) - t\right| = 0 \quad \text{for } \mathcal{W}\text{-almost all } \psi \in \mathfrak{P}(\mathbb{R})$$

and every $T \in [0,\infty)$.

Hint: Set

$$\bar{V}_n(M,\psi) = \sum_{m=1}^{M} \left[\left(\psi\left(\frac{m}{n}\right) - \psi\left(\frac{m-1}{n}\right)\right)^2 - \frac{1}{n}\right], \quad M \in \mathbb{Z}^+,$$

use Kolmogorov's Inequality to show that

$$\mathcal{W}\left(\max_{1\le M\le nT} \left|\bar{V}_n(M)\right| \ge \epsilon\right) \le \frac{2T}{(\epsilon n)^2}, \quad n \in \mathbb{Z}^+,$$

and apply the Borel–Cantelli Lemma.

4.1.13 Exercise: As a further confirmation that Wiener paths have unbounded variation, again let $N = 1$ and consider the three Riemann sums

$$S_n^-(\psi) = \sum_{m=1}^{n} \psi\left(\frac{m-1}{n}\right)\left(\psi\left(\frac{m}{n}\right) - \psi\left(\frac{m-1}{n}\right)\right),$$

$$S_n(\psi) = \sum_{m=1}^{n} \psi\left(\frac{2m-1}{2n}\right)\left(\psi\left(\frac{m}{n}\right) - \psi\left(\frac{m-1}{n}\right)\right),$$

and

$$S_n^+(\psi) = \sum_{m=1}^{n} \psi\left(\frac{m}{n}\right)\left(\psi\left(\frac{m}{n}\right) - \psi\left(\frac{m-1}{n}\right)\right).$$

Whenever ψ has bounded variation on the interval $[0,1]$, all three of these Riemann sums tend to $\frac{\psi(1)^2}{2}$. As an application of (4.1.12), show that, for $\mathcal{W}$-almost every $\psi \in \mathfrak{P}(\mathbb{R})$,

$$\lim_{n\to\infty} S_n^{\pm}(\psi) = \frac{\psi(1)^2 \pm 1}{2} \quad \text{and} \quad \lim_{n\to\infty} S_n(\psi) = \frac{\psi(1)^2}{2}.$$

This observation can be taken as the starting point for Itô's theory of stochastic integration.

§4.2: Gaussian Aspects of Wiener's Measure.

Having discussed Wiener's measure on the basis of its independent increment property, we (like Wiener himself) will now look at it as a Gaussian measure; and the point of view which we will adopt is the one introduced by I. Segal and developed further by L. Gross.†

In order to get started, we begin with a somewhat fanciful presentation of Wiener's measure. Namely, given $\ell \in \mathbb{Z}^+$, $0 = t_0 < t_1 < \cdots < t_\ell < \infty$, and a set $A \in \left(\mathcal{B}_{\mathbb{R}^N}\right)^\ell$, suppose that we first rewrite the determining property (cf. (3.3.21)) of $\mathcal{W}^{(N)}$ as

$$\mathcal{W}^{(N)}\Big(\Big\{\psi : \big(\psi(t_1),\dots,\psi(t_\ell)\big) \in A\Big\}\Big) = C\big(t_1,\dots,t_\ell\big)^N \times \int_A \exp\left[-\sum_{k=1}^{\ell} \frac{t_k - t_{k-1}}{2}\left(\frac{|\mathbf{y}_k - \mathbf{y}_{k-1}|}{t_k - t_{k-1}}\right)^2\right] d\mathbf{y}_1 \cdots d\mathbf{y}_\ell$$

where $\mathbf{y}_0 \equiv \mathbf{0}$, and then rename the variables $\mathbf{y}_k$ as "$\psi(t_k)$" so that the preceding expression becomes

$$\mathcal{W}^{(N)}\Big(\Big\{\psi : \big(\psi(t_1),\dots,\psi(t_\ell)\big) \in A\Big\}\Big) = C\big(t_1,\dots,t_\ell\big)^N \times \int_A \exp\left[-\sum_{k=1}^{\ell} \frac{t_k - t_{k-1}}{2}\left(\frac{|\psi(t_k) - \psi(t_{k-1})|}{t_k - t_{k-1}}\right)^2\right] d\psi(t_1) \cdots d\psi(t_\ell).$$

† See I.E. Segal's "Distributions in Hilbert space and canonical systems of operators," *T.A.M.S.* **88** (1958) and L. Gross's "Abstract Wiener spaces," *Proc. 5th Berkeley Symp. on Prob. & Stat.*, *2* (1965). A good exposition of this topic can be found in H.-H. Kuo's *Gaussian Measures in Banach Spaces*, publ. by Springer–Verlag Math. Lec. Notes., no. **463**.

Obviously nothing very significant has happened yet since nothing very exciting has been done yet. However, if we now close our eyes, suspend our disbelief, and *pass to the limit* as ℓ tends to infinity and the t_k's become dense, we arrive at the remarkable expression

$$\mathcal{W}^{(N)}(A) = C^N \int_A \exp\left[-\frac{1}{2}\int_{[0,\infty)} \left|\dot{\psi}(t)\right|^2 dt\right] d\psi \tag{4.2.1}$$

for any $A \in \mathcal{B}_{\mathfrak{P}(\mathbb{R}^N)}$. Of course, when we return to reality and take a look at (4.2.1), we see that it is riddled with flaws. Indeed, not even one of the ingredients on the right-hand side (4.2.1) makes sense! In the first place, the constant C must be ∞. Secondly, since the image of the "measure $d\psi$" under

$$\psi \in \mathfrak{P}(\mathbb{R}^N) \longmapsto \left(\psi(t_1) \dots, \psi(t_\ell)\right) \in \left(\mathbb{R}^N\right)^\ell$$

is Lebesgue's measure for every $\ell \in \mathbb{Z}^+$ and $0 < t_1 \cdots < t_\ell$, "$d\psi$" must be the nonexistent *translation invariant measure* on the infinite dimensional space $\mathfrak{P}(\mathbb{R}^N)$. Finally, if it has any rigorous meaning at all, (4.2.1) certainly seems to be saying that $\mathcal{W}^{(N)}\left(\mathbf{H}(\mathbb{R}^N)\right) = 1$ where

$$\mathbf{H}(\mathbb{R}^N) \equiv \Big\{\psi \in \mathfrak{P}(\mathbb{R}^N) : \psi(t) = \int_0^t \dot{\psi}(s)\, ds,\ t \in [0,\infty),$$
$$\text{with } \dot{\psi} \in L^2\left([0,\infty); \mathbb{R}^N\right)\Big\}.$$

However, even this is entirely wrong, because

$$\psi \in \mathbf{H}(\mathbb{R}^N) \implies \operatorname{var}_{[0,T]}\left(\psi(\,\cdot\,)\right) \le T^{\frac{1}{2}} \|\psi\|_{\mathbf{H}(\mathbb{R}^N)}, \quad T \in [0,\infty),$$

where

$$\|\psi\|_{\mathbf{H}(\mathbb{R}^N)} \equiv \left\|\dot{\psi}\right\|_{L^2([0,\infty);\mathbb{R}^N)},$$

and yet (cf. Theorem 4.1.10 and Exercise 4.1.11) we know that $\mathcal{W}^{(N)}$-almost no ψ has bounded variation on any open interval. That is, although (4.2.1) would seem to be predicting that $\mathbf{H}(\mathbb{R}^N)$ should have full $\mathcal{W}^{(N)}$-measure, $\mathbf{H}(\mathbb{R}^N)$ is in fact a set of $\mathcal{W}^{(N)}$-measure 0.

At this point it would appear that (4.2.1) does not have very much to recommend it. On the other hand, it is such an intuitively appealing formula that one is reluctant to simply abandon it; and, for this reason, we are going to take another look at it from a slightly different perspective. Notice that $\mathbf{H}(\mathbb{R}^N)$ becomes a separable, real Hilbert space under the norm $\|\cdot\|_{\mathbf{H}(\mathbb{R}^N)}$; the corresponding inner product being

$$\left(\varphi, \psi\right)_{\mathbf{H}(\mathbb{R}^N)} = \left(\dot{\varphi}, \dot{\psi}\right)_{L^2([0,\infty);\mathbb{R}^N)}.$$

Next, we rewrite (4.2.1) in the form

$$\mathcal{W}^{(N)}(d\mathbf{h}) = C^N \exp\left[-\frac{\|\mathbf{h}\|^2_{\mathbf{H}(\mathbb{R}^N)}}{2}\right] d\mathbf{h} \tag{4.2.2}$$

and thereby come to the conclusion that (4.2.1) is the statement that $\mathcal{W}^{(N)}$ *is the standard Gaussian measure for* $\mathbf{H}(\mathbb{R}^N)$. (It should be remarked that, in the preceding statement, we have said *for* $\mathbf{H}(\mathbb{R}^N)$ instead of *on* $\mathbf{H}(\mathbb{R}^N)$ since we already know that $\mathbf{H}(\mathbb{R}^N)$ is invisible to $\mathcal{W}^{(N)}$.) In order to give some substance to this interpretation, suppose that we use it to guess what the *characteristic function of* $\mathcal{W}^{(N)}$ looks like. That is, we want to see whether (4.2.2) can be used to guess the value of

$$\widehat{\mathcal{W}^{(N)}}(\mathbf{h}) \equiv \int_{\mathfrak{P}(\mathbb{R}^N)} \exp\left[\sqrt{-1}\,(\boldsymbol{\psi},\mathbf{h})_{\mathbf{H}(\mathbb{R}^N)}\right] \mathcal{W}^{(N)}(d\boldsymbol{\psi})$$

for $\mathbf{h} \in \mathbf{H}(\mathbb{R}^N)$, where, at least for the moment, we ignore the problem of giving a rigorous meaning to $(\boldsymbol{\psi},\mathbf{h})_{\mathbf{H}(\mathbb{R}^N)}$ for $\boldsymbol{\psi}$'s which are not in $\mathbf{H}(\mathbb{R}^N)$. But, as soon as one poses the problem in this way, the answer is immediate: namely, by analogy with what we know about Gaussian measures in finite dimensions, we are compelled to guess that

$$\widehat{\mathcal{W}^{(N)}}(\mathbf{h}) = \exp\left[-\frac{\|\mathbf{h}\|^2_{\mathbf{H}(\mathbb{R}^N)}}{2}\right], \quad \mathbf{h} \in \mathbf{H}(\mathbb{R}^N). \tag{4.2.3}$$

With these heuristic preliminaries in place, we will now see what can be done to make mathematics out of them. From the point of view adopted by Segal's school, this means that we want to find a separable Banach space Θ which, on the one hand, is small enough that $\mathbf{H}(\mathbb{R}^N)$ is embedded as a dense subspace while, at the same time, it is large enough to support a measure for which (4.2.3) (properly interpreted) holds. To make all this more precise, we will use the following rather simple application of elementary Banach space theory.

4.2.4 Theorem. *Let* Θ *with norm* $\|\cdot\|_\Theta$ *be a separable, real Banach space, and use*

$$(\theta,\lambda) \in \Theta \times \Theta^* \longmapsto \langle \theta,\lambda\rangle \in \mathbb{R}$$

to denote the duality relation between Θ *and its dual space* Θ^*. *Then* $\mathcal{B}_\Theta$ *coincides with the* σ*-algebra generated by the maps* $\theta \in \Theta \longmapsto \langle\theta,\lambda\rangle$ *as* λ *runs over* Θ^*. *In particular, if, for* $\mu \in \mathbf{M}_1(\Theta)$, *we define its* **characteristic function** $\hat{\mu} : \Theta^* \longrightarrow \mathbb{C}$ *by*

$$\hat{\mu}(\lambda) = \int_\Theta \exp\left[\sqrt{-1}\,\langle\theta,\lambda\rangle\right] \mu(d\theta), \quad \lambda \in \Theta^*,$$

then $\hat{\mu}$ is a continuous function of weak convergence on Θ^*, and $\hat{\mu}$ uniquely determines μ in the sense that if ν is a second element of $\mathbf{M}_1(\Theta)$ and $\hat{\mu} = \hat{\nu}$ then $\mu = \nu$.*

Next, suppose that H is a separable Hilbert space which is continuously embedded as a dense subspace of Θ. Then, for each $\lambda \in \Theta^*$, there is a unique $h_\lambda \in H$ with the property that

$$\big(h, h_\lambda\big)_H = \langle h, \lambda \rangle, \quad h \in H;$$

the mapping $\lambda \in \Theta^* \longmapsto h_\lambda \in H$ is continuous from the weak topology on Θ^* into the weak topology on H; and $\{h_\lambda : \lambda \in \Theta^*\}$ is dense in H. Moreover, there is at most one $\mathcal{W}_H \in \mathbf{M}_1(\Theta)$ with the property that*

$$\widehat{\mathcal{W}_H}(\lambda) = \exp\left[-\frac{\|h_\lambda\|_H^2}{2}\right], \quad \lambda \in \Theta^*, \tag{4.2.5}$$

and in order for $\mathcal{W}_H$ to exist it is necessary that the inclusion mapping taking H into Θ be compact. Finally, if $\mathcal{W}_H$ exists, then there is a unique isometric mapping $h \in H \longmapsto \mathcal{I}(h) \in L^2(\mathcal{W}_H;\mathbb{R})$ with the property that

$$\big[\mathcal{I}(h_\lambda)\big](\theta) = \langle \theta, \lambda \rangle, \ \theta \in \Theta, \quad \text{for each } \lambda \in \Theta^*.$$

In fact, each $\mathcal{I}(h)$ is, under $\mathcal{W}_H$, an $\mathfrak{N}(0, \|h\|_H^2)$ random variable; and so, if

$$\mathcal{F}_S \equiv \sigma\big(\mathcal{I}(h) : h \in S\big) \quad \text{for nonempty } S \subseteq H,$$

then $\mathcal{F}_S$ is independent of $\mathcal{F}_T$ under $\mathcal{W}_H$ if and only if $S \perp T$ in H.

PROOF: Since it is clear that each of the maps $\theta \in \Theta \longmapsto \langle \theta, \lambda \rangle \in \mathbb{R}$ is continuous and therefore $\mathcal{B}_\Theta$-measurable, the first assertion will follow as soon as we show that the norm $\|\cdot\|_\Theta$ can be expressed a measurable function of these maps. But, because Θ is separable, we know that Θ^* is separable with respect to the weak* topology and therefore that we can find a sequence $\{\lambda_n\}_1^\infty \subseteq \Theta^*$ so that

$$\|\theta\|_\Theta = \sup_{n \in \mathbb{Z}^+} \langle \theta, \lambda_n \rangle, \quad \theta \in \Theta.$$

Turning to the properties of $\hat{\mu}$, note that its continuity with respect to weak* convergence is an immediate consequence of Lebesgue's Dominated Convergence Theorem. Furthermore, in view of the preceding, we will know that $\hat{\mu}$ completely determines μ as soon as we show that, for each $n \in \mathbb{Z}^+$ and $\Lambda = \big(\lambda_1, \dots, \lambda_n\big) \in \big(\Theta^*\big)^n$, $\hat{\mu}$ determines the marginal distribution $\mu_\Lambda \in \mathbf{M}_1(\mathbb{R}^N)$ of

$$\theta \in \Theta \longmapsto \Big(\langle \theta, \lambda_1 \rangle, \dots, \langle \theta, \lambda_n \rangle\Big) \in \mathbb{R}^n$$

under μ. But this is clear (cf. Lemma 2.2.8), since

$$\widehat{\mu_\Lambda}(\boldsymbol{\xi}) = \hat{\mu}\left(\sum_1^n \xi_m \lambda_m\right) \quad \text{for } \boldsymbol{\xi} \in \mathbb{R}^n.$$

Next, suppose that H is given. Because H is continuously embedded in Θ, there exists a $C \in (0, \infty)$ such that

$$\left|\langle h, \lambda\rangle\right| \le C\|h\|_H\|\lambda\|_{\Theta^*}, \quad h \in H.$$

Hence, because H is densely embedded in Θ, the Riesz Representation Theorem for Hilbert spaces guarantees both the existence and the uniqueness of h_λ. In fact, $\|h_\lambda\|_H \le C\|\lambda\|_{\Theta^*}$. In particular, if $\{\lambda_\alpha;\, \alpha \in I\}$ is a net in Θ^* which is weak* convergent to λ, then $\{h_{\lambda_\alpha} : \alpha \in I\}$ is a bounded subset of H and is therefore relatively compact in the weak topology. But, because h_λ is the only possible weak limit of any subnet of $\{h_{\lambda_\alpha} : \alpha \in I\}$, this means that the whole net converges to h_λ. Hence, we have now proved the required continuity property of $\lambda \in \Theta^* \longmapsto h_\lambda \in H$. As for the density of $L \equiv \{h_\lambda : \lambda \in \Theta^*\}$, suppose that $h \perp L$. We would then know that $\langle h, \lambda\rangle = 0$ for all $\lambda \in \Theta^*$, and so h would have to be 0, first as an element of Θ and therefore also as an element of H.

Now assume that $\mathcal{W}_H$ exists. To prove that the inclusion map must be compact, note that, because $\widehat{\mathcal{W}_H}$ is continuous with respect to the weak* topology, (4.2.5) implies that $\lambda \in \Theta^* \longmapsto \|h_\lambda\|_H \in \mathbb{R}$ must also be continuous with respect to the weak* topology. But, after combining this with the continuity statement derived in the preceding paragraph, this implies that $\lambda \in \Theta^* \longmapsto h_\lambda \in H$ is continuous from the weak* topology on Θ^* into the strong topology on H; and, therefore, $A \equiv \{h_\lambda : \|\lambda\|_{\Theta^*} \le 1\}$ is relatively compact with the respect to the strong topology on H. In particular, this means that if $\{g_\alpha : \alpha \in I\} \subseteq H$ converges weakly to 0 in H, then

$$\|g_\alpha\|_\Theta = \sup_{\|\lambda\|_{\Theta^*} \le 1} \langle g_\alpha, \lambda\rangle = \sup_{h \in A} \big(g_\alpha, h\big)_H \longrightarrow 0;$$

and so we have now proved that the embedding is a compact map.

Turning to the map $\mathcal{I}$, recall that, under $\mathcal{W}_H$, $\langle\,\cdot\,, \lambda\rangle$ is an $\mathfrak{N}(0, \|h_\lambda\|_H^2)$ random variable for each $\lambda \in \Theta^*$, and conclude that $\mathcal{I}$ is isometric on $\{h_\lambda : \lambda \in \Theta^*\}$. Hence, because $\{h_\lambda : \lambda \in \Theta^*\}$ is dense in H, the existence as well as the uniqueness of $\mathcal{I}$ are clear. In addition, if h is any element of H and we choose $\{\lambda_n\}_1^\infty$ so that $h_{\lambda_n} \longrightarrow h$ in $\mathbf{H}$, then

$$\mathbb{E}^{\mathcal{W}_H}\left[\exp\Big(\sqrt{-1}\,\xi\,\mathcal{I}(h)\Big)\right] = \lim_{n\to\infty} \widehat{\mathcal{W}_H}(\xi\, h_{\lambda_n}) = \exp\left[-\frac{\xi^2\,\|h\|_{H^2}^2}{2}\right], \quad \xi \in \mathbb{R},$$

and therefore $\mathcal{I}(h)$ is an $\mathfrak{N}(0, \|h\|_H^2)$ random variable under $\mathcal{W}_H$.

Given the preceding, the final assertion is based on a general fact about linear families of Gaussian random variables (cf. Exercise 4.2.39 below). Namely, for $g_1, \dots, g_m \in S$ and $h_1, \dots, h_n \in T$, we have that

$$\begin{aligned}
&\mathbb{E}^{\mathcal{W}_H}\left[\exp\Big(\sqrt{-1}\sum_{i=1}^m \xi_i\,\mathcal{I}(g_i)\Big)\,\exp\Big(\sqrt{-1}\sum_{j=1}^n \eta_j\,\mathcal{I}(h_j)\Big)\right]\\
&\quad = \widehat{\mathcal{W}_H}\left(\sum_{i=1}^m \xi_i\,g_i + \sum_{j=1}^n \eta_j h_j\right)\\
&\quad = \exp\left[-\frac{1}{2}\left\|\sum_{i=1}^m \xi_i\,g_i + \sum_{j=1}^n \eta_j h_j\right\|_H^2\right]\\
&\quad = \exp\left[-\frac{1}{2}\left\|\sum_{i=1}^m \xi_i\,g_i\right\|_H^2\right]\,\exp\left[-\frac{1}{2}\left\|\sum_{j=1}^n \eta_j\,h_j\right\|_H^2\right]\\
&\quad = \mathbb{E}^{\mathcal{W}_H}\left[\exp\Big(\sqrt{-1}\sum_{i=1}^m \xi_i\,\mathcal{I}(g_i)\Big)\right]\,\mathbb{E}^{\mathcal{W}_H}\left[\exp\Big(\sqrt{-1}\sum_{j=1}^n \eta_j\,\mathcal{I}(h_j)\Big)\right]
\end{aligned}$$

for all $(\xi_1, \dots, \xi_m) \in \mathbb{R}^m$ and $(\eta_1, \dots, \eta_n) \in \mathbb{R}^N$ if and only if $(g_i, h_j)_H = 0$ for all $1 \le i \le m$ and $1 \le j \le n$. Hence, if $\mathcal{F}_S$ is independent of $\mathcal{F}_T$, then certainly $S \perp T$. On the other hand, if $S \perp T$, then the preceding in conjunction with elementary Fourier analysis leads immediately to the statement that

$$\begin{aligned}
&\mathbb{E}^{\mathcal{W}_H}\Big[F\big(\mathcal{I}(g_1), \dots, \mathcal{I}(g_m)\big)\,G\big(\mathcal{I}(h_1), \dots, \mathcal{I}(h_n)\big)\Big]\\
&\qquad = \mathbb{E}^{\mathcal{W}_H}\Big[F\big(\mathcal{I}(g_1), \dots, \mathcal{I}(g_m)\big)\Big]\,\mathbb{E}^{\mathcal{W}_H}\Big[G\big(\mathcal{I}(h_1), \dots, \mathcal{I}(h_n)\big)\Big]
\end{aligned}$$

for all bounded measurable $F : \mathbb{R}^m \longrightarrow \mathbb{R}$ and $G : \mathbb{R}^n \longrightarrow \mathbb{R}$. □

If H is a separable Hilbert space which is embedded as a dense subspace of the separable Banach space Θ and if $\mathcal{W}_H \in \mathbf{M}_1(\Theta)$ satisfies (4.2.5), the triple $(H, \Theta, \mathcal{W}_H)$ is called an **abstract Wiener space**. As we will see (cf. Remark 4.2.29), although the Hilbert space H is *canonical*, the Banach space Θ is not. Nonetheless, L. Gross proved that every H admits a Θ on which there exists a $\mathcal{W}_H$ for which $(H, \Theta, \mathcal{W}_H)$ is an abstract Wiener space. Finally, the isometry $h \in H \longmapsto \mathcal{I}(h) \in L^2(\mathcal{W}_H)$ was introduced by Paley and Wiener and will be called the **Paley–Wiener map**.

4.2.6 Remark. The central issue being discussed in Theorem 4.2.4, as well as the paragraph which follows it, is that of understanding on what Θ's $\mathcal{W}_H$ can exist. In particular, when H is finite dimensional with dimension d, the whole issue disappears since we can then identify H with $\mathbb{R}^d$ and take $\mathcal{W}_H = {\gamma_1}^d$. That is, in this case there is no choice but to take $\Theta = H$. However, when H is infinite dimensional, the situation is entirely different. In fact, because bounded subsets of H are relatively compact if and only if H is finite dimensional, we know from Theorem 4.2.4 that *the only time when we can take* $\Theta = H$ *is when* H *is finite dimensional.* The problem is, of course, that although $\mathcal{W}_H$ always exists (cf. Exercise 4.2.30) on H as a *finitely additive* measure defined on the *algebra* of subsets generated by the maps $h \in H \longmapsto (h, g)_H \in \mathbb{R}$ as g runs over H, *when* H *is infinite dimensional,* $\mathcal{W}_H$ *cannot be extended to* $\mathcal{B}_H$ *as a countably additive measure.*

At first this sort of issue may look a little unfamiliar and might be written off as the sort of pathology which one encounters only in infinite dimensional situations. However, this is not the case at all. Indeed, consider the problem of putting a translation invariant probability measure on the countable set $\mathbb{Q}_1$ of rational $q \in [0, 1)$. That is, suppose one attempts to construct a $\mu \in \mathbf{M}_1(\mathbb{Q}_1)$ with the property that μ is invariant with respect rational translation (i.e., addition) modulo 1. To this end, one might start by taking $\mathcal{A}$ to be the algebra over $\mathbb{Q}_1$ which is generated by the collection

$$\big\{[p, q) \cap \mathbb{Q}_1 : p,\, q \in \mathbb{Q}_1 \text{ with } p \le q\big\}.$$

It is then an elementary matter to see that μ exists as the one and only finitely additive measure on $\mathcal{A}$ such that $\mu\big([p, q)\big) = q - p$ for all $p,\, q \in \mathbb{Q}_1$ with $p \le q$. On the other hand, it is equally elementary to see that μ cannot be extended to $\sigma(\mathcal{A})$ (i.e., the set of all subsets of $\mathbb{Q}_1$) as a countably additive measure. In fact, because μ would have to assign measure 0 to each point, countable additivity would mean that $1 = \mu(\mathbb{Q}_1) = 0$. Hence, in this *very* finite dimensional setting, $\mathbb{Q}_1$ plays the rôle of H, the interval $[0, 1]$ plays that of Θ, and Lebesgue's measure is $\mathcal{W}_H$.

The notion of an abstract Wiener space provides us with a context in which to complete our interpretation of (4.2.3). However, we must first find an appropriate separable Banach $\Theta(\mathbb{R}^N)$; and, because we already know that $\mathcal{W}^{(N)}$ lives on $\mathfrak{P}(\mathbb{R}^N)$ (which is not itself a Banach space), we should look for a suitable subspace of $\mathfrak{P}(\mathbb{R}^N)$. Actually, because of Exercise 3.3.31, we need not look very far. Namely, set

$$\Theta(\mathbb{R}^N) = \left\{\boldsymbol{\theta} \in \mathfrak{P}(\mathbb{R}^N) : \boldsymbol{\theta}(0) = 0 \quad \text{and} \quad \lim_{t \to \infty} \frac{|\boldsymbol{\theta}(t)|}{t} = 0\right\},$$

and define

$$\|\boldsymbol{\psi}\|_{\Theta(\mathbb{R}^N)} = \sup_{t \in [0,\infty)} \frac{|\boldsymbol{\psi}(t)|}{1 + t} \in [0, \infty] \quad \text{for all } \boldsymbol{\psi} \in \mathfrak{P}(\mathbb{R}^N).$$

At the same time, let $\Lambda(\mathbb{R}^N)$ be the space of $\mathbb{R}^N$-valued Borel measures $\boldsymbol{\lambda}$ on $(0,\infty)$ satisfying

$$(4.2.7) \qquad \|\boldsymbol{\lambda}\|_{\Lambda(\mathbb{R}^N)} \equiv \int_{(0,\infty)} (1+t)\,|\boldsymbol{\lambda}|(dt) < \infty,$$

where $|\boldsymbol{\lambda}|$ denotes the total variation measure determined by $\boldsymbol{\lambda}$. Finally we will say that $\boldsymbol{\lambda} \in \Lambda(\mathbb{R}^N)$ is **simple** if there exist an $n \in \mathbb{Z}^+$, $\mathbf{v}_1, \dots, \mathbf{v}_n \in \mathbb{R}^N$, and $0 < t_1 < \cdots < t_n$ such that

$$\boldsymbol{\lambda} = \sum_{m=1}^{n} \mathbf{v}_m \delta_{t_m}.$$

4.2.8 Lemma. *The map*

$$\boldsymbol{\psi} \in \mathfrak{P}(\mathbb{R}^N) \longmapsto \|\boldsymbol{\psi}\|_{\Theta(\mathbb{R}^N)} \in [0,\infty]$$

is lower semicontinuous, and the pair $\big(\Theta(\mathbb{R}^N), \|\cdot\|_{\Theta(\mathbb{R}^N)}\big)$ is a separable Banach space which is continuously embedded as a dense, measurable subset of $\mathfrak{P}(\mathbb{R}^N)$. In particular, $\mathcal{B}_{\Theta(\mathbb{R}^N)}$ coincides with $\mathcal{B}_{\mathfrak{P}(\mathbb{R}^N)}[\Theta(\mathbb{R}^N)] = \{A \cap \Theta(\mathbb{R}^N) : A \in \mathcal{B}_{\mathfrak{P}(\mathbb{R}^N)}\}$; and the dual space $\Theta(\mathbb{R}^N)^*$ of $\Theta(\mathbb{R}^N)$ can be identified with the $\Lambda(\mathbb{R}^N)$ via the duality relation given by

$$\langle \boldsymbol{\theta}, \boldsymbol{\lambda} \rangle \equiv \int_{(0,\infty)} \boldsymbol{\theta}(t) \cdot \boldsymbol{\lambda}(dt), \quad \boldsymbol{\theta} \in \Theta(\mathbb{R}^N),$$

in which case (cf. (4.2.7)) $\|\boldsymbol{\lambda}\|_{\Lambda(\mathbb{R}^N)}$ is the norm of $\boldsymbol{\lambda}$ as an element of $\Theta(\mathbb{R}^N)^*$. In addition, $\mathbf{H}(\mathbb{R}^N)$ is continuously embedded as a dense, measurable subset of $\Theta(\mathbb{R}^N)$; and, if $\boldsymbol{\lambda} \in \Lambda(\mathbb{R}^N)$, then, for all $\boldsymbol{\lambda} \in \Lambda(\mathbb{R}^N)$:

$$(4.2.9) \qquad \mathbf{h}_{\boldsymbol{\lambda}}(t) = \int_0^t \boldsymbol{\lambda}\big([s,\infty)\big)\,ds, \quad t \in [0,\infty),$$

is the unique element of $\mathbf{H}(\mathbb{R}^N)$ satisfying

$$(4.2.10) \qquad \langle \mathbf{h}, \boldsymbol{\lambda} \rangle = \big(\mathbf{h}, \mathbf{h}_{\boldsymbol{\lambda}}\big)_{\mathbf{H}(\mathbb{R}^N)}, \quad \mathbf{h} \in \mathbf{H}(\mathbb{R}^N).$$

Finally, for each $\boldsymbol{\lambda} \in \Lambda(\mathbb{R}^N)$, there is a sequence $\{\boldsymbol{\lambda}_n\}_1^\infty$ of simple elements of $\Lambda(\mathbb{R}^N)$ with the properties that $\boldsymbol{\lambda}_n \longrightarrow \boldsymbol{\lambda}$ in the weak topology and $\mathbf{h}_{\boldsymbol{\lambda}_n} \longrightarrow \mathbf{h}_{\boldsymbol{\lambda}}$ (strongly) in $\mathbf{H}(\mathbb{R}^N)$.*

PROOF: To see that $\|\cdot\|_{\Theta(\mathbb{R}^N)}$ is lower semicontinuous on $\mathfrak{P}(\mathbb{R}^N)$, simply notice that

$$F_n(\boldsymbol{\psi}) \equiv \sup_{t\in[0,n]} \frac{|\boldsymbol{\psi}(t)|}{1+t} \nearrow \|\boldsymbol{\psi}\|_{\Theta(\mathbb{R}^N)} \quad \text{as} \quad n \to \infty,$$

and observe that F_n is continuous for each $n \in \mathbb{Z}^+$. Hence, $\|\cdot\|_{\Theta(\mathbb{R}^N)}$ is lower semicontinuous and

$$\Theta(\mathbb{R}^N) = \left\{\boldsymbol{\psi} \in \mathfrak{P}(\mathbb{R}^N) : \boldsymbol{\psi}(0) = \mathbf{0} \text{ and } \|\boldsymbol{\psi}\|_{\Theta(\mathbb{R}^N)} < \infty\right\} \in \mathcal{B}_{\mathfrak{P}(\mathbb{R}^N)}.$$

Moreover, there can be no doubt that the inclusion map from $\Theta(\mathbb{R}^N)$ into $\mathfrak{P}(\mathbb{R}^N)$ is continuous or that $\Theta(\mathbb{R}^N)$ is dense in $\mathfrak{P}(\mathbb{R}^N)$.

In order to analyze the space $\left(\Theta(\mathbb{R}^N), \|\cdot\|_{\Theta(\mathbb{R}^N)}\right)$, define

$$F : \Theta(\mathbb{R}^N) \longrightarrow C_0\left(\mathbb{R}; \mathbb{R}^N\right) \equiv \left\{f \in C\left(\mathbb{R}; \mathbb{R}^N\right) : \lim_{|s| \to \infty} |f(s)| = 0\right\}$$

by

$$\left[F(\boldsymbol{\theta})\right](s) = \frac{\boldsymbol{\theta}\left(e^s\right)}{1 + e^s}, \qquad s \in \mathbb{R}.$$

As is well-known, $C_0\left(\mathbb{R}; \mathbb{R}^N\right)$ with the uniform norm is a separable Banach space; and it is obvious that F is an isometry from $\Theta(\mathbb{R}^N)$ onto $C_0\left(\mathbb{R}; \mathbb{R}^N\right)$. Moreover, by the Riesz Representation Theorem for $C_0\left(\mathbb{R}; \mathbb{R}^N\right)$, one knows that the dual of $C_0\left(\mathbb{R}; \mathbb{R}^N\right)$ is isometric to the space $\mathbf{M}\left(\mathbb{R}; \mathbb{R}^N\right)$ of totally finite, $\mathbb{R}^N$-valued measures on $\left(\mathbb{R}; \mathcal{B}_{\mathbb{R}}\right)$ with the norm given by total variation. Hence, the identification of $\Theta(\mathbb{R}^N)^*$ with $\Lambda(\mathbb{R}^N)$ reduces to the obvious interpretation of the adjoint map F^T as a mapping from $\mathbf{M}\left(\mathbb{R}; \mathbb{R}^N\right)$ onto $\Lambda(\mathbb{R}^N)$.

Finally, turning to the relationship between $\Theta(\mathbb{R}^N)$ and $\mathbf{H}(\mathbb{R}^N)$, first note that

$$\|\mathbf{h}\|_{\Theta(\mathbb{R}^N)} \leq \sup_{t \in [0,\infty)} \frac{t^{\frac{1}{2}}}{1+t} \|\mathbf{h}\|_{\mathbf{H}(\mathbb{R}^N)} \leq \|\mathbf{h}\|_{\mathbf{H}(\mathbb{R}^N)}, \qquad \mathbf{h} \in \mathbf{H}(\mathbb{R}^N).$$

Hence $\mathbf{H}(\mathbb{R}^N)$ is certainly continuously embedded in $\Theta(\mathbb{R}^N)$. Moreover, to see that it is dense, simply use the obvious fact that $C_c^\infty\left((0,\infty); \mathbb{R}^N\right)$ is already dense in $\Theta(\mathbb{R}^N)$; and to see that it is measurable as a subset of $\Theta(\mathbb{R}^N)$, note that the map $\boldsymbol{\theta} \in \Theta(\mathbb{R}^N) \longmapsto \Psi(\boldsymbol{\theta}) \in [0,\infty]$ given by

$$\Psi(\boldsymbol{\theta}) = \sup\left\{\int_{[0,\infty)} \left(\boldsymbol{\theta}(t), \dot{\boldsymbol{\varphi}}(t)\right)_{\mathbb{R}^N} dt \right.$$
$$\left. : \boldsymbol{\varphi} \in C_c\left((0,\infty); \mathbb{R}^N\right) \text{ and } \|\boldsymbol{\varphi}\|_{L^2([0,\infty);\mathbb{R}^N)} \leq 1\right\}$$

is a lower-semicontinuous function with the property that $\mathbf{H}(\mathbb{R}^N) = \left\{\boldsymbol{\theta} \in \Theta(\mathbb{R}^N) : \Psi(\boldsymbol{\theta}) < \infty\right\}$.

Finally, let $\boldsymbol{\lambda} \in \Lambda(\mathbb{R}^N) \longmapsto \mathbf{h}_{\boldsymbol{\lambda}} \in \mathbf{H}(\mathbb{R}^N)$ be defined as in (4.2.9). It is then an easy application of integration by parts to check that (4.2.10) holds. In addition, if $\boldsymbol{\lambda} \in \Lambda(\mathbb{R}^N)$ is given and

$$\boldsymbol{\lambda}_n \equiv \sum_{m=1}^{n^2} \boldsymbol{\lambda}\left(\left[\tfrac{m-1}{n}, \tfrac{m}{n}\right)\right) \delta_{\frac{m}{n}} \quad \text{for } n \in \mathbb{Z}^+,$$

then it is clear that $\{\boldsymbol{\lambda}_n\}_1^\infty$ is weak* convergent to $\boldsymbol{\lambda}$ while $\mathbf{h}_{\boldsymbol{\lambda}_n} \longrightarrow \mathbf{h}_{\boldsymbol{\lambda}}$ in $\mathbf{H}(\mathbb{R}^N)$. □

We now know that $\mathbf{H}(\mathbb{R}^N)$ is embedded in $\Theta(\mathbb{R}^N)$ as a dense subspace. At the same time, $\Theta(\mathbb{R}^N)$ is a measurable subset of $\mathfrak{P}(\mathbb{R}^N)$, and, by Exercise 3.3.31, it has $\mathcal{W}^{(N)}$-measure 1. Thus, all that remains is to check (4.2.5).

4.2.11 Theorem. *For any $\mu \in \mathbf{M}_1\left(\mathfrak{P}(\mathbb{R}^N)\right)$, the following are equivalent:*

(i) $\mu = \mathcal{W}^{(N)}$.

(ii) *For every $n \in \mathbb{Z}^+$, $\mathbf{v}_1, \dots, \mathbf{v}_n \in \mathbb{R}^N$, and $0 < t_1 < \cdots < t_n < \infty$, the random variable*

$$\psi \in \mathfrak{P}(\mathbb{R}^N) \longmapsto \sum_{m=1}^{n} \left(\mathbf{v}_m, \psi(t_m)\right)_{\mathbb{R}^N} \in \mathbb{R}$$

under μ is Gaussian with mean-value 0 and variance

$$\sum_{k,\,\ell=1}^{n} t_k \wedge t_\ell \left(\mathbf{v}_k, \mathbf{v}_\ell\right)_{\mathbb{R}^N}.$$

(iii) $\mu\left(\Theta(\mathbb{R}^N)\right) = 1$ *and, for every $\boldsymbol{\lambda} \in \Lambda(\mathbb{R}^N)$, the random variable $\boldsymbol{\theta} \in \Theta(\mathbb{R}^N) \longmapsto \langle \boldsymbol{\theta}, \boldsymbol{\lambda} \rangle \in \mathbb{R}$ under μ is Gaussian with mean-value 0 and variance*

$$\iint_{(0,\infty)^2} s \wedge t\, \boldsymbol{\lambda}(ds) \cdot \boldsymbol{\lambda}(dt).$$

In particular, if we restrict $\mathcal{W}^{(N)}$ to $\Theta(\mathbb{R}^N)$, then

$$\widehat{\mathcal{W}^{(N)}}(\boldsymbol{\lambda}) = \exp\left[-\frac{\|\mathbf{h}_{\boldsymbol{\lambda}}\|^2_{\mathbf{H}(\mathbb{R}^N)}}{2}\right] \quad \text{for all } \boldsymbol{\lambda} \in \Lambda(\mathbb{R}^N). \tag{4.2.12}$$

PROOF: We begin by showing that

$$\iint_{[0,\infty)^2} s \wedge t\, \boldsymbol{\lambda}(ds) \cdot \boldsymbol{\lambda}(dt) = \int_{[0,\infty)} \left|\boldsymbol{\lambda}((s,\infty))\right|^2 ds = \|\mathbf{h}_{\boldsymbol{\lambda}}\|^2_{\mathbf{H}(\mathbb{R}^N)} \tag{4.2.13}$$

for every $\boldsymbol{\lambda} \in \Lambda(\mathbb{R}^N)$, and, from the definition of $\mathbf{h}_{\boldsymbol{\lambda}}$ in (4.2.9), this comes down to checking the first equality. But

$$\begin{aligned}\iint\limits_{[0,\infty)^2} s \wedge t\, \boldsymbol{\lambda}(ds) \cdot \boldsymbol{\lambda}(dt) &= \iint\limits_{s \le t} s\, \boldsymbol{\lambda}(ds) \cdot \boldsymbol{\lambda}(dt) + \iint\limits_{t<s} t\, \boldsymbol{\lambda}(ds) \cdot \boldsymbol{\lambda}(dt) \\ &= \int_{[0,\infty)} \left(\int_0^t s\, \boldsymbol{\lambda}(ds) \right) \cdot \boldsymbol{\lambda}(dt) + \int_{[0,\infty)} t \boldsymbol{\lambda}((t,\infty)) \cdot \boldsymbol{\lambda}(dt) \\ &= \int_{[0,\infty)} \left(\int_0^t \boldsymbol{\lambda}((s,\infty))\, ds \right) \cdot \boldsymbol{\lambda}(dt) = \int_{[0,\infty)} \left| \boldsymbol{\lambda}(t,\infty)) \right|^2 dt,\end{aligned}$$

where, at the end, we have integrated by parts twice.

To see that (**i**) $\Longrightarrow$ (**ii**), first observe (cf. Theorem 3.3.20) that $\mu = \mathcal{W}^{(N)}$ if and only if $\mu(\boldsymbol{\psi}(0) = \mathbf{0}) = 1$ and, for every $n \in \mathbb{Z}^+$ and $0 = t_0 < \cdots < t_n$, the functions $\boldsymbol{\psi} \in \mathfrak{P}(\mathbb{R}^N) \longmapsto \boldsymbol{\psi}(t_m) - \boldsymbol{\psi}(t_{m-1})$, $1 \le m \le n$, under μ are mutually independent $\mathbb{R}^N$-valued, Gaussian random variables with mean-value $\mathbf{0}$ and covariance $(t_m - t_{m-1})\mathbf{I}$. Next, given $\mathbf{v}_1, \dots, \mathbf{v}_n \in \mathbb{R}^N$, define $\mathbf{V}_m = \sum_{k=m}^n \mathbf{v}_m$ for $1 \le m \le n$, use summation by parts to check that

$$\sum_{m=1}^n \big(\mathbf{v}_m, \boldsymbol{\psi}(t_m)\big)_{\mathbb{R}^N} = \big(\mathbf{V}_1, \boldsymbol{\psi}(0)\big)_{\mathbb{R}^N} + \sum_{m=1}^n \big(\mathbf{V}_m, \boldsymbol{\psi}(t_m) - \boldsymbol{\psi}(t_{m-1})\big)_{\mathbb{R}^N},$$

and conclude that all the functions of the form

$$\boldsymbol{\psi} \in \mathfrak{P}(\mathbb{R}^N) \longmapsto \sum_{m=1}^n \big(\mathbf{v}_m, \boldsymbol{\psi}(t_m)\big)_{\mathbb{R}^N}$$

under μ are Gaussian with mean-value 0 and (cf. (4.2.13) with $\boldsymbol{\lambda} = \sum_1^n \mathbf{v}_m \delta_{t_m}$) variance

$$\sum_{m=1}^n (t_m - t_{m-1}) \big|\mathbf{V}_m\big|^2 = \sum_{k,\,\ell=1}^n t_k \wedge t_\ell \big(\mathbf{v}_k, \mathbf{v}_\ell\big)_{\mathbb{R}^N}$$

if and only if $\mu = \mathcal{W}^{(N)}$.

Knowing (4.2.13) and the implication (**i**) $\Longrightarrow$ (**ii**), we have (4.2.12) immediately for simple $\boldsymbol{\lambda}$'s and then, by the last part of Lemma 4.2.8, for all $\boldsymbol{\lambda} \in \Lambda(\mathbb{R}^N)$. Finally, in order to complete the proof, we use (4.2.13) and (4.2.12) to check that (**iii**) is equivalent to $\hat{\mu} = \widehat{\mathcal{W}^{(N)}}$, which, by Theorem 4.2.4, is equivalent to $\mu = \mathcal{W}^{(N)}$. $\square$

With Theorem 4.2.11, we have completed our interpretation of (4.2.3). Namely, we have shown that

$$\Big(\mathbf{H}(\mathbb{R}^N), \Theta(\mathbb{R}^N), \mathcal{W}_{\mathbf{H}(\mathbb{R}^N)}\Big) \quad \text{with } \mathcal{W}_{\mathbf{H}(\mathbb{R}^N)} \equiv \mathcal{W}^{(N)} \restriction \mathcal{B}_{\Theta(\mathbb{R}^N)}$$

is an abstract Wiener space; and it is in this sense that we have given meaning to (4.2.3). Moreover, because there can be no harm done by doing so, we will, whenever convenient, think of $\mathcal{W}^{(N)}$ as an element of $\mathbf{M}_1(\Theta(\mathbb{R}^N))$ rather than $\mathbf{M}_1(\mathfrak{P}(\mathbb{R}^N))$; thereby making the identification $\mathcal{W}^{(N)} = \mathcal{W}_{\mathbf{H}(\mathbb{R}^N)}$.

As our first application of these considerations, we present the following general procedure for deducing invariance properties of $\mathcal{W}_H$.

4.2.14 Lemma. *Let $(H, \Theta, \mathcal{W}_H)$ be an abstract Wiener space, suppose that $\mathbf{L}$ is a continuous linear mapping from Θ into itself, and use $\mathbf{L}^T : \Theta^* \longrightarrow \Theta^*$ to denote the corresponding adjoint map. Then $\mathcal{W}_H$ is $\mathbf{L}$-invariant (i.e., $\mathcal{W}_H = \mathbf{L}^*\mathcal{W}_H$) if and only if*

$$\left\|h_{\mathbf{L}^T\lambda}\right\|_H = \|h_\lambda\|_H \quad \text{for all } \lambda \in \Theta^*.$$

In particular, if $\mathbf{L}$ is a bounded linear operator on $\Theta(\mathbb{R}^N)$, then $\mathcal{W}^{(N)}$ is $\mathbf{L}$-invariant if and only if

$$\iint\limits_{(0,\infty)^2} s \wedge t\, \mathbf{L}^T\boldsymbol{\lambda}(ds) \cdot \mathbf{L}^T\boldsymbol{\lambda}(dt)$$
$$= \iint\limits_{(0,\infty)^2} s \wedge t\, \boldsymbol{\lambda}(ds) \cdot \boldsymbol{\lambda}(dt), \quad \boldsymbol{\lambda} \in \Lambda(\mathbb{R}^N).$$

PROOF: Because $\mathcal{W}_H$ is uniquely determined by (4.2.5), there is nothing to do. □

It should be recognized that Lemma 4.2.14 represents a vast generalization of the rotation invariance of $\mathcal{W}^{(N)}$ which was developed in Exercise 3.3.28. Indeed, both the rotation invariance presented there as well as the scaling and incremental time-shift invariance described at the beginning of Section 4.1 (cf. (4.1.1) and Lemma 4.1.4) can all be seen as instances of the general result in the last part of Lemma 4.2.14; and another (perhaps more intriguing) example is the subject of Exercise 4.2.31 below. However, rather than pursuing the consequences of Lemma 4.2.14 further here, we will next apply Theorem 4.2.11 to obtain a famous *quasi-invariance* result about the way in which $\mathcal{W}^{(N)}$ transforms under translation in the directions of $\mathbf{H}(\mathbb{R}^N)$. Since, like the one in Lemma 4.2.14, the result applies equally well to an arbitrary abstract Wiener space, we will formulate it in the abstract setting. Thus, let $(H, \Theta, \mathcal{W}_H)$ be an abstract Wiener space, and, for each $h \in H$, define the translation operator $T_h : \Theta \longrightarrow \Theta$ by $T_h\theta = \theta + h$, $\theta \in \Theta$.

4.2.15 Lemma (Cameron & Martin). *Referring to the preceding, let $h \in H$ be given, define $\mathcal{I}(h) \in L^2(\mathcal{W}_H)$ as in the last part of Theorem 4.2.4, and set*

$$R_h(\theta) = \exp\left[\left[\mathcal{I}(h)\right](\theta) - \tfrac{\|h\|_H^2}{2}\right] \quad \text{for } \theta \in \Theta. \tag{4.2.16}$$

Then, for all $F \in B(\Theta;\mathbb{R})$,

$$\mathbb{E}^{\mathcal{W}_H}\big[F \circ T_h\big] = \mathbb{E}^{\mathcal{W}_H}\big[F\, R_h\big], \quad h \in H.$$

In other words, for every $h \in H$, ${T_h}^*\mathcal{W}_H$ *is absolutely continuous with respect to* $\mathcal{W}_H$, *and* R_h *is its Radon–Nikodym derivative. (Cf. Exercise 5.2.38 for the converse statement.)*

PROOF: We first note that, because $\mathcal{I}(h)$ is an $\mathfrak{N}\big(0, \|h\|_H^2\big)$ random variable under $\mathcal{W}_H$,

$$A \in \mathcal{B}_\Theta \longmapsto \mu(A) \equiv \mathbb{E}^{\mathcal{W}_H}\big[R_h,\, A\big] \in \mathbb{R}$$

determines an element of $\mathbf{M}_1(\Theta)$. Hence, all we have to do is check that $\hat{\mu} = \widehat{{T_h}^*\mathcal{W}_H}$. But, clearly,

$$\widehat{{T_h}^*\mathcal{W}_H}(\lambda) = \exp\Big[\sqrt{-1}\,\langle h, \lambda\rangle - \tfrac{1}{2}\|h_\lambda\|_H^2\Big], \quad \lambda \in \Theta^*.$$

At the same time, for any $g,\, h \in H$,

$$\mathbb{E}^{\mathcal{W}_H}\Big[\exp\big(\zeta\mathcal{I}(g) + z\mathcal{I}(h)\big)\Big] = \exp\left[\frac{\zeta^2\|g\|_H^2}{2} + \zeta\, z\big(g, h\big)_H + \frac{z^2\|h\|_H^2}{2}\right]$$

for all $\zeta,\, z \in \mathbb{C}$. (To see this, one reduces to the case in which $g \perp h$ and then applies analytic continuation in order to handle ζ's and z's which are not pure imaginary.) In particular, by taking $g = h_\lambda$, $\zeta = \sqrt{-1}$, and $z = 1$, we arrive at

$$\begin{aligned}\hat{\mu}(\lambda) &= \int_\Theta \exp\Big[\sqrt{-1}\,\langle \theta, \lambda\rangle + \big[\mathcal{I}(h)\big](\theta) - \tfrac{1}{2}\|h\|_H^2\Big]\, \mathcal{W}_H(d\theta)\\ &= \exp\Big[\sqrt{-1}\,\langle h, \lambda\rangle - \tfrac{1}{2}\|h_\lambda\|_H^2\Big] = \widehat{{T_h}^*\mathcal{W}_H}(\lambda). \quad \square\end{aligned}$$

4.2.17 Remark. It is significant that the formula in (4.2.16) is exactly what one would have to predict if one were to take the formula (cf. (4.2.2))

$$(*) \qquad \mathcal{W}_H(dh) = C \exp\left(-\frac{\|h\|_H^2}{2}\right) dh$$

seriously. Indeed, suppose that $h = h_\lambda$ for some $\lambda \in \Theta^*$. Then,

$$\begin{aligned}\int_\Theta F(\theta + h)\, \mathcal{W}_H(d\theta) &= C\int_H F(g+h)e^{-\frac{\|g\|_H^2}{2}}\, dg\\ &= C\int_H F(g)e^{-\frac{\|g-h\|_H^2}{2}}\, dg = C\int_H F(g)\exp\Big[(g,h)_H - \tfrac{1}{2}\|h\|_H^2\Big]\, \mathcal{W}_H(dg)\\ &= \int_\Theta F(\theta)\exp\Big[\langle\theta,\lambda\rangle - \tfrac{1}{2}\|h\|_H^2\Big]\, \mathcal{W}_H(d\theta) = \int_\Theta F(\theta)R_h(\theta)\, \mathcal{W}_H(d\theta).\end{aligned}$$

Hence, since $\mathcal{I}(h)$ is the extension of $\langle\,\cdot\,,\lambda\rangle$ for $h \in H$ which cannot be expressed as h_λ for any $\lambda \in \Theta^*$, (4.2.16) can be viewed as confirmation that (*) *is correct even if it makes no sense.* Actually, a useful metatheorem is that formulae which are guessed on the basis of (*) are likely to be true if, like the preceding, (*) enters only into intermediate steps and does not appear in the final formula.

However one chooses to view the formula in (4.2.16), it was discovered first (for $\mathcal{W}^{(1)}$) by Cameron and Martin, and, for this reason, (4.2.16) is known as the **Cameron–Martin formula** and H is called the **Cameron–Martin subspace**. In many ways, their discovery provides the single most compelling reason for considering Wiener's measure to be the Lebesgue measure on path-space. Indeed, when applied to $\mathcal{W}^{(N)}$, their result shows that Wiener's measure is just about as translation-invariant as a probability measure can be on an infinite dimensional space. In fact, no probability measure on an infinite dimensional space can be quasi-invariant under translation in all directions (we will see in Exercise 5.2.38 that translation of $\mathcal{W}^{(N)}$ in directions outside of $\mathbf{H}(\mathbb{R}^N)$ results in a measure which is singular to $\mathcal{W}^{(N)}$), and so it is significant that $\mathcal{W}^{(N)}$ is quasi-invariant in a dense set of directions. Furthermore, for the purpose of doing calculus on path-space, it is extremely important that the Radon–Nikodym derivative R_h be as tractable an expression as it turns out to be. For example, R_h is much more than integrable. In fact, because $\mathcal{I}(h)$ is $\mathfrak{N}(0,\|h\|_H^2)$, it is easy to check that

$$\|R_h\|_{L^p(\mathcal{W}_H)} = \exp\left[\tfrac{p-1}{2}\,\|h\|_H^2\right] \quad \text{for all } p \in [1,\infty).$$

As yet, we have not taken advantage of the relationship, developed in Theorem 4.2.4, between orthogonality and independence. However, as the following dramatically demonstrates, this relationship can be exploited to prove highly nontrivial results.

4.2.18 Theorem. *Given $T \in (0,\infty)$, define the* **pinned path** $\psi \in \mathfrak{P}(\mathbb{R}^N) \longmapsto \tilde{\psi}_T \in \mathfrak{P}(\mathbb{R}^N)$ *so that*

$$\tilde{\psi}_T(t) = \psi(t) - \tfrac{t\wedge T}{T}\psi(T), \quad t \in [0,\infty). \tag{4.2.19}$$

Then $\{\tilde{\psi}_T(t) : t \in [0,\infty)\}$ is independent of $\psi(T)$ under $\mathcal{W}^{(N)}$. In addition, $\tilde{\psi}_T \restriction [0,T]$ is **reversible** *under $\mathcal{W}^{(N)}$ in the sense that the processes*

$$\psi \in \mathfrak{P}(\mathbb{R}^N) \longmapsto \tilde{\psi}_T(\,\cdot\,) \restriction [0,T] \in C\big([0,T];\mathbb{R}^N\big)$$
$$\psi \in \mathfrak{P}(\mathbb{R}^N) \longmapsto \tilde{\psi}_T(T-\,\cdot\,) \restriction [0,T] \in C\big([0,T];\mathbb{R}^N\big)$$

have the same distribution under $\mathcal{W}^{(N)}$. Finally, if

$$\ell_{T,\mathbf{y}}(t) \equiv \tfrac{t\wedge T}{T}\mathbf{y}, \quad (t,\mathbf{y}) \in [0,\infty)\times\mathbb{R}^N,$$

then for every bounded, $\mathcal{B}_T$-measurable (cf. (4.1.3)) function F on $\mathfrak{P}(\mathbb{R}^N)$ and $\Gamma \in \mathcal{B}_{\mathbb{R}^N}$:

$$\int_{\{\psi:\psi(T)\in\Gamma\}} F(\psi)\,\mathcal{W}^{(N)}(d\psi) = \int_\Gamma \left(\int_{\mathfrak{P}(\mathbb{R}^N)} F\big(\tilde{\psi}_T + \ell_{T,\mathbf{y}}\big)\,\mathcal{W}^{(N)}(d\psi) \right) {\gamma_T}^N(d\mathbf{y}). \tag{4.2.20}$$

PROOF: In order to prove the independence statement, we think of $\mathcal{W}^{(N)}$ as $\mathcal{W}_{\mathbf{H}(\mathbb{R}^N)}$ and define (cf. (4.2.7))

$$t \in [0,\infty) \longmapsto \boldsymbol{\lambda}_{t,k} = \mathbf{e}_k \delta_t \in \Lambda(\mathbb{R}^N), \quad \text{for } 1 \le k \le N,$$

where $(\mathbf{e}_1, \dots, \mathbf{e}_N)$ is an orthonormal basis for $\mathbb{R}^N$. It is then an easy matter to check that

$$\sum_{k=1}^N \big[\mathcal{I}(\mathbf{h}_{\boldsymbol{\lambda}_{T,k}})\big](\boldsymbol{\theta}) = \big(\boldsymbol{\theta}(T), \mathbf{e}_k\big)_{\mathbb{R}^N} \qquad (\text{a.s.}, \mathcal{W}^{(N)}),$$

$$\mathbf{h}_{\boldsymbol{\lambda}_{T,k}} \perp S_T \equiv \left\{ \boldsymbol{\lambda}_{t,\ell} - \tfrac{t\wedge T}{T} \boldsymbol{\lambda}_{T,\ell} : t \in [0,\infty) \text{ and } 1 \le \ell \le N \right\},$$

and (cf. Theorem 4.2.4)

$$\mathcal{F}_{S_T} = \sigma\Big(\boldsymbol{\theta}(t) - \tfrac{t\wedge T}{T} \boldsymbol{\theta}(T) : t \in [0,\infty) \Big).$$

Hence, the required independence is now an easy application of Theorem 4.2.4. Moreover, given this independence result, (4.2.20) is derived by writing

$$\psi(t) = \tilde{\psi}_T(t) + \tfrac{t\wedge T}{T} \psi(T), \quad t \in [0,\infty),$$

then computing the left-hand side of (4.2.20) using the independence just established, and finally remembering that the distribution of $\psi \in \mathfrak{P}(\mathbb{R}^N) \longmapsto \psi(T) \in \mathbb{R}^N$ under $\mathcal{W}^{(N)}$ is ${\gamma_T}^N$.

We turn now to the proof of the reversibility statement, and for this purpose we again work via characteristic functions. Thus, let μ and ν on $C\big([0,T];\mathbb{R}^N\big)$ be the distributions under $\mathcal{W}^{(N)}$ of the processes $\psi \longmapsto \tilde{\psi}_T \restriction [0,T]$ and $\psi \longmapsto \tilde{\psi}_T(T - \cdot) \restriction [0,T]$. What we must show is that $\hat{\mu}(\boldsymbol{\lambda}) = \hat{\nu}(\boldsymbol{\lambda})$ for every finite, $\mathbb{R}^N$-valued measure $\boldsymbol{\lambda}$ on $[0,T]$. To this end, let such a $\boldsymbol{\lambda}$ be given, and note that

$$\langle \tilde{\psi}_T, \boldsymbol{\lambda} \rangle = \langle \psi, \tilde{\boldsymbol{\lambda}}_T \rangle \quad \text{where } \tilde{\boldsymbol{\lambda}}_T \equiv \boldsymbol{\lambda} - \sum_{k=1}^N \langle \ell_{T,\mathbf{e}_k}, \boldsymbol{\lambda} \rangle \mathbf{e}_k \delta_T.$$

Hence, by part (**iii**) of Theorem 4.2.11,

$$\hat{\mu}(\boldsymbol{\lambda}) = \widehat{\mathcal{W}^{(N)}}(\tilde{\boldsymbol{\lambda}}_T) = \exp\left[-\iint_{[0,T]^2}\left(s\wedge t - \tfrac{st}{T}\right)\boldsymbol{\lambda}(ds)\cdot\boldsymbol{\lambda}(dt)\right].$$

On the other hand, it is clear that $\hat{\nu}(\boldsymbol{\lambda}) = \hat{\mu}(\check{\boldsymbol{\lambda}})$, where $\check{\boldsymbol{\lambda}}$ is determined by

$$\langle \boldsymbol{\psi}, \check{\boldsymbol{\lambda}}\rangle = \langle \check{\boldsymbol{\psi}}, \boldsymbol{\lambda}\rangle, \quad \boldsymbol{\psi} \in C([0,T];\mathbb{R}^N), \quad \text{with } \check{\boldsymbol{\psi}}(\,\cdot\,) \equiv \boldsymbol{\psi}(T - \,\cdot\,).$$

Hence, the desired equality comes from the elementary fact that

$$s\wedge t - \tfrac{st}{T} = (T-s)\wedge(T-t) - \tfrac{(T-s)(T-t)}{T}, \quad (s,t) \in [0,T]. \quad \square$$

Before closing this section, we want to use the ideas developed here to explain one of the methods which Wiener himself proposed† for the construction of his measure $\mathcal{W}^{(N)}$; and, because of Exercise 3.3.28, we need only consider the case when $N = 1$. Furthermore, as we will see in the following lemma, it suffices to construct the restriction $\mathcal{W}_1$ of $\mathcal{W}$ to the $\Theta_1 \equiv \{\theta \restriction [0,1] : \theta \in \Theta(\mathbb{R})\}$ with the usual uniform norm.

4.2.21 Lemma. *Define* $\Psi : {\Theta_1}^{\mathbb{N}} \longrightarrow \mathfrak{P}(\mathbb{R})$ *by*

$$[\Psi(\mathbf{p})](t) = \sum_{k=0}^{\infty} p_k(T_k(t)), \quad (t,\mathbf{p}) \in [0,\infty) \times {\Theta_1}^{\mathbb{N}},$$

where $T_k(t) \equiv (t-k)^+ \wedge 1$ *for* $k \in \mathbb{N}$ *and* $t \in [0,\infty)$*. Then, for any* $\mu \in \mathbf{M}_1(\Theta_1)$*, the following are equivalent:*

(**i**) *For every* $f \in L^2([0,1];\mathbb{R})$,

$$p \in \Theta_1 \longmapsto (f,p)_{L^2([0,1];\mathbb{R})} \in \mathbb{R}$$

under μ *is Gaussian with mean-value* 0 *and variance*

$$\mathbf{V}(f) \equiv \iint_{[0,1]^2} s\wedge t\, f(s)\, f(t)\, dsdt.$$

(**ii**) $\mathcal{W} = \Psi^*\mu^{\mathbb{N}}$*, and therefore* $\mu = \mathcal{W}_1$.

† Wiener's article "Differential-space," *J. Math. & Phys.* **2**, contains at least three approaches to the construction of his measure. However, to the average mortal, none of these three seems to be complete.

PROOF: Obviously, if $\mathcal{W} = \Psi^*\mu^{\mathrm{N}}$, then $\mu = \mathcal{W}_1$ and therefore the property in (i) follows by taking $\lambda(dt) = \mathbf{1}_{(0,1]}(t)f(t)\,dt$ in part (iii) of Theorem 4.2.11. Thus, all that we have to do is check that (i) implies (ii).

Assume that μ has the property in (i). In order to prove that it has the property in (ii), we first check that, for $0 < t_1 < \cdots < t_n \le 1$ and $v_1, \dots, v_m \in \mathbb{R}$, $p \in \Theta_1 \longmapsto \sum_{m=1}^n v_m p(t_m)$ under μ is Gaussian with mean-value 0 and variance $\sum_{\ell,m=1}^n t_\ell \wedge t_m v_\ell v_m$. To this end, set

$$f_\epsilon = \epsilon^{-1} \sum_{m=1}^n \mathbf{v}_m \mathbf{1}_{(t_m-\epsilon,t_m]} \quad \text{for} \quad 0 < \epsilon < t_1,$$

note that $\left(f_\epsilon, p\right)_{L^2([0,1];\mathbb{R})} \longrightarrow \sum_1^n v_m p(t_m)$ for each $p \in \Theta_1$ while $\mathbf{V}(f_\epsilon) \longrightarrow \sum_1^n t_\ell \wedge t_m\, v_\ell v_m$ as $\epsilon \searrow 0$, and deduce

$$\int_{\Theta_1} \exp\left[\sqrt{-1}\sum_{m=1}^n v_m\, p(t_m)\right] \mu(dp) = \exp\left[-\tfrac{1}{2}\sum_{\ell,m=1}^n t_\ell \wedge t_m\, v_\ell v_m\right].$$

Next, set $t_0 = 0$, apply the preceding to see that

$$\int_{\Theta_1} \exp\left[\sqrt{-1}\sum_{m=1}^n \xi_m \big(p(t_m) - p(t_{m-1})\big)\right] \mu(dp)$$
$$= \exp\left[-\frac{1}{2}\sum_{m=1}^n (t_m - t_{m-1})\xi_m^2\right]$$

for all $\xi_1, \dots, \xi_n \in \mathbb{R}$, and conclude that the functions

$$p \in \Theta_1 \longmapsto p(t_m) - p(t_{m-1}), \quad 1 \le m \le n,$$

are mutually independent, $\mathfrak{N}\big(0, t_m - t_{m-1}\big)$ random variables under μ.

In order to complete the proof that (i) $\Longrightarrow$ (ii), we must still check that, for all $0 = t_0 < t_1 < \cdots < t_n < \infty$, the functions

$$\mathbf{p} \in {\Theta_1}^{\mathrm{N}} \longmapsto \big[\Psi(\mathbf{p})\big](t_m) - \big[\Psi(\mathbf{p})\big](t_{m-1}), \quad 1 \le m \le n,$$

are mutually independent, Gaussian random variables under μ^{N} having mean-value 0 and variance $t_m - t_{m-1}$. But,

$$\big[\Psi(\mathbf{p})\big](t_m) - \big[\Psi(\mathbf{p})\big](t_{m-1}) = \sum_{k=0}^\infty \Big(p_k\big(T_k(t_m)\big) - p_k\big(T_k(t_{m-1})\big)\Big),$$

and, by the preceding paragraph, the functions

$$\mathbf{p} \in \Theta_1{}^{\mathbb{N}} \longmapsto p_k\big(T_k(t_m)\big) - p_k\big(T_k(t_{m-1})\big), \quad 1 \le m \le n \text{ and } k \in \mathbb{N},$$

are mutually independent, $\mathfrak{N}\big(0, T_k(t_m) - T_k(t_{m-1})\big)$ random variables under $\mu^{\mathbb{N}}$. Hence, since

$$t_m - t_{m-1} = \sum_{k=0}^{\infty}\Big(T_k(t_m) - T_k(t_{m-1})\Big), \quad \text{for each } 1 \le m \le n,$$

the required conclusion is obvious. □

We will now explain how Wiener went about constructing $\mathcal{W}_1$ on Θ_1; and, in order to understand Wiener's idea, we will not only assume for the moment that $\mathcal{W}_1$ exists but will even pretend that it lives on

$$\mathbf{H}_1 \equiv \big\{h \restriction [0,1] : h \in \mathbf{H}(\mathbb{R})\big\}.$$

Next, we choose an orthonormal basis $\{h_n\}_0^\infty$ for $\mathbf{H}_1$ and express the path p in terms of this basis:

$$p \sim \sum_{k=0}^{\infty} \big(p, h_k\big)_{\mathbf{H}} h_k. \tag{4.2.22}$$

Of course, because $\mathcal{W}_1$ does not really even see $\mathbf{H}_1$, we should be somewhat wary about the interpretation of "$\sim$" in (4.2.22): *it certainly cannot mean that the convergence is taking place in* $\mathbf{H}_1$. On the other hand, the interpretation of $\big(p, h_k\big)_{\mathbf{H}_1}$ is easy; namely, we adopt (cf. Theorem 4.2.4)

$$\big(p, h_k\big)_{\mathbf{H}_1} = \big[\mathcal{I}(h_k)\big](p)$$

as its meaning and therefore replace (4.2.22) by

$$p \sim \sum_{k=0}^{\infty} \big[\mathcal{I}(h_k)\big](p)\, h_k. \tag{4.2.23}$$

In particular, if the convergence on the right-hand side of (4.2.23) were taking place in Θ_1 for $\mathcal{W}_1$-almost every $p \in \Theta_1$, one would then have a representation of $\mathcal{W}_1$ as the image of $\gamma_1{}^{\mathbb{N}}$.

Wiener's idea is to turn the preceding line of reasoning around. That is, he started with the measure $\gamma_1{}^{\mathbb{N}}$ on $\mathbb{R}^{\mathbb{N}}$ and, by making a judicious choice of orthonormal basis $\{h_n\}_0^\infty$, he was able to show that, for $\gamma_1{}^{\mathbb{N}}$-almost every $\mathbf{x} \in \mathbb{R}^{\mathbb{N}}$, the sequence of functions

$$\big[\Sigma_n(\mathbf{x})\big](t) \equiv \sum_{k=0}^{n} x_k\, h_k(t) \tag{4.2.24}$$

converges in Θ_1 (*not* $\mathbf{H}_1$) to some $\Sigma(\mathbf{x})$. In fact, Wiener chose to take

$$(4.2.25)\qquad h_0(t) = t \wedge 1 \quad \text{and} \quad h_n(t) = \frac{\sqrt{2}\,\sin\bigl(n\pi(t\wedge 1)\bigr)}{n\pi} \quad \text{for } n \in \mathbb{Z}^+.$$

That this constitutes an orthonormal basis in $\mathbf{H}_1$ is equivalent to the well-known fact that

$$\{1\} \cup \bigl\{\sqrt{2}\cos(n\pi\,\cdot\,) : n \in \mathbb{Z}^+\bigr\}$$

is an orthonormal basis in $L^2\bigl([0,1];\mathbb{R}\bigr)$. Moreover, this choice has the advantage that it facilitates computations in $L^2\bigl([0,1];\mathbb{R}\bigr)$. In fact, had Wiener been trying to show that $\mathcal{W}_1$ lives on $L^2\bigl([0,1];\mathbb{R}\bigr)$ instead of Θ_1, he would have been done. Indeed, when the h_n's are given by (4.2.25), it is clear that

$$\sup_{n>m} \bigl\|\Sigma_n(\mathbf{x}) - \Sigma_m(\mathbf{x})\bigr\|^2_{L^2([0,1];\mathbb{R})} = \frac{2}{\pi^2}\sum_{k=m+1}^{\infty} \frac{x_k^2}{k^2},$$

and therefore that

$$\int_{\mathbb{R}^{\mathbb{N}}} \sup_{n>m} \bigl\|\Sigma_n(\mathbf{x}) - \Sigma_m(\mathbf{x})\bigr\|^2_{L^2([0,1];\mathbb{R})}\, \gamma_1{}^{\mathbb{N}}(d\mathbf{x}) \longrightarrow 0 \quad \text{as } m \to \infty.$$

In other words, there is no problem about convergence in $L^2\bigl([0,1];\mathbb{R}\bigr)$: as elements of $L^2\bigl([0,1];\mathbb{R}\bigr)$, the $\Sigma_n(\mathbf{x})$'s converge both $\gamma_1{}^{\mathbb{N}}$-almost surely as well as in $\gamma_1{}^{\mathbb{N}}$-square mean to a random variable $\mathbf{x} \longmapsto \Sigma(\mathbf{x}) \in L^2\bigl([0,1];\mathbb{R}\bigr)$. In particular, for every $f \in L^2\bigl([0,1];\mathbb{R}\bigr)$, $\mathbf{x} \longmapsto \bigl(\Sigma(\mathbf{x}), f\bigr)_{L^2([0,1];\mathbb{R})}$ is Gaussian under $\gamma_1{}^{\mathbb{N}}$ with mean-value 0 and (cf. part (**i**) of Lemma 4.2.21) variance

$$\begin{aligned}\int_{\mathbb{R}^{\mathbb{N}}} \bigl(\Sigma(\mathbf{x}), f\bigr)^2_{L^2([0,1];\mathbb{R})}\, \gamma_1{}^{\mathbb{N}}(d\mathbf{x}) &= \sum_{n=0}^{\infty} \bigl(f, h_n\bigr)^2_{L^2([0,1];\mathbb{R})} \\ &= \sum_{n=0}^{\infty} \bigl(h_f, h_n\bigr)^2_{\mathbf{H}_1} = \|h_f\|^2_{\mathbf{H}} = \mathbf{V}(f),\end{aligned}$$

where h_f is the element of $\mathbf{H}_1$ determined by

$$\dot{h}_f(t) = \int_{t\wedge 1}^{1} f(s)\, ds.$$

Hence, if μ on $L^2\bigl([0,1];\mathbb{R}\bigr)$ is the image $\Sigma^*\gamma_1{}^{\mathbb{N}}$ of $\gamma_1{}^{\mathbb{N}}$ under Σ, then μ is very nearly the measure described in (**i**) of Lemma 4.2.21; its only problem is that it lives on $L^2\bigl([0,1];\mathbb{R}\bigr)$ instead of Θ_1. Put into the language of the Segal school, what we have already shown is that

$$\Bigl(\mathbf{H}_1, L^2\bigl([0,1];\mathbb{R}\bigr), \Sigma^*\gamma_1{}^{\mathbb{N}}\Bigr)$$

is an abstract Wiener space; and what we still need to show is that $L^2([0,1];\mathbb{R})$ can be replaced by Θ_1. Equivalently, what remains is to improve the sense in which $\Sigma_n(\mathbf{x})$ tends to $\Sigma(\mathbf{x})$ from a statement about $L^2([0,1];\mathbb{R})$-convergence to one about uniform convergence.

As an harmonic analyst, Wiener viewed the problem just posed as a question about the convergence of trigonometric series with random coefficients; and his own solution to this problem remains one of the more baffling computations of classical harmonic analysis! Fortunately, we have the machinery to greatly simplify Wiener's proof. Thus, choose the orthonormal basis $\{h_n\}_0^\infty$ described in (4.2.25), and define $\mathbf{x} \in \mathbb{R}^{\mathbb{N}} \longmapsto \Sigma_n(\mathbf{x}) \in \Theta_1$ by (4.2.24). Because we already know that there exists a measurable $\mathbf{x} \in \mathbb{R}^{\mathbb{N}} \longmapsto \Sigma(\mathbf{x}) \in L^2([0,1];\mathbb{R})$ to which $\{\Sigma_n(\mathbf{x})\}_0^\infty$ converges in $L^2([0,1];\mathbb{R})$ for ${\gamma_1}^{\mathbb{N}}$-almost every $\mathbf{x} \in \mathbb{R}^{\mathbb{N}}$, it is sufficient for us to show that

$$\text{(4.2.26)} \qquad \sup_{n\in\mathbb{N}} \sup_{0\le s<t\le 1} \frac{\left|[\Sigma_n(\mathbf{x})](t) - [\Sigma_n(\mathbf{x})](s)\right|}{(t-s)^{\frac{1}{8}}} < \infty$$

for ${\gamma_1}^{\mathbb{N}}$-almost every $\mathbf{x} \in \mathbb{R}^{\mathbb{N}}$. Indeed, we will then know that, for ${\gamma_1}^{\mathbb{N}}$-almost every $\mathbf{x} \in \mathbb{R}^{\mathbb{N}}$, the sequence $\{\Sigma_n(\mathbf{x})\}_0^\infty$ is equicontinuous and therefore (since we already know $\Sigma_n(\mathbf{x}) \longrightarrow \Sigma(\mathbf{x})$ in $L^2([0,1];\mathbb{R})$ for ${\gamma_1}^{\mathbb{N}}$-almost every $\mathbf{x} \in \mathbb{R}^{\mathbb{N}}$) that

$$(t,\mathbf{x}) \in [0,1] \times \mathbb{R}^{\mathbb{N}} \longmapsto [\Sigma(\mathbf{x})](t) \in \mathbb{R}$$

can be modified in such a way that $\Sigma_n(\mathbf{x}) \longrightarrow \Sigma(\mathbf{x})$ in Θ_1 for ${\gamma_1}^{\mathbb{N}}$-almost every $\mathbf{x} \in \mathbb{R}^{\mathbb{N}}$. But, by exactly the same reasoning as we used in the derivation of Kolmogorov's criterion (Theorem 3.3.16) from Lemma 3.3.13, (4.2.26) will follow as soon as we check that

$$\text{(4.2.27)} \qquad \int_{\mathbb{R}^{\mathbb{N}}} \sup_{n\in\mathbb{Z}^+} \left|[\Sigma_n(\mathbf{x})](t) - [\Sigma_n(\mathbf{x})](s)\right|^4 {\gamma_1}^{\mathbb{N}}(d\mathbf{x}) \le C(t-s)^2,$$

for some $C \in (0,\infty)$ and all $s,\ t \in [0,1]$. To this end, we use (1.4.26) to see that

$$\begin{aligned} &\int_{\mathbb{R}^{\mathbb{N}}} \sup_{n\in\mathbb{Z}^+} \left|[\Sigma_n(\mathbf{x})](t) - [\Sigma_n(\mathbf{x})](s)\right|^4 {\gamma_1}^{\mathbb{N}}(d\mathbf{x}) \\ &\qquad \le 4 \sup_{n\in\mathbb{Z}^+} \int_{\mathbb{R}^{\mathbb{N}}} \left|[\Sigma_n(\mathbf{x})](t) - [\Sigma_n(\mathbf{x})](s)\right|^4 {\gamma_1}^{\mathbb{N}}(d\mathbf{x}). \end{aligned}$$

Furthermore, for each $n \in \mathbb{Z}^+$ and $0 \le s < t \le 1$, $\mathbf{x} \in \mathbb{R}^{\mathbb{N}} \longmapsto [\Sigma_n(\mathbf{x})](t) - [\Sigma_n(\mathbf{x})](s)$ under ${\gamma_1}^{\mathbb{N}}$ is a Gaussian random variable with mean-value 0 and variance

$$\frac{2}{\pi^2} \sum_{k=1}^{n} \Delta_k(s,t) \le \frac{2}{\pi^2} \sum_{k=1}^{\infty} \Delta_k(s,t)$$

where

$$\Delta_k(s,t) \equiv \frac{\big(\sin(k\pi t) - \sin(k\pi s)\big)^2}{k^2}.$$

Hence, we now see that

$$\int_{\mathbb{R}^{\mathbb{N}}} \sup_{n\in\mathbb{Z}^+} \Big|[\Sigma_n(\mathbf{x})](t) - [\Sigma_n(\mathbf{x})](s)\Big|^4 \gamma_1{}^{\mathbb{N}}(d\mathbf{x}) \le \frac{3}{\pi^4}\left(\sum_{k=1}^{\infty}\Delta_k(s,t)\right)^2.$$

Finally, by splitting the sum on the right according to whether

$$k \le (t-s)^{-1} \quad \text{or} \quad k > (t-s)^{-1},$$

it is easy to find a $C \in (0,\infty)$ for which (4.2.27) holds.

In view of the preceding considerations, we have now proved the following remarkable result.

4.2.28 Theorem (Wiener). *For each $n \in \mathbb{N}$, define $\mathbf{x} \in \mathbb{R}^{\mathbb{N}} \longmapsto \Sigma_n(\mathbf{x}) \in \Theta_1$ by (4.2.24), where the h_n's are given by (4.2.25). Then there is a measurable $\mathbf{x} \in \mathbb{R}^{\mathbb{N}} \longmapsto \Sigma(\mathbf{x}) \in \Theta_1$ such that $\Sigma_n(\mathbf{x}) \longrightarrow \Sigma(\mathbf{x})$ in Θ_1 for $\gamma_1{}^{\mathbb{N}}$-almost every $\mathbf{x} \in \mathbb{R}^{\mathbb{N}}$; and $\mathcal{W}_1 = \Sigma^*\gamma_1{}^{\mathbb{N}}$. (Cf. Theorem 5.3.32 and Remark 5.3.36 for more information.)*

4.2.29 Remark. The preceding discussion brings out a point which we alluded to earlier: the Banach space Θ of an abstract Wiener space *is not canonical.* Indeed, as we have just seen, when $H = \mathbf{H}_1$ there is no *a priori* reason dictated by the space $\mathbf{H}_1$ itself to take the associated Banach space to be Θ_1; in fact, for technical reasons, it would have been more convenient to take it to be $L^2\big([0,1];\mathbb{R}\big)$. On the other hand, if we were to settle for $L^2\big([0,1];\mathbb{R}\big)$ instead of Θ_1 we would have given up a great deal. For example, all sorts of functions which are obviously measurable on Θ_1 are not so clearly measurable (or even well-defined) on $L^2\big([0,1];\mathbb{R}\big)$. Thus, as a general rule, it is best to take the Banach space to be as close to the Hilbert space as possible. In this connection, there is no reason why we should have stopped at Θ_1, since, according to Theorem 4.1.6, we could have taken the subspace of Θ_1 consisting of those paths whose modulus of continuity is dominated by a multiple of the function ω described in that theorem.

Exercises

4.2.30 Exercise: Let H be a separable Hilbert space, and, for each $n \in \mathbb{Z}^+$ and subset $\{g_1,\dots,g_n\} \subseteq H$, let $\mathcal{A}(g_1,\dots,g_n)$ denote the σ-algebra over H generated by the mapping

$$h \in H \longmapsto \big((h,g_1)_H,\dots,(h,g_n)_H\big) \in \mathbb{R}^n.$$

Check that

$$\mathcal{A} = \bigcup \{\mathcal{A}(g_1, \dots, g_n) : n \in \mathbb{Z}^+ \text{ and } g_1, \dots, g_n \in H\}$$

is an algebra which generates $\mathcal{B}_H$.

(i) Show that there *always* exists a *finitely additive* $\mathcal{W}_H$ on $\mathcal{A}$ which is uniquely determined by the properties that it is σ-additive on $\mathcal{A}(g_1, \dots, g_n)$ for every $n \in \mathbb{Z}^+$ and $\{g_1, \dots, g_n\} \subseteq H$ and

$$\int_H \exp\left[\sqrt{-1}\,(h, g)_H\right] \mathcal{W}_H(dh) = \exp\left[-\frac{\|g\|_H^2}{2}\right], \quad g \in H.$$

(ii) Although Theorem 4.2.4 provides a proof that $\mathcal{W}_H$ admits no σ-additive extension to $\mathcal{B}_H$ unless H is finite dimensional (in which case no extension is necessary), there is a far more direct route to the same conclusion. Namely, by a simple computation with independent Gaussian random variables, show that if $\mathcal{W}_H$ admitted such an extension and if H is infinite dimensional, then

$$\begin{aligned}\int_{\mathbf{H}} \exp\left[-\|h\|_{\mathbf{H}}^2\right] \mu(dh) &= \lim_{n\to\infty} \prod_{m=1}^{n} \int_{\mathbf{H}} \exp\left[-(\mathbf{e}_m, h)_{\mathbf{H}}^2\right] \mu(dh) \\ &= \lim_{n\to\infty} 3^{-\frac{n}{2}} = 0,\end{aligned}$$

where $\{\mathbf{e}_n : n \in \mathbb{Z}^+\}$ is an orthonormal basis for $\mathbf{H}$. In other words, $\mathcal{W}_H$ would have to be concentrated at the origin, which is utter nonsense.

Obviously, what this argument is exploiting is the fact that in order to be invariant under the entire rotation group on the infinite dimensional H, it is impossible for $\mathcal{W}_H$ to be nearly concentrated on a compact subset of H.

4.2.31 Exercise: In this exercise we will prove yet another invariance property of $\mathcal{W}^{(N)}$. Namely, define $\mathfrak{I} : \Theta(\mathbb{R}^N) \longrightarrow \Theta(\mathbb{R}^N)$ so that

$$[\mathfrak{I}(\boldsymbol{\theta})](t) = \begin{cases} t\boldsymbol{\theta}\left(\frac{1}{t}\right) & \text{if} \quad t \in (0, \infty) \\ \mathbf{0} & \text{if} \quad t = 0 \end{cases}.$$

Show that $\mathfrak{I}$ is an isometry on both $\Theta(\mathbb{R}^N)$ and $\mathbf{H}(\mathbb{R}^N)$, and use Lemma 4.2.14 to derive the **Wiener time-inversion invariance property** $\mathfrak{I}^*\mathcal{W}^{(N)} = \mathcal{W}^{(N)}$.

4.2.32 Exercise: As an application of time-inversion invariance (cf. Exercise 4.2.31), prove that, for each $s \in [0, \infty)$,

$$\overline{\lim_{\substack{t\to s \\ t>0,\, t\neq s}}} \frac{|\boldsymbol{\theta}(t) - \boldsymbol{\theta}(s)|}{\beta(t-s)} = 1 \quad \text{for } \mathcal{W}^{(N)}\text{-almost every } \boldsymbol{\theta} \in \Theta(\mathbb{R}^N), \tag{4.2.33}$$

where

$$\beta(\delta) \equiv \sqrt{2|\delta| \log_{(2)} \tfrac{1}{|\delta|}} \quad \text{when} \quad |\delta| \le \tfrac{1}{9}.$$

For the reader who is concerned with what might look like an inconsistency between (4.2.33) and the Lévy's modulus of continuity which was mentioned after the proof of Theorem 4.1.6, we point out that (4.2.33) deals with only a *single time* s, whereas Lévy's result is a statement about *all* $s \in [0, \infty)$.

Hint: First note that, by (4.1.8) and time-inversion invariance,

$$\varlimsup_{\delta \searrow 0} \frac{|\boldsymbol{\theta}(\delta)|}{\beta(\delta)} = \varlimsup_{t \nearrow \infty} \frac{\big|[\mathfrak{I}(\boldsymbol{\theta})](t)\big|}{\sqrt{2t \log_{(2)} t}} = 1$$

for $\mathcal{W}^{(N)}$-almost every $\boldsymbol{\theta} \in \Theta(\mathbb{R}^N)$; and therefore (4.2.22) has been proved when $s = 0$. To handle general $s \in (0, \infty)$, one can first use scaling invariance (cf. (4.1.2)) to see that it is sufficient to consider the case when $s = 1$. Second, using Lemma 4.1.4 and the result already proved for $s = 0$, check that

$$\varlimsup_{t \searrow 1} \frac{|\boldsymbol{\theta}(t) - \boldsymbol{\theta}(1)|}{\beta(t-1)} = \varlimsup_{t \searrow 0} \frac{|\delta_1 \boldsymbol{\theta}(t)|}{\beta(t)} = 1$$

for $\mathcal{W}^{(N)}$-almost every $\boldsymbol{\theta} \in \Theta(\mathbb{R}^N)$. Finally, by another application of time-inversion invariance, conclude that

$$\begin{aligned}\varlimsup_{t \nearrow 1} \frac{|\boldsymbol{\theta}(t) - \boldsymbol{\theta}(1)|}{\beta(t-1)} &= \varlimsup_{t \searrow 1} \frac{\big|[\mathfrak{I}(\boldsymbol{\theta})](t) - t\boldsymbol{\theta}(1)\big|}{t\beta(1 - \frac{1}{t})} \\ &= \varlimsup_{t \searrow 1} \frac{|\boldsymbol{\theta}(t) - \boldsymbol{\theta}(1)|}{\beta(t-1)} = 1\end{aligned}$$

for $\mathcal{W}^{(N)}$-almost every $\boldsymbol{\theta} \in \Theta(\mathbb{R}^N)$.

4.2.34 Exercise: Take $T = 1$ in Theorem 4.2.18.

(i) Set

$$\mathbf{H}_0\big([0,1];\mathbb{R}^N\big) = \big\{\mathbf{h} \restriction [0,1] : \mathbf{h} \in \mathbf{H}(\mathbb{R}^N) \text{ and } \mathbf{h}(1) = \mathbf{0}\big\}$$

and

$$\Theta_0\big([0,1];\mathbb{R}^N\big) = \big\{\boldsymbol{\theta} \restriction [0,1] : \boldsymbol{\theta} \in \Theta(\mathbb{R}^N) \text{ and } \boldsymbol{\theta}(1) = \mathbf{0}\big\}$$

and note that each of these is a closed subspace of its parent space. Show that the triple

$$\Big(\mathbf{H}_0\big([0,1];\mathbb{R}^N\big), \Theta_0\big([0,1];\mathbb{R}^N\big), \mathcal{W}_{\mathbf{H}_0([0,1];\mathbb{R}^N)}\Big)$$

is an abstract Wiener space when $\mathcal{W}_{\mathbf{H}_0([0,1];\mathbb{R}^N)}$ is the distribution of (cf. (4.2.19))

$$\boldsymbol{\psi} \in \mathfrak{P}(\mathbb{R}^N) \longmapsto \tilde{\boldsymbol{\psi}}_1 \restriction [0,1] \in \Theta_0\big([0,1];\mathbb{R}^N\big)$$

under $\mathcal{W}^{(N)}$.

(ii) Take $N = 1$, and show that Theorem 4.2.28 can be interpreted as the statement that, for $\mathcal{W}$-almost every $\psi \in \mathfrak{P}(\mathbb{R})$,

$$\tilde{\psi}_1(t) = -2\sum_{n=1}^{\infty} a_n(\psi)\, \sin\bigl(n\pi(t)\bigr), \quad t \in [0,1], \tag{4.2.35}$$

where

$$a_n(\psi) \equiv \int_{[0,1]} \tilde{\psi}_1(t)\, \sin(n\pi t)\, dt, \quad n \in \mathbb{Z}^+$$

and the convergence is uniform. Note, in particular, that another proof of the reversibility assertion in Theorem 4.2.18 can be based on the representation in (4.2.35).

4.2.36 Exercise: Use Theorem 4.2.28 to see that the distribution of $\psi \in \mathfrak{P}(\mathbb{R}) \longmapsto \int_0^1 \psi(t)^2\, dt$ under $\mathcal{W}$ is the same as that of

$$\mathbf{x} \in \mathbb{R}^{\mathbb{N}} \longmapsto \frac{x_0^2}{3} + \frac{1}{\pi^2}\sum_{n=1}^{\infty} \frac{x_n^2 + \sqrt{8}\, x_n x_0}{n^2} \in [0,\infty]$$

under ${\gamma_1}^{\mathbb{N}}$ (the convergence on the right is both ${\gamma_1}^{\mathbb{N}}$-almost everywhere and in $L^p({\gamma_1}^{\mathbb{N}};\mathbb{R})$ for each $p \in [1,\infty)$.) In particular, for $\alpha \in (0,\infty)$, conclude that

$$\begin{aligned}
&\int_{\mathfrak{P}(\mathbb{R})} \exp\left[-\alpha \int_0^1 \psi(t)^2\, dt\right] \mathcal{W}(d\theta) \\
&= \lim_{n\to\infty} \int_{\mathbb{R}} e^{-\frac{\alpha x_0^2}{3}} \left(\prod_{m=1}^{n} \int_{\mathbb{R}} \exp\left[-\frac{\alpha\bigl(x_m^2 + \sqrt{8}\, x_m x_0\bigr)}{n^2\pi^2}\right] \gamma(dx_m)\right) \gamma(dx_0) \\
&= \left[\left(\prod_{n=1}^{\infty}\left(1 + \frac{2\alpha}{n^2\pi^2}\right)\right)\left(1 + 4\alpha\sum_{n=1}^{\infty}\frac{1}{n^2\pi^2 + 2\alpha}\right)\right]^{-\frac{1}{2}},
\end{aligned}$$

and, after recalling Euler's product formula

$$\sinh z = z\prod_{n=1}^{\infty}\left(1 + \frac{z^2}{n^2\pi^2}\right), \quad z \in \mathbb{C},$$

deduce that, for all $\alpha \in (0,\infty)$,

$$\int_{\mathfrak{P}(\mathbb{R})} \exp\left[-\alpha \int_0^1 \psi(t)^2\, dt\right] \mathcal{W}(d\theta) = \left[\cosh\sqrt{2\alpha}\right]^{-\frac{1}{2}}. \tag{4.2.37}$$

This is a famous calculation to which we will return in part **(ii)** of Exercise 4.3.52.

Hint: Use Euler's product formula to see that

$$\frac{d}{dt}\log\frac{\sinh t}{t} = 2t\sum_{n=1}^{\infty}\frac{1}{n^2\pi^2+t^2} \quad \text{for} \quad t\in\mathbb{R}.$$

4.2.38 Exercise: The reader might well have asked himself whether Wiener could have simplified his and our lives by choosing a different basis from the one described in (4.2.25). Indeed, because the convergence for which Wiener was looking is uniform and not L^2 convergence, it is not at all clear why trigonometric functions make the best choice. That he would have been better off with a cleverer basis was suggested by P. Lévy and implemented by Z. Ciesielski, whom we will follow in this exercise.

(i) Define the function $\eta:\mathbb{R}\longrightarrow\{-1,0,1\}$ so that

$$\eta = \begin{cases} 0 & \text{on} \quad (-\infty,0)\cup[1,\infty) \\ 1 & \text{on} \quad \left[0,\frac{1}{2}\right) \\ -1 & \text{on} \quad \left[\frac{1}{2},1\right). \end{cases}$$

Next, set $f_0 = \mathbf{1}_{[0,1)}$ and

$$f_{2^n+k}(t) = 2^{\frac{n}{2}}\eta\left(2^n t - k\right) \quad \text{for} \quad t\in[0,\infty),\ n\in\mathbb{N},\ \text{and } 0\le k<2^n.$$

Show that $\left\{f_m \restriction [0,1]\right\}_0^\infty$ is an orthonormal basis in $L^2\big([0,1];\mathbb{R}\big)$.

Hint: The orthonormality is similar to the argument given for the Rademacher functions in Section 1.1. To prove the completeness, first use a dimension counting argument to show that, for each $n\in\mathbb{N}$, $\left\{f_m : 0\le m<2^n\right\}$ has the same span as

$$\left\{\mathbf{1}_{\left[\frac{k}{2^n},\frac{k+1}{2^n}\right)} : 0\le k<2^n\right\}.$$

Once this has been done, suppose that $f\in L^2\big([0,1];\mathbb{R}\big)$ is perpendicular to the span of the f_m's, and conclude that $f\perp C\big([0,1];\mathbb{R}\big)$.

(ii) For each $m\in\mathbb{N}$, let h_m be the element of $\mathbf{H}_1$ with $\dot{h}_m = f_m$, where f_m is defined as in (i). Note that $\{h_m\}_0^\infty$ is an orthonormal basis in $\mathbf{H}_1$, and show that

$$\sup_{t\in[0,1]}\left|\sum_{k=0}^{2^n-1} a_k\, h_{2^n+k}(t)\right| \le 2^{-\frac{n}{2}}\max_{0\le k<2^n}|a_k| \le 2^{-\frac{n}{2}}\left(\sum_{k=0}^{2^n-1} a_k^4\right)^{\frac{1}{4}}$$

for all $\{a_k\}_0^{2^n-1}\subseteq\mathbb{R}$.

(iii) Let $(\Omega, \mathcal{F}, P)$ be a probability space and $\{X_m\}_0^\infty$ a sequence of random variables for which

$$M \equiv \sup_{m\in\mathbb{N}} \mathbb{E}^P\left[X_m^4\right]^{\frac{1}{4}} < \infty.$$

For $n \in \mathbb{N}$, define $\omega \in \Omega \longmapsto \Sigma_n(\omega) \in \Theta_1$ by

$$\left[\Sigma_n(\omega)\right](t) = \sum_{m=0}^{2^n-1} X_m(\omega)\, h_m(t), \quad t \in [0,1],$$

where the h_m's are those in **(ii)**. Using the estimate in **(ii)**, show that

$$\mathbb{E}^P\left[\sup_{n\geq m} \left\|\Sigma_n - \Sigma_m\right\|_{\Theta_1}\right] \leq M \sum_{n=m}^{\infty} 2^{-\frac{n}{4}} \longrightarrow 0 \quad \text{as } m \to \infty;$$

and conclude that there is a measurable $\Sigma : \Omega \longrightarrow \Theta_1$ to which $\left\{\Sigma_n(\omega)\right\}_0^\infty$ converges in Θ_1 (i.e., uniformly) for P-almost every $\omega \in \Omega$.

(iv) Continuing in the setting of **(iii)**, show that if the X_m's are mutually independent, $\mathfrak{N}(0,1)$ random variables under P, then $\mathcal{W}_1 = \Sigma^* P$.

4.2.39 Exercise: As we mentioned in the proof of Theorem 4.2.4, the last part of that theorem is a particular case of a general phenomenon. Namely, let $(\Omega, \mathcal{F}, P)$ be a probability space. A subset $\mathfrak{G} \subseteq L^2(P;\mathbb{R})$ is called a **centered Gaussian family** if $\mathfrak{G}$ is a linear subspace and all of its elements are Gaussian random variables with mean-value 0.

(i) Show that the closure $\overline{\mathfrak{G}}$ in $L^2(P;\mathbb{R})$ of a centered Gaussian family $\mathfrak{G}$ is again a centered Gaussian family and that $\overline{\mathfrak{G}}$ is closed under convergence in P-measure. Further, show that if L is a closed linear subspace of a centered Gaussian family $\mathfrak{G}$ and Π_L denotes the orthogonal projection operator onto L, then, for every $X \in \overline{\mathfrak{G}}$, $X - \Pi_L X$ is independent of L (i.e., of the σ-algebra $\sigma(L)$ over Ω generated by the elements of L).

(ii) Let $\Omega = \mathfrak{P}(\mathbb{R}^N)$ and $\mathcal{F} = \mathcal{B}_{\mathfrak{P}(\mathbb{R}^N)}$, and suppose that P is an element of $\mathbf{M}_1\left(\mathfrak{P}(\mathbb{R}^N)\right)$ with the properties that $\psi(t) \in L^2(P;\mathbb{R}^N)$ and $\mathbb{E}^P\left[\psi(t)\right] = \mathbf{0}$ for every $t \in [0,\infty)$. Next, define the **covariance function** $\mathbf{C}_P : [0,\infty)^2 \longrightarrow \mathbb{R}^N \otimes \mathbb{R}^N$ so that

$$\left[\mathbf{C}_P(s,t)\right](\boldsymbol{\xi},\boldsymbol{\eta}) = \mathbb{E}^P\left[\left(\boldsymbol{\xi},\psi(s)\right)_{\mathbb{R}^N}\left(\boldsymbol{\eta},\psi(t)\right)_{\mathbb{R}^N}\right]$$

for $(s,t) \in [0,\infty)^2$ and $(\boldsymbol{\xi},\boldsymbol{\eta}) \in \left(\mathbb{R}^N\right)^2$; and let $\mathfrak{G}(\mathbb{R}^N)$ denote the span of

$$\left\{\left(\boldsymbol{\xi},\psi(t)\right)_{\mathbb{R}^N} : t \in [0,\infty) \text{ and } \boldsymbol{\xi} \in \mathbb{R}^N\right\}.$$

One says that P is a **centered Gaussian measure** on $\mathfrak{P}(\mathbb{R}^N)$ if $\mathfrak{G}(\mathbb{R}^N)$ is a centered Gaussian family under P. Show that P is a centered Gaussian measure if and only if

$$\mathbb{E}^P\left[\exp\left(\sum_{m=1}^{n}\big(\boldsymbol{\xi}_m,\boldsymbol{\psi}(t_m)\big)_{\mathbb{R}^N}\right)\right]=\exp\left[-\frac{1}{2}\sum_{\ell,m=1}^{n}\big[\mathbf{C}_P(t_\ell,t_m)\big](\boldsymbol{\xi}_\ell,\boldsymbol{\xi}_m)\right]$$

for all $n\in\mathbb{Z}^+$, $0\le t_1<\cdots<t_n$, and $\boldsymbol{\xi}_1,\dots,\boldsymbol{\xi}_n\in\mathbb{R}^N$. In particular, conclude that if P and Q are centered Gaussian measures on $\mathfrak{P}(\mathbb{R}^N)$, then $P=Q$ if and only if $\mathbf{C}_P=\mathbf{C}_Q$. In other words, *if P is a centered Gaussian measure on $\mathfrak{P}(\mathbb{R}^N)$, then P is completely determined by its covariance function $\mathbf{C}_P$.*

(iii) Define $F:\Theta(\mathbb{R}^N)\longrightarrow C\big([0,1];\mathbb{R}^N\big)$ by

$$\big[F(\boldsymbol{\theta})\big](t)=\begin{cases}(1-t)\boldsymbol{\theta}\left(\frac{t}{1-t}\right) & \text{if } t\in[0,1)\\ \mathbf{0} & \text{if } t=0,\end{cases}$$

and use the preceding considerations to check that (cf. **(i)** in Exercise 4.2.34)

$$F^*\mathcal{W}^{(N)}=\mathcal{W}_{\mathbf{H}_0([0,1];\mathbb{R}^N)}.$$

4.2.40 Exercise: In this exercise, we will discuss another famous example of an abstract Wiener space which was introduced originally by Ornstein and Uhlenbeck.†

(i) Define the function $\mathbf{U}:\Theta(\mathbb{R}^N)\longrightarrow\Theta(\mathbb{R}^N)$ so that

$$\big[\mathbf{U}(\boldsymbol{\theta})\big](t)=\boldsymbol{\theta}(t)-\tfrac{1}{2}\int_0^t e^{-\frac{t-s}{2}}\boldsymbol{\theta}(s)\,ds\quad\text{for}\quad t\in[0,\infty),$$

and let $\mathcal{U}^{(N)}\in\mathbf{M}_1\big(\Theta(\mathbb{R}^N)\big)$ denote the distribution of $\boldsymbol{\theta}\in\Theta(\mathbb{R}^N)\longmapsto\mathbf{U}(\boldsymbol{\theta})\in\Theta(\mathbb{R}^N)$ under $\mathcal{W}^{(N)}$. Referring to the notation in Theorem 4.2.11, show that

$$\widehat{\mathcal{U}^{(N)}}(\boldsymbol{\lambda})=\exp\left[-\tfrac{1}{2}\iint_{[0,\infty)^2}u(s,t)\,\boldsymbol{\lambda}(ds)\cdot\boldsymbol{\lambda}(dt)\right],\quad \boldsymbol{\lambda}\in\Lambda(\mathbb{R}^N),\tag{4.2.41}$$

where

$$u(s,t)\equiv e^{-\frac{|t-s|}{2}}-e^{-\frac{|t+s|}{2}},\quad\text{for } (s,t)\in[0,\infty)^2.$$

† In their article "On the theory of Brownian motion," *Phys. Reviews* **36** (3), L.S. Ornstein & G. Uhlenbeck introduced this process in an attempt to reconcile some of the more disturbing properties of Wiener paths with physical reality.

Conclude that (cf. part (**ii**) of Exercise 4.2.39) $\mathcal{U}^{(N)}$ is a centered Gaussian measure on $\mathfrak{P}(\mathbb{R}^N)$ with covariance $\mathbf{C}(s,t) = u(s,t)\mathbf{I}_{\mathbb{R}^N}$; and, in particular, that, for any $t \in (0,\infty)$ and $f \in B(\mathbb{R};\mathbb{R})$

$$\mathbb{E}^{\mathcal{U}^{(N)}}\left[f(\boldsymbol{\theta}(t))\right] = \int_{\mathbb{R}} f(\mathbf{y})\,(\gamma_{1-e^{-t}})^N\,(d\mathbf{y}).$$

Hint: Note that $\widehat{\mathcal{U}^{(N)}}(\boldsymbol{\lambda}) = \widehat{\mathcal{W}^{(N)}}(\check{\boldsymbol{\lambda}})$, where

$$\check{\boldsymbol{\lambda}}(dt) = \boldsymbol{\lambda}(dt) - \left(\frac{1}{2}\int_{[t,\infty)} e^{-\frac{\tau-t}{2}}\,\boldsymbol{\lambda}(d\tau)\right) dt;$$

and apply (4.2.12) and (4.2.13).

(**ii**) Define the Hilbert norm

$$\|\mathbf{h}\|_{\mathbf{H}^U(\mathbb{R}^N)} = \sqrt{\|\mathbf{h}\|^2_{\mathbf{H}(\mathbb{R}^N)} + \tfrac{1}{4}\|\mathbf{h}\|^2_{L^2([0,\infty);\mathbb{R}^N)}}$$

on

$$\mathbf{H}^U(\mathbb{R}^N) \equiv \mathbf{H}(\mathbb{R}^N) \cap L^2\big([0,\infty);\mathbb{R}^N\big).$$

Note that $\mathbf{H}^U(\mathbb{R}^N)$ is continuously embedded as a dense, measurable subspace of $\Theta(\mathbb{R}^N)$, and show that, for each $\boldsymbol{\lambda} \in \Lambda(\mathbb{R}^N)$, the function (cf. the notation in (4.2.41))

$$t \in [0,\infty) \longmapsto \mathbf{h}^u_{\boldsymbol{\lambda}}(t) \equiv \int_{[0,\infty)} u(s,t)\,\boldsymbol{\lambda}(ds) \in \mathbb{R}$$

is the unique element of $\mathbf{H}^U(\mathbb{R}^N)$ with the property that

$$\langle \mathbf{h}, \boldsymbol{\lambda}\rangle = \left(\mathbf{h}, \mathbf{h}^u_{\boldsymbol{\lambda}}\right)_{\mathbf{H}^U(\mathbb{R}^N)} \quad \text{for all } \mathbf{h} \in \mathbf{H}^U(\mathbb{R}^N).$$

Finally, check that

$$\|\mathbf{h}^u_{\boldsymbol{\lambda}}\|^2_{\mathbf{H}^U(\mathbb{R}^N)} = \iint_{[0,\infty)^2} u(s,t)\,\boldsymbol{\lambda}(ds)\cdot\boldsymbol{\lambda}(dt),$$

and conclude that

$$\left(\mathbf{H}^U(\mathbb{R}^N), \Theta(\mathbb{R}^N), \mathcal{U}^{(N)}\right)$$

is an abstract Wiener space. The process $\{\boldsymbol{\theta}(t) : t \in [0,\infty)\}$ under $\mathcal{U}^{(N)}$ is called the **Ornstein–Uhlenbeck process**, and it has played an important rôle in both the mathematical and the physical development of diffusion theory.

(iii) For each $s \in (0,\infty)$, define $\boldsymbol{\delta}^U_s : \Theta(\mathbb{R}^N) \longrightarrow \Theta(\mathbb{R}^N)$ by

$$\left[\boldsymbol{\delta}^U_s \boldsymbol{\theta}\right](t) = \boldsymbol{\theta}(s+t) - e^{-\frac{t}{2}}\boldsymbol{\theta}(s), \quad t \in [0,\infty).$$

Show that the σ-algebras

$$\sigma\big(\boldsymbol{\delta}^U_s \boldsymbol{\theta}(t) : t \in [0,\infty)\big) \quad \text{and} \quad \sigma\big(\boldsymbol{\theta}(t) : t \in [0,s]\big)$$

are independent under $\mathcal{U}^{(N)}$ and that $\mathcal{U}^{(N)}$ is $\boldsymbol{\delta}^U_s$-invariant (i.e., $\boldsymbol{\delta}^U_s{}^*\mathcal{U}^{(N)} = \mathcal{U}^{(N)}$.)

4.2.42 Exercise: Referring to Exercise 4.2.40, notice that, as distinguished from what happens with $\mathcal{W}^{(N)}$, the distribution of $\boldsymbol{\theta}(T)$ under $\mathcal{U}^{(N)}$ *settles down* as $T \to \infty$. In fact, it tends to $\gamma_1{}^N$. In this exercise will we see that the same phenomenon can be seen at a path-space level. In order to explain what we have in mind, we first have to introduce the two-sided path-space $C(\mathbb{R};\mathbb{R}^N)$, which we turn into a Polish space by giving it the topology of uniform convergence on compacts. Next, for each $T \in [0,\infty)$,

$$(\mathbf{x},\boldsymbol{\theta}) \in \mathbb{R}^N \times \Theta(\mathbb{R}^N) \longmapsto \boldsymbol{\psi}_T(\mathbf{x},\boldsymbol{\theta}) \in C(\mathbb{R};\mathbb{R}^N)$$

is the mapping defined so that

$$\left[\boldsymbol{\psi}_T(\mathbf{x},\boldsymbol{\theta})\right](t) = e^{-\frac{(t+T)^+}{2}}\mathbf{x} + \boldsymbol{\theta}\big((t+T)^+\big), \quad t \in \mathbb{R}.$$

Finally, define $\Psi_T : \Theta(\mathbb{R}^N) \longrightarrow C(\mathbb{R};\mathbb{R}^N)$ by $\Psi_T(\boldsymbol{\theta}) = \boldsymbol{\psi}_T(\mathbf{0},\boldsymbol{\theta})$, and set $\mu_T = \Psi_T{}^*\mathcal{U}^{(N)} \in \mathbf{M}_1\big(C(\mathbb{R};\mathbb{R}^N)\big)$.

(i) Let $L \in [0,\infty)$ be given, and suppose that $F : C(\mathbb{R};\mathbb{R}^N) \longrightarrow \mathbb{R}$ is a bounded function which is $\sigma\big(\boldsymbol{\psi}(t) : t \geq -L\big)$-measurable. After noting that

$$\left[\Psi_T(\boldsymbol{\theta})\right](t) = \left[\boldsymbol{\psi}_L\big(\boldsymbol{\theta}(T-L), \boldsymbol{\delta}^U_{T-L}\boldsymbol{\theta}\big)\right](t), \quad t \in [-L,\infty) \text{ and } T \geq L,$$

use part **(iii)** of Exercise 4.2.40 to see that

$$\mathbb{E}^{\mu_T}[F] = \iint\limits_{\mathbb{R}^N \times \Theta(\mathbb{R}^N)} F\big(\boldsymbol{\psi}_L(\mathbf{x},\boldsymbol{\theta})\big)\,\big(\gamma_{1-e^{L-T}}\big)^N(d\mathbf{x})\,\mathcal{U}^{(N)}(d\boldsymbol{\theta}) \quad \text{for } T \geq L$$

and (with the help of Lemma 3.3.3 and Exercise 3.3.26) conclude that $\mu_T \Longrightarrow \mathcal{U}^{(N)}_S$ in $\mathbf{M}_1\big(C(\mathbb{R};\mathbb{R}^N)\big)$ as $T \to \infty$, where

$$\mathbb{E}^{\mathcal{U}^{(N)}_S}[F] = \iint\limits_{\mathbb{R}^N \times \Theta(\mathbb{R}^N)} F\big(\boldsymbol{\psi}_L(\mathbf{x},\boldsymbol{\theta})\big)\,\gamma_1{}^N(d\mathbf{x})\,\mathcal{U}^{(N)}(d\boldsymbol{\theta})$$

for every bounded, $\sigma\big(\boldsymbol{\psi}(t) : t \geq -L\big)$-measurable F.

(ii) For $n \in \mathbb{Z}^+$ and $-\infty < t_1 < \cdots < t_n < \infty$, show that the distribution of

$$\psi \in C(\mathbb{R};\mathbb{R}^N) \longmapsto (\psi(t_1),\dots,\psi(t_n)) \in (\mathbb{R}^N)^n$$

under $\mathcal{U}_S^{(N)}$ is Gaussian with mean-value $\mathbf{0}$ and covariance

$$\left(\left(e^{-\frac{|t_i-t_j|}{2}}\right)\right)_{1\le i,j\le n} \otimes \mathbf{I}.$$

In particular, conclude that $\mathcal{U}_S^{(N)}$ is **stationary** in the sense that it is invariant under the time-shift $\psi \in C(\mathbb{R};\mathbb{R}^N) \longmapsto \psi(\,\cdot + T) \in C(\mathbb{R};\mathbb{R}^N)$ for every $T \in \mathbb{R}$. In fact, show that it is **reversible** in the sense that $\psi \in C(\mathbb{R};\mathbb{R}^N) \longmapsto \psi(-\,\cdot\,) \in C(\mathbb{R};\mathbb{R}^N)$ has the same $\mathcal{U}_S^{(N)}$-distribution as ψ itself.

(iii) First show that

$$\lim_{t\to\infty} \frac{|\psi(t)|}{t} = 0 \quad \text{for } \mathcal{U}_S^{(N)}\text{-almost every } \psi \in C(\mathbb{R};\mathbb{R}^N),$$

and then, as an application of reversibility, that

$$\lim_{|t|\to\infty} \frac{|\psi(t)|}{|t|} = 0 \quad \text{for } \mathcal{U}_S^{(N)}\text{-almost every } \psi \in C(\mathbb{R};\mathbb{R}^N).$$

Next, check that

$$\Theta_S(\mathbb{R}^N) \equiv \left\{\boldsymbol{\theta} \in C(\mathbb{R};\mathbb{R}^N) : \lim_{|t|\to\infty} \frac{|\boldsymbol{\theta}(t)|}{|t|} = 0\right\}$$

is a Borel measurable subset of $C(\mathbb{R};\mathbb{R}^N)$ which becomes a separable Banach space under the norm

$$\|\boldsymbol{\theta}\|_{\Theta_S(\mathbb{R}^N)} \equiv \sup_{t\in\mathbb{R}} \frac{|\psi(t)|}{1+|t|}.$$

Finally, when $\mathbf{H}_S(\mathbb{R}^N)$ denotes the space of $\mathbf{h} \in L^2(\mathbb{R};\mathbb{R}^N) \cap \Theta_S(\mathbb{R}^N)$ with the property that

$$\mathbf{h}(t) - \mathbf{h}(s) = \int_s^t \dot{\mathbf{h}}(\tau)\,d\tau, \quad s < t,$$

for some $\dot{\mathbf{h}} \in L^2(\mathbb{R};\mathbb{R}^N)$ and $\mathbf{H}_S(\mathbb{R}^N)$ is given the Hilbert norm

$$\|\mathbf{h}\|_{\mathbf{H}_S(\mathbb{R}^N)} \equiv \sqrt{\|\dot{\mathbf{h}}\|^2_{L^2(\mathbb{R};\mathbb{R}^N)} + \tfrac{1}{4}\|\mathbf{h}\|^2_{L^2(\mathbb{R};\mathbb{R}^N)}},$$

show that

$$\left(\mathbf{H}_S(\mathbb{R}^N), \Theta_S(\mathbb{R}^N), \mathcal{W}_{\mathbf{H}_S(\mathbb{R}^N)}\right) \quad \text{with } \mathcal{W}_{\mathbf{H}_S(\mathbb{R}^N)} \equiv \mathcal{U}_S^{(N)} \restriction \Theta_S(\mathbb{R}^N)$$

is an abstract Wiener space.

§4.3: Markov Aspects of Wiener's Measure.

In the preceding section, we concentrated on properties of $\mathcal{W}$ which derive from the fact that (cf. Exercise 4.2.39) $\{\psi(t) : t \in [0,\infty)\}$ is a centered Gaussian family in $L^2(\mathcal{W};\mathbb{R})$. In this section, we will focus on the properties (cf. Lemma 4.1.4) which stem from equation (4.1.5).

We will begin by generalizing (4.1.5) to cover certain situations in which the time displacement s depends nicely on the path ψ. To be more precise, call the function $\tau : \mathfrak{P}(\mathbb{R}^N) \longrightarrow [0,\infty]$ a $\{\mathcal{B}_t : t \geq 0\}$**-stopping time**† (cf. (4.1.3)) if $\{\tau \leq t\} \in \mathcal{B}_t$ for every $t \in [0,\infty)$. In other words, a stopping time is a *time* which depends on the path in such a way that *one can determine whether it has occurred during the interval* $[0,t]$ *by observing the path during that interval.* (That is, a knowledge of history is necessary but clairvoyance is not.) Thus, for example, if F is a closed subset of $\mathbb{R}^N$ and one defines the **first entrance time into** F by

$$\tau_F(\psi) \equiv \inf\{t \geq 0 : \psi(t) \in F\}, \quad \psi \in \mathfrak{P}(\mathbb{R}^N),$$

(we take $\inf \emptyset = \infty$) then, because

$$\{\tau_F > t\} = \bigcup_{n=1}^{\infty} \bigcap_{s \in \mathbb{Q}\cap[0,t]} \{\psi : |\psi(s) - F| \geq \tfrac{1}{n}\} \in \mathcal{B}_t, \quad t \in [0,\infty),$$

τ_F is a stopping time. By contrast, the *last exit time*

$$\psi \in \mathfrak{P}(\mathbb{R}^N) \longmapsto \inf\{s \geq 0 : \psi(t) \notin F \text{ for all } t > s\},$$

is not a stopping time!

Given a $\{\mathcal{B}_t : t \in [0,\infty)\}$-stopping time τ, we will use the notation

$$\mathcal{B}_\tau \equiv \Big\{A \subseteq \mathfrak{P}(\mathbb{R}^N) : A \cap \{\tau \leq t\} \in \mathcal{B}_t \text{ for all } t \in [0,\infty)\Big\}. \tag{4.3.1}$$

(Cf. Exercise 4.3.45 for an alternative description.) Check that this notation is consistent in the sense that $\mathcal{B}_\tau = \mathcal{B}_t$ if $\tau \equiv t$. More generally, it is an elementary matter to check that $\mathcal{B}_\tau$ is always a sub σ-algebra of $\mathcal{B}_{\mathfrak{P}(\mathbb{R}^N)}$. The following lemma contains a few additional elementary facts about stopping times and their associated σ-algebras.

4.3.2 Lemma. *If τ is a $\{\mathcal{B}_t : t \in [0,\infty)\}$-stopping time, then τ itself is a $\mathcal{B}_\tau$-measurable function. Moreover, if σ is a second $\{\mathcal{B}_t : t \in [0,\infty)\}$-stopping time, then $\sigma + \tau$, $\sigma \vee \tau$, and $\sigma \wedge \tau$ are all $\{\mathcal{B}_t : t \in [0,\infty)\}$-stopping times, and*

† A precise formulation of this notion evolved through several metamorphoses. The elegant one given here seems to be due to E.B. Dynkin and is taken from his 1956 article "Infinitesimal operators of Markov processes," *Theory of Prob. & Appl.* **1**.

$A \cap \{\sigma \le \tau\} \in \mathcal{B}_\tau$ whenever $A \in \mathcal{B}_\sigma$. In particular, if $\sigma \le \tau$, then $\mathcal{B}_\sigma \subseteq \mathcal{B}_\tau$. Finally, if σ is a $\{\mathcal{B}_t : t \in [0,\infty)\}$-stopping time, $F \subseteq \mathbb{R}^N$ is closed, and

$$\tau_F^\sigma(\psi) \equiv \begin{cases} \inf\{t \ge \sigma(\psi) : \psi(t) \in F\} & \text{if } \sigma(\psi) < \infty \\ \infty & \text{otherwise,} \end{cases}$$

then τ_F^σ is again a $\{\mathcal{B}_t : t \in [0,\infty)\}$-stopping time.

PROOF: It is obvious that τ is $\mathcal{B}_\tau$-measurable and that both $\sigma \vee \tau$ and $\sigma \wedge \tau$ are stopping times. To see that $\sigma + \tau$ is a stopping time, simply observe that

$$\{\sigma + \tau > t\} = \bigcup_{s \in \mathbb{Q} \cap [-1,t]} \{\sigma > s\} \cap \{\tau > t - s\} \in \mathcal{B}_t, \quad t \in [0,\infty).$$

To finish the proof, we next note that

$$\{\sigma \le \tau \le t\} = \{\sigma \wedge t \le \tau \wedge t\} \cap \{\sigma \vee \tau \le t\} \in \mathcal{B}_t, \quad t \in [0,\infty),$$

because, as is easily checked, both $\sigma \wedge t$ and $\tau \wedge t$ are $\mathcal{B}_t$-measurable. Hence, if $A \in \mathcal{B}_\sigma$, then

$$\big(A \cap \{\sigma \le \tau\}\big) \cap \{\tau \le t\} = \big(A \cap \{\sigma \le t\}\big) \cap \{\sigma \le \tau \le t\} \in \mathcal{B}_t, \quad t \in [0,\infty),$$

and therefore $A \cap \{\sigma \le \tau\} \in \mathcal{B}_\tau$.

Finally let σ, F, and τ_F^σ be as in the last part of the statement. When σ is constant, the argument given above for the case $\sigma \equiv 0$ works again and shows that τ_F^s is a $\{\mathcal{B}_t : t \in [0,\infty)\}$-stopping time for every $s \in [0,\infty)$. To handle the general case, note that

$$\{\tau_F^\sigma > t\} = \{\sigma > t\} \cup \{\sigma \le t < \tau_F^\sigma\}$$

and

$$\{\sigma \le t < \tau_F^\sigma\} = \bigcup_{s \in [0,t) \cap \mathbb{Q}} \{s \le \sigma \le t < \tau_F^s\}. \quad \square$$

The generalization of (4.1.5) for which we are looking is contained in the following.

4.3.3 Theorem. *For any $\{\mathcal{B}_t : t \in [0,\infty)\}$-stopping time τ and any bounded function $F : \mathfrak{P}(\mathbb{R}^N)^2 \longrightarrow \mathbb{R}$ which is $\mathcal{B}_\tau \times \mathcal{B}_{\mathfrak{P}(\mathbb{R}^N)}$-measurable:*

$$\begin{aligned} &\int_{\{\psi:\tau(\psi)<\infty\}} F\big(\psi, \delta_{\tau(\psi)}(\psi)\big)\, \mathcal{W}^{(N)}(d\psi) \\ &= \int_{\{\varphi:\tau(\varphi)<\infty\}} \left(\int_{\mathfrak{P}(\mathbb{R}^N)} F(\varphi,\psi)\, \mathcal{W}^{(N)}(d\psi) \right) \mathcal{W}^{(N)}(d\varphi). \end{aligned} \tag{4.3.4}$$

PROOF: When τ is identically equal to some $s \in [0,\infty)$, (4.3.4) is precisely (4.1.5). More generally, if τ takes only a finite number of finite values $0 \le s_1 < \cdots s_n < \infty$, then, because $\{\tau = s_k\} \in \mathcal{B}_{s_k}$,

$$\begin{aligned}
&\int_{\{\psi:\tau(\psi)<\infty\}} F(\psi, \boldsymbol{\delta}_{\tau(\psi)}(\psi))\, \mathcal{W}^{(N)}(d\psi) \\
&\quad = \sum_{k=1}^{n} \int_{\{\psi:\tau(\psi)=s_k\}} F(\psi, \boldsymbol{\delta}_{s_k}(\psi))\, \mathcal{W}^{(N)}(d\psi) \\
&\quad = \sum_{k=1}^{n} \int_{\{\varphi:\tau(\varphi)=s_k\}} \left(\int_{\mathfrak{P}(\mathbb{R}^N)} F(\varphi, \psi)\, \mathcal{W}^{(N)}(d\psi) \right) \mathcal{W}^{(N)}(d\varphi) \\
&\quad = \int_{\{\varphi:\tau(\varphi)<\infty\}} \left(\int_{\mathfrak{P}(\mathbb{R}^N)} F(\varphi, \psi)\, \mathcal{W}^{(N)}(d\psi) \right) \mathcal{W}^{(N)}(d\varphi).
\end{aligned}$$

Now let τ be general, and define τ_n for $n \in \mathbb{Z}^+$ so that

$$\tau_n = \begin{cases} \frac{k}{2^n} & \text{if } \frac{k-1}{2^n} \le \tau \le \frac{k}{2^n} < 2^n \\ \infty & \text{if } \tau \ge 2^n. \end{cases}$$

Note that, because $\tau_n \searrow \tau$, $\mathcal{B}_\tau \subseteq \bigcap_1^\infty \mathcal{B}_{\tau_n}$ and $\boldsymbol{\delta}_{\tau_n(\psi)}\psi \longrightarrow \boldsymbol{\delta}_{\tau(\psi)}\psi$ if $\tau(\psi) < \infty$. Hence, if $\psi \in \mathfrak{P}(\mathbb{R}^N) \longmapsto F(\varphi, \psi)$ is continuous for each $\varphi \in \mathfrak{P}(\mathbb{R}^N)$, we then have that

$$\begin{aligned}
&\int_{\{\psi:\tau(\psi)<\infty\}} F(\psi, \boldsymbol{\delta}_{\tau(\psi)}(\psi))\, \mathcal{W}^{(N)}(d\psi) \\
&\quad = \lim_{n\to\infty} \int_{\{\psi:\tau_n(\psi)<\infty\}} F(\psi, \boldsymbol{\delta}_{\tau_n(\psi)}(\psi))\, \mathcal{W}^{(N)}(d\psi) \\
&\quad = \int_{\{\varphi:\tau(\varphi)<\infty\}} \left(\int_{\mathfrak{P}(\mathbb{R}^N)} F(\varphi, \psi)\, \mathcal{W}^{(N)}(d\psi) \right) \mathcal{W}^{(N)}(d\varphi).
\end{aligned}$$

Finally, since (again by a standard measure theoretic argument) it suffices to prove (4.3.4) when F is a continuous function of its second variable, the proof is now complete. □

In order to demonstrate as quickly as possible just how powerful a tool (4.3.4) can be, let $N = 1$ and, for $a \in (0,\infty)$, set

$$\tau_a(\psi) = \inf\{t \ge 0 : \psi(t) \ge a\}.$$

As the first entrance time into $[a,\infty)$, τ_a is a $\{\mathcal{B}_t : t \in [0,\infty)\}$-stopping time. Moreover, if

$$F(\varphi,\psi) = \begin{cases} 1 & \text{when} \quad t > \tau_a(\varphi) \text{ and } \psi(t-\tau_a(\varphi)) > 0 \\ 0 & \text{otherwise,} \end{cases}$$

then F satisfies the conditions of Theorem 4.3.3,

$$\mathbf{1}_{(0,\infty)}\big(\tau_a(\psi)\big)\, F\big(\psi, \boldsymbol{\delta}_{\tau_a(\psi)}\psi\big) = \mathbf{1}_{(a,\infty)}\big(\psi(t)\big),$$

and, for each $\varphi \in \mathfrak{P}(\mathbb{R}^N)$,

$$\int_{\mathfrak{P}(\mathbb{R})} F(\varphi,\psi)\, \mathcal{W}(d\psi) = \tfrac{1}{2}\, \mathbf{1}_{[0,t)}\big(\tau_a(\varphi)\big).$$

Thus, by (4.3.4), we conclude that

$$\mathcal{W}\big(\tau_a < t\big) = 2\mathcal{W}\big(\psi(t) > a\big) = \sqrt{\tfrac{2}{\pi t}} \int_{(a,\infty)} \exp\left[-\frac{y^2}{2t}\right] dy;$$

and, after taking left limits with respect to a and right limits with respect to t, we get the **reflection principle** :

$$\mathcal{W}\left(\left\{\psi : \max_{s\in[0,t]} \psi(s) \ge a\right\}\right) = \sqrt{\tfrac{2}{\pi}} \int_{at^{-\frac{1}{2}}}^{\infty} \exp\left[-\frac{y^2}{2}\right] dy. \tag{4.3.5}$$

Notice that (4.3.5) represents a tightening of the reasoning on which Lévy's Inequality (1.4.16) was based. To be precise, consider the partial sums S_n of independent, symmetric random variables on a probability space $(\Omega, \mathcal{F}, P)$, and set

$$n_a(\omega) = \inf\{n \ge 1 : S_n(\omega) \ge a\} \quad \text{for} \quad a \in (0,\infty).$$

Then the reasoning in Theorem 1.4.15 can be summarized as:

$$\begin{aligned} P\big(S_n \ge a\big) &= P\Big(\{\omega : S_n(\omega) \ge a \ \&\ n_a(\omega) \le n\}\Big) \\ &\ge \int_{\{\omega : n_a(\omega) \le n\}} P\Big(S_n - S_{n_a(\omega)}(\omega) \ge 0\Big)\, P(d\omega) \ge \tfrac{1}{2} P\Big(\{\omega : n_a(\omega) \le n\}\Big); \end{aligned}$$

the origin of the first inequality being that we only know $S_{n_a(\omega)}(\omega) \ge a$ (not necessarily $= a$) and the second that

$$P\Big(S_n - S_{n_a(\omega)}(\omega) \ge 0\Big) \ge \frac{1}{2},$$

(not necessarily $= \frac{1}{2}$). Obviously, it is continuity of paths which accounts for our ability to erase the first of these and continuity of the distribution function which removes the second.

Although the precise reflection principle does not hold for partial sums of random variables, the Invariance Principle shows that, in some sense, it holds in the limit. Namely, we have from (4.3.5) and Theorem 3.3.20 the following famous result due to Erdös and Kac, which (as was made manifest by Donsker's original proof[†]) is equivalent to the Invariance Principle itself. To be precise, given a probability space $(\Omega, \mathcal{F}, P)$ and a sequence $\{X_n\}_1^\infty$ of independent, uniformly square P-integrable random variables with mean-value 0 and variance 1, set $S_n = \sum_1^n X_\ell$, $n \in \mathbb{Z}^+$, and define the polygonal paths $t \in [0,\infty) \longmapsto [\mathbf{S}_n(\omega)](t) \in \mathbb{R}$ for $\omega \in \Omega$ as we did in Section 3.3 (cf. (3.3.9)). Then, because

$$\max_{1\le m\le n} \frac{S_m(\omega)}{n^{\frac{1}{2}}} = \sup_{t\in[0,1]} [\mathbf{S}_n(\omega)](t), \quad \omega \in \Omega,$$

we have

$$\lim_{n\to\infty} P\left(\sup_{1\le m\le n} \frac{S_m}{n^{\frac{1}{2}}} \ge a\right) = \sqrt{\tfrac{2}{\pi}} \int_a^\infty e^{-\frac{x^2}{2}}\,dx, \quad a \in (0,\infty), \tag{4.3.6}$$

as an essentially immediate consequence of Theorem 3.3.20, (**vii**) in Theorem 3.1.5, and (4.3.5).

As a second application of (4.3.5), note that, by Exercise 3.3.28, for $\mathbf{e} \in \mathbb{R}^N$ and $a \in (0,\infty)$:

$$\begin{aligned}\mathcal{W}^{(N)}&\left(\left\{\psi : \max_{s\in[0,t]} \left(\mathbf{e}, \psi(s)\right)_{\mathbb{R}^N} \ge a\right\}\right)\\ &= \sqrt{\tfrac{2}{\pi}} \int_{at^{-\frac{1}{2}}}^\infty \exp\left[-\frac{y^2}{2}\right] dy \longrightarrow 1\end{aligned}$$

as $t \to \infty$. Thus, if G is an open subset of $\mathbb{R}^N$ and ζ^G is the **first exit time** from G (i.e., the first entrance time into $G\complement$), then

$$\mathcal{W}^{(N)}\left(\left\{\psi : \zeta^G(\psi) < \infty\right\}\right) = 1 \quad \text{if} \quad G \subseteq \left\{\mathbf{y} : (\mathbf{e}, \mathbf{y})_{\mathbb{R}^N} \le a\right\}$$

for some $\mathbf{e} \in \mathbb{R}^N$ and $a \in (0,\infty)$; in particular, this will be the case if G is bounded.

Our next goal is to reformulate (4.3.4) as a Markov property, and for this purpose it will be convenient to introduce two new families of transformations on $\mathfrak{P}(\mathbb{R}^N)$. The first of these is the family $\{\mathbf{T}_\mathbf{x} : \mathbf{x} \in \mathbb{R}^N\}$ of **translations**

† See "Four papers in probability," *Mem. Am. Math. Soc.* **6**.

given by $[\mathbf{T_x}\psi](t) = \mathbf{x} + \psi(t)$, $t \in [0,\infty)$. Clearly, $(\mathbf{x},\psi) \in \mathbb{R}^N \times \mathfrak{P}(\mathbb{R}^N) \longmapsto \mathbf{T_x}\psi \in \mathfrak{P}(\mathbb{R}^N)$ is a continuous map, and therefore, if

$$\mathcal{W}_{\mathbf{x}}^{(N)} \equiv \mathbf{T_x}^*\mathcal{W}^{(N)}, \quad \mathbf{x} \in \mathbb{R}^N,$$

then $\mathbf{x} \in \mathbb{R}^N \longmapsto \mathcal{W}_{\mathbf{x}}^{(N)} \in \mathbf{M}_1\big(\mathfrak{P}(\mathbb{R}^N)\big)$ is also continuous. Hence, if $F : \mathbb{R}^N \times \mathfrak{P}(\mathbb{R}^N) \longrightarrow \mathbb{R}$ is a bounded measurable (continuous) function, then $\mathbf{x} \in \mathbb{R}^N \longmapsto \mathbb{E}^{\mathcal{W}_{\mathbf{x}}^{(N)}}\big[F(\mathbf{x},\,\cdot\,)\big] \in \mathbb{R}$ is measurable (continuous). The second family of transformations which we will want are the **time-shifts** $\boldsymbol{\Sigma}_s$, $s \in [0,\infty)$, given by $\big[\boldsymbol{\Sigma}_s\psi\big](t) = \psi(s+t)$, $t \in [0,\infty)$. Once again, it is an easy matter to check that $(s,\psi) \in [0,\infty) \times \mathfrak{P}(\mathbb{R}^N) \longmapsto \boldsymbol{\Sigma}_s\psi \in \mathfrak{P}(\mathbb{R}^N)$ is continuous.

The technical key to converting (4.3.4) into a Markov statement is contained in the following lemma.

Caution: In the statement of this lemma, as well as in what follows, we use $\varphi(\tau)$ in place of the more precise but uglier $\varphi\big(\tau(\varphi)\big)$.

4.3.7 Lemma. *If τ is an $\big\{\mathcal{B}_t : t \in [0,\infty)\big\}$-stopping time, then the function $\psi \in \mathfrak{P}(\mathbb{R}^N) \longmapsto \psi(\tau\wedge t) \in \mathbb{R}^N$ is $\mathcal{B}_{\tau\wedge t}$-measurable for each $t \in [0,\infty)$. Moreover, if F is a $\mathcal{B}_\tau \times \mathcal{B}_{\mathfrak{P}(\mathbb{R}^N)}$-measurable function on $\mathfrak{P}(\mathbb{R}^N)^2$ into some measurable space, then the function*

$$(\varphi,\psi) \in \mathfrak{P}(\mathbb{R}^N)^2 \longmapsto \mathbf{1}_{[0,\infty)}\big(\tau(\varphi)\big)\, F\Big(\varphi, \mathbf{T}_{\varphi(\tau)}\psi\Big)$$

is also $\mathcal{B}_\tau \times \mathcal{B}_{\mathfrak{P}(\mathbb{R}^N)}$-measurable.

PROOF: Clearly, everything comes down to showing that $\psi \in \mathfrak{P}(\mathbb{R}^N) \longmapsto \psi(\tau\wedge t) \in \mathbb{R}^N$ is $\mathcal{B}_{\tau\wedge t}$-measurable. But, by Lemma 4.3.2, for any $t \in [0,\infty)$, $\psi \in \mathfrak{P}(\mathbb{R}^N) \longmapsto \tau(\psi)\wedge t$ is $\mathcal{B}_t$-measurable and so $\psi \in \mathfrak{P}(\mathbb{R}^N) \longmapsto \psi(\tau\wedge t)$ is also. Clearly the required result follows immediately from this. □

4.3.8 Theorem. *Let τ be a $\big\{\mathcal{B}_t : t \in [0,\infty)\big\}$-stopping time and $F : \mathfrak{P}(\mathbb{R}^N)^2 \longrightarrow \mathbb{R}$ a bounded $\mathcal{B}_\tau \times \mathcal{B}_{\mathfrak{P}(\mathbb{R}^N)}$-measurable function. Then, for all $\mathbf{x} \in \mathbb{R}^N$,*

$$\begin{aligned} &\int_{\{\psi:\tau(\psi)<\infty\}} F\big(\psi, \boldsymbol{\Sigma}_{\tau(\psi)}\psi\big)\, \mathcal{W}_{\mathbf{x}}^{(N)}(d\psi) \\ &\qquad = \int_{\{\varphi:\tau(\varphi)<\infty\}} \left(\int_{\mathfrak{P}(\mathbb{R}^N)} F(\varphi,\psi)\, \mathcal{W}_{\varphi(\tau)}^{(N)}(d\psi)\right) \mathcal{W}_{\mathbf{x}}^{(N)}(d\varphi). \end{aligned} \tag{4.3.9}$$

In particular, if $f \in B(\mathbb{R}^N;\mathbb{R})$ *and* $F : \mathfrak{P}(\mathbb{R}^N) \longrightarrow \mathbb{R}$ *is bounded and* $\mathcal{B}_\tau$*-measurable, then*

$$\int_{\{\psi:\tau(\psi)<\infty\}} F(\psi) f\big(\psi(t+\tau)\big)\, \mathcal{W}_{\mathbf{x}}^{(N)}(d\psi)$$

$$= \int_{\{\psi:\tau(\psi)<\infty\}} F(\psi) \left(\int_{\mathbb{R}^N} f\big(\psi(\tau)+\mathbf{y}\big)\, {\gamma_t}^N(d\mathbf{y})\right) \mathcal{W}_{\mathbf{x}}^{(N)}(d\psi)$$

for all $(t,\mathbf{x}) \in (0,\infty) \times \mathbb{R}^N$.

PROOF: Define

$$G(\varphi,\psi) = \mathbf{1}_{[0,\infty)}\big(\tau(\varphi)\big)\, F\Big(\mathbf{T}_{\mathbf{x}}\varphi, \mathbf{T}_{\mathbf{x}+\varphi(\tau)}\psi\Big).$$

Then G is $\mathcal{B}_\tau \times \mathcal{B}_{\mathfrak{P}(\mathbb{R}^N)}$-measurable, and (4.3.9) for F is equivalent to (4.3.4) for G. □

Although, as its proof makes evident, (4.3.9) is really just another way of writing (4.3.4), it is often the more useful of the two expressions. Indeed, (4.3.9) brings out a slightly different intuitive picture. Namely, it says that, under $\mathcal{W}_{\mathbf{x}}^{(N)}$, for any stopping time τ, *the distribution of the path* $\boldsymbol{\Sigma}_{\psi(\tau)}\psi$ *is unaffected by the past of* $\tau(\psi)$ *and depends only on the position* $\psi(\tau)$ *of* ψ *at time* $\tau(\psi)$. Such a property is called a **Markov property**, and, because it holds for stopping times (not just constant times), the one in Theorem 4.3.8 is called the **strong Markov property** of Wiener's measure (cf. Exercise 4.3.55).

From an analytic point of view, the Markov property for Wiener's measure is an expression of the fact that the Gauss measures $\big\{{\gamma_s}^N : s \in (0,\infty)\big\}$ form a semigroup; that is $\gamma_{t+s}^N = {\gamma_s}^N \bigstar {\gamma_t}^N$ for all $s,\, t \in (0,\infty)$. To be more precise, after further aggravating our abuse of notation by introducing

$$\gamma_s^N(\mathbf{x}) \equiv \prod_1^N \gamma_s(x_k), \quad s \in (0,\infty) \text{ and } \mathbf{x} \in \mathbb{R}^N, \tag{4.3.10}$$

to stand for the density of the Gauss measure ${\gamma_s}^N$, we define the operators $P_s : B(\mathbb{R}^N;\mathbb{R}) \longrightarrow B(\mathbb{R}^N;\mathbb{R})$ by $P_s f(\mathbf{x}) = \big[\gamma_s^N \bigstar f\big](\mathbf{x})$. Clearly, the aforementioned semigroup property for the measures ${\gamma_s}^N$ becomes the statement that $\{P_s : s > 0\}$ is a **semigroup of operators**: $P_{s+t} = P_s \circ P_t$. Moreover, P_s tends to the identity operation $\mathbf{I}$ in the sense that, for $f \in C_{\mathrm{b}}(\mathbb{R}^N;\mathbb{R})$, $P_s f \longrightarrow f$ uniformly on compacts as $s \searrow 0$. In fact, if $\boldsymbol{\Delta}$ is used to denote the **standard Laplacian** $\sum_1^N \frac{\partial^2}{\partial x_k^2}$ for $\mathbb{R}^N$, then $\frac{P_s - \mathbf{I}}{s} \longrightarrow \frac{1}{2}\boldsymbol{\Delta}$ in the sense that, for $f \in C_{\mathrm{b}}^2(\mathbb{R}^N;\mathbb{R})$,

$$\frac{P_s f - f}{s} \longrightarrow \tfrac{1}{2}\boldsymbol{\Delta} f \quad \text{uniformly on compacts as } s \searrow 0.$$

(The most instructive derivation of this fact is to use Taylor's Theorem and write

$$\begin{aligned}\big[P_s f\big](\mathbf{x}) - f(\mathbf{x}) &= \sum_{k=1}^{N} \frac{\partial f}{\partial x_k}(\mathbf{x}) \int_{\mathbb{R}^N} y_k \, {\gamma_s}^N(d\mathbf{y}) \\ &\quad + \frac{1}{2} \sum_{k,\ell=1}^{N} \frac{\partial^2 f}{\partial x_k \partial x_\ell}(\mathbf{x}) \int_{\mathbb{R}^N} y_k y_\ell \, {\gamma_s}^N(d\mathbf{y}) + sE(s,\mathbf{x}) \\ &= \frac{s}{2}\big[\boldsymbol{\Delta} f\big](\mathbf{x}) + sE(s,\mathbf{x}),\end{aligned}$$

where

$$|E(s,\mathbf{x})| \le \frac{1}{2s} \sum_{k,\ell=1}^{N} \iint_{[0,1]\times\mathbb{R}^N} |\mathbf{y}|^2 \left| \frac{\partial^2 f}{\partial x_k \partial x_\ell}(\mathbf{x}+\theta\mathbf{y}) - \frac{\partial^2 f}{\partial x_k \partial x_\ell}(\mathbf{x}) \right| d\theta {\gamma_s}^N(d\mathbf{y})$$

tends to 0 uniformly on compacts as $s \searrow 0$.) Thus, since P_s maps $B(\mathbb{R}^N;\mathbb{R})$ into $C^2_{\mathrm{b}}(\mathbb{R}^N;\mathbb{R})$, we see that $\{P_s : s > 0\}$ is the **heat flow semigroup** in the sense that, at least when $f \in C_{\mathrm{b}}(\mathbb{R}^N;\mathbb{R})$, $s \in (0,\infty) \longmapsto P_s f \in B(\mathbb{R}^N;\mathbb{R})$ is the unique (cf. Theorem 4.3.15) bounded mapping $s \in (0,\infty) \longmapsto u(s) \in B(\mathbb{R}^N;\mathbb{R})$ which is a classical (i.e., smooth) solution to the **Cauchy initial value problem** for the **heat equation**

$$\tfrac{\partial u}{\partial s}(s) = \tfrac{1}{2}\boldsymbol{\Delta} u(s) \quad \text{with} \quad \lim_{s\searrow 0} u(s) = f \text{ pointwise.}$$

Alternatively, in the somewhat formal but suggestive notation of operator theory, "$P_s = e^{\frac{s}{2}\boldsymbol{\Delta}}$."

In the remainder of this section, we will use the Markov property to show how Wiener's measure can be used to represent solutions to certain perturbations of the heat equation. Specifically, let $V : \mathbb{R}^N \longmapsto \mathbb{R}$ be a measurable function which is bounded above, and consider the Cauchy initial value problem

$$\tag{4.3.11} \tfrac{\partial u}{\partial s} = \tfrac{1}{2}\boldsymbol{\Delta} u + Vu \text{ in } (0,T] \times \mathbb{R}^N \quad \text{with } u(0,\,\cdot\,) = f.$$

Following R. Feynman and M. Kac,† we will show that, under very general circumstances, the *reasonable* solution to (4.3.11) is given at $(t,\mathbf{x})$ by

$$\tag{4.3.12} \big[P_t^V f\big](\mathbf{x}) \equiv \int_{\mathfrak{P}(\mathbb{R}^N)} f\big(\psi(t)\big) \exp\left[\int_0^t V\big(\psi(s)\big)\,ds\right] \mathcal{W}^{(N)}(d\psi).$$

That is, "$P_t^V = e^{t(\frac{1}{2}\boldsymbol{\Delta}+V)}$."

† More accurately, we will follow Kac, who himself followed Feynman. Indeed, this formula grows out of Feynman's *path-integral* approach to solving Schrödinger's equation in terms of integrals involving expressions like the one in (4.2.1), only with a $\sqrt{-1}$ in the exponent! After hearing Feynman lecture on his method, Kac realized that one could transfer Feynman's ideas from the Schrödinger to the heat context and thereby arrive at a mathematically rigorous but far less exciting theory.

To make all this more precise, we begin by checking that $\{P_t^V : t > 0\}$ determines a semigroup on $B(\mathbb{R}^N;\mathbb{R})$. But clearly, for each $f \in B(\mathbb{R}^N;\mathbb{R})$, $(t,\mathbf{x}) \in (0,\infty) \times \mathbb{R}^N \longmapsto [P_t^V f](\mathbf{x}) \in \mathbb{R}$ is a measurable function which satisfies the estimate

$$\|P_t^V f\|_{\mathrm{u}} \leq \|f\|_{\mathrm{u}} \exp\left[t \sup_{\mathbf{x}\in\mathbb{R}^N} V(\mathbf{x})\right], \quad t \in (0,\infty). \tag{4.3.13}$$

In addition, if

$$M(t,\psi) = \exp\left[\int_0^t V(\psi(s))\,ds\right], \quad (t,\psi) \in [0,\infty) \times \mathfrak{P}(\mathbb{R}^N),$$

Then $M(s+t,\psi) = M(s,\psi)\,M(t,\Sigma_s\psi)$, and therefore, by (4.3.9),

$$\begin{aligned}
[P_{s+t}^V f](\mathbf{x}) &= \int_{\mathfrak{P}(\mathbb{R}^N)} M(s,\psi)\, f([\Sigma_s\psi](t))\, M(t,\Sigma_s\psi)\ \mathcal{W}_{\mathbf{x}}^{(N)}(d\psi) \\
&= \int_{\mathfrak{P}(\mathbb{R}^N)} M(s,\varphi) \left(\int_{\mathfrak{P}(\mathbb{R}^N)} f(\psi(t))\, M(t,\psi)\, \mathcal{W}_{\varphi(s)}^{(N)}(d\psi)\right) \mathcal{W}_{\mathbf{x}}^{(N)}(d\varphi) \\
&= [P_s^V \circ P_t^V f](\mathbf{x}).
\end{aligned}$$

Hence, we have now proved the semigroup property.

What remains is to establish the connection with (4.3.11) by justifying the equation "$P_t^V = e^{t(\frac{1}{2}\Delta+V)}$," and the first step in this program is to eliminate technical differentiability questions by converting (4.3.11) into an integral equation. To this end, let $t \in (0,T]$ be given, suppose that u is a bounded classical (i.e., smooth) solution to (4.3.11), consider

$$s \in (0,t) \longmapsto w(s) \equiv \gamma_{t-s}^N \bigstar u(s) \in B(\mathbb{R}^N),$$

and note that

$$\dot{w}(s) = \gamma_{t-s}^N \bigstar \left(-\tfrac{1}{2}\Delta u(s) + \tfrac{\partial u}{\partial t}(s)\right) = \gamma_{t-s}^N \bigstar (V u(s)).$$

Hence, after integrating in s over $(0,t)$, we arrive at

$$u(t) = \gamma_t^N \bigstar f + \int_0^t \gamma_{t-s}^N \bigstar (V\,u(s))\,ds, \quad t \in [0,T]. \tag{4.3.14}$$

In other words, if u is a bounded, classical solution to (4.3.11), then it is a solution to the integral equation in (4.3.14); and so (4.3.14) represents a generalized statement of the problem in (4.3.11), one in which all smoothness requirements can be ignored.

With these preliminaries, we are now ready to present the following version of Feynman and Kac's result.

4.3.15 Theorem (Feynman–Kac Formula). *Let $V:\mathbb{R}^N \longrightarrow \mathbb{R}$ be a bounded, measurable function. Then, for each $f \in B(\mathbb{R}^N;\mathbb{R})$ and $T\in(0,\infty)$, $t\in(0,T] \longmapsto P_t^V f \in B(\mathbb{R}^N;\mathbb{R})$ is the unique bounded, measurable $t\in(0,T]\longmapsto u(t)\in B(\mathbb{R}^N;\mathbb{R})$ which satisfies (4.3.14). In particular,*

$$P_t^V f - f = \int_0^t P_s^V\left(\tfrac{1}{2}\mathbf{\Delta} f + Vf\right)ds \tag{4.3.16}$$

for $t\in(0,\infty)$ and $f\in C_{\mathrm{b}}^2(\mathbb{R}^N;\mathbb{R})$.

Finally, let $T\in(0,\infty)$ be given, and suppose that $\{u_n\}_1^\infty$ is a sequence of functions from $C_{\mathrm{b}}^{1,2}\big((0,T)\times\mathbb{R}^N;\mathbb{R}\big)\cap C_{\mathrm{b}}\big([0,T]\times\mathbb{R}^N;\mathbb{R}\big)$ with the properties that

$$\sup_{n\in\mathbb{Z}^+}\|u_n\|_{\mathrm{u}}\vee\left\|\frac{\partial u_n}{\partial t}-\tfrac{1}{2}\mathbf{\Delta}u_n - Vu_n\right\|_{\mathrm{u}}<\infty,$$

$$u_n(t,\mathbf{x})\longrightarrow u(t,\mathbf{x})\quad \text{for each } (t,\mathbf{x})\in[0,T]\times\mathbb{R}^N,$$

and

$$\left[\tfrac{\partial u_n}{\partial t}-\tfrac{1}{2}\mathbf{\Delta}u_n - V\,u_n\right](t,\mathbf{x})\longrightarrow 0$$

for Lebesgue-almost every $(t,\mathbf{x})\in[0,T]\times\mathbb{R}^N$. Then

$$u(t,\mathbf{x}) = \big[P_t^V f\big](\mathbf{x}),\quad (t,\mathbf{x})\in[0,T]\times\mathbb{R}^N, \tag{4.3.17}$$

where $f\equiv u(0,\,\cdot\,)$; and so (4.3.17) holds whenever $u\in C_{\mathrm{b}}^{1,2}\big((0,T)\times\mathbb{R}^N;\mathbb{R}\big)\cap C_{\mathrm{b}}\big([0,T]\times\mathbb{R}^N;\mathbb{R}\big)$ is a solution to Cauchy initial value problem in (4.3.11).

PROOF: We begin by noting that, for each $f\in B(\mathbb{R}^N;\mathbb{R})$ and $T\in(0,\infty)$, there is at most one bounded, measurable $t\in(0,T]\longrightarrow B(\mathbb{R}^N;\mathbb{R})$ satisfying (4.3.14); and, by linearity, this will follow as soon as we check that the only bounded solution to (4.3.14) when $f\equiv 0$ is $u\equiv 0$. But if u satisfies (4.3.14) with $f\equiv 0$, then

$$U(t)\le\|V\|_{\mathrm{u}}\int_0^t U(s)\,ds,\quad t\in[0,T],$$

where $U(t)\equiv\sup_{s\in[0,t]}\|u(t)\|_{\mathrm{u}}$ for each $t\in[0,T]$. Thus, by induction, we see that

$$U(t)\le\frac{\big(\|V\|_{\mathrm{u}}\,t\big)^n}{n!}U(T),\quad n\in\mathbb{N}\text{ and } t\in[0,T],$$

from which the conclusion $U(T)=0$ is obvious.

Next, we show that $t\in(0,T]\longmapsto P_t^V f$ satisfies (4.3.14). For this purpose, set $w(t)=P_t^V f$, and note that, for each $0\le s<t$:

$$\begin{aligned}\big[\gamma_{t-s}^N\bigstar\big(Vw(s)\big)\big](\mathbf{x}) &= \int_{\mathfrak{P}(\mathbb{R}^N)} V\big(\varphi(t-s)\big)\,\big[w(s)\big]\big(\varphi(t-s)\big)\,\mathcal{W}_{\mathbf{x}}^{(N)}(d\varphi)\\ &= \int_{\mathfrak{P}(\mathbb{R}^N)}\left(\int_{\mathfrak{P}(\mathbb{R}^N)} F(\varphi,\psi)\,\mathcal{W}_{\varphi(t-s)}^{(N)}(d\psi)\right)\mathcal{W}_{\mathbf{x}}^{(N)}(d\varphi),\end{aligned}$$

where

$$F(\varphi,\psi) \equiv f(\psi(s))\,V(\varphi(t-s))\,\exp\left[\int_0^s V(\psi(\sigma))\,d\sigma\right].$$

Hence, since F is $\mathcal{B}_{t-s}\times\mathcal{B}_{\mathfrak{P}(\mathbb{R}^N)}$-measurable and

$$F(\psi,\Sigma_{t-s}\psi) = f(\psi(t))\,V(\psi(t-s))\,\exp\left[\int_{t-s}^t V(\psi(\sigma))\,d\sigma\right],$$

(4.3.9) plus Fubini's Theorem leads to

$$\begin{aligned}
&\int_0^t \left[\gamma_{t-s}^N\bigstar\big(Vw(s)\big)\right](\mathbf{x})\,ds \\
&= \int_{\mathfrak{P}(\mathbb{R}^N)} f(\psi(t))\bigg(\int_0^t V(\psi(t-s)) \\
&\qquad\qquad \times\exp\left[\int_{t-s}^t V(\psi(\sigma))\,d\sigma\right]ds\bigg)\,\mathcal{W}^{(N)}(d\psi) \\
&= \int_{\mathfrak{P}(\mathbb{R}^N)} f(\psi(t))\left(\exp\left[\int_0^t V(\psi(\sigma))\,d\sigma\right]-1\right)\mathcal{W}^{(N)}_{\mathbf{x}}(d\psi) \\
&= \big[w(t)\big](\mathbf{x}) - \big[\gamma_t^N\bigstar f\big](\mathbf{x}).
\end{aligned}$$

In other words, w satisfies (4.3.14).

To prove (4.3.16), let $f\in C_{\mathrm{b}}^2(\mathbb{R}^N;\mathbb{R})$ be given, and observe that the class of V's for which (4.3.16) holds must be closed under bounded, pointwise convergence. Hence, it suffices to prove (4.3.16) for $V\in C_{\mathrm{b}}(\mathbb{R}^N;\mathbb{R})$. But, for $\delta>0$,

$$P_{t+\delta}^V f - P_t^V f = P_t^V\left(P_\delta^V f - f\right);$$

and

$$\frac{P_\delta^V f - f}{\delta} = \frac{\gamma_\delta^N\bigstar f - f}{\delta} + \frac{1}{\delta}\int_0^\delta \gamma_{\delta-s}^N\bigstar\left(V\,P_s^V f\right)ds.$$

As we already know (cf. the discussion of the heat equation given earlier) the first term on the right tends boundedly and pointwise to $\frac{1}{2}\Delta f$ as $\delta\searrow 0$, and clearly, when $V\in C_{\mathrm{b}}(\mathbb{R}^N;\mathbb{R})$, the second term tends boundedly and pointwise to Vf. In other words, at least when V is continuous, we have now shown that

$$\frac{d\,P_t^V f}{dt} = P_t^V\left(\tfrac{1}{2}\Delta f + V f\right),\quad t\in[0,\infty),$$

and therefore that (4.3.16) holds first for continuous V's and then for general ones.

Finally, let $T \in (0,\infty)$, $\{u_n\}_1^\infty \cup \{u\}$, and f be as in the last assertion, and set

$$u_n(t) = u_n(t,\,\cdot\,), \quad g_n(s) = \frac{\partial u_n}{\partial s} - \tfrac{1}{2}\Delta u_n(s) - V\,u_n(s),$$

and $f_n = u_n(0)$. Then, for $t \in (0,T]$,

$$\frac{d}{ds}\gamma_{t-s}^N \bigstar u_n(s) = \gamma_{t-s}^N \bigstar g_n(s) + \gamma_{t-s}^N \bigstar \big(V\,u_n(s)\big), \quad s \in (0,t),$$

and therefore

$$u_n(t) - \gamma_t^N \bigstar f = \int_0^t \gamma_{t-s}^N \bigstar g_n(s)\,ds + \int_0^t \gamma_{t-s}^N \bigstar \big(V\,u_n(s)\big)\,ds.$$

Thus, after letting $n \to \infty$ and using the assumptions about the convergence of g_n to 0 and u_n to u, we conclude that u satisfies (4.3.14) during the time interval $(0,T]$. □

Warning: In the future, we will seldom give such a detailed account of how (4.3.9) applies. In particular, we will seldom provide the explicit expression for the function $F(\varphi,\psi)$ to which we are applying (4.3.9) and, instead, will merely use a phrase like *by the (strong) Markov property* to indicate that such an application has been made.

In the terminology of functional analysis, the conclusion drawn in (4.3.16) says that $C_b^2(\mathbb{R}^N;\mathbb{R})$ is contained in the *domain of the generator of* $\{P_t^V : t \in [0,\infty)\}$ and that $\frac{1}{2}\Delta f + Vf$ is what one gets by acting the generator on an $f \in C_b^2(\mathbb{R}^N;\mathbb{R})$. Because in Schrödinger's model of quantum mechanics (cf. the footnote following (4.3.11)), $\frac{1}{2}\Delta + V$ is the Hamiltonian corresponding to the *potential energy* V, we will call $\{P_t^V : t \in (0,\infty)\}$ the **Feynman–Kac semigroup with potential V.**

4.3.18 Corollary. *Assume that $V \le -\epsilon$ for some $\epsilon > 0$, and let $\{f_n\}_1^\infty \subseteq C_b^2(\mathbb{R}^N;\mathbb{R})$ be a uniformly bounded sequence of functions for which*

$$\tfrac{1}{2}\Delta f_n + V\,f_n \longrightarrow -g \in B(\mathbb{R}^N;\mathbb{R})$$

boundedly and (Lebesque) almost everywhere. Then,

$$\lim_{n\to\infty} f_n(\mathbf{x}) = \int_0^\infty \big[P_t^V g\big](\mathbf{x})\,dt \quad \textit{for each } \mathbf{x} \in \mathbb{R}^N.$$

PROOF: Set $g_n = -\frac{1}{2}\Delta f_n - V\,f_n$. Then, by (4.3.16),

$$f_n - P_T^V f_n = \int_0^T P_t^V g_n\,dt, \quad T \in (0,\infty).$$

Moreover, by (4.3.13), $\|P_t^V f\|_{\mathrm{u}} \le Me^{-\epsilon t}$, $t \in [0,\infty)$, for some $M < \infty$. Thus, the desired conclusion is reached by first letting $T \nearrow \infty$ and then sending $n \to \infty$. □

The following application, which is due to Kac himself,† of Corollary 4.3.18 gives dramatic evidence of the power that the Feynman–Kac formula has to compute certain nontrivial (i.e., ones involving functionals which require sampling the path at infinitely many times) Wiener integrals.

4.3.19 Theorem (The Arcsine Law). ‡ *For every $T \in (0,\infty)$ and $\alpha \in [0,1]$,*

$$\mathcal{W}\left(\left\{\psi \in \mathfrak{P}(\mathbb{R}) : \frac{1}{T}\int_0^T \mathbf{1}_{[0,\infty)}\big(\psi(t)\big)\,dt \le \alpha\right\}\right) = \frac{2}{\pi}\arcsin\big(\sqrt{\alpha}\big).$$

In fact, if $(\Omega, \mathcal{F}, P)$ is a probability space and $\{X_n\}_1^\infty$ is a sequence of independent, uniformly square P-integrable random variables with mean-value 0 and variance 1, then, for every $\alpha \in [0,1]$,

$$\lim_{n\to\infty} P\left(\left\{\omega : \frac{N_n(\omega)}{n} \le \alpha\right\}\right) = \frac{2}{\pi}\arcsin\big(\sqrt{\alpha}\big), \tag{4.3.20}$$

where $N_n(\omega)$ is the number of $m \in \mathbb{Z}^+ \cap [0,n]$ for which $S_m(\omega) \equiv \sum_{\ell=1}^m X_\ell(\omega) \ge 0$.

PROOF: First note that, by Wiener scaling-invariance, it suffices to prove the first part when $T = 1$. Next, set

$$F(\alpha) = \mathcal{W}\left(\left\{\psi \in \mathfrak{P}(\mathbb{R}) : \int_0^1 \mathbf{1}_{[0,\infty)}\big(\psi(s)\big)\,ds \le \alpha\right\}\right), \quad \alpha \in [0,\infty),$$

and let μ denote the element of $\mathbf{M}_1\big([0,\infty)\big)$ for which F is the distribution function. We are going to compute $F(\alpha)$ by looking at the double Laplace transform

$$G(\lambda) \equiv \int_{(0,\infty)} e^{-\lambda t}\, g(t)\,dt, \quad \lambda \in (0,\infty),$$

where

$$g(t) \equiv \int_{[0,\infty)} e^{-t\alpha}\,\mu(d\alpha), \quad t \in (0,\infty);$$

and, by another application of the Wiener scaling property, we see that

$$\begin{aligned} G(\lambda) &= \int_0^\infty \left(\int_{\mathfrak{P}(\mathbb{R})} \exp\left[-\int_0^t \Big(\lambda + \mathbf{1}_{[0,\infty)}\big(\psi(s)\big)\Big)\,ds\right]\mathcal{W}(d\psi)\right)dt \\ &= \int_0^\infty \big[P_t^{V_\lambda}\mathbf{1}\big](0)\,dt \quad \text{where } V_\lambda \equiv -\lambda - \mathbf{1}_{[0,\infty)}. \end{aligned}$$

† See Kac's "On some connections between probability theory and differential and integral equations," *Proc. 2nd Berkeley Symp. on Prob. & Stat.* (1951), where he gives several other intriguing applications as well.

‡ The first part of this theorem was discovered by P. Lévy

At this point, the idea is to calculate $G(\lambda)$ with the help of Corollary 4.3.18. Thus, we seek as good a solution $x \in \mathbb{R} \longmapsto f_\lambda(x) \in \mathbb{R}$ to the equation $\frac{1}{2}f''(x) - \big(\lambda + \mathbf{1}_{[0,\infty)}\big)f = -1$ as we can find. By considering this equation separately on the left and right half lines and then matching, in so far as possible, at 0, we find that the best choice of bounded f_λ will be to take

$$f_\lambda(x) = \begin{cases} A_\lambda \exp\Big[-\sqrt{2(1+\lambda)}\,x\Big] + \frac{1}{1+\lambda} & \text{if } \ x \in [0,\infty) \\ B_\lambda \exp\Big[\sqrt{2\lambda}\,x\Big] + \frac{1}{\lambda} & \text{if } \ x \in (-\infty, 0), \end{cases}$$

where

$$A_\lambda = \left(\frac{1}{\lambda(1+\lambda)}\right)^{\frac{1}{2}} - \frac{1}{1+\lambda} \quad \text{and} \quad B_\lambda = \left(\frac{1}{\lambda(1+\lambda)}\right)^{\frac{1}{2}} - \frac{1}{\lambda}.$$

(The choice of sign in the exponent is dictated by our desire to have f_λ bounded.) Notice that, although f_λ has a discontinuous second derivative at 0, f'_λ is nonetheless uniformly Lipschitz continuous everywhere. Hence, by taking $\rho \in C^\infty_c\big(\mathbb{R};[0,\infty)\big)$ with Lebesgue integral 1 and setting

$$f_{\lambda,n}(x) = n\int_{\mathbb{R}} f_\lambda(x-y)\rho(ny)\,dy, \quad n \in \mathbb{Z}^+,$$

we see that: $f_{\lambda,n} \in C^\infty_b(\mathbb{R};\mathbb{R})$ for each $n \in \mathbb{Z}^+$, $f_{\lambda,n} \longrightarrow f_\lambda$ uniformly on $\mathbb{R}$ as $n \to \infty$, $\sup_{n\in\mathbb{Z}^+} \|f_{\lambda,n}\|_{C^2_b(\mathbb{R};\mathbb{R})} < \infty$, and, as $n \to \infty$,

$$\frac{1}{2}f''_{\lambda,n} - \big(\lambda + \mathbf{1}_{[0,\infty)}\big)f_{\lambda,n} \longrightarrow -1 \quad \text{on } \mathbb{R}\setminus\{0\};$$

and so, by Corollary 4.3.18, we know that

$$G(\lambda) = f_\lambda(0) = \left(\frac{1}{\lambda(1+\lambda)}\right)^{\frac{1}{2}}.$$

Given the preceding expression for $G(\lambda)$, the rest of the calculation is easy. Indeed, since

$$\int_0^\infty t^{-\frac{1}{2}}e^{-\lambda t}\,dt = \sqrt{\frac{\pi}{\lambda}},$$

the multiplication rule for Laplace transforms tells us that

$$g(t) = \frac{1}{\pi}\int_0^t \frac{e^{-s}}{\sqrt{s(t-s)}}\,ds = \frac{1}{\pi}\int_0^1 \frac{e^{-t\alpha}}{\sqrt{\alpha(1-\alpha)}}\,d\alpha;$$

and so we now find that

$$F(\alpha) = \frac{1}{\pi}\int_0^{\alpha\wedge 1} \frac{1}{\sqrt{\beta(1-\beta)}}\,d\beta = \frac{2}{\pi}\arcsin\bigl(\sqrt{\alpha\wedge 1}\bigr).$$

Obviously, what we would like is to get the second assertion as a consequence of the Invariance Principle (cf. Theorem 3.3.20). Indeed, thinking of $\frac{N_n(\omega)}{n}$ as a Riemann approximation to (cf. the notation in Section 3.3)

$$\int_0^1 \mathbf{1}_{[0,\infty)}\Bigl(\bigl[\mathbf{S}_n(\omega)\bigr](t)\Bigr)\,dt,$$

it is reasonable to hope, on the basis of the Invariance Principle, that the distribution of $\omega \longmapsto A_n(\omega) \equiv \frac{N_n(\omega)}{n}$ under P tends to that of

$$\psi \in \mathfrak{P}(\mathbb{R}) \longmapsto F(\psi) \equiv \int_0^1 \mathbf{1}_{[0,\infty)}\bigl(\psi(t)\bigr)\,dt$$

under $\mathcal{W}$. However, the argument here is a little less straightforward than it was in the derivation of (4.3.6) because the functions with which we are dealing now are somewhat more complicated. To get around this difficulty, define

$$F^f(\psi) = \int_0^1 f\bigl(\psi(t)\bigr)\,dt \quad\text{and}\quad F^f_n(\psi) = \frac{1}{n}\sum_{m=1}^{n} f\Bigl(\psi\bigl(\tfrac{m}{n}\bigr)\Bigr), \quad n \in \mathbb{Z}^+,$$

for $f \in B(\mathbb{R};\mathbb{R})$ and $\psi \in \mathfrak{P}(\mathbb{R})$. Then, so long as $f \in C_b(\mathbb{R};\mathbb{R})$, it is clear that $F^f_n \longrightarrow F^f$ uniformly fast on compact subsets of $\mathfrak{P}(\mathbb{R})$; and therefore, by Lemma 3.1.10 and Theorem 3.3.20, the distribution of

$$\omega \in \Omega \longmapsto A^f_n(\omega) \equiv \frac{1}{n}\sum_{m=1}^{n} f\left(\frac{S_m(\omega)}{n^{\frac{1}{2}}}\right)$$

under P tends to that of $\psi \in \mathfrak{P}(\mathbb{R}) \longmapsto F^f(\psi)$ under $\mathcal{W}$. Next, for each $\delta \in (0,\infty)$, choose continuous functions $f_{\pm\delta}$ so that $\mathbf{1}_{(\delta,\infty)} \le f_\delta \le \mathbf{1}_{[0,\infty)}$ and $\mathbf{1}_{[0,\infty)} \le f_{-\delta} \le \mathbf{1}_{[-\delta,\infty)}$, and conclude that

$$\overline{\lim_{n\to\infty}}\, P\Bigl(\frac{N_n}{n} \le \alpha\Bigr) \le \mathcal{W}\Bigl(F^{f_\delta} \le \alpha\Bigr)$$

and

$$\underline{\lim_{n\to\infty}}\, P\Bigl(\frac{N_n}{n} < \alpha\Bigr) \ge \mathcal{W}\Bigl(F^{f_{-\delta}} < \alpha\Bigr)$$

for every $\delta > 0$. Passing to the limit as $\delta \searrow 0$, we arrive at

$$\varlimsup_{n\to\infty} P\left(\frac{N_n}{n} \le \alpha\right) \le \mathcal{W}\left(\left\{\psi : \int_0^1 \mathbf{1}_{(0,\infty)}(\psi(t))\,dt \le \alpha\right\}\right)$$

and

$$\varliminf_{n\to\infty} P\left(\frac{N_n}{n} < \alpha\right) \ge \mathcal{W}\left(\left\{\psi : \int_0^1 \mathbf{1}_{[0,\infty)}(\psi(t))\,dt < \alpha\right\}\right).$$

Finally, since

$$\int_{\mathfrak{P}(\mathbb{R})} \left(\int_0^1 \mathbf{1}_{\{0\}}(\psi(t))\,dt\right) \mathcal{W}(d\psi) = \int_0^1 \mathcal{W}(\psi(t) = 0)\,dt = 0,$$

and $\alpha \in [0,1] \longmapsto \arcsin\big(\sqrt{\alpha}\big)$ is continuous, (4.3.20) follows.† □

The renown of the Arcsine Law stems, in large part, from the following counterintuitive deduction which can be drawn from it. Namely, given $\delta \in (0,1)$, set

$$I(\alpha,\delta) = \big\{t \in [0,1] : |\alpha - t| \wedge |1-\alpha-t| < \delta\big\} \quad \text{for} \quad \alpha \in [0,1],$$

and guess which α maximizes $\lim_{n\to\infty} P\left(\frac{N_n}{n} \in I(\alpha,\delta)\right)$ for a fixed δ. Because of The Law of Large Numbers (in more common parlance, "The Law of Averages"), most people are inclined to guess that the maximum should occur at $\alpha = \frac{1}{2}$. Thus, it is surprising that, since

$$\alpha \in [0,1] \longmapsto \frac{1}{\sqrt{\alpha(1-\alpha)}} \in [0,\infty]$$

is convex and has its minimum at $\frac{1}{2}$, the Arcsine Law makes the exact opposite prediction! The point is, of course, that the sequence of partial sums $\big\{S_n(\omega)\big\}_1^\infty$ is most likely to make long excursions above and below 0, but tends to spend relatively little time in a neighborhood of 0. In other words, although one may be correct to feel that "my luck has got to change," one had better be prepared to wait a long time.

---

† S. Sternberg pointed out that the arcsine distribution is familiar to people studying iterated maps and is important to them because it is the one and only nonatomic probability distribution on $[0,1]$ which is invariant under $x \in [0,1] \longmapsto 4x(1-x) \in [0,1]$. He has asked whether a derivation of the preceding result about Wiener's measure can be based on this observation. Taking $T_+ = \int_0^1 \mathbf{1}_{[0,\infty)}(\psi(s))\,ds$ and $S = \int_0^1 \operatorname{sgn}(\psi(s))\,ds$, what Sternberg is asking is whether there is a *pure thought* way to check that T_+ and S^2 have the same distribution under $\mathcal{W}$. I have posed this problem to several experts but, as yet, none of them has come up with a satisfactory solution.

Before closing this discussion, we will develop some refinements of Feynman and Kac's formula. The first of these refinements is the extension of their formula to semibounded V's. Indeed, it was clear from the outset that at least the semigroup $\{P_t^V : t > 0\}$ is well defined even if V is only bounded above. Thus, one suspects that the Feynman–Kac formula should extend to such V's. In fact, the only problem presented by V's which are unbounded below comes when one wants to interpret the equation in (4.3.14). The author learned the argument which follows from M. Nagasawa.

4.3.21 Theorem. *Let $W : \mathbb{R}^N \longrightarrow [0,\infty)$ be a given measurable function. Then for each $f \in B(\mathbb{R}^N;[0,\infty))$ and $U \in B(\mathbb{R}^N;\mathbb{R})$, the measurable function $t \in (0,\infty) \longmapsto P_t^{U-W} f \in B(\mathbb{R}^N;[0,\infty))$ given by*

$$\text{(4.3.22)} \qquad \begin{aligned} &\left[P_t^{U-W} f\right](\mathbf{x}) \\ &\qquad \equiv \int_{\mathfrak{P}(\mathbb{R}^N)} f(\psi(t)) \exp\left[\int_0^t (U-W)(\psi(s))\,ds\right] \mathcal{W}_{\mathbf{x}}^{(N)}(d\psi) \end{aligned}$$

is the unique nonnegative, measurable $t \in (0,\infty) \longmapsto u(t) \in B(\mathbb{R}^N;[0,\infty))$ satisfying

$$\text{(4.3.23)} \qquad u(t) = P_t^U f - \int_0^t P_{t-s}^U\big(W\,u(s)\big)\,ds, \quad t \in [0,\infty).$$

PROOF: First suppose that u is given by (4.3.22) and let u_n, $n \in \mathbb{Z}^+$, be the function determined by the right-hand side of (4.3.22) after W has been replaced by $W \wedge n$. Then, by exactly the same argument we used to show that the function given in (4.3.12) is a solution to (4.3.14), one sees that u_n solves (4.3.23) with $W \wedge n$ replacing W. Hence, because $u_n \searrow u \geq 0$, (4.3.23) follows from the Monotone Convergence Theorem.

Next, suppose that u is a nonnegative solution to (4.3.23). Our proof that u is given by (4.3.22) will rely on our showing that for each $W' \in B(\mathbb{R}^N;[0,\infty))$ satisfying $W' \leq W$:

$$\text{(4.3.24)} \qquad u(t) = P_t^{U-W'} f - \int_0^t P_{t-s}^{U-W'}\big((W-W')u(s)\big)\,ds, \quad t \in [0,\infty).$$

Indeed, once we know (4.3.24), we get (4.3.22) by taking $W' = W_n \equiv W \wedge n$ in (4.3.24), noting that

$$\int_0^t P_{t-s}^{U-W_n}\big((W-W_n)\big)u(s)\,ds \leq \int_0^t P_{t-s}^U\big(Wu(s)\big)\,ds \leq \left\|P_t^U f\right\|_{\mathrm{u}} < \infty$$

while

$$P_t^{U-W_n} f \searrow P_t^{U-W} f \quad \text{and} \quad P_{t-s}^{U-W_n}\big((W - W_n)\big)u(s) \searrow 0 \quad \text{pointwise},$$

and then applying the Monotone Convergence Theorem.

In order to prove (4.3.24) from (4.3.23), first note that

$$\begin{aligned}\int_0^t P_{t-s}^{U-W'}\big(W' u(s)\big)\, ds = \int_0^t P_{t-s}^{U-W'}\big(W' P_s^U f\big)\, ds \\ - \int_0^t P_{t-s}^{U-W'}\left(W' \int_0^s P_{s-\sigma}^U\big(W u(\sigma)\big)\, d\sigma\right) ds.\end{aligned}$$

Next, set

$$M(\xi, \eta, \psi) = \exp\left[-\int_\xi^\eta W'\big(\psi(\sigma)\big)\, d\sigma\right]$$

and

$$N(\xi, \psi) = \exp\left[\int_0^\xi U\big(\psi(\sigma)\big)\, d\sigma\right]$$

for $\xi, \eta \in [0, \infty)$ and $\psi \in \mathfrak{P}(\mathbb{R}^N)$. Then, by the Markov property and Tonelli's Theorem:

$$\begin{aligned}&\int_0^t \left[P_{t-s}^{U-W'}\big(W' P_s^U f\big)\right](\mathbf{x})\, ds \\ &\quad = -\int_0^t \left(\int_{\mathfrak{P}(\mathbb{R}^N)} f\big(\psi(t)\big) N(t, \psi) \frac{\partial}{\partial s} M(t-s, t, \psi)\, \mathcal{W}_{\mathbf{x}}^{(N)}(d\psi)\right) ds \\ &\quad = \int_{\mathfrak{P}(\mathbb{R}^N)} f\big(\psi(t)\big) N(t, \psi)\big(1 - M(0, t, \psi)\big)\, \mathcal{W}_{\mathbf{x}}^{(N)}(d\psi) \\ &\quad = \big[P_t^U f\big](\mathbf{x}) - \big[P_t^{U-W'} f\big](\mathbf{x}).\end{aligned}$$

Similarly,

$$\begin{aligned}&\int_0^t \left[P_{t-s}^{U-W'}\left(W' \int_0^s P_{s-\sigma}^U\big(W u(\sigma)\big)\, d\sigma\right](\mathbf{x})\right) ds \\ &\quad = \iint\limits_{0 \le \sigma \le s \le t} \left(\int_{\mathfrak{P}(\mathbb{R}^N)} \big[W u(\sigma)\big]\big(\psi(t-\sigma)\big) N(t-\sigma, \psi)\right. \\ &\qquad\qquad \left. \times \frac{\partial}{\partial s} M(0, t-s, \psi)\, \mathcal{W}_{\mathbf{x}}^{(N)}(d\psi)\right) d\sigma ds\end{aligned}$$

$$= \int_0^t \left(\int_{\mathfrak{P}(\mathbb{R}^N)} [Wu(\sigma)]\big(\psi(t-\sigma)\big) N(t-\sigma,\psi) \right.$$

$$\left. \times \big(1 - M(0, t-\sigma, \psi)\big)\, \mathcal{W}_{\mathbf{x}}^{(N)}(d\psi) \right) d\sigma$$

$$= \int_0^t \Big[P_{t-\sigma}^{U}\big(Wu(\sigma)\big)\Big](\mathbf{x})\, d\sigma - \int_0^t \Big[P_{t-\sigma}^{U-W'}\big(W\,u(\sigma)\big)\Big](\mathbf{x})\, d\sigma.$$

Thus, by combining these two with (4.3.23), we arrive at (4.3.24). □

4.3.25 Corollary. *Suppose that $V : \mathbb{R}^N \longrightarrow \mathbb{R}$ is a measurable function which is bounded above, and define $\{P_t^V : t > 0\}$ accordingly. Then, for each $f \in B(\mathbb{R}^N; [0,\infty))$ and $T \in (0,\infty)$, $t \in (0,T] \longmapsto P_t^V f \in B(\mathbb{R}^N; [0,\infty))$ is the unique measurable mapping*

$$t \in (0,T] \longmapsto u(t) \in B\big(\mathbb{R}^N; [0,\infty)\big)$$

which satisfies the relation

$$\text{(4.3.26)} \qquad u(t) = P_t^{V^+} f - \int_0^t P_{t-s}^{V^+}\big(V^- u(s)\big)\, ds, \quad t \in [0,T].$$

In particular, if $u \in C_{\mathrm{b}}^{1,2}\big((0,T) \times \mathbb{R}^N; \mathbb{R}\big) \cap C_{\mathrm{b}}\big([0,T] \times \mathbb{R}^N; [0,\infty)\big)$ solves the Cauchy initial value problem (4.3.11), then $u(t,\mathbf{x}) = \big[P_t^V f\big](\mathbf{x})$ for all $(t,\mathbf{x}) \in [0,T] \times \mathbb{R}^N$.

PROOF: The characterization of $t \in [0,\infty) \longmapsto P_t^V f \in B\big(\mathbb{R}^N; [0,\infty)\big)$ for nonnegative $f \in B(\mathbb{R}^N;\mathbb{R})$ in terms of (4.3.26) is an immediate application of Theorem 4.3.21 when one takes $U = V^+$ and $W = V^-$. Moreover, if $u \in C_{\mathrm{b}}^{1,2}\big((0,\infty) \times \mathbb{R}^N; \mathbb{R}\big)$ is a nonnegative solution to the Cauchy initial value problem in (4.3.17), then, by (4.3.16) with V^+ replacing V,

$$\begin{aligned} -\int_0^t P_{t-s}^{V^+}\big(V^- u(s)\big)\, ds &= \int_0^t P_{t-s}^{V^+}\left(\tfrac{\partial u}{\partial s} - \tfrac{1}{2}\boldsymbol{\Delta} u - V^+ u\right)(s)\, ds \\ &= \int_0^t \frac{d}{ds} P_{t-s}^{V^+}\big(u(s)\big)\, ds = u(t) - P_t^{V^+} f, \end{aligned}$$

and so $t \in [0,\infty) \longmapsto u(t,\,\cdot\,)$ is a nonnegative solution to (4.3.26). □

The second refinement which we want to make concerns the operators P_t^V for V's of the sort in Corollary 4.3.25. Obviously,

$$\big[P_t^V f\big](\mathbf{x}) = \int_{\mathbb{R}^N} f(\mathbf{y})\, P^V(t,\mathbf{x},d\mathbf{y}),$$

where $P^V(t,\mathbf{x},\,\cdot\,)$ is the finite Borel measure on $\mathbb{R}^N$ given by

$$P^V(t,\mathbf{x},\Gamma) = \mathbb{E}^{\mathcal{W}_{\mathbf{x}}^{(N)}}\left[\exp\left(\int_0^t V(\psi(s))\,ds\right),\ \psi(t)\in\Gamma\right]$$

for $\Gamma \in \mathcal{B}_{\mathbb{R}^N}$. Moreover, the measurability properties of the P_t^V's translate immediately into the measurability of

$$(t,\mathbf{x}) \in [0,\infty)\times\mathbb{R}^N \longmapsto P^V(t,\mathbf{x},\Gamma) \in [0,\infty)$$

for each $\Gamma \in \mathcal{B}_{\mathbb{R}^N}$; and the semigroup property $P_{s+t}^V = P_s^V \circ P_t^V$ becomes the **Chapman–Kolmogorov equation**

$$P^V(s+t,\mathbf{x},\Gamma) = \int_{\mathbb{R}^N} P^V(t,\mathbf{y},\Gamma)\,P^V(s,\mathbf{x},d\mathbf{y}),\quad \Gamma\in\mathcal{B}_{\mathbb{R}^N} \tag{4.3.27}$$

for all $(s,t) \in (0,\infty)$. Thus far, all this represents is an alternative formulation of things we already know from the first part of Corollary 4.3.25. However, the result which follows provides us with significant new information.

4.3.28 Theorem. *Let $V : \mathbb{R}^N \longmapsto \mathbb{R}$ be a measurable function which is bounded above, and define $(t,\mathbf{x}) \longmapsto P^V(t,\mathbf{x},\,\cdot\,) \in \mathbf{M}(\mathbb{R}^N)$ as in the preceding discussion. Next, set (cf. Theorem 4.2.18)*

$$r^V(t,\mathbf{x},\mathbf{y}) = \mathbb{E}^{\mathcal{W}^{(N)}}\left[\exp\left(\int_0^t V\left((1-\tfrac{s}{t})\mathbf{x} + \tilde{\psi}_t(s) + \tfrac{s}{t}\mathbf{y}\right)ds\right)\right] \tag{4.3.29}$$

for $(t,\mathbf{x},\mathbf{y}) \in (0,\infty)\times\mathbb{R}^N\times\mathbb{R}^N$. Then, for each $(t,\mathbf{x}) \in (0,\infty)\times\mathbb{R}^N$,

$$\begin{aligned} &P^V(t,\mathbf{x},d\mathbf{y}) = p^V(t,\mathbf{x},\mathbf{y})\,d\mathbf{y}\\ &\qquad\qquad \text{where } p^V(t,\mathbf{x},\mathbf{y}) \equiv \gamma_t^N(\mathbf{y}-\mathbf{x})r^V(t,\mathbf{x},\mathbf{y}). \end{aligned} \tag{4.3.30}$$

Moreover,

$$p^V(t,\mathbf{x},\mathbf{y}) = p^V(t,\mathbf{y},\mathbf{x}),\quad (t,\mathbf{x},\mathbf{y}) \in (0,\infty)\times\mathbb{R}^N\times\mathbb{R}^N;$$

and therefore

$$\int_{\mathbb{R}^N} g(\mathbf{x})\,[P_t^V f](\mathbf{x})\,d\mathbf{x} = \int_{\mathbb{R}^N} f(\mathbf{x})\,[P_t^V g](\mathbf{x})\,d\mathbf{x}$$

for all nonnegative, measurable f and g on $\mathbb{R}^N$. Finally, when V is continuous as well as bounded above, $(t,\mathbf{x},\mathbf{y}) \in (0,\infty)\times\mathbb{R}^N\times\mathbb{R}^N \longmapsto p^V(t,\mathbf{x},\mathbf{y}) \in (0,\infty)$ is continuous.

PROOF: To prove (4.3.30), let $f \in B(\mathbb{R}^N;\mathbb{R})$ be given and apply (4.2.20) to the function $F : \mathfrak{P}(\mathbb{R}^N) \longmapsto \mathbb{R}$ given by

$$F(\psi) = \exp\left[\int_0^t V(\psi(s))\,ds\right] f(\psi(t)), \quad \psi \in \mathfrak{P}(\mathbb{R}^N),$$

in order to see that

$$\int_{\mathbb{R}^N} f(\mathbf{y})\,P^V(t,\mathbf{x},d\mathbf{y}) = \int_{\mathbb{R}^N} f(\mathbf{y})\,r^V(t,\mathbf{x},\mathbf{y})\gamma_t^N(\mathbf{y}-\mathbf{x})\,d\mathbf{y}.$$

As for the symmetry assertion, simply note that, because $\psi \longmapsto \tilde{\psi}_T \restriction [0,T]$ is reversible (cf. Theorem 4.2.18), $r^V(t,\mathbf{x},\mathbf{y}) = r^V(t,\mathbf{y},\mathbf{x})$. Finally, if V is continuous, then the continuity of r^V on $(0,\infty)\times\mathbb{R}^N\times\mathbb{R}^N$ becomes an easy application of Lebesgue's Dominated Convergence Theorem. □

From a probabilistic standpoint, Feynman–Kac semigroups are flawed by their failure to leave the function $\mathbf{1}$ invariant, with the consequence that the corresponding measures $P^V(t,\mathbf{x},\,\cdot\,)$ cannot be interpreted as the distribution of paths at time t. From the Schrödinger standpoint, this failure is a reflection of the fact that $\mathbf{1}$ is not *the ground-state* of the quantum mechanical system with Hamiltonian $\frac{1}{2}\boldsymbol{\Delta}+V$. On the other hand, from the physical standpoint (cf. the footnote following (4.3.11)), there is nothing sacrosanct about any particular representation of the Hamiltonian: any unitarily equivalent representation is just as good. Thus, physicists will often force $\mathbf{1}$ to be the ground-state by performing what they call a transformation to the **ground-state representation**, and, as a consequence, they produce a situation which is amenable to a nice probabilistic interpretation. In order to avoid difficulties, we will restrict our attention here to V's for which it is easy to find the ground-state. In fact, we will cheat by starting with a candidate and producing the potential for which it is the ground-state. To be precise, let $U \in C^2(\mathbb{R}^N;\mathbb{R})$ be given, set

$$V^U = \tfrac{1}{2}\boldsymbol{\Delta}U - \tfrac{1}{2}|\nabla U|^2, \tag{4.3.31}$$

and observe that e^{-U} is the ground-state for $\frac{1}{2}\boldsymbol{\Delta}+V^U$ in the sense that $\frac{1}{2}\boldsymbol{\Delta}e^{-U}+V^Ue^{-U}=0$. In particular, this means that the map taking a function φ to $e^U\varphi$ transforms a solution to (4.3.11) with initial condition e^Uf into a solution to

$$\frac{\partial w}{\partial t} = \tfrac{1}{2}\boldsymbol{\Delta}w - \nabla U\cdot\nabla w \text{ in } [0,\infty)\times\mathbb{R}^N \text{ with } w(0,\,\cdot\,) = f. \tag{4.3.32}$$

As distinguished from the operator $\frac{1}{2}\boldsymbol{\Delta}+V^U$, the operator $\frac{1}{2}\boldsymbol{\Delta}-\nabla U\cdot\nabla$ annihilates constants. Thus, there is reason to hope that the corresponding semigroup "$e^{t(\frac{1}{2}\boldsymbol{\Delta}-\nabla U\cdot\nabla)}$" will preserve the function $\mathbf{1}$ and, if everything works out well, will admit a nice probabilistic interpretation.

In order to implement the preceding program, let $U \in C^2(\mathbb{R}^N;\mathbb{R})$ be given, take V^U as in (4.3.31), and define $R^U : [0,\infty) \times \mathfrak{P}(\mathbb{R}^N) \longrightarrow [0,\infty)$ by

$$R^U(t,\psi) = \exp\left[U\big(\psi(0)\big) - U\big(\psi(t)\big) + \int_0^t V^U\big(\psi(s)\big)\,ds\right]. \tag{4.3.33}$$

Finally, define $Q_t^U f$ for $t \in [0,\infty)$ and measurable $f : \mathbb{R}^N \longrightarrow [0,\infty]$ to be the function given by

$$\big[Q_t^U f\big](\mathbf{x}) = \mathbb{E}^{\mathcal{W}_{\mathbf{x}}^{(N)}}\Big[f\big(\psi(t)\big)\,R^U(t)\Big], \quad \mathbf{x} \in \mathbb{R}^N. \tag{4.3.34}$$

Obviously, $(t,\mathbf{x}) \in [0,\infty) \times \mathbb{R}^N \longmapsto \big[Q_t^U f\big](\mathbf{x}) \in [0,\infty]$ is measurable for each nonnegative, measurable f on $\mathbb{R}^N$.

4.3.35 Lemma. *Assume that both $-U$ and the associated function V^U in (4.3.31) are bounded above. If $w \in C^{1,2}\big((0,T)\times\mathbb{R}^N;\mathbb{R}\big) \cap C\big([0,T]\times\mathbb{R}^N;[0,\infty)\big)$ satisfies (4.3.32) and if $e^{-U}w \in C_{\mathrm{b}}^{1,2}\big([0,T]\times\mathbb{R}^N;[0,\infty)\big)$, then*

$$w(t,\mathbf{x}) = \big[Q_t^U f\big](\mathbf{x}), \quad (t,\mathbf{x}) \in [0,T] \times \mathbb{R}^N.$$

In particular, if $e^{-U} \in C_{\mathrm{b}}^2\big(\mathbb{R}^N;[0,\infty)\big)$, then $Q_t^U\mathbf{1} = \mathbf{1}$ for all $t \in [0,\infty)$.

PROOF: Set $u(t) = e^{-U}w(t,\,\cdot\,)$ and observe that

$$\tfrac{\partial u}{\partial t} = \tfrac{1}{2}\boldsymbol{\Delta} u + V^U u \quad \text{on } (0,T)\times\mathbb{R}^N.$$

Hence, by the final part of Corollary 4.3.25,

$$w(t,\,\cdot\,) = e^U u(t) = e^U\,P_t^{V^U}\big(u(0,\,\cdot\,)\big) = Q_t^U f \quad \text{for } t \in [0,T].$$

In particular, when $e^{-U} \in C_{\mathrm{b}}^2(\mathbb{R}^N;\mathbb{R})$, we can apply this to $w = \mathbf{1}$. □

4.3.36 Theorem. *Let $U \in C^2(\mathbb{R}^N;\mathbb{R})$ be given, define V^U and R^U from U as in (4.3.31) and (4.3.33), and assume that*

$$\big(-U(\mathbf{x})\big) \vee V^U(\mathbf{x}) \le C\big(1+|\mathbf{x}|\big), \quad \mathbf{x} \in \mathbb{R}^N, \tag{4.3.37}$$

for some $C \in (0,\infty)$. Then, for each $\mathbf{x} \in \mathbb{R}^N$, there is a unique $Q_{\mathbf{x}}^U \in \mathbf{M}_1\big(\mathfrak{P}(\mathbb{R}^N)\big)$ with the property that

$$Q_{\mathbf{x}}^U(A) = \mathbb{E}^{\mathcal{W}_{\mathbf{x}}^{(N)}}\big[R^U(t),\,A\big] \quad \text{for all } t \in [0,\infty) \text{ and } A \in \mathcal{B}_t. \tag{4.3.38}$$

Moreover, the map $\mathbf{x} \in \mathbb{R}^N \longmapsto Q_{\mathbf{x}}^U \in \mathbf{M}_1\big(\mathfrak{P}(\mathbb{R}^N)\big)$ is continuous. Finally, for every $\{\mathcal{B}_t : t \in [0,\infty)\}$-stopping time τ and every bounded $F : \mathfrak{P}(\mathbb{R}^N)^2 \longrightarrow \mathbb{R}$ which is $\mathcal{B}_\tau \times \mathcal{B}_{\mathfrak{P}(\mathbb{R}^N)}$-measurable,

$$\begin{aligned}&\int_{\{\psi:\tau(\psi)<\infty\}} F\big(\psi, \boldsymbol{\Sigma}_{\tau(\psi)}\psi\big)\, Q_{\mathbf{x}}^U(d\psi)\\ &\qquad = \int_{\{\varphi:\tau(\varphi)<\infty\}} \left(\int_{\mathfrak{P}(\mathbb{R}^N)} F(\varphi,\psi)\, Q_{\varphi(\tau)}^U(d\psi)\right) Q_{\mathbf{x}}^U(d\varphi).\end{aligned} \tag{4.3.39}$$

PROOF: We begin with some preparations. Let $T \in (0,\infty)$ be given, and suppose that τ is a $\{\mathcal{B}_t : t \in [0,\infty)\}$-stopping time which is dominated by T. Given a bounded, $\mathcal{B}_\tau \times \mathcal{B}_{\mathfrak{P}(\mathbb{R}^N)}$-measurable F for which $\psi \in \mathfrak{P}(\mathbb{R}^N) \longmapsto F\big(\psi, \boldsymbol{\Sigma}_{\tau(\psi)}\psi\big) \in \mathbb{R}$ is $\mathcal{B}_T$-measurable, set

$$\tilde{F}(\varphi,\psi) = R^U\big(\tau(\varphi),\varphi\big)\, R^U\big(T-\tau(\varphi),\psi\big)\, F(\varphi,\psi), \quad (\varphi,\psi) \in \mathfrak{P}(\mathbb{R}^N)^2,$$

and observe that $\tilde{F}$ is $\mathcal{B}_\tau \times \mathcal{B}_T$-measurable and that

$$R^U(T,\psi)F\big(\psi, \boldsymbol{\Sigma}_{\tau(\psi)}\psi\big) = \tilde{F}\big(\psi, \boldsymbol{\Sigma}_{\tau(\psi)}\psi\big).$$

Hence, by (4.3.9),

$$\begin{aligned}&\int F\big(\psi, \boldsymbol{\Sigma}_{\tau(\psi)}\psi\big)\, R^U(T,\psi)\, \mathcal{W}_{\mathbf{x}}^{(N)}(d\psi)\\ &\qquad = \int R^U\big(\tau(\varphi),\varphi\big)\\ &\qquad\qquad \times \left(\int F(\varphi,\psi)\, R^U\big(T-\tau(\varphi),\psi\big)\mathcal{W}_{\varphi(\tau)}^{(N)}(d\psi)\right) \mathcal{W}_{\mathbf{x}}^{(N)}(d\varphi).\end{aligned} \tag{4.3.40}$$

The first step in the proof that $Q_{\mathbf{x}}^U$ exists is to check that

$$\mathbb{E}^{\mathcal{W}_{\mathbf{x}}^{(N)}}\big[R^U(\tau)\big] = 1, \quad \mathbf{x} \in \mathbb{R}^N \tag{4.3.41}$$

for all bounded $\{\mathcal{B}_t : t \in [0,\infty)\}$-stopping times τ. To this end, take $F = \mathbf{1}$ in (4.3.40) and thereby obtain (cf. (4.3.34))

$$\big[Q_T^U\mathbf{1}\big](\mathbf{x}) = \int R^U\big(\tau(\varphi),\varphi\big)\, \big[Q_{T-\tau(\varphi)}^U\mathbf{1}\big]\big(\varphi(\tau)\big)\, \mathcal{W}_{\mathbf{x}}^{(N)}(d\varphi)$$

for any τ which is dominated by T. In particular, if $U \in C^2_{\mathrm{b}}(\mathbb{R}^N;[0,\infty))$, then the last part of Lemma 4.3.35 says that $Q^U_t\mathbf{1} = \mathbf{1}$ and therefore (4.3.41) holds in this case. Next, to handle the general case, choose $\eta \in C^\infty(\mathbb{R};[0,1])$ so that

$$\eta(\mathbf{x}) = \begin{cases} 1 & \text{if } \;|\mathbf{x}| \le 1 \\ 0 & \text{if } \;|\mathbf{x}| \ge 2, \end{cases}$$

and, for $n \in \mathbb{Z}^+$, define $U_n \in C^2_{\mathrm{b}}(\mathbb{R}^N;\mathbb{R})$ by $U_n(\mathbf{x}) = \eta\left(\frac{\mathbf{x}}{n}\right) U(\mathbf{x})$. At the same time, given a τ bounded by T, set

$$\tau_n = \tau \wedge \zeta^{B_{\mathbb{R}^N}(\mathbf{0},n)} \quad \text{where } \zeta^{B_{\mathbb{R}^N}(\mathbf{0},n)}(\psi) = \inf\left\{t : \psi(t) \notin B_{\mathbb{R}^N}(\mathbf{0},n)\right\}$$

is the first exit time from $B_{\mathbb{R}^N}(\mathbf{0},n)$. Then, because U_n and U coincide on $B_{\mathbb{R}^N}(\mathbf{0},n)$, (4.3.41) for $U = U_n$ and $\tau = \tau_n$ immediately implies (4.3.41) for U and $\tau = \tau_n$. Moreover, by (4.3.37), we know that, for all $0 \le t \le T$:

$$R^U(t,\psi) \le \exp\left[U(\psi(0)) + C(2+T)\left(1 + \sup_{t\in[0,T]} |\psi(t)|\right)\right], \tag{4.3.42}$$

and therefore (because $\zeta^{B_{\mathbb{R}^N}(\mathbf{0},n)} \nearrow \infty$), (4.3.41) for U and τ follows easily from (3.3.30) and Lebesgue's Dominated Convergence Theorem.

With the preceding in hand, we can now prove the existence and uniqueness of $Q^U_{\mathbf{x}}$ as an application of Exercise 3.3.26. Namely, define $\mu_n \in \mathbf{M}_1(\mathfrak{P}(\mathbb{R}^N))$ for $n \in \mathbb{Z}^+$ so that

$$\mu_n(A) = \mathbb{E}^{\mathcal{W}^{(N)}_{\mathbf{x}}}\left[R^U(n), A\right] \quad \text{for all } A \in \mathcal{B}_{\mathfrak{P}(\mathbb{R}^N)}.$$

Given $n \in \mathbb{Z}^+$ and $A \in \mathcal{B}_n$, we take $T = n+1$, $\tau = n$, and $F = \mathbf{1}_A$ in (4.3.40) and conclude (from (4.3.41) with $\tau \equiv 1$) that

$$\mu_{n+1}(A) = \int_A R^U(n,\varphi)\left(\int R^U(1,\psi)\,\mathcal{W}^{(N)}_{\varphi(n)}(d\psi)\right)\mathcal{W}^{(N)}_{\mathbf{x}}(d\varphi) = \mu_n(A).$$

Hence, the hypotheses in Exercise 3.3.26 are trivially satisfied by $\{\mu_n\}_1^\infty$, and therefore both the existence and uniqueness of $Q^U_{\mathbf{x}}$ are assured. Moreover, again by (4.3.42), (3.3.30), and Lebesgue's Dominated Convergence Theorem, the map $\mathbf{x} \in \mathbb{R}^N \longmapsto \mathbb{E}^{Q^U_{\mathbf{x}}}[F] \in \mathbb{R}$ is continuous for any $F \in C_{\mathrm{b}}(\mathfrak{P}(\mathbb{R}^N);\mathbb{R})$ which is $\mathcal{B}_T$-measurable for some $T \in [0,\infty)$; and so the continuity of $\mathbf{x} \in \mathbb{R}^N \longmapsto Q^U_{\mathbf{x}} \in \mathbf{M}_1(\mathfrak{P}(\mathbb{R}^N))$ can also be seen as an application of Exercise 3.3.26.

To complete the proof, note that (4.3.39) and (4.3.40) coincide when $\tau \le T$ and $\psi \in \mathfrak{P}(\mathbb{R}^N) \longmapsto F(\psi, \Sigma_{\tau(\psi)}\psi) \in \mathbb{R}$ is $\mathcal{B}_T$-measurable; and starting from this case, the general result follows from elementary measuretheoretic considerations. □

Put into the jargon of probability theory, the property of the family $\{Q^U_{\mathbf{x}} : \mathbf{x} \in \mathbb{R}^N\}$ expressed by (4.3.39) is the statement that, like the family $\{\mathcal{W}^{(N)}_{\mathbf{x}} : \mathbf{x} \in \mathbb{R}^N\}$, this family has the strong Markov property (cf. Exercise 4.3.55).

4.3.43 Corollary. *Let everything be as in Theorem 4.3.36, and define the operators Q^U_t, $t \in [0,\infty)$, accordingly, as in (4.3.34). Then each Q^U_t admits a unique extension as a contraction on $B(\mathbb{R}^N;\mathbb{R})$ and $Q^U_{s+t} = Q^U_s \circ Q^U_t$ for all $s, t \in [0,\infty)$. Furthermore, there exists a continuous $(t,\mathbf{x}) \in [0,\infty) \times \mathbb{R}^N \longmapsto Q^U(t,\mathbf{x},\cdot\,) \in \mathbf{M}_1(\mathbb{R}^N)$ with the property*

$$\big[Q^U_t f\big](\mathbf{x}) = \int_{\mathbb{R}^N} f(\mathbf{y})\, Q^U(t,\mathbf{x},d\mathbf{y}), \quad (t,\mathbf{x}) \in [0,\infty) \times \mathbb{R}^B$$

for each $f \in B(\mathbb{R}^N;\mathbb{R})$; and the Chapman–Kolmogorov equation

$$Q^U(s+t,\mathbf{x},\cdot\,) = \int_{\mathbb{R}^N} Q^U(t,\mathbf{y},\cdot\,)\, Q^U(s,\mathbf{x},d\mathbf{y})$$

holds for all $s, t \in [0,\infty)$ and $\mathbf{x} \in \mathbb{R}^N$. Finally, if $r^{V^U}(t,\mathbf{x},\mathbf{y})$ denotes the right-hand side of (4.3.29) with $V = V^U$ and $q^U : (0,\infty) \times \mathbb{R}^N \times \mathbb{R}^N \longrightarrow (0,\infty)$ is defined by

$$q^U(t,\mathbf{x},\mathbf{y}) = e^{U(\mathbf{x})}\, \gamma^N_t(\mathbf{y}-\mathbf{x})\, r^{V^U}(t,\mathbf{x},\mathbf{y})\, e^{U(\mathbf{y})}, \tag{4.3.44}$$

then q^U is continuous, $(\mathbf{x},\mathbf{y}) \in \mathbb{R}^N \times \mathbb{R}^N \longmapsto q^U(t,\mathbf{x},\mathbf{y})$ is symmetric for each $t \in (0,\infty)$, and

$$Q^U(t,\mathbf{x},d\mathbf{y}) = q^U(t,\mathbf{x},\mathbf{y})\, \mu^U(d\mathbf{y}) \quad \text{where } \mu^U(d\mathbf{y}) \equiv e^{-2U(\mathbf{y})}\, d\mathbf{y}.$$

In particular,

$$q^U(s+t,\mathbf{x},\mathbf{y}) = \int_{\mathbb{R}^N} q^U(s,\mathbf{x},\boldsymbol{\xi})\, q^U(t,\boldsymbol{\xi},\mathbf{y})\, \mu^U(d\boldsymbol{\xi}).$$

PROOF: Seeing as

$$\big[Q^U_t f\big](\mathbf{x}) = \mathbb{E}^{Q^U_{\mathbf{x}}}\big[f(\psi(t))\big],$$

the first assertion which requires any comment is the one about the semigroup property of the operators Q^U_t. But, by (4.3.39) with $\tau \equiv s$ and $F(\varphi,\psi) = f(\psi(t))$,

$$\begin{aligned}
\big[Q^U_{s+t} f\big](\mathbf{x}) &= \int_{\mathfrak{P}(\mathbb{R}^N)} f(\psi(s+t))\, Q^U(d\psi) \\
&= \int_{\mathfrak{P}(\mathbb{R}^N)} \left(\int_{\mathfrak{P}(\mathbb{R}^N)} f(\psi(t))\, Q^U_{\varphi(s)}(d\psi) \right) Q^U(d\varphi) \\
&= \big[Q^U_s \circ Q^U_t f\big](\mathbf{x}).
\end{aligned}$$

Next, let q^U be defined on $(0,\infty)\times\mathbb{R}^N\times\mathbb{R}^N$ as in (4.3.44). To see that q^U is continuous, it suffices to check that r^{V^U} is continuous. But, by (4.3.37) and (4.3.26),

$$\begin{aligned}\sup_{t\in[0,T]}\int_0^t V^U\Bigl(\bigl(1-\tfrac{s}{t}\bigr)\mathbf{x}+\tilde{\psi}_t(s)+\tfrac{s}{t}\mathbf{y}\Bigr)\,ds\\ \le CT\left(1+|\mathbf{x}|+|\mathbf{y}|+2\sup_{s\in[0,T]}|\psi(s)|\right),\end{aligned}$$

and so the desired continuity follows from (3.3.30) and Lebesgue's Dominated Convergence Theorem. Finally, both the symmetry of q^U and the equation

$$\bigl[Q_t^U f\bigr](\mathbf{x})=\int_{\mathbb{R}^N} f(\mathbf{y})\,q^U(t,\mathbf{x},\mathbf{y})\,\mu^U(d\mathbf{y})$$

are easy corollaries of Theorem 4.2.18; and therefore all the remaining assertions are immediate from what we have already proved. □

With Theorem 4.3.36 and Corollary 4.3.43, we have certainly shown that *transforming to the ground-state* is, from a probabilistic standpoint, a good idea which provides us with an interesting source of strong Markov processes. However, as yet, we do not know much about these processes. For example, we do not know how properties of U are reflected in the distributional properties of their paths. For this reason, we will return to the study of these processes in Section 7.5, where we will interpret their paths as *Wiener paths which have been perturbed by a conservative force field.*

Exercises

4.3.45 Exercise: As was mentioned in the footnote following their introduction, our presentation of stopping times and their associated σ-algebras is based on definitions made originally by E.B. Dynkin. Although his definitions are very elegant and have served us well, his description of the σ-algebra $\mathcal{B}_\tau$ is a little opaque and does not fully capture the intuition on which it is based. For this reason, it may be helpful to develop the following alternative description of $\mathcal{B}_\tau$ as the σ-algebra $\sigma\bigl(\psi(t\wedge\tau):t\in[0,\infty)\bigr)$ over $\mathfrak{P}(\mathbb{R}^N)$ generated by the maps $\bigl\{\psi(t\wedge\tau),\,t\in[0,\infty)\bigr\}$. Since, by Lemma 4.3.7, it is clear that $\sigma\bigl(\psi(t\wedge\tau):t\in[0,\infty)\bigr)\subseteq\mathcal{B}_\tau$, we will only deal with the opposite inclusion.

(i) Suppose that $f:[0,\infty]\times\mathfrak{P}(\mathbb{R}^N)\longmapsto\mathbb{R}$ is a $\mathcal{B}_{[0,\infty]}\times\mathcal{B}_\tau$-measurable function with the property that $\psi\in\mathfrak{P}(\mathbb{R}^N)\longmapsto f(t,\psi)\in\mathbb{R}$ is $\mathcal{B}_t$-measurable for each

$t \in [0,\infty)$. Show that there exists a measurable $F_f : [0,\infty] \times \mathbb{R}^{\mathbb{Z}^+} \longrightarrow \mathbb{R}$ and a sequence $\{s_n\}_1^\infty \subseteq [0,\infty)$ such that

$$f(t,\psi) = F_f\big(t, \psi(s_1 \wedge t), \dots, \psi(s_n \wedge t), \dots\big)$$

for all $(t,\psi) \in [0,\infty) \times \mathfrak{P}(\mathbb{R}^N)$.

(ii) Apply part **(i)** to show that if $f : \mathfrak{P}(\mathbb{R}^N) \longrightarrow \mathbb{R}$ is $\mathcal{B}_\tau$-measurable, then there exists a measurable F_f and $\{s_n\}_1^\infty \subseteq [0,\infty)$ such that

$$\mathbf{1}_{\{t\}}\big(\tau(\psi)\big)\, f(\psi) = F_f\big(t, \psi(s_1 \wedge t), \dots, \psi(s_n \wedge t), \dots\big).$$

First use this with $f = \mathbf{1}$ to see that $\tau(\psi) = t \in [0,\infty)$ and $\varphi(s) = \psi(s)$ for $s \in [0,t]$ imply $\tau(\varphi) = t$; and conclude that τ itself is $\sigma\big(\psi(t \wedge \tau) : t \in [0,\infty)\big)$-measurable. Finally, complete the proof by letting f be any $\mathcal{B}_\tau$-measurable function and noting that

$$f(\psi) = F_f\big(\tau(\psi), \psi(s_1 \wedge \tau), \dots, \psi(s_n \wedge t), \dots\big).$$

4.3.46 Exercise: For $a \in (0,\infty)$, let $\nu_a \in \mathbf{M}_1(\mathbb{R})$ denote the distribution of

$$\psi \longmapsto \tau_a(\psi) = \inf\Big\{t \in [0,\infty) : \psi(t) \ge a\Big\}$$

under $\mathcal{W}$. From (4.3.5), we know that

$$\nu_a(dt) = \frac{a\mathbf{1}_{(0,\infty)}(t)}{\sqrt{2\pi}\, t^{\frac{3}{2}}} \exp\left[-\frac{a^2}{2t}\right] dt.$$

(i) Noting that

$$\tau_{a+b}(\psi) = \tau_a(\psi) + \tau_b\big(\boldsymbol{\delta}_{\tau_a(\psi)}\psi\big) \quad \text{when} \quad \psi(0) = 0,$$

use Theorem 4.3.3 to see that

$$\nu_{a+b} = \nu_a \bigstar \nu_b, \quad a,\, b \in (0,\infty), \tag{4.3.47}$$

Next, use Wiener scaling invariance to see that τ_a has the same distribution under $\mathcal{W}$ as $a^2\tau_1$, and combine this with the preceding to conclude that, for $\lambda \in (0,\infty)$,

$$f_a(\lambda) \equiv \int_{(0,\infty)} e^{-\lambda t}\, \nu_a(dt) = \exp\Big[-c\, a\, \lambda^{\frac{1}{2}}\Big] \quad \text{where} \quad c \equiv -\log\big[f_1(1)\big] > 0.$$

Finally, use

$$-c = \lim_{\alpha\to\infty} \frac{1}{\alpha} \log\Big[f_1(\alpha^2)\Big] = \lim_{\alpha\to\infty} \frac{1}{\alpha} \log\left[\int_{(0,\infty)} \exp\left[-\alpha\left(\frac{1}{2t}+t\right)\right] \frac{dt}{t^{\frac{3}{2}}}\right]$$

together with Exercise 1.3.17 to see that $c = \sqrt{2}$ and therefore that

$$\int_{(0,\infty)} e^{-\lambda t}\,\nu_a(dt) = \exp\Big[-a\sqrt{2\lambda}\Big], \quad a,\,\lambda \in (0,\infty). \tag{4.3.48}$$

As a consequence of (4.3.47), we see that $\big\{\nu_a : a \in (0,\infty)\big\}$ is a convolution semigroup and therefore that each of the distributions ν_a is infinitely divisible in the sense discussed in Section 3.2. Moreover, starting from (4.3.48), extending both sides analytically to the open right half-plane in $\mathbb{C}$, and passing to the limit as $\mathfrak{Re}(\lambda) \searrow 0$, come to

$$\widehat{\nu_a}(\xi) = \exp\left[\frac{a}{\sqrt{2\pi}} \int_{(0,\infty)} \left(e^{\sqrt{-1}\,\xi y} - 1\right) \frac{dy}{y^{\frac{3}{2}}}\right],$$

which is, of course, a special case of the Lévy–Khinchine formula derived in Section 3.2. Because they are concentrated on a half-line and have the scaling property summarized in (4.3.48), the distributions ν_a are called the **one-sided stable laws of order $\frac{1}{2}$**.

4.3.49 Exercise: Given $N \in \mathbb{Z}^+\setminus\{1\}$, let $\mathbb{R}^N_+$ denote the open upper half-space $\big\{(\mathbf{x},y) \in \mathbb{R}^{N-1}\times\mathbb{R} : y > 0\big\}$ and denote by

$$\psi \in \mathfrak{P}(\mathbb{R}^N) \longmapsto \zeta(\psi) = \inf\Big\{t \in [0,\infty) : \psi(t) \notin \mathbb{R}^N_+\Big\}$$

the first exit time from $\mathbb{R}^N_+$. Clearly, for $y \in (0,\infty)$, the distribution of ζ under $\mathcal{W}^{(N)}_{(\mathbf{x},y)}$ is the probability measure ν_y in Exercise 4.3.46. Noting that the Nth coordinate $\psi(\zeta)_N$ of $\psi(\zeta)$ is $\mathcal{W}^{(N)}$-almost surely 0, think of the distribution $P^{(N-1)}_{(\mathbf{x},y)}$ of $\psi \longmapsto \psi(\zeta)$ under $\mathcal{W}^{(N)}_{(\mathbf{x},y)}$ as being an element of $\mathbf{M}_1\big(\mathbb{R}^{N-1}\big)$, and use the independence provided by Exercise 3.3.28 to see that $P^{(N-1)}_{(\mathbf{x},y)}$ admits the density $\boldsymbol{\eta} \in \mathbb{R}^{N-1} \longmapsto p^{(N-1)}_y(\boldsymbol{\eta}-\mathbf{x}) \in (0,\infty)$ with respect to Lebesgue's measure on $\mathbb{R}^{N-1}$, where

$$\begin{aligned} p^{(N-1)}_y(\mathbf{x}) &= \int_{(0,\infty)} \gamma^{N-1}_t(\mathbf{x})\,\nu_y(dt) \\ &= \frac{2}{\omega_{N-1}} \frac{y}{\big(y^2+|\mathbf{x}|^2\big)^{\frac{N}{2}}}. \end{aligned} \tag{4.3.50}$$

(The constant

$$\omega_{N-1} = \frac{2\pi^{\frac{N}{2}}}{\Gamma(\frac{N}{2})}$$

in (4.3.50) is the surface area of the unit sphere $\mathbf{S}^{N-1}$ in $\mathbb{R}^N$.) Next, by the same sort of reasoning as was used to get (4.3.47), show that

$$p^{(N-1)}_{y_1+y_2} = p^{(N-1)}_{y_1} \bigstar p^{(N-1)}_{y_2} \quad \text{for all} \quad (y_1, y_2 \in (0,\infty).$$

Thus, once again, we are dealing with a convolution semigroup of infinitely divisible distributions. In addition, these distributions also possess a scaling property: the distribution of $\boldsymbol{\eta} \in \mathbb{R}^{N-1} \longmapsto y\,\boldsymbol{\eta}$ under $P^{(N-1)}_{(\mathbf{0},1)}$ is $P^{(N-1)}_{(\mathbf{0},y)}$. Finally, use the first line of (4.3.50) together with (4.3.48) to see that

$$\widehat{P^{(N-1)}_{(\mathbf{0},y)}}(\boldsymbol{\xi}) = \exp\bigl[-y\,|\boldsymbol{\xi}|\bigr], \quad y \in (0,\infty) \text{ and } \boldsymbol{\xi} \in \mathbb{R}^{N-1}.$$

The measures $P^{(N-1)}_{(\mathbf{x},y)}$ are called **Cauchy distributions** by probabilists, and, because they arise as the in connection with the harmonic analysis of $\mathbb{R}^N_+$, harmonic analysts call the densities $p^{(N-1)}_y$ **Poisson's kernel** (cf. Section 8.3, especially (8.2.19), for more information about the connection with harmonic analysis). Finally, the first equality in (4.3.50) is a classic example of **subordination**.

4.3.51 Exercise: In this exercise, we will make a somewhat subtle application of the fact that Wiener paths are continuous.
(**i**) For, $r \in (0,\infty)$, set

$$\sigma_r(\boldsymbol{\psi}) = \inf\bigl\{t \in [0,\infty) : |\boldsymbol{\psi}(t) - \boldsymbol{\psi}(0)| \geq r\bigr\},$$

and show that, for any $\mathbf{z} \in \mathbb{R}^N$, the distribution of $\boldsymbol{\psi} \in \mathfrak{P}(\mathbb{R}^N) \longmapsto \boldsymbol{\psi}(\tau_r)$ under $\mathcal{W}^{(N)}$ is uniform on the sphere $\mathbf{S}^{N-1}(\mathbf{z}, r) \equiv \{\boldsymbol{\eta} \in \mathbb{R}^N : |\boldsymbol{\eta} - \mathbf{z}| = r\}$.

Hint: Use the rotation invariance coming from Exercise 3.3.28 to show that the distribution in question must be rotation invariant.

(**ii**) Consider the position of a point which evolves in $\mathbb{R}^N_+$ (cf. Exercise 4.3.49) according to the following random prescription. At time 0 the point is at $\mathbf{Z}_0 = (\mathbf{X}_0, Y_0) = (\mathbf{0}, 1)$, where it remains for one time unit, when it moves to a randomly chosen point $\mathbf{Z}_1 = (\mathbf{X}_1, Y_1)$ on the sphere $\mathbf{S}^{N-1}(\mathbf{Z}_0, Y_0)$, where it again rests for a unit time before moving to a randomly chosen point $\mathbf{Z}_2 = (\mathbf{X}_2, Y_2)$ on $\mathbf{S}^{N-1}(\mathbf{Z}_1, Y_1)$, etc. Interpreting this to mean that:

$$\mathbf{Z}_0 = (\mathbf{0}, 1) \quad \text{and} \quad \mathbf{Z}_n = \mathbf{Z}_{n-1} + Y_{n-1}\boldsymbol{\alpha}_n, \quad n \in \mathbb{Z}^+,$$

where $\{\boldsymbol{\alpha}_n\}_1^\infty$ are independent, uniformly distributed $\mathbf{S}^{N-1}$-valued random variables, the problem is to show that, with probability 1, $\mathbf{Z}_n \longrightarrow (\mathbf{X}_\infty, 0)$, where $\mathbf{X}_\infty$ has Cauchy distribution $P^{(N-1)}_{(\mathbf{0},1)}$. The principle underlying this example is sometimes called the **method of balayage**; see Exercise 8.2.37 for a generalization.

Hint: Define $\tau_0 \equiv 0$ and, for $n \in \mathbb{Z}^+$,

$$\tau_n(\psi) = \inf\left\{t \in \left[\tau_{n-1}, \infty\right) : \left|\psi(t) - \psi(\tau_{n-1})\right| \geq \psi(\tau_{n-2})_N\right\}$$

if $\tau_{m-1}(\psi) < \infty$ and $\tau_n(\psi) = \infty$ otherwise. Show (cf. Lemma 4.3.2) that the τ_n's are $\{\mathcal{B}_t : t \in [0, \infty)\}$-stopping times and that, when $\psi(0)_N \geq 0$, $\tau_n(\psi) \leq \zeta(\psi)$ (cf. Exercise 4.3.49). In particular, conclude first that $\mathcal{W}^{(N)}_{(\mathbf{0},1)}\left(\tau_n < \infty\right) = 1$ and second that the distribution of $\left\{\psi(\tau_n) : n \in \mathbb{N}\right\}$ under $\mathcal{W}^{(N)}_{(\mathbf{0},1)}$ is the same as the distribution of the family $\left\{\mathbf{Z}_n : n \in \mathbb{N}\right\}$ defined above. Finally, complete the proof by showing that $\tau_n(\psi) \nearrow \zeta(\psi)$ if $\psi(0)_N = 1$ and $\zeta(\psi) < \infty$, since otherwise there would exist an $r_\psi > 0$, $T_\psi \in (0, \infty)$, and $0 = t_0 < \cdots < t_n < \cdots \leq T_\psi$ such that

$$\inf_{n \in \mathbb{Z}^+} \left|\psi(t_n) - \psi(t_{n-1})\right| \geq r_\psi .$$

4.3.52 Exercise: Refer to the setting in Theorem 4.3.36 and Corollary 4.3.43. There are only a few cases in which it is possible to write down an expression for the quantity $q^U(t, \mathbf{x}, \mathbf{y})$ (cf. (4.3.44)) which is tractable. In fact, among the only ones for which such an expression is known are U's of the form

$$U_\alpha(\mathbf{x}) = \frac{\alpha|\mathbf{x}|^2}{4} + \frac{N}{4}\log\frac{2\pi}{\alpha} \quad \text{and therefore} \quad V^{U_\alpha}(\mathbf{x}) = -\alpha^2\frac{|\mathbf{x}|^2}{8} + \frac{N\alpha}{4},$$

where $\alpha \in (0, \infty)$. In what follows, q_α will be used to denote the function q^{U_α}.

(**i**) To find an expression for q_α, begin by observing that if

$$\tfrac{\partial u}{\partial t} = \tfrac{1}{2}\mathbf{\Delta} u \quad \text{in } (0, \infty) \times \mathbb{R}^N$$

and

$$w_\alpha(t, \mathbf{x}) \equiv u\left(\frac{1 - e^{-\alpha t}}{\alpha}, e^{-\frac{\alpha t}{2}}\mathbf{x}\right)$$

then

$$\tfrac{\partial w_\alpha}{\partial t} = \tfrac{1}{2}\mathbf{\Delta} w_\alpha - \nabla U_\alpha \cdot \nabla w_\alpha .$$

Next, as an application of Lemma 4.3.35, conclude that

$$\begin{aligned} q_\alpha(t, \mathbf{x}, \mathbf{y}) &= \left(\frac{2\pi}{\alpha}\right)^{\frac{N}{2}} \gamma^N_{\frac{1-e^{-\alpha t}}{\alpha}}\left(\mathbf{y} - e^{-\frac{\alpha t}{2}}\mathbf{x}\right) e^{\frac{\alpha|\mathbf{y}|^2}{2}} \\ &= \left(\frac{1}{1 - e^{-\alpha t}}\right)^{\frac{N}{2}} \exp\left[-\frac{\alpha}{2}\,\frac{e^{-\alpha t}|\mathbf{x}|^2 - 2e^{-\frac{\alpha t}{2}}(\mathbf{x}, \mathbf{y})_{\mathbb{R}^N} + e^{-\alpha t}|\mathbf{y}|^2}{1 - e^{-\alpha t}}\right]. \end{aligned} \tag{4.3.53}$$

In particular, when $N = 1$ and $\alpha = 1$, check that this is consistent with the result obtained in part (**i**) of Exercise 2.3.41.

(ii) Take $N = 1$ and $\alpha = 1$ in the preceding, and starting from (4.3.53), show that

$$\mathbb{E}^{\mathcal{W}_x}\left[-\frac{1}{8}\int_0^t |\psi(s)|^2\,ds\right] = e^{-\frac{t+|x|^2}{4}}\int_{\mathbb{R}} \gamma_{1-e^{-t}}\left(y - e^{-\frac{t}{2}}x\right)e^{\frac{y^2}{4}}\,dy$$
$$= \left(\cosh\tfrac{t}{2}\right)^{-\frac{1}{2}}\exp\left[-\tfrac{|x|^2}{4}\tanh\tfrac{t}{2}\right];$$

and use this together with Wiener scaling to give another derivation of (4.2.37).

4.3.54 Exercise: Take $\alpha = 1$ in Exercise 4.3.52 and use $Q_{\mathbf{x}}$ to denote $Q_{\mathbf{x}}^{U_1} \in \mathbf{M}_1\big(\mathfrak{P}(\mathbb{R}^N)\big)$ for each $\mathbf{x} \in \mathbb{R}^N$.

(i) By combining (4.3.53) with the Markov property (cf. (4.3.40)) for $\{Q_{\mathbf{x}} : \mathbf{x} \in \mathbb{R}^N\}$, show that, for each $s \in [0,\infty)$, $t \in (0,\infty)$, and $\mathbf{x} \in \mathbb{R}^N$,

$$\psi \in \mathfrak{P}(\mathbb{R}^N) \longmapsto \psi(s+t) - e^{-\frac{t}{2}}\psi(s) \in \mathbb{R}^N$$

under $Q_{\mathbf{x}}$ is an $\mathbb{R}^N$-valued random variable which is independent of $\mathcal{B}_s$ and is Gaussian with mean-value $\mathbf{0}$ and covariance $(1 - e^{-t})\,\mathbf{I}$.

(ii) Using the result in part (i) and proceeding by induction on $n \in \mathbb{Z}^+$, show that for any $n \in \mathbb{Z}^+$, $0 \le t_1 \cdots < t_n$, and $\boldsymbol{\xi}_1, \dots, \boldsymbol{\xi}_n \in \mathbb{R}^N$,

$$\psi \in \mathfrak{P}(\mathbb{R}^N) \longmapsto \sum_{m=1}^{n} \big(\boldsymbol{\xi}_m, \psi(t_m)\big)_{\mathbb{R}^N} \in \mathbb{R}$$

under $Q_{\mathbf{0}}$ is Gaussian with mean-value $\mathbf{0}$ and variance

$$\sum_{k,\ell=1}^{n}\left(e^{-\frac{|t_\ell - t_k|}{2}} - e^{-\frac{|t_\ell + t_k|}{2}}\right)\big(\boldsymbol{\xi}_\ell, \boldsymbol{\xi}_k\big)_{\mathbb{R}^N}.$$

In particular, conclude that (cf. part (ii) of Exercise 4.2.39) $\big\{(\boldsymbol{\xi}, \psi(t)) : \boldsymbol{\xi} \in \mathbb{R}^N \,\&\, t \in [0,\infty)\big\}$ under $Q_{\mathbf{0}}$ is a centered Gaussian family with covariance function

$$\big[\mathbf{C}(s,t)\big](\boldsymbol{\xi}, \boldsymbol{\eta}) = \left(e^{-\frac{|t-s|}{2}} - e^{-\frac{|t+s|}{2}}\right)(\boldsymbol{\xi}, \boldsymbol{\eta})_{\mathbb{R}^N}.$$

Finally, after referring to Exercise 4.2.40, use this to identify $Q_{\mathbf{0}}$ with the image of the Ornstein–Uhlenbeck measure $\mathcal{U}^{(N)}$ under the trivial embedding $\boldsymbol{\theta} \in \Theta(\mathbb{R}^N) \longmapsto \boldsymbol{\theta} \in \mathfrak{P}(\mathbb{R}^N)$

4.3.55 Exercise: In this exercise we will introduce and discuss the abstract version of the (strong) Markov property. To this end, let E be a Polish space and define the σ-algebras $\mathcal{B}_s$, $s \in [0,\infty)$, and the notion of a $\{\mathcal{B}_t : t \in [0,\infty)\}$-**stopping time** $\tau : \mathfrak{P}(E) \longrightarrow [0,\infty]$ and the associated σ-algebra $\mathcal{B}_\tau$ by analogy with their definitions when $E = \mathbb{R}^N$ (cf. (4.1.3) and (4.3.1)). Obviously, the results in Lemma 4.3.2 and Exercise 4.3.45 are unaffected by the replacement of $\mathbb{R}^N$ with more general E. Next, given a collection of measures $P_x \in \mathbf{M}_1(\mathfrak{P}(E))$, we will say that $\{P_x : x \in E\}$ is a **Markov family** if $x \in E \longmapsto P_x \in \mathbf{M}_1(\mathfrak{P}(E))$ is measurable, $P_x(\psi(0) = x) = 1$ for every $x \in E$, and, for every $(s,x) \in [0,\infty) \times E$ and bounded, $\mathcal{B}_s \times \mathcal{B}_{\mathfrak{P}(E)}$-measurable $F : \mathfrak{P}(E)^2 \longmapsto \mathbb{R}$,

$$\begin{aligned}\int_{\mathfrak{P}(E)} & F(\psi, \mathbf{\Sigma}_s\psi)\, P_x(d\psi) \\ &= \int_{\mathfrak{P}(E)} \left(\int_{\mathfrak{P}(E)} F(\varphi,\psi)\, P_{\varphi(s)}(d\psi) \right) P_x(d\varphi),\end{aligned}$$

where $\mathbf{\Sigma}_s : \mathfrak{P}(E) \longrightarrow \mathfrak{P}(E)$ is the **time-shift** mapping defined by $[\mathbf{\Sigma}_s\psi](t) = \psi(t+s)$, $t \in [0,\infty)$. If, in addition, for every $\{\mathcal{B}_t : t \in [0,\infty)\}$-stopping time τ, every bounded $\mathcal{B}_\tau \times \mathcal{B}_{\mathfrak{P}(\mathbb{R}^N)}$-measurable $F : \mathfrak{P}(E)^2 \longmapsto \mathbb{R}$, and every $x \in E$:

$$\begin{aligned}\int_{\{\psi:\tau(\psi)<\infty)\}} & F(\psi, \mathbf{\Sigma}_{\tau(\psi)}\psi)\, P_x(d\psi) \\ &= \int_{\{\varphi:\tau(\varphi)<\infty\}} \left(\int_{\mathfrak{P}(E)} F(\varphi,\psi)\, P_{\varphi(\tau)}(d\psi) \right) P_x(d\varphi),\end{aligned} \tag{4.3.56}$$

then we say that $\{P_x : x \in E\}$ is a **strong Markov family**.

(**i**) Given a measurable map $x \in E \longmapsto P_x \in \mathbf{M}_1(\mathfrak{P}(E))$ with the property that $P_x(\Psi(0) = x) = 1$ for every $x \in E$, set

$$P(t,x,\Gamma) = P_x(\{\psi : \psi(t) \in \Gamma\}), \quad (t,x,\Gamma) \in (0,\infty) \times E \times \mathcal{B}_E,$$

check that **transition probability function** $(t,x) \in [0,\infty)\times E \longmapsto P(t,x,\cdot\,) \in \mathbf{M}_1(E)$ is measurable, and show that $\{P_x : x \in E\}$ is a Markov family if and only if, for every $n \in \mathbb{Z}^+$, $0 \le t_0 < \cdots < t_n < \infty$, and $\Gamma_0, \ldots, \Gamma_n \in \mathcal{B}_E$:

$$\begin{aligned}P_x & \Big(\{\psi : \psi(t_0) \in \Gamma_0, \ldots, \psi(t_n) \in \Gamma_n\} \Big) \\ &= \int_{A(t_0,\ldots,t_{n-1};\Gamma_0,\ldots,\Gamma_{n-1})} P(t_n - t_{n-1}, \psi(t_{n-1}), \Gamma_n)\, P_x(d\psi),\end{aligned} \tag{4.3.57}$$

where

$$A(t_0,\dots,t_{n-1};\Gamma_0,\dots,\Gamma_{n-1}) \equiv \{\psi : \psi(t_0)\in\Gamma_0,\dots,\psi(t_{n-1})\in\Gamma_{n-1}\}.$$

In particular, if $\{P_x : x \in E\}$ is a Markov family, show that $P(t,x,\,\cdot\,)$ satisfies the **Chapman–Kolomogorov equation**

$$P(s+t,x,\Gamma) = \int_E P(t,y,\Gamma)\,P(s,x,dy) \tag{4.3.58}$$

for all $(s,t,x,\Gamma) \in (0,\infty)^2 \times E \times \mathcal{B}_E$.

(ii) Suppose that $(t,x) \in [0,\infty) \times E \longmapsto P(s,x,\,\cdot\,) \in \mathbf{M}_1(E)$ is a measurable map satisfying (4.3.58), and show that there is at most one Markov family $\{P_x : x \in E\}$ for which (4.3.57) can hold.

(iii) Following Dynkin and Yushkevich,† show that if $\{P_x : x \in E\}$ is a Markov family and $x \in E \longmapsto P_x \in \mathbf{M}_1(\mathfrak{P}(E))$ is continuous, then $\{P_x : x \in E\}$ is a strong Markov family.

Hint: Note that one need only prove (4.3.56) when

$$\psi \in \mathfrak{P}(E) \longmapsto F(\varphi,\psi) \in \mathbb{R}$$

is continuous for each $\varphi \in \mathfrak{P}(E)$, and apply the approximation procedure used in the proof of Theorem 4.3.3.

† See E.B. Dynkin & A.A. Yushkevich's "Strong Markov processes," *Theory of Prob. & Appl.* **1** (1956). In fact, see everything to which Dynkin's name is attached in that volume.

Chapter V:
Conditioning and Martingales

§5.1: Conditioning.

Up to this point we have been dealing with random variables which are either themselves mutually independent or are built out of other random variables which are. For this reason, it has not been necessary for us to make explicit use of the concept of *conditioning*, although, as we will see shortly, this concept has been lurking silently in the background.

Let $(\Omega, \mathcal{F}, P)$ be a probability space, and suppose that $A \in \mathcal{F}$ is a set of positive P-measure. For reasons which are most easily understand when Ω is finite and P is uniform, the ratio

$$(5.1.1) \qquad P(B|A) \equiv \frac{P(A \cap B)}{P(A)}, \quad B \in \mathcal{F},$$

is called the **conditional probability of B given A**. As one learns in an elementary course, the introduction of conditional probabilities makes many calculations much simpler; in particular, conditional probabilities help to clarify dependence relations between the events represented by A and B. For example, B is independent of A precisely when $P(B|A) = P(B)$ or, in words, *when the condition that A occurs does not change the probability that B occurs.* Thus, it is unfortunate that the naïve definition of conditioning as described in (5.1.1) does not cover many important situations. For example, suppose that X and Y are random variables and that one wants to talk about the conditional probability that $Y \le b$ given that $X = a$. Obviously, unless one is very lucky and $P(X = a) > 0$, (5.1.1) is not going to do the job. Hence, it is of great importance to generalize the concept of conditional probability to include situations when the event on which one is conditioning has P-measure 0, and in this section we will present Kolmogorov's elegant solution to this problem.

In order to appreciate the idea behind Kolmogorov's solution, imagine someone told you the conditional probability that the event B occurs given that the event A occurs. Obviously, since you have no way of saying anything about the probability of B when A does not occur, she has provided you with incomplete information about B. Thus, before you are satisfied, you should demand to know what is the conditional probability of B given that A does not occur. Of course, this second piece of information is relevant only if A *is not certain*, in which case $P(A) < 1$ and therefore $P(B|A\complement)$ is well defined. More generally, suppose that $\mathcal{P} = \{A_1, \ldots, A_N\}$ (N here may be either finite or countably infinite) is a partition of Ω into elements of $\mathcal{F}$ of positive P-measure. Then, in order to have complete information about the probability of $B \in \mathcal{F}$ relative to $\mathcal{P}$, one has to know the entire list of the numbers $P(B|A_n)$, $1 \le n \le N$. Next, suppose that we attempt to describe this list in a way which does not depend explicitly on the positivity of the numbers $P(A_n)$. For this purpose, consider the function

$$\omega \in \Omega \longmapsto f(\omega) \equiv \sum_{n=1}^{N} P(B|A_n)\,\mathbf{1}_{A_n}(\omega).$$

Obviously, f is not only $\mathcal{F}$-measurable, it is measurable with respect to the σ-algebra $\sigma(\mathcal{P})$ over Ω generated by $\mathcal{P}$. In particular (because the only $\sigma(\mathcal{P})$-measurable set of P-measure 0 is empty), f is uniquely determined by its P-integrals $\mathbb{E}^P[f, A]$ over sets $A \in \sigma(\mathcal{P})$. Moreover, because, for each $A \in \sigma(\mathcal{P})$ and n, either $A_n \subseteq A$ or $A \cap A_n = \emptyset$, we have that

$$\mathbb{E}^P[f, A] = \sum_{n=1}^{N} P(B|A_n)\,P(A \cap A_n) = \sum_{\{n: A_n \subseteq A\}} P(A_n \cap B) = P(A \cap B).$$

Hence, the function f is uniquely determined by the property that

$$\mathbb{E}^P[f, A] = P(A \cap B) \quad \text{for every} \quad A \in \sigma(\mathcal{P}).$$

The beauty of this description is that it makes perfectly good sense even if some of the A_n's have P-measure 0, only in that case the description would not determine f pointwise but merely up to a $\sigma(\mathcal{P})$-measurable P-**null set** (i.e., a set of P-measure 0), which is the very least one should expect to pay for *dividing by* 0.

With the preceding discussion in mind, one ought to find the following formulation reasonable. Namely, given a sub-σ-algebra $\Sigma \subseteq \mathcal{F}$ and a $(-\infty, \infty]$-valued random variable X for which $X^-(\equiv -(X \wedge 0))$ is P-integrable , we will say that the random variable X_Σ is a **conditional expectation of X given Σ** if X_Σ is $(-\infty, \infty]$-valued and Σ-measurable, $(X_\Sigma)^-$ is P-integrable, and

$$\mathbb{E}^P[X_\Sigma, A] = \mathbb{E}^P[X, A] \quad \text{for every } A \in \Sigma. \tag{5.1.2}$$

Obviously, having made this definition, our first order of business is to show that such an X_Σ always exists and to discover in what sense it is uniquely determined. The latter problem is dealt with by the following lemma.

5.1.3 Lemma. *Let Σ be a sub-σ-algebra of $\mathcal{F}$, and suppose that X_Σ and Y_Σ are a pair of $(-\infty,\infty]$-valued Σ-measurable random variables for which X_Σ^- and Y_Σ^- are both P-integrable. Then*

$$\mathbb{E}^P\big[X_\Sigma,\, A\big] \le \mathbb{E}^P\big[Y_\Sigma,\, A\big] \quad \textit{for every } A \in \Sigma,$$

if and only if $X_\Sigma \le Y_\Sigma$ (a.s., P).

PROOF: Without loss in generality, we may and will assume that $\Sigma = \mathcal{F}$ and will therefore drop the subscript Σ; and, since the "if" implication is completely trivial, we will only discuss the minimally less trivial "only if" assertion. Thus, suppose that P-integrals of Y dominate those of X and yet that $X > Y$ on a set of positive P-measure. We could then choose an $M \in [1,\infty)$ so that $P(A) \vee P(B) > 0$ where

$$A \equiv \big\{X \le M \text{ and } Y \le X - \tfrac{1}{M}\big\} \quad \text{and} \quad B \equiv \big\{X = \infty \text{ and } Y \le M\big\}.$$

But if $P(A) > 0$, then

$$\mathbb{E}^P\big[X,\, A\big] \le \mathbb{E}^P\big[Y,\, A\big] \le \mathbb{E}^P\big[X,\, A\big] - \tfrac{1}{M}P(A),$$

which, because $\mathbb{E}^P\big[X,\, A\big]$ is a finite number, is impossible. At the same time, if $P(B) > 0$, then

$$\infty = \mathbb{E}^P\big[X,\, B\big] \le \mathbb{E}^P\big[Y,\, B\big] \le M < \infty,$$

which is also impossible. □

5.1.4 Theorem. *Let Σ be a sub-σ-algebra of $\mathcal{F}$ and X a $(-\infty,\infty]$-valued random variable for which X^- is P-integrable. Then there exists a conditional expectation value X_Σ of X. Moreover, if Y is a second $(-\infty,\infty]$-valued random variable and $Y \ge X$* (a.s., P), *then Y^- is P-integrable and $Y_\Sigma \ge X_\Sigma$* (a.s., P) *for any Y_Σ which is a conditional expectation value of Y given Σ. In particular, if $X = Y$* (a.s., P), *then $\big\{Y_\Sigma \ne X_\Sigma\big\}$ is a Σ-measurable, P-null set.*†

PROOF: In view of Lemma 5.1.3, it suffices for us to handle the initial existence statement. To this end, let $\mathcal{G}$ denote the class of X satisfying $\mathbb{E}^P\big[X^-\big] < \infty$ for which an X_Σ exists, and let $\mathcal{G}^+$ denote the nonnegative elements of $\mathcal{G}$. If $\{X_n\}_1^\infty \subseteq \mathcal{G}^+$ is nondecreasing and, for each $n \in \mathbb{Z}^+$, $\big(X_n\big)_\Sigma$ denotes a conditional expectation of X_n given Σ, then $0 \le \big(X_n\big)_\Sigma \le \big(X_{n+1}\big)_\Sigma$ (a.s., P), and therefore we can arrange that $0 \le \big(X_n\big)_\Sigma \le \big(X_{n+1}\big)_\Sigma$ everywhere. In particular,

† Kolmogorov himself, and most authors ever since, have obtained the existence of conditional expectation values as a consequence of the Radon–Nikodym Theorem. Because I find projections more intuitively appealing, I prefer the approach given here.

if X and X_Σ are the pointwise limits of the X_n's and $\left(X_n\right)_\Sigma$'s, respectively, then the Monotone Convergence Theorem guarantees that X_Σ is a conditional expectation of X given Σ. Hence, we now know that $\mathcal{G}^+$ is closed under nondecreasing, pointwise limits, and therefore we will know that $\mathcal{G}^+$ contains all nonnegative random variables X as soon as we show that $\mathfrak{G}$ contains all bounded X's. But if X is bounded (and is therefore an element of $L^2(P;\mathbb{R})$) and $\mathbf{L}_\Sigma = L^2(\Omega,\Sigma,P;\mathbb{R})$ is the subspace of $L^2(P;\mathbb{R})$ consisting of its Σ-measurable elements, then the orthogonal projection X_Σ of X onto $\mathbf{L}_\Sigma$ is a Σ-measurable random variable which is square P-integrable and satisfies (5.1.2).

So far we have proved that $\mathcal{G}^+$ contains all nonnegative X's. Furthermore, if X is nonnegative, then (by Lemma 5.1.3) $X_\Sigma \geq 0$ (a.s., P) and so X_Σ is P-integrable precisely when X itself is. In particular, we can arrange to make X_Σ take its values in $[0,\infty)$ when X is nonnegative and P-integrable. Finally, to see that $X \in \mathfrak{G}$ for every X with $\mathbb{E}^P\left[X^-\right] < \infty$, simply consider X^+ and X^- separately, apply the preceding to show that $\left(X^\pm\right)_\Sigma \geq 0$ (a.s., P) and that $\left(X^-\right)_\Sigma$ is P-integrable, and check that the random variable

$$X_\Sigma \equiv \begin{cases} \left(X^+\right)_\Sigma - \left(X^-\right)_\Sigma & \text{when } \left(X^\pm\right)_\Sigma \geq 0 \text{ and } \left(X^-\right)_\Sigma < \infty \\ 0 & \text{otherwise} \end{cases}$$

is a conditional expectation of X given Σ. □

5.1.5 Convention. Because it is determined only up to a Σ-measurable P-null set, one cannot, in general, talk about *the* conditional expectation of X as a *function*. Instead, the best that one can do is say that **the conditional expectation of** X is the equivalence class of Σ-measurable X_Σ's which satisfy (5.1.2), and we will adopt the notation $\mathbb{E}^P\left[X\middle|\Sigma\right]$ to denote this equivalence class. On the other hand, because one is usually interested only in P-integrals of conditional expectations, it has become common practice to ignore, for the most part, the distinction between the equivalence class $\mathbb{E}^P\left[X\middle|\Sigma\right]$ and the members of that equivalence class. Thus (just as one would when dealing with the Lebesgue spaces) we will abuse notation by using $\mathbb{E}^P\left[X\middle|\Sigma\right]$ to denote a generic element of the equivalence class $\mathbb{E}^P\left[X\middle|\Sigma\right]$, and will be more precise only when $\mathbb{E}^P\left[X\middle|\Sigma\right]$ contains some particularly distinguished member. For example, recall the random variables $\mathbf{T}_n$ entering the definition of the simple Poisson process $\{\mathbf{N}(t) : t \in (0,\infty)\}$ (cf. (3.2.23)). Then, by (3.2.25), it is clear that we can take

$$\mathbb{E}^P\left[\mathbf{1}_{\{n\}}(\mathbf{N}(t))\,\middle|\,\sigma(\mathbf{T}_1,\dots,\mathbf{T}_n)\right] = \mathbf{1}_{[0,t]}(\mathbf{T}_n)e^{-(t-\mathbf{T}_n)},$$

and we would be foolish to take any other representative. More generally, we will always take nonnegative representatives of $\mathbb{E}^P\left[X\middle|\Sigma\right]$ when X itself is nonnegative and $\mathbb{R}$-valued representatives when X is P-integrable. Finally, for historical reasons, it is usual to distinguish the case when X is the indicator function $\mathbf{1}_B$ of a set $B \in \mathcal{F}$ and to call $\mathbb{E}^P[\mathbf{1}_B|\Sigma]$ the **conditional probability of B given Σ** and to write $P(B|\Sigma)$ instead of $\mathbb{E}^P[\mathbf{1}_B|\Sigma]$. Of course, representatives of $P(B|\Sigma)$ will always be assumed to take their values in $[0,1]$.

Once one has established the existence and uniqueness of conditional expectations, there is a long list of more or less obvious properties which one can easily verify. The following theorem contains some of the more important items which might appear on such a list.

5.1.6 Theorem. *Let Σ be a sub-σ-algebra of $\mathcal{F}$. If X is a P-integrable random variable and $\mathcal{C} \subseteq \Sigma$ is a π-system (cf. Exercise 1.1.12) which generates Σ, then*

$$\begin{gathered} Y = \mathbb{E}^P[X|\Sigma] \quad (\text{a.s.}, P) \iff \\ Y \in L^1(\Omega, \Sigma, P) \text{ and } \mathbb{E}^P[Y, A] = \mathbb{E}^P[X, A] \text{ for } A \in \mathcal{C} \cup \{\Omega\}. \end{gathered} \tag{5.1.7}$$

Moreover, if X is any $(-\infty,\infty]$-valued random variable which satisfies $\mathbb{E}^P[X^-] < \infty$, then each of the following relations holds P-almost surely:

$$\left|\mathbb{E}^P[X|\Sigma]\right| \le \mathbb{E}^P[|X||\Sigma]; \tag{5.1.8}$$

$$\mathbb{E}^P[X|\mathcal{T}] = \mathbb{E}^P\Big[\mathbb{E}^P[X|\Sigma]\,\Big|\mathcal{T}\Big] \tag{5.1.9}$$

when $\mathcal{T}$ is a sub-σ-algebra of Σ; and, when X is $\mathbb{R}$-valued and P-integrable,

$$\mathbb{E}^P[-X|\Sigma] = -\mathbb{E}^P[X|\Sigma]. \tag{5.1.10}$$

Next, let Y be a second $(-\infty,\infty]$-valued random variable with $\mathbb{E}^P[Y^-] < \infty$. Then, P-almost surely:

$$\mathbb{E}^P[\alpha X + \beta Y|\Sigma] = \alpha\mathbb{E}^P[X|\Sigma] + \beta\mathbb{E}^P[Y|\Sigma] \tag{5.1.11}$$

for each α, $\beta \in [0,\infty)$, and

$$\mathbb{E}^P[YX|\Sigma] = Y\,\mathbb{E}^P[X|\Sigma] \tag{5.1.12}$$

if Y is Σ-measurable and $(XY)^-$ is and P-integrable.
Finally, suppose that $\{X_n\}_1^\infty$ is a sequence of $(-\infty,\infty]$-valued random variables. Then, P-almost surely:

$$\mathbb{E}^P[X_n|\Sigma] \nearrow \mathbb{E}^P[X|\Sigma] \tag{5.1.13}$$

if $\mathbb{E}^P[X_1^-] < \infty$ and $X_n \nearrow X$ (a.s., P); and

$$\mathbb{E}^P\left[\left.\varliminf_{n\to\infty} X_n\right|\Sigma\right] \le \varliminf_{n\to\infty} \mathbb{E}^P[X_n|\Sigma] \tag{5.1.14}$$

if $X_n \ge 0$ (a.s., P) for each $n \in \mathbb{Z}^+$.

PROOF: To prove (5.1.7), note that the set of $A \in \Sigma$ for which $\mathbb{E}^P[X, A] = \mathbb{E}^P[Y, A]$ is (cf. Exercise 1.1.12) a λ-system which contains $\mathcal{C}$ and therefore Σ. Next, clearly (5.1.8) is just an application of Lemma 5.1.3, while (5.1.9) through (5.1.11) are all expressions of uniqueness. (5.1.12) is also an expression of uniqueness when Y is the indicator function of a element of Σ, follows for general nonnegative X's and Y's by taking monotone limits of the case when Y is simple, and is completed by considering $(XY)^+$ and $(XY)^-$ separately. Finally, (5.1.13) is an immediate application of the Monotone Convergence Theorem; whereas (5.1.14) comes from the conjunction of

$$\mathbb{E}^P\left[\inf_{n\geq m} X_n \middle| \Sigma\right] \leq \inf_{n\geq m} \mathbb{E}^P\left[X_n \middle| \Sigma\right] \quad \text{(a.s., } P), \quad m \in \mathbb{Z}^+,$$

with (5.1.13). □

It probably will have occurred to most readers that the properties expressed by (5.1.8), (5.1.10), (5.1.11), and (5.1.13) give strong evidence that, for fixed $\omega \in \Omega$, $X \longmapsto \mathbb{E}^P\left[X\middle|\Sigma\right](\omega)$ behaves like an integral (in the sense of Daniell) and therefore ought to be expressible in terms of integration with respect to a probability measure P_ω. Indeed, if one could actually talk about $X \longmapsto \mathbb{E}^P\left[X\middle|\Sigma\right](\omega)$ for a fixed (as opposed to P-almost every) $\omega \in \Omega$, then there is no doubt that such a P_ω would have to exist. Thus, it is reasonable to ask whether there are circumstances in which one can gain sufficient control over all the P-null sets involved to really make sense out of $X \longmapsto \mathbb{E}^P\left[X\middle|\Sigma\right](\omega)$ for fixed $\omega \in \Omega$. One answer to this question is contained in the following theorem.

5.1.15 Theorem. *Suppose that Ω is a Polish space and that $\mathcal{F} = \mathcal{B}_\Omega$. Then, for every sub-$\sigma$-algebra Σ of $\mathcal{F}$, there is a P-almost surely unique Σ-measurable map $\omega \in \Omega \longmapsto P^\Sigma_\omega \in \mathbf{M}_1(\Omega)$ with the property that*

$$(5.1.16) \qquad P(A \cap B) = \int_A P^\Sigma_\omega(B)\, P(d\omega) \quad \textit{for all } A \in \Sigma \textit{ and } B \in \mathcal{F}.$$

In particular, for each $(-\infty, \infty]$-valued random variable X which is bounded below,

$$\omega \in \Omega \longmapsto \mathbb{E}^{P^\Sigma_\omega}\left[X\right]$$

is a conditional expectation value of X given Σ. Finally, if Σ is countably generated, then $\omega \in \Omega \longmapsto P^\Sigma_\omega$ can be chosen so that

$$(5.1.17) \qquad P^\Sigma_\omega(A) = \mathbf{1}_A(\omega) \quad \textit{for all } \omega \in \Omega \textit{ and } A \in \Sigma.$$

PROOF: To prove the uniqueness, suppose $\omega \in \Omega \longmapsto Q^\Sigma_\omega \in \mathbf{M}_1(\Omega)$ were a second such mapping. We would then know that, for each $B \in \mathcal{F}$, $Q^\Sigma_\omega(B) = P^\Sigma_\omega(B)$ for P-almost every $\omega \in \Omega$. Hence, since $\mathcal{F}$ (as the Borel field over a second countable topological space) is countably generated, we could find one Σ-measurable P-null set off of which $Q_\omega^\Sigma = P_\omega^\Sigma$. Similarly, to prove the final assertion when Σ is countably generated, note (cf. (5.1.12)) that, for each $A \in \Sigma$, $P_\omega^\Sigma(A) = \mathbf{1}_A(\omega) = \delta_\omega(A)$ for P-almost every $\omega \in \Omega$. Hence, once again countability allows us to choose one Σ-measurable P-null set $\mathcal{N}$ such that $P_\omega^\Sigma \restriction \Sigma = \delta_\omega \restriction \Sigma$ if $\omega \notin \mathcal{N}$. Finally, let $\mathcal{C} \subseteq \Sigma$ be a countable generating set for Σ, and, for each $\omega \in \Omega$, set

$$A_\Sigma(\omega) \equiv \bigcap \Big\{ A \ni \omega : \ A \in \mathcal{C} \text{ or } A\complement \in \mathcal{C} \Big\},$$

and note that $A_\Sigma(\omega)$ is the **atom** in Σ containing ω (i.e., for every $A \in \Sigma$, either $\omega \in A$ and $A_\Sigma(\omega) \subseteq A$ or $\omega \notin A$ and $A_\Sigma(\omega) \cap A = \emptyset$.) Thus, if we modify $\omega \in \Omega \longmapsto P_\omega^\Sigma$ by taking

$$P_\omega^\Sigma(B) = \begin{cases} 1 & \text{if} \quad B \supseteq A_\Sigma(\omega) \\ 0 & \text{otherwise} \end{cases} \qquad \text{for} \quad \omega \in \mathcal{N} \text{ and } B \in \mathcal{B}_\Omega,$$

then, by the preceding, it is easy to check that $\omega \in \Omega \longmapsto P_\omega^\Sigma$ is still a conditional probability distribution of P given Σ and that now (5.1.17) holds.

We next turn to the question of existence. For this purpose, first choose (cf. (**ii**) of Lemma 3.1.4) ρ to be a totally bounded metric for Ω, and let $\mathcal{U} = U_b^\rho(\Omega; \mathbb{R})$ be the space of bounded, ρ-uniformly continuous, $\mathbb{R}$-valued functions on Ω. Then (cf. (**iii**) of Lemma 3.1.4) $\mathcal{U}$ is a separable Banach space with respect to the uniform norm. In particular, we can choose a sequence $\{f_n\}_0^\infty \subseteq \mathcal{U}$ so that $f_0 = \mathbf{1}$, the functions $f_0, \dots, f_n$ are linearly independent for each $n \in \mathbb{Z}^+$, and the linear span $\mathcal{S}$ of $\{f_n : n \in \mathbb{N}\}$ is dense in $\mathcal{U}$. Set $g_0 = \mathbf{1}$, and, for each $n \in \mathbb{Z}^+$, let g_n be some fixed representative of $\mathbb{E}^P\left[f_n \middle| \Sigma\right]$. Next, set

$$\mathfrak{R} = \Big\{ \boldsymbol{\alpha} \in \mathbb{R}^{\mathbb{N}} : \exists m \in \mathbb{N} \ \alpha_n = 0 \text{ for all } n \geq m \Big\}$$

and define

$$f_{\boldsymbol{\alpha}} = \sum_{n=0}^\infty \alpha_n f_n \quad \text{and} \quad g_{\boldsymbol{\alpha}} = \sum_{n=0}^\infty \alpha_n g_n$$

for $\boldsymbol{\alpha} \in \mathfrak{R}$. Because of the linear independence of the f_n's, we know that $f_{\boldsymbol{\alpha}} = f_{\boldsymbol{\beta}}$ if and only if $\boldsymbol{\alpha} = \boldsymbol{\beta}$. Hence, for each $\omega \in \Omega$, we can define the (not necessarily continuous) linear functional $\Lambda_\omega : \mathcal{S} \longrightarrow \mathbb{R}$ so that

$$\Lambda_\omega\big(f_{\boldsymbol{\alpha}}\big) = g_{\boldsymbol{\alpha}}(\omega), \qquad \boldsymbol{\alpha} \in \mathfrak{R}.$$

Clearly, $\Lambda_\omega(\mathbf{1}) = \mathbf{1}$ for all $\omega \in \Omega$. On the other hand, we cannot say that Λ_ω is always nonnegative as a linear functional on $\mathcal{S}$. In fact, the best we can do is extract a Σ-measurable P-null set $\mathcal{N}$ so that Λ_ω is a nonnegative linear functional on $\mathcal{S}$ whenever $\omega \notin \mathcal{N}$. To this end, let $\mathbb{Q}$ denote the rational reals and set

$$\mathfrak{Q}^+ = \left\{\boldsymbol{\alpha} \in \mathfrak{R} \cap \mathbb{Q}^{\mathbb{N}} : f_{\boldsymbol{\alpha}} \geq 0\right\}.$$

Since $g_{\boldsymbol{\alpha}} \geq 0$ (a.s., P) for every $\boldsymbol{\alpha} \in \mathfrak{Q}^+$ and $\mathfrak{Q}^+$ is countable,

$$\mathcal{N} \equiv \left\{\omega \in \Omega : \exists \boldsymbol{\alpha} \in \mathfrak{Q}^+ \quad g_{\boldsymbol{\alpha}}(\omega) < 0\right\}$$

is a Σ-measurable, P-null set. In addition, it is obvious that, for every $\omega \notin \mathcal{N}$, $\Lambda_\omega(f) \geq 0$ whenever f is a nonnegative element of $\mathcal{S}$. In particular, for $\omega \notin \mathcal{N}$,

$$\|f\|_{\mathrm{u}} \pm \Lambda_\omega(f) = \Lambda_\omega\big(\|f\|_{\mathrm{u}}\mathbf{1} \pm f\big) \geq 0, \quad f \in \mathcal{S},$$

and therefore Λ_ω admits a unique extension as a nonnegative, linear functional on $\mathcal{U}$ which takes $\mathbf{1}$ to 1. Furthermore, it is an easy matter to check that, for every $f \in \mathcal{U}$, the function

$$g(\omega) = \begin{cases} \Lambda_\omega(f) & \text{for} \quad \omega \notin \mathcal{N} \\ \mathbb{E}^P[f] & \text{for} \quad \omega \in \mathcal{N} \end{cases}$$

is a conditional expectation value of f given Σ.

At this point, all that remains is to show that, for P-almost every $\omega \notin \mathcal{N}$, Λ_ω is given by integration with respect to a $P_\omega \in \mathbf{M}_1(\Omega)$. In particular, by the Riesz Representation Theorem, there is nothing more to do in the case when Ω is compact. To treat the case when Ω is not compact, we want to use Lemma 3.1.7. For this purpose, first choose (cf. the last part of Lemma 3.1.7) a nondecreasing sequence of sets $K_n \subset\subset \Omega$, $n \in \mathbb{Z}^+$, with the property that $P(K_n\complement) \leq \frac{1}{2^n}$. Next, define

$$\eta_{m,n}(\omega) = \frac{m\,\rho(\omega, K_n)}{1 + m\,\rho(\omega, K_n)} \quad \text{for} \quad k,\, n \in \mathbb{Z}^+.$$

Clearly, $\eta_{m,n} \in \mathcal{U}$ for each pair (m, n) and $0 \leq \eta_{m,n} \nearrow \mathbf{1}_{K_n\complement}$ as $m \to \infty$ for each $n \in \mathbb{Z}^+$. Thus, by the Monotone Convergence Theorem, for each $n \in \mathbb{Z}^+$,

$$\begin{aligned} \int_{\mathcal{N}\complement} \sup_{m \in \mathbb{Z}^+} \Lambda_\omega\big(\eta_{m,n}\big)\, P(d\omega) &= \lim_{m\to\infty} \int_{\mathcal{N}\complement} \Lambda_\omega\big(\eta_{m,n}\big)\, P(d\omega) \\ &= \lim_{m\to\infty} \mathbb{E}^P\big[\eta_{m,n}\big] \leq \frac{1}{2^n}; \end{aligned}$$

and so, by the Borel–Cantelli Lemma, we can find a Σ-measurable P-null set $\mathcal{N}' \supseteq \mathcal{N}$ such that

$$M(\omega) \equiv \sup_{n \in \mathbb{Z}^+} n \left(\sup_{m \in \mathbb{Z}^+} \Lambda_\omega\big(\eta_{m,n}\big)\right) < \infty \quad \text{for every } \omega \notin \mathcal{N}'.$$

Hence, if $\omega \notin \mathcal{N}'$, then, for every $f \in \mathcal{U}$ and $n \in \mathbb{Z}^+$,

$$\begin{aligned}\big|\Lambda_\omega(f)\big| &\le \big|\Lambda_\omega\big((1-\eta_{m,n})\,f\big)\big| + \big|\Lambda_\omega\big(\eta_{m,n}\,f\big)\big| \\ &\le \big\|(1-\eta_{m,n})\,f\big\|_{\mathrm{u}} + \frac{M(\omega)}{n}\,\|f\|_{\mathrm{u}}\end{aligned}$$

for all $m \in \mathbb{Z}^+$. But $\big\|(1-\eta_{m,n})\,f\big\|_{\mathrm{u}} \longrightarrow \|f\|_{\mathrm{u},K_n}$ as $m \to \infty$, and so we now see that the condition in (3.1.8) is satisfied by Λ_ω for every $\omega \notin \mathcal{N}'$. In other words, we have shown that, for each $\omega \notin \mathcal{N}'$, there is a unique $P^\Sigma_\omega \in \mathbf{M}_1(\Omega)$ such that $\Lambda_\omega(f) = \mathbb{E}^{P^\Sigma_\omega}\big[f\big]$ for all $f \in \mathcal{U}$. Finally, if we complete the definition of the map $\omega \in \Omega \longmapsto P^\Sigma_\omega$ by taking $P^\Sigma_\omega = P$ for $\omega \in \mathcal{N}'$, then this map is Σ-measurable and

$$\mathbb{E}^P\big[f,\,A\big] = \int_\Omega \mathbb{E}^{P^\Sigma_\omega}\big[f\big]\,P(d\omega), \quad A \in \Sigma,$$

first for all $f \in \mathcal{U}$ and thence for all bounded $\mathcal{F}$-measurable f's. □

Given a measurable space $(\Omega, \mathcal{F})$ and a sub-σ-algebra Σ, a **Σ-measurable transition probability** is a map $(\omega, B) \in \Omega \times \mathcal{F} \longmapsto P(\omega, B) \in [0,1]$ with the properties that $B \in \mathcal{F} \longmapsto P(\omega, B) \in [0,1]$ is a probability measure for each $\omega \in \Omega$ and $\omega \in \Omega \longmapsto P(\omega, B) \in [0,1]$ is Σ-measurable for each $B \in \mathcal{F}$. In particular, if P is a probability measure on $(\Omega, \mathcal{F})$, then a **conditional probability distribution of P given Σ** is a Σ-measurable transition $(\omega, B) \longmapsto P^\Sigma_\omega(B)$ probability for which (5.1.16) holds. If, in addition, (5.1.17) holds, then the conditional probability distribution is said to be **regular**. Notice that, although they may not always exist, conditional probability distributions are always unique up to a Σ-measurable P-null set. Moreover, Theorem 5.1.15 says that they will always exist if Ω is Polish and $\mathcal{F} = \mathcal{B}_\Omega$. Finally, whenever a conditional probability distribution of P given Σ exists, the argument leading to (5.1.17) when Σ is countably generated is completely general and shows that a regular version can be found.

For various applications, it is convenient to have two extensions of the basic theory developed above. In particular, as we will now show, the theory is not restricted to probability (or even finite) measures and can be applied to random variables which take their values in a separable Banach space. Thus, from now on, μ will be an arbitrary (nonnegative) measure on $(\Omega, \mathcal{F})$ and $\big(E, \|\cdot\|_E\big)$ will be a separable Banach space; and we begin by reviewing a few elementary facts about μ-integration for E-valued random variables.

A function $\mathbf{X} : \Omega \longrightarrow E$ is said to be **μ-simple** if $\mathbf{X}$ is $\mathcal{F}$-measurable, $\mu\big(\mathbf{X} \ne \mathbf{0}\big) < \infty$, and $\mathbf{X}$ takes on only a finite number of distinct values, in which case its integral with respect to μ is the element of E given by:

$$\mathbb{E}^\mu\big[\mathbf{X}\big] = \int_\Omega \mathbf{X}(\omega)\,\mu(d\omega) \equiv \sum_{\mathbf{x} \in E \setminus \{\mathbf{0}\}} \mathbf{x}\,\mu\big(\mathbf{X} = \mathbf{x}\big).$$

Notice that another description of $\mathbb{E}^\mu[\mathbf{X}]$ is as the unique element of E with the property that

$$\left\langle \mathbb{E}^\mu[\mathbf{X}], \lambda \right\rangle = \mathbb{E}^\mu\left[\langle \mathbf{X}, \lambda \rangle\right] \quad \text{for all } \lambda \in E^* \tag{5.1.18}$$

(we use E^* to denote the dual of E and $\langle \mathbf{x}, \lambda \rangle$ to denote the action of $\lambda \in E^*$ on $\mathbf{x} \in E$), and therefore the mapping taking μ-simple $\mathbf{X}$ to $\mathbb{E}^\mu[\mathbf{X}]$ is linear. Next, because E is separable and therefore there exists a sequence $\{\lambda_n\}_1^\infty$ from the unit ball in E^* with the property that

$$\sup_{n \in \mathbb{Z}^+} \langle \mathbf{x}, \lambda_n \rangle = \|\mathbf{x}\|_E \quad \text{for all} \quad \mathbf{x} \in E, \tag{5.1.19}$$

we know that $\omega \in \Omega \longmapsto \|\mathbf{X}(\omega)\|_E \in \mathbb{R}$ is $\mathcal{F}$-measurable if $\mathbf{X} : \Omega \longrightarrow E$ is $\mathcal{F}$-measurable. In particular, for $\mathcal{F}$-measurable $\mathbf{X} : \Omega \longrightarrow E$, we can set

$$\|\mathbf{X}\|_{L^p(\mu;E)} = \begin{cases} \mathbb{E}^\mu\left[\|\mathbf{X}\|_E^p\right]^{\frac{1}{p}} & \text{if } \; p \in [1,\infty) \\ \inf\left\{M : \mu(\|\mathbf{X}\|_E > M) = 0\right\} & \text{if } \; p = \infty \end{cases}$$

and write $\mathbf{X} \in L^p(\mu; E)$ when $\|\mathbf{X}\|_{L^p(\mu;E)} < \infty$. Also, we will say the $\mathbf{X} : \Omega \longrightarrow E$ is **μ-integrable** if $\mathbf{X} \in L^1(\mu; E)$; and we will say that $\mathbf{X}$ is **μ-locally integrable** if $\mathbf{1}_A \mathbf{X}$ is μ-integrable for every $A \in \mathcal{F}$ with $\mu(A) < \infty$.

The definition of μ-integration for E-valued $\mathbf{X}$ is completed in the following lemma.

5.1.20 Lemma. *For each μ-integrable $\mathbf{X} : \Omega \longrightarrow E$ there is a unique element $\mathbb{E}^\mu[\mathbf{X}] \in E$ for which (5.1.18) holds. In particular, the mapping $\mathbf{X} \in L^1(\mu; E) \longmapsto \mathbb{E}^\mu[\mathbf{X}] \in E$ is linear and satisfies*

$$\left\|\mathbb{E}^\mu[\mathbf{X}]\right\|_E \le \mathbb{E}^\mu\left[\|\mathbf{X}\|_E\right]. \tag{5.1.21}$$

Finally, if $\mathbf{X} \in L^p(\mu; E)$ where $p \in [1,\infty)$, then there exists a sequence $\{\mathbf{X}_n\}_1^\infty$ of E-valued, μ-simple functions with the property that $\|\mathbf{X}_n - \mathbf{X}\|_{L^p(\mu;E)} \longrightarrow 0$.

PROOF: Clearly uniqueness, linearity, and (5.1.21) all follow immediately from the characterization of $\mathbb{E}^\mu[\mathbf{X}]$ given in (5.1.18). Thus, all that remains is to prove existence and the final approximation assertion. In fact, once the approximation assertion is proved, then existence will follow immediately from the observation that, by (5.1.21), $\mathbb{E}^\mu[\mathbf{X}]$ can be taken equal to $\lim_{n\to\infty} \mathbb{E}^\mu[\mathbf{X}_n]$ if $\|\mathbf{X} - \mathbf{X}_n\|_{L^1(\mu;E)} \longrightarrow 0$.

To prove the approximation assertion, we begin with the case when μ is finite and

$$M = \sup_{\omega \in \Omega} \|\mathbf{X}(\omega)\|_E < \infty.$$

Next, choose a dense sequence $\{\mathbf{x}_\ell\}_1^\infty$ in E, set $A_{0,n} = \emptyset$, and, for $\ell \in \mathbb{Z}^+$,

$$A_{\ell,n} = \left\{\omega : \|\mathbf{X}(\omega) - \mathbf{x}_\ell\|_E < \tfrac{1}{n}\right\} \quad \text{for } n \in \mathbb{Z}^+.$$

Then, for each $n \in \mathbb{Z}^+$ there exists an $L_n \in \mathbb{Z}^+$ with the property that

$$\mu\left(\Omega \setminus \bigcup_{\ell=1}^{L_n} A_{\ell,n}\right) < \frac{1}{n^p}.$$

Hence, if $\mathbf{X}_n : \Omega \longrightarrow E$ is defined so that

$$\mathbf{X}_n(\omega) = \mathbf{x}_\ell \quad \text{if } 1 \le \ell \le L_n \text{ and } \mathbf{x}_\ell \in A_{\ell,n} \setminus \bigcup_{k=0}^{\ell-1} A_{k,n}$$

and $\mathbf{X}_n(\omega) = \mathbf{0}$ when $\omega \notin \bigcup_1^{L_n} A_{\ell,n}$, then $\mathbf{X}_n$ is μ-simple and

$$\|\mathbf{X} - \mathbf{X}_n\|_{L^p(\mu;E)} \le \frac{M + \mu(E)}{n}.$$

In order to handle the general case, let $\mathbf{X} \in L^p(\mu; E)$ and $n \in \mathbb{Z}^+$ be given. We can then find an $r_n \in (0,1]$ with the property that

$$r_n^p\, \mu(\Omega(r_n)) \le \int_{\Omega(r_n)\complement} \|\mathbf{X}(\omega)\|_E^p\, \mu(d\omega) \le \frac{1}{(2n)^p},$$

where

$$\Omega(r) \equiv \left\{\omega : r \le \|\mathbf{X}(\omega)\|_E \le \tfrac{1}{r}\right\} \quad \text{for } r \in (0,1].$$

At the same time, by applying the preceding to the restrictions of μ and $\mathbf{X}$ to $\Omega(r_n)$, we can find a μ-simple $\mathbf{X}_n : \Omega(r_n) \longrightarrow E$ with the property

$$\left(\int_{\Omega(r_n)} \|\mathbf{X}(\omega) - \mathbf{X}_n(\omega)\|_E^p\, \mu(d\omega)\right)^{\frac{1}{p}} \le \frac{1}{2n}.$$

Hence, after extending $\mathbf{X}_n$ to Ω by taking it to be $\mathbf{0}$ off of $\Omega(r_n)$, we arrive at a μ-simple $\mathbf{X}_n$ for which $\|\mathbf{X} - \mathbf{X}_n\|_{L^p(\mu;E)} \le \frac{1}{n}$. □

Given an $\mathcal{F}$-measurable $\mathbf{X} : \Omega \longrightarrow E$ and a $B \in \mathcal{F}$ for which $\mathbf{1}_B\,\mathbf{X} \in L^1(\mu; E)$, we will use the notation

$$\mathbb{E}^\mu\big[\mathbf{X},\, B\big] \quad \text{or} \quad \int_B \mathbf{X}\, d\mu \quad \text{or} \quad \int_B \mathbf{X}(\omega)\, \mu(d\omega)$$

all to denote the quantity $\mathbb{E}^\mu\big[\mathbf{1}_B\,\mathbf{X}\big]$. Also, when discussing the spaces $L^p(\mu; E)$, we will adopt the usual convention of blurring the distinction between a particular $\mathcal{F}$-measurable $\mathbf{X} : \Omega \longrightarrow E$ belonging to $L^p(\mu; E)$ and the equivalence class of those $\mathcal{F}$-measurable $\mathbf{Y}$'s which differ from $\mathbf{X}$ on a μ-null set. Thus, with this convention, $\|\cdot\|_{L^p(\mu;E)}$ becomes a bona fide norm (not just a seminorm) on $L^p(\mu; E)$ with respect to which $L^p(\mu; E)$ becomes a Banach space.

5.1.22 Theorem. *Let $(\Omega, \mathcal{F}, \mu)$ be a σ-finite measure space and $\mathbf{X} : \Omega \longrightarrow E$ a μ-locally integrable function. Then*

$$\mu(\mathbf{X} \neq \mathbf{0}) = 0 \iff \mathbb{E}^{\mu}\big[\mathbf{X},\, A\big] = \mathbf{0} \text{ for } A \in \mathcal{F} \text{ with } \mu(A) < \infty. \tag{5.1.23}$$

Next, assume that Σ is a sub-σ-algebra for which $\mu \restriction \Sigma$ is σ-finite. Then for each μ-locally integrable $\mathbf{X} : \Omega \longrightarrow E$ there is a μ-almost everywhere unique μ-locally integrable, Σ-measurable $\mathbf{X}_\Sigma : \Omega \longrightarrow E$ such that

$$\mathbb{E}^{\mu}\big[\mathbf{X}_\Sigma,\, A\big] = \mathbb{E}^{\mu}\big[\mathbf{X},\, A\big] \quad \text{for every } A \in \Sigma \text{ with } \mu(A) < \infty. \tag{5.1.24}$$

In particular, if $\mathbf{Y} : \Omega \longrightarrow E$ is a second μ-locally integrable function, then, for all α, $\beta \in \mathbb{R}$,

$$\big(\alpha\mathbf{X} + \beta\mathbf{Y}\big)_\Sigma = \alpha\mathbf{X}_\Sigma + \beta\mathbf{Y}_\Sigma \quad \text{(a.e., } \mu).$$

Finally,

$$\big\|\mathbf{X}_\Sigma\big\|_E \leq \big(\|\mathbf{X}\|_E\big)_\Sigma \quad \text{(a.e., } \mu). \tag{5.1.25}$$

Hence, not only does (5.1.24) continue to hold for any $A \in \Sigma$ with $\mathbf{1}_A\mathbf{X} \in L^1(\mu; E)$; but also, for each $p \in [1, \infty]$, the mapping $\mathbf{X} \in L^p(\mu; E) \longmapsto \mathbf{X}_\Sigma \in L^p(\mu; E)$ is a linear contraction.

PROOF: Clearly, it is only necessary to prove the "$\Longleftarrow$" part of (5.1.23). Thus, suppose that $\mu(\mathbf{X} \neq \mathbf{0}) > 0$. Then (cf. (5.1.19)) there exists an $\epsilon > 0$ and a $\lambda \in E^*$ with the property that $\mu\big(\langle \mathbf{X}, \lambda\rangle \geq \epsilon\big) > 0$; from which it follows (by σ-finiteness) that there is an $A \in \mathcal{F}$ for which $\mu(A) < \infty$ and

$$\Big\langle \mathbb{E}^{\mu}\big[\mathbf{X},\, A\big], \lambda\Big\rangle = \mathbb{E}^{\mu}\Big[\langle \mathbf{X}, \lambda\rangle,\, A\Big] \neq 0.$$

We turn next to the uniqueness and other properties of $\mathbf{X}_\Sigma$. But it is obvious that uniqueness is an immediate consequence of (5.1.23) and that linearity follows from uniqueness. As for (5.1.25), notice that if $\lambda \in E^*$ and $\|\lambda\|_{E^*} \leq 1$, then

$$\begin{aligned}\mathbb{E}^{\mu}\Big[\langle \mathbf{X}_\Sigma, \lambda\rangle,\, A\Big] &= \mathbb{E}^{\mu}\Big[\langle \mathbf{X}, \lambda\rangle,\, A\Big] \\ &\leq \mathbb{E}^{\mu}\Big[\|\mathbf{X}\|_E,\, A\Big] = \mathbb{E}^{\mu}\Big[\big(\|\mathbf{X}\|_E\big)_\Sigma,\, A\Big]\end{aligned}$$

for every $A \in \Sigma$ with $\mu(A) < \infty$. Hence, at least when μ is a probability measure, Theorem 5.1.4 implies that

$$\langle \mathbf{X}_\Sigma, \lambda\rangle \leq \big(\|\mathbf{X}\|_E\big)_\Sigma \quad \text{(a.e., } \mu)$$

for each element λ from the unit ball in E^*; and so, because E is separable,

(5.1.25) follows in this case from (5.1.19). To handle μ's which are not probability measures, note that either $\mu(\Omega) = 0$, in which case everything is trivial, or $\mu(\Omega) \in (0,\infty)$, in which case we can renormalize μ to make it a probability measure, or $\mu(\Omega) = \infty$, in which case we can use the σ-finiteness of $\mu \restriction \Sigma$ to reduce ourselves to the countable, disjoint union of the preceding cases.

Finally, to prove the existence of $\mathbf{X}_\Sigma$, we proceed as in the last part of the preceding paragraph to reduce ourselves to the case when μ is a probability measure P. Next, suppose that $\mathbf{X}$ is P-simple, let R denote its range, and note that

$$\mathbf{X}_\Sigma \equiv \sum_{\mathbf{x}\in R} \mathbf{x}\, P(\mathbf{X} = \mathbf{x}\,|\,\Sigma)$$

has the required properties. In order to handle general $\mathbf{X} \in L^1(P;E)$, we use the approximation result in Lemma 5.1.20 to find a sequence $\{\mathbf{X}_n\}_1^\infty$ of P-simple functions which tend to $\mathbf{X}$ in $L^1(P;E)$. Then, since

$$(\mathbf{X}_n)_\Sigma - (\mathbf{X}_m)_\Sigma = \left(\mathbf{X}_n - \mathbf{X}_m\right)_\Sigma \quad \text{(a.s., } P)$$

and therefore, by (5.1.25),

$$\left\|(\mathbf{X}_n)_\Sigma - (\mathbf{X}_m)_\Sigma\right\|_{L^1(P;E)} \le \left\|\mathbf{X}_n - \mathbf{X}_m\right\|_{L^1(P;E)},$$

we know that there exists a Σ-measurable $\mathbf{X}_\Sigma \in L^1(P;E)$ to which the sequence $\left\{(\mathbf{X}_n)_\Sigma\right\}_1^\infty$ converges; and clearly $\mathbf{X}_\Sigma$ has the required properties. □

Referring to the setting in the second part of Theorem 5.1.22, we will extend the convention introduced in Convention 5.1.5 and call the μ-equivalence class of $\mathbf{X}_\Sigma$'s satisfying (5.1.24) the **μ-conditional expectation of $\mathbf{X}$ given Σ**, will use $\mathbb{E}^\mu[\mathbf{X}|\Sigma]$ to denote this μ-equivalence class, and will, in general, ignore the distinction between the equivalence class and a generic representative of that class. In addition, if $\mathbf{X} : \Omega \longrightarrow E$ is μ-locally integrable, then, just as in Theorem 5.1.6, the following are essentially immediate consequences of uniqueness:

(5.1.26) $$\mathbb{E}^\mu\left[Y\mathbf{X}\middle|\Sigma\right] = Y\,\mathbb{E}^\mu\left[\mathbf{X}\middle|\Sigma\right] \quad \text{(a.e., } \mu) \quad \text{for } Y \in L^\infty(\Omega,\Sigma,\mu;\mathbb{R}),$$

and

(5.1.27) $$\mathbb{E}^\mu\left[\mathbf{X}\middle|\mathcal{T}\right] = \mathbb{E}^\mu\left[\mathbb{E}^\mu\left[\mathbf{X}\middle|\Sigma\right]\middle|\mathcal{T}\right] \quad \text{(a.e., } \mu)$$

whenever $\mathcal{T}$ is a sub-σ-algebra of Σ for which $\mu \restriction \mathcal{T}$ is σ-finite.

Exercises

5.1.28 Exercise: As the proof of existence in Theorem 5.1.6 makes clear, the operation $X \in L^2(P;\mathbb{R}) \longmapsto \mathbb{E}^P[X|\Sigma]$ is just the operation of orthogonal projection from $L^2(P;\mathbb{R})$ onto the space $L^2(\Omega,\Sigma,P;\mathbb{R})$ of Σ-measurable elements of $L^2(P;\mathbb{R})$. For this reason, one might be inclined to think that the concept of conditional expectation is basically a Hilbert space notion. However, as we will show in this exercise, that inclination is not entirely well-founded. The point is that, although conditional expectation is definitely an orthogonal projection, not every orthogonal projection is a conditional expectation!

(i) Let $\mathbf{L}$ be a closed linear subspace of $L^2(P;\mathbb{R})$ and let $\Sigma_{\mathbf{L}} = \sigma(X : X \in \mathbf{L})$ be the σ-algebra over Ω generated by $X \in \mathbf{L}$. Show that $\mathbf{L} = L^2(\Omega,\Sigma_{\mathbf{L}},P;\mathbb{R})$ if and only if $\mathbf{1} \in \mathbf{L}$ and $X^+ \in \mathbf{L}$ whenever $X \in \mathbf{L}$.

Hint: To prove the "if" assertion, let $X \in \mathbf{L}$ be given, and show that

$$X_n \equiv \left[n\big(X - \alpha\mathbf{1}\big)^+ \wedge \mathbf{1}\right] \in \mathbf{L} \quad \text{for every } \alpha \in \mathbb{R} \text{ and } n \in \mathbb{Z}^+.$$

Conclude that $X_n \nearrow \mathbf{1}_{(\alpha,\infty)} \circ X$ must be an element of $\mathbf{L}$.

(ii) Let Π be an orthogonal projection operator on $L^2(P;\mathbb{R})$, set $\mathbf{L} = \text{Range}(\Pi)$, and let $\Sigma = \Sigma_{\mathbf{L}}$, where $\Sigma_{\mathbf{L}}$ is defined as in part (**i**). Show that $\Pi X = \mathbb{E}^P[X|\Sigma]$ (a.s., P) for all $X \in L^2(P;\mathbb{R})$ if and only if $\Pi\mathbf{1} = \mathbf{1}$ and

$$\Pi\big(X\,\Pi Y\big) = (\Pi X)(\Pi Y) \quad \text{for all} \quad X, Y \in L^\infty(P;\mathbb{R}). \tag{5.1.29}$$

Hint: Assume that $\Pi\mathbf{1} = \mathbf{1}$ and that (5.1.29) holds. Given $X \in L^\infty(P;\mathbb{R})$, use induction to show that

$$\|\Pi X\|^n_{L^{2n}(P)} \le \|X\|^{n-1}_{L^\infty(P)}\|X\|_{L^2(P)} \quad \text{and} \quad \big(\Pi X\big)^n = \Pi\big(X(\Pi X)^{n-1}\big)$$

for all $n \in \mathbb{Z}^+$. Conclude that $\|\Pi X\|_{L^\infty(P)} \le \|X\|_{L^\infty(P)}$ and that $\big(\Pi X\big)^n \in \mathbf{L}$, $n \in \mathbb{Z}^+$, for every $X \in L^\infty(P;\mathbb{R})$. Next, using the preceding together with Weierstrass's Approximation Theorem, show that $(\Pi X)^+ \in \mathbf{L}$, first for $X \in L^\infty(P;\mathbb{R})$ and then for all $X \in L^2(P;\mathbb{R})$. Finally, apply (**i**) to arrive at $\mathbf{L} = L^2\big(\Omega,\Sigma,P;\mathbb{R}\big)$.

(iii) Just in case the situation is not completely clarified by part (**ii**), consider once again a closed linear subspace $\mathbf{L}$ of $L^2(P;\mathbb{R})$ and let $\Pi_{\mathbf{L}}$ be the orthogonal projection mapping onto $\mathbf{L}$. Given $X \in L^2(P;\mathbb{R})$, recall that $\Pi_{\mathbf{L}}X$ is characterized as the unique element of $\mathbf{L}$ for which $X - \Pi_{\mathbf{L}}X \perp \mathbf{L}$, and show that $\mathbb{E}^P[X|\Sigma_{\mathbf{L}}]$ is the unique element of $L^2(\Omega,\Sigma_{\mathbf{L}},P;\mathbb{R})$ with the property that

$$X - \mathbb{E}^P\big[X\big|\Sigma_L\big] \perp f\big(Y_1,\dots,Y_n\big)$$

for all $n \in \mathbb{Z}^+$, $f \in C_b\big(\mathbb{R}^n;\mathbb{R}\big)$, and $Y_1,\dots,Y_n \in \mathbf{L}$. In particular, $\Pi_{\mathbf{L}}X = \mathbb{E}^P[X|\Sigma_{\mathbf{L}}]$ if and only if $X - \Pi_{\mathbf{L}}X$ is perpendicular not only to all *linear* functions of the Y's in $\mathbf{L}$ but even to all *nonlinear* ones.

(**iv**) There is an important, nontrivial situation in which conditioning and projecting turn out to be the same thing. Namely, let $\mathfrak{G} \subseteq L^2(P;\mathbb{R})$ be a centered Gaussian family (cf. Exercise 4.2.39), and show that for any closed linear subspace $\mathbf{L}$ of $\mathfrak{G}$ and any $X \in \mathfrak{G}$, $\Pi_{\mathbf{L}} X = \mathbb{E}^P\left[X \middle| \Sigma_{\mathbf{L}}\right]$.

(**v**) Because most projections are not conditional expectations, it is an unfortunate fact of life that, for the most part, partial sums of Fourier series cannot be interpreted as conditional expectations. Be that as it may, there are special cases in which such an interpretation is possible. To see this, take $\Omega = [0,1)$, $\mathcal{F} = \mathcal{B}_{[0,1)}$, and P to be the restriction of Lebesgue's measure to $[0,1)$. Next, for each $n \in \mathbb{N}$, take $\mathcal{F}_n$ to be the σ-algebra generated by those $\mathcal{F}$-measurable functions f on $[0,1)$ which are periodic with period 2^{-n-1}. Thus, an $\mathcal{F}$-measurable f is $\mathcal{F}_n$-measurable if and only if $f(x) = f(y)$ for all x, $y \in [0,1)$ with the property that $|y-x| = 2^{-n-1}$. Using elementary Fourier analysis, show that, for each $n \in \mathbb{N}$,

$$\{\mathbf{1}\} \cup \left\{ e^{\sqrt{-1}2^m \pi t} : |m| > n+1 \right\}$$

constitutes an orthonormal basis for $L^2(\Omega, \mathcal{F}_n, P; \mathbb{C})$, and conclude that, for every $f \in L^2(P;\mathbb{C})$:

$$\mathbb{E}^P\left[f \mid \mathcal{F}_n\right] = \mathbb{E}^P[f] + \sum_{|m|>n+1}^{\infty} c_m e^{\sqrt{-1}2^m \pi t} \text{ where } c_m \equiv \int_0^1 f(t) e^{-\sqrt{-1}2^m \pi t}\, dt$$

and the convergence is in $L^2([0,1];\mathbb{C})$. (See part (**v**) of Exercise 5.2.40 and Exercise 6.3.29 for a continuation of this exercise.)

5.1.30 Exercise: In this exercise we will show how to formulate the (strong) Markov property (cf. Exercise 4.3.55) in the language of conditional expectations. For this purpose, suppose that E is a Polish space and that $x \in E \longmapsto P_x \in \mathbf{M}_1(\mathfrak{P}(E))$ is a measurable map for which $P_x(\psi(0) = x) = 1$, $x \in E$.

(**i**) If $(t,x) \in [0,\infty) \times E \longmapsto P(t,x,\cdot\,) \in \mathbf{M}_1(E)$ is defined as in part (**i**) of Exercise 4.3.55, show that $\{P_x : x \in E\}$ is a Markov family if and only if, for every $(s,x) \in [0,\infty) \times E$, $t \in (0,\infty)$, and $f \in B(E;\mathbb{R})$,

$$\varphi \in \mathfrak{P}(E) \longmapsto \int_E f(y)\, P\big(t, \varphi(s), dy\big) \in \mathbb{R}$$

is a representative of the P_x-conditional expectation of

$$\psi \in P(E) \longmapsto f\big(\psi(s+t)\big) \in \mathbb{R} \text{ given } \mathcal{B}_s.$$

More generally, recall the time-shift transformation $\mathbf{\Sigma}_s$ and, under the assumption that $\{P_x : x \in E\}$ is a Markov family, show that, for every $(s,x) \in [0,\infty) \times E$ and $F \in B\big(\mathfrak{P}(E);\mathbb{R}\big)$,

$$\varphi \in \mathfrak{P}(E) \longmapsto \mathbb{E}^{P_{\varphi(s)}}\left[F\right] \in \mathbb{R}$$

is a representative of the conditional expectation $\mathbb{E}^{P_x}\left[F \circ \mathbf{\Sigma}_s \middle| \mathcal{B}_s\right]$.

(ii) Given $F \in B(\mathfrak{P}(E);\mathbb{R})$ and a $\{\mathcal{B}_t : t \in [0,\infty)\}$-stopping time τ, define

$$\psi \in \mathfrak{P}(E) \longmapsto F \circ \mathbf{\Sigma}_\tau(\psi) \equiv \begin{cases} F\big(\mathbf{\Sigma}_{\tau(\psi)}\psi\big) & \text{if } \tau(\psi) < \infty \\ 0 & \text{otherwise,} \end{cases}$$

and show that $\{P_x : x \in E\}$ is a strong Markov family if and only if, for every $x \in E$, $F \in B(\mathcal{P}(E);\mathbb{R})$, and $\{\mathcal{B}_t : t \in [0,\infty)\}$-stopping time τ,

$$\varphi \in \mathfrak{P}(E) \longmapsto \begin{cases} \mathbb{E}^{P_{\varphi(\tau)}}[F] & \text{if } \tau(\varphi) < \infty \\ 0 & \text{otherwise} \end{cases}$$

is a representative of the conditional expectation $\mathbb{E}^{P_x}\big[F \circ \mathbf{\Sigma}_\tau \big| \mathcal{B}_\tau\big]$.

Notice that the preceding provides an extremely intuitive formulation of the Markov property. Indeed, it says that $\{P_x : x \in E\}$ is a (strong) Markov family if, given *the history of the past up to any (stopping) time τ, the distribution of paths in the future of that time depends only on the position $\psi(\tau)$ of the path at time τ and will be the same as the distribution of paths which issue from $\psi(\tau)$ initially.*

5.1.31 Exercise: The abstract statement that conditional probability distributions may exist is less interesting than the fact that there are nontrivial circumstances in which they can be computed. In this and the next exercises we will develop two important examples based on the results which we obtained in Sections 4.2 and 4.3.

Let $P \in \mathbf{M}_1\Big(C\big([0,1];\mathbb{R}^N\big)\Big)$ denote the distribution of $\boldsymbol{\psi} \in \mathfrak{P}(\mathbb{R}^N) \longmapsto \boldsymbol{\psi} \restriction [0,1]$ under Wiener's measure $\mathcal{W}^{(N)}$, and let Σ be the σ-algebra over $C\big([0,1];\mathbb{R}^N\big)$ generated by $\boldsymbol{\psi} \in C\big([0,1];\mathbb{R}^N\big) \longmapsto \boldsymbol{\psi}(1) \in \mathbb{R}^N$. Referring to Theorem 4.3.18, define $P^\Sigma_{\mathbf{y}} \in \mathbf{M}_1\Big(C\big([0,1];\mathbb{R}^N\big)\Big)$ for $\mathbf{y} \in \mathbb{R}^N$ to be the distribution of

$$\boldsymbol{\psi} \in C\big([0,\infty);\mathbb{R}^N\big) \longmapsto \big(\tilde{\boldsymbol{\psi}}_1 + \ell_{1,\mathbf{y}}\big) \restriction [0,1]$$

under $\mathcal{W}^{(N)}$, and interpret (4.3.20) (with $T = 1$) as the statement that

$$\boldsymbol{\varphi} \in C\big([0,\infty);\mathbb{R}^N\big) \longmapsto P^\Sigma_{\boldsymbol{\varphi}(1)}$$

is a regular conditional probability distribution of P given Σ. In other words, *$P^\Sigma_{\mathbf{y}}$ is the conditional distribution of $\boldsymbol{\psi} \in \mathfrak{P}(\mathbb{R}^N) \longmapsto \boldsymbol{\psi} \restriction [0,1]$ under $\mathcal{W}^{(N)}$ given that $\boldsymbol{\psi}(1) = \mathbf{y}$.* In particular, $P^\Sigma_{\mathbf{0}}$ is the distribution of **Wiener loops**.

5.1.32 Exercise: Our second example comes from rephrasing the contents of Exercises 4.3.55 and 5.1.30 as a statement about regular conditional probabilities. However, before we can do so, we will need to introduce a construction which enables us to *splice* measures together. To begin, define

$$\varphi \underset{s}{\otimes} \psi(t) = \begin{cases} \varphi(t) & \text{if } t \in [0,s) \\ \varphi(s) & \text{if } t \ge s \text{ and } \psi(0) \ne \varphi(s) \\ \psi(t-s) & \text{if } t \ge s \text{ and } \psi(0) = \varphi(s) \end{cases} \tag{5.1.33}$$

for $s \in [0,\infty]$, $t \in [0,\infty)$, and $(\varphi,\psi) \in \mathfrak{P}(E)^2$, and check that $(s,\varphi,\psi) \in [0,\infty] \times \mathfrak{P}(E)^2 \longmapsto \varphi \underset{s}{\otimes} \psi \in \mathfrak{P}(E)$ is measurable.

(i) Given a $\{\mathcal{B}_t : t \in [0,\infty)\}$-stopping time τ, show that

$$(\varphi,\psi) \in \mathfrak{P}(E)^2 \longmapsto \varphi \underset{\tau}{\otimes} \psi \equiv \varphi \underset{\tau(\varphi)}{\otimes} \psi \in \mathfrak{P}(E)$$

is $\mathcal{B}_\tau \times \mathcal{B}_{\mathfrak{P}(E)}$-measurable. Next, let $x \in E \longmapsto P_x \in \mathbf{M}_1(\mathfrak{P}(E))$ be a measurable map satisfying $P_x(\psi(0) = x) = 1$, $x \in E$, and, for $\varphi \in \mathfrak{P}(E)$, define $\delta_\varphi \underset{\tau}{\otimes} P. \in \mathbf{M}_1(\mathfrak{P}(E))$ by

$$\delta_\varphi \underset{\tau}{\otimes} P.(A) = P_{\varphi(\tau)}\Big(\big\{\psi : \varphi \underset{\tau}{\otimes} \psi \in A\big\}\Big) \tag{5.1.34}$$

for $A \in \mathcal{B}_{\mathfrak{P}(E)}$. Show that $\varphi \in \mathfrak{P}(E) \longmapsto \delta_\varphi \underset{\tau}{\otimes} P. \in \mathbf{M}_1(\mathfrak{P}(E))$ is $\mathcal{B}_\tau$-measurable, and check that, for $P \in \mathbf{M}_1(\mathfrak{P}(E))$, the **spliced measure**

$$P \underset{\tau}{\otimes} P. \equiv \int_{\mathfrak{P}(E)} \delta_\varphi \underset{\tau}{\otimes} P.\, P(d\varphi) \tag{5.1.35}$$

is the unique $Q \in \mathbf{M}_1(\mathfrak{P}(E))$ with the properties that $Q \restriction \mathcal{B}_\tau = P \restriction \mathcal{B}_\tau$ and

$$\varphi \in \mathfrak{P}(E) \longmapsto \delta_\varphi \underset{\tau}{\otimes} P.$$

is a regular conditional probability distribution of Q given $\mathcal{B}_\tau$.

(ii) Referring to part (i), show that the measure $P \underset{\tau}{\otimes} P.$ is the unique $Q \in \mathbf{M}_1(\mathfrak{P}(E))$ with the property that, for every bounded, $\mathcal{B}_\tau \times \mathcal{B}_{\mathfrak{P}(E)}$-measurable $F : \mathfrak{P}(E)^2 \longrightarrow \mathbb{R}$,

$$\begin{aligned} &\int_{\{\psi:\tau(\psi)<\infty\}} \int_{\mathfrak{P}(E)} F\big(\psi, \Sigma_{\tau(\psi)}\psi\big)\, Q(d\psi) \\ &\qquad = \int_{\{\varphi:\tau(\varphi)<\infty\}} \left(\int_{\mathfrak{P}(E)} F(\varphi,\psi)\, P_{\varphi(\tau)}(d\psi)\right) P(d\varphi). \end{aligned}$$

(iii) As a consequence of (i) and (ii), show that $\{P_x : x \in E\}$ is a Markov family if and only if, for every $(s,x) \in [0,\infty) \times E$, $P_x = P_x \underset{s}{\otimes} P.$ and that it is a strong Markov family if and only if $P_x = P_x \underset{\tau}{\otimes} P.$ for every $x \in E$ and $\{\mathcal{B}_t : t \in [0,\infty)\}$-stopping time τ.

5.1.36 Exercise: Let $(\Omega, \mathcal{F}, \mu)$ be a measure space and Σ a sub-σ-algebra of $\mathcal{F}$ with the property that $\mu \restriction \Sigma$ is σ-finite. Next, let E be a separable Hilbert space, $p \in [1,\infty]$, $\mathbf{X} \in L^p(\mu; E)$, and $\mathbf{Y}$ a Σ-measurable element of $L^{p'}(\mu; E)$ (p' is the Hölder conjugate of p). Show that

$$\mathbb{E}^\mu\Big[\big(\mathbf{Y},\mathbf{X}\big)_E\Big|\Sigma\Big] = \Big(\mathbf{Y}, \mathbb{E}^\mu\big[\mathbf{X}\big|\Sigma\big]\Big)_E \quad \mu\text{-almost surely.}$$

Hint: First observe that it suffices to check that

$$\mathbb{E}^\mu\Big[\big(\mathbf{Y},\mathbf{X}\big)_E\Big] = \mathbb{E}^\mu\Big[\Big(\mathbf{Y}, \mathbb{E}^\mu\big[\mathbf{X}\big|\Sigma\big]\Big)_E\Big].$$

Next, choose an orthonormal basis $\{\mathbf{e}_n\}_1^\infty$ for E and justify the steps in

$$\mathbb{E}^\mu\Big[\big(\mathbf{Y},\mathbf{X}\big)_E\Big] = \sum_1^\infty \mathbb{E}^\mu\Big[\big(\mathbf{Y},\mathbf{e}_n\big)_E\big(\mathbf{e}_n,\mathbf{X}\big)_E\Big]$$

$$= \sum_1^\infty \mathbb{E}^\mu\Big[\big(\mathbf{Y},\mathbf{e}_n\big)_E\,\mathbb{E}^\mu\Big[\big(\mathbf{e}_n,\mathbf{X}\big)_E\Big|\Sigma\Big]\Big] = \mathbb{E}^\mu\Big[\Big(\mathbf{Y},\mathbb{E}^\mu\big[\mathbf{X}\big|\Sigma\big]\Big)_E\Big].$$

§5.2: Discrete Parameter Martingales.

In this section we will introduce an interesting and useful class of stochastic processes which unifies and simplifies several branches of probability theory as well as other branches of analysis. From the analytic point of view, what we will be doing is developing an abstract version of differentiation theory (cf. Corollary 5.2.7 and Theorem 5.2.26).

Although we will want to make some extensions later (cf. Section 5.3), we start with the following setting. $(\Omega, \mathcal{F}, P)$ is a probability space and $\{\mathcal{F}_n : n \in \mathbb{N}\}$ is a nondecreasing sequence of sub-σ-algebra's of $\mathcal{F}$. Given a measurable space $(E, \mathcal{B})$, we say that the family $\{X_n : n \in \mathbb{N}\}$ of E-valued random variables is $\{\mathcal{F}_n : n \in \mathbb{N}\}$**-progressively measurable** if X_n is $\mathcal{F}_n$-measurable for each

$n \in \mathbb{N}$. Next, a family $\{X_n : n \in \mathbb{N}\}$ of $(-\infty, \infty]$-valued random variables is said to be a ***P*-submartingale with respect to** $\{\mathcal{F}_n : n \in \mathbb{N}\}$ if it is $\{\mathcal{F}_n : n \in \mathbb{N}\}$-progressively measurable and

$$\text{(5.2.1)} \qquad \mathbb{E}^P\big[X_n^-\big] < \infty \quad \text{and} \quad X_n \le \mathbb{E}^P\big[X_{n+1}\big|\mathcal{F}_n\big] \quad \text{(a.s., } P)$$

for each $n \in \mathbb{N}$; and it is said to be a ***P*-martingale with respect to** $\{\mathcal{F}_n : n \in \mathbb{N}\}$ if $\{X_n : n \in \mathbb{N}\}$ is an $\{\mathcal{F}_n : n \in \mathbb{N}\}$-progressively measurable family of $\mathbb{R}$-valued, P-integrable random variables satisfying

$$\text{(5.2.2)} \qquad X_n = \mathbb{E}^P\big[X_{n+1}\big|\mathcal{F}_n\big] \quad \text{(a.s., } P)$$

for each $n \in \mathbb{N}$. In the future, we will abbreviate these statements by saying that the triple $(X_n, \mathcal{F}_n, P)$ is a submartingale or martingale.

5.2.3 Examples. The most trivial example of a submartingale is provided by a nondecreasing sequence $\{a_n\}_1^\infty$. That is, if $X_n \equiv a_n$, $n \in \mathbb{Z}^+$, then $(X_n, \mathcal{F}, P)$ is a submartingale on any probability space $(\Omega, \mathcal{F}, P)$ relative to any nondecreasing $\{\mathcal{F}_n : n \in \mathbb{N}\}$. More interesting examples are those given below.[†]

(i) Let $\{Y_n\}_1^\infty$ be a sequence of mutually independent $(-\infty, \infty]$-valued random variables with $\mathbb{E}^P\big[Y_n^-\big] < \infty$, $n \in \mathbb{N}$, set $\mathcal{F}_0 = \{\emptyset, \Omega\}$, $\mathcal{F}_n = \sigma\big(Y_1, \dots, Y_n\big)$ for $n \in \mathbb{Z}^+$, and define

$$X_n = \sum_{m=1}^{n} Y_m \; (\equiv 0 \text{ if } n = 0) \quad \text{for } n \in \mathbb{N}.$$

Then, because $\mathbb{E}^P\big[Y_{n+1}\big|\mathcal{F}_n\big] = \mathbb{E}^P\big[Y_{n+1}\big]$ (a.s., P) and therefore

$$\mathbb{E}^P\big[X_{n+1}\big|\mathcal{F}_n\big] = X_n + \mathbb{E}^P\big[Y_{n+1}\big] \quad \text{(a.s., } P)$$

for every $n \in \mathbb{N}$, we see that $(X_n, \mathcal{F}_n, P)$ is a submartingale if and only if $\mathbb{E}^P\big[Y_n\big] \ge 0$ for all $n \in \mathbb{Z}^+$. In fact, if the Y_n's are $\mathbb{R}$-valued and P-integrable, then the same line of reasoning shows that $(X_n, \mathcal{F}_n, P)$ is a martingale if and only if $\mathbb{E}^P\big[Y_n\big] = 0$ for all $n \in \mathbb{Z}^+$. Finally, if $\{Y_n\}_1^\infty \subseteq L^2(P)$ and $\mathbb{E}^P[Y_n] = 0$ for each $n \in \mathbb{Z}^+$, then

$$\mathbb{E}^P\big[X_{n+1}^2 \,\big|\, \mathcal{F}_n\big] = X_n^2 + \mathbb{E}^P\big[Y_{n+1}^2 \,\big|\, \mathcal{F}_n\big] \ge X_n^2 \quad \text{(a.s., } P),$$

and so $(X_n^2, \mathcal{F}_n, P)$ is a submartingale.

[†] For a much more interesting and complete list of examples, the reader might want to consult J. Neveu's *Discrete-parameter Martingales*, publ. in 1975 by North–Holland.

(ii) If X is an $\mathbb{R}$-valued, P-integrable random variable and $\{\mathcal{F}_n : n \in \mathbb{N}\}$ is any nondecreasing sequence of sub-σ-algebras of $\mathcal{F}$, then, by (5.1.9),

$$\left(\mathbb{E}^P[X|\mathcal{F}_n], \mathcal{F}_n, P\right)$$

is a martingale.

(iii) If $(X_n, \mathcal{F}_n, P)$ is a martingale, then, by (5.1.8), $(|X_n|, \mathcal{F}_n, P)$ is a submartingale.

In view of (**i**) in the Examples 5.2.3, we see that partial sums of independent random variables with mean-value 0 are a source of martingales and that their squares are a source of submartingales. Hence, it is reasonable to ask whether some of the important facts about such partial sums will continue to be true for all martingales; and perhaps the single most important indication that the answer may be "yes" is contained in the following generalization of Kolmogorov's Inequality (cf. Theorem 1.4.5).

5.2.4 Theorem (Doob's Inequality). *Assume that* $(X_n, \mathcal{F}_n, P)$ *is a submartingale. Then, for every* $N \in \mathbb{Z}^+$ *and* $\alpha \in (0,\infty)$*:*

$$P\left(\max_{0\le n\le N} X_n \ge \alpha\right) \le \frac{1}{\alpha}\mathbb{E}^P\left[X_N, \max_{0\le n\le N} X_n \ge \alpha\right]. \tag{5.2.5}$$

In particular, if the X_n*'s are nonnegative, then, for each* $p \in (1,\infty)$*,*

$$\mathbb{E}^P\left[\sup_{n\in\mathbb{N}} X_n^p\right]^{\frac{1}{p}} \le \frac{p}{p-1}\sup_{n\in\mathbb{N}} \mathbb{E}^P\left[X_n^p\right]^{\frac{1}{p}}. \tag{5.2.6}$$

PROOF: To prove (5.2.5), set $A_0 = \{X_0 \ge \alpha\}$ and

$$A_n = \left\{X_n \ge \alpha \text{ but } \max_{0\le m<n} X_m < \alpha\right\} \quad \text{for} \quad n \in \mathbb{Z}^+.$$

Then the A_n's are mutually disjoint and $A_n \in \mathcal{F}_n$ for each $n \in \mathbb{N}$. Thus,

$$\begin{aligned} P\left(\max_{0\le n\le N} X_n \ge \alpha\right) &= \sum_{n=0}^N P(A_n) \le \sum_{n=0}^N \frac{\mathbb{E}^P[X_n, A_n]}{\alpha} \\ &\le \sum_{n=0}^N \frac{\mathbb{E}^P[X_N, A_n]}{\alpha} = \frac{1}{\alpha}\mathbb{E}^P\left[X_N, \max_{0\le n\le N} X_n \ge \alpha\right]. \end{aligned}$$

Now assume that the X_n's are nonnegative. Given (5.2.5), (5.2.6) becomes an easy application of Exercise 1.4.20. □

Doob's inequality is an example of what analysts call a **weak-type inequality**. To be more precise, it is a *weak-type* 1–1 inequality. (The terminology derives from the fact that such an inequality follows immediately from an L^1-norm, or *strong-type* 1–1, inequality between the objects under consideration; but, in general, it is strictly weaker. (See Theorem 6.1.20 for more about the relationship between weak and strong type inequalities.) In order to demonstrate how powerful such a result can be, we will now present the following preliminary convergence result for martingales; and, because it is an argument to which we will return several times, the reader would do well to become comfortable with the line of reasoning which allows one to pass from a *weak-type inequality*, like Doob's, to almost sure convergence results.

5.2.7 Corollary. *Let X be an $\mathbb{R}$-valued random variable and $p \in [1,\infty)$. If $X \in L^p(P;\mathbb{R})$, then for any nondecreasing sequence $\{\mathcal{F}_n : n \in \mathbb{N}\}$ of sub-σ-algebras of $\mathcal{F}$:*

$$\mathbb{E}^P\big[X\big|\mathcal{F}_n\big] \longrightarrow \mathbb{E}^P\left[X\middle|\bigvee_0^\infty \mathcal{F}_n\right] \quad \text{(a.s., } P\text{) and in } L^p(P;\mathbb{R}) \tag{5.2.8}$$

as $n \to \infty$. In particular, if X is $\bigvee_0^\infty \mathcal{F}_n$-measurable, then $\mathbb{E}^P\big[X\big|\mathcal{F}_n\big] \longrightarrow X$ (a.s., P) and in $L^p(P;\mathbb{R})$.

PROOF: Without loss in generality, we will assume that $\mathcal{F} = \bigvee_0^\infty \mathcal{F}_n$.

Given $X \in L^1(P;\mathbb{R})$, set $X_n = \mathbb{E}^P\big[X\big|\mathcal{F}_n\big]$ for $n \in \mathbb{N}$. The key to our proof will be the inequality

$$P\left(\sup_{n\in\mathbb{N}} |X_n| \geq \alpha\right) \leq \frac{1}{\alpha}\mathbb{E}^P\left[|X|, \sup_{n\in\mathbb{N}}|X_n| \geq \alpha\right], \quad \alpha \in (0,\infty); \tag{5.2.9}$$

and, since, by (5.1.8), $|X_n| \leq \mathbb{E}^P\big[|X|\,\big|\mathcal{F}_n\big]$ (a.s., P), while proving (5.2.9) we may and will assume that X and all the X_n's are nonnegative. But then, by (5.2.5),

$$\begin{aligned} P\left(\sup_{0\leq n\leq N} X_n \geq \alpha\right) &\leq \frac{1}{\alpha}\mathbb{E}^P\left[X_N, \sup_{0\leq n\leq N} X_n \geq \alpha\right] \\ &= \frac{1}{\alpha}\mathbb{E}^P\left[X, \sup_{0\leq n\leq N} X_n \geq \alpha\right] \end{aligned}$$

for all $N \in \mathbb{Z}^+$; and therefore (5.2.9) follows when $N \to \infty$.

As our first application of (5.2.9), we note that $\{X_n\}_0^\infty$ is uniformly P-integrable. Indeed, because $|X_n| \leq \mathbb{E}^P\big[|X|\,\big|\mathcal{F}_n\big]$, we have from (5.2.9) that

$$\begin{aligned} \sup_{n\in\mathbb{N}} \mathbb{E}^P\Big[|X_n|, |X_n| \geq \alpha\Big] &\leq \sup_{n\in\mathbb{N}} \mathbb{E}^P\Big[|X|, |X_n| \geq \alpha\Big] \\ &\leq \mathbb{E}^P\left[|X|, \sup_{n\in\mathbb{N}}|X_n| \geq \alpha\right] \longrightarrow 0 \end{aligned}$$

as $\alpha \to \infty$. Thus, we will know that the convergence in (5.2.8) takes place in $L^1(P;\mathbb{R})$ as soon as we show that it happens P-almost surely. In addition, if $X \in L^p(P;\mathbb{R})$ for some $p \in (1,\infty)$, then, by (5.2.6) and Exercise 1.4.20, we see that $\{|X_n|^p : n \in \mathbb{N}\}$ is uniformly P-integrable and, therefore, that $X_n \longrightarrow X$ in $L^p(\mu)$ as soon as it does (a.s., P). In other words, everything comes down to checking the P-almost sure convergence.

To prove the P-almost sure convergence, let $\mathcal{G}$ be the set of $X \in L^1(P;\mathbb{R})$ for which $X_n \longrightarrow X$ (a.s., P). Clearly, $X \in \mathcal{G}$ if $X \in L^1(P;\mathbb{R})$ is $\mathcal{F}_n$-measurable for some $n \in \mathbb{N}$; and, therefore, $\mathcal{G}$ is dense in $L^1(P;\mathbb{R})$. Thus, all that remains is to prove that $\mathcal{G}$ is closed in $L^1(P;\mathbb{R})$. But if $\{X^{(k)}\}_1^\infty \subseteq \mathcal{G}$ and $X^{(k)} \longrightarrow X$ in $L^1(P;\mathbb{R})$, then, by (5.2.9),

$$\begin{aligned}
&P\left(\sup_{n\ge N}\left|X_n - X\right| \ge 3\alpha\right)\\
&\le P\left(\sup_{n\ge N}\left|X_n - X_n^{(k)}\right| \ge \alpha\right) + P\left(\sup_{n\ge N}\left|X_n^{(k)} - X^{(k)}\right| \ge \alpha\right)\\
&\qquad + P\left(\left|X^{(k)} - X\right| \ge \alpha\right)\\
&\le \frac{2}{\alpha}\left\|X - X^{(k)}\right\|_{L^1(P)} + P\left(\sup_{n\ge N}\left|X_n^{(k)} - X^{(k)}\right| \ge \alpha\right)
\end{aligned}$$

for every $N \in \mathbb{Z}^+$, $\alpha \in (0,\infty)$, and $k \in \mathbb{Z}^+$. Hence, by first letting $N \to \infty$ and then $k \to \infty$, we see that

$$\lim_{N\to\infty} P\left(\sup_{n\ge N}\left|X_n - X\right| \ge 3\alpha\right) = 0 \quad \text{for every } \alpha \in (0,\infty);$$

and this proves that $X \in \mathcal{G}$. □

Before moving on to more sophisticated convergence results, we will spend a little time showing that Corollary 5.2.7 is already interesting. In order to introduce our main application, recall our preliminary discussion of conditioning when we were attempting to explain Kolmogorov's idea at the beginning of Section 5.1. As we said there, the most easily understood situation occurs when one conditions with respect to a sub-σ-algebra Σ which is generated by a countable partition $\mathcal{P}$. Indeed, in that case one can easily verify that

$$\text{(5.2.10)} \qquad \mathbb{E}^P\left[X\middle|\Sigma\right] = \sum_{A\in\mathcal{P}} \frac{\mathbb{E}^P[X,\, A]}{P(A)}\,\mathbf{1}_A,$$

where it is understood that

$$\frac{\mathbb{E}^P[X,\, A]}{P(A)} \equiv 0 \quad \text{when} \quad P(A) = 0.$$

Unfortunately, even when $\mathcal{F}$ is countably generated, Σ need not be (cf. Exercise 1.1.8). Furthermore, just because Σ is countably generated, it will be seldom true that its generators can be chosen to form a countable partition. (For example, as soon as Σ contains an uncountable number of atoms, such a partition cannot exist.) Nonetheless, if Σ is any countably generated σ-algebra, then we can find a sequence $\{\mathcal{P}_n\}_0^\infty$ of finite partitions with the properties that

$$\Sigma = \sigma\left(\bigcup_0^\infty \mathcal{P}_n\right) \quad \text{and} \quad \sigma(\mathcal{P}_{n-1}) \subseteq \sigma(\mathcal{P}_n), \quad n \in \mathbb{Z}^+. \tag{5.2.11}$$

In fact, simply choose a countable generating sequence $\{A_n\}_0^\infty$ for Σ and take $\mathcal{P}_n$ to be the collection of distinct sets of the form $B_0 \cap \cdots \cap B_n$, where $B_m \in \{A_m, A_m\complement\}$ for each $0 \le m \le n$.

5.2.12 Theorem. *Let Σ be a countably generated sub-σ-algebra of $\mathcal{F}$ and choose $\{\mathcal{P}_n\}_0^\infty$ to be a sequence of finite partitions satisfying (5.2.11). Next, given $p \in [1,\infty)$ and a random variable $X \in L^p(P;\mathbb{R})$, define X_n for $n \in \mathbb{N}$ by the right-hand side of (5.2.10) with $\mathcal{P} = \mathcal{P}_n$. Then $X_n \longrightarrow \mathbb{E}^P[X|\Sigma]$ both P-almost surely and in $L^p(P;\mathbb{R})$. In particular, even if Σ is not countably generated, for each separable, closed subspace $\mathbf{L}$ of $L^p(P;\mathbb{R})$ there exists a sequence of finite partitions $\mathcal{P}_n$, $n \in \mathbb{N}$, such that*

$$\sum_{A \in \mathcal{P}_n} \frac{\mathbb{E}^P[X, A]}{P(A)} \mathbf{1}_A \longrightarrow \mathbb{E}^P[X|\Sigma] \quad (\text{a.s.}, P) \text{ and in } L^p(P;\mathbb{R}) \tag{5.2.13}$$

for every $X \in \mathbf{L}$.

PROOF: To prove the first part, simply set $\mathcal{F}_n = \sigma(\mathcal{P}_n)$, then identify X_n as $\mathbb{E}^P[X|\mathcal{F}_n]$, and finally apply Corollary 5.2.7. As for the second part, let $\Sigma(\mathbf{L})$ be the σ-algebra generated by

$$\left\{\mathbb{E}^P[X|\Sigma] : X \in \mathbf{L}\right\},$$

note that $\Sigma(\mathbf{L})$ is countably generated and that

$$\mathbb{E}^P[X|\Sigma] = \mathbb{E}^P[X|\Sigma(\mathbf{L})] \quad (\text{a.s.}, P) \quad \text{for each } X \in \mathbf{L},$$

and apply the first part with Σ replaced by $\Sigma(\mathbf{L})$. □

5.2.14 Corollary (Jensen's Inequality). *Let C be a closed, convex subset of $\mathbb{R}^N$, $\mathbf{X}$ a C-valued, P-integrable random variable, and Σ a sub-σ-algebra of $\mathcal{F}$. Then there is a C-valued representative $\mathbf{X}_\Sigma$ of*

$$\mathbb{E}^P[\mathbf{X}|\Sigma] \equiv \begin{bmatrix} \mathbb{E}^P[X_1|\Sigma] \\ \vdots \\ \mathbb{E}^P[X_N|\Sigma] \end{bmatrix}.$$

In addition, if $g : C \longrightarrow [0,\infty)$ is continuous and concave, then

$$\mathbb{E}^P\big[g(\mathbf{X})\big|\Sigma\big] \le g\big(\mathbf{X}_\Sigma\big|\Sigma\big) \quad (\text{a.s.}, P). \tag{5.2.15}$$

PROOF: By the classical Jensen's Inequality, $Y \equiv g(\mathbf{X})$ is P-integrable. Hence, by the second part of Theorem 5.2.12, we can find finite partitions $\mathcal{P}_n$, $n \in \mathbb{N}$, so that

$$\mathbf{X}_n \equiv \sum_{A\in\mathcal{P}_n} \frac{\mathbb{E}^P\big[\mathbf{X}, A\big]}{P(A)}\mathbf{1}_A \longrightarrow \mathbb{E}^P\big[\mathbf{X}\big|\Sigma\big]$$

and

$$Y_n \equiv \sum_{A\in\mathcal{P}_n} \frac{\mathbb{E}^P\big[g(\mathbf{X}), A\big]}{P(A)}\mathbf{1}_A \longrightarrow \mathbb{E}^P\big[g(\mathbf{X})\big|\Sigma\big]$$

P-almost surely. Furthermore, again by the classical Jensen's Inequality,

$$\frac{\mathbb{E}^P\big[\mathbf{X}, A\big]}{P(A)} \in C \quad\text{and}\quad \frac{\mathbb{E}^P\big[g(\mathbf{X}), A\big]}{P(A)} \le g\left(\frac{\mathbb{E}^P\big[\mathbf{X}, A\big]}{P(A)}\right)$$

for all $A \in \mathcal{F}$ with $P(A) > 0$. Hence, if $\Lambda \in \Sigma$ denotes the set of ω for which

$$\lim_{n\to\infty} \begin{bmatrix} \mathbf{X}_n(\omega) \\ Y_n(\omega) \end{bmatrix} \in \mathbb{R}^{N+1}$$

exists, $\mathbf{v}$ is a fixed element of C,

$$\mathbf{X}_\Sigma(\omega) \equiv \begin{cases} \lim_{n\to\infty} \mathbf{X}_n(\omega) & \text{if } \ \omega \in \Lambda \\ \mathbf{v} & \text{if } \ \omega \notin \Lambda, \end{cases}$$

and

$$Y(\omega) \equiv \begin{cases} \lim_{n\to\infty} Y_n(\omega) & \text{if } \ \omega \in \Lambda \\ 0 & \text{if } \ \omega \notin \Lambda, \end{cases}$$

then $\mathbf{X}_\Sigma$ is a C-valued representative of $\mathbb{E}^P\big[\mathbf{X}\big|\Sigma\big]$, Y is a representative of $\mathbb{E}^P\big[g(\mathbf{X})\big|\Sigma\big]$, and $Y(\omega) \le g\big(\mathbf{X}_\Sigma(\omega)\big)$ for every $\omega \in \Omega$. □

5.2.16 Corollary. *Let I be a nontrivial, closed interval in $\mathbb{R}\cup\{+\infty\}$ (i.e., either $I \subset \mathbb{R}$ is bounded on the right or $I \cap \mathbb{R}$ is unbounded on the right and I includes the point $+\infty$). Then every I-valued random variable X with P-integrable negative part admits an I-valued representative of $\mathbb{E}^P[X|\Sigma]$. Furthermore, given a continuous, convex function $f : I \longrightarrow \mathbb{R}\cup\{+\infty\}$,*

$$f\Big(\mathbb{E}^P\big[X\big|\Sigma\big]\Big) \le \mathbb{E}^P\big[f(X)\big|\Sigma\big] \quad (\text{a.s.}, P) \tag{5.2.17}$$

if either f is bounded above and X is P-integrable or f is bounded below and to the left (i.e., f is bounded on each interval of the form $I \cap (-\infty, a]$ with $a \in I \cap \mathbb{R}$). In particular, for each $p \in [1, \infty)$,

$$\left\| \mathbb{E}^P [X \mid \Sigma] \right\|_{L^p(P)} \le \|X\|_{L^p(P)}.$$

Finally, if $(X_n, \mathcal{F}_n, P)$ is an I-valued martingale and f is as above or if $(X_n, \mathcal{F}_n, P)$ is an I-valued submartingale and f is bounded below and nondecreasing (as well as continuous and convex), then $(f(X_n), \mathcal{F}_n, P)$ is a submartingale.

PROOF: In view of Corollary 5.2.14, we know that an I-valued representative of $\mathbb{E}^P[X|\Sigma]$ exists when X is P-integrable, and the general case follows after a trivial truncation procedure. In order to prove (5.2.17), first assume that f is bounded above by some $M < \infty$ and that $X \in L^1(P)$. Then (5.2.17) is an immediate consequence of (5.2.15) with $g = M - f$. To handle the case when f is bounded below and to the left, first observe that either f is nonincreasing everywhere, or there is an $a \in I \cap \mathbb{R}$ with the property that f is nonincreasing to the left of a and nondecreasing to the right of a. Next, let an I-valued X with $X^- \in L^1(P)$ be given, and set $X_n = X \wedge n$. Then there exists an $m \in \mathbb{Z}^+$ such that X_n is I-valued for all $n \ge m$; and clearly, by the preceding, we know that

$$f\Big(\mathbb{E}^P[X_n|\Sigma]\Big) \le \mathbb{E}^P[f(X_n)|\Sigma] \quad \text{(a.s., } P\text{)} \quad \text{for all } n \ge m.$$

Moreover, in the case when f is nonincreasing, $\{f(X_n) : n \ge m\}$ is bounded and nonincreasing; and, in the other case, $\{f(X_n) : n \ge m \vee a\}$ is bounded below and nondecreasing. Hence, in both cases, (5.2.17) follows from the preceding after an application of the version of the Monotone Convergence Theorem in (5.1.13).

To complete the proof, simply note that in either of the two cases given, the results just proved justify:

$$\mathbb{E}^P[f(X_n)|\mathcal{F}_{n-1}] \ge f\Big(\mathbb{E}^P[X_n|\mathcal{F}_{n-1}]\Big) \ge f(X_{n-1})$$

P-almost surely. □

Our next goal is to show that, even when they are not given in the form covered by Corollary 5.2.7, *martingales want to converge.* If for no other reason, such a result has got to be more difficult because one does not know ahead of time what, if it exists, the limit ought to be. Thus, the reasoning will have to be more subtle than that used in the proof of Corollary 5.2.7. We will follow Doob and base our argument on the idea that, in some sense, a martingale has got to be *nearly constant* and that a submartingale is the sum of a martingale and a nondecreasing process. Although the first of these observations will take a little time to develop, the second one is nearly trivial and is covered by the following simple lemma.

5.2.18 Lemma (Doob's Decomposition). *Let $(X_n, \mathcal{F}_n, P)$ be a P-integrable submartingale. Then there is a P-almost surely unique sequence $\{I_n : n \in \mathbb{N}\}$ of random variables with the properties that $I_0 \equiv 0$,*

$$I_{n-1} \le I_n \in L^1(\Omega, \mathcal{F}_{n-1}, P; [0,\infty)) \quad \text{for each } n \in \mathbb{Z}^+,$$

and $(X_n - I_n, \mathcal{F}_n, P)$ is a martingale.

PROOF: To prove the existence, set $I_0 \equiv 0$ and

$$I_n = I_{n-1} + \mathbb{E}^P[X_n - X_{n-1} | \mathcal{F}_{n-1}] \vee 0 \quad \text{for} \quad n \in \mathbb{Z}^+.$$

To prove the uniqueness, suppose that $\{J_n : n \in \mathbb{N}\}$ is a second such sequence and set $\Delta_n = I_n - J_n$, $n \in \mathbb{N}$. Then $\Delta_0 \equiv 0$, Δ_n is $\mathcal{F}_{n-1}$-measurable for each $n \in \mathbb{Z}^+$, and $(\Delta_n, \mathcal{F}_n, P)$ is a martingale. Hence

$$\Delta_n = \mathbb{E}^P[\Delta_n | \mathcal{F}_{n-1}] = \Delta_{n-1} \quad \text{(a.s., } P) \quad \text{for each} \quad n \in \mathbb{Z}^+,$$

and so, by induction, $\Delta_n = 0$ (a.s., P) for all $n \in \mathbb{N}$. □

In order to make mathematics out of the idea that a martingale *is nearly constant*, we will first adjust the notion of a stopping time to the present setting. Namely, we will say that the function $\tau : \Omega \longrightarrow \mathbb{N} \cup \{\infty\}$ is an $\{\mathcal{F}_n : n \in \mathbb{N}\}$-**stopping time** if $\{\omega : \tau(\omega) = n\} \in \mathcal{F}_n$ for each $n \in \mathbb{N}$. In addition, given an $\{\mathcal{F}_n : n \in \mathbb{N}\}$-stopping time τ, let $\mathcal{F}_\tau$ be the σ-algebra of $A \in \mathcal{F}$ such that $A \cap \{\tau = n\} \in \mathcal{F}_n$, $n \in \mathbb{Z}^+$. Notice that these definitions are completely consistent with the ones introduced earlier (at the beginning of Section 4.3) and that the obvious variant of Lemma 4.3.2 holds equally well here. In addition, if $\{X_n : n \in \mathbb{N}\}$ is $\{\mathcal{F}_n : n \in \mathbb{N}\}$-progressively measurable, then the random variable X_τ given by $X_\tau(\omega) = X_{\tau(\omega)}(\omega)$ is $\mathcal{F}_\tau$-measurable on $\{\tau < \infty\}$.

5.2.19 Theorem (Hunt). *Let $(X_n, \mathcal{F}_n, P)$ be a P-integrable submartingale. Given bounded $\{\mathcal{F}_n : n \in \mathbb{N}\}$-stopping times σ and τ satisfying $\sigma \le \tau$,*

$$X_\sigma \le \mathbb{E}^P[X_\tau | \mathcal{F}_\sigma] \quad \text{(a.s., } P), \tag{5.2.20}$$

and the inequality can be replaced by equality when $(X_n, \mathcal{F}_n, P)$ is a martingale. (Cf. Exercise 5.2.32 for unbounded stopping times.)

PROOF: Choose $\{I_n : n \in \mathbb{N}\}$ as in Lemma 5.2.18, and set $Y_n = X_n - I_n$, $n \in \mathbb{N}$. Then, because $I_\sigma \le I_\tau$ and I_σ is $\mathcal{F}_\sigma$-measurable:

$$\mathbb{E}^P[X_\tau | \mathcal{F}_\sigma] \ge \mathbb{E}^P[Y_\tau + I_\sigma | \mathcal{F}_\sigma] = \mathbb{E}^P[Y_\tau | \mathcal{F}_\sigma] + I_\sigma,$$

and so it suffices to prove that equality holds in (5.2.20) when $(X_n, \mathcal{F}_n, P)$ is a martingale. To this end, choose $N \in \mathbb{Z}^+$ to be an upper bound for τ, let $A \in \mathcal{F}_\sigma$ be given, and note that

$$\begin{aligned}\mathbb{E}^P\big[X_N,\, A\big] &= \sum_{n=0}^{N} \mathbb{E}^P\big[X_N,\, A \cap \{\sigma = n\}\big] \\ &= \sum_{n=0}^{N} \mathbb{E}^P\big[X_n,\, A \cap \{\sigma = n\}\big] = \mathbb{E}^P\big[X_\sigma,\, A\big];\end{aligned}$$

and similarly (since $A \in \mathcal{F}_\sigma \subseteq \mathcal{F}_\tau$), $\mathbb{E}^P\big[X_N,\, A\big] = \mathbb{E}^P\big[X_\tau,\, A\big]$. □

The following easy consequence of Hunt's Theorem becomes more interesting when it is interpreted from the point of view of a gambler who is trying to *beat the system.* Namely, let $(X_n, \mathcal{F}_n, P)$ be a martingale with mean-value 0, and think of X_n as the gambler's fortune after n plays of a *fair game.* With this model in mind, it is natural to interpret an $\{\mathcal{F}_n : n \in \mathbb{N}\}$-stopping time τ as a *strategy.* That is, τ can be considered as a feasible (i.e., one which does not require the power of prophesy) scheme which the gambler can use to determine when he should stop playing. When couched in these terms, the next result predicts that *there is no strategy with which the gambler can alter his expected take!*

5.2.21 Corollary (Doob's Stopping Time Theorem). *For any P-integrable submartingale (martingale) $(X_n, \mathcal{F}_n, P)$ and any $\{\mathcal{F}_n : n \in \mathbb{N}\}$-stopping time τ, $(X_{n\wedge\tau}, \mathcal{F}_n, P)$ is again a P-integrable submartingale (martingale).*

PROOF: Let $A \in \mathcal{F}_{n-1}$. Then, since $A \cap \{\tau > n-1\} \in \mathcal{F}_{(n-1)\wedge\tau}$, (5.2.20) implies that

$$\begin{aligned}&\mathbb{E}^P\big[X_{n\wedge\tau},\, A\big] \\ &\quad = \mathbb{E}^P\big[X_{n\wedge\tau},\, A \cap \{\tau \le n-1\}\big] + \mathbb{E}^P\big[X_{n\wedge\tau},\, A \cap \{\tau > n-1\}\big] \\ &\quad \ge \mathbb{E}^P\big[X_\tau,\, A \cap \{\tau \le n-1\}\big] + \mathbb{E}^P\big[X_{(n-1)\wedge\tau},\, A \cap \{\tau > n-1\}\big] \\ &\quad = \mathbb{E}^P\big[X_{(n-1)\wedge\tau},\, A\big];\end{aligned}$$

and, in the case of martingales, the inequality in the preceding can be replaced by an equality. □

Given a sequence $\{x_n\}_0^\infty \subseteq \mathbb{R}$ and $-\infty < a < b < \infty$, we say that $\{x_n\}_0^\infty$ **upcrosses the interval** $[a, b]$ **at least** N **times** if there exist integers $0 \le m_1 < n_1 < \cdots < m_N < n_N$ such that $x_{m_i} \le a$ and $x_{n_i} \ge b$ for each $1 \le i \le N$ and that it **upcrosses** $[a, b]$ **precisely** N **times** if it upcrosses $[a, b]$ at least N but does not upcross $[a, b]$ at least $N+1$ times. Notice that $\underline{\lim}_{n\to\infty} x_n < \overline{\lim}_{n\to\infty} x_n$ if and only if there exist rational numbers $a < b$ such that $\{x_n\}$ upcrosses $[a, b]$ at least N times for every $N \in \mathbb{Z}^+$.

5.2.22 Doob's Martingale Convergence Theorem.[†] *Let $(X_n, \mathcal{F}_n, P)$ be a P-integrable submartingale, and, for $-\infty < a < b < \infty$, let $U_{[a,b]}(\omega)$ denote the precise number of times that $\{X_n(\omega)\}_0^\infty$ upcrosses $[a,b]$. Then*

$$\mathbb{E}^P\left[U_{[a,b]}\right] \le \sup_{n\in\mathbb{N}} \frac{\mathbb{E}^P\left[(X_n - a)^+\right]}{b-a}. \tag{5.2.23}$$

In particular, if

$$\sup_{n\in\mathbb{N}} \mathbb{E}^P\left[X_n^+\right] < \infty, \tag{5.2.24}$$

then there exists a P-integrable random variable X to which $\{X_n\}_1^\infty$ converges P-almost surely. (See Exercises 5.2.27 and 5.2.31 for other derivations.)

PROOF: Set $Y_n = \frac{(X_n-a)^+}{b-a}$ and note that (by Corollary 5.2.16) $(Y_n, \mathcal{F}_n, P)$ is a P-integrable submartingale. Next, set $\tau_0 = 0$, and, for $k \in \mathbb{Z}^+$, define

$$\sigma_k = \inf\{n \ge \tau_{k-1} : X_n \le a\} \quad \text{and} \quad \tau_k = \inf\{n \ge \sigma_k : X_n \ge b\}.$$

Proceeding by induction, it is an easy matter to check that all the σ_k's and τ_k's are $\{\mathcal{F}_n : n \in \mathbb{N}\}$-stopping times. Moreover, if $N \in \mathbb{Z}^+$ and $U_{[a,b]}^{(N)}(\omega)$ is the precise number of times $\{X_{n\wedge N}(\omega)\}_0^\infty$ upcrosses $[a,b]$, then

$$\begin{aligned} U_{[a,b]}^{(N)} &\le \sum_{k=1}^N \left(Y_{N\wedge\tau_k} - Y_{N\wedge\sigma_k}\right) \\ &= Y_N - Y_0 - \sum_{k=1}^N \left(Y_{N\wedge\sigma_k} - Y_{N\wedge\tau_{k-1}}\right) \\ &\le Y_N - \sum_{k=1}^N \left(Y_{N\wedge\sigma_k} - Y_{N\wedge\tau_{k-1}}\right). \end{aligned}$$

Hence, since $\tau_{k-1} \le \sigma_k$ and therefore, by (5.2.20),

$$\mathbb{E}^P\left[Y_{N\wedge\sigma_k} - Y_{N\wedge\tau_{k-1}}\right] \ge 0 \quad \text{for all} \quad k \in \mathbb{Z}^+,$$

we see that $\mathbb{E}^P\left[U_{[a,b]}^{(N)}\right] \le \mathbb{E}^P\left[Y_N\right]$; and, clearly (5.2.23) follows from this after one lets $N \to \infty$.

[†] In the notes to Chapter VII of his *Stochastic Processes*, publ. by J. Wiley in 1953, Doob gives a thorough account of the relationship between his convergence result and earlier attempts in the same direction. In particular, he points out that, in 1946, S. Anderson and B. Jessen formulated and proved a closely related convergence theorem.

Given (5.2.23), the convergence result is easy. Namely, if (5.2.24) is satisfied, then (5.2.23) implies that there is a set Λ of full P-measure such that $U_{[a,b]}(\omega) < \infty$ for all rational $a < b$ and $\omega \in \Lambda$; and so, by the remark preceding the statement of this theorem, for each $\omega \in \Lambda$, $\{X_n(\omega)\}_0^\infty$ converges to some $X(\omega) \in [-\infty, \infty]$. Hence, we will be done as soon as we show that $\mathbb{E}^P[|X|, \Lambda] < \infty$. But

$$\mathbb{E}^P[|X_n|] = 2\mathbb{E}^P[X_n^+] - \mathbb{E}^P[X_n] \le 2\mathbb{E}^P[X_n^+] - \mathbb{E}^P[X_0], \quad n \in \mathbb{N},$$

and therefore Fatou's Lemma shows that X is P-integrable on Λ. $\square$

The inequality in (5.2.23) is quite famous and is known as **Doob's upcrossing inequality**.

5.2.25 Corollary. *Let $(X_n, \mathcal{F}_n, P)$ be a martingale. Then there exists an $X \in L^1(P;\mathbb{R})$ such that $X_n = \mathbb{E}^P[X|\mathcal{F}_n]$ (a.s., P) for each $n \in \mathbb{N}$ if and only if the sequence $\{X_n\}_0^\infty$ is uniformly P-integrable. In addition, if $p \in (1,\infty]$, then there is an $X \in L^p(P;\mathbb{R})$ such that $X_n = \mathbb{E}^P[X|\mathcal{F}_n]$ (a.s., P) for each $n \in \mathbb{N}$ if and only if $\{X_n\}_0^\infty$ is a bounded subset of $L^p(P;\mathbb{R})$.*

PROOF: Because of Corollary 5.2.7 and (5.2.6), we need only check the first "if" assertion in the first assertion. But, if $\{X_n\}_0^\infty$ is uniformly P-integrable, then (5.2.24) holds and therefore $X_n \longrightarrow X$ (a.s., P) for some P-integrable X. Moreover, uniform integrability together with almost sure convergence implies convergence in $L^1(P;\mathbb{R})$, and therefore, by (5.1.8), for each $m \in \mathbb{N}$,

$$X_m = \lim_{n\to\infty} \mathbb{E}^P[X_n|\mathcal{F}_m] = \mathbb{E}^P[X|\mathcal{F}_m] \quad \text{(a.s., } P). \quad \square$$

Just as Corollary 5.2.7 led us to an intuitively appealing way to construct conditional expectations, so Doob's Theorem gives us an appealing approximation procedure for Radon–Nikodym derivatives.

5.2.26 Theorem (Jessen). *Let P and Q be a pair of probability measures on the measurable space $(\Omega, \mathcal{F})$ and $\{\mathcal{F}_n : n \in \mathbb{N}\}$ a nondecreasing sequence of sub-σ-algebras whose union generates $\mathcal{F}$. For each $n \in \mathbb{N}$, let $Q_{n,\mathrm{a}}$ and $Q_{n,\mathrm{s}}$ denote, respectively, the absolutely continuous and singular parts of $Q_n \equiv Q \restriction \mathcal{F}_n$ with respect to $P_n \equiv P \restriction \mathcal{F}_n$, and set $X_n = \frac{dQ_{n,\mathrm{a}}}{dP_n}$. Also, let Q_a be the absolutely continuous part of Q with respect to P, and set $Y = \frac{dQ_\mathrm{a}}{dP}$. Then $X_n \longrightarrow Y$ (a.s., P). In particular, $Q \perp P$ if and only if $X_n \longrightarrow 0$ (a.s., P). Moreover, if $Q_n \ll P_n$ for each $n \in \mathbb{N}$, then $Q \ll P$ if and only if $\{X_n\}_0^\infty$ is uniformly P-integrable, in which case $X_n \longrightarrow Y$ in $L^1(P;\mathbb{R})$ as well as P-almost surely. Finally, if $Q_n \sim P_n$ (i.e., $P_n \ll Q_n$ as well as $Q_n \ll P_n$) for each $n \in \mathbb{N}$ and $B \equiv \{\varliminf_{n\to\infty} X_n = 0\}$, then $Q_\mathrm{a}(A) = Q(A \cap B\complement)$ for all $A \in \mathcal{F}$, and therefore $Q(B) = 0 \implies Q \ll P$.*

PROOF: Without loss in generality, we will assume throughout that all the X_n's as well as $Y \equiv \frac{dQ_a}{dP}$ take values in $[0,\infty)$; and clearly, $\mathbb{E}^P[X_n]$, $n \in \mathbb{N}$, and $\mathbb{E}^P[Y]$ are all dominated by 1.

We first note that

$$Q_{n,s}(A) = \sup\left\{Q(A \cap B) : B \in \mathcal{F}_n \text{ and } P(B) = 0\right\} \quad \text{for} \quad A \in \mathcal{F}_n.$$

Hence, $Q_{n,s} \restriction \mathcal{F}_{n-1} \geq Q_{n-1,s}$ for each $n \in \mathbb{Z}^+$, and so

$$\mathbb{E}^P\big[X_n,\, A\big] = Q_{n,a}(A) \leq Q_{n-1,a}(A) = \mathbb{E}^P\big[X_{n-1},\, A\big]$$

for all $n \in \mathbb{Z}^+$ and $A \in \mathcal{F}_{n-1}$. In other words, $(-X_n, \mathcal{F}_n, P)$ is a nonpositive submartingale. Moreover, in the case when $Q_n \ll P_n$, $n \in \mathbb{N}$, the same argument shows that $(X_n, \mathcal{F}_n, P)$ is a nonnegative martingale. Thus, in any case, there is a nonnegative, P-integrable random variable X with the property that $X_n \longrightarrow X$ (a.s., P). In order to identify X as Y, we use Fatou's Lemma to see that, for any $m \in \mathbb{N}$ and $A \in \mathcal{F}_m$:

$$\mathbb{E}^P\big[X,\, A\big] \leq \varliminf_{n\to\infty} \mathbb{E}^P\big[X_n,\, A\big] = \varliminf_{n\to\infty} Q_{n,a}(A) \leq Q(A);$$

and therefore $\mathbb{E}^P\big[X,\, A\big] \leq Q(A)$ for every $A \in \mathcal{F}$. In particular, by choosing $B \in \mathcal{F}$ so that $Q_a(A) = Q(A \cap B)$, $A \in \mathcal{F}$, and $P(B\complement) = 0$, we arrive at

$$\mathbb{E}^P\big[X,\, A\big] \leq Q(A \cap B) = Q_a(A) = \mathbb{E}^P\big[Y,\, A\big] \quad \text{for all } A \in \mathcal{F};$$

which means that $X \leq Y$ (a.s., P). On the other hand, if $Y_n = \mathbb{E}^P\big[Y \big| \mathcal{F}_n\big]$ for $n \in \mathbb{N}$, then

$$\mathbb{E}^P\big[Y_n,\, A\big] = Q_a(A) \leq Q_{n,a}(A) = \mathbb{E}^P\big[X_n,\, A\big] \quad \text{for all} \quad A \in \mathcal{F}_n,$$

and therefore $Y_n \leq X_n$ (a.s., P) for each $n \in \mathbb{N}$. But, since $Y_n \longrightarrow Y$ and $X_n \longrightarrow X$ P-almost surely, this means that $Y \leq X$ (a.s., P).

Next, assume that $Q_n \ll P_n$ for each $n \in \mathbb{N}$ and therefore that $(X_n, \mathcal{F}_n, P)$ is a nonnegative martingale. If $\{X_n\}_0^\infty$ is uniformly P-integrable, then $X_n \longrightarrow Y$ in $L^1(P;\mathbb{R})$ and therefore $Q_s(\Omega) = 1 - \mathbb{E}^P[Y] = 0$. Hence, $Q \ll P$ when $\{X_n\}_0^\infty$ is uniformly P-integrable. Conversely, if $Q \ll P$, then it is easy to see that $X_n = \mathbb{E}^P\big[Y \big| \mathcal{F}_n\big]$ for each $n \in \mathbb{N}$, and therefore (by Corollary 5.2.7) that $\{X_n\}_0^\infty$ is uniformly P-integrable.

Finally, assume that $Q_n \sim P_n$ for each $n \in \mathbb{N}$. Then, the X_n's can be chosen to take their values in $(0,\infty)$ and

$$Y_n \equiv \frac{1}{X_n} = \frac{dP_n}{dQ_n}.$$

Hence, if P_a and P_s are the absolutely continuous and singular parts of P relative to Q and if $Y \equiv \varliminf_{n\to\infty} Y_n$, then $Y = \frac{dP_a}{dQ}$ and so $P_a(A) = \mathbb{E}^Q[Y, A]$ for all $A \in \mathcal{F}$. Thus, when $C \in \mathcal{F}$ is chosen so that $P_s(C) = 0 = Q(C\complement)$, then, since $Y = \frac{1}{X}$ (a.s., P) on $B\complement$, it is becomes an easy step from the above to

$$Q(A \cap B\complement) = \mathbb{E}^{P_a}\left[Y^{-1}, A \cap B\complement\right] = \mathbb{E}^P\left[X, A \cap C\right] = Q_a(A \cap C) = Q_a(A)$$

for all $A \in \mathcal{F}$. □

Exercises

5.2.27 Exercise: In this exercise we will present a quite independent derivation of the convergence assertion in Doob's Martingale Convergence Theorem. The key observations here are first that, given Doob's Inequality (cf. (5.2.5)), the result is nearly trivial for martingales having two bounded moments and, second, that everything can be reduced to that case.

(i) Let $(M_n, \mathcal{F}_n, P)$ be a martingale for which

$$\sup_{n\in\mathbb{N}} \mathbb{E}^P\left[M_n^2\right] < \infty. \tag{5.2.28}$$

Note that

$$\mathbb{E}^P\left[M_n^2\right] - \mathbb{E}^P\left[M_{m-1}^2\right] = \mathbb{E}^P\left[\left(M_n - M_{m-1}\right)^2\right] \quad \text{for} \quad 1 \le m \le n; \tag{5.2.29}$$

and starting from (5.2.28) and (5.2.29), show that there is an $M \in L^2(P;\mathbb{R})$ such that $M_n \longrightarrow M$ in $L^2(P;\mathbb{R})$. Next, by applying (5.2.5) to the submartingale $\left(\left(M_{n\vee m} - M_m\right)^2, \mathcal{F}_n, P\right)$, show that, for every $\epsilon > 0$,

$$P\left(\sup_{n\ge m} |M - M_n| \ge \epsilon\right) \le \frac{1}{\epsilon^2}\mathbb{E}^P\left[\left(M - M_m\right)^2\right] \longrightarrow 0 \quad \text{as} \quad m \to \infty,$$

and conclude that $M_n \longrightarrow M$ (a.s., P).

(ii) Let $(X_n, \mathcal{F}_n, P)$ be a nonnegative submartingale with the property that $\sup_{n\in\mathbb{N}} \mathbb{E}^P\left[X_n^2\right] < \infty$, define the sequence $\{I_n : n \in \mathbb{N}\}$ as in Lemma 5.2.18, and set $M_n = X_n - I_n$, $n \in \mathbb{N}$. Then $(M_n, \mathcal{F}_n, P)$ is a martingale, and clearly both M_n and I_n are square P-integrable for each $n \in \mathbb{N}$. In fact, check that

$$\begin{aligned}
\mathbb{E}^P\left[M_n^2 - M_{n-1}^2\right] &= \mathbb{E}^P\left[\left(M_n - M_{n-1}\right)\left(X_n + X_{n-1}\right)\right] \\
&= \mathbb{E}^P\left[X_n^2 - X_{n-1}^2\right] - \mathbb{E}^P\left[\left(I_n - I_{n-1}\right)\left(X_n + X_{n-1}\right)\right] \\
&\le \mathbb{E}^P\left[X_n^2 - X_{n-1}^2\right],
\end{aligned}$$

and therefore that

$$\mathbb{E}^P\big[M_n^2\big] \le \mathbb{E}^P\big[X_n^2\big] \quad\text{and}\quad \mathbb{E}^P\big[I_n^2\big] \le 2\mathbb{E}^P\big[X_n^2\big] \quad\text{for every } n \in \mathbb{N}.$$

Finally, show that there exist $M \in L^2(P;\mathbb{R})$ and $I \in L^2\big(P;[0,\infty)\big)$ such that $M_n \longrightarrow M$, $I_n \nearrow I$, and, therefore, $X_n \longrightarrow X \equiv M + I$ both P-almost surely and in $L^2(P;\mathbb{R})$.

(iii) Let $\big(X_n, \mathcal{F}_n, P\big)$ be a nonnegative martingale, set $Y_n = e^{-X_n}$, $n \in \mathbb{N}$, use Corollary 5.2.16 to see that $\big(Y_n, \mathcal{F}_n, P\big)$ is a uniformly bounded, nonnegative, submartingale, and apply part **(ii)** to conclude that $\big\{X_n\big\}_0^\infty$ converges P-almost surely to a nonnegative $X \in L^1(P;\mathbb{R})$.

(iv) Let $\big(X_n, \mathcal{F}_n, P\big)$ be a martingale for which

$$\sup_{n\in\mathbb{N}} \mathbb{E}^P\Big[\big|X_n\big|\Big] < \infty. \tag{5.2.30}$$

For each $m \in \mathbb{N}$, define

$$Y_{n,m}^{\pm} = \mathbb{E}^P\Big[X_{n\vee m}^{\pm}\Big|\mathcal{F}_m\Big] \vee 0, \quad n \in \mathbb{N}.$$

Show that $Y_{n+1,m}^{\pm} \ge Y_{n,m}^{\pm}$ (a.s., P), define $Y_m^{\pm} = \underline{\lim}_{n\to\infty} Y_{n,m}^{\pm}$, check that both $\big(Y_m^+, \mathcal{F}_m, P\big)$ and $\big(Y_m^-, \mathcal{F}_m, P\big)$ are nonnegative martingales with

$$\mathbb{E}^P\big[Y_0^+ + Y_0^-\big] \le \sup_{n\in\mathbb{N}} \mathbb{E}^P\big[|X_n|\big],$$

and note that $X_m = Y_m^+ - Y_m^-$ (a.s., P) for each $m \in \mathbb{N}$. In other words, *every martingale* $\big(X_n, \mathcal{F}_n, P\big)$ *satisfying* (5.2.30) *admits a* **Hahn decomposition** *as the difference of two nonnegative martingales whose sum has expectation value dominated by the left-hand side of* (5.2.30). Finally, use this observation together with **(iii)** to see that every such martingale converges P-almost surely to some $X \in L^1(P;\mathbb{R})$.

(v) By combining the final assertion in **(iv)** together with Doob's Decomposition in Lemma 5.2.18, give another proof of the convergence assertion in Theorem 5.2.22.

5.2.31 Exercise: In this exercise we will develop yet another way to reduce Doob's Martingale Convergence Theorem to the case of L^2-bounded martingales. The technique here is due to R. Gundy and derives from the ideas introduced by Calderón and Zygmund in connection with their famous work on weak-type 1–1 estimates for singular integrals (cf. Theorem 5.3.26).

(i) Let $\{Z_n : n \in \mathbb{N}\}$ be a $\{\mathcal{F}_n : n \in \mathbb{N}\}$-progressively measurable, $[0, R]$-valued sequence with the property that $(-Z_n, \mathcal{F}_n, P)$ is a submartingale. Next, choose $\{I_n : n \in \mathbb{N}\}$ for $(-Z_n, \mathcal{F}_n, P)$ as in Lemma 5.2.18, note that I_n's can be chosen so that $0 \le I_n - I_{n-1} \le R$ for all $n \in \mathbb{Z}^+$, and set $M_n = Z_n + I_n$, $n \in \mathbb{N}$. Check that $(M_n, \mathcal{F}_n, P)$ is a nonnegative martingale with $M_n \le (n+1)R$ for each $n \in \mathbb{N}$. Next, show that

$$\begin{aligned}\mathbb{E}^P\left[M_n^2 - M_{n-1}^2\right] &= \mathbb{E}^P\left[(M_n - M_{n-1})(Z_n + Z_{n-1})\right]\\ &= \mathbb{E}^P\left[Z_n^2 - Z_{n-1}^2\right] + \mathbb{E}^P\left[(I_n - I_{n-1})(Z_n + Z_{n-1})\right]\\ &\le \mathbb{E}^P\left[Z_n^2 - Z_{n-1}^2\right] + 2R\,\mathbb{E}^P\left[I_n - I_{n-1}\right],\end{aligned}$$

and conclude that

$$\mathbb{E}^P\left[I_n^2\right] \le \mathbb{E}^P\left[M_n^2\right] \le 3R\mathbb{E}^P\left[Z_0\right], \quad n \in \mathbb{N}.$$

(ii) Let $(X_n, \mathcal{F}_n, P)$ be a nonnegative martingale. Show that, for each $R \in (0, \infty)$,

$$X_n = M_n^{(R)} - I_n^{(R)} + \Delta_n^{(R)}, \quad n \in \mathbb{N},$$

where $(M_n^{(R)}, \mathcal{F}_n, P)$ is a nonnegative martingale satisfying

$$\mathbb{E}^P\left[\left(M_n^{(R)}\right)^2\right] \le 3R\,\mathbb{E}^P\left[X_0\right], \quad n \in \mathbb{N},$$

$\{I_n^{(R)} : n \in \mathbb{N}\}$ is a nondecreasing sequence of random variables with the properties that $I_0^{(R)} \equiv 0$, $I_n^{(R)}$ is $\mathcal{F}_{n-1}$-measurable and

$$\mathbb{E}^P\left[\left(I_n^{(R)}\right)^2\right] \le 3R\,\mathbb{E}^P\left[X_0\right] \quad \text{for } n \in \mathbb{Z}^+,$$

and $\{\Delta_n^{(R)} : n \in \mathbb{N}\}$ is a $\{\mathcal{F}_n : n \in \mathbb{N}\}$-progressively measurable sequence with the property that

$$P\left(\exists n \in \mathbb{N}\ \Delta_n^{(R)} \ne 0\right) \le \frac{1}{R}\mathbb{E}^P\left[X_0\right].$$

Hint: Set $Z_n^{(R)} = X_n \wedge R$ and $\Delta_n^{(R)} = X_n - Z_n^{(R)}$ for $n \in \mathbb{N}$, apply part **(i)** to $\{Z_n^{(R)} : n \in \mathbb{N}\}$, and use Doob's inequality to estimate the probability that $\Delta_n^{(R)} \ne 0$ for some $n \in \mathbb{N}$.

(iii) Let $(X_n, \mathcal{F}_n, P)$ be any P-integrable martingale. Show that, for each $R \in (0, \infty)$,

$$X_n = M_n^{(R)} + V_n^{(R)} + \Delta_n^{(R)}, \quad n \in \mathbb{N},$$

where $(M_n^{(R)}, \mathcal{F}_n, P)$ is a martingale satisfying

$$\mathbb{E}^P\Big[\big(M_n^{(R)}\big)^2\Big] \le 12\, R\mathbb{E}^P\big[|X_n|\big],$$

$\{V_n^{(R)} : n \in \mathbb{N}\}$ is a sequence of random variables satisfying

$$V_0^{(R)} \equiv 0 \quad \text{and} \quad V_n^{(R)} \text{ is } \mathcal{F}_{n-1}\text{-measurable}$$

and

$$\mathbb{E}^P\left[\left(\sum_1^n \big|V_m^{(R)} - V_{m-1}^{(R)}\big|\right)^2\right] \le 12R\mathbb{E}^P\big[|X_n|\big]$$

for $n \in \mathbb{Z}^+$, and $\{\Delta_n :\in \mathbb{N}\}$ is an $\{\mathcal{F}_n : n \in \mathbb{N}\}$-progressively measurable sequence satisfying

$$P\Big(\exists 0 \le m \le n\ \Delta_m^{(R)} \ne 0\Big) \le \frac{2}{R}\mathbb{E}^P\big[|X_n|\big].$$

The preceding representation is called the **Calderón–Zygmund decomposition of the martingale** $(X_n, \mathcal{F}_n, P)$.

Hint: Use part **(iv)** of Exercise 5.2.27 to reduce the present case to the one just treated in **(ii)**.

(iv) Let $(X_n, \mathcal{F}_n, P)$ martingale which satisfies (5.2.30), and use part **(iii)** above together with part **(i)** of Exercise 5.2.27 to show that, for each $R \in (0, \infty)$, $\{X_n\}_0^\infty$ converges off of a set whose P-measure is no more than $\frac{1}{R}$ times the supremum over $n \in \mathbb{N}$ of $\mathbb{E}^P\big[|X_n|\big]$. In particular, when combined with Lemma 5.2.18, the preceding line of reasoning leads to yet another proof of the convergence result in Theorem 5.2.22.

5.2.32 Exercise: In this exercise we want to extend Hunt's Theorem (cf. Theorem 5.2.19) to allow for unbounded stopping times. To this end, let $(X_n, \mathcal{F}_n, P)$ be a uniformly P-integrable submartingale on the probability space $(\Omega, \mathcal{F}, P)$, and set $M_n = X_n - I_n$, $n \in \mathbb{N}$, where $\{I_n : n \in \mathbb{N}\}$ is the sequence discussed in Lemma 5.2.18. After checking that $(M_n, \mathcal{F}_n, P)$ is a uniformly P-integrable martingale, show that, for any $\{\mathcal{F}_n : n \in \mathbb{N}\}$-stopping time τ:

$$X_\tau = \mathbb{E}^P\big[M_\infty \big| \mathcal{F}_\tau\big] + I_\tau \quad \text{(a.s., } P\text{)},$$

where, X_∞, M_∞, and I_∞ are, respectively, the P-almost sure limits of $\{X_n\}_0^\infty$, $\{M_n\}_0^\infty$, and $\{I_n\}_0^\infty$. In particular, if σ and τ are a pair of $\{\mathcal{F}_n : n \in \mathbb{N}\}$-stopping times and $\sigma \le \tau$, conclude that

$$X_\sigma \le \mathbb{E}^P\big[X_\tau \big| \mathcal{F}_\sigma\big] \quad (\text{a.s.}, P). \tag{5.2.33}$$

5.2.34 Exercise: There are times when submartingales converge even though they are not bounded in $L^1(P;\mathbb{R})$. For example, suppose that $(X_n, \mathcal{F}_n, P)$ is a P-submartingale for which there exists a function $\rho : \mathbb{R} \longmapsto \mathbb{R}$ with the properties that $\rho(R) \ge R$ for all R and $X_{n+1} \le \rho(X_n)$ (a.e., P) for each $n \in \mathbb{N}$.

(i) Set $\tau_R(\omega) = \inf\{n \in \mathbb{N} : X_n(\omega) \ge R\}$ for $R \in (0,\infty)$, and note that

$$\sup_{n\in\mathbb{N}} X_{n\wedge\tau_R} \le X_0 \vee \rho(R) \quad (\text{a.e.}, P).$$

In particular, if X_0 is P-integrable, show that $\{X_n(\omega)\}_0^\infty$ converges in $\mathbb{R}$ for P-almost every ω for which $\{X_n(\omega)\}_0^\infty$ is bounded above.

Hint: After observing that

$$\sup_{n\in\mathbb{N}} \mathbb{E}^P\big[X^+_{n\wedge\tau_R}\big] < \infty \quad \text{for every} \quad R \in (0,\infty),$$

conclude that, for each $R \in (0,\infty)$, $\{X_n\}_0^\infty$ converges P-almost everywhere on $\{\tau_R = \infty\}$.

(ii) Let $\{Y_n\}_1^\infty$ be a sequence of mutually independent, P-integrable random variables, assume that

$$\mathbb{E}^P\big[Y_n\big] \ge 0 \text{ for } n \in \mathbb{N} \quad \text{and} \quad \sup_{n\in\mathbb{N}} \|Y_n^+\|_{L^\infty(P)} < \infty,$$

and set $S_n = \sum_1^n Y_m$. Show that $\{S_n(\omega)\}_1^\infty$ is either P-almost surely unbounded above or P-almost surely convergent in $\mathbb{R}$.

(iii) Let $\{\mathcal{F}_n : n \in \mathbb{N}\}$ be a nondecreasing sequence of sub-σ-algebras and A_n an element of $\mathcal{F}_n$ for each $n \in \mathbb{N}$. Show that the set of $\omega \in \Omega$ for which either

$$\sum_{n=0}^\infty \mathbf{1}_{A_n}(\omega) < \infty \text{ but } \sum_{n=1}^\infty P\big(A_n \,\big|\, \mathcal{F}_{n-1}\big)(\omega) = \infty$$

or

$$\sum_{n=0}^\infty \mathbf{1}_{A_n}(\omega) = \infty \text{ but } \sum_{n=1}^\infty P\big(A_n \,\big|\, \mathcal{F}_{n-1}\big)(\omega) < \infty$$

has P-measure 0. In particular, note that this gives another derivation of the Borel–Cantelli Lemma (cf. Lemma 1.1.4).

5.2.35 Exercise: For each $n \in \mathbb{N}$, let $(E_n, \mathcal{B}_n)$ be a measurable space and μ_n and ν_n a pair of probability measures on $(E_n, \mathcal{B}_n)$ with the property that $\nu_n \ll \mu_n$. Prove **Kakutani's Theorem** which says that (cf. Exercise 1.1.14) either $\prod_{n\in\mathbb{N}} \nu_n \perp \prod_{n\in\mathbb{N}} \mu_n$ or $\prod_{n\in\mathbb{N}} \nu_n \ll \prod_{n\in\mathbb{N}} \mu_n$.

Hint: Set

$$\Omega = \prod_{n\in\mathbb{N}} E_n, \quad \mathcal{F} = \prod_{n\in\mathbb{N}} \mathcal{B}_n, \quad P = \prod_{n\in\mathbb{N}} \mu_n, \quad \text{and } Q = \prod_{n\in\mathbb{N}} \nu_n.$$

Next, take $\mathcal{F}_n = \pi_n^{-1}\left(\prod_0^n \mathcal{B}_m\right)$, where π_n is the natural projection from Ω onto $\prod_0^n E_m$, set $P_n = P \restriction \mathcal{F}_n$ and $Q_n = Q \restriction \mathcal{F}_n$, and note that

$$X_n(\mathbf{x}) \equiv \frac{dQ_n}{dP_n}(\mathbf{x}) = \prod_0^n f_m(x_m), \quad \mathbf{x} \in \Omega,$$

where $f_n \equiv \frac{d\nu_n}{d\mu_n}$. In particular, when $\nu_n \sim \mu_n$ for each $n \in \mathbb{N}$, use Kolmogorov's 0–1 Law (cf. Theorem 1.1.2) to see that $Q(B) \in \{0,1\}$, where $B \equiv \left\{\varliminf_{n\to\infty} X_n = 0\right\}$, and combine this with the last part of Theorem 5.2.26 to conclude that $Q \not\perp P \implies Q \ll P$. Finally, to remove the assumption that $\nu_n \sim \mu_n$ for all n's, define $\tilde{\nu}_n$ on $(E_n, \mathcal{B}_n)$ by

$$\tilde{\nu}_n = \left(1 - 2^{-n-1}\right)\nu_n + 2^{-n-1}\mu_n,$$

check that $\tilde{\nu}_n \sim \mu_n$ and $Q \ll \tilde{Q} \equiv \prod_{n\in\mathbb{N}} \tilde{\nu}_n$, and use the preceding to complete the proof.

5.2.36 Excercise: Let $(\Omega, \mathcal{F})$ be a measurable space and Σ a sub-σ-algebra of $\mathcal{F}$. Given a pair of probability measures P and Q on $(\Omega, \mathcal{F})$, let X_Σ and Y_Σ be nonnegative Radon–Nikodym derivatives of $P_\Sigma \equiv P \restriction \Sigma$ and $Q_\Sigma \equiv Q \restriction \Sigma$, respectively, with respect to $\left(P_\Sigma + Q_\Sigma\right)$, and define

$$(P,Q)_\Sigma = \int X_\Sigma^{\frac{1}{2}} Y_\Sigma^{\frac{1}{2}}\, d(P+Q). \tag{5.2.37}$$

(i) Show that if R is any σ-finite measure on (Ω, Σ) with the property that $P_\Sigma \ll R$ and $Q_\Sigma \ll R$, then the number $(P,Q)_\Sigma$ in (5.2.37) is equal to

$$\int \left(\frac{dP_\Sigma}{dR}\right)^{\frac{1}{2}} \left(\frac{dQ_\Sigma}{dR}\right)^{\frac{1}{2}} dR.$$

Also, check that $P_\Sigma \perp Q_\Sigma$ if and only if $(P,Q)_\Sigma = 0$.

(ii) Suppose that $\{\mathcal{F}_n : n \in \mathbb{N}\}$ is a nondecreasing sequence of sub-σ-algebras of $\mathcal{F}$, and show that

$$(P,Q)_{\mathcal{F}_n} \longrightarrow (P,Q)_{\bigvee_0^\infty \mathcal{F}_n}.$$

(iii) Referring to part **(ii)**, assume that $Q \restriction \mathcal{F}_n \ll P \restriction \mathcal{F}_n$ for each $n \in \mathbb{N}$, let X_n be a nonnegative Radon–Nikodym derivative of $Q \restriction \mathcal{F}_n$ with respect to $P \restriction \mathcal{F}_n$, and show that $Q \restriction \bigvee_0^\infty \mathcal{F}_n$ is singular to $P \restriction \bigvee_0^\infty \mathcal{F}_n$ if and only if

$$\mathbb{E}^P\left[\sqrt{X_n}\right] \longrightarrow 0 \quad \text{as} \quad n \to \infty.$$

(iv) Let $\{\sigma_n\}_0^\infty \subseteq (0,\infty)$, and, for each $n \in \mathbb{N}$, let μ_n and ν_n be Gaussian measures on $\mathbb{R}$ with variance σ_n^2. If a_n and b_n are, respectively, the mean value of μ_n and ν_n, show that

$$\prod_{n\in\mathbb{N}} \nu_n \sim \prod_{n\in\mathbb{N}} \mu_n \quad \text{or} \quad \prod_{n\in\mathbb{N}} \nu_n \perp \prod_{n\in\mathbb{N}} \mu_n$$

depending on whether $\sum_0^\infty \sigma_n^{-2}(b_n - a_n)^2$ converges or diverges.

5.2.38 Exercise: In this exercise we will complete the program initiated in Lemma 4.2.15. Namely, we will show here that *the only directions in which Wiener's measure $\mathcal{W}^{(N)}$ is translation quasi-invariant are those in the Cameron–Martin subspace* $\mathbf{H}(\mathbb{R}^N)$. We begin by making a few preparations.

(i) Choose $\{\mathbf{f}_n\}_0^\infty \subseteq C^1([0,\infty);\mathbb{R}^N)$ so that it forms an orthonormal basis for $L^2([0,\infty);\mathbb{R}^N)$, and assume that

$$\int_{[0,\infty)} (1+t)\left|\dot{\mathbf{f}}_n(t)\right| dt < \infty \quad \text{for each} \quad n \in \mathbb{N}.$$

(For example, one could start with a countable set $\{\varphi_k : k \in \mathbb{N}\} \subseteq C_c^\infty((0,\infty);\mathbb{R}^N)$ which is dense in $L^2([0,\infty);\mathbb{R}^N)$ and apply the Gram–Schmidt orthogonalization procedure to construct $\mathbf{f}_n$'s which are elements of $C_c^\infty((0,\infty);\mathbb{R}^N)$.) Next, for each $n \in \mathbb{N}$, define $\mathbf{h}_n \in \mathbf{H}(\mathbb{R}^N)$ and $\boldsymbol{\lambda}_n \in \Theta(\mathbb{R}^N)^*$ by

$$\mathbf{h}_n(t) = \int_0^t \mathbf{f}_n(s)\,ds \quad \text{and} \quad \langle \boldsymbol{\theta}, \boldsymbol{\lambda}_n \rangle = -\int_{[0,\infty)} \left(\boldsymbol{\theta}(t), \dot{\mathbf{f}}_n(t)\right)_{\mathbb{R}^N} dt.$$

Given $\boldsymbol{\theta} \in \Theta(\mathbb{R}^N)$, show that $\boldsymbol{\theta} \in \mathbf{H}(\mathbb{R}^N)$ if and only if

$$\sum_{n=0}^\infty \langle \boldsymbol{\theta}, \boldsymbol{\lambda}_n \rangle^2 < \infty. \tag{5.2.39}$$

Hint: To check the "if" assertion, suppose that (5.2.39) holds and set

$$\mathbf{h} = \sum_{n=0}^{\infty} \langle \boldsymbol{\theta}, \boldsymbol{\lambda}_n \rangle \, \mathbf{h}_n \quad \text{and} \quad \boldsymbol{\xi} = \boldsymbol{\theta} - \mathbf{h}.$$

Clearly $\mathbf{h} \in \mathbf{H}(\mathbb{R}^N)$ and $\langle \boldsymbol{\xi}, \boldsymbol{\lambda}_n \rangle = 0$ for every $n \in \mathbb{N}$. Thus, one will know that $\boldsymbol{\theta} = \mathbf{h} \in \mathbf{H}(\mathbb{R}^N)$ as soon as one shows that

$$\sum_{m=0}^{n} \langle \mathbf{h}_m, \boldsymbol{\lambda} \rangle \langle \boldsymbol{\theta}, \boldsymbol{\lambda}_m \rangle \longrightarrow \langle \boldsymbol{\theta}, \boldsymbol{\lambda} \rangle \quad \text{for each } \boldsymbol{\lambda} \in \Theta(\mathbb{R}^N)^*.$$

(ii) Let $\boldsymbol{\eta} \in \Theta(\mathbb{R}^N)$ be given and define $T_{\boldsymbol{\eta}}\Theta(\mathbb{R}^N) \longrightarrow \Theta(\mathbb{R}^N)$ by $T_{\boldsymbol{\eta}}\boldsymbol{\theta} = \boldsymbol{\eta} + \boldsymbol{\theta}$. Next, referring to part **(i)**, let $\mathcal{F}_n$ be the σ-algebra over $\Theta(\mathbb{R}^N)$ generated by the functions $\boldsymbol{\theta} \in \Theta(\mathbb{R}^N) \longmapsto \langle \boldsymbol{\theta}, \boldsymbol{\lambda}_m \rangle \in \mathbb{R}$, $0 \le m \le n$. Show that $T^*_{\boldsymbol{\eta}}\mathcal{W}^{(N)} \restriction \mathcal{F}_n \ll \mathcal{W}^{(N)} \restriction \mathcal{F}_n$ and that

$$\boldsymbol{\theta} \in \Theta(\mathbb{R}^N) \longmapsto X_n(\boldsymbol{\theta}) \equiv \exp\left[\sum_{m=0}^{n} \left(\langle \boldsymbol{\eta}, \boldsymbol{\lambda}_m \rangle \langle \boldsymbol{\theta}, \boldsymbol{\lambda}_m \rangle - \frac{1}{2} \langle \boldsymbol{\eta}, \boldsymbol{\lambda}_m \rangle^2 \right)\right]$$

is a corresponding Radon–Nikodym for each $n \in \mathbb{N}$. Further, show that, for any $p \in (0, \infty)$,

$$\mathbb{E}^P\left[X_n^p\right]^{\frac{1}{p}} = \exp\left[\frac{p-1}{2} \sum_{0}^{n} \langle \boldsymbol{\eta}, \boldsymbol{\lambda}_m \rangle^2\right].$$

In particular (cf. part **(iii)** of Exercise 5.2.36), conclude that $T^*_{\boldsymbol{\eta}}\mathcal{W}^{(N)} \perp \mathcal{W}^{(N)}$ unless $\boldsymbol{\eta} \in \mathbf{H}(\mathbb{R}^N)$. An alternative proof can be based on part **(iv)** of the preceding exercise with $\sigma_n = 1$, $a_n = 0$, and $b_n = \langle \boldsymbol{\eta}, \lambda_n \rangle$.

5.2.40 Exercise: Again let $(\Omega, \mathcal{F}, P)$ be a probability space, only this time suppose that $\{\mathcal{F}_n : n \in \mathbb{N}\}$ is a sequence of sub-σ-algebras which is *nonincreasing*. Given a sequence $\{X_n : n \in \mathbb{N}\}$ of $(-\infty, \infty]$-valued random variables, we say that the triple $(X_n, \mathcal{F}_n, P)$ is a **reversed submartingale** or a **reversed martingale** according to whether $(X_{N-n\wedge N}, \mathcal{F}_{N-n\wedge N}, P)$ is a submartingale or martingale for every $N \in \mathbb{N}$.

(i) Let I be an interval and $f : I \longrightarrow \mathbb{R}$ be as in Corollary 5.2.16. If $(X_n, \mathcal{F}_n, P)$ is an integrable I-valued reversed submartingale and f is nondecreasing or if $(X_n, \mathcal{F}_n, P)$ is an I-valued reversed martingale, show that $(f(X_n), \mathcal{F}_N, P)$ is a reversed submartingale. In particular, $(|X_n|, \mathcal{F}_n, P)$ is a reversed submartingale if $(X_n, \mathcal{F}_n, P)$ is a reversed martingale.

(ii) Given a reversed submartingale $(X_n, \mathcal{F}_n, P)$, show that

$$P\left(\sup_{n\in\mathbb{N}} X_n \geq \alpha\right) \leq \frac{1}{\alpha}\mathbb{E}^P\left[X_0,\, \sup_{n\in\mathbb{N}} X_n \geq \alpha\right], \quad \alpha \in (0,\infty). \tag{5.2.41}$$

In particular, if $(X_n, \mathcal{F}_n, P)$ is a nonnegative reversed submartingale and $X_0 \in L^p(P;\mathbb{R})$ for some $p \in [1,\infty]$, conclude from (5.2.41) that $\{X_n : n \in \mathbb{N}\}$ is uniformly P-integrable and that

$$\left\|\sup_{n\in\mathbb{N}} X_n\right\|_{L^p(P)} \leq \frac{p}{p-1}\|X_0\|_{L^p(P)} \quad \text{when} \quad p \in (1,\infty]. \tag{5.2.42}$$

(iii) Given a reversed submartingale $(X_n, \mathcal{F}_n, P)$ with $X_0 \in L^1(P;\mathbb{R})$, show that there is a P-almost surely unique, $\mathcal{F}_\infty \equiv \bigcap_0^\infty \mathcal{F}_n$-measurable $X : \Omega \longrightarrow [-\infty,\infty)$ such that $X_n \longrightarrow X$ (a.s., P). Further, show that $X \in L^1(P;\mathbb{R})$ if $\sup_{n\in\mathbb{N}} \mathbb{E}^P[X_n^-] < \infty$. Finally, if $X_0 \in L^p(P;\mathbb{R})$ for some $p \in [1,\infty)$ and $(X_n, \mathcal{F}_n, P)$ is either nonnegative or a reversed martingale, show that $X_n \longrightarrow X$ in $L^p(P;\mathbb{R})$.

Hint: Given $a < b$, let $D_{[a,b]}$ be the precise number of times that $\{X_n : n \in \mathbb{N}\}$ **downcrosses** $[a,b]$ (i.e., the precise number of times that $\{-X_n : n \in \mathbb{N}\}$ upcrosses $[-b,-a]$), and prove the **downcrossing inequality**

$$\mathbb{E}^P\left[D_{[a,b]}\right] \leq \frac{\mathbb{E}^P\left[(X_0 - a)^+\right]}{b-a}.$$

(iv) As an application of the preceding, consider a sequence $\{X_n : n \in \mathbb{Z}^+\}$ of random variables which are **exchangeable** in the sense that for any finite permutation σ of $\mathbb{Z}^+$ (i.e., any isomorphism σ of $\mathbb{Z}^+$ onto itself which leaves all but a finite set invariant) the distribution of

$$\omega \in \Omega \longmapsto \mathbf{X}_\sigma(\omega) \equiv \left(X_{\sigma(1)}(\omega), \dots, X_{\sigma(n)}(\omega), \dots\right) \in \mathbb{R}^{\mathbb{Z}^+}$$

under P is the same as the distribution of

$$\omega \in \Omega \longmapsto \mathbf{X}(\omega) \equiv \left(X_1(\omega), \dots, X_n(\omega), \dots\right) \in \mathbb{R}^{\mathbb{Z}^+}.$$

Next, for each $n \in \mathbb{Z}^+$, let Σ_n be the set of permutations $\sigma : \mathbb{Z}^+ \longrightarrow \mathbb{Z}^+$ with the property that $\sigma(m) = m$ for all $m > n$, and set $\mathcal{F}_n = \mathbf{X}^{-1}(\mathcal{B}_n)$ where $\mathcal{B}_n$ denotes the σ-algebra of $\Gamma \in \mathcal{B}_{\mathbb{R}^{\mathbb{Z}^+}}$ which are invariant under each $\sigma \in \Sigma_n$ (i.e., for each $\sigma \in \Sigma_n$ and $\mathbf{x} \in \mathbb{R}^{\mathbb{Z}^+}$, $\mathbf{x} \in \Gamma$ if and only if $\sigma(\mathbf{x}) \in \Gamma$). Assuming that X_1 is P-integrable, show that

$$\mathbb{E}^P\left[X_1 \middle| \mathcal{F}_n\right] = \overline{S}_n \equiv \frac{1}{n}\sum_{m=1}^{n} X_m,$$

and conclude that $\{\overline{S}_n\}_1^\infty$ converges both P-almost surely as well as in $L^1(P;\mathbb{R})$. In honor of its author, this result is often called **De Finetti's Theorem**. In particular, when the X_n's are mutually independent and identically distributed, use this to give another proof of The Strong Law of Large Numbers (cf. Theorem 1.4.11).

(v) Let $(\Omega, \mathcal{F}, P)$ and $\{\mathcal{F}_n : n \in \mathbb{N}\}$ be as in part **(v)** of Exercise 5.1.28. Next, let $f \in L^2([0,1);\mathbb{C})$ be given, and define $\{c_m : |m| > 1\} \subseteq \mathbb{C}$ as in Exercise 5.1.28; and, after first noting that $\{\mathcal{F}_n : n \in \mathbb{N}\}$ is nonincreasing, use the result obtained there together with part **(iii)** here to see that,

$$\mathbb{E}^P[f \mid \mathcal{F}_0] = \mathbb{E}^P[f] + \sum_{|m|>1} c_m e^{\sqrt{-1}2^m \pi t},$$

where the convergence is both almost everywhere as well as in $L^2([0,1];\mathbb{C})$.

The preceding almost everywhere convergence result for lacunary trigonometric series was discovered by Kolmogorov. However, ever since L. Carleson's definitive theorem on the almost every convergence of the Fourier series of an arbitrary square integrable function, the interest in this result of Kolmogorov is mostly historical.

§5.3: Some Extensions.

It turns out that many of the results obtained in Section 5.2 admit easy extensions to both infinite measures and Banach space valued random variables. Furthermore, in many applications, these extensions play a useful, and occasionally essential, rôle. Throughout the discussion which follows, $(\Omega, \mathcal{F}, \mu)$ will be a measure space and $\{\mathcal{F}_n : n \in \mathbb{N}\}$ will be a nondecreasing sequence of sub-σ-algebras with the property that $\mu \restriction \mathcal{F}_0$ is σ-finite. In particular, this means that the conditional expectation of a μ-locally integrable random variable given $\mathcal{F}_n$ is well-defined (cf. Theorem 5.1.22) even if the random variable takes values in a separable Banach space E. Thus, we will say that the sequence $\{\mathbf{X}_n;\, n \in \mathbb{N}\}$ of E-valued random variables is a **μ-martingale with respect to $\{\mathcal{F}_n : n \in \mathbb{N}\}$**, or, more briefly, that the triple $(\mathbf{X}_n, \mathcal{F}_n, \mu)$ is a **martingale** if $\{\mathbf{X}_n : n \in \mathbb{N}\}$ is $\{\mathcal{F}_n : n \in \mathbb{N}\}$-progressively measurable, each $\mathbf{X}_n$ is μ-locally integrable, and

$$\mathbf{X}_{n-1} = \mathbb{E}^{\mu}\big[\mathbf{X}_n \big| \mathcal{F}_{n-1}\big] \quad \text{(a.e., } \mu\text{)} \quad \text{for each } n \in \mathbb{Z}^+. \tag{5.3.1}$$

Furthermore, when $E = \mathbb{R}$, we will say that $\{X_n : n \in \mathbb{N}\}$ is a **μ-submartingale with respect to $\{\mathcal{F}_n : n \in \mathbb{N}\}$** (equivalently, the triple $(X_n, \mathcal{F}_n, \mu)$ is a **submartingale**) if $\{X_n : n \in \mathbb{N}\}$ is $\{\mathcal{F}_n : n \in \mathbb{N}\}$-progressively measurable, each X_n is μ-locally integrable, and

$$X_{n-1} \le \mathbb{E}^{\mu}\big[X_n \big| \mathcal{F}_{n-1}\big] \quad \text{(a.e., } \mu\text{)} \quad \text{for each } n \in \mathbb{Z}^+. \tag{5.3.2}$$

Without any real effort, we can now prove the following variant of each of the main results in Section 5.2.

5.3.3 Theorem. *Let $(X_n, \mathcal{F}_n, \mu)$ be a μ-submartingale. Then, for each $N \in \mathbb{N}$ and $A \in \mathcal{F}_0$ on which X_N is μ-integrable:*

$$\mu\left(\max_{0\le n\le N} X_n \ge \alpha \,\&\, A\right) \le \frac{1}{\alpha}\mathbb{E}^\mu\left[X_N, \left\{\max_{0\le n\le N} X_n \ge \alpha\right\}\cap A\right] \tag{5.3.4}$$

for all $\alpha \in (0,\infty)$. In particular, if the X_n's are all nonnegative, then, for every $p \in (1,\infty)$ and $A \in \mathcal{F}_0$:

$$\mathbb{E}^\mu\left[\sup_{n\in\mathbb{N}} |X_n|^p,\ A\right]^{\frac{1}{p}} \le \frac{p}{p-1}\sup_{n\in\mathbb{N}} \mathbb{E}^\mu\left[|X_n|^p,\ A\right]^{\frac{1}{p}}. \tag{5.3.5}$$

Furthermore, for any bounded $\{\mathcal{F}_n : n \in \mathbb{N}\}$-stopping times $\sigma \le \tau$,

$$X_\sigma \le \mathbb{E}^\mu\left[X_\tau \middle| \mathcal{F}_\sigma\right] \quad (\text{a.e.}, \mu); \tag{5.3.6}$$

and, for any $a < b$ and $A \in \mathcal{F}_0$,

$$\mathbb{E}^\mu\left[U_{[a,b]},\ A\right] \le \sup_{n\in\mathbb{N}} \frac{\mathbb{E}^\mu\left[(X_n - a)^+,\ A\right]}{b-a}, \tag{5.3.7}$$

where $U_{[a,b]}(\omega)$ denotes the precise number of times that $\{X_n(\omega)\}_0^\infty$ upcrosses $[a,b]$ (cf. the discussion preceding Theorem 5.2.22). In particular, for each stopping time τ, $(X_{n\wedge\tau}, \mathcal{F}_n, \mu)$ is a submartingale; and, when $(X_n, \mathcal{F}_n, \mu)$ is itself a martingale, then the inequality in (5.3.6) is an equality and $(X_{n\wedge\tau}, \mathcal{F}_n, \mu)$ is again a martingale. Finally,

$$\begin{aligned}&\sup_{n\in\mathbb{N}} \mathbb{E}^\mu\left[X_n^+,\ A\right] < \infty \text{ for every } A \in \mathcal{F}_0 \text{ with } \mu(A) < \infty \\ &\qquad\qquad \Longrightarrow X_n \longrightarrow X \quad (\text{a.e.}, \mu),\end{aligned} \tag{5.3.8}$$

where X is $\bigvee_0^\infty \mathcal{F}_n$-measurable and μ-locally integrable; and, for each $p \in (1,\infty)$, the convergence is in $L^p(\mu)$ if and only if $\{X_n\}_0^\infty$ is bounded in $L^p(\mu)$. In fact, in the case of martingales, there is a $\bigvee_0^\infty \mathcal{F}_n$-measurable X such that

$$X_n = \mathbb{E}^\mu\left[X \middle| \mathcal{F}_n\right] \quad (\text{a.e.}, \mu) \quad \text{for all } n \in \mathbb{N}; \tag{5.3.9}$$

if and only if $\{X_n\}_0^\infty$ is uniformly μ-integrable on each $A \in \mathcal{F}_0$ with $\mu(A) < \infty$; the X in (5.3.9) is μ-integrable if and only if $X_n \longrightarrow X$ in $L^1(\mu)$; and, for each $p \in (1,\infty)$, $X \in L^p(\mu)$ if and only if $\{X_n : n \in \mathbb{N}\}$ is bounded in $L^p(\mu;\mathbb{R})$, in which case, $X_n \longrightarrow X$ in $L^p(\mu;\mathbb{R})$.

PROOF: Obviously, there is no problem unless $\mu(\Omega) = \infty$. However, even then, each of these results follows immediately from its counterpart in Section 5.2 once one makes the following trivial observation. Namely, given $\Omega' \in \mathcal{F}_0$ with $\mu(\Omega') \in (0, \infty)$, set

$$\mathcal{F}' = \mathcal{F}[\Omega'], \quad \mathcal{F}'_n = \mathcal{F}_n[\Omega'], \quad X'_n = X_n \restriction \Omega', \quad \text{and } P = \frac{\mu \restriction \mathcal{F}'}{\mu\Omega')}.$$

Then $(X'_n, \mathcal{F}'_n, P')$ is a submartingale or martingale depending on whether the original $(X_n, \mathcal{F}_n, \mu)$ was a submartingale or martingale. Hence, when $\mu(\Omega) = \infty$, we simply choose a sequence $\{\Omega_k\}_1^\infty$ of mutually disjoint, μ-finite elements of $\mathcal{F}_0$ so that $\Omega = \bigcup_1^\infty \Omega_k$, work on each Ω_k separately, and, at the end, sum the results. □

Before moving on to see what can be said about martingales with values in a Banach space, we will spend a little time seeing how Theorem 5.3.3 can be applied to give simple proofs of basic facts in real analysis; and we will begin with the following derivation of a famous estimate proved originally by Hardy and Littlewood. In order to describe their result, define the **Hardy–Littlewood maximal function** Mf for $f \in L^1(\mathbb{R}^N; \mathbb{R})$ by

$$\mathbf{M}f(\mathbf{x}) = \sup_{Q \ni \mathbf{x}} \frac{1}{|Q|} \int_Q |f(\mathbf{y})|\, d\mathbf{y}, \quad \mathbf{x} \in \mathbb{R}^N, \tag{5.3.10}$$

where Q is used to denote a generic **cube**

$$Q = \prod_{j=1}^N [a_j, a_j + r) \quad \text{with } \mathbf{a} \in \mathbb{R}^N \text{ and } r > 0 \tag{5.3.11}$$

and we have introduced the notation $|\Gamma|$ to denote the Lebesgue measure of $\Gamma \in \mathbb{R}^N$. As is easily checked, $\mathbf{M}f : \mathbb{R}^N \longrightarrow [0, \infty]$ is lower semicontinuous and therefore certainly Borel measurable. Furthermore, if we restrict our attention to *nicely meshed* families of cubes, then it is easy to relate $\mathbf{M}f$ to martingales. More precisely, for each $n \in \mathbb{Z}$, the **nth standard diadic partition of $\mathbb{R}^N$** is the partition $\mathcal{P}_n$ of $\mathbb{R}^N$ into the cubes

$$C_n(\mathbf{k}) \equiv \prod_{i=1}^N \left[\frac{k_i}{2^n}, \frac{k_i + 1}{2^n}\right), \quad \mathbf{k} \in \mathbb{Z}^N. \tag{5.3.12}$$

These partitions are nicely meshed in the sense that the $(n+1)$st is a refinement of the nth. Equivalently, if $\mathcal{F}_n$ denotes the σ-algebra over $\mathbb{R}^N$ generated by the partition $\mathcal{P}_n$, then $\mathcal{F}_n \subseteq \mathcal{F}_{n+1}$. Moreover, if $f \in L^1(\mathbb{R}^N; \mathbb{R})$ and

$$X_n^f(\mathbf{x}) \equiv 2^{nN} \int_{C_n(\mathbf{k})} |f(\mathbf{y})|\, d\mathbf{y} \quad \text{for } \mathbf{x} \in C_n(\mathbf{k}) \text{ and } \mathbf{k} \in \mathbb{Z}^N, \tag{5.3.13}$$

then

$$X_n^f = \mathbb{E}^{\text{Leb}}\big[|f|\big|\mathcal{F}_n\big] \quad \text{(a.e., Leb)} \tag{5.3.14}$$

for each $n \in \mathbb{Z}$. In particular, for each $m \in \mathbb{Z}$,

$$\left(X_{m+n}^f, \mathcal{F}_{m+n}, \text{Leb}\right), \quad n \in \mathbb{N},$$

is a nonnegative martingale; and so, by applying (5.3.4) for each $m \in \mathbb{Z}$ and then letting $m \searrow -\infty$, we see that

$$\left|\left\{\mathbf{x} : \mathbf{M}^{(\mathbf{0})}f \geq \alpha\right\}\right| \leq \frac{1}{\alpha} \int_{\{\mathbf{M}^{(\mathbf{0})}f \geq \alpha\}} f(\mathbf{y})\, dy, \quad \alpha \in (0,\infty), \tag{5.3.15}$$

where

$$\mathbf{M}^{(\mathbf{0})}f(\mathbf{x}) = \sup\left\{\frac{1}{|Q|}\int_Q f(\mathbf{y})\, d\mathbf{y} : \mathbf{x} \in Q \in \bigcup_{n\in\mathbb{Z}} \mathcal{P}_n\right\}.$$

At first sight, one might hope that it should be possible to pass directly from (5.3.15) to analogous estimates on the level sets of $\mathbf{M}f$. However, the passage from (5.3.15) to control on $\mathbf{M}f$ is not so easy as it might at first appear: the "sup" in (5.3.10) involves many more cubes than the one in the definition of $\mathbf{M}^{(\mathbf{0})}f$; and it is for this reason we will have to introduce additional families of meshed partitions. Namely, for each $\boldsymbol{\eta} \in \{0,1\}^N$ set

$$\mathcal{P}_n(\boldsymbol{\eta}) = \left\{\frac{(-1)^n\boldsymbol{\eta}}{3\,2^n} + C_n(\mathbf{k}) : \mathbf{k} \in \mathbb{Z}^N\right\},$$

where $C_n(\mathbf{k})$ is the cube described in (5.3.12). It is then an easy matter to check that, for each $\boldsymbol{\eta} \in \{0,1\}^N$, $\{\mathcal{P}_n(\boldsymbol{\eta}) : n \in \mathbb{Z}\}$ is a family of meshed partitions of $\mathbb{R}^N$. Furthermore, if

$$\left[\mathbf{M}^{(\boldsymbol{\eta})}f\right](\mathbf{x}) = \sup\left\{\frac{1}{|Q|}\int_Q |f(\mathbf{y})|\, d\mathbf{y} : \mathbf{x} \in Q \in \bigcup_{n\in\mathbb{Z}} \mathcal{P}_n(\boldsymbol{\eta})\right\}, \quad \mathbf{x} \in \mathbb{R}^N,$$

then exactly the same argument which (when $\boldsymbol{\eta} = \mathbf{0}$) led us to (5.3.15) can now be used to get

$$\left|\left\{\mathbf{x} \in \mathbb{R}^N : \left[\mathbf{M}^{(\boldsymbol{\eta})}f\right](\mathbf{x}) \geq \alpha\right\}\right| \leq \frac{1}{\alpha} \int_{\{\mathbf{M}^{(\boldsymbol{\eta})}f \geq \alpha\}} |f(\mathbf{y})|\, d\mathbf{y}, \tag{5.3.16}$$

for each $\boldsymbol{\eta} \in \{0,1\}^N$ and $\alpha \in (0,\infty)$. Finally, if Q is given by (5.3.11) and $r \leq \frac{1}{3\,2^n}$, then it is possible to find an $\boldsymbol{\eta} \in \{0,1\}^N$ and a $C \in \mathcal{P}_n(\boldsymbol{\eta})$ for which $Q \subseteq C$. (To see this, first reduce to the case when $N = 1$.) Hence,

$$\max_{\boldsymbol{\eta}\in\{0,1\}^N} \mathbf{M}^{(\boldsymbol{\eta})} f \leq \mathbf{M} f \leq 6^N \max_{\boldsymbol{\eta}\in\{0,1\}^N} \mathbf{M}^{(\boldsymbol{\eta})} f; \tag{5.3.17}$$

and so, after combining (5.3.16) and (5.3.17), we arrive at the following version of the **Hardy–Littlewood inequality**

$$\left|\left\{\mathbf{x} \in \mathbb{R}^N : \mathbf{M}f(\mathbf{x}) \geq \alpha\right\}\right| \leq \frac{(12)^N}{\alpha} \int_{\mathbb{R}^N} |f(\mathbf{y})|\, d\mathbf{y}. \tag{5.3.18}$$

At the same time, because (cf. Exercise 1.4.20) (1.5.16) implies that

$$\max_{\boldsymbol{\eta}\in\{0,1\}^N} \left\|\mathbf{M}^{(\boldsymbol{\eta})} f\right\|_{L^p(\mathbb{R}^N)} \leq \frac{p}{p-1} \|f\|_{L^p(\mathbb{R}^N)}, \quad p \in (1,\infty],$$

we can also use (5.3.17) to get the useful estimate

$$\left\|\mathbf{M}f\right\|_{L^p(\mathbb{R}^N)} \leq \frac{(12)^N p}{p-1} \|f\|_{L^p(\mathbb{R}^N)}, \quad p \in (1,\infty]. \tag{5.3.19}$$

In this connection, notice that there is no hope of getting this estimate when $p = 1$, since it is clear that

$$\varliminf_{|\mathbf{x}|\to\infty} |\mathbf{x}|^N\, \mathbf{M}f(\mathbf{x}) > 0$$

whenever f does not vanish Leb-almost everywhere.

The inequality in (5.3.18) plays the same rôle in classical analysis as Doob's inequality plays in martingale theory. For example, by essentially the same argument as we used to pass from Doob's inequality to Corollary 5.2.7, we obtain the following famous **Lebesgue Differentiation Theorem**.

5.3.20 Theorem. *For each $f \in L^1(\mathbb{R}^N;\mathbb{R})$,*

$$\lim_{B\searrow\{\mathbf{x}\}} \frac{1}{|B|} \int_B \left|f(\mathbf{y}) - f(\mathbf{x})\right| d\mathbf{y} = 0 \quad \textit{for Leb-almost every } \mathbf{x} \in \mathbb{R}^N, \tag{5.3.21}$$

where, for each $\mathbf{x} \in \mathbb{R}^N$, the limit is taken over balls B which contain $\mathbf{x}$ and tend to $\mathbf{x}$ in the sense that their radii shrink to 0. In particular,

$$f(\mathbf{x}) = \lim_{B\searrow\{\mathbf{x}\}} \frac{1}{|B|} \int_B f(\mathbf{y})\, d\mathbf{y} \quad \textit{for Leb-almost every } \mathbf{x} \in \mathbb{R}^N. \tag{5.3.22}$$

PROOF: We begin with the observation that, for each $f \in L^1(\mathbb{R}^N;\mathbb{R})$,

$$\tilde{\mathbf{M}}f(\mathbf{x}) \equiv \sup_{B \ni \mathbf{x}} \frac{1}{|B|} \int_B |f(\mathbf{y})|\, d\mathbf{y} \le \kappa_N \mathbf{M}f(\mathbf{x}), \quad \mathbf{x} \in \mathbb{R}^N$$

where $\kappa_n = \frac{2^N}{\mathbf{\Omega}_N}$ with $\mathbf{\Omega}_N = \big|B_{\mathbb{R}^N}(\mathbf{0},1)\big|$. Second, we remark that (5.3.21) for *every* $\mathbf{x} \in \mathbb{R}^N$ is trivial when $f \in C_c(\mathbb{R}^N;\mathbb{R})$. Hence, all that remains is to check that if $f_n \longrightarrow f$ in $L^1(\mathbb{R}^N;\mathbb{R})$ and if (5.3.21) holds for each f_n, then it holds for f. To this end, let $\epsilon > 0$ be given and observe that, because of the preceding and (5.3.18),

$$\begin{aligned}
&\left|\left\{\mathbf{x} : \varlimsup_{B \searrow \{\mathbf{x}\}} \frac{1}{|B|}\int_B |f(\mathbf{y}) - f(\mathbf{x})|\, d\mathbf{y} \ge \epsilon\right\}\right| \\
&\quad \le \left|\left\{\mathbf{x} : \tilde{\mathbf{M}}(f - f_n) \ge \frac{\epsilon}{3}\right\}\right| \\
&\qquad + \left|\left\{\mathbf{x} : \varlimsup_{B \searrow \{\mathbf{x}\}} \frac{1}{|B|}\int_B |f_n(\mathbf{y}) - f_n(\mathbf{x})|\, d\mathbf{y} \ge \frac{\epsilon}{3}\right\}\right| \\
&\qquad + \left|\left\{\mathbf{x} : |f_n(\mathbf{x}) - f(\mathbf{x})| \ge \tfrac{\epsilon}{3}\right\}\right| \\
&\quad \le \frac{3}{\epsilon}\big(1 + (12)^N \kappa_N\big)\|f - f_n\|_{L^1(\mathbb{R}^N)}
\end{aligned}$$

for every $n \in \mathbb{Z}^+$. Hence, after letting $n \to \infty$, we see that (5.3.21) also holds for f. □

Although applications like Lebesgue's Differentiation Theorem incline one to think that (5.3.18) is most interesting because of what it says about averages over small cubes, its implications for large cubes are also significant. In fact, as we will see in Section 6.1, it allows us to prove Birkhoff's Individual Ergodic Theorem (cf. Theorem 6.1.8), which may be viewed as *differentiation at infinity.* The link between ergodic theory and the Hardy–Littlewood Inequality is found in the following deterministic version of the Maximal Ergodic Lemma (cf. Lemma 6.1.2). Namely, let $\{a_{\mathbf{k}} : \mathbf{k} \in \mathbb{Z}^N\}$ be a summable subset of $[0,\infty)$ and set

$$\overline{S}_n(\mathbf{k}) = \frac{1}{(2n)^N}\sum_{\mathbf{j} \in Q_n} a_{\mathbf{j}+\mathbf{k}}, \quad n \in \mathbb{N} \text{ and } \mathbf{k} \in \mathbb{Z}^N,$$

where $Q_n = \{\mathbf{j} \in \mathbb{Z}^N : -n \le j_i < n \text{ for } 1 \le i \le N\}$. By applying (5.3.18) and (5.3.19) to the function f given by (cf. (5.3.12))

$$f(\mathbf{x}) = a_{\mathbf{k}} \quad \text{when } \mathbf{x} \in C_0(\mathbf{k}),$$

we see that

$$\left|\left\{\mathbf{k} \in \mathbb{Z}^N : \sup_{n\in\mathbb{Z}^+} \overline{S}_n(\mathbf{k}) \geq \alpha\right\}\right| \leq \frac{(12)^N}{\alpha} \sum_{\mathbf{k}\in\mathbb{Z}^+} a_{\mathbf{k}}, \quad \alpha \in (0,\infty), \tag{5.3.23}$$

(on the left-hand side of (5.3.23), we have used $|\Gamma|$ to denote the cardinality of the set Γ) and

$$\left(\sum_{\mathbf{k}\in\mathbb{Z}^N} \sup_{n\in\mathbb{Z}^+} |\overline{S}_n(\mathbf{k})|^p\right)^{\frac{1}{p}} \leq \frac{(12)^N p}{p-1} \left(\sum_{\mathbf{k}\in\mathbb{Z}^N} |a_{\mathbf{k}}|^p\right)^{\frac{1}{p}} \tag{5.3.24}$$

for each $p \in (1,\infty]$. The inequality in (5.3.23) is called **Hardy's Inequality**. Actually, Hardy was drawn to this line of research by his passion for the game of cricket. What Hardy wanted to find is the optimal order in which to arrange batters to maximize the average score per inning. Thus, he worked with a nonnegative sequence $\{a_k\}_0^\infty$ in which a_k represented the expected number of runs scored by player k, and what he showed is that, for each $\alpha \in (0,\infty)$,

$$\left|\left\{k \in \mathbb{N} : \sup_{n\in\mathbb{Z}^+} \overline{S}_n(k) \geq \alpha\right\}\right|$$

is maximized when $\{a_n\}_0^\infty$ is nonincreasing; from which it is an easy application of Markov's inequality to prove that

$$\left|\left\{k \in \mathbb{N} : \sup_{n\in\mathbb{Z}^+} \overline{S}_n(k) \geq \alpha\right\}\right| \leq \frac{1}{\alpha}\sum_0^\infty a_k, \quad \alpha \in (0,\infty).$$

Although this sharpened result can also be obtained as a corollary the *Sunrise Lemma*,† Hardy's approach remains the most appealing.

The preceding development for the Hardy–Littlewood maximal function contains all the ingredients which are required to introduce another basic tool of real analysis: the Calderón–Zygmund decomposition of a function $f \in L^1(\mathbb{R}^N;\mathbb{R})$. Indeed, everything that one needs to carry out their decomposition is contained in the following simple application of (5.3.15) and Theorem 5.3.20.

5.3.25 Lemma. *Let $f \in L^1(\mathbb{R}^N;\mathbb{R})$ be given. For each $R \in (0,\infty)$ there is an at most countable collection $\mathcal{Q}(R)$ of mutually disjoint cubes with the properties that*

$$|f| \leq R \quad \text{(a.e., Leb) off of } \bigcup \mathcal{Q}(R), \qquad \left|\bigcup \mathcal{Q}(R)\right| \leq \frac{\|f\|_{L^1(\mathbb{R}^N)}}{R},$$

† See Theorem 5.2.2 in Deuschel and Stroock, *Large Deviations*, Academic Press Pure Math Series **137** (1989).

and

$$\frac{1}{|Q|}\int_Q |f(\mathbf{y})|\,d\mathbf{y} \le 2^N R \quad \text{for each } Q \in \mathcal{Q}(R).$$

PROOF: Define X_n^f as in (5.3.14) and set

$$B_n = \left\{\mathbf{x} : \sup_{m<n} X_m^f(\mathbf{x}) \le R \text{ and } X_n^f(\mathbf{x}) > R\right\} \quad \text{for } n \in \mathbb{Z}.$$

Clearly $B_m \cap B_n = \emptyset$ for $m \ne n$; and, because $B_n \in \sigma(\mathcal{P}_n)$, either $B_n = \emptyset$ or there is a subset $\mathcal{Q}_n(R) \subseteq \mathcal{P}_n$ whose union is B_n. In particular, $\mathcal{Q}(R) \equiv \bigcup_{n\in\mathbb{Z}} \mathcal{Q}_n(R)$ is an at most countable set of mutually disjoint cubes whose union is $B \equiv \bigcup_{n\in\mathbb{Z}} B_n$.

To see that $\mathcal{Q}(R)$ has the required properties, we first note that, because

$$\sup_{\mathbf{x}\in\mathbb{R}^N} X_n^f(\mathbf{x}) \le 2^{-nN}\|f\|_{L^1(\mathbb{R}^N)} \longrightarrow 0 \quad \text{as } n \to -\infty,$$

$$B = \left\{\sup_{n\in\mathbb{Z}} X_n^f(\mathbf{x}) > R\right\}.$$

In particular, by Theorem 5.3.20,

$$|f(\mathbf{x})| \le \sup_{n\in\mathbb{Z}} X_n^f(\mathbf{x}) \le R \quad \text{for Leb-almost every } \mathbf{x} \notin B.$$

Furthermore, by (5.3.15), $|B| \le \frac{\|f\|_{L^1(\mathbb{R}^N)}}{R}$. Finally, if $Q \in \mathcal{Q}(R)$, $\mathbf{x} \in Q$, and Q' is the element of $\mathcal{P}_{n-1}$ which contains Q, then

$$\frac{1}{|Q|}\int_Q |f(\mathbf{y})|\,d\mathbf{y} \le \frac{2^N}{|Q'|}\int_{Q'} f(\mathbf{y})\,d\mathbf{y} = 2^N X_{n-1}^f(\mathbf{x}) \le 2^N R. \quad \square$$

5.3.26 Theorem (Calderón–Zygmund Decomposition). *Let E be a separable Banach space and $\mathbf{f} \in L^1(\mathbb{R}^N; E)$. Then, for each $R \in (0,\infty)$, there exists an at most countable collection $\mathcal{Q}(R)$ of mutually disjoint cubes Q, a function $\mathbf{g} \in L^2(\mathbb{R}^N; E)$, and functions $\{\mathbf{h}_Q : Q \in \mathcal{Q}(R)\} \subseteq L^1(\mathbb{R}^N; E)$ such that*

$$\left|\bigcup \mathcal{Q}(R)\right| \le \frac{\|\mathbf{f}\|_{L^1(\mathbb{R};E)}}{R} \quad \text{and} \quad \|\mathbf{g}\|^2_{L^2(\mathbb{R}^N;E)} \le 2^N R\,\|\mathbf{f}\|_{L^1(\mathbb{R}^N;E)},$$

$$\mathbf{h}_Q \equiv \mathbf{0} \text{ off } Q \quad \text{and} \quad \int_Q \mathbf{h}_Q(\mathbf{y})\,d\mathbf{y} = \mathbf{0} \quad \text{for each } Q \in \mathcal{Q}(R),$$

$$\sum_{Q\in\mathcal{Q}(R)} \|\mathbf{h}_Q\|_{L^1(\mathbb{R}^N)} \le 2\|\mathbf{f}\|_{L^1(\mathbb{R}^N)},$$

and

$$\mathbf{f} = \mathbf{g} + \sum_{Q \in \mathcal{Q}(R)} \mathbf{h}_Q.$$

PROOF: Choose $\mathcal{Q}(R)$ for $\|f\|_E$ and R as in Lemma 5.3.25, set $B = \bigcup \mathcal{Q}(R)$, define

$$\mathbf{g} = \begin{cases} \mathbf{f} & \text{off of } B \\ \frac{1}{|Q|} \int_Q \mathbf{f}(\mathbf{y})\, d\mathbf{y} & \text{on } Q \in \mathcal{Q}(R), \end{cases}$$

and take

$$\mathbf{h}_Q = \left(\mathbf{f} - \frac{1}{|Q|} \int_Q \mathbf{f}(\mathbf{y})\, d\mathbf{y} \right) \mathbf{1}_Q \quad \text{for } Q \in \mathcal{Q}(R).$$

In view of Lemma 5.3.25, all that we have to do is check that the required estimates hold for $\|\mathbf{g}\|_{L^2(\mathbb{R}^N;E)}$ and the $\|\mathbf{h}_Q\|_{L^1(\mathbb{R}^N;E)}$. But, obviously, $\|\mathbf{g}\|_{L^1(\mathbb{R}^N;E)} \leq \|\mathbf{f}\|_{L^1(\mathbb{R}^N;E)}$, and so

$$\sum_{Q \in \mathcal{Q}(R)} \|\mathbf{h}_Q\|_{L^1(\mathbb{R}^N;E)} = \|\mathbf{f} - \mathbf{g}\|_{L^1(\mathbb{R}^N;E)} \leq 2\|\mathbf{f}\|_{L^1(\mathbb{R}^N;E)}.$$

In addition, by Lemma 5.3.25, $\|\mathbf{g}\|_{L^\infty(\mathbb{R}^N;E)} \leq 2^N R$. Hence, since

$$\|\mathbf{g}\|^2_{L^2(\mathbb{R}^N;E)} \leq \|\mathbf{g}\|_{L^1(\mathbb{R}^N;E)} \|\mathbf{g}\|_{L^\infty(\mathbb{R}^N;E)},$$

we are done. □

Since they take us somewhat far afield, we will postpone the applications of Theorem 5.3.26 until Section 6.2.

We turn next to Banach space valued martingales. Actually, everything except the easiest aspects of this topic becomes extremely complicated and technical very quickly, and, for this reason, we will restrict our attention to those results which do not involve any deep properties of the geometry of Banach spaces. In fact, the only general theory with which we will deal is contained in the following.

5.3.27 Theorem. *Let E be a separable Banach space and $(\mathbf{X}_n, \mathcal{F}_n, \mu)$ an E-valued martingale. Then $(\|\mathbf{X}_n\|_E, \mathcal{F}_n, \mu)$ is a nonnegative submartingale and therefore, for each $N \in \mathbb{Z}^+$ and all $\alpha \in (0,\infty)$,*

$$\mu\left(\sup_{0 \leq n \leq N} \|\mathbf{X}_n\|_E \geq \alpha \right) \leq \frac{1}{\alpha} \mathbb{E}^\mu \left[\|\mathbf{X}_N\|_E, \sup_{0 \leq n \leq N} \|\mathbf{X}_n\|_E \geq \alpha \right]. \tag{5.3.28}$$

In particular, for each $p \in (1,\infty]$,

$$\left\| \sup_{n \in \mathbb{N}} \|\mathbf{X}_n\|_E \right\|_{L^p(\mu)} \leq \frac{p}{p-1} \sup_{n \in \mathbb{N}} \|\mathbf{X}_n\|_{L^p(\mu;E)}. \tag{5.3.29}$$

Finally, if $\mathbf{X} \in L^p(\mu; E)$ *for some* $p \in [1, \infty)$*, then*

$$\text{(5.3.30)} \qquad \mathbb{E}^\mu\left[\mathbf{X}\middle|\mathcal{F}_n\right] \longrightarrow \mathbb{E}^\mu\left[\mathbf{X}\middle|\bigvee_0^\infty \mathcal{F}_n\right] \quad \textit{both (a.e., } \mu\textit{) and in } L^p(\mu; E).$$

PROOF: The fact $\left(\|\mathbf{X}_n\|_E, \mathcal{F}_n, \mu\right)$ is a submartingale is an easy application of the inequality in (5.1.25); and, given this fact, the inequalities in (5.3.28) and (5.3.29) follow from (5.3.4) and (5.3.5), respectively.

While proving the convergence statements, we may and will assume that $\mathcal{F} = \bigvee_0^\infty \mathcal{F}_n$. Now let $\mathbf{X} \in L^p(\mu; E)$ be given, and set $\mathbf{X}_n = \mathbb{E}^\mu[\mathbf{X}|\mathcal{F}_n]$, $n \in \mathbb{N}$. Because of (5.3.28) and (5.3.29), we know (cf. the proofs of Corollary 5.2.7 and Theorem 5.3.20) that the set of $\mathbf{X}$ for which $\mathbf{X}_n \longrightarrow \mathbf{X}$ (a.e., μ) is a closed subset of $L^p(\mu; E)$. Moreover, if $\mathbf{X}$ is μ-simple, then the μ-almost everywhere convergence of $\mathbf{X}_n$ to $\mathbf{X}$ follows easily from the $\mathbb{R}$-valued result. Hence, we now know that $\mathbf{X}_n \longrightarrow \mathbf{X}$ (a.s, μ) for each $\mathbf{X} \in L^1(\mu; E)$. In addition, because of (5.3.29), when $p \in (1, \infty)$, the convergence in $L^p(\mu; E)$ follows by Lebesgue's Dominated Convergence Theorem. Finally, to prove the convergence in $L^1(\mu; E)$ when $\mathbf{X} \in L^1(\mu; E)$, note that, by Fatou's Lemma,

$$\|\mathbf{X}\|_{L^1(\mu;E)} \le \varliminf_{n\to\infty} \|\mathbf{X}_n\|_{L^1(\mu;E)};$$

whereas (5.1.25) guarantees that

$$\|\mathbf{X}\|_{L^1(\mu;E)} \ge \varlimsup_{n\to\infty} \|\mathbf{X}_n\|_{L^1(\mu;E)}.$$

Hence, because

$$\Big| \|\mathbf{X}_n\|_E - \|\mathbf{X}\|_E - \|\mathbf{X}_n - \mathbf{X}\|_E \Big| \le 2\|\mathbf{X}\|_E,$$

the convergence in $L^1(\mu; E)$ is again an application of Lebesgue's Dominated Convergence Theorem. □

As an application of the preceding, we will close this section with a proof that the sort of procedure discussed in Theorem 4.2.28 and Exercise 4.2.38 can be carried out in complete generality (cf. Remark 5.3.36 below). That is, let H be an infinite dimensional, separable, real Hilbert space, and suppose that $(\Theta, H, \mathcal{W}_H)$ is an abstract Wiener space (cf. the discussion preceding Remark 4.2.6) with the property that $\theta \in \Theta \longmapsto \|\theta\|_\Theta^p \in [0, \infty)$ is $\mathcal{W}_H$-integrable for every $p \in [1, \infty)$. Actually, by a beautiful result proved by X. Fernique, this latter integrability assumption will *always* be satisfied, since his theorem says that there will always exist an $\alpha \in (0, \infty)$ with the property that

$$\int_\Theta \exp\left[\alpha\|\theta\|_\Theta^2\right] \mathcal{W}_H(d\theta) < \infty.$$

Next, choose an orthonormal basis $\{h_n\}_0^\infty$ for H, and define (cf. the last part of Theorem 4.2.4)

$$\text{(5.3.31)} \qquad \mathbf{X}_n(\theta) = \sum_{m=0}^{n} \big[\mathcal{I}(h_m)\big](\theta)\, h_m, \quad n \in \mathbb{N} \text{ and } \theta \in \Theta.$$

5.3.32 Theorem. *Referring to the preceding discussion, one has that*

$$\text{(5.3.33)} \qquad \mathbf{X}_n(\theta) \xrightarrow{\Theta} \theta \quad \text{both (a.s., } \mathcal{W}_H) \text{ and in } L^p(\mathcal{W}_H;\Theta)$$

for every $p \in [1,\infty)$. *In particular, there is a measurable function* $F : \mathbb{R}^{\mathbb{N}} \longrightarrow \Theta$ *such that*

$$\sum_{m=0}^{n} x_m\, h_m \xrightarrow{\Theta} F(\mathbf{x}) \quad \text{for } {\gamma_1}^{\mathbb{N}}\text{-almost every } \mathbf{x} \in \mathbb{R}^{\mathbb{N}} \text{ and } \mathcal{W}_H = F^* {\gamma_1}^{\mathbb{N}}.$$

PROOF: Let $\mathcal{F}_n$ denote the σ-algebra over Θ generated by the maps $\mathcal{I}(h_m)$, $0 \le m \le n$, set $\mathcal{F}_\infty = \bigvee_0^\infty \mathcal{F}_n$, and, let $\mathbf{X}$ be a representative of the $\mathcal{W}_H$-conditional expectation of θ given $\mathcal{F}_\infty$. Our proof of (5.3.33) will involve two steps: the identification of $\mathbf{X}_n$ as a representative of the $\mathcal{W}_H$-conditional expectation of θ given $\mathcal{F}_n$ and a proof that $\mathbf{X}(\theta) = \theta$ for $\mathcal{W}_H$-almost every $\theta \in \Theta$. Indeed, in conjunction with Theorem 5.3.25, the first of these shows that $\mathbf{X}_n \longrightarrow \mathbf{X}$ both $\mathcal{W}_H$-almost surely and in $L^p(\mathcal{W}_H;\Theta)$ for every $p \in [1,\infty)$, and so (5.3.33) will follow immediately after one combines these two steps.

Since $\mathbf{X}_n$ is obviously $\mathcal{F}_n$-measurable, to prove that $\mathbf{X}_n$ is a representative of $\mathbb{E}^{\mathcal{W}_H}[\theta|\mathcal{F}_n]$ it suffices to show that, for each $\lambda \in \Theta^*$, $\langle \mathbf{X}_n, \lambda\rangle$ is a representative of $\mathbb{E}^{\mathcal{W}_H}\big[\langle\,\cdot\,,\lambda\rangle\big|\mathcal{F}_n\big]$. But, by the last part of Theorem 4.2.4, for any $\lambda \in \Theta^*$,

$$\langle\,\cdot\,,\lambda\rangle - \langle \mathbf{X}_n,\lambda\rangle = \mathcal{I}(h_\lambda) - \sum_{m=0}^{n} \big(h_\lambda, h_m\big)_H \mathcal{I}(h_m)$$

is independent of $\mathcal{F}_n$ under $\mathcal{W}_H$, and therefore (cf. (**iv**) of Exercise 5.1.28) $\langle \mathbf{X}_n,\lambda\rangle$ is a $\mathcal{W}_H$-conditional expectation of $\langle\,\cdot\,,\lambda\rangle$ given $\mathcal{F}_n$.

To see that $\mathbf{X}(\theta) = \theta$ for $\mathcal{W}_H$-almost every $\theta \in \Theta$, we need only check that $\mathcal{B}_\Theta$ is contained in the $\mathcal{W}_H$-completion of $\mathcal{F}_\infty$. To this end, recall that $\mathcal{B}_\Theta$ is generated by the maps $\theta \in \Theta \longmapsto \langle \theta,\lambda\rangle$ as λ runs over Θ^*. Thus, we need only show that each $\langle\,\cdot\,,\lambda\rangle$ can be expressed as the $\mathcal{W}_H$-almost sure limit of $\mathcal{F}_\infty$-measurable functions. But if $\lambda \in \Theta^*$, then, again by the last part of Theorem 4.2.4,

$$\mathbb{E}^{\mathcal{W}_H}\Big[\big|\langle\,\cdot\,,\lambda\rangle - \langle \mathbf{X}_n,\lambda\rangle\big|^2\Big] = \|h_\lambda\|_H^2 - \sum_0^n \big(h_\lambda, h_m\big)_H^2 \longrightarrow 0.$$

Hence, because $\big(\langle \mathbf{X}_n, \lambda\rangle, \mathcal{F}_n, P\big)$ is a martingale, $\langle \mathbf{X}_n, \lambda\rangle \longrightarrow \langle\,\cdot\,,\lambda\rangle$ $\mathcal{W}_{\mathbf{H}}$-almost surely.

The second part of the theorem is a more or less immediate consequence of the first part. Namely, define

$$F_n(\mathbf{x}) = \sum_{m=0}^{n} x_m\, h_m \quad \text{for } n \in \mathbb{N} \text{ and } \mathbf{x} \in \mathbb{R}^{\mathbb{N}}.$$

Then, the map

$$\mathbf{x} \in \mathbb{R}^{\mathbb{N}} \longmapsto \big(F_0(\mathbf{x}), \dots, F_n(\mathbf{x}), \dots\big) \in \Theta^{\mathbb{N}}$$

has the same distribution under ${\gamma_1}^{\mathbb{N}}$ as

$$\theta \in \Theta \longmapsto \big(\mathbf{X}_0(\theta), \dots, \mathbf{X}_n(\theta), \dots\big) \in \Theta^{\mathbb{N}}$$

has under $\mathcal{W}_H$. In particular, by the preceding, this means that there is a measurable $F : \mathbb{R}^{\mathbb{N}} \longrightarrow \Theta$ such that

$$F(\mathbf{x}) = \lim_{n\to\infty} F_n(\mathbf{x}) \quad \text{for } {\gamma_1}^{\mathbb{N}}\text{-almost every } \mathbf{x} \in \mathbb{R}^{\mathbb{N}},$$

and $\mathbf{x} \in \mathbb{R}^{\mathbb{N}} \longmapsto F(\mathbf{x}) \in \Theta$ has the same distribution under ${\gamma_1}^{\mathbb{N}}$ as θ has under $\mathcal{W}_H$. □

5.3.34 Corollary. *Let $\mathbf{H}(\mathbb{R})$ be the Hilbert space described in (4.2.2) and $\Theta(\mathbb{R})$ the Banach space in (4.2.12) when $N = 1$. Then, for every choice of orthonormal basis $\{\mathbf{h}_n\}_0^\infty$ for $\mathbf{H}(\mathbb{R})$,*

$$\left\| \sum_{m=0}^{n} [\mathcal{I}(\mathbf{h}_m)](\boldsymbol{\theta})\, \mathbf{h}_m - \theta \right\|_{\Theta(\mathbb{R})} \longrightarrow 0 \quad \textit{both (a.s., } \mathcal{W}\textit{) and in } L^p(\mathcal{W};\mathbb{R})$$

for every $p \in [1,\infty)$. In particular, for each choice of $\{\mathbf{h}_m\}_0^\infty$, there is a measurable $F : \mathbb{R}^{\mathbb{N}} \longrightarrow \Theta(\mathbb{R})$ such that $\mathcal{W} = F^* {\gamma_1}^{\mathbb{N}}$ and

$$\left\| \sum_{m=0}^{n} x_m\, \mathbf{h}_m - F(\mathbf{x}) \right\|_{\Theta(\mathbb{R})} \longrightarrow 0 \quad \textit{both for } {\gamma_1}^{\mathbb{N}}\textit{-almost every } \mathbf{x} \in \mathbb{R}^{\mathbb{N}}$$

and in $L^p({\gamma_1}^{\mathbb{N}};\mathbb{R})$ for every $p \in [1,\infty)$.

PROOF: All that we have to do is verify that all powers of $\|\theta\|_{\Theta(\mathbb{R})}$ are $\mathcal{W}$-integrable; and we will do this by checking (in this case, by hand) the estimate of Fernique alluded to above. But, by (3.3.30),

$$\begin{aligned}\mathcal{W}(\|\theta\|_{\Theta(\mathbb{R})} \geq R) &\leq \sum_{n=1}^{\infty} \mathcal{W}\left(\sup_{t\in[0,n]} |\theta(t)| \geq nR\right)\\ &\leq 2\sum_{n=1}^{\infty} \exp\left[-n\frac{R^2}{2}\right] = 2\frac{e^{-\frac{R^2}{2}}}{1-e^{-\frac{R^2}{2}}},\end{aligned}$$

and therefore

$$\int_{\Theta(\mathbb{R})} \exp\left[\alpha\|\theta\|^2_{\Theta(\mathbb{R})}\right] \mathcal{W}(d\theta) < \infty \tag{5.3.35}$$

for some $\alpha \in (0,\infty)$. In particular, this certainly means that all powers of $\|\theta\|_{\Theta(\mathbb{R})}$ are $\mathcal{W}$-integrable. □

5.3.36 Remark: It should be pointed out that although Corollary 5.3.32 might seem to say that Wiener's task would have been essentially trivial had he had Theorem 5.3.27 available to him, the fact is that Corollary 5.3.34 would have been no help to him. Indeed, Wiener was trying to prove that $\mathcal{W}$ exists at all, whereas Corollary 5.3.34 only says that, once one knows it exists, there are lots of ways to represent $\mathcal{W}$ as the image of $\gamma_1{}^{\mathbb{N}}$.

Exercises

5.3.37 Exercise: In this exercise, we will develop Jensen's inequality in the Banach space setting. Thus, $(\Omega, \mathcal{F}, P)$ will be a probability space, C will be a closed, convex subset of the separable Banach space E, and $\mathbf{X}$ will be a C-valued element of $L^1(P;E)$.

(i) Show that there exists a sequence $\{\mathbf{X}_n\}_1^\infty$ of C-valued, P-simple functions which tend to $\mathbf{X}$ both P-almost surely and in $L^1(P;E)$.

(ii) Show that $\mathbb{E}^P[\mathbf{X}] \in C$ and that

$$\mathbb{E}^P\left[g(\mathbf{X})\right] \leq g(\mathbb{E}^P[\mathbf{X}])$$

for every continuous, concave $g : C \longrightarrow [0,\infty)$.

(iii) Given a sub-σ-algebra Σ of $\mathcal{F}$, follow the argument in Corollary 5.2.14 to show that there exists a sequence $\{\mathcal{P}_n\}_0^\infty$ of finite, Σ-measurable partitions with the property that

$$\sum_{A\in\mathcal{P}_n} \frac{\mathbb{E}^P[\mathbf{X},\, A]}{P(A)} \mathbf{1}_A \longrightarrow \mathbb{E}^P[\mathbf{X}|\Sigma] \quad \text{both } P\text{-almost surely and in } L^1(P;E).$$

In particular, conclude that there is a representative $\mathbf{X}_\Sigma$ of $\mathbb{E}^P[\mathbf{X}|\Sigma]$ which is C-valued and that

$$\mathbb{E}^P\big[g(\mathbf{X})\big|\Sigma\big] \le g(\mathbf{X}_\Sigma) \quad \text{(a.s., } P)$$

for each continuous, convex $g : C \longrightarrow [0,\infty)$.

5.3.38 Exercise: Again let $(\Omega, \mathcal{F}, P)$ be a probability space and E a separable, real Banach space. Further, suppose that $\{\mathcal{F}_n\}_0^\infty$ is a *nonincreasing* sequence of sub-σ-algebras of $\mathcal{F}$, and set $\mathcal{F}_\infty = \bigcap_0^\infty \mathcal{F}_n$. Finally, let $\mathbf{X} \in L^1(P;E)$.

(i) Show that

$$\mathbb{E}^P\big[\mathbf{X}\big|\mathcal{F}_n\big] \longrightarrow \mathbb{E}^P[\mathbf{X}|\mathcal{F}_\infty] \quad \text{both } P\text{-almost surely and in } L^p(P;E)$$

for any $p \in [1,\infty)$ with $\mathbf{X} \in L^p(P;E)$.

Hint: Use (5.3.28) and the approximation result in Theorem 5.1.20 to reduce to the case when $\mathbf{X}$ is P-simple. When $\mathbf{X}$ is P-simple, get the result as an application of **(iii)** in Exercise 5.2.40.

(ii) Using part **(i)** and following the line of reasoning given in part **(iv)** of Exercise 5.2.40, give another proof of The Strong Law of Large Numbers for E-valued random variables. (See Exercise 3.1.21 for an entirely different approach.)

5.3.39 Exercise: As we saw in the proof of Theorem 5.3.20, the Hardy–Littlewood maximal function can be used to dominate other quantities of interest. As further confirmation of its importance, we will use it in this exercise to prove the analogue of Theorem 5.3.20 for a large class of approximate identities. That is, let $\psi \in L^1(\mathbb{R}^N;\mathbb{R})$ with $\int_{\mathbb{R}^N} \psi\, d\mathbf{x} = 1$ be given, and set

$$\psi_t(\mathbf{x}) = t^{-N}\psi\left(\tfrac{\mathbf{x}}{t}\right), \quad t \in (0,\infty) \text{ and } \mathbf{x} \in \mathbb{R}^N. \tag{5.3.40}$$

Then $\{\psi_t : t > 0\}$ forms an **approximate identity** in the sense that, as tempered distributions, $\psi_t \longrightarrow \delta_\mathbf{0}$ as $t \searrow 0$. In fact, because

$$\|\psi_t \bigstar f\|_{L^p(\mathbb{R}^N)} \le \|\psi\|_{L^1(\mathbb{R}^N)}\, \|f\|_{L^p(\mathbb{R}^N)}, \quad t \in (0,\infty) \text{ and } p \in [1,\infty]$$

and

$$\psi_t\bigstar f(\mathbf{x}) = \int_{\mathbb{R}^N} \psi(\mathbf{y})\, f(\mathbf{x} - t\mathbf{y})\, d\mathbf{y},$$

it is easy to see that, for each $p \in [1,\infty)$,

$$\lim_{t\searrow 0} \left\|\psi_t\bigstar f - f\right\|_{L^p(\mathbb{R}^N)} = 0$$

first for $f \in C_c(\mathbb{R}^N;\mathbb{R})$ and then for all $f \in L^p(\mathbb{R}^N;\mathbb{R})$.

The purpose of this exercise is to sharpen the preceding under the assumption that

$$\text{(5.3.41)}\qquad \begin{gathered}\psi(\mathbf{x}) = \alpha\big(|\mathbf{x}|\big), \quad \mathbf{x} \in \mathbb{R}^N \setminus \{\mathbf{0}\} \quad \text{for some } \alpha \in C^1\big((0,\infty);\mathbb{R}\big) \text{ with} \\ A \equiv \int_{(0,\infty)} r^N |\alpha'(r)|\, dr < \infty.\end{gathered}$$

Notice that when α is nonnegative and nonincreasing, integration by parts shows that $A = N$.

(i) Let $f \in C_c(\mathbb{R}^N;\mathbb{R})$ be given, and set

$$\tilde{f}(r,\mathbf{x}) = \frac{1}{|B_{\mathbb{R}^N}(\mathbf{x},r)|} \int_{B_{\mathbb{R}^N}(\mathbf{x},r)} f(\mathbf{y})\, d\mathbf{y} \quad \text{for } r \in (0,\infty) \text{ and } \mathbf{x} \in \mathbb{R}^N.$$

Using integration by parts and the hypotheses in (5.3.41), show that

$$\psi_t\bigstar f(\mathbf{x}) = -\tfrac{1}{N} \int_{(0,\infty)} r^N \alpha'(r)\, \tilde{f}(tr,\mathbf{x})\, dr,$$

and conclude that

$$\left|\psi_t\bigstar f(\mathbf{x})\right| \le \tfrac{A}{N}\, \tilde{\mathbf{M}}f(\mathbf{x}),$$

where $\tilde{\mathbf{M}}f$ is the quantity introduced at the beginning of the proof of Theorem 5.3.20. In particular, conclude that there is a constant $K_N \in (0,\infty)$, depending only on $N \in \mathbb{Z}^+$, such that

$$\text{(5.3.42)}\qquad \mathbf{M}_\psi f(\mathbf{x}) \equiv \sup_{t\in(0,\infty)} \left|\psi_t\bigstar f(\mathbf{x})\right| \le K_N A\, \mathbf{M}f(\mathbf{x}), \quad \mathbf{x} \in \mathbb{R}^N.$$

(ii) Starting from (5.3.42), show that

$$\text{(5.3.43)}\qquad \left|\{\mathbf{x} : M_\psi f(\mathbf{x}) \ge R\}\right| \le \frac{(12)^N K_N A \|f\|_{L^1(\mathbb{R}^N)}}{R}, \quad f \in L^1(\mathbb{R}^N;\mathbb{R}),$$

and that for $p \in (1,\infty]$,

$$\text{(5.3.44)}\qquad \left\|M_\psi f\right\|_{L^p(\mathbb{R}^N)} \le \frac{(12)^N K_N A p}{p-1} \|f\|_{L^p(\mathbb{R}^N)}, \quad f \in L^p(\mathbb{R}^N;\mathbb{R}).$$

Finally, proceeding as in the proof of Theorem 5.3.20, use (5.3.43) to prove that, for Leb-almost every $\mathbf{x} \in \mathbb{R}^N$:

$$\tag{5.3.45} \begin{aligned} &\overline{\lim_{t\searrow 0}} \left|\psi_t \star f(\mathbf{x}) - f(\mathbf{x})\right| \\ &\leq \overline{\lim_{t\searrow 0}} \int_{\mathbb{R}^N} \left|\psi_t(\mathbf{y})\big(f(\mathbf{x}-\mathbf{y}) - f(\mathbf{x})\big)\right| d\mathbf{y} = 0. \end{aligned}$$

Two of the most familiar examples to which the preceding applies are the Gauss kernel (cf. (1.3.5)) γ^N (notice that the notation γ_t used previously corresponds to $\gamma_{\sqrt{t}}$ in the convention used here) and the Poisson kernel (cf. (4.3.50)) $p^{(N)}$. In both these cases, $A = N$.

5.3.46 Exercise: Let E be a separable Hilbert space and $(\mathbf{X}_n, \mathcal{F}, P)$ an E-valued martingale on some probability space $(\Omega, \mathcal{F}, P)$ satisfying the condition

$$\tag{5.3.47} \sup_{n\in\mathbb{Z}^+} \mathbb{E}^P\left[\|\mathbf{X}_n\|_E^2\right] < \infty.$$

Proceeding as in (**i**) of Exercise 5.2.27, first prove that there is a $\bigvee_1^\infty \mathcal{F}_n$-measurable $\mathbf{X} \in L^2(P; E)$ to which $\{\mathbf{X}_n\}_1^\infty$ converges in $L^2(P; E)$, next check that

$$\mathbf{X}_n = \mathbb{E}^P\left[\mathbf{X}\middle|\mathcal{F}_n\right] \quad \text{(a.s., } P\text{) for each } n \in \mathbb{Z}^+,$$

and finally apply the last part of Theorem 5.3.27 to see that $\mathbf{X}_n \longrightarrow \mathbf{X}$ P-almost surely.

Chapter VI:
Some Applications of Martingale Theory

§6.1: The Individual Ergodic Theorem.

This chapter contains various applications of the ideas and results in Chapter 5, the first being an application of Hardy's Inequality (cf. (5.3.23)) to the derivation of Birkhoff's Individual Ergodic Theorem.†

The setting in which we will prove the Ergodic Theorem will be the following. $(\Omega, \mathcal{F}, \mu)$ will be a σ-finite measure space on which there exits a semigroup $\{\mathbf{\Sigma^k} : \mathbf{k} \in \mathbb{N}^N\}$ of measurable, **μ-measure preserving transformations.** That is, for each $\mathbf{k} \in \mathbb{N}^N$, $\mathbf{\Sigma^k}$ is an $\mathcal{F}$-measurable map from Ω into itself, $\mathbf{\Sigma^0}$ is the identity map, $\mathbf{\Sigma}^{\mathbf{k}+\boldsymbol{\ell}} = \mathbf{\Sigma^k} \circ \mathbf{\Sigma}^{\boldsymbol{\ell}}$ for all $\mathbf{k}$, $\boldsymbol{\ell} \in \mathbb{N}^N$, and

$$\mu(\Gamma) = \mu\big((\mathbf{\Sigma^k})^{-1}(\Gamma)\big) \quad \text{for all } \mathbf{k} \in \mathbb{N} \text{ and } \Gamma \in \mathcal{F}.$$

Further, E will be a separable Banach space with norm $\|\cdot\|_E$; and, given a function $f : \Omega \longrightarrow E$, we will be considering the averages

$$\mathbf{A}_n F(\omega) \equiv \frac{1}{n^N} \sum_{\mathbf{k} \in Q_n^+} F \circ \mathbf{\Sigma^k}(\omega), \quad n \in \mathbb{Z}^+, \tag{6.1.1}$$

where Q_n^+ is the cube $\{\mathbf{k} \in \mathbb{N}^N : \|\mathbf{k}\| < n\}$ and $\|\mathbf{k}\| \equiv \max_{1 \le j \le N} k_j$. Our goal (cf. Theorem 6.1.8 below) is to show that for each $p \in [1,\infty)$ and $f \in L^p(\mu; E)$, $\{\mathbf{A}_n f\}_1^\infty$ converges μ-almost everywhere. In fact, when either μ is finite or $p \in (1,\infty)$, we will show that the convergence is also in $L^p(\mu; E)$.

The key to proving this result is contained in the following weak-type inequality.

6.1.2 Lemma (Maximal Ergodic Lemma). *For each $n \in \mathbb{Z}^+$, $\mathbf{A}_n$ is a contraction on $L^p(\mu; E)$ for every $p \in [1,\infty)$. Moreover, for each $F \in L^p(\mu; E)$:*

$$\mu\left(\sup_{n \ge 1} \|\mathbf{A}_n F\|_E \ge \lambda\right) \le \frac{(24)^N}{\lambda} \|F\|_{L^1(\mu;E)}, \quad \lambda \in (0,\infty), \tag{6.1.3}$$

† The statement which we prove here is due to N. Wiener. The idea of using Hardy's Inequality was suggested to P. Hartman by J. von Neumann and appears for the first time in Hartman's "On the ergodic theorem," *Am. J. Math.* **69**: 193–199 (1947).

or

$$\left\|\sup_{n\geq 1}\|\mathbf{A}_n F\|_E\right\|_{L^p(\mu)} \leq \frac{(24)^N p}{p-1}\|F\|_{L^p(\mu;E)}, \tag{6.1.4}$$

depending on whether $p=1$ *or* $p\in(1,\infty)$.

PROOF: First observe that it suffices to prove all of these assertions in the case when $E=\mathbb{R}$ and F is nonnegative. Thus, we will restrict ourselves to this case. But then $F\circ\mathbf{\Sigma}^{\mathbf{k}}$ has the same distribution as F itself, and so the first assertion is trivial. To prove (6.1.3) and (6.1.4), let $n\in\mathbb{Z}^+$ be given, apply (5.3.23) and (5.2.24) to

$$a_{\mathbf{k}}(\omega)\equiv\begin{cases} F\circ\mathbf{\Sigma}^{\mathbf{k}}(\omega) & \text{if } \mathbf{k}\in Q_{2n}^+ \\ 0 & \text{if } \mathbf{k}\notin Q_{2n}^+,\end{cases}$$

and conclude that

$$\begin{aligned} C_n(\omega) &\equiv \left|\left\{\mathbf{k}\in Q_n^+ :\ \max_{1\leq m\leq n}\mathbf{A}_m\big(F\circ\mathbf{\Sigma}^{\mathbf{k}}\big)(\omega)\geq\lambda\right\}\right| \\ &\leq \frac{(12)^N}{\lambda}\sum_{\mathbf{k}\in Q_{2n}^+} F\circ\mathbf{\Sigma}^{\mathbf{k}}(\omega)\end{aligned}$$

and

$$\sum_{\mathbf{k}\in Q_n^+}\max_{1\leq m\leq n}\left|\mathbf{A}_m\big(F\circ\mathbf{\Sigma}^{\mathbf{k}}\big)(\omega)\right|^p \leq \left(\frac{(12)^N p}{p-1}\right)^p\sum_{\mathbf{k}\in Q_{2n}^+}\left|F\circ\mathbf{\Sigma}^{\mathbf{k}}(\omega)\right|^p.$$

Hence, by Tonelli's Theorem,

$$\begin{aligned}\sum_{\mathbf{k}\in Q_n^+}\mu\left(\max_{1\leq m\leq n}\mathbf{A}_m\big(F\circ\mathbf{\Sigma}^{\mathbf{k}}\big)\leq\lambda\right) &= \int C_n(\omega)\,\mu(d\omega) \\ &\leq \frac{(12)^N}{\lambda}\sum_{\mathbf{k}\in Q_{2n}^+}\int F\circ\mathbf{\Sigma}^{\mathbf{k}} f\,d\mu\end{aligned}$$

and, similarly,

$$\sum_{\mathbf{k}\in Q_n^+}\int\max_{1\leq m\leq n}\mathbf{A}_m\big(F\circ\mathbf{\Sigma}^{\mathbf{k}}\big)^p\,d\mu \leq \left(\frac{(12)^N p}{p-1}\right)^p\sum_{\mathbf{k}\in Q_{2n}^+}\int F\circ\mathbf{\Sigma}^{\mathbf{k}}\,d\mu.$$

Finally, since the distributions of $\max_{1\leq m\leq n}\mathbf{A}_m\big(F\circ\mathbf{\Sigma}^{\mathbf{k}}\big)$ and $F\circ\mathbf{\Sigma}^{\mathbf{k}}$ do not depend on $\mathbf{k}\in\mathbb{N}^N$, the preceding lead immediately to

$$\mu\left(\max_{1\leq m\leq n}\mathbf{A}_m F\leq\lambda\right)\leq\frac{(24)^N}{\lambda}\|F\|_{L^1(\mu)}$$

and

$$\left\|\max_{1\leq m\leq n}\mathbf{A}_m F\right\|_{L^p(\mu)}\leq\frac{2^{\frac{N}{p}}(12)^N p}{p-1}\|F\|_{L^p(\mu)}$$

for all $n\in\mathbb{Z}^+$. Thus, (6.1.3) and (6.1.4) follow after one lets $n\to\infty$. □

Given (6.1.3) and (6.1.4), we adopt again the strategy used in the proof of Corollary 5.2.7. That is, we must find a dense subset of each L^p-space on which the desired convergence results can be checked by hand, and for this purpose we will have to introduce the notion of invariance.

A set $\Gamma \in \mathcal{F}$ is said to be **invariant**, and we write $\Gamma \in \mathfrak{I}$, if $\Gamma = (\boldsymbol{\Sigma}^{\mathbf{k}})^{-1}(\Gamma)$ for every $\mathbf{k} \in \mathbb{N}^N$. As is easily checked, $\mathfrak{I}$ is a sub-σ-algebra of $\mathcal{F}$. In addition, it is clear that $\Gamma \in \mathcal{F}$ is invariant if $\Gamma = (\boldsymbol{\Sigma}^{\mathbf{e}_j})^{-1}(\Gamma)$ for each $1 \le j \le N$. Finally, if $\overline{\mathfrak{I}}$ is the μ-completion of $\mathfrak{I}$ relative to $\mathcal{F}$ in the sense that $\Gamma \in \overline{\mathfrak{I}}$ if and only if $\Gamma \in \mathcal{F}$ and there is $\tilde{\Gamma} \in \mathfrak{I}$ such that $\mu(\Gamma \Delta \tilde{\Gamma}) = 0$ ($A \Delta B \equiv (A \setminus B) \cup (B \setminus A)$ is the **symmetric difference** of the sets A and B), then an $\mathcal{F}$-measurable $F : \Omega \longrightarrow E$ is $\Gamma \in \overline{\mathfrak{I}}$ if and only if $F = F \circ \boldsymbol{\Sigma}^{\mathbf{k}}$ (a.e., μ) for each $\mathbf{k} \in \mathbb{N}^N$. Indeed, one need only check this equivalence for indicator functions of sets. But if $\Gamma \in \mathcal{F}$ and $\mu(\Gamma \Delta \tilde{\Gamma}) = 0$ for some $\tilde{\Gamma} \in \mathfrak{I}$, then

$$\mu\Big(\Gamma \Delta (\boldsymbol{\Sigma}^{\mathbf{k}})^{-1}(\Gamma)\Big) \le \mu\big((\boldsymbol{\Sigma}^{\mathbf{k}})^{-1}(\Gamma \Delta \tilde{\Gamma})\big) + \mu(\Gamma \Delta \tilde{\Gamma}) = 0,$$

and so $\Gamma \in \overline{\mathfrak{I}}$. Conversely, if $\Gamma \in \overline{\mathfrak{I}}$, set

$$\tilde{\Gamma} = \bigcup_{\mathbf{k} \in \mathbb{N}^N} (\boldsymbol{\Sigma}^{\mathbf{k}})^{-1}(\Gamma),$$

and check that $\tilde{\Gamma} \in \mathfrak{I}$ and $\mu(\Gamma \Delta \tilde{\Gamma}) = 0$.

6.1.5 Lemma. *Let $\mathfrak{I}(E)$ be the subspace of $\overline{\mathfrak{I}}$-measurable elements of $L^2(\mu; E)$. Then, $\mathfrak{I}^2(E)$ is a closed linear subspace of $L^2(\mu; E)$. Moreover, if $\Pi_{\mathfrak{I}(\mathbb{R})}$ denotes orthogonal projection from $L^2(\mu; \mathbb{R})$ onto $\mathfrak{I}(\mathbb{R})$, then there exists a unique linear contraction $\Pi_{\mathfrak{I}(E)} : L^2(\mu; E) \longrightarrow \mathfrak{I}(E)$ with the property that $\Pi_{\mathfrak{I}(E)}(\mathbf{a}f) = \mathbf{a}\Pi_{\mathfrak{I}(\mathbb{R})} f$ for $\mathbf{a} \in E$ and $f \in L^2(\mu; \mathbb{R})$. Finally, for each $F \in L^2(\mu; E)$,*

$$\mathbf{A}_n F \longrightarrow \Pi_{\mathfrak{I}(E)} F \quad \text{(a.e., } \mu) \text{ and in } L^2(\mu; E). \tag{6.1.6}$$

PROOF: We begin with the case when $E = \mathbb{R}$. The first step is to identify the orthogonal complement $\mathfrak{I}(\mathbb{R})^\perp$ of $\mathfrak{I}(\mathbb{R})$. To this end, let $\mathcal{N}$ denote the subspace of $L^2(\mu; \mathbb{R})$ consisting of elements having the form $g - g \circ \boldsymbol{\Sigma}^{\mathbf{e}_j}$ for some $g \in L^2(\mu; \mathbb{R}) \cap L^\infty(\mu; \mathbb{R})$ and $1 \le j \le N$. Given $f \in \mathfrak{I}(\mathbb{R})$, observe that

$$\big(f, g - g \circ \boldsymbol{\Sigma}^{\mathbf{e}_j}\big)_{L^2(\mu)} = \big(f, g\big)_{L^2(\mu)} - \big(f \circ \boldsymbol{\Sigma}^{\mathbf{e}_j}, g \circ \boldsymbol{\Sigma}^{\mathbf{e}_j}\big)_{L^2(\mu)} = 0.$$

Hence, $\mathcal{N} \subseteq \mathfrak{I}(\mathbb{R})^\perp$. On the other hand, if $f \in L^2(\mu; \mathbb{R})$ and $f \perp \mathcal{N}$, then it is clear that $f \perp f - f \circ \boldsymbol{\Sigma}^{\mathbf{e}_j}$ for each $1 \le j \le N$ and therefore that

$$\begin{aligned}
&\big\|f - f \circ \boldsymbol{\Sigma}^{\mathbf{e}_j}\big\|^2_{L^2(\mu;\mathbb{R})} \\
&\quad = \|f\|^2_{L^2(\mu;\mathbb{R})} - 2\big(f, f \circ \boldsymbol{\Sigma}^{\mathbf{e}_j}\big)_{L^2(\mu;\mathbb{R})} + \big\|f \circ \boldsymbol{\Sigma}^{\mathbf{e}_j}\big\|^2_{L^2(\mu;\mathbb{R})} \\
&\quad = 2\Big(\|f\|^2_{L^2(\mu)} - \big(f, f \circ \boldsymbol{\Sigma}^{\mathbf{e}_j}\big)_{L^2(\mu)}\Big) = 2\big(f, f - f \circ \boldsymbol{\Sigma}^{\mathbf{e}_j}\big)_{L^2(\mu)} = 0.
\end{aligned}$$

Thus, for each $1 \le j \le N$, $f = f \circ \mathbf{\Sigma}^{\mathbf{e}_j}$ μ-almost everywhere; and, by induction on $\|\mathbf{k}\|$, one concludes that $f = f \circ \mathbf{\Sigma}^{\mathbf{k}}$ μ-almost everywhere for all $\mathbf{k} \in \mathbb{N}^N$. In other words, we have now shown that $\mathfrak{I}(\mathbb{R}) = \mathcal{N}^{\perp}$, or, equivalently, that $\overline{\mathcal{N}} = \mathfrak{I}(\mathbb{R})^{\perp}$.

Continuing with $E = \mathbb{R}$, next note that if $f \in \mathfrak{I}^2(\mathbb{R})$ then $\mathbf{A}_n f = f$ (a.e., μ) for each $n \in \mathbb{Z}^+$. Hence, (6.1.6) is completely trivial in this case. On the other hand, if $g \in L^2(\mu;\mathbb{R}) \cap L^\infty(\mu;\mathbb{R})$ and $f = g - g \circ \mathbf{\Sigma}^{\mathbf{e}_j}$, then

$$n^N \mathbf{A}_n f = \sum_{\{\mathbf{k} \in Q_n^+ : k_j = 0\}} g \circ \mathbf{\Sigma}^{\mathbf{k}} - \sum_{\{\mathbf{k} \in Q_n^+ : k_j = n\}} g \circ \mathbf{\Sigma}^{\mathbf{k}+\mathbf{e}_j},$$

and so, with $p \in \{2, \infty\}$,

$$\big\|\mathbf{A}_n f\big\|_{L^p(\mu;\mathbb{R})} \le \frac{4\|g\|_{L^p(\mu;\mathbb{R})}}{n} \longrightarrow 0 \quad \text{as } n \to \infty.$$

Hence, in this case also, (6.1.6) is easy. Finally, to complete the proof for $E = \mathbb{R}$, simply note that, by (6.1.4) with $p = 2$ and $E = \mathbb{R}$, the set of $f \in L^2(\mu;\mathbb{R})$ for which (6.1.6) holds is a closed linear subspace of $L^2(\mu;\mathbb{R})$ and that we have already verified (6.1.6) for $f \in \mathfrak{I}(\mathbb{R})$ and f from a dense subspace of $\mathfrak{I}(\mathbb{R})^{\perp}$.

Turning to general E's, first note $\Pi_{\mathfrak{I}(E)} F$ is well-defined for μ-simple F's. Indeed, if $F = \sum_1^\ell \mathbf{a}_i \mathbf{1}_{\Gamma_i}$ for some $\{\mathbf{a}_i\}_1^\ell \subseteq E$ and $\{\Gamma_i\}_1^\ell$ of mutually disjoint elements of $\mathcal{F}$ with finite μ-measure, then

$$\Pi_{\mathfrak{I}(E)} F = \sum_1^\ell \mathbf{a}_i \Pi_{\mathfrak{I}(\mathbb{R})} \mathbf{1}_{\Gamma_i}$$

and so

$$\begin{aligned}\big\|\Pi_{\mathfrak{I}(E)} F\big\|_{L^2(\mu;E)}^2 &\le \int \left(\sum_1^\ell \|\mathbf{a}_i\|_E \Pi_{\mathfrak{I}(\mathbb{R})} \mathbf{1}_{\Gamma_i}\right)^2 d\mu \\ &= \left\|\Pi_{\mathfrak{I}(\mathbb{R})}\left(\sum_1^\ell \|\mathbf{a}_i\|_E \mathbf{1}_{\Gamma_i}\right)\right\|_{L^2(\mu;\mathbb{R})}^2 \le \sum_1^\ell \|\mathbf{a}_i\|_E^2 \mu(\Gamma_i) = \|F\|_{L^2(\mu;E)}^2.\end{aligned}$$

Thus, since the space of μ-simple functions is dense in $L^2(\mu;E)$, it is clear that $\Pi_{\mathfrak{I}(E)}$ not only exists but is also unique.

Finally, to check (6.1.6) for general E's, note that (6.1.6) for E-valued, μ-simple F's is an immediate consequence of (6.1.6) for $E = \mathbb{R}$. Thus, we already know (6.1.6) for a dense subspace of $L^2(\mu;E)$; and so the rest is a simple application of (6.1.4). □

For general $p \in [1,\infty)$, let $\mathfrak{I}^p(E)$ denote the subspace of $\overline{\mathfrak{I}}$-measurable elements of $L^p(\mu;E)$. Clearly $\mathfrak{I}^p(E)$ is closed for every $p \in [1,\infty)$. Moreover, since

$$\mu(E) < \infty \implies \Pi_{\mathfrak{I}(E)}F = \mathbb{E}^{\mu}\big[F\big|\mathfrak{I}\big], \tag{6.1.7}$$

$\Pi_{\mathfrak{I}(E)}$ extends automatically as a linear contraction from $L^p(\mu;E)$ onto $\mathfrak{I}^p(E)$ for each $p \in [1,\infty)$, the extension being given by the right-hand side of (6.1.7). However, when $\mu(E) = \infty$, there is a problem. Namely, because $\mu \upharpoonright \mathfrak{I}$ will seldom be σ-finite, it will not be possible to condition μ with respect to $\mathfrak{I}$. Be that as it may, (6.1.6) provides an extension of $\Pi_{\mathfrak{I}(E)}$. Namely, from (6.1.6) and Fatou's Lemma, it is clear that, for each $p \in [1,\infty)$,

$$\big\|\Pi_{\mathfrak{I}(E)}F\big\|_{L^p(\mu;E)} \le \|F\|_{L^p(\mu;E)}, \quad F \in L^p(\mu;E) \cap L^2(\mu;E);$$

and therefore the desired extension follows by continuity.

6.1.8 The Individual Ergodic Theorem. *For each $p \in [1,\infty)$ and $F \in L^p(\mu;E)$:*

$$\mathbf{A}_nF \longrightarrow \Pi_{\mathfrak{I}(E)}F \quad (\text{a.e.}, \mu). \tag{6.1.9}$$

Moreover, if either $p \in (1,\infty)$ or $\mu(E) < \infty$, then the convergence in (6.1.8) *is also in $L^p(\mu;E)$. Finally, if $\mu(\Gamma) \in \{0,\mu(E)\}$ for every $\Gamma \in \mathfrak{I}$, then* (6.1.9) *can be replaced by*

$$\lim_{n\to\infty} \mathbf{A}_nF = \begin{cases} \dfrac{\mathbb{E}^{\mu}[F]}{\mu(E)} & \text{if } \mu(E) \in (0,\infty) \\ \mathbf{0} & \text{if } \mu(E) = \infty \end{cases} \quad (\text{a.e.}, \mu), \tag{6.1.10}$$

and the convergence is in $L^p(\mu;E)$ when either $p \in (1,\infty)$ or $\mu(E) < \infty$.

PROOF: As we said above, the proof is now an easy application of the strategy used to prove Corollary 5.2.7. Namely, by (6.1.3), the set of $F \in L^1(\mu;E)$ for which (6.1.9) holds is closed and, by (6.1.6), it includes $L^1(\mu;E) \cap L^\infty(\mu;E)$. Hence, (6.1.9) is proved for $p = 1$. On the other hand, when $p \in (1,\infty)$, (6.1.4) applies and shows first that the set of $F \in L^p(\mu;E)$ for which (6.1.9) holds is closed in $L^p(\mu;E)$ and second that μ-almost everywhere convergence already implies convergence in $L^p(\mu;E)$. Hence, (6.1.9) has been proved, and the convergence is in $L^p(\mu;E)$ when $p \in (1,\infty)$. In addition, when $\mu(\Gamma) \in \{0,\mu(\Omega)\}$ for all $\Gamma \in \mathfrak{I}$, it is clear that the only elements of $\mathfrak{I}^p(E)$ are μ-almost everywhere constant, which, in the case when $\mu(E) < \infty$ means (cf. (6.1.7)) that

$$\Pi_{\mathfrak{I}(E)}F = \frac{\mathbb{E}^{\mu}[F]}{\mu(E)},$$

and, when $\mu(E) = \infty$, means that $\mathfrak{I}^p(E) = \{\mathbf{0}\}$ for all $p \in [1,\infty)$.

In view of the preceding, all that remains is to discuss the $L^1(\mu;E)$-convergence in the case when $p = 1$ and $\mu(\Omega) < \infty$. To this end, observe that, because the $\mathbf{A}_n$'s are all contractions in $L^1(\mu;E)$, it suffices to prove $L^1(\mu;E)$-convergence for E-valued, μ-simple F's. But $L^1(\mu;E)$-convergence for such F's reduces to showing that $\mathbf{A}_n f \longrightarrow \Pi_{\mathfrak{I}(\mathbb{R})} f$ in $L^1(\mu;\mathbb{R})$ for nonnegative $f \in L^\infty(\mu;\mathbb{R})$. Finally, if $f \in L^1\big(\mu;[0,\infty)\big)$, then

$$\big\|\mathbf{A}_n f\big\|_{L^1(\mu)} = \|f\|_{L^1(\mu)} = \big\|\Pi_{\mathfrak{I}(\mathbb{R})} f\big\|_{L^1(\mu;\mathbb{R})}, \quad n \in \mathbb{Z}^+,$$

where, in the last equality we used (6.1.7). □

We say that semigroup $\{\boldsymbol{\Sigma}^{\mathbf{k}} : \mathbf{k} \in \mathbb{N}^N\}$ is **ergodic** on $(\Omega, \mathcal{F}, \mu)$ if, in addition to being μ-measure preserving, $\mu(\Gamma) \in \{0, \mu(\Omega)\}$ for every invariant $\Gamma \in \mathcal{F}$.

6.1.11 Classic Example: In order to get a feeling for what the Ergodic Theorem is saying, take μ to be Lebesgue's measure on the interval $[0,1)$ and, for a given $\alpha \in (0,1)$, define $\boldsymbol{\Sigma}_\alpha : [0,1) \longrightarrow [0,1)$ so that

$$\boldsymbol{\Sigma}_\alpha(\omega) \equiv \omega + \alpha - [\omega + \alpha] = \omega + \alpha \bmod 1. \tag{6.1.12}$$

If α is rational and m is the smallest element of $\mathbb{Z}^+$ with the property that $m\alpha \in \mathbb{Z}^+$, then it is clear that, for any F on $[0,1)$, $F \circ \boldsymbol{\Sigma}_\alpha = F$ if and only if F has period $\frac{1}{m}$. Hence, if $F \in L^2\big([0,1);\mathbb{C}\big)$ and

$$c_\ell(F) \equiv \int_{[0,1)} F(\omega) e^{-\sqrt{-1}\,2\pi\ell\omega}\, d\omega, \quad \ell \in \mathbb{Z}, \tag{6.1.13}$$

then elementary Fourier analysis leads to the conclusion that, in this case:

$$\lim_{n\to\infty} \mathbf{A}_n F(\omega) = \sum_{\ell\in\mathbb{Z}} c_{m\ell}(F) e^{\sqrt{-1}\,2m\ell\pi\omega}$$
$$\text{for Lebesgue-almost every } \omega \in [0,1).$$

On the other hand, if α is irrational, then $\{\boldsymbol{\Sigma}_\alpha^k : k \in \mathbb{N}\}$ is μ-ergodic on $[0,1)$. To see this, suppose that $F \in \mathfrak{I}(\mathbb{C})$. Then (cf. (6.1.13) and use Parseval's identity)

$$0 = \big\|F - F \circ \boldsymbol{\Sigma}_\alpha\big\|^2_{L^2([0,1);\mathbb{C})} = \sum_{\ell\in\mathbb{Z}} \big|c_\ell(F) - c_\ell(F\circ\boldsymbol{\Sigma}_\alpha)\big|^2.$$

But, clearly,

$$c_\ell(F \circ \boldsymbol{\Sigma}_\alpha) = e^{\sqrt{-1}\,2\pi\ell\alpha} c_\ell(F), \quad \ell \in \mathbb{Z},$$

and so (because α is irrational) $c_\ell(F) = 0$ for each $\ell \neq 0$. In other words, the only elements of $\mathfrak{I}(\mathbb{C})$ are μ-almost everywhere constant. Thus, for each irrational $\alpha \in (0,1)$, $p \in [1,\infty)$, and $F \in L^p\big([0,1);E\big)$:

$$\lim_{n\to\infty} \mathbf{A}_n F = \int_{[0,1)} F(\omega)\, d\omega \tag{6.1.14}$$
$$\text{Lebesgue-almost everywhere and in } L^p(\mu;E).$$

Finally, notice that the situation changes radically when we replace $[0,1)$ by $[0,\infty)$ and again take μ to be Lebesgue's measure. If we continue to define $\boldsymbol{\Sigma}_\alpha(\omega)$ as in (6.1.12), only now for all $\omega \in [0,\infty)$, then it is clear that no choice of α leads to ergodicity: $[0,1)$ will always be a nontrivial invariant set. In addition, for any $p \in [1,\infty)$ and $F \in L^p(\mu;E)$, the limit behavior of $\mathbf{A}_n F$ in the present setting will be the same as the limit behavior of $\mathbf{A}_n\big(\mathbf{1}_{[0,1)}F\big)$ in the preceding. On the other hand, if we define $\boldsymbol{\Sigma}_\alpha(\omega) = \omega + \alpha$, then every invariant set which has nonzero measure will have infinite measure, and so, now, every choice of $\alpha \in (0,1)$ will give rise to an ergodic system. In particular, we will have, for each $p \in [1,\infty)$ and $F \in L^p(\mu;E)$,

$$\lim_{n\to\infty} \mathbf{A}_n F = 0 \quad \text{Lebesgue-almost everywhere;}$$

and the convergence is in $L^p(\mu;E)$ when $p \in (1,\infty)$.

For applications to probability theory, it is useful to reformulate these considerations in terms of stationary families of random variables. Thus, let $(\Omega,\mathcal{F},P)$ be a probability space and $(E,\mathcal{B})$ a measurable space (E need not be a Banach space). Given a family $\mathfrak{F} = \{X_{\mathbf{k}} : \mathbf{k} \in \mathbb{N}^N\}$ of E-valued random variables on $(\Omega,\mathcal{F},P)$, we say that $\mathfrak{F}$ is **P-stationary** (or simply **stationary**) if, for each $\boldsymbol{\ell} \in \mathbb{N}^N$, the family

$$\mathfrak{F}_{\boldsymbol{\ell}} \equiv \big\{X_{\mathbf{k}+\boldsymbol{\ell}} : \mathbf{k} \in \mathbb{N}^N\big\} \tag{6.1.15}$$

has the same (joint) distribution under P as $\mathfrak{F}$ itself. Clearly, one can test for stationarity by checking that distribution of $\mathfrak{F}_{\mathbf{e}_j}$ is the same as that of $\mathfrak{F}$ for each $1 \le j \le N$. In order to apply the considerations of this section to stationary families, note that all questions about the properties of $\mathfrak{F}$ can be phrased in terms of the following **canonical setting** . Namely, set $\mathbf{E} = E^{\mathbb{N}^N}$ and define μ on $\big(\mathbf{E},\mathcal{B}^{\mathbb{N}^N}\big)$ to be the image measure $\mathfrak{F}^*P$. In other words, for each $\Gamma \in \mathcal{B}^{\mathbb{N}^N}$, $\mu(\Gamma) = P\big(\mathfrak{F} \in \Gamma\big)$. Next, for each $\boldsymbol{\ell} \in \mathbb{N}^N$, define $\boldsymbol{\Sigma}^{\boldsymbol{\ell}} : \mathbf{E} \longrightarrow \mathbf{E}$ to be the **natural shift** transformation given by

$$\boldsymbol{\Sigma}^{\boldsymbol{\ell}}(\mathbf{x})_{\mathbf{k}} = x_{\mathbf{k}+\boldsymbol{\ell}} \quad \text{for all } \mathbf{k} \in \mathbb{N}^N \text{ and } \mathbf{x} \in \mathbf{E}. \tag{6.1.16}$$

Obviously, *stationarity of $\mathfrak{F}$ is equivalent to the statement that $\{\boldsymbol{\Sigma}^{\mathbf{k}} : \mathbf{k} \in \mathbb{N}^N\}$ is μ-measure preserving.* Moreover, if $\mathfrak{I}$ is the σ-algebra of **shift invariant** elements $\Gamma \in \mathcal{B}^{\mathbb{N}^N}$ (i.e., $\Gamma = \big(\boldsymbol{\Sigma}^{\mathbf{k}}\big)^{-1}(\Gamma)$ for all $\mathbf{k} \in \mathbb{N}^N$), then, for any Banach space B, any $p \in [1,\infty)$, and any $F \in L^p(\mu;B)$:

$$\lim_{n\to\infty} \frac{1}{n^N} \sum_{\mathbf{k}\in Q_n^+} F \circ \mathfrak{F}_{\mathbf{k}} = \mathbb{E}^P\Big[F \circ \mathfrak{F} \,\Big|\, \mathfrak{F}^{-1}(\mathfrak{I})\Big] \tag{6.1.17}$$

$$\text{(a.s., } P\text{) and in } L^p(P;B).$$

In particular, when $\{\boldsymbol{\Sigma}^{\mathbf{k}} : \mathbf{k} \in \mathbb{N}^N\}$ is ergodic on $\left(\mathbf{E}, \mathcal{B}^{\mathbb{N}^N} \mu\right)$, we say that the family $\mathfrak{F}$ is **ergodic** and conclude that (6.1.17) can be replaced by

$$\lim_{n\to\infty} \frac{1}{n^N} \sum_{\mathbf{k}\in Q_n^+} F \circ \mathfrak{F}_{\mathbf{k}} = \mathbb{E}^P\big[F \circ \mathfrak{F}\big] \quad \text{(a.s.,}, P) \text{ and in } L^p(P; B). \tag{6.1.18}$$

So far we have discussed *one-sided* stationary families, that is families indexed by $\mathbb{N}^N$. However, for various reasons (cf. Theorem 6.1.20 below) it is useful to know that *one can usually embed a one-sided stationary family into a two-sided one.* In terms of the *semigroup* of shifts, this corresponds to the trivial observation that the semigroup $\{\boldsymbol{\Sigma}^{\mathbf{k}} : \mathbf{k} \in \mathbb{N}^N\}$ on $\mathbf{E} = E^{\mathbb{N}^N}$ can be viewed as a subsemigroup of the *group* of shifts $\{\boldsymbol{\Sigma}^{\mathbf{k}} : \mathbf{k} \in \mathbb{Z}^N\}$ on $\hat{\mathbf{E}} = E^{\mathbb{Z}^N}$. With these comments in mind, we have the following.

6.1.19 Lemma. *Assume that E is a Polish space and that $\mathfrak{F} = \{X_{\mathbf{k}} : \mathbf{k} \in \mathbb{N}^N\}$ is a stationary family of E-valued random variables on the probability space $(\Omega, \mathcal{F}, P)$. Then there exists a probability space $(\hat{\Omega}, \hat{\mathcal{F}}, \hat{P})$ and a family $\hat{\mathfrak{F}} = \{\hat{X}_{\mathbf{k}} : \mathbf{k} \in \mathbb{Z}^N\}$ with the property that, for each $\boldsymbol{\ell} \in \mathbb{Z}^N$,*

$$\hat{\mathfrak{F}}_{\boldsymbol{\ell}} \equiv \{X_{\mathbf{k}+\boldsymbol{\ell}} : \mathbf{k} \in \mathbb{N}^N\}$$

has the same distribution under $\hat{P}$ as $\mathfrak{F}$ has under P.

PROOF: When formulated correctly, this theorem is an essentially trivial application of Kolmogorov's Extension Theorem (cf. Exercise 3.1.18). Namely, for $n \in \mathbb{N}$, set

$$\Lambda_n = \{\mathbf{k} \in \mathbb{Z}^N : k_j \geq -n \text{ for } 1 \leq j \leq N\},$$

and take

$$E_0 = E^{\Lambda_0} \quad \text{and} \quad E_n = E^{\Lambda_n \setminus \Lambda_{n-1}} \text{ for } n \in \mathbb{Z}^+.$$

Clearly $\mathbf{E}_n \equiv E^{\Lambda_n} = \prod_0^n E_m$. Moreover, if $\mathbf{n} \equiv (n, \dots, n)$ and $\mu_{[0,n]}$ is defined on $\mathbf{E}_n$ by

$$\mu_{[0,n]}(\Gamma) = P\big(\boldsymbol{\Sigma}^{-\mathbf{n}}\mathfrak{F} \in \Gamma\big) \quad \text{for } \Gamma \in \mathcal{B}_{\mathbf{E}_n},$$

then it is a trivial consequence of stationarity that

$$\mu_{[0,n+1]}\big(E_{n+1} \times \Gamma\big) = \mu_{[0,n]}(\Gamma) \quad \text{for all } n \in \mathbb{N} \text{ and } \Gamma \in \mathcal{B}_{\mathbf{E}_n}.$$

Hence, by Kolmogorov's Extension Theorem, there is a unique Borel probability measure $\hat{P}$ on $\hat{\Omega} \equiv E^{\mathbb{Z}^N} = \prod_0^\infty E_n$ such that

$$\hat{P}\big(E^{\mathbb{Z}^N \setminus \Lambda_n} \times \Gamma\big) = \mu_{[0,n]}(\Gamma) \quad \text{for all } n \in \mathbb{N} \text{ and } \Gamma \in \mathcal{B}_{\mathbf{E}_n}.$$

In other words, if we define $\hat{X}_{\mathbf{k}}(\hat{\omega}) = \hat{\omega}_{\mathbf{k}}$ for $\mathbf{k} \in \mathbb{Z}^N$ and $\hat{\omega} \in \Omega$, then we are done. □

As an example of the advantage which Lemma 6.1.19 affords, we present the following beautiful observation made by M. Kac.

6.1.20 Theorem. *Let $(E, \mathcal{B})$ be a measurable space and $\{X_k : k \in \mathbb{N}\}$ a stationary sequence of E-valued random variables on the probability space $(\Omega, \mathcal{F}, P)$. Given $\Gamma \in \mathcal{B}$, define the return time*

$$\rho_\Gamma(\omega) = \inf\{k \geq 1 : X_k(\omega) \in \Gamma\}.$$

Then,

$$\mathbb{E}^P\big[\rho_\Gamma,\, X_0 \in \Gamma\big] = P\big(X_k \in \Gamma \text{ for some } k \in \mathbb{N}\big). \tag{6.1.21}$$

In particular, if $\{X_k : k \in \mathbb{N}\}$ is ergodic, then

$$P\big(X_0 \in \Gamma\big) > 0 \implies \mathbb{E}^P\big[\rho_\Gamma,\, X_0 \in \Gamma\big] = 1. \tag{6.1.22}$$

PROOF: Set $U_k = \mathbf{1}_\Gamma \circ X_k$ for $k \in \mathbb{N}$. Then $\{U_k : k \in \mathbb{N}\}$ is a stationary sequence of $\{0,1\}$-valued random variables. Hence, by Lemma 6.1.19, we can find a probability space $(\hat{\Omega}, \hat{\mathcal{F}}, \hat{P})$ on which there is a family $\{\hat{U}_k : k \in \mathbb{Z}\}$ of $\{0,1\}$-valued random variables with the property that, for every $n \in \mathbb{Z}$, $(\hat{U}_n, \dots, \hat{U}_{n+k}, \dots)$ has the same distribution under $\hat{P}$ as $(U_0, \dots, U_k, \dots)$ has under P. In particular,

$$P\big(\rho_\Gamma \geq 1,\, X_0 \in \Gamma\big) = \hat{P}\big(\hat{U}_0 = 1\big)$$
$$P\big(\rho_\Gamma \geq n+1,\, X_0 \in \Gamma\big) = \hat{P}\big(\hat{U}_{-n} = 1, \hat{U}_{-n+1} = 0, \dots, \hat{U}_0 = 0\big), \quad n \in \mathbb{Z}^+.$$

Thus, if

$$\lambda_\Gamma(\hat{\omega}) \equiv \sup\big\{k \in \mathbb{N} : U_{-k}(\hat{\omega}) = 1\big\},$$

then

$$P\big(\rho_\Gamma \geq n,\, X_0 \in \Gamma\big) = \hat{P}\big(\lambda_\Gamma = n-1\big), \quad n \in \mathbb{Z}^+;$$

and so

$$\mathbb{E}^P\big[\rho_\Gamma,\, X_0 \in \Gamma\big] = \hat{P}\big(\lambda_\Gamma < \infty\big).$$

Finally, note that

$$\hat{P}\big(\lambda_\Gamma > n\big) = \hat{P}\big(\hat{U}_{-n} = 0, \dots, \hat{U}_0 = 0\big) = P\big(X_0 \notin \Gamma, \dots, X_n \notin \Gamma\big),$$

from which it is clear that

$$\hat{P}\big(\lambda_\Gamma < \infty\big) = P\big(\exists k \in \mathbb{N}\ X_k \in \Gamma\big).$$

To handle the case when $\{X_k : k \in \mathbb{N}\}$ is ergodic and $\Gamma \in \mathcal{B}$ satisfies $P(X_0 \in \Gamma) > 0$, simply note that, by (6.1.18), $\sum_0^\infty \mathbf{1}_\Gamma(X_k) = \infty$ P-almost surely and therefore that the right-hand side of (6.1.21) is 1. □

We turn now to the setting of continuously parameterized semigroups of transformations. Thus, again $(\Omega, \mathcal{F}, \mu)$ is a σ-finite measure space and $\{\mathbf{\Sigma}^{\mathbf{t}} : \mathbf{t} \in [0,\infty)^N\}$ is a measurable semigroup of μ-measure preserving transformations on Ω. That is, $\mathbf{\Sigma}^{\mathbf{0}}$ is the identity, $\mathbf{\Sigma}^{\mathbf{s}+\mathbf{t}} = \mathbf{\Sigma}^{\mathbf{s}} \circ \mathbf{\Sigma}^{\mathbf{t}}$,

$$(\mathbf{t}, \omega) \in [0,\infty)^N \times \Omega \longmapsto \mathbf{\Sigma}^{\mathbf{t}}(\omega) \in \Omega \quad \text{is } \mathcal{B}_{[0,\infty)^N} \times \mathcal{F}\text{-measurable},$$

and $\left(\mathbf{\Sigma}^{\mathbf{t}}\right)^* \mu = \mu$ for every $\mathbf{t} \in [0,\infty)^N$. Next, given an $\mathcal{F}$-measurable F with values in some separable Banach space E, let $\mathfrak{G}(F)$ be the set of $\omega \in \Omega$ with the property that

$$\int_{[0,T)^N} \left\| F \circ \mathbf{\Sigma}^{\mathbf{t}}(\omega) \right\|_E d\mathbf{t} < \infty \quad \text{for all } T \in (0,\infty).$$

Clearly,

$$\omega \in \mathfrak{G}(F) \implies \mathbf{\Sigma}^{\mathbf{t}}(\omega) \in \mathfrak{G}(F) \quad \text{for every } \mathbf{t} \in [0,\infty)^N. \tag{6.1.23}$$

In addition, if $F \in L^p(\mu; E)$ for some $p \in [1,\infty)$, then

$$\int_\Omega \left(\int_{[0,T)^N} \left\| F \circ \mathbf{\Sigma}^{\mathbf{t}}(\omega) \right\|_E^p d\mathbf{t} \right) \mu(d\omega) = T^N \|F\|_{L^p(\mu;E)}^p < \infty,$$

and so

$$F \in \bigcup_{p \in [1,\infty)} L^p(\mu; E) \implies \mu\Big(\mathfrak{G}(F)\complement\Big) = 0.$$

Next, for each $T \in (0,\infty)$, define

$$\mathcal{A}_T F(\omega) = \begin{cases} T^{-N} \int_{[0,T)^N} F \circ \mathbf{\Sigma}^{\mathbf{t}}(\omega)\, d\mathbf{t} & \text{if } \omega \in \mathfrak{G}(F) \\ 0 & \text{if } \omega \notin \mathfrak{G}(F). \end{cases} \tag{6.1.24}$$

Note that, as a consequence of (6.1.23),

$$\left(\mathcal{A}_T F\right) \circ \mathbf{\Sigma}^{\mathbf{t}} = \mathcal{A}_T\left(F \circ \mathbf{\Sigma}^{\mathbf{t}}\right) \quad \text{for all } \mathbf{t} \in [0,\infty)^N. \tag{6.1.25}$$

Finally, use $\hat{\mathfrak{J}}$ to denote the σ-algebra of $\Gamma \in \mathcal{F}$ with the property that $\Gamma = (\mathbf{\Sigma}^{\mathbf{t}})^{-1}(\Gamma)$ for each $t \in [0,\infty)^N$, and say that $\{\mathbf{\Sigma}^{\mathbf{t}} : \mathbf{t} \in [0,\infty)^N\}$ is **ergodic** if $\mu(\Gamma) \in \{0, \mu(\Omega)\}$ for every $\Gamma \in \hat{\mathfrak{J}}$.

6.1.26 Theorem. *Let $(\Omega, \mathcal{F}, \mu)$ be a σ-finite measure space and $\{\boldsymbol{\Sigma}^{\mathbf{t}} : \mathbf{t} \in [0,\infty)^N\}$ be a measurable semigroup of μ-measure preserving transformations on Ω. Then, for each separable Banach space E, $p \in [1,\infty)$, and $T \in (0,\infty)$, $\mathcal{A}_T$ is a contraction on $L^p(\mu; E)$. Next, set $\Pi_{\hat{\mathfrak{I}}(E)} = \Pi_{\mathfrak{I}(E)} \circ \mathcal{A}_1$, where $\Pi_{\mathfrak{I}(E)}$ is defined in terms of $\{\boldsymbol{\Sigma}^{\mathbf{k}} : \mathbf{k} \in \mathbb{N}^N\}$ as in Theorem 6.1.8. Then, for each $p \in [1,\infty)$ and $F \in L^p(\mu; E)$:*

$$\lim_{T\to\infty} \mathcal{A}_T F = \Pi_{\hat{\mathfrak{I}}(E)} F \quad \text{(a.e., } \mu). \tag{6.1.27}$$

Moreover, if $p \in (1,\infty)$ or $\mu(E) < \infty$, then the convergence is also in $L^p(\mu; E)$. In fact, if $\mu(\Omega) < \infty$, then

$$\lim_{T\to\infty} \mathcal{A}_T F = \mathbb{E}^{\mu}\big[F \,\big|\, \hat{\mathfrak{I}}\big] \quad \text{(a.e., } \mu) \text{ and in } L^p(\mu : E). \tag{6.1.28}$$

Finally, if $\{\boldsymbol{\Sigma}^{\mathbf{t}} : \mathbf{t} \in [0,\infty)^N\}$ is ergodic, then (6.1.27) can be replaced by

$$\lim_{T\to\infty} \mathcal{A}_T F = \frac{\mathbb{E}^{\mu}[F]}{\mu(\Omega)}, \tag{6.1.29}$$

*where it is understood that the ratio is **0** when the denominator is infinite.*

PROOF: The first step is the observation that

$$\mu\left(\sup_{T>0} \big\|\mathcal{A}_T F\big\|_E \ge \lambda\right) \le \frac{(24)^N}{\lambda} \|F\|_{L^1(\mu;E)}, \quad \lambda \in (0,\infty), \tag{6.1.30}$$

and

$$\left\| \sup_{T>0} \big\|\mathcal{A}_T F\big\|_E \right\|_{L^p(\mu;E)} \le \frac{(24)^N p}{p-1} \|F\|_{L^p(\mu;E)} \quad \text{for } p \in (1,\infty). \tag{6.1.31}$$

Indeed (because of (6.1.25)), (6.1.30) is derived from (5.3.18) in precisely the same way as we derived (6.1.3) from (5.2.23), and (6.1.31) comes from (5.3.19) just as (6.1.4) came from (5.3.24).

Given (6.1.30) and (6.1.31), we know that it suffices to prove (6.1.27) in the case when F is a uniformly bounded element of $L^1(\mu; E)$. But in that case, set $\hat{F} = \mathcal{A}_1 F$ and observe that

$$\big\|T^N \mathcal{A}_T F(\omega) - n^N \mathbf{A}_n \hat{F}(\omega)\big\|_E \le \int_{[0,n+1)^N \setminus [0,n)^N} \|F \circ \boldsymbol{\Sigma}^{\mathbf{t}}(\omega)\|_E \, d\mathbf{t}$$

for $n \le T \le n+1$, and conclude that

$$\lim_{n\to\infty} \left\| \sup_{n \le T \le n+1} \big\|\mathcal{A}_T F - \mathbf{A}_n \hat{F}\big\|_E \right\|_{L^p(\mu;E)} = 0 \quad \text{for every } p \in [1,\infty].$$

Hence, (6.1.27) follows from (6.1.9). As for (6.1.28), all that we have to do is check that

$$\Pi_{\hat{\mathfrak{J}}(E)} F = \mathbb{E}^{\mu}\big[F \big| \hat{\mathfrak{J}}(E)\big] \quad \text{(a.e., } \mu)$$

when $\mu(\Omega) < \infty$. However, from (6.1.27), it is easy to see that $\Pi_{\hat{\mathfrak{J}}(E)} F$ is measurable with respect to the μ-completion of $\hat{\mathfrak{J}}$; and so it suffices to show that

$$\mathbb{E}^{\mu}\big[F, \Gamma\big] = \mathbb{E}^{\mu}\big[\mathcal{A}_1 F, \Gamma\big] \quad \text{for all } \Gamma \in \hat{\mathfrak{J}}.$$

But, if $\Gamma \in \hat{\mathfrak{J}}$, then

$$\begin{aligned}\mathbb{E}^{\mu}\big[\mathcal{A}_1 F, \Gamma\big] &= \int_{[0,1)^N} \mathbb{E}^{\mu}\big[F \circ \boldsymbol{\Sigma}^{\mathbf{t}}, \Gamma\big]\, d\mathbf{t} \\ &= \int_{[0,1)^N} \mathbb{E}^{\mu}\big[F \circ \boldsymbol{\Sigma}^{\mathbf{t}}, \big(\boldsymbol{\Sigma}^{\mathbf{t}}\big)^{-1}(\Gamma)\big]\, d\mathbf{t} = \mathbb{E}^{\mu}[F, \Gamma].\end{aligned}$$

Finally, assume that $\{\boldsymbol{\Sigma}^{\mathbf{t}} : \mathbf{t} \in [0,\infty)^N\}$ is μ-ergodic. When $\mu(\Omega) < \infty$, (6.1.29) follows immediately from (6.1.28); and when $\mu(\Omega) = \infty$, it follows from the fact that $\Pi_{\hat{\mathfrak{J}}(E)} F$ is measurable with respect to the μ-completion of $\hat{\mathfrak{J}}$. □

Exercises

6.1.32 Exercise: Assume that $\mu(\Omega) < \infty$ and that $\{\boldsymbol{\Sigma}^{\mathbf{k}} : \mathbf{k} \in \mathbb{N}^N\}$ is ergodic. Given a nonnegative $\mathcal{F}$-measurable function f, show that

$$\begin{aligned}\overline{\lim_{n\to\infty}}\, \mathbf{A}_n f < \infty \text{ on a set of positive } \mu\text{-measure} &\implies f \in L^1(\mu;\mathbb{R}) \\ &\implies \lim_{n\to\infty} \mathbf{A}_n f = \mathbb{E}^{\mu}[f] \quad \text{(a.e., } \mu).\end{aligned}$$

6.1.33 Exercise: Let $\mathfrak{F} = \{X_{\mathbf{k}} : \mathbf{k} \in \mathbb{N}^N\}$ be a stationary family of random variables on the probability space $(\Omega, \mathcal{F}, P)$ with values in the measurable space $(E, \mathcal{B})$; and let $\mathfrak{J}$ denote the σ-algebra of shift invariant $\Gamma \in \mathcal{B}_E^{\mathbb{N}^N}$.

(i) Let

$$\mathcal{T} \equiv \bigcap_{n \geq 0} \sigma\big(X_{\mathbf{k}} : k_j \geq n \text{ for all } 1 \leq j \leq N\big)$$

be the tail field determined by $\{X_{\mathbf{k}} : \mathbf{k} \in \mathbb{N}^N\}$. Show that $\mathcal{F}^{-1}(\mathfrak{J}) \subseteq \mathcal{T}$, and conclude that $\{X_{\mathbf{k}} : \mathbf{k} \in \mathbb{N}^N\}$ is ergodic if $\mathcal{T}$ is P-trivial (i.e., $P(\Gamma) \in \{0,1\}$ for all $\Gamma \in \mathcal{T}$.)

(ii) By combining (i), Kolmogorov's 0–1 Law, and the Individual Ergodic Theorem, give another derivation of the Strong Law of Large Numbers for independent, identically distributed, integrable random variables with values in a separable Banach space.

6.1.34 Exercise: Let $\{X_k : k \in \mathbb{N}\}$ be a stationary, ergodic sequence of $\mathbb{R}$-valued, integrable random variables on $(\Omega, \mathcal{F}, P)$. Using the reasoning suggested in Exercise 1.4.31, show that

$$\mathbb{E}^P\big[X_1\big] = 0 \implies \varliminf_{n\to\infty} \left|\sum_{k=0}^{n-1} X_k\right| < \infty.$$

6.1.35 Exercise: Given an irrational $\alpha \in (0,1)$ and an $\epsilon \in (0,1)$, let $N_n(\alpha, \epsilon)$ be the number of $1 \le m \le n$ with the property that

$$\left|\alpha - \frac{\ell}{m}\right| \le \frac{\epsilon}{2m} \quad \text{for some } \ell \in \mathbb{Z}.$$

As an application of the considerations in Example 6.1.11, show that

$$\varlimsup_{n\to\infty} \frac{N_n(\alpha,\epsilon)}{n} \ge \epsilon.$$

Hint: Let $\delta \in \big(0, \frac{\epsilon}{2}\big)$ be given, take f equal to the indicator function of $[0,\delta) \cup (1-\delta, 1)$, and observe that

$$N_n(\alpha,\epsilon) \ge \sum_{k=1}^{n} f \circ \mathbf{\Sigma}_\alpha^k(\omega)$$

so long as $0 \le \omega \le \frac{\epsilon}{2} - \delta$.

§6.2: Singular Integrals & Square Functions in Analysis.

The topic of this section is somewhat out of place in a book on probability theory. However, the material here is not only interesting in its own right, it has been included to indicate the kind of delicate cancellation properties which underlie the most challenging applications of martingale theory. At the same time, the considerations here serve as an introduction to the ones in Section 6.3.[†]

We will begin with singular integral operators. In general, an operator involves *singular integrals* whenever its expression in terms of its kernel fails to be Lebesgue integrable. Thus, for example, the Fourier transform on $L^2(\mathbb{R}^N;\mathbb{C})$ involves singular integrals. However, ever since the ground-breaking work of Calderón and Zygmund, the term *singular integral operator* has been reserved for operators whose kernels have a very specific type of singularity and equally special cancellation properties. The archetypical example of such an operator is the **Hilbert transform** given by

$$\text{``}\,[Hf](x) = \frac{1}{\pi}\int_{\mathbb{R}} \frac{f(\xi)}{x-\xi}\,d\xi, \quad x\in\mathbb{R}.\text{''}$$

The reason for the quotation marks is obvious: the indicated integral cannot be interpreted in the sense of Lebesgue. On the other hand, if one adopts the Riemannian attitude that such integrals should be defined by first excising the singularity and then passing to a limit, and if one assumes, for instance, that $f\in C^1(\mathbb{R};\mathbb{C})$ satisfies

$$\lim_{|x|\to\infty}\big|f(x)\log|x|\big| = 0 \quad\text{and}\quad \int_{\mathbb{R}}\big|f'(x)\log|x|\big|\,dx<\infty,$$

then one comes to the conclusion that

$$[Hf](x) = -\frac{1}{\pi}\int_{\mathbb{R}}\log\big(|x-\xi|\big)\,f'(\xi)\,d\xi.$$

That is, in the language of Schwartz's theory of tempered distributions, the Hilbert transform acts on test functions by convolution with respect to the tempered distribution $\frac{1}{\pi x} = \frac{1}{\pi}\big(\log|x|\big)'$.

However one chooses to describe the Hilbert transform, it should be emphasized that its kernel $\frac{1}{\pi(x-\xi)}$ has two important properties: it fails to be integrable because of *logarithmic divergences at the diagonal and infinity* and it has *strong cancellation properties.* These two features are common to all the operators studied by Calderón and Zygmund, and, as will we see, the damage done by the first is balanced by the good done by the second in such a way that the resulting operator possesses unexpectedly good mapping properties on the Lebesgue spaces. However, before getting any further into their theory, we will first introduce the Riesz transformations, which are the higher dimensional analogs of the Hilbert transform. Namely, for $N\in\mathbb{Z}^+$ and $1\le j\le N$, set

$$q_j^{(N)}(\mathbf{x}) = \frac{2}{\omega_N}\frac{x_j}{|\mathbf{x}|^{N+1}}, \quad \mathbf{x}\in\mathbb{R}^N\setminus\{\mathbf{0}\}, \tag{6.2.1}$$

[†] Most of this material is a diluted version of material in E.M. Stein's beautiful *Singular Integrals and Differentiability Properties of Functions*, publ. by Princeton Univ. Press. (1970)

where ω_N is the surface area of the unit sphere $\mathbf{S}^N$ in $\mathbb{R}^{N+1}$. We have already seen that, as the distributional derivative of a tempered, locally integrable function, $q_1^{(1)}$ is a tempered distribution; and a similar argument proves that $q_j^{(N)}$ is a tempered distribution for each $N \geq 2$ and $1 \leq j \leq N$. In fact, if we define

$$q_{j,y}^{(N)} \in C_b^\infty(\mathbb{R}^N;\mathbb{R}) \cap L^2(\mathbb{R}^N;\mathbb{R}) \quad \text{for } y \in (0,\infty)$$

by

$$q_{j,y}^{(N)}(\mathbf{x}) = \frac{2}{\omega_N} \frac{x_j}{\left(y^2 + |\mathbf{x}|^2\right)^{\frac{N+1}{2}}}, \quad \mathbf{x} \in \mathbb{R}^N, \tag{6.2.2}$$

then it is an easy matter to check that

$$q_{j,y}^{(N)} \xrightarrow{\mathcal{S}'(\mathbb{R}^N;\mathbb{R})} q_j^{(N)} \quad \text{as } y \searrow 0, \tag{6.2.3}$$

where the indicated convergence is meant in the sense of tempered distributions. Thus, if we define the operators $Q_{j,y}^{(N)} : L^1(\mathbb{R}^N\mathbb{C}) \longrightarrow C_b(\mathbb{R}^N;\mathbb{C})$ by

$$Q_{j,y}^{(N)} f = q_{j,y}^{(N)} \star f, \quad 1 \leq j \leq N \text{ and } y \in (0,\infty) \tag{6.2.4}$$

(the convolution being defined here in the usual Lebesgue sense), then we know that, for f from Schwartz's test-function class $\mathcal{S}(\mathbb{R}^N;\mathbb{C})$ of infinitely differentiable functions having rapidly decreasing derivatives of all orders,

$$Q_{j,y}^{(N)} f \longrightarrow Q_j^{(N)} f \equiv q_j^{(N)} \star f \quad \text{uniformly on compacts,} \tag{6.2.5}$$

where the convolution on the right is taken in the sense of tempered distributions acting on $\mathcal{S}(\mathbb{R}^N;\mathbb{C})$. Obviously, $Q_1^{(1)}$ is just the Hilbert transform H described before; and, when $N \geq 2$, the operators $Q_j^{(N)}$ is called the jth **Riesz transform**.

The first step toward a better understanding of the Riesz transforms is to see how they act on $L^2(\mathbb{R}^N;\mathbb{C})$; and, for this purpose, we begin by computing the Fourier transform of $q_j^{(N)}$.

6.2.6 Lemma. *For each $y \in (0,\infty)$,*

$$\widehat{q_{j,y}^{(N)}}(\boldsymbol{\xi}) = \frac{\sqrt{-1}\,\xi_j}{|\boldsymbol{\xi}|} e^{-y|\boldsymbol{\xi}|}, \quad \boldsymbol{\xi} \in \mathbb{R}^N \setminus \{\mathbf{0}\}; \tag{6.2.7}$$

and therefore,

$$\widehat{q_j^{(N)}}(\boldsymbol{\xi}) = \frac{\sqrt{-1}\,\xi_j}{|\boldsymbol{\xi}|}, \quad \boldsymbol{\xi} \in \mathbb{R}^N \setminus \{\mathbf{0}\}, \tag{6.2.8}$$

is a representative of the Fourier transform (in the sense of tempered distributions) of $q_j^{(N)}$. In particular,

$$\delta_{\mathbf{0}} = -\sum_{j=1}^N q_j^{(N)} \star q_j^{(N)} \quad \text{and} \quad q_{j,s+t}^{(N)} = p_s^{(N)} \star q_{j,t}^{(N)}, \; s,\, t \in (0,\infty), \tag{6.2.9}$$

where $p_y^{(N)}$, $y \in (0,\infty)$, is the Poisson kernel defined in (4.3.50).

PROOF: Because of (6.2.3), (6.2.8) follows immediately from (6.2.7). In addition, by taking Fourier transforms throughout, the first part of (6.2.9) follows from (6.2.8); and to prove the second part, one again uses Fourier transforms together with the computations in (6.2.7) and Exercise 4.3.49. Thus, all that remains is to prove (6.2.7). To this end, note that

$$q_{j,y}^{(N)}(\mathbf{x}) = \frac{x_j\, p_y^{(N)}(\mathbf{x})}{y}$$

and therefore, because

$$\widehat{x_j\, u}(\boldsymbol{\xi}) = -\sqrt{-1}\,\frac{\partial \hat{u}}{\partial \xi_j}(\boldsymbol{\xi}) \quad \text{for } u \in \mathcal{S}'(\mathbb{R}^N;\mathbb{C}),$$

that (cf. Exercise 4.3.49)

$$\widehat{q_{j,y}^{(N)}}(\boldsymbol{\xi}) = -\frac{\sqrt{-1}}{y}\,\frac{\partial \widehat{p_y^{(N)}}}{\partial \xi_j}(\boldsymbol{\xi}) = \frac{\sqrt{-1}\,\xi_j}{|\boldsymbol{\xi}|}\,e^{-y|\boldsymbol{\xi}|}. \quad \square$$

As an immediate consequence of (6.2.8), we see that each of the operators $Q_j^{(N)}$ determines a unique extension as a skew-adjoint contraction from $L^2(\mathbb{R}^N;\mathbb{C})$ to itself. Moreover, because of (6.2.7), we know that $Q_j^{(N)}$ is the limit, in the strong operator norm, of the operators $Q_{j,y}^{(N)}$ as $y \searrow 0$. In fact, by (6.2.9), we see that

$$Q_{j,y}^{(N)} f = p_y^{(N)} \star Q_j^{(N)} f, \quad y \in (0,\infty),$$

and therefore, by part (**ii**) of Exercise 5.3.39, we have that

$$\text{(6.2.10)} \qquad \left\| \sup_{y>0} Q_{j,y}^{(N)} f \right\|_{L^2(\mathbb{R}^N)} \le 2N(12)^N K_N \|f\|_{L^2(\mathbb{R}^N)}, \quad f \in L^2(\mathbb{R}^N;\mathbb{C}).$$

Hence, since $q_{j,y}^{(N)} \star f \longrightarrow Q_j^{(N)} f$ pointwise for each $f \in \mathcal{S}(\mathbb{R}^N;\mathbb{C})$, we see (cf. the argument given in the proof of Theorem 5.3.20) that

$$\text{(6.2.11)} \qquad q_{j,y}^{(N)} \star f \longrightarrow Q_j^{(N)} f \quad \text{as } y \searrow 0 \quad \text{(a.e., Leb)}$$

for each $f \in L^2(\mathbb{R}^N;\mathbb{C})$. On the other hand, the situation is not so clear when it comes to the other Lebesgue spaces. Indeed, because $p_1^{(N)} \in L^1(\mathbb{R}^N;\mathbb{R}) \cap C_b(\mathbb{R}^N;\mathbb{R})$ and yet, by (6.2.9),

$$Q_j^{(N)} p_1^{(N)} = q_{j,1}^{(N)} \notin L^1(\mathbb{R}^N;\mathbb{C}),$$

we know that $Q_j^{(N)}$ *cannot* be extended as a bounded operator from the space $L^1(\mathbb{R}^N;\mathbb{C})$ into itself. Thus, if we are going to give a meaning to $Q_j^{(N)} f$ for every $f \in L^1(\mathbb{R}^N;\mathbb{C})$, then we are going to have to use a line of reasoning which is subtler than the one which worked in the case of $L^2(\mathbb{R}^N;\mathbb{C})$; and that subtler reasoning is contained in the following variation on Calderón and Zygmund's basic application of their decomposition theorem (cf. Theorem 5.3.26).

6.2.12 Theorem. *Let E and F be a pair of separable Banach spaces, and use $\mathrm{Hom}(E;F)$ to denote the Banach space of all bounded, linear operators from E into F. Next, let*

$$\mathbf{x} \in \mathbb{R}^N \longmapsto \mathbf{k}(\mathbf{x}) \in \mathrm{Hom}(E;F)$$

be a bounded, continuous mapping; and define

$$\mathbf{K} : L^1(\mathbb{R}^N;E) \longrightarrow C_b(\mathbb{R}^N;E)$$

so that, for each $\mathbf{f} \in L^1(\mathbb{R}^N;E)$,

$$[\mathbf{K}f](\mathbf{x}) = \int_{\mathbb{R}^N} \mathbf{k}(\mathbf{x}-\boldsymbol{\xi})\,\mathbf{f}(\boldsymbol{\xi})\,d\boldsymbol{\xi}, \quad \mathbf{x} \in \mathbb{R}^N.$$

Finally, assume that there exist A, $B \in (0,\infty)$ with the properties that

$$\|\mathbf{Kf}\|_{L^2(\mathbb{R}^N;F)} \le A\|\mathbf{f}\|_{L^2(\mathbb{R}^N;E)}, \quad \mathbf{f} \in L^1(\mathbb{R}^N;E) \cap L^2(\mathbb{R}^N;E), \tag{6.2.13}$$

and

$$\|\mathbf{k}(\mathbf{x}) - \mathbf{k}(\mathbf{x}-\mathbf{a})\|_{E\to F} \le \frac{B|\mathbf{a}|}{|\mathbf{x}|^{N+1}} \quad \text{for all } \mathbf{a} \in \mathbb{R}^N \setminus \{\mathbf{0}\} \text{ and } |\mathbf{x}| \ge 2|\mathbf{a}|. \tag{6.2.14}$$

(The norm $\|\cdot\|_{E\to F}$ is the operator norm on $\mathrm{Hom}(E;F)$.) Then

$$\left|\{\mathbf{x} \in \mathbb{R}^N : \|[\mathbf{Kf}](\mathbf{x})\|_F \ge R\}\right| \le \frac{C\|\mathbf{f}\|_{L^1(\mathbb{R}^N;E)}}{R} \quad \text{for } R \in (0,\infty) \text{ and } \mathbf{f} \in L^1(\mathbb{R}^N;E), \tag{6.2.15}$$

where the constant $C \in (0,\infty)$ depends only on A, B, and N.

PROOF: We begin by observing that, by (6.2.14) and a trivial change of variables, for each $\mathbf{c} \in \mathbb{R}^N$ and $\boldsymbol{\xi} \in \mathbb{R}^N \setminus \{\mathbf{c}\}$,

$$\int_{\{\mathbf{x}:|\mathbf{x}-\boldsymbol{\xi}|\ge 2|\mathbf{x}-\mathbf{c}|\}} \left\|\mathbf{k}(\mathbf{x}-\boldsymbol{\xi}) - \mathbf{k}(\mathbf{x}-\mathbf{c})\right\|_{E\to F} d\mathbf{x} \le B', \tag{6.2.16}$$

where $B' = B\boldsymbol{\Omega}_N$.

Now, let $\mathbf{f} \in L^1(\mathbb{R}^N;E)$ and $R \in (0,\infty)$ be given, and choose the cubes $\mathcal{Q}(R)$ and the functions $\mathbf{g}$ and $\{\mathbf{h}_Q : Q \in \mathcal{Q}(R)\}$ accordingly, as in Theorem 5.3.26. Next, for each $Q \in \mathcal{Q}(R)$, choose $\mathbf{c}(Q) \in \mathbb{R}^N$ and $r(Q) > 0$ so that

$$Q = \prod_{j=1}^N \left[c(Q)_j - \tfrac{r(Q)}{2}, c(Q)_j + \tfrac{r(Q)}{2}\right),$$

and define

$$\hat{Q} = \prod_{j=1}^{N} \Big[c(Q)_j - \sqrt{N}\, r(Q), c(Q)_j + \sqrt{N}\, r(Q) \Big].$$

Finally, set $\hat{B} = \bigcup \{\hat{Q} : Q \in \mathcal{Q}(R)\}$ and $\mathbf{h} = \mathbf{f} - \mathbf{g}$. Then,

$$\begin{aligned} &\big|\{\mathbf{x} : \|[\mathbf{Kf}](\mathbf{x})\|_F \geq 2R\}\big| \\ &\quad \leq \big|\{\mathbf{x} : \|[\mathbf{Kg}](\mathbf{x})\|_F \geq R\}\big| + \big|\{\mathbf{x} : \|[\mathbf{Kh}](\mathbf{x})\|_F \geq R\}\big| \\ &\quad \leq \frac{\|\mathbf{Kg}\|^2_{L^2(\mathbb{R}^N;F)}}{R^2} + |\hat{B}| + \frac{1}{R} \int_{\hat{B}\complement} \|\mathbf{Kh}(\mathbf{x})\|_F \, d\mathbf{x}. \end{aligned}$$

Because (cf. Theorem 5.3.26)

$$\|\mathbf{Kg}\|^2_{L^2(\mathbb{R}^N;F)} \leq A^2 \|\mathbf{g}\|^2_{L^2(\mathbb{R}^N;E)} \leq 2^N A^2 R \|\mathbf{f}\|_{L^1(\mathbb{R}^N;E)}$$

while

$$|\hat{B}| \leq \left(2\sqrt{N}\right)^N \sum_{Q \in \mathcal{Q}(R)} |Q| \leq \frac{\left(2\sqrt{N}\right)^N \|\mathbf{f}\|_{L^1(\mathbb{R}^N;E)}}{R},$$

all that remains is to estimate

$$\int_{\hat{B}\complement} \|\mathbf{Kh}(\mathbf{x})\|_F \, d\mathbf{x} \leq \sum_{Q \in \mathcal{Q}(R)} \int_{\hat{Q}\complement} \|\mathbf{Kh}_Q(\mathbf{x})\|_F \, d\mathbf{x}$$

in terms of $\|\mathbf{f}\|_{L^1(\mathbb{R}^N;E)}$. To this end, note that, because $\int_Q \mathbf{h}_Q(\boldsymbol{\xi})\, d\boldsymbol{\xi} = \mathbf{0}$,

$$\begin{aligned} &\int_{\hat{Q}\complement} \|\mathbf{Kh}_Q(\mathbf{x})\|_F \, d\mathbf{x} \\ &\quad = \int_{\hat{Q}\complement} \left\| \int_Q \Big(\mathbf{k}(\mathbf{x} - \boldsymbol{\xi}) - \mathbf{k}\big(\mathbf{x} - \mathbf{c}(Q)\big) \Big) \mathbf{h}_Q(\boldsymbol{\xi})\, d\boldsymbol{\xi} \right\|_F d\mathbf{x} \\ &\quad \leq \int_Q \|\mathbf{h}_Q(\boldsymbol{\xi})\|_E \left(\int_{\hat{Q}\complement} \big\|\mathbf{k}(\mathbf{x} - \boldsymbol{\xi}) - \mathbf{k}\big(\mathbf{x} - \mathbf{c}(Q)\big)\big\|_{E \to F} \, d\mathbf{x} \right) d\boldsymbol{\xi} \\ &\quad \leq B' \|\mathbf{h}_Q\|_{L^1(\mathbb{R}^N;E)}, \end{aligned}$$

where, at the final step, we have used (6.2.16). Hence, because

$$\sum_{Q \in \mathcal{Q}(R)} \|\mathbf{h}_Q\|_{L^1(\mathbb{R}^N;E)} \leq 2\|\mathbf{f}\|_{L^1(\mathbb{R}^N;E)},$$

we are done. □

As our first application of Theorem 6.2.12, we will continue our analysis of the Riesz transforms.

6.2.17 Theorem. *There is a $C \in (0,\infty)$, depending only on $N \in \mathbb{Z}^+$, such that*

$$\left|\left\{\mathbf{x} \in \mathbb{R}^N : \sup_{y>0} \left|\left[Q^{(N)}_{j,y} f(\mathbf{x})\right]\right| \geq R\right\}\right| \leq \frac{C\|f\|_{L^1(\mathbb{R}^N)}}{R} \tag{6.2.18}$$

for every $1 \leq j \leq N$, $f \in L^1(\mathbb{R}^N;\mathbb{C})$, and $R \in (0,\infty)$. Hence, for each $f \in L^1(\mathbb{R}^N;\mathbb{C})$, there is a measurable function $Q^{(N)}_j f : \mathbb{R}^N \longrightarrow \mathbb{C}$ with the property that, for Leb-almost every $\mathbf{x} \in \mathbb{R}^N$, the mapping

$$y \in [0,\infty) \longmapsto \begin{cases} [Q^{(N)}_j f](\mathbf{x}) & \text{if } \; y = 0 \\ [Q^{(N)}_{j,y} f](\mathbf{x}) & \text{if } \; y > 0 \end{cases} \in \mathbb{C}$$

is continuous and tends to 0 as $y \nearrow \infty$.

PROOF: Given the estimate in (6.2.18), the second assertion follows easily from the fact that it is true when $f \in \mathcal{S}(\mathbb{R}^N;\mathbb{C})$. To prove the first assertion, let F be the separable Banach space $C_0([0,\infty);\mathbb{C})$ of continuous functions on $[0,\infty)$ which vanish at infinity, and, for each $t \in (0,\infty)$, define $\mathbf{k}_t : \mathbb{R}^N \longrightarrow F$ by

$$\big[\mathbf{k}_t(\mathbf{x})\big](y) = q^{(N)}_{j,t+y}(\mathbf{x}), \quad y \in [0,\infty).$$

Clearly, $\mathbf{k}_t$ is a bounded, continuous function. Furthermore, a simple computation shows that there is a $B \in (0,\infty)$ for which

$$\big\|\mathbf{k}_t(\mathbf{x}) - \mathbf{k}_t(\mathbf{x}-\mathbf{a})\big\|_F \leq \frac{B|\mathbf{a}|}{|\mathbf{x}|^{N+1}}, \quad t \in (0,\infty),\ \mathbf{a} \in \mathbb{R}^N \setminus \{\mathbf{0}\}, \text{ and } |\mathbf{x}| \geq 2|\mathbf{a}|.$$

At the same time, by (6.2.10), we know that there is an $A \in (0,\infty)$ for which

$$\|\mathbf{k}_t f\|_{L^2(\mathbb{R}^N;F)} \leq A\|f\|_{L^2(\mathbb{R}^N)}, \quad t \in (0,\infty) \text{ and } f \in L^2(\mathbb{R}^N;\mathbb{C}).$$

Hence, by Theorem 6.2.12 (with $E = \mathbb{C}$), we see that there is a $C \in (0,\infty)$ with the property that

$$\big|\{\mathbf{x} : \|\mathbf{k}_t f\|_F \geq R\}\big| \leq \frac{C\|f\|_{L^1(\mathbb{R}^N)}}{R}, \quad R \in (0,\infty),$$

for all $t \in (0,\infty)$ and $f \in L^1(\mathbb{R}^N;\mathbb{C})$; and clearly this leads directly to (6.2.18). □

Before moving on, we want to show that Theorem 6.2.12 in conjunction with elementary interpolation theory leads to estimates for $\mathbf{K}$ as a mapping from $L^p(\mathbb{R}^N;E)$ to $L^p(\mathbb{R}^N;F)$ with $p\in(1,2)$. The interpolation theory which we will need is due to Marcinkiewicz and is very closely related to Exercise 1.4.20, in that they both turn on the familiar identity

$$\|f\|^p_{L^p(m)} = p\int_{(0,\infty)} t^{p-1}\,m\big(|f|>t\big)\,dt$$

for $p\in[1,\infty)$ and m-integrable f.

6.2.19 Theorem (Marcinkiewicz). *Let E and F be a pair of separable Banach spaces and $(X,\mathcal{A},\mu)$ and $(Y,\mathcal{B},\nu)$ a pair of measure spaces. Further, suppose that $\mathbf{K}$ is a linear operator which takes an $\mathbf{f}\in L^1(\mu;E)\cap L^\infty(\mu;E)$ into a ν-almost surely unique, $\mathcal{B}$-measurable function $\mathbf{Kf}:Y\longrightarrow F$. Finally, suppose that there exist $1\le p_1<p_2\le\infty$ and $M_1,\ M_2\in(0,\infty)$ with the properties that, for $i\in\{1,2\}$, $R\in(0,\infty)$, and $\mathbf{f}\in L^1(\mu;E)\cap L^\infty(\mu;E)$,*

$$\Big(\nu\big(\{y\in Y:\ \|\mathbf{Kf}(y)\|_F\ge R\}\big)\Big)^{\frac{1}{p_i}} \le \frac{M_i\|\mathbf{f}\|_{L^{p_i}(\mu;F)}}{R}, \tag{6.2.20}$$

where, when $p_2=\infty$ and $i=2$, the interpretation of (6.2.20) is given by the convention that $a^{\frac{1}{\infty}}$ equals 0 or 1 according to whether $a=0$ or $a>0$. Then, for each $p\in(p_1,p_2)$ and $\mathbf{f}\in L^1(\mu;E)\cap L^\infty(\mu;E)$:

$$\|\mathbf{Kf}\|_{L^p(\nu;F)}\le C_p M_1^\theta\,M_2^{1-\theta}\,\|\mathbf{f}\|_{L^p(\mu;E)}, \tag{6.2.21}$$

where

$$C_p\le 2\left(1+\frac{p_1}{p-p_1}+\frac{\mathbf{1}_{[1,\infty)}(p_2)p_2}{p_2-p}\right)^{\frac1p}$$

is universal and $\theta\in(0,1)$ is determined by the relation $\frac1p=\frac{\theta}{p_1}+\frac{1-\theta}{p_2}$.

PROOF: We begin with the case in which $M_1=M_2=1$. Define

$$\Phi(s)=\mu\big(\{x:\ \|\mathbf{f}(\mathbf{x})\|_F>s\}\big)\quad\text{and}\quad\Psi(t)=\nu\big(\{y:\ \|\mathbf{Kf}(y)\|_F>t\}\big)$$

for $s,\ t\in(0,\infty)$. Next, for $t\in(0,\infty)$, set

$$\mathbf{f}_t=\mathbf{f}\,\mathbf{1}_{[0,t]}\big(2\|\mathbf{f}\|_E\big),$$

and note that

$$\nu\big(\{y:\ \big\|\mathbf{K}(\mathbf{f}-\mathbf{f}_t)(y)\big\|_F>\tfrac t2\}\big)\le\left(\frac{2\|\mathbf{f}-\mathbf{f}_t\|_{L^{p_1}(\mu;E)}}{t}\right)^{p_1}.$$

At the same time,

$$\nu\big(\{y : \|\mathbf{K}\mathbf{f}_t(y)\|_F > \tfrac{t}{2}\}\big) \le \left(\frac{2\|\mathbf{f}_t\|_{L^{p_2}(\mu;E)}}{t}\right)^{p_2} \quad \text{when } p_2 < \infty$$

and

$$\nu\big(\{y : \|\mathbf{K}\mathbf{f}_t(y)\|_F > \tfrac{t}{2}\}\big) = 0 \quad \text{when } p_2 = \infty.$$

Hence,

$$\begin{aligned}\Psi(t) \le \frac{p_1 2^{p_1}}{t^{p_1}} \int_0^\infty s^{p_1-1}\, \Phi\left(s \vee \tfrac{t}{2}\right) ds \\ + \mathbf{1}_{[1,\infty)}(p_2)\, \frac{p_2 2^{p_2}}{t^{p_2}} \int_0^{\frac{t}{2}} s^{p_2-1}\, \Phi(s)\, ds;\end{aligned}$$

and so,

$$\begin{aligned}\frac{1}{p}\|\mathbf{K}\mathbf{f}\|^p_{L^p(\nu;F)} &\le p_1 2^{p_1} \int_0^\infty t^{p-p_1-1} \left(\int_0^\infty s^{p_1-1}\, \Phi\left(s \vee \tfrac{t}{2}\right) ds\right) dt \\ &\quad + \mathbf{1}_{[1,\infty)}(p_2)\, p_2 2^{p_2} \int_0^\infty t^{p-p_2-1} \left(\int_0^{\frac{t}{2}} s^{p_1-1}\, \Phi(s)\, ds\right) dt \\ &= p_1 2^p \int_0^\infty t^{p-p_1-1} \left(\int_0^\infty s^{p_1-1}\, \Phi\big(s \vee t\big)\, ds\right) dt \\ &\quad + \mathbf{1}_{[1,\infty)}(p_2)\, p_2 2^p \int_0^\infty t^{p-p_2-1} \left(\int_0^t s^{p_2-1}\, \Phi(s)\, ds\right) dt \\ &= 2^p \int_0^\infty t^{p-1}\, \Phi(t)\, dt + \frac{2^p p_1}{p-p_1} \int_0^\infty s^{p-1}\, \Phi(s)\, ds \\ &\quad + \mathbf{1}_{[1,\infty)}(p_2) \frac{2^p p_2}{p_2-p} \int_0^\infty s^{p-1}\, \Phi(s)\, ds \\ &= \frac{2^p}{p}\left(1 + \frac{p_1}{p-p_1} + \frac{p_2 \mathbf{1}_{[1,\infty)}(p_2)}{p_2-p}\right) \|\mathbf{f}\|^p_{L^p(\mu;E)};\end{aligned}$$

which means that (6.2.21) holds with the required sort of C_p when $M_1 = M_2 = 1$.

To complete the proof, we must still show how to reduce the general case to the one when $M_1 = M_2 = 1$. To this end, replace ν by $\frac{1}{\alpha}\nu$ and $\mathbf{K}$ by $\frac{1}{\beta}\mathbf{K}$, where

$$\alpha^{\frac{1}{p_1}}\beta = M_1 \quad \text{and} \quad \alpha^{\frac{1}{p_2}}\beta = M_2.$$

(6.2.20) for μ, ν, and $\mathbf{K}$ then implies (6.2.20) with $M_1 = M_2 = 1$ for μ, $\frac{1}{\alpha}\nu$, and $\frac{1}{\beta}\mathbf{K}$; and therefore the result just proved implies (6.2.21). □

6.2.22 Corollary. *Let everything be as in Theorem 6.2.12. Then for each $p \in (1,2]$ there exists a $C_p \in (0,\infty)$, which depends only on A, B, and N in addition to p, such that $\overline{\lim}_{p\searrow 1}(p-1)C_p < \infty$ and*

$$\|\mathbf{Kf}\|_{L^p(\mathbb{R}^N;F)} \le C_p \|\mathbf{f}\|_{L^p(\mathbb{R}^N;E)}, \quad \mathbf{f} \in L^1(\mathbb{R}^N;E). \tag{6.2.23}$$

If, in addition, both E and F are reflexive, then $\sup_{p\in(1,2]}(p-1)C_p < \infty$ and (6.2.23) continues to hold for all $p \in (2,\infty)$ with $C_p = C_{p'}$ and $p' = \frac{p}{p-1}$, the Hölder conjugate of p.

PROOF: In view of Theorems 6.2.12 and 6.2.19, there is nothing to do when $p \in (1,2]$. Thus, we suppose that E and F are reflexive and consider $p \in (2,\infty)$. To this end, let $\mathbf{k}(\mathbf{x})^* : F^* \longrightarrow E^*$ be the adjoint of $\mathbf{k}(\mathbf{x})$ for each $\mathbf{x} \in \mathbb{R}^N$, and note that the adjoint $\mathbf{K}^* : L^2(\mathbb{R}^N;F^*) \longrightarrow L^2(\mathbb{R}^N;E^*)$ of $\mathbf{K}$ is given by

$$[\mathbf{K}^*\mathbf{g}](\mathbf{x}) = \int_{\mathbb{R}^N} \mathbf{k}(\boldsymbol{\xi}-\mathbf{x})^*\mathbf{g}(\boldsymbol{\xi})\,d\boldsymbol{\xi}, \quad \mathbf{x} \in \mathbb{R}^N,$$

for $\mathbf{g} \in L^1(\mathbb{R}^N;F^*) \cap L^2(\mathbb{R}^N;F^*)$. Hence Theorems 6.2.12 and 6.2.19 apply to $\mathbf{K}^*$ with the same constants A and B; and so

$$\|\mathbf{K}^*\mathbf{g}\|_{L^{p'}(\mathbb{R}^N;E^*)} \le C_{p'}\|\mathbf{g}\|_{L^{p'}(\mathbb{R}^N;F^*)}, \quad \mathbf{g} \in L^1(\mathbb{R}^N;F^*).$$

Thus, by duality, (6.2.23) holds with $C_p = C_{p'}$. Finally, to see that $(p-1)C_p$ is bounded for $p \in (1,2]$, it suffices to check that the C_p's remain bounded for p's in an open interval around 2. But, because we know that $C_{\frac{1}{4}} = C_{\frac{4}{3}} < \infty$, this follows by another application of Theorem 6.2.19. □

With the preceding at hand, we can now complete our analysis of the Riesz transforms.

6.2.24 Theorem. *For each $p \in (1,\infty)$ there is a constant $C_p \in (0,\infty)$, satisfying $\sup_{p\in(1,2]}(p-1)C_p < \infty$, with the property that, for all $1 \le j \le N$ and $f \in L^p(\mathbb{R}^N;\mathbb{C})$:*

$$\left\| \sup_{y>0} \left| Q^{(N)}_{j,y} f \right| \right\|_{L^p(\mathbb{R}^N)} \le C_p \|f\|_{L^p(\mathbb{R}^N)}. \tag{6.2.25}$$

In particular, for each $p \in (1,\infty)$ and $f \in L^p(\mathbb{R}^N;\mathbb{C})$, there is a unique $Q^{(N)}_j f \in L^p(\mathbb{R}^N;\mathbb{C})$ with the property that, as $y \searrow 0$,

$$Q^{(N)}_{j,y} f \longrightarrow Q^{(N)}_j f \quad \text{both Leb-almost everywhere and in } L^p(\mathbb{R}^N;\mathbb{C}).$$

Finally, for each $p \in (1,\infty)$, there exists a $c_p \in [1,\infty)$ with the properties that $c_p = c_{p'}$, $\sup_{p\in(1,2]}(p-1)c_p < \infty$, and, for all $f \in L^p(\mathbb{R}^N;\mathbb{C})$:

$$\frac{1}{c_{p'}}\|f\|_{L^p(\mathbb{R}^N)} \le \big\|\mathbf{Q}^{(N)}f\big\|_{L^p(\mathbb{R}^N;\mathbb{C}^N)} \le c_p\|f\|_{L^p(\mathbb{R}^N)}, \tag{6.2.26}$$

where $\mathbf{Q}^{(N)} : L^p(\mathbb{R}^N;\mathbb{C}) \longrightarrow L^p(\mathbb{R}^N;\mathbb{C}^N)$ is given by

$$\mathbf{Q}^{(N)}f \equiv \left(Q_1^{(N)}f,\dots,Q_N^{(N)}f\right).$$

PROOF: Using the same notation as in the proof of Theorem 6.2.17 together with the estimates in (6.2.10) and (6.2.18), we see from the first part of Corollary 6.2.22 that, when $p \in (1,2]$, there is a C_p satisfying $\overline{\lim}_{p\searrow 1}(p-1)C_p < \infty$ such that

$$\left\|\sup_{y>0}\big|Q_{j,y}^{(N)}f\big|\right\|_{L^p(\mathbb{R}^N)} = \sup_{t>0}\big\|\mathbf{K}_tf\big\|_{L^p(\mathbb{R}^N;F)} \le C_p\|f\|_{L^p(\mathbb{R}^N)}.$$

However, because $C_0\big([0,\infty);\mathbb{C}\big)$ is not reflexive, we cannot apply Corollary 6.2.22 directly to get (6.2.25) for $p \in (2,\infty)$. On the other hand, we can use Corollary 6.2.22 (with $E = \mathbb{C}$ and $F = \mathbb{C}^N$) and the estimates in (6.2.10) and (6.2.18) to show that, for each $p \in (1,\infty)$, there is a $c_p \in (0,\infty)$ such that $c_p = c_{p'}$, $\sup_{p\in(1,2]}(p-1)c_p < \infty$, and

$$\big\|\mathbf{Q}^{(N)}f\big\|_{L^p(\mathbb{R}^N;\mathbb{C}^N)} \le \sup_{y>0}\big\|\mathbf{Q}_y^{(N)}f\big\|_{L^p(\mathbb{R}^N;\mathbb{C}^N)} \le c_p\|f\|_{L^p(\mathbb{R}^N)},$$

where

$$\mathbf{Q}_y^{(N)}f \equiv \left(Q_{1,y}^{(N)}f,\dots,Q_{N,y}^{(N)}f\right).$$

Hence, the right-hand inequality in (6.2.26) holds; and, as an application of the second part of (6.2.9) and part (**ii**) of Exercise 5.3.39, we also see that (6.2.25) holds with a C_p which is an appropriate multiple of c_p.

In view of the preceding, all that remains is to prove the left-hand inequality in (6.2.26). To this end, note that, by (6.2.8), $\mathbf{Q}^{(N)}$ is an isometry from $L^2(\mathbb{R}^N;\mathbb{C})$ into $L^2(\mathbb{R}^N;\mathbb{C}^N)$. In particular, (6.2.26) holds with $c_2 = 1$. Hence, for any $p \in (1,\infty)$ and pair $f,\, g \in L^2(\mathbb{R}^N;\mathbb{C})$, we have that

$$\begin{aligned}
\big(f,g\big)_{L^2(\mathbb{R}^N)} &= \big(\mathbf{Q}^{(N)}f,\mathbf{Q}^{(N)}g\big)_{L^2(\mathbb{R}^N;\mathbb{C}^N)} \\
&\le \big\|\mathbf{Q}^{(N)}f\big\|_{L^p(\mathbb{R}^N;\mathbb{C}^N)}\,\big\|\mathbf{Q}^{(N)}g\big\|_{L^{p'}(\mathbb{R}^N;\mathbb{C})} \\
&\le c_{p'}\big\|\mathbf{Q}^{(N)}f\big\|_{L^p(\mathbb{R}^N;\mathbb{C}^N)}\,\|g\|_{L^{p'}(\mathbb{R}^N)};
\end{aligned}$$

which is equivalent to the left-hand side of (6.2.26). □

We turn now to a topic which may be viewed as a harmonic analytic precursor of the results in the next section. Namely, given a function $f \in L^2(\mathbb{R}^N;\mathbb{C})$, define the **Littlewood–Paley g-function of f** to be the function

$$\mathbf{x} \in \mathbb{R}^N \longmapsto [g(f)](\mathbf{x}) \equiv \left(\int_{(0,\infty)} y\left|\nabla u_f(\mathbf{x}),y)\right|^2 dy\right)^{\frac{1}{2}} \in [0,\infty], \tag{6.2.27}$$

where

$$(\mathbf{x},y) \in \mathbb{R}^{N+1}_+ \longmapsto u_f(\mathbf{x},y) = p_y^{(N)}\bigstar f(\mathbf{x}) \in \mathbb{C}$$

is the harmonic extension of f to the upper half-space $\mathbb{R}^{N+1}_+ = \mathbb{R}^N \times (0,\infty)$ and ∇ is the Euclidean gradient on $\mathbb{R}^{N+1}_+$. Notice that the Fourier transform of $\nabla u_f(\,\cdot\,,y)$ as a function of $\mathbf{x} \in \mathbb{R}^N$ is

$$\boldsymbol{\xi} \in \mathbb{R}^N \longmapsto -\big(\sqrt{-1}\,\boldsymbol{\xi},|\boldsymbol{\xi}|\big)\, e^{-y|\boldsymbol{\xi}|}\hat{f}(\boldsymbol{\xi}) \in \mathbb{C}^{N+1};$$

and therefore, by Fubini's Theorem and Parseval's identity:

$$\begin{aligned}\|g(f)\|^2_{L^2(\mathbb{R}^N)} &= \int_{(0,\infty)} y\left(\int_{\mathbb{R}^N} \left|\nabla u_f(\mathbf{x},y)\right|^2 d\mathbf{x}\right) dy\\ &= 2(2\pi)^{-N}\int_{(0,\infty)} y\left(\int_{\mathbb{R}^N} |\boldsymbol{\xi}|^2 e^{-2y|\boldsymbol{\xi}|}\left|\hat{f}(\boldsymbol{\xi})\right|^2 d\boldsymbol{\xi}\right) dy\\ &= \frac{(2\pi)^{-N}}{2}\int_{\mathbb{R}^N} \left|\hat{f}(\boldsymbol{\xi})\right|^2 d\boldsymbol{\xi};\end{aligned}$$

which means, after another application of Parseval, that

$$\|g(f)\|^2_{L^2(\mathbb{R}^N)} = \frac{\|f\|^2_{L^2(\mathbb{R}^N)}}{2}, \quad f \in L^2(\mathbb{R}^N;\mathbb{C}). \tag{6.2.28}$$

Thus, once again, we have a situation in which Fourier analysis makes it clear that everything is fine in L^2, and, once again we will use Corollary 6.2.22 to get information about $p \neq 2$. For this purpose, note that

$$\nabla u_f(\mathbf{x},y) = \int_{\mathbb{R}^N} \dot{\mathbf{p}}_y(\mathbf{x}-\boldsymbol{\xi})\, f(\boldsymbol{\xi})\, d\boldsymbol{\xi}, \quad (\mathbf{x},y) \in \mathbb{R}^{N+1}_+,$$

where

$$\begin{aligned}\dot{\mathbf{p}}_y(\mathbf{x}) &= \left(\tfrac{\partial p_y}{\partial x_1}(\mathbf{x}),\dots,\tfrac{\partial p_y}{\partial x_N}(\mathbf{x}),\tfrac{\partial p_y}{\partial y}(\mathbf{x})\right)\\ &= -\tfrac{2}{\Omega_{N+1}}\big(y^2+|\mathbf{x}|^2\big)^{-\frac{N+3}{2}}\big(yx_1,\dots,yx_N,|\mathbf{x}|^2-Ny^2\big)\end{aligned} \tag{6.2.29}$$

and $\boldsymbol{\Omega}_{N+1}$ is the volume of the unit ball in $\mathbb{R}^{N+1}$. Next, define the Borel measure ν on $(0,\infty)$ by $\nu(dy) = y\,dy$, let F be the separable Hilbert space $\left(L^2(\nu;\mathbb{C})\right)^{N+1}$, and define

$$\mathbf{x} \in \mathbb{R}^N \longmapsto \mathbf{k}_t(\mathbf{x}) \in F \text{ by } \big[\mathbf{k}_t(\mathbf{x})\big](y) = \dot{\mathbf{p}}_{t+y}(\mathbf{x}) \quad \text{for } t \in (0,\infty).$$

Clearly

$$\mathbf{k}_t \in C_b(\mathbb{R}^N;F) \cap \bigcap_{p\in(1,\infty)} L^p(\mathbb{R}^N;F)$$

and, for $f \in \bigcup_{p\in(1,\infty)} L^p(\mathbb{R}^N;\mathbb{C})$ and $\mathbf{x} \in \mathbb{R}^N$,

$$\nabla u_f(\mathbf{x}, t+y) = \big[\mathbf{K}_t f(\mathbf{x})\big](y) \quad \text{where } \mathbf{K}_t f(\mathbf{x}) \equiv \int_{\mathbb{R}^N} \mathbf{k}_t(\mathbf{x}-\boldsymbol{\xi})\, f(\boldsymbol{\xi})\, d\boldsymbol{\xi}.$$

In particular,

$$\nabla u_f(\mathbf{x}, y) = \lim_{t\searrow 0} \big[\mathbf{K}_t f(\mathbf{x})\big](y), \quad (\mathbf{x}, y) \in \mathbb{R}^{N+1}_+,$$

and so, by Fatou's Lemma,

$$\big[g(f)\big](\mathbf{x}) \leq \varliminf_{t\searrow 0} \big\|\mathbf{K}_t f(\mathbf{x})\big\|_F;$$

which, by another application of Fatou, means that

$$\|g(f)\|_{L^p(\mathbb{R}^N)} \leq \varliminf_{t\searrow 0} \big\|\mathbf{K}_t f\big\|_{L^p(\mathbb{R}^N;F)}, \quad f \in L^p(\mathbb{R}^N;\mathbb{C}), \tag{6.2.30}$$

for each $p \in (1,\infty)$. With these preliminaries, we are now in a position to prove the following estimates discovered originally (in the case when $N = 1$) by Littlewood and Paley.

6.2.31 Theorem. *For each $N \in \mathbb{Z}^+$ and $p \in (1,\infty)$ there is a $C_p \in (0,\infty)$ with the properties that $C_p = C_{p'}$, $\sup_{p\in(1,2]}(p-1)C_p < \infty$, and*

$$\frac{1}{2C_p}\|f\|_{L^p(\mathbb{R}^N)} \leq \|g(f)\|_{L^p(\mathbb{R}^N)} \leq C_p\|f\|_{L^p(\mathbb{R}^N)}, \quad f \in L^p(\mathbb{R}^N;\mathbb{C}). \tag{6.2.32}$$

PROOF: In view of (6.2.30), the right-hand side of (6.2.32) will follow as soon as we check that the estimates in (6.2.13) and (6.2.14) (with $E = \mathbb{C}$) are satisfied by $\mathbf{K}_t$ and its kernel $\mathbf{k}_t$, with constants A and B which do not depend on $t \in (0,\infty)$. But, as is easily checked, $\big\|\mathbf{K}_t f(\mathbf{x})\big\|_F \leq \big[g(f)\big](\mathbf{x})$ for all $t \in (0,\infty)$ and $\mathbf{x} \in \mathbb{R}^N$;

and therefore, by (6.2.28), $\mathbf{K}_t$ satisfies (6.2.13) with $A = 2^{-\frac{1}{2}}$. At the same time, an elementary computation leads to the existence of a $c_N \in (0,\infty)$ for which

$$\max_{1\le j\le N}\left|\frac{\partial \dot{\mathbf{p}}_y}{\partial x_j}(\mathbf{x})\right| \le \frac{c_N\left(y^3+|\mathbf{x}|^3\right)}{\left(y^2+|\mathbf{x}|^2\right)^{\frac{N+5}{2}}}, \quad (\mathbf{x},y)\in\mathbb{R}^{N+1}_+;$$

and from this it is easy to see that the $\mathbf{k}_t$'s satisfy (6.2.14) with a $B \in (0,\infty)$ which is independent of $t \in (0,\infty)$. Hence, by Corollary 6.2.22 and the remark with which we started this proof, the right-hand side of (6.2.32) holds with C_p's which satisfy the required conditions.

To prove the left-hand side of (6.2.32), we use essentially to same line of reasoning as we used in the derivation of the left-hand side of (6.2.26). Namely, given φ, $\psi \in L^1(\mathbb{R}^N;\mathbb{C})\cap L^\infty(\mathbb{R}^N;\mathbb{C})$, we know, from (6.2.28) and the right-hand side of (6.2.32), that

$$\bigl(\varphi,\psi\bigr)_{L^2(\mathbb{R}^N)} \le 2\bigl(g(\varphi),g(\psi)\bigr)_{L^2(\mathbb{R}^N)} \le 2C_p\bigl\|g(\varphi)\bigr\|_{L^p(\mathbb{R}^N)}\,\|\psi\|_{L^{p'}(\mathbb{R}^N)}.$$

Hence,

$$\|\varphi\|_{L^p(\mathbb{R}^N)} \le 2C_p\bigl\|g(\varphi)\bigr\|_{L^p(\mathbb{R}^N)}, \quad \varphi\in L^1(\mathbb{R}^N;\mathbb{C})\cap L^\infty(\mathbb{R}^N;\mathbb{C});$$

and so, again by the right-hand side of (6.2.32), we can conclude that the left-hand side must hold for each $f \in L^p(\mathbb{R}^N;\mathbb{C})$. □

6.2.33 Corollary. *Given $f \in \bigcup_{p\in(1,\infty)} L^p(\mathbb{R}^N;\mathbb{C})$, define*

$$\text{(6.2.34)}\qquad \begin{aligned}\bigl[g_v(f)\bigr](\mathbf{x}) &= \left(\int_0^\infty y\left|\frac{\partial u_f}{\partial y}(\mathbf{x},y)\right|^2 dy\right)^{\frac{1}{2}}\\ \bigl[g_h(f)\bigr](\mathbf{x}) &= \sqrt{\bigl[g(f)\bigr](\mathbf{x})^2-\bigl[g_v(f)\bigr](\mathbf{x})^2}.\end{aligned}$$

Then for each $p \in (1,\infty)$

$$\text{(6.2.35)}\qquad \begin{aligned}\frac{1}{4C_p}\|f\|_{L^p(\mathbb{R}^N)} &\le \|g_h\|_{L^p(\mathbb{R}^N)} \wedge \|g_v(f)\|_{L^p(\mathbb{R}^N)}\\ &\le \|g_h\|_{L^p(\mathbb{R}^N)} \vee \|g_v(f)\|_{L^p(\mathbb{R}^N)} \le C_p\|f\|_{L^p(\mathbb{R}^N)},\end{aligned}$$

where the C_p's are the same as they were in Theorem 6.2.31.

PROOF: Seeing as both $g_v(f)$ and $g_h(f)$ are dominated pointwise by $g(f)$, the right-hand side of (6.2.35) is an immediate consequence of the right-hand side of (6.2.32). To prove the left-hand side, note that the Fourier transform of $\mathbf{x}\in\mathbb{R}^N \longmapsto \frac{\partial u_f}{\partial y}(\mathbf{x},y)\in\mathbb{C}$ is

$$\boldsymbol{\xi}\in\mathbb{R}^N \longmapsto -|\boldsymbol{\xi}|\,e^{-y|\boldsymbol{\xi}|}\,\hat{f}(\boldsymbol{\xi})\in\mathbb{C},$$

and, proceeding as in the derivation of (6.2.28), conclude that

$$\|g_v(f)\|^2_{L^2(\mathbb{R}^N)} = \frac{\|f\|^2_{L^2(\mathbb{R}^N)}}{4}; \tag{6.2.36}$$

which, since $g(f)^2 = g_v(f)^2 + g_h(f)^2$, means that

$$\|g_h(f)\|^2_{L^2(\mathbb{R}^N)} = \frac{\|f\|^2_{L^2(\mathbb{R}^N)}}{4}$$

as well. Given these and the right-hand side of (6.2.35), the derivation of the left-hand side is now identical to the derivation of the left-hand side of (6.2.32) from (6.2.28) and the right-hand side of (6.2.32). □

The quantities $g(f)$, $g_v(f)$, and $g_h(f)$ are all examples of **square functions**, and (6.2.32) and (6.2.35) are important examples the relation between the L^p-norm of a function and the L^p-norm of related square functions. An indication of the power of such relations is given in Exercises 6.2.41 and 6.2.41 below.

Exercises

6.2.37 Exercise: As mentioned earlier, when $N = 1$, the Riesz transform (of which there is only one) is called the Hilbert transform and has a beautiful interpretation in terms of complex analysis on the complex upper half-plane

$$\mathbb{C}_+ = \{x + \sqrt{-1}\,y : (x, y) \in \mathbb{R}^2_+\}.$$

The origin of this interpretation comes from the easily verified relation

$$\frac{\sqrt{-1}}{\pi z} = p_y^{(1)}(x) + \sqrt{-1}\,q_y^{(1)}(x), \quad z = x + \sqrt{-1}\,y \in \mathbb{C}_+. \tag{6.2.38}$$

(i) Let $f \in \bigcup_{p\in[1,\infty)} L^p(\mathbb{R};\mathbb{R})$ be given, and set $u_f(x, y) = p_y^{(1)}\bigstar f(x)$ and $v_f(x, y) = q_y^{(1)}\bigstar f$ for $(x, y) \in \mathbb{R}^2_+$, and define $F_f : \mathbb{C}_+ \longrightarrow \mathbb{C}$ by

$$F_f(z) = u_f(x, y) + \sqrt{-1}\,v_f(x, y), \quad z = x + \sqrt{-1}\,y \in \mathbb{C}_+. \tag{6.2.39}$$

Show that F_f is the unique holomorphic function F on $\mathbb{C}_+$ with the properties that

$$\lim_{y\nearrow\infty} \sup_{x\in\mathbb{R}} \left|F_f(x + \sqrt{-1}\,y)\right| = 0$$

and

$$f(x) = \lim_{y \searrow 0} \mathfrak{Re}\Big(F\big(x + \sqrt{-1}\,y\big)\Big) \quad \text{for Leb-almost every } x \in \mathbb{R}.$$

In particular, conclude that *the Hilbert transform Hf of f is characterized as the imaginary part of the boundary value of F_f* in the sense that

$$\lim_{y \searrow 0} \mathfrak{Im}\Big(F_f\big(x + \sqrt{-1}\,y\big)\Big) = Hf(x) \quad \text{for Leb-almost every } x \in \mathbb{R}.$$

(ii) From part **(i)**, we know that u_f and v_f are conjugate harmonic functions. Using this fact together with the Cauchy–Riemann equations, note (cf. (6.2.27)) that $g(f) = g(Hf)$ and use this in conjunction with (6.2.32) to give another proof that H is bounded as an operator on $L^p(\mathbb{R};\mathbb{R})$ into itself for each $p \in (1,\infty)$.

(iii) The original proof that H is bounded on $L^p(\mathbb{R};\mathbb{R})$ was based on (6.2.39). The approach taken went as follows. Given an $m \in \mathbb{Z}^+$, use Cauchy's integral formula to justify

$$\int_{\mathbb{R}} F_f\left(x + \sqrt{-1}\,y\right)^{2m} dx = 0, \quad y \in (0,\infty) \text{ and } f \in \mathcal{S}(\mathbb{R};\mathbb{C}). \tag{6.2.40}$$

When $m = 1$, note that this leads immediately to

$$\begin{aligned}\big\|v_f(\cdot,y)\big\|_{L^2(\mathbb{R})}^2 &= \big\|u_f(\cdot,y)\big\|_{L^2(\mathbb{R})}^2 - \mathfrak{Re}\left(\int_{\mathbb{R}} F_f\left(x + \sqrt{-1}\,y\right)^2 dx\right) \\ &= \big\|u_f(\cdot,y)\big\|_{L^2(\mathbb{R})}^2 \le \|f\|_{L^2(\mathbb{R})}^2.\end{aligned}$$

More generally, show that for each $m \in \mathbb{Z}^+$ there exists a $C_m \in (0,\infty)$ with the property that

$$\big(\mathfrak{Im}(\zeta)\big)^{2m} \le C_{2m}^{2m}\Big(\mathfrak{Re}(\zeta)^{2m} + (-1)^m \mathfrak{Re}\big(\zeta^{2m}\big)\Big) \quad \zeta \in \mathbb{C}$$

or, equivalently,

$$\big(\sin\theta\big)^{2m} \le C_m^{2m}\Big(\big(\cos\theta\big)^{2m} + (-1)^m \cos(2m\theta)\Big), \quad \theta \in \mathbb{R};$$

and use this together with (6.2.40) to prove that

$$\big\|u_f(\cdot,y)\big\|_{L^{2m}(\mathbb{R})} \le C_m \big\|u_f(\cdot,y)\big\|_{L^{2m}(\mathbb{R})}, \quad y \in (0,\infty).$$

To handle other $p \in [2,\infty)$, one proceeds via interpolation; and to handle $p \in (1,2)$, one works by duality.

6.2.41 Exercise: Even when $N > 1$, the relation between the square functions $g_v(f)$ and $g_h(f)$ can be used to give another proof that the Riesz transforms are bounded on $L^p(\mathbb{R}^N;\mathbb{C})$ for each $p \in (1,\infty)$. Indeed, given f, set $\mathbf{f} = (f_1,\dots,f_N) = \mathbf{Q}^{(N)}f$ and define

$$\mathbf{U}_{\mathbf{f}}(\mathbf{x},y) = \big(u_{f_1}(\mathbf{x},y),\dots,u_{f_N}(\mathbf{x},y)\big), \quad (\mathbf{x},y) \in \mathbb{R}^{N+1}_+.$$

Working via Fourier transform, check that

$$\frac{\partial \mathbf{U}_{\mathbf{f}}}{\partial y} = \left(\frac{\partial u_f}{\partial x_1},\dots,\frac{\partial u_f}{\partial x_N}\right),$$

which means that $\sum_1^N g_v(f_j)^2 = g_h(f)^2$. Hence, by (6.2.35), we get (6.2.26) with $c_p \le 4C_p^2$.

6.2.42 Exercise: Perhaps the most useful fact that comes out of the considerations in this section is the remarkable† observation that the right-hand side of (6.2.26) leads to

$$\max_{1\le i,j\le N}\left\|\frac{\partial^2 f}{\partial x_i\partial x_j}\right\|_{L^p(\mathbb{R}^N)} \le C_p^2\|\boldsymbol{\Delta} f\|_{L^p(\mathbb{R}^N)}, \quad f \in C^2_c(\mathbb{R}^N;\mathbb{C}); \tag{6.2.43}$$

where $\boldsymbol{\Delta}$ is the Euclidean Laplacian on $\mathbb{R}^N$. To prove (6.2.43), recall that

$$\widehat{\frac{\partial f}{\partial x_j}}(\boldsymbol{\xi}) = -\sqrt{-1}\,\xi_j\hat{f}(\boldsymbol{\xi}), \quad 1 \le j \le N \text{ and } \boldsymbol{\xi} \in \mathbb{R}^N,$$

and conclude that

$$\frac{\partial^2 f}{\partial x_i\partial x_j} = Q_i^{(N)} \circ Q_j^{(N)}(\boldsymbol{\Delta} f).$$

§6.3 Burkholder's Inequality.

In the preceding section, we adopted the Calderón–Zygmund point of view toward operators like the Riesz transforms. That is, we took the attitude that the kernel associated with these operators is the primary object. An alternative

† The fact that (6.2.43) is *false* when $N \ge 2$ and $p \in \{1,\infty\}$ has kept bread on the table of many analysts.

approach to these operators is to present them as Fourier *multipliers*. Thus, for example, we could have given

$$\widehat{Q_j^{(N)}f}(\boldsymbol{\xi}) = \sqrt{-1}\,\frac{\xi_j}{|\boldsymbol{\xi}|}\hat{f}(\boldsymbol{\xi}), \quad \boldsymbol{\xi} \in \mathbb{R}^N \quad \text{and } f \in L^2(\mathbb{R}^N;\mathbb{C}),$$

as the primary description of $Q_j^{(N)}$. The reason why one might want the multiplier description is obvious: it is ideally suited for analysis in $L^2(\mathbb{R}^N;\mathbb{C})$. On the other hand, it has the disadvantage that it provides little information about the possibility for analysis in any of the other Lebesgue spaces. Indeed, given $m \in L^\infty(\mathbb{R}^N;\mathbb{C})$, Parseval's identity makes it obvious that

$$\begin{aligned} f \in \mathcal{S}(\mathbb{R}^N;\mathbb{C}) \longmapsto \big[K_m f\big](\mathbf{x}) \\ \equiv (2\pi)^{-N} \int_{\mathbb{R}^N} \exp\left[\sqrt{-1}\,(\mathbf{x},\boldsymbol{\xi})_{\mathbb{R}^N}\right] m(\boldsymbol{\xi})\,\hat{f}(\boldsymbol{\xi})\,d\boldsymbol{\xi} \in \mathbb{C} \end{aligned}$$

determines a unique, bounded, normal operator K_m on $L^2(\mathbb{R}^N;\mathbb{C})$ with norm equal to $\|m\|_{L^\infty(\mathbb{R}^N)}$. On the other hand, it is not at all clear which m's from $L^\infty(\mathbb{R}^N;\mathbb{C})$ determine operators K_m that are bounded on $L^p(\mathbb{R}^N;\mathbb{C})$ when $p \neq 2$. In fact, even the trivial case when $m = \hat{k}$ for some $k \in L^1(\mathbb{R}^N;\mathbb{C})$ and therefore

$$\big\|K_m f\big\|_{L^p(\mathbb{R}^N)} \le \|k\|_{L^1(\mathbb{R}^N)}\,\|f\|_{L^p(\mathbb{R}^N)}, \quad p \in [1,\infty],$$

is not so easy to spot by simply staring at the multiplier m; and the problem becomes only harder when the multiplier, like $\boldsymbol{\xi} \longmapsto \sqrt{-1}\,\frac{\xi_j}{|\boldsymbol{\xi}|}$, cannot possibly be the Fourier transform of $k \in L^1(\mathbb{R}^N;\mathbb{C})$.

The problem of deciding which Fourier multipliers determine bounded operations on which Lebesgue spaces has received a great deal of attention over the years (Marcinkiewicz provided what remains one of the key results in this topic); and a completely satisfactory answer has yet to be found. Be that as it may, there are nontrivial multiplier problems in other contexts which turn out to have simple solutions; and, as Burkholder was the first to discover, martingale theory provides one such example. Before explaining what Burkholder did, it may be helpful to abstract the procedure of defining an operation in terms of a *multiplier*; and, for our purposes, the following description will do. Namely, let $(A,\mathcal{A},\alpha)$ and $(B,\mathcal{B},\beta)$ be a pair of σ-finite measure spaces and E and F a pair of separable Hilbert spaces. Further, assume that $\mathcal{U}$ is a unitary mapping from $L^2(\alpha;E)$ onto $L^2(\beta;F)$. Then, for each $m \in L^\infty(\beta;\mathbb{R})$ or, when F is complex, $m \in L^\infty(\beta;\mathbb{C})$, we can define the bounded, normal (self-adjoint when m is real) operator K_m on $L^2(\alpha;E)$ by

$$K_m f = \mathcal{U}^*\big(m\,\mathcal{U}f\big), \quad f \in L^2(\alpha;E),$$

in which case K_m is said to be **operator determined by the multiplier** m **relative to** $\mathcal{U}$. When $A = B = \mathbb{R}^N$, $E = F = \mathbb{C}$, $\alpha = \text{Leb}$, $\beta = \frac{\alpha}{(2\pi)^N}$, and $\mathcal{U}$ is the Fourier transform, we are in the setting discussed in the preceding paragraph. When $A = \mathbb{R}$, $B = \mathbb{N}$, $\alpha = \gamma$ is the standard Gauss measure, β is counting measure, $E = F = \mathbb{C}$, and

$$(\mathcal{U}f)(n) = (n!)^{-\frac{1}{2}}\big(f, H_n\big)_{L^2(\gamma)}, \quad n \in \mathbb{N},$$

with H_n the nth Hermite polynomial described in (2.3.1), we are in the setting of Section 2.3. Finally, when $(A, \mathcal{A}, \alpha)$ and $(B, \mathcal{B}, \beta)$ are the same probability space $(\Omega, \mathcal{F}, P)$, $E = \mathbb{R}$, $F = \ell^2(\mathbb{N}; \mathbb{R})$ (the sequence space of square summable, real-valued sequences indexed by $\mathbb{N}$), and $\mathcal{U}$ is defined by

$$\big[\mathcal{U}f\big](n,\omega) = \begin{cases} \mathbb{E}^P\big[f\big|\mathcal{F}_0\big](\omega) & \text{if } n = 0 \\ \mathbb{E}^P\big[f\big|\mathcal{F}_n\big](\omega) - \mathbb{E}^P\big[f\big|\mathcal{F}_{n-1}\big](\omega) & \text{if } n \in \mathbb{Z}^+, \end{cases}$$

for some nondecreasing family $\{\mathcal{F}_n : n \in \mathbb{N}\}$ of σ-algebras satisfying $\mathcal{F} = \bigvee_0^\infty \mathcal{F}_n$, then we are in the setting studied by Burkholder.

As our first derivation of Burkholder's result, we will give an argument which parallels the one developed by Calderón and Zygmund's to handle singular integral operators.

6.3.1 Theorem (Burkholder). *Let $(\Omega, \mathcal{F}, P)$ be a probability space, $\{\mathcal{F}_n : n \in \mathbb{N}\}$ a nondecreasing sequence of σ-algebras satisfying $\mathcal{F} = \bigvee_0^\infty \mathcal{F}_n$, and $\{\sigma_n : n \in \mathbb{N}\}$ an $\{\mathcal{F}_n : n \in \mathbb{N}\}$-progressively measurable sequence of $[-1,1]$-valued functions. Then, for each $f \in L^2(P;\mathbb{R})$ there is a unique $Kf \in L^2(P;\mathbb{R})$ with the property that*

$$K_n f \equiv \mathbb{E}^P\big[Kf\big|\mathcal{F}_n\big] = X_0 + \sum_{m=1}^{n} \sigma_{m-1}\big(X_m - X_{m-1}\big), \quad n \in \mathbb{N},$$

where $X_n \equiv \mathbb{E}^P[f|\mathcal{F}_n]$, $n \in \mathbb{N}$. Moreover,

$$P\Big(\sup_{n\in\mathbb{N}} \big|K_n f\big| \geq R\Big) \leq \frac{98}{R}\|f\|_{L^1(P)}, \quad R \in (0,\infty), \tag{6.3.2}$$

and

$$\Big\|\sup_{n\in\mathbb{N}} \big|K_n f\big|\Big\|_{L^2(P)} \leq 2\|f\|_{L^2(P)}, \quad f \in L^2(P;\mathbb{R}) \tag{6.3.3}$$

Finally, for each $p \in (1,\infty)$ there exists a universal (i.e., independent of everything except p) constant $C_p \in [1,\infty)$ with the property that

$$\Big\|\sup_{n\in\mathbb{N}} \big|K_n f\big|\Big\|_{L^p(P)} \leq C_p\|f\|_{L^p(P)}, \quad f \in L^2(P;\mathbb{R}) \cap L^p(P;\mathbb{R}). \tag{6.3.4}$$

In fact, there is a universal $C \in (0,\infty)$ such that $C_p \leq C\frac{p^2}{p-1}$ for all $p \in (1,\infty)$.

PROOF: Because, by an easy limit argument, it is possible to obtain the general case from the one in which only finitely many $\mathcal{F}_n$'s differ from $\mathcal{F}$, we will assume throughout that there is an $N \in \mathbb{Z}^+$ such that $\mathcal{F} = \mathcal{F}_n$ for all $n \geq N$. In particular, the existence of Kf presents no difficulty in this case.

We begin by observing that, because $(K_n f, \mathcal{F}_n, P)$ is a martingale, the left-hand side of (6.3.3) is dominated by $2\|Kf\|_{L^2(P)}$ and

$$\begin{aligned}\|Kf\|^2_{L^2(P)} &= \left\|K_0 f\right\|^2_{L^2(P)} + \sum_{n\in\mathbb{Z}^+} \left\|K_n f - K_{n-1} f\right\|^2_{L^2(P)} \\ &\leq \left\|X_0\right\|^2_{L^2(P)} + \sum_{n\in\mathbb{Z}^+} \left\|X_n - X_{n-1}\right\|^2_{L^2(P)} = \|f\|^2_{L^2(P)}.\end{aligned}$$

Hence, (6.3.3) presents no problem. To prove (6.3.2), let $R \in (0,\infty)$ be given and, referring to part **(iii)** of Exercise 5.2.31, use Gundy's Calderón –Zygmund decomposition of martingales to write $X_n = M_n^{(R)} + V_n^{(R)} + \Delta_n^{(R)}$ where: $(M_n^{(R)}, \mathcal{F}_n, P)$ is a martingale satisfying

$$\mathbb{E}^P\left[\left(M_n^{(R)}\right)^2\right] \leq 12R\|f\|_{L^1(P)}, \quad n \in \mathbb{N};$$

$V_0^{(R)} \equiv 0$, $V_n^{(R)}$ is $\mathcal{F}_{n-1}$-measurable for each $n \in \mathbb{Z}^+$, and

$$\mathbb{E}^P\left[\left(\sum_{n=1}^\infty \left|V_n^{(R)} - V_{n-1}^{(R)}\right|\right)^2\right] \leq 12R\|f\|_{L^1(P)};$$

and $\{\Delta_n : n \in \mathbb{N}\}$ is an $\{\mathcal{F}_n : n \in \mathbb{N}\}$-progressively measurable sequence with the property that

$$P\left(\exists n \in \mathbb{N}\ \Delta_n^{(R)} \neq 0\right) \leq \tfrac{2}{R}\,\|f\|_{L^2(P)}.$$

We can then write $K_n f = N_n^{(R)} + W_n^{(R)} + E_n^{(R)}$, where

$$N_n^{(R)} = M_0^{(R)} + \sum_{m=1}^n \sigma_{m-1}\left(M_m^{(R)} - M_{m-1}^{(R)}\right)$$

$$W_n^{(R)} = \sum_{m=1}^n \sigma_{m-1}\left(V_m^{(R)} - V_{m-1}^{(R)}\right)$$

and

$$E_n^{(R)} = \Delta_0^{(R)} + \sum_{m=1}^n \sigma_{m-1}\left(\Delta_m^{(R)} - \Delta_{m-1}^{(R)}\right).$$

In particular, since

$$P\Big(\exists n \in \mathbb{N}\ E_n^{(R)} \neq 0\Big) \leq \tfrac{2}{R}\,\|f\|_{L^1(P)},$$

the left-hand side of (6.3.2) is dominated by

$$\frac{2}{R}\,\|f\|_{L^1(P)} + P\left(\sup_{n\in\mathbb{N}} \big|N_n^{(R)}\big| \geq \frac{R}{2}\right) + P\left(\sup_{n\in\mathbb{N}} \big|W_n^{(R)}\big| \geq \frac{R}{2}\right).$$

But $\big(N_n^{(R)}, \mathcal{F}_n, P\big)$ is a martingale, and therefore

$$P\left(\sup_{n\in\mathbb{N}} \big|N_n^{(R)}\big| \geq \frac{R}{2}\right) \leq \frac{4}{R^2} \sup_{n\in\mathbb{N}} \mathbb{E}^P\Big[\big(N_n^{(R)}\big)^2\Big]$$

and

$$\begin{aligned}\sup_{n\in\mathbb{N}} \mathbb{E}^P\Big[\big(N_n^{(R)}\big)^2\Big] &= \mathbb{E}^P\Big[\big(N_0^{(R)}\big)^2\Big] + \sum_{n=1}^{\infty} \mathbb{E}^P\Big[\big(N_n^{(R)} - N_{n-1}^{(R)}\big)^2\Big] \\ &\leq \mathbb{E}^P\Big[\big(M_0^{(R)}\big)^2\Big] + \sum_{n=1}^{\infty} \mathbb{E}^P\Big[\big(M_n^{(R)} - M_{n-1}^{(R)}\big)^2\Big] \\ &= \sup_{n\in\mathbb{N}} \mathbb{E}^P\Big[\big(M_n^{(R)}\big)^2\Big] \leq 12R\,\|f\|_{L^1(P)}.\end{aligned}$$

Hence,

$$P\left(\sup_{n\in\mathbb{N}} \big|N_n^{(R)}\big| \geq \frac{R}{2}\right) \leq \frac{48}{R}\,\|f\|_{L^1(P)}.$$

At the same time,

$$\sup_{n\in\mathbb{N}} \big|W_n^{(R)}\big| \leq \sum_{n=1}^{\infty} \big|W_n^{(R)} - W_{n-1}^{(R)}\big| \leq \sum_{n=1}^{\infty} \big|V_n^{(R)} - V_{n-1}^{(R)}\big|,$$

and so

$$\begin{aligned}&P\left(\sup_{n\in\mathbb{N}} \big|W_n^{(R)}\big| \geq \frac{R}{2}\right) \\ &\qquad \leq \frac{4}{R^2}\mathbb{E}^P\left[\left(\sum_{n=1}^{\infty} \big|V_n^{(R)} - V_{n-1}^{(R)}\big|\right)^2\right] \leq \frac{48}{R}\,\|f\|_{L^1(P)}.\end{aligned}$$

Thus, after combining these, we arrive at the estimate in (6.3.2).

Given (6.3.2) and (6.3.3), (6.3.4) becomes an application of the Marcinkiewicz Interpolation Theorem (cf. Theorem 6.2.19). Indeed, because we have assumed that $\mathcal{F}_n = \mathcal{F}$ for $n \geq N$,

$$\sup_{n\in\mathbb{N}} |K_n f| = \|\mathbf{K}f\|_F,$$

where

$$f \in L^2(P;\mathbb{R}) \longmapsto \mathbf{K}f \equiv \big(K_0 f, \dots, K_N f\big) \in L^2(P;E)$$

and $F = \mathbb{R}^{N+1}$ with the norm $\|x\|_F = \max_{0\leq n\leq N} |x_n|$. But, with this notation, (6.3.2) and (6.3.3) become, respectively,

$$P\big(\|\mathbf{K}f\|_F \geq R\big) \leq \frac{98}{R}\|f\|_{L^1(P)} \quad \text{and} \quad \big\|\mathbf{K}f\big\|_{L^2(P;F)} \leq 2\|f\|_{L^2(P)};$$

and therefore, by Theorem 6.2.19, we know that there exists a universal $C \in (0,\infty)$ such that

$$\begin{aligned} \big\|K_N f\big\|_{L^p(P)} &\leq \bigg\|\sup_{n\in\mathbb{N}}\big|K_n f\big|\bigg\|_{L^p(P)} \\ &\leq \frac{C}{(p-1)\wedge(2-p)}\|f\|_{L^p(P)}, \quad p\in(1,2); \end{aligned} \tag{6.3.5}$$

and this completes the proof of (6.3.4) for $p \in \big(1,\frac{3}{2}\big)$. To handle large p's, note that K_N is self-adjoint on $L^2(P;\mathbb{R})$ and therefore that, by duality and (6.3.5),

$$\big\|K_N f\big\|_{L^{p'}(P)} \leq \frac{C}{(p-1)\wedge(2-p)}\|f\|_{L^{p'}(P)} \quad \text{for } p \in (1,2).$$

But $\big(K_n f, \mathcal{F}_n, P\big)$ is a martingale, and therefore, by (5.2.6), the preceding implies that

$$\begin{aligned} \big\|\mathbf{K}f\big\|_{L^p(P;E)} &= \bigg\|\sup_{n\in\mathbb{N}}\big|K_n f\big|\bigg\|_{L^p(P)} \\ &\leq \frac{2C}{(p'-1)\wedge(2-p')}\|f\|_{L^p(P)}, \quad p\in(2,\infty). \end{aligned}$$

Thus, we now have (6.3.4) for $p \in (3,\infty)$. Finally, to get (6.3.4) for $p \in \big[\frac{3}{2},3\big]$, note that, by the results already obtained for $p_1 = \frac{5}{4}$ and $p_2 = 5$,

$$P\big(\|\mathbf{K}f\|_F \geq R\big) \leq \left(\frac{8C\,\|f\|_{L^{p_j}(P)}}{R}\right)^{\frac{1}{p_j}}, \quad j \in \{1,2\},$$

and so another application of Theorem 6.2.19 completes the proof. □

Theorem 6.3.1 contains the essence of Burkholder's basic result: it says that *the multiplier problem associated with martingales is trivial in the sense that any uniformly bounded multiplier preserves $L^p(P;\mathbb{R})$ for every $p \in (1,\infty)$*. However, there are several more or less immediate reformulations which lend themselves better to applications.

6.3.6 Corollary. *Let $(X_n, \mathcal{F}_n, P)$ be a martingale which is bounded in $L^1(P)$, let $\{\sigma_n : n \in \mathbb{N}\}$ be an $\{\mathcal{F}_n : n \in \mathbb{N}\}$-progressively measurable sequence of $[-1,1]$-valued functions, and set*

$$Y_n = X_0 + \sum_{m=1}^{n} \sigma_{m-1}\big(X_m - X_{m-1}\big), \quad n \in \mathbb{N}. \tag{6.3.7}$$

Then

$$P\left(\sup_{n\in\mathbb{N}} |Y_n| \geq R\right) \leq \frac{98}{R} \sup_{n\in\mathbb{N}} \mathbb{E}^P\big[|X_n|\big], \quad R \in (0,\infty), \tag{6.3.8}$$

and, with the same constants C_p, $p \in (1,\infty)$, as in (6.3.4),

$$\mathbb{E}^P\left[\sup_{n\in\mathbb{N}} |Y_m|^p\right]^{\frac{1}{p}} \leq C_p \sup_{n\in\mathbb{N}} \mathbb{E}^P\Big[|X_n|^p\Big]^{\frac{1}{p}}, \quad p \in (1,\infty). \tag{6.3.9}$$

Finally, there exists a $\bigvee_0^\infty \mathcal{F}_n$-measurable $Y : \Omega \longrightarrow \mathbb{R}$ to which $\{Y_n\}_0^\infty$ converges P-almost surely; and, when $p \in (1,\infty)$ and $\{X_n\}_0^\infty$ is bounded in $L^p(P)$, the convergence takes place in $L^p(P)$ as well.

Proof: By taking $f = X_N$ and applying (6.3.2) and (6.3.3), one gets (6.3.8) and (6.3.9) first for the martingales

$$\big(X_{n\wedge N}, \mathcal{F}_n, P\big) \quad \text{and} \quad \big(Y_{n\wedge N}, \mathcal{F}_n, P\big)$$

and then for the original martingales after letting $N \nearrow \infty$. Moreover, once one has (6.3.9), the only assertion which requires comment is the one about the P-almost sure convergence of $\{Y_n\}_0^\infty$. To handle this problem, let $R \in (0,\infty)$ be given, use $X_n = M_n^{(R)} + V_n^{(R)} + \Delta_n^{(R)}$ to denote Gundy's decomposition (cf. Exercise 5.2.31) of $(X_n, \mathcal{F}_n, P)$ at level R, and define $\{N_n^{(R)}\}_0^\infty$, $\{W_n^{(R)}\}_0^\infty$, and $\{E_n^{(R)}\}_0^\infty$ accordingly, as in the proof of Theorem 6.3.1. Because $\big(N_n^{(R)}, \mathcal{F}_n, P\big)$ is an $L^2(P)$-bounded martingale and

$$\sum_1^\infty \big|W_n^{(R)} - W_{n-1}^{(R)}\big| < \infty \quad \text{(a.s., } P\text{)},$$

we know that $\lim_{n\to\infty}\left(N_n^{(R)}+W_n^{(R)}\right)\in\mathbb{R}$ exists P-almost surely. At the same time, $Y_n=N_n^{(R)}+W_n^{(R)}$, $n\in\mathbb{N}$, on $A(R)\equiv\left\{\omega: E_n^{(R)}(\omega)=0,\ n\in\mathbb{N}\right\}$ and

$$P\big(A(R)\complement\big)\le\frac{2}{R}\sup_{n\in\mathbb{N}}\|X_n\|_{L^1(P)}.$$

Hence, we now know that $\left\{Y_n\right\}_0^\infty$ fails to converge in $\mathbb{R}$ on a set of arbitrarily small P-measure. □

As we have already discussed, the preceding results are martingale analogs of the singular integral results obtained in Section 6.2; and as we saw in part **(ii)** of Exercise 6.2.37 and again in Exercise 6.2.41, such results are intimately connected to inequalities involving square functions. In the context of martingales, this relationship turns out to be an equivalence. To be precise, we have the following.

6.3.10 Theorem (Burkholder). *Let $(\Omega,\mathcal{F},P)$ be a probability space, $\{\mathcal{F}_n : n\in\mathbb{N}\}$ a nondecreasing sequence of sub-σ-algebras of $\mathcal{F}$, and $(X_n,\mathcal{F}_n,P)$ a martingale. Then, for each $p\in(1,\infty)$,*

$$\frac{p-1}{Bp^3}\mathbb{E}^P\left[\sup_{n\in\mathbb{N}}|X_n-X_0|^p\right]^{\frac{1}{p}}$$

$$\le\mathbb{E}^P\left[\left(\sum_{n=1}^{\infty}(X_n-X_{n-1})^2\right)^{\frac{p}{2}}\right]^{\frac{1}{p}} \tag{6.3.11}$$

$$\le\frac{Bp^2}{p-1}\sup_{n\in\mathbb{N}}\mathbb{E}^P\left[|X_n-X_0|^p\right]^{\frac{1}{p}},$$

where $B\in(0,\infty)$ is universal.

PROOF: Without loss in generality, we will assume that $X_0\equiv 0$ and that there is an $N\in\mathbb{Z}^+$ with the property that $\mathcal{F}_n=\mathcal{F}$ for all $n\ge N$. In particular, this means that $X_n=X_N$ whenever $n\ge N$; and so all extrema and sums in (6.3.11) involve only N quantities.

Our proof of (6.3.11) will turn on a special choice of the sequence $\{\sigma_n : n\in\mathbb{N}\}$ in Corollary 6.3.6. To guarantee that this choice is possible, first augment the underlying probability space by taking

$$\tilde\Omega=\Omega\times\{-1,1\}^{\mathbb{N}},\quad \tilde{\mathcal{F}}=\mathcal{F}\times\mathcal{B},\quad \text{and } \tilde P=P\times\mu,$$

where $\mathcal{B}$ is the Borel field over $\{-1,1\}^{\mathbb{N}}$ and μ is the Bernoulli measure $\frac{1}{2}(\delta_{-1}+\delta_1)^{\mathbb{N}}$ on $(\{-1,1\}^{\mathbb{N}},\mathcal{B})$. Next, we take $\tilde{\mathcal{F}}_n=\mathcal{F}_n\times\mathcal{B}$ for each $n\in\mathbb{N}$, and define

$$\tilde X_n(\tilde\omega)=X_n(\omega)\quad\text{for } (\omega,\boldsymbol{\sigma})\in\tilde\Omega.$$

It is then an easy matter to check that

$$\tilde{\omega} \in \tilde{\Omega} \longmapsto \left(\tilde{X}_0(\tilde{\omega}), \dots, \tilde{X}_n(\tilde{\omega}), \dots\right) \in \mathbb{R}^{\mathbb{N}}$$

has the same distribution under $\tilde{P}$ as

$$\omega \in \Omega \longmapsto \left(X_0(\omega), \dots, X_n(\omega), \dots\right) \in \mathbb{R}^{\mathbb{N}}$$

has under P. At the same time, it is clear that $\left(\tilde{X}_n, \tilde{\mathcal{F}}_n, \tilde{P}\right)$ is a martingale; and, obviously, if $\sigma_n(\tilde{\omega}) \equiv \sigma_n$ for $\tilde{\omega} = (\omega, \boldsymbol{\sigma})$, then the sequence $\left\{\sigma_n : n \in X_n : n \in \mathbb{N}\right\}$ is $\left\{\tilde{\mathcal{F}}_n : n \in \mathbb{N}\right\}$-progressively measurable. Thus, if we define

$$Y_n = \sum_{m=1}^{n} \sigma_{m-1}\left(\tilde{X}_m - \tilde{X}_{m-1}\right), \quad n \in \mathbb{N},$$

then (6.3.9) holds and leads to

$$\left[\int_\Omega \left(\int_{\{-1,1\}^{\mathbb{N}}} \left| \sum_{m=1}^{N} \sigma_{m-1}\left(X_m(\omega) - X_{m-1}(\omega)\right)\right|^p \mu(d\boldsymbol{\sigma}) \right) P(d\omega)\right]^{\frac{1}{p}}$$
$$\leq C_p \mathbb{E}^P\left[|X_N|^p\right]^{\frac{1}{p}}.$$

Hence, by Khinchine's Inequality (2.2.29) (with $\beta = 1$) applied to the μ-integral for each $\omega \in \Omega$, we obtain

$$\mathbb{E}^P\left[\left(\sum_{m=1}^{N} \left(X_m - X_{m-1}\right)^2\right)^{\frac{p}{2}}\right]^{\frac{1}{p}} \leq B_p \mathbb{E}^P\left[|X_N|^p\right]^{\frac{1}{p}},$$

with (cf. part (**iii**) of Exercise 2.2.26) B_p's which are bounded by a multiple of the C_p's in Theorem 6.3.1.

To prove the left-hand side of (6.3.11), note that (because $\sigma_n \in \{-1,1\}$ for every $n \in \mathbb{N}$)

$$X_n = \sum_{m=1}^{n} \sigma_m\left(Y_m - Y_{m-1}\right), \quad n \in \mathbb{N}.$$

Hence, by (6.3.9) with the rôles of the X_n's and Y_N's reversed, we now get

$$\mathbb{E}^P\left[\max_{n \in \mathbb{N}} |X_n|^p\right]^{\frac{1}{p}} \leq C_p \mathbb{E}^P\left[|Y_N|^p\right]^{\frac{1}{p}}$$

$$= C_p \left[\int_\Omega \left(\int_{\{-1,1\}^{\mathbb{N}}} \left| \sum_{m=1}^{N} \sigma_{m-1}\left(X_m(\omega) - X_{m-1}(\omega)\right)\right|^p \mu(d\boldsymbol{\sigma}) \right) P(d\omega)\right]^{\frac{1}{p}}$$

$$\leq B_p \mathbb{E}^P\left[\left(\sum_{m=1}^{N} \left(X_m - X_{m-1}\right)^2\right)^{\frac{p}{2}}\right]^{\frac{1}{p}},$$

where now (again see part (**iii**) of Exercise 2.2.26)

$$B_p = C_p K_p^{\frac{1}{p}} \le \frac{Cp^{\frac{5}{2}}}{p-1},$$

and we have used Stirling's Formula (cf. (1.3.23)) to get the final estimate. □

The inequality in (6.3.11) is called **Burkholder's inequality**, and it turns out to be one of two or three most useful tools in any analysis which involves martingales. Indeed, what it enables one to do is parlay *local* comparisons of two martingales into a *global* comparison. That is, although we derived it on the basis of (6.3.9) (in which the relative sizes of martingales are compared only if they are related by (6.3.7)), (6.3.11) itself enables one to compare the sizes of any pair of martingales for which one can compare corresponding increments. To be precise, we have the following.

6.3.12 Corollary. *Let $(\Omega, \mathcal{F}, P)$ and $\{\mathcal{F}_n : n \in \mathbb{N}\}$ be as in Theorem 6.3.10, and let $(X_n, \mathcal{F}_n, P)$ and $(Y_n, \mathcal{F}_n, P)$ be a pair of martingales. If*

$$\text{(6.3.13)} \qquad |X_0| \le |Y_0| \quad \text{and} \quad |X_n - X_{n-1}| \le |Y_n - Y_{n-1}|, \quad n \in \mathbb{Z}^+,$$

P-almost surely, then, with the same $B \in (0,\infty)$ as in (6.3.11),

$$\text{(6.3.14)} \qquad \|X_n\|_{L^p(P)} \le \left\| \max_{0 \le m \le n} |X_m| \right\|_{L^p(P)} \le \frac{3B^2 p^{\frac{9}{2}}}{(p-1)^2} \|Y_n\|_{L^p(P)}$$

for each $p \in (1,\infty)$ and $n \in \mathbb{N}$.

PROOF: When $X_0 = Y_0 \equiv 0$, one gets

$$\left\| \max_{0 \le m \le n} |X_m| \right\|_{L^p(P)} \le \frac{B^2 p^{\frac{9}{2}}}{(p-1)^2} \|Y_n\|_{L^p(P)}$$

as an immediate consequence of (6.3.11). In the general case, one combines this special case with Minkowski's inequality and thereby gets

$$\begin{aligned}
\left\| \max_{0 \le m \le n} |X_m| \right\|_{L^p(P)} &\le \|X_0\|_{L^p(P)} + \left\| \max_{0 \le m \le n} |X_m - X_0| \right\|_{L^p(P)} \\
&\le \|Y_0\|_{L^p(P)} + \frac{B^2 p^{\frac{9}{2}}}{(p-1)^2} \|Y_n - Y_0\|_{L^p(P)} \\
&\le \|Y_n\|_{L^p(P)} + \frac{2B^2 p^{\frac{9}{2}}}{(p-1)^2} \|Y_n\|_{L^p(P)} \\
&\le \frac{3B^2 p^{\frac{9}{2}}}{(p-1)^2} \|Y_n\|_{L^p(P)}. \quad \square
\end{aligned}$$

The route which we have been following thus far is more or less the same one as Burkholder mapped out when he first proved his inequality; and, for most purposes, the results to which it has brought us are adequate. Nonetheless, there are two directions in which one wants to improve these results. First, and most important, one would like to obtain a statement of (6.3.11) for vector valued martingales in such a way that the constants do not depend on dimension. (Notice that it is Gundy's version of the Calderón–Zygmund decomposition of martingales which weds the argument given to $\mathbb{R}$-valued martingales.) Secondly, one suspects that the application of Khinchine's inequality leads to less than optimal constants. In addition to these two objections, one can also say, with hindsight, that the problem here is essentially easier than the ones in Section 6.2 and should therefore yield to an essentially simpler argument. Quite recently, Burkholder has discovered *the right argument*: it removes the two objections mentioned and it is completely elementary. Unfortunately, it is also completely opaque. Indeed, his new argument is nothing but an elementary verification that he has got the right answer; it gives no hint about how he came to that answer.†

The strategy of Burkholder's new approach is to first derive a generalized version of Corollary 6.3.12. Namely, the main step in his new strategy is to prove the elegant statement which follows.

6.3.15 Theorem (Burkholder). *Let $(\Omega, \mathcal{F}, P)$ be a probability space, $\{\mathcal{F}_n : n \in \mathbb{N}\}$ a nondecreasing sequence of sub-σ-algebras of $\mathcal{F}$, and E and F a pair of separable Hilbert spaces. Next, suppose that $(\mathbf{X}_n, \mathcal{F}_n, P)$ and $(\mathbf{Y}_n, \mathcal{F}_n, P)$ are, respectively, E- and F-valued martingales. If*

$$\|\mathbf{X}_0\|_E \leq \|\mathbf{Y}_0\|_F \text{ and } \|\mathbf{X}_n - \mathbf{X}_{n-1}\|_E \leq \|\mathbf{Y}_n - \mathbf{Y}_{n-1}\|_F, \; n \in \mathbb{Z}^+, \tag{6.3.16}$$

P-almost surely, then, for each $p \in (1, \infty)$,

$$\begin{aligned} \|\mathbf{X}_n\|_{L^p(P;E)} \leq B_p \|\mathbf{Y}_n\|_{L^p(P;F)}, \quad n \in \mathbb{N}, \\ \text{where } B_p = (p-1) \vee \frac{1}{p-1} \end{aligned} . \tag{6.3.17}$$

Before giving the proof of Theorem 6.3.15, we will show that the desired form of Burkholder's inequality is a trivial corollary.

6.3.18 Corollary (Burkholder). *Let $(\Omega, \mathcal{F}, P)$ and $\{\mathcal{F}_n : n \in \mathbb{N}\}$ be as in Theorem 6.3.15, and let $(\mathbf{X}_n, \mathcal{F}_n, P)$ be a martingale with values in the separable*

---

† For those who want to know the secret behind this proof, Burkholder has written an explanation in his article "Explorations in martingale theory and its applications" for the 1989 Saint-Flour Ecole d'Eté lectures published by Springer–Verlag, LNM **1464** (1991).

Hilbert space E. Then, for each $p \in (1,\infty)$ and $n \in \mathbb{N}$,

$$\begin{aligned}\frac{1}{B_p}\sup_{n\in\mathbb{N}}\left\|\mathbf{X}_n - \mathbf{X}_0\right\|_{L^p(P;E)} \\ \le \mathbb{E}^P\left[\left(\sum_1^\infty \left\|\mathbf{X}_n - \mathbf{X}_{n-1}\right\|_E^2\right)^{\frac{p}{2}}\right]^{\frac{1}{p}} \\ B_p \sup_{n\in\mathbb{N}}\left\|\mathbf{X}_n - \mathbf{X}_0\right\|_{L^p(P;E)},\end{aligned} \tag{6.3.19}$$

with B_p as in (6.2.17).

PROOF: Let $F = \ell^2(\mathbb{N};E)$ be the separable Hilbert space of sequences

$$\mathbf{y} = \left(\mathbf{x}_0, \dots, \mathbf{x}_n, \dots\right) \in E^{\mathbb{N}}$$

satisfying

$$\|\mathbf{y}\|_F \equiv \left(\sum_0^\infty \|\mathbf{x}_n\|^2\right)^{\frac{1}{2}} < \infty;$$

and define

$$\mathbf{Y}_n(\omega) = (\mathbf{X}_0(\omega), \mathbf{X}_1(\omega) - \mathbf{X}_0(\omega), \dots, \mathbf{X}_n(\omega) - \mathbf{X}_{n-1}(\omega), \mathbf{0}, \mathbf{0}, \dots) \in F$$

for $\omega \in \Omega$ and $n \in \mathbb{N}$. Obviously, $\left(\mathbf{Y}_n, \mathcal{F}_n, P\right)$ is an F-valued martingale. Moreover,

$$\|\mathbf{X}_0\|_E = \|\mathbf{Y}_0\|_F \quad \text{and} \quad \|\mathbf{X}_n - \mathbf{X}_{n-1}\|_E = \|\mathbf{Y}_n - \mathbf{Y}_{n-1}\|_F, \quad n \in \mathbb{N};$$

and therefore the left-hand side of (6.3.19) is implied by (6.3.17) while the right-hand side also follows from (6.3.17) when the rôles of the $\mathbf{X}_n$ and $\mathbf{Y}_n$'s are reversed. □

We now turn to the proof of (6.3.17), which, as we said before, is both elementary and mysterious. The heart of the proof lies in the computations contained in the following two lemmas.

6.3.20 Lemma. *Let $p \in (1,\infty)$ be given, set*

$$\alpha_p = \begin{cases} p^{2-p}(p-1)^{p-1} & \text{if } \ p \in [2,\infty) \\ p^{2-p} & \text{if } \ p \in (1,2], \end{cases}$$

and define $u : E \times F \longrightarrow \mathbb{R}$ by

$$u(\mathbf{x},\mathbf{y}) = \begin{cases} \left(\|\mathbf{x}\|_E - (p-1)\|\mathbf{y}\|_F\right)\left(\|\mathbf{x}\|_E + \|\mathbf{y}\|_F\right)^{p-1} & \text{if } \ p \in [2,\infty) \\ \left(\|\mathbf{x}\|_E - \frac{\|\mathbf{y}\|_F}{p-1}\right)\left(\|\mathbf{x}\|_E + \|\mathbf{y}\|_F\right)^{p-1} & \text{if } \ p \in (1,2]. \end{cases}$$

Then (cf. (6.2.17))

$$\|\mathbf{x}\|_E^p - \big(B_p\|\mathbf{y}\|_F\big)^p \le \alpha_p\, u(\mathbf{x},\mathbf{y}), \quad (\mathbf{x},\mathbf{y}) \in E \times F.$$

PROOF: By embedding both E and F into $E \times F$ if necessary, we may and will assume the $E = F$. Next, observe that it suffices to show that for all $(\mathbf{x},\mathbf{y}) \in E^2$ satisfying $\|\mathbf{x}\|_E + \|\mathbf{y}\|_E = 1$:

$$(6.3.21) \quad \|\mathbf{x}\|_E^p - \big((p-1)\|\mathbf{y}\|_E\big)^p \begin{cases} \le (p-1)^{p-1}p^{2-p}\big(\|\mathbf{x}\|_E - (p-1)\|\mathbf{y}\|_E\big) \\ \ge (p-1)^{p-1}p^{2-p}\big(\|\mathbf{x}\|_E - (p-1)\|\mathbf{y}\|_E\big). \end{cases}$$

depending on whether $p \in (2,\infty)$ or $p \in (1,2)$. Indeed, when $p = 2$ there is nothing to do, when $p \in (2,\infty)$, (6.3.21) is precisely the result desired, and, when $p \in (1,2)$, (6.3.21) gives the desired result after one divides through by $(p-1)^{p-1}$ and reverses the rôles of $\mathbf{x}$ and $\mathbf{y}$.

We begin the verification of (6.3.21) by checking that

$$p^{2-p}\,(p-1)^{p-1} \begin{cases} > 1 & \text{if} \quad p \in (2,\infty) \\ < 1 & \text{if} \quad p \in (1,2). \end{cases}$$

To this end, set $f(p) = (p-1)\log(p-1) - (p-2)\log p$ for $p \in (1,\infty)$. Then f is strictly convex on $(1,2)$ and strictly concave on $(2,\infty)$. Thus, $f \restriction (1,2)$ cannot achieve a maximum and therefore, since $\lim_{p\searrow 1} f(p) = 0 = f(2)$, $f < 0$ on $(1,2)$. Similarly, $f \restriction (2,\infty)$ cannot achieve a minimum, and therefore, since $f(2) = 0$ while $\lim_{p\nearrow\infty} f(p) = \infty$, we have that $f > 0$ on $(2,\infty)$.

Next, observe that proving (6.3.21) comes down to checking that

$$\Phi(s) \equiv p^{2-p}\,(p-1)^{p-1}\,(1-ps) - (1-s)^p + (p-1)^p\, s^p \begin{cases} \ge 0 & \text{if } p \in (2,\infty) \\ \le 0 & \text{if } p \in (1,2), \end{cases}$$

for $s \in [0,1]$. To this end, note that, by the preceding paragraph, $\Phi(0) > 0$ when $p \in (2,\infty)$ and $\Phi(0) < 0$ when $p \in (1,2)$. Also, for $s \in (0,1)$,

$$\Phi'(s) = p\Big[(p-1)^p\, s^{p-1} + (1-s)^{p-1} - p^{2-p}\,(p-1)^{p-1}\Big]$$

and

$$\Phi''(s) = p(p-1)\Big[(p-1)^p\, s^{p-2} - (1-s)^{p-2}\Big].$$

In particular, we see that:

$$\Phi\left(\tfrac{1}{p}\right) = \Phi'\left(\tfrac{1}{p}\right) = 0.$$

In addition, depending on whether $p \in (2,\infty)$ or $p \in (1,2)$: $\lim_{s\searrow 0} \Phi''(0)$ is negative or positive, Φ'' is strictly increasing or decreasing on $(0,1)$, and $\lim_{s\nearrow 1} \Phi''(1)$

is positive or negative. Hence, there exists a unique $t \in (0,1)$ with the property that

$$\Phi'' \restriction (0,t)\begin{cases} <0 & \text{if } p\in(2,\infty)\\ >0 & \text{if } p\in(1,2)\end{cases} \quad\text{and}\quad \Phi'' \restriction (t,1)\begin{cases} >0 & \text{if } p\in(2,\infty)\\ <0 & \text{if } p\in(1,2)\end{cases}.$$

Moreover, from the equation $\Phi''(t)=0$, it is easy to see that $t\in\left(1,\frac{1}{p}\right)$. Now suppose that $p\in(2,\infty)$ and consider Φ on each of the intervals $\left[\frac{1}{p},1\right]$, $\left[t,\frac{1}{p}\right]$, and $\left[0,\frac{1}{p}\right]$ separately. Because both Φ and Φ' vanish at $\frac{1}{p}$ while $\Phi''>0$ on $\left(\frac{1}{p},1\right)$, it is clear that $\Phi>0$ on $\left(\frac{1}{p},1\right]$. Next, because $\Phi'\left(\frac{1}{p}\right)=0$ and $\Phi''\restriction\left(t,\frac{1}{p}\right)>0$, we know that Φ is strictly decreasing on $\left(t,\frac{1}{p}\right)$ and therefore that $\Phi\restriction\left[t,\frac{1}{p}\right)>\Phi\left(\frac{1}{p}\right)=0$. Finally, because $\Phi''\restriction(0,t)<0$ while $\Phi(0)\wedge\Phi(t)\geq 0$, we also know that $\Phi\restriction(0,t)>0$. The argument when $p\in(1,2)$ is similar; only this time all the signs are reversed. □

6.3.22 Lemma. *Again let $p\in(1,\infty)$ be given, and define $u: E\times F\longrightarrow\mathbb{R}$ as in Lemma 6.3.20. In addition, define the functions v and w on $E\times F\setminus\{\mathbf{0},\mathbf{0}\}$ by*

$$v(\mathbf{x},\mathbf{y})=p\big(\|\mathbf{x}\|_E+\|\mathbf{y}\|_F\big)^{p-2}\big(\|\mathbf{x}\|_E+(2-p)\|\mathbf{y}\|_F\big)$$

and

$$w(\mathbf{x},\mathbf{y})=p(1-p)\big(\|\mathbf{x}\|_E+\|\mathbf{y}\|_F\big)^{p-2}\|\mathbf{y}\|_F.$$

Then, for $(\mathbf{x},\mathbf{k})\in E^2$ and $(\mathbf{y},\mathbf{h})\in F^2$ satisfying

$$\min_{t\in[0,1]}\big(\|\mathbf{x}+t\mathbf{k}\|_E\wedge\|\mathbf{y}+t\mathbf{h}\|_F\big)>0\quad\text{and}\quad\|\mathbf{k}\|_E\leq\|\mathbf{h}\|_F,$$

$$\begin{aligned} &u(\mathbf{x}+\mathbf{k},\mathbf{y}+\mathbf{h})-u(\mathbf{x},\mathbf{y})\\ &\qquad\leq v(\mathbf{x},\mathbf{y})\,\mathfrak{Re}\left(\tfrac{\mathbf{x}}{\|\mathbf{x}\|_E},\mathbf{k}\right)_E+w(\mathbf{x},\mathbf{y})\,\mathfrak{Re}\left(\tfrac{\mathbf{y}}{\|\mathbf{y}\|_F},\mathbf{h}\right)_E \end{aligned}\tag{6.3.23}$$

when $p\in[2,\infty)$ and

$$\begin{aligned} &(p-1)\big[u(\mathbf{x}+\mathbf{k},\mathbf{y}+\mathbf{h})-u(\mathbf{x},\mathbf{y})\big]\\ &\qquad\leq -w(\mathbf{y},\mathbf{x})\,\mathfrak{Re}\left(\tfrac{\mathbf{x}}{\|\mathbf{x}\|_E},\mathbf{k}\right)_E-v(\mathbf{y},\mathbf{x})\,\mathfrak{Re}\left(\tfrac{\mathbf{y}}{\|\mathbf{y}\|_F},\mathbf{h}\right)_E \end{aligned}\tag{6.3.24}$$

when $p\in(1,2]$.

PROOF: Just as in the proof of Lemma 6.3.20, we may and will assume that $E = F$. Next, set

$$\begin{aligned}\Phi(t) &= \Phi\big(t; (\mathbf{x},\mathbf{k}), (\mathbf{y},\mathbf{h})\big)\\ &\equiv \big(\|\mathbf{x}+t\mathbf{k}\|_E - (p-1)\|\mathbf{y}+t\mathbf{h}\|_E\big)\big(\|\mathbf{x}+t\mathbf{k}\|_E + \|\mathbf{y}+t\mathbf{h}\|_E\big)^{p-1},\end{aligned}$$

and observe that

$$u\big(\mathbf{x}+t\mathbf{k}, \mathbf{y}+t\mathbf{h}\big) = \begin{cases} \Phi\big(t; (\mathbf{x},\mathbf{k}), (\mathbf{y},\mathbf{h})\big) & \text{if } \ p \in [2,\infty) \\ -(p-1)^{-1}\,\Phi\big(t; (\mathbf{y},\mathbf{h}), (\mathbf{x},\mathbf{k})\big) & \text{if } \ p \in (1,2). \end{cases}$$

Hence, it suffices for us to prove that

$$\Phi'(t) = v(\mathbf{x}+t\mathbf{k}, \mathbf{y}+t\mathbf{h})\left(\tfrac{\mathbf{x}+t\mathbf{k}}{\|\mathbf{x}+t\mathbf{k}\|_E}, \mathbf{k}\right)_E + w(\mathbf{x}+t\mathbf{h}, \mathbf{y}+t\mathbf{h})\left(\tfrac{\mathbf{y}+t\mathbf{h}}{\|\mathbf{y}+t\mathbf{h}\|_E}, \mathbf{h}\right)_E$$

and that

$$\Phi''\big(t; (\mathbf{x},\mathbf{k}), (\mathbf{y},\mathbf{h})\big)\begin{cases} \le 0 & \text{if } p \in [2,\infty) \text{ and } \|\mathbf{k}\|_E \le \|\mathbf{h}\|_E \\ \ge 0 & \text{if } p \in (1,2] \ \text{ and } \|\mathbf{k}\|_E \ge \|\mathbf{h}\|_E. \end{cases}$$

To prove the preceding, set $\mathbf{x}(t) = \mathbf{x}+t\mathbf{k}$, $\mathbf{y}(t) = \mathbf{y}+t\mathbf{h}$, $\Psi(t) = \|\mathbf{x}(t)\|_E + \|\mathbf{y}(t)\|_E$,

$$a(t) = \frac{\mathfrak{Re}\big(\mathbf{x}(t),\mathbf{k}\big)_E}{\|\mathbf{x}(t)\|_E}, \quad \text{and} \quad b(t) = \frac{\mathfrak{Re}\big(\mathbf{y}(t),\mathbf{h}\big)_E}{\|\mathbf{y}(t)\|_E}.$$

One then has that

$$\begin{aligned}\Phi'(t) &= p\Psi(t)^{p-2}\Big[\Big(\|\mathbf{x}(t)\|_E + (2-p)\|\mathbf{y}(t)\|_E\Big)\,a(t) + (1-p)\|\mathbf{y}(t)\|_E\,b(t)\Big]\\ &= p\Big[(1-p)\Psi(t)^{p-2}\,\|\mathbf{y}(t)\|_E\,\big(a(t)+b(t)\big) + \Psi(t)^{p-1}\,a(t)\Big].\end{aligned}$$

In particular, the first expression in the preceding establishes the required form for $\Phi'(t)$. In addition, from the second expression, we see that

$$\begin{aligned}-\frac{\Phi''(t)}{p} &= (p-1)(p-2)\,\Psi(t)^{p-3}\,\|\mathbf{y}(t)\|_E\,\big(a(t)+b(t)\big)^2\\ &\qquad + (p-1)\Psi(t)^{p-2}\Big[b(t)\big(a(t)+b(t)\big) + \tfrac{\|\mathbf{y}(t)\|_E}{\|\mathbf{x}(t)\|_E}\,a_\perp(t)^2 + b_\perp(t)^2\Big]\\ &\qquad - \Psi(t)^{p-2}\Big[(p-1)\big(a(t)+b(t)\big)\,a(t) + \Psi(t)\,\tfrac{a_\perp(t)^2}{\|\mathbf{x}(t)\|_E}\Big]\\ &= (p-1)(p-2)\,\Psi(t)^{p-3}\,\|\mathbf{y}(t)\|_E\,\big(a(t)+b(t)\big)^2\\ &\qquad + (p-1)\Psi(t)^{p-2}\big(\|\mathbf{h}\|_E^2 - \|\mathbf{k}\|_E^2\big)\\ &\qquad + (p-2)\Psi(t)^{p-1}\,\tfrac{a_\perp(t)^2}{\|\mathbf{x}(t)\|_E},\end{aligned}$$

where

$$a_\perp(t) = \sqrt{\|\mathbf{k}\|_E^2 - a(t)^2} \quad \text{and} \quad b_\perp(t) = \sqrt{\|\mathbf{h}\|_E^2 - b(t)^2}.$$

Hence the required properties of $\Phi''(t)$ have also been established. □

PROOF OF THEOREM 6.3.15: Again we will assume that $E = F$. In addition, we set $\mathbf{K}_n = \mathbf{X}_n - \mathbf{X}_{n-1}$ and $\mathbf{H}_n = \mathbf{Y}_n - \mathbf{Y}_{n-1}$ for $n \in \mathbb{Z}^+$, and we will assume that there is an $\epsilon > 0$ with the property that

$$\left\|\mathbf{X}_0(\omega) - \operatorname{span}\{\mathbf{K}_n(\omega)\} : n \in \mathbb{Z}^+\}\right\|_E \geq \epsilon$$

and

$$\left\|\mathbf{Y}_0(\omega) - \operatorname{span}\{\mathbf{H}_n(\omega)\} : n \in \mathbb{Z}^+\}\right\|_E \geq \epsilon$$

for all $\omega \in \Omega$. Indeed, if this is not already the case, then we can replace E by $\mathbb{R} \times E$ or, when E is complex, $\mathbb{C} \times E$, $\mathbf{X}_n(\omega)$ and $\mathbf{Y}_n(\omega)$, respectively, by

$$\mathbf{X}_n^{(\epsilon)}(\omega) \equiv \big(\epsilon, \mathbf{X}_n(\omega)\big) \quad \text{and} \quad \mathbf{Y}_n^{(\epsilon)}(\omega) \equiv \big(\epsilon, \mathbf{Y}_n(\omega)\big),$$

for each $n \in \mathbb{N}$; and, clearly, (6.3.17) for each $\mathbf{X}_n^{(\epsilon)}$ and $\mathbf{Y}_n^{(\epsilon)}$ implies (6.3.17) for $\mathbf{X}_n$ and $\mathbf{Y}_n$ after one lets $\epsilon \searrow 0$. Finally, because there is nothing to do when the right-hand side of (6.3.17) is infinite, we let $p \in (1, \infty)$ be given, and assume that $\mathbf{Y}_n \in L^p(P; E)$ for each $n \in \mathbb{N}$. In particular, if u is the function defined in Lemma 6.3.20 and v and w are those defined in Lemma 6.3.22, then

$$u(\mathbf{X}_n, \mathbf{Y}_n) \in L^1(P; \mathbb{R}) \quad \text{and} \quad v(\mathbf{X}_n, \mathbf{Y}_n),\ w(\mathbf{X}_n, \mathbf{Y}_n) \in L^{p'}(P; \mathbb{R})$$

for all $n \in \mathbb{N}$.

Note that, by Lemma 6.3.20, it suffices for us to show that

$$A_n \equiv \mathbb{E}^P\Big[u\big(\mathbf{X}_n, \mathbf{Y}_n\big)\Big] \leq 0, \quad n \in \mathbb{N}. \tag{6.3.25}$$

Since $u\big(\mathbf{X}_0, \mathbf{Y}_0\big) \leq 0$ P-almost surely, there is no question that $A_0 \leq 0$. Next, assume that $A_n \leq 0$, and, depending on whether $p \in [2, \infty)$ or $p \in (1, 2]$, use (6.3.23) or (6.3.24) to see that

$$\begin{aligned} A_{n+1} \leq & \mathbb{E}^P\left[v(\mathbf{X}_n, \mathbf{Y}_n)\left(\tfrac{\mathbf{X}_n}{\|\mathbf{X}_n\|_E}, \mathbf{K}_{n+1}\right)_E\right] \\ & + \mathbb{E}^P\left[w(\mathbf{X}_n, \mathbf{Y}_n)\left(\tfrac{\mathbf{Y}_n}{\|\mathbf{Y}_n\|_E}, \mathbf{H}_{n+1}\right)_E\right] \end{aligned}$$

or

$$\begin{aligned} A_{n+1} \leq & -\mathbb{E}^P\left[w(\mathbf{Y}_n, \mathbf{X}_n)\left(\tfrac{\mathbf{X}_n}{\|\mathbf{X}_n\|_E}, \mathbf{K}_{n+1}\right)_E\right] \\ & - \mathbb{E}^P\left[v(\mathbf{Y}_n, \mathbf{X}_n)\left(\tfrac{\mathbf{Y}_n}{\|\mathbf{Y}_n\|_E}, \mathbf{H}_{n+1}\right)_E\right]. \end{aligned}$$

But $v(\mathbf{X}_n, \mathbf{Y}_n)\frac{\mathbf{X}_n}{\|\mathbf{X}_n\|_E}$ is $\mathcal{F}_n$-measurable, $\mathbb{E}^P[\mathbf{K}_n|\mathcal{F}_n] = \mathbf{0}$, and therefore (cf. Exercise 5.1.36)

$$\mathbb{E}^P\left[v(\mathbf{X}_n, \mathbf{Y}_n)\left(\tfrac{\mathbf{X}_n}{\|\mathbf{X}_n\|_E}, \mathbf{K}_{n+1}\right)_E\right] = 0.$$

Since the same reasoning shows that each of the other terms on the right-hand side vanishes, we have now proved that $A_{n+1} \le 0$. □

Exercises

6.3.26 Exercise: Because it arises repeatedly in the theory of stochastic integration, one of the most frequent applications of Burkholder's inequality is to situations in which $(\mathbf{X}_n, \mathcal{F}_n, P)$ is an E-valued martingale for which one has an estimate of the form

$$K_p \equiv \sup_{m\in\mathbb{Z}^+} \left\| \mathbb{E}^P\Big[\|\mathbf{X}_m - \mathbf{X}_{m-1}\|_E^{2p} \big| \mathcal{F}_{m-1}\Big]^{\frac{1}{2p}} \right\|_{L^\infty(P)} < \infty \tag{6.3.27}$$

for some $p \in [1, \infty)$. To see how (6.3.27) gets used, let F be a second separable Hilbert space and suppose that $\{\boldsymbol{\sigma}_m : m \in \mathbb{N}\}$ is an $\{\mathcal{F}_n : n \in \mathbb{N}\}$-progressively measurable sequence of maps $\boldsymbol{\sigma}_m : \Omega \longrightarrow \mathrm{Hom}(E; F)$ for which

$$\sigma_m \equiv \mathbb{E}^P\Big[\|\boldsymbol{\sigma}_m\|_{E\to F}^{2p}\Big]^{\frac{1}{2p}} < \infty, \quad m \in \mathbb{N},$$

set

$$\mathbf{Y}_n = \sum_{m=1}^{n} \boldsymbol{\sigma}_{m-1}\big(\mathbf{X}_m - \mathbf{X}_{m-1}\big) \quad \text{for } n \in \mathbb{N},$$

and show that

$$\big\|\mathbf{Y}_n\big\|_{L^{2p}(P;F)} \le (2p-1)\, K_p \left(\sum_{m=0}^{n-1} \sigma_m^2\right)^{\frac{1}{2}}. \tag{6.3.28}$$

6.3.29 Exercise: Let $\{c_m : |m| > 1\} \subseteq \mathbb{C}$ be square summable sequence, and determine $f \in L^2\big([-1,1];\mathbb{C}\big)$ by the Fourier series

$$f = \sum_{|m|>1} c_m e^{2^m \pi t}.$$

Using part (**v**) of Exercise 5.1.28 and Burkholder's inequality, show that, for each $p \in (1, \infty)$,

$$\begin{aligned} &(p-1)\wedge\frac{1}{p-1}\|f\|_{L^p([-1,1])} \\ &\le \left(\int_{[-1,1]} \left(\sum_{m=2}^{\infty} \Big|c_m e^{2^m \pi t} + c_{-m} e^{-2^m \pi t}\Big|^2\right)^{\frac{p}{2}} dt\right)^{\frac{1}{p}} \\ &\le (p-1)\vee\frac{1}{p-1}\|f\|_{L^p([-1,1])}. \end{aligned} \tag{6.3.30}$$

Apart from the constants, the inequalities in (6.3.30) are a very special case of a famous theorem proved by Littlewood and Paley generalizing Parseval's identity when $p \neq 2$.

6.3.31 Exercise: In connection with the preceding exercise, it is interesting to note that there is an orthonormal basis for $L^2\big([0,1);\mathbb{R}\big)$ which, as distinguished from the trigonometric functions, can be completely understood in terms of martingale analysis. Namely, recall the Rademacher functions $\{R_n : n \in \mathbb{Z}^+\}$ introduced in Section 1.1. Next, use $\mathfrak{F}$ to denote the set of all finite subsets F of $\mathbb{Z}^+$, and define the **Walsh function** W_F for $F \in \mathfrak{F}$ by

$$W_F = \begin{cases} \mathbf{1} & \text{if } F = \emptyset \\ \prod_{m\in F} R_m & \text{if } F \neq \emptyset. \end{cases} \tag{6.3.32}$$

Finally, set $A_0 = \emptyset$ and $A_n = \{1,\dots,n\}$ for $n \in \mathbb{Z}^+$.

(i) For each $n \in \mathbb{N}$, let $\mathcal{F}_n$ be the σ-algebra generated by the partition

$$\left\{\left[\tfrac{k}{2^n}, \tfrac{k+1}{2^n}\right) : 0 \le k < 2^n\right\},$$

and show that $\{W_F : F \subseteq A_n\}$ is an orthonormal basis for the subspace $L^2\big([0,1),\mathcal{F}_n,\text{Leb};\mathbb{R}\big)$; and conclude from this that $\{W_F : F \in \mathfrak{F}\}$ forms an orthonormal basis for $L^2\big([0,1);\mathbb{R}\big)$.

(ii) Let $f \in L^1\big([0,1);\mathbb{R}\big)$ be given, and set

$$X_n^f = \sum_{F\subseteq A_n} \left(\int_{[0,1)} f(t)\,W_F(t)\,dt\right) W_F \quad \text{for } n \in \mathbb{N}. \tag{6.3.33}$$

Using the result in (i), show that $X_n^f = E^{\text{Leb}}[f|\mathcal{F}_n]$ and therefore that $\big(X_n^f, \mathcal{F}_n, \text{Leb}\big)$ is a martingale. In particular, $X_n^f \longrightarrow f$ both (a.e., Leb) and in $L^1\big([0,1],\mathbb{R}\big)$.

(iii) Show that for each $p \in (1,\infty)$ and $f \in L^1\big([0,1);\mathbb{R}\big)$:

$$\begin{aligned}
&(p-1)\wedge(p-1)^{-1}\|f\|_{L^p([0,1))}\\
&\quad\le \left[\int_{[0,1)} \left(\sum_{n=0}^{\infty}\left[\sum_{|F|=n}\left(\int_{[0,1)} f(s)\,W_F(s)\,ds\right) W_F(t)\right]^2\right)^{\frac{p}{2}} dt\right]^{\frac{1}{p}}\\
&\quad\le (p-1)\vee(p-1)^{-1}\|f\|_{L^p([0,1))}.
\end{aligned}$$

Chapter VII:
Continuous Martingales and Elementary Diffusion Theory

§7.1: Continuous Parameter Martingales.

It turns out that many of the ideas and results introduced in Section 5.2 can be easily transferred to the setting of processes depending on a *continuous time parameter*. Thus, let $(\Omega, \mathcal{F})$ be a measurable space and $\{\mathcal{F}_t : t \in [0,\infty)\}$ a nondecreasing family of sub-σ-algebras. We will say that a function X on $[0,\infty) \times \Omega$ into a measurable space $(E, \mathcal{B})$ is **$\{\mathcal{F}_t : t \in [0,\infty)\}$-progressively measurable** if $X \upharpoonright [0,T] \times \Omega$ is $\mathcal{B}_{[0,T]} \times \mathcal{F}_T$-measurable for every $T \in [0,\infty)$. When E is a metric space, we say that $X : [0,\infty)\times\Omega \longrightarrow E$ is **right-continuous** or **left-continuous** depending on whether

$$X(s,\omega) = \lim_{t \searrow s} X(t,\omega) \quad \text{or} \quad X(s,\omega) = \lim_{t \nearrow s} X(t,\omega)$$

for every $(s,\omega) \in [0,\infty) \times \Omega$; and, of course, if X is both left- and right-continuous, we say that it is **continuous**.

7.1.1 Remark. The reader might have been expecting a slightly different definition of progressive measurability here. Namely, he might have thought that one would say that X is $\{\mathcal{F}_t : t \in [0,\infty)\}$-progressively measurable if it is $\mathcal{B}_{[0,\infty)} \times \mathcal{F}$-measurable and $\omega \in \Omega \longmapsto X(t,\omega) \in E$ is $\mathcal{F}_t$-measurable for each $t \in [0,\infty)$. Indeed, in extrapolating from the discrete parameter setting, this would be the first definition at which one would arrive. In fact, it was the notion with which Doob and Itô originally worked; and such functions were said by them to be **$\{\mathcal{F}_t : t \in [0,\infty)\}$-adapted**. However, it came to be realized that there are various problems with the notion of adaptedness. For example, even if X is $\{\mathcal{F}_t : t \in [0,\infty)\}$-adapted and $f : E \longrightarrow \mathbb{R}$ is a bounded, $\mathcal{B}$-measurable function, the function

$$(t,\omega) \longmapsto Y(t,\omega) \equiv \int_0^t f\big(X(s,\omega)\big)\, ds \in \mathbb{R}$$

need not be $\{\mathcal{F}_t : t \in [0,\infty)\}$-adapted. On the other hand, if X is $\{\mathcal{F}_t : t \in [0,\infty)\}$-progressively measurable, then Y will be also.

The following simple lemma should help to explain the virtue of progressive measurability and its relationship to adaptedness.

7.1.2 Lemma. *Let $\mathcal{PM}$ denote the set of $A \subseteq [0,\infty) \times \Omega$ such that $\mathbf{1}_A$ is an $\{\mathcal{F}_t : t \in [0,\infty)\}$-progressively measurable function. Then $\mathcal{PM}$ is a sub-σ-algebra of $\mathcal{B}_{[0,\infty)} \times \mathcal{F}$ and X is $\{\mathcal{F}_t : t \in [0,\infty)\}$-progressively measurable if and only if it is $\mathcal{PM}$-measurable. Furthermore, if E is a metric space and $X : [0,\infty) \times \Omega \longrightarrow E$ is a left-continuous function which is $\{\mathcal{F}_t : t \in [0,\infty)\}$-adapted, then X is $\{\mathcal{F}_t : t \in [0,\infty)\}$-progressively measurable. Finally, if*

$$\mathcal{F}_{s+} \equiv \bigcap_{t>s} \mathcal{F}_t \quad \text{for} \quad s \in [0,\infty), \tag{7.1.3}$$

then every right-continuous, $\{\mathcal{F}_t : t \in [0,\infty)\}$-adapted function is $\{\mathcal{F}_{t+} : t \in [0,\infty)\}$-progressively measurable

PROOF: First note that

$$A \in \mathcal{PM} \iff \big([0,T] \times \Omega\big) \cap A \in \mathcal{B}_{[0,T]} \times \mathcal{F}_T \quad \text{for all } T \in [0,\infty);$$

from which it is clear that $\mathcal{PM}$ is a σ-algebra. Furthermore, for any $X : [0,\infty) \times \Omega \longrightarrow E$, $T \in [0,\infty)$, and $\Gamma \in \mathcal{B}_E$,

$$\begin{aligned}&\big\{(t,\omega) \in [0,T] \times \Omega : X(t,\omega) \in \Gamma\big\} \\ &\quad = \big([0,T] \times \Omega\big) \cap \big\{(t,\omega) \in [0,\infty) \times \Omega : X(t,\omega) \in \Gamma\big\},\end{aligned}$$

and so X is $\{\mathcal{F}_t : t \in [0,\infty)\}$-progressively measurable if and only if it is $\mathcal{PM}$-measurable. Hence, the first assertion has been proved.

Next, suppose that X is a left-continuous, $\{\mathcal{F}_t : t \in [0,\infty)\}$-adapted function, and define

$$X_n(t,\omega) = X\left(\tfrac{[2^n t]}{2^n}, \omega\right), \quad (t,\omega) \in [0,\infty) \times \Omega \text{ and } n \in \mathbb{N}.$$

Obviously, X_n is $\{\mathcal{F}_t : t \in [0,\infty)\}$-progressively measurable for every $n \in \mathbb{N}$ and $X_n(t,\omega) \longrightarrow X(t,\omega)$ as $n \to \infty$ for every $\omega \in \Omega$. Hence, by the first part of this lemma, X is $\{\mathcal{F}_t : t \in [0,\infty)\}$-progressively measurable.

Finally, suppose that X is a right-continuous, $\{\mathcal{F}_t : t \in [0,\infty)\}$-adapted function and set

$$X_n(t,\omega) = X\left(\tfrac{[2^n t]+1}{2^n}, \omega\right), \quad (t,\omega) \in [0,\infty) \times \Omega \text{ and } n \in \mathbb{N}.$$

Given $\epsilon > 0$, X_n is $\{\mathcal{F}_{t+\epsilon} : t \in [0,\infty)\}$-progressively measurable so long as $\frac{1}{2^n} < \epsilon$; and therefore, since $X_n \longrightarrow X$ pointwise, X is $\{\mathcal{F}_{t+\epsilon} : t \in [0,\infty)\}$-progressively measurable for every $\epsilon > 0$, which is, of course, the same as saying that X is $\{\mathcal{F}_{t+} : t \in [0,\infty)\}$-progressively measurable. □

Given a probability space $(\Omega, \mathcal{F}, P)$ and a nondecreasing family of sub-σ-algebras $\{\mathcal{F}_t : t \in [0,\infty)\}$, we say that $X : [0,\infty) \times \Omega \longrightarrow (-\infty,\infty]$ is a **P-submartingale with respect to $\{\mathcal{F}_t : t \in [0,\infty)\}$** or, equivalently, that $\big(X(t), \mathcal{F}_t, P\big)$ is a **submartingale** if X is a right-continuous, $\{\mathcal{F}_t : t \in [0,\infty)\}$-progressively measurable function with the properties that $X(t)^-$ is P-integrable for every $t \in [0,\infty)$ and

$$X(s) \le \mathbb{E}^P\big[X(t)\big|\mathcal{F}_s\big] \quad \text{(a.s., } P) \quad \text{for all} \quad 0 \le s \le t < \infty;$$

and when both $\big(X(t), \mathcal{F}_t, P\big)$ and $\big(-X(t), \mathcal{F}_t, P\big)$ are submartingales, we say either that X is a **P-martingale with respect to $\{\mathcal{F}_t : t \in [0,\infty)\}$** or simply that $\big(X(t), \mathcal{F}_t, P\big)$ is a **martingale**. Finally, if $Z : [0,\infty) \times \Omega \longrightarrow \mathbb{C}$ is a right-continuous, $\{\mathcal{F}_t : t \in [0,\infty)\}$-progressively measurable function, then we say that $\big(Z(t), \mathcal{F}_t, P\big)$ is a **(complex) martingale** if both $\big(\mathfrak{Re}\, Z(t), \mathcal{F}_t, P\big)$ and $\big(\mathfrak{Im}\, Z(t), \mathcal{F}_t, P\big)$ are.

7.1.4 Basic Example. We give here an example which, all by itself, justifies the introduction of continuous parameter martingales. Namely, recall the path-space $\mathfrak{P}(\mathbb{R}^N)$ and the sub-σ-algebras $\{\mathcal{B}_t : t \in [0,\infty)\}$ defined in (4.1.3), and let $\mathcal{W}_{\mathbf{x}}^{(N)}$, $\mathbf{x} \in \mathbb{R}^N$, be the shifts of Wiener's measure defined in the discussion preceding Lemma 4.3.7. From (4.1.5), we see that, for each $0 \le s < t$, $\mathbf{x} \in \mathbb{R}^N$, and $\boldsymbol{\xi} \in \mathbb{R}^N$,

$$\mathbb{E}^{\mathcal{W}_{\mathbf{x}}^{(N)}}\Big[e_{\boldsymbol{\xi}}\big(\psi(t) - \psi(s)\big)\,\Big|\,\mathcal{B}_s\Big] = \int_{\mathbb{R}^N} e_{\boldsymbol{\xi}}(\mathbf{y})\,\gamma_{t-s}^N(d\mathbf{y}) = e^{-\frac{|\boldsymbol{\xi}|^2}{2}(t-s)}$$

$\mathcal{W}_{\mathbf{x}}^{(N)}$-almost surely, where we have introduced the notation

$$e_{\boldsymbol{\xi}}(\mathbf{y}) \equiv \exp\Big[\sqrt{-1}\,(\boldsymbol{\xi}, \mathbf{y})_{\mathbb{R}^N}\Big] \quad \text{for } (\boldsymbol{\xi}, \mathbf{y}) \in \mathbb{R}^N \times \mathbb{R}^N. \tag{7.1.5}$$

Hence, after shifting factors a little, we come to the conclusion that

$$\Big(\exp\Big[\sqrt{-1}\,\big(\boldsymbol{\xi}, \psi(t)\big)_{\mathbb{R}^N} + \tfrac{|\boldsymbol{\xi}|^2}{2}t\Big], \mathcal{B}_t, \mathcal{W}_{\mathbf{x}}^{(N)}\Big) \quad \text{is a martingale} \tag{7.1.6}$$

for every $\boldsymbol{\xi} \in \mathbb{R}^N$. As we will see in Corollary 7.1.20, (7.1.6) is the source for a great many other martingales. Moreover, together with the initial condition $\mathcal{W}_{\mathbf{x}}^{(N)}\big(\psi(0) = \mathbf{x}\big) = 1$, (7.1.6) completely characterizes $\mathcal{W}_{\mathbf{x}}^{(N)}$. That is, *if* $P \in \mathbf{M}_1(\mathfrak{P}(\mathbb{R}^N))$, $P\big(\psi(0) = \mathbf{x}\big) = 1$, *and, for every* $\boldsymbol{\xi} \in \mathbb{R}^N$,

$$\Big(\exp\Big[\sqrt{-1}\,\big(\boldsymbol{\xi}, \psi(t)\big)_{\mathbb{R}^N} + \tfrac{|\boldsymbol{\xi}|^2}{2}t\Big], \mathcal{B}_t, P\Big) \quad \textit{is a martingale,}$$

then $P = \mathcal{W}_{\mathbf{x}}^{(N)}$. To see this, simply reverse the preceding line of reasoning and conclude that

$$\mathbb{E}^P\Big[e_\xi\big(\psi(t)-\psi(s)\big)\,\Big|\,\mathcal{B}_s\Big] = e^{-\frac{|\xi|^2}{2}(t-s)} \quad (\text{a.s.}, P).$$

Now, proceeding by induction on $n \in \mathbb{Z}^+$, one sees that, for any $0 \le t_0 < t_1 < \cdots < t_n$, $\{\psi(t_m)-\psi(t_{m-1})\}_1^n$ are mutually independent, $\mathbb{R}^N$-valued random variables under P with distribution $\gamma^N_{t_m - t_{m-1}}$, and from this and the condition $P(\psi(0)=\mathbf{x}) = 1$, it is clear (cf. Theorem 3.3.20) that $P = \mathcal{W}_{\mathbf{x}}^{(N)}$.

We will now run through some of the results from Section 5.2 which transfer immediately to the continuous parameter setting.

7.1.7 Lemma. *Let the interval I and the function $f : I \longrightarrow \mathbb{R}\cup\{\infty\}$ be as in Corollary 5.2.16. If either $\big(X(t),\mathcal{F}_t,P\big)$ is an I-valued martingale or f is nondecreasing and $\big(X(t),\mathcal{F}_t,P\big)$ is an I-valued submartingale, then $\big(f\circ X(t),\mathcal{F}_t,P\big)$ is a submartingale.*

PROOF: The fact that the parameter is continuous makes no difference here, and so this result is already covered by the argument in Corollary 5.2.16. □

7.1.8 Theorem (Doob's Inequality). *Let $\big(X(t),\mathcal{F}_t,P\big)$ be a submartingale. Then, for every $\alpha\in(0,\infty)$ and $T\in[0,\infty)$,*

$$P\left(\sup_{t\in[0,T]} X(t)\ge\alpha\right) \le \frac{1}{\alpha}\mathbb{E}^P\left[X(T),\ \sup_{t\in[0,T]} X(t)\ge\alpha\right]. \tag{7.1.9}$$

In particular, for nonnegative submartingales and $T\in[0,\infty)$,

$$\mathbb{E}^P\left[\sup_{t\in[0,T]} X(t)^p\right]^{\frac{1}{p}} \le \frac{p}{p-1}\mathbb{E}^P\big[X(T)^p\big]^{\frac{1}{p}}, \quad p\in(0,\infty). \tag{7.1.10}$$

PROOF: For each $T\in(0,\infty)$ and $N\in\mathbb{N}$, apply Theorem 5.2.4 to the discrete parameter submartingale

$$\left(X\left(\tfrac{nT}{2^N}\right),\mathcal{F}_{\frac{nT}{2^N}},P\right),$$

and observe that

$$\sup\left\{X\left(\tfrac{nT}{2^N}\right) : 0\le n\le 2^N\right\} \nearrow \sup_{t\in[0,T]} X(t) \quad \text{as } N\to\infty. \quad \square$$

In order to state the analogs of Hunt's and Doob's stopping time results (cf. Theorem 5.2.19 and Corollary 5.2.21), we must first make sure that it is clear what a stopping time is in the present context. Thus, call a function $\tau : \Omega \longrightarrow [0,\infty]$ an $\{\mathcal{F}_t : t \in [0,\infty)\}$**-stopping time** if $\{\tau \le t\} \in \mathcal{F}_t$ for every $t \in [0,\infty)$; set $\mathcal{F}_\tau = \{A \in \mathcal{F} : A \cap \{\tau \le t\} \in \mathcal{F}_t \text{ for every } t \in [0,\infty)\}$; note that $\mathcal{F}_\tau$ is a sub-σ-algebra of $\mathcal{F}$; and check that all but the final part of Lemma 4.3.2 continue to hold when "$\mathcal{B}$" is replaced by "$\mathcal{F}$" throughout. As for the final part of Lemma 4.3.2, let (E,ρ) be a Polish space, $X : [0,\infty) \times \Omega \longrightarrow E$ a right-continuous, $\{\mathcal{F}_t : t \in [0,\infty)\}$-progressively measurable function, and σ a $\{\mathcal{F}_t : t \in [0,\infty)\}$-stopping time, and check that for every closed F in E

$$\tau_F^\sigma \equiv \inf\left\{t \ge \sigma : \inf_{s\in[\sigma,t]} \rho\big(X(t),F\big) = 0\right\} \tag{7.1.11}$$

is an $\{\mathcal{F}_t : t \in [0,\infty)\}$-stopping time. (Of course, when X is continuous, one can eliminate the infimum over $s \in [0,t]$ without altering the definition of τ_F^σ.) The key to extending Hunt and Doob's results is contained in the following application of Theorem 7.1.8.

7.1.12 Lemma. *Suppose that X is an $\{\mathcal{F}_t : t \in [0,\infty)\}$-progressively measurable function into some measurable space $(E,\mathcal{B})$ and that τ is an $\{\mathcal{F}_t : t \in [0,\infty)\}$-stopping time. Then, the function $\omega \in \{\tau < \infty\} \longmapsto X(\tau(\omega),\omega) \in E$ is $\mathcal{F}_\tau[\{\tau < \infty\}]$-measurable. Moreover, if $\big(X(t),\mathcal{F}_t,P\big)$ is a martingale or a nonnegative, P-integrable submartingale, then, for each $T \in [0,\infty)$, the set*

$$\Big\{X(\tau) : \tau \text{ is an } \{\mathcal{F}_t : t \in [0,\infty)\}\text{-stopping time and } \tau \le T\Big\}$$

is uniformly P-integrable.

PROOF: To prove the first assertion, note that for any $t \in [0,\infty)$ and $\Gamma \in \mathcal{B}$,

$$A(t,\Gamma) \equiv \Big\{(s,\omega) \in [0,t] \times \Omega : X(s,\omega) \in \Gamma\Big\} \in \mathcal{B}_{[0,t]} \times \mathcal{F}_t,$$

and therefore

$$\Big\{\omega : X\big(\tau(\omega),\omega\big) \in \Gamma\Big\} \cap \{\tau \le t\} = \Big\{\omega : \big(\tau(\omega),\omega\big) \in A(t,\Gamma)\Big\} \in \mathcal{F}_t.$$

Turning to the second assertion, first note that, by Lemma 7.1.7, it suffices to consider nonnegative submartingales. Thus, let $\big(X(t),\mathcal{F}_t,P\big)$ be a nonnegative, P-integrable submartingale, and define

$$\tau_n(\omega) = \begin{cases} \frac{[2^n\tau(\omega)]+1}{2^n} & \text{if } \tau(\omega) < \infty \\ \infty & \text{if } \tau(\omega) = \infty \end{cases} \tag{7.1.13}$$

for $\{\mathcal{F}_t : t \in [0,\infty)\}$-stopping times τ and $n \in \mathbb{N}$; and let $\mathcal{T}(T)$ denote the set of $\{\mathcal{F}_t : t \in [0,\infty)\}$-stopping times dominated by T. It is then clear that, for each $N \in \mathbb{Z}^+$ and $\tau \in \mathcal{T}(2^N)$,

$$\tau_n \le 2^N + 1 \quad \text{for all } n \in \mathbb{N} \quad \text{and } X(\tau) = \lim_{n\to\infty} X(\tau_n).$$

Hence, it suffices for us to show that, for each $N \in \mathbb{Z}^+$,

$$\lim_{R\nearrow\infty} \sup_{n\in\mathbb{N}} \sup_{\tau\in\mathcal{T}(2^N)} \mathbb{E}^P\Big[X(\tau_n),\, X(\tau_n) \ge R\Big] = 0.$$

But, by Theorem 5.2.19 applied to discrete parameter process

$$\left(X\left(\frac{m}{2^n}\right), \mathcal{F}_{\frac{m}{2^n}}, P\right),$$

we have

$$\begin{aligned}\mathbb{E}^P\Big[X(\tau_n),\, X(\tau_n) \ge R\Big] &\le \mathbb{E}^P\Big[X(2^N+1),\, X(\tau_n) \ge R\Big]\\ &\le \mathbb{E}^P\left[X(2^N+1),\, \sup_{t\in[0,2^N+1]} X(t) \ge R\right],\end{aligned}$$

and, by (7.1.9),

$$P\left(\sup_{t\in[0,2^N+1]} X(t) \ge R\right) \longrightarrow 0 \quad \text{as} \quad R \nearrow \infty. \quad \square$$

7.1.14 Theorem (Hunt). *Let σ and τ be a pair of bounded $\{\mathcal{F}_t : t \in [0,\infty)\}$-stopping times and assume that $\sigma \le \tau$. If $(X(t), \mathcal{F}_t, P)$ is a nonnegative, P-integrable submartingale, then $X(\sigma) \le \mathbb{E}^P[X(\tau)|\mathcal{F}_\sigma]$ (a.s., P); and if $(X(t), \mathcal{F}_t, P)$ is a martingale, then $X(\sigma) = \mathbb{E}^P[X(\tau)|\mathcal{F}_\sigma]$ (a.s., P).*

PROOF: For $n \in \mathbb{N}$, define σ_n and τ_n from σ and τ by the prescription in (7.1.13), note that $\mathcal{F}_\sigma \subseteq \mathcal{F}_{\sigma_n}$, and apply Theorem 5.2.19 to conclude that, for each $n \in \mathbb{N}$ and $A \in \mathcal{F}_\sigma$,

$$\mathbb{E}^P\Big[X(\sigma_n),\, A\Big] \le \mathbb{E}^P\Big[X(\tau_n),\, A\Big] \quad \text{or} \quad \mathbb{E}^P\Big[X(\sigma_n),\, A\Big] = \mathbb{E}^P\Big[X(\tau_n),\, A\Big]$$

depending on whether $(X(t), \mathcal{F}_t, P)$ is a nonnegative, P-integrable submartingale or a martingale. Finally, use Lemma 7.1.12 and right-continuity to justify a passage to the limit as $n \to \infty$. $\square$

Just as Theorem 5.2.19 led immediately to Corollary 5.2.21, so the preceding implies the following.

7.1.15 Corollary (Doob's Stopping Time Theorem). *Depending on whether $\big(X(t),\mathcal{F}_t,P\big)$ is a nonnegative, P-integrable submartingale or a martingale, $\big(X(t\wedge\tau),\mathcal{F}_t,P\big)$ is again a nonnegative, P-integrable submartingale or a martingale for every $\{\mathcal{F}_t : t\in[0,\infty)\}$-stopping time τ.*

Before stating the analog of Theorem 5.2.22 for the continuous parameter case, note that there is no problem adapting the convergence criterion used in the proof there to the setting here. Namely, given a function $f:[0,\infty)\longmapsto\mathbb{R}$, $\lim_{t\nearrow\infty} f(t)$ exists in $[-\infty,\infty]$ if and only if for every pair of rational numbers $-\infty<a<b<\infty$ there is an $s\in[0,\infty)$ such that either $f(t)\le b$ for all $t\ge s$ or $f(t)\ge a$ for all $t\ge s$.

7.1.16 Doob's Martingale Convergence Theorem. *Let $\big(X(t),\mathcal{F}_t,P\big)$ be a P-integrable submartingale. If*

$$\sup_{t\in[0,\infty)} \mathbb{E}^P\big[X(t)^+\big]<\infty,$$

then there exists an $\mathcal{F}_\infty\equiv\bigvee_{t\ge0}\mathcal{F}_t$-measurable $X=X(\infty)\in L^1(P;\mathbb{R})$ to which $X(t)$ converges P-almost surely as $t\to\infty$. Moreover, when $\big(X(t),\mathcal{F}_t,P\big)$ is either a nonnegative submartingale or a martingale, the convergence takes place in $L^1(P)$ if and only if the family $\{X(t):t\in[0,\infty)\}$ is uniformly P-integrable, in which case

$$\Big\{X(\tau) : \tau \text{ is an } \{\mathcal{F}_t : t\in[0,\infty)\}\text{-stopping time}\Big\}$$

is uniformly P-integrable and, for every $\{\mathcal{F}_t : t\in[0,\infty)\}$-stopping time τ,

$$X(\tau)\le\mathbb{E}^P\big[X\,\big|\,\mathcal{F}_\tau\big] \quad\text{or}\quad X(\tau)=\mathbb{E}^P\big[X\,\big|\,\mathcal{F}_\tau\big],$$

depending on whether $\big(X(t),\mathcal{F}_t,P\big)$ is a submartingale or a martingale. Finally, again when $\big(X(t),\mathcal{F}_t,P\big)$ is either a nonnegative submartingale or a martingale, for each $p\in(1,\infty)$ the family $\{|X(t)|^p : t\in[0,\infty)\}$ is uniformly P-integrable if and only if $\sup_{t\in[0,\infty)}\|X(t)\|_{L^p(P)}<\infty$, in which case $X(t)\longrightarrow X$ in $L^p(P)$.

PROOF: To prove the initial convergence assertion, note that, by Theorem 5.2.22 applied to the discrete parameter process $\big(X(n),\mathcal{F}_n,P\big)$, there is an $\bigvee_{n\in\mathbb{N}}\mathcal{F}_n$-measurable $X\in L^1(P)$ to which $X(n)$ converges P-almost surely. Hence, we need only check that $\lim_{t\to\infty}X(t)$ exists in $[-\infty,\infty]$ P-almost surely. To this end, define $U^{(n)}_{[a,b]}(\omega)$ for $n\in\mathbb{N}$ and $a<b$ to be the precise number of times that the sequence

$$\left\{X\left(\tfrac{m}{2^n},\omega\right) : m\in\mathbb{N}\right\}$$

upcrosses the interval $[a,b]$ (cf. the paragraph preceding Theorem 5.2.22), observe that $U^{(n)}_{[a,b]}(\omega)$ is nondecreasing as n increases, and set

$$U_{[a,b]}(\omega) = \lim_{n\to\infty} U^{(n)}_{[a,b]}(\omega).$$

Note that if $U_{[a,b]}(\omega) < \infty$, then (by right-continuity), there is an $s \in [0,\infty)$ such that either $X(t,\omega) \le b$ for all $t \ge s$ or $X(t,\omega) \ge a$ for all $t \ge s$. Hence, we will know that $X(t,\omega)$ converges in $[-\infty,\infty]$ for P-almost every $\omega \in \Omega$ as soon as we show that $\mathbb{E}^P\left[U_{[a,b]}\right] < \infty$ for every pair $a < b$. But, by (5.2.23), we know that

$$\sup_{n\in\mathbb{N}} \mathbb{E}^P\left[U^{(n)}_{[a,b]}\right] \le \sup_{t\in[0,\infty)} \frac{\mathbb{E}^P\left[(X(t)-a)^+\right]}{b-a} < \infty,$$

and so the required estimate follows from the Monotone Convergence Theorem.

Now assume that $\left(X(t),\mathcal{F}_t,P\right)$ is either a nonnegative submartingale or a martingale. Given the preceding, it is clear that $X(t) \longrightarrow X$ in $L^1(P)$ if $\left\{X(t) : t \in [0,\infty)\right\}$ is uniformly P-integrable. Conversely, suppose that $X(t) \longrightarrow X$ in $L^1(P;\mathbb{R})$. Then one can use Theorem 7.1.14 to easily check that

$$|X(\tau)| \le \mathbb{E}^P\left[|X| \,\middle|\, \mathcal{F}_\tau\right] \quad \text{(a.s.}, P) \tag{7.1.17}$$

for all bounded $\left\{\mathcal{F}_t : t \in [0,\infty)\right\}$-stopping times τ. But given (7.1.17) for bounded τ's, one can prove it for arbitrary τ's by replacing τ with $\tau \wedge n$ and using Corollary 5.2.7 to pass to the limit as $n \to \infty$. Thus, we now know (7.1.17) for all $\left\{\mathcal{F}_t : t \in [0,\infty)\right\}$-stopping times. In particular, after combining (7.1.17) with (7.1.9), we see first that

$$P\left(\sup_{t\in[0,\infty)} |X(t)| \ge \alpha\right) \le \frac{1}{\alpha}\mathbb{E}^P\left[|X|,\ \sup_{t\in[0,\infty)} |X(t)| \ge \alpha\right] \tag{7.1.18}$$

for every $\alpha \in (0,\infty)$ and then that

$$\left\{X(\tau) : \tau \text{ is an } \left\{\mathcal{F}_t : t \in [0,\infty)\right\}\text{-stopping time}\right\}$$

is uniformly P-integrable. Thus, when $\left(X(t),\mathcal{F}_t,P\right)$ is a martingale which converges to $X = X(\infty)$ in $L^1(P)$, we can replace (7.1.17) by

$$X(\tau) = \mathbb{E}^P\left[X \,\middle|\, \mathcal{F}_\tau\right]$$

for any $\left\{\mathcal{F}_t : t \in [0,\infty)\right\}$-stopping time.

Finally, the assertions about $p \in (1,\infty)$ are now easy applications of the results already proved and the reasoning used in proof of the analogous results for the discrete parameter case (cf. Corollary 5.2.25). The details are left to the reader. □

The reader will have undoubtedly noticed that the one result from the Section 5.2 for which we have not given a counterpart here is Doob's Decomposition for submartingales (cf. Lemma 5.2.18). The reason for our not doing so is that, although it is, perhaps, the most trivial result which we proved there, its counterpart here (known as the **Doob–Meyer Decomposition Theorem**) is highly nontrivial and requires a good many preparations which would lead us too far afield. (Cf. Exercise 7.1.32 below for an important special case.) Hence, we will content ourselves with the following simple observation about the relationship between martingales and functions of bounded variation.

7.1.19 Theorem. *Suppose $V : [0,\infty)\times\Omega \longrightarrow \mathbb{C}$ is a right-continuous, $\{\mathcal{F}_t : t \in [0,\infty)\}$-progressively measurable function, and let $|V|(t,\omega) \in [0,\infty]$ denote the total variation of $V(\,\cdot\,,\omega)$ on the interval $[0,t]$. Then $|V| : [0,\infty)\times\Omega \longrightarrow [0,\infty]$ is a nondecreasing, $\{\mathcal{F}_t : t \in [0,\infty)\}$-progressively measurable function which is right-continuous on each interval $[0,t)$ for which $|V|(t,\omega) < \infty$. Next, suppose that $\big(X(t),\mathcal{F}_t,P\big)$ is a $\mathbb{C}$-valued martingale with the property that, for each $(t,\omega) \in (0,\infty)\times\Omega$, the product $|X(\,\cdot\,,\omega)|\,|V|(t,\omega)$ is bounded on $[0,t]$. Define*

$$B(t,\omega) = \begin{cases} \int_{(0,t]} X(s,\omega)\,V(ds,\omega) & \text{if } \ |V|(t,\omega) < \infty \\ 0 & \text{otherwise} \end{cases}$$

where, in the case when $|V|(t,\omega) < \infty$, the integral is the Lebesgue integral of $X(\,\cdot\,,\omega)$ on $[0,t]$ with respect to the $\mathbb{C}$-valued measure induced by $V(\,\cdot\,,\omega)$. If

$$\mathbb{E}^P\left[\sup_{t\in[0,T]} |X(t)|\Big(|V|(T) + \big|V(0)\big|\Big)\right] < \infty \quad \text{for all } T \in (0,\infty),$$

then $\big(X(t)V(t) - B(t),\mathcal{F}_t,P\big)$ is a martingale.

PROOF: Without loss in generality, we will assume that both X and V are $\mathbb{R}$-valued. To see that $|V|$ is $\{\mathcal{F}_t : t \in [0,\infty)\}$-progressively measurable, simply observe that, by right-continuity,

$$|V|(t,\omega) = \sup_{n\in\mathbb{N}} \sum_{k=0}^{[2^n t]} \left|V\left(\tfrac{k+1}{2^n}\wedge t,\omega\right) - V\left(\tfrac{k}{2^n},\omega\right)\right|;$$

and to see that $|V|(\,\cdot\,,\omega)$ is right-continuous on $[0,t)$ whenever $|V|(t,\omega) < \infty$, recall that the magnitude of the jumps (from the right and left) of the variation of a function coincide with those of the function itself.

We now turn to the second part. Certainly B is $\{\mathcal{F}_t : t \in [0,\infty)\}$-progressively measurable. In addition, because

$$\sup_{s\in[0,t]} |X(s,\omega)|\,|V|(t,\omega) < \infty, \quad (t,\omega) \in [0,\infty)\times\Omega,$$

for any $\omega \in \Omega$ one has that

$$B(t,\omega) = 0 \quad \text{or} \quad B(t,\omega) = \int_{(0,t]} X(s,\omega)\, V(ds,\omega) \quad \text{for all } t \in [0,\infty);$$

and so, in either case, $B(\,\cdot\,,\omega)$ is right-continuous and $B(t,\omega) - B(s,\omega)$ can be computed as

$$\lim_{n\to\infty} \sum_{k=[2^n s]}^{[2^n t]} X\left(\tfrac{k+1}{2^n} \wedge t, \omega\right)\Big(V\left(\tfrac{k+1}{2^n} \wedge t,\omega\right) - V\left(\tfrac{k}{2^n} \vee s,\omega\right)\Big).$$

In fact, under the stated integrability condition, the convergence in the preceding takes place in $L^1(P;\mathbb{R})$ for every $t \in [0,\infty)$; and therefore, for any $0 \le s \le t < \infty$ and $A \in \mathcal{F}_s$:

$$\begin{aligned}
&\mathbb{E}^P\big[B(t) - B(s),\, A\big] \\
&= \lim_{n\to\infty} \sum_{k=[2^n s]}^{[2^n t]} \mathbb{E}^P\Big[X\left(\tfrac{k+1}{2^n} \wedge t,\omega\right)\Big(V\left(\tfrac{k+1}{2^n} \wedge t,\omega\right) - V\left(\tfrac{k}{2^n} \vee s,\omega\right)\Big),\, A\Big] \\
&= \lim_{n\to\infty} \sum_{k=[2^n s]}^{[2^n t]} \mathbb{E}^P\Big[X(t+1)\Big(V\left(\tfrac{k+1}{2^n} \wedge t,\omega\right) - V\left(\tfrac{k}{2^n} \vee s,\omega\right)\Big),\, A\Big] \\
&= \mathbb{E}^P\Big[X(t+1)\big(V(t) - V(s)\big),\, A\Big] = \mathbb{E}^P\Big[X(t)V(t) - X(s)V(s),\, A\Big];
\end{aligned}$$

and clearly this is equivalent to the asserted martingale property. □

We next use Theorem 7.1.19 to produce some of the additional martingales advertised in Example 7.1.4.

7.1.20 Corollary. *Suppose that* $\mathbf{Y} : [0,\infty) \times \mathfrak{P}(\mathbb{R}^N) \longmapsto \mathbb{C}^M$ *is a* $\{\mathcal{B}_t : t \in [0,\infty)\}$*-progressively measurable function with the property that* $\mathbf{Y}(\,\cdot\,,\boldsymbol{\psi}) \in C^1\big([0,\infty);\mathbb{R}^M\big)$ *for each* $\boldsymbol{\psi} \in \mathfrak{P}(\mathbb{R}^N)$ *and*

$$\mathbb{E}^{\mathcal{W}^{(N)}_{\mathbf{x}}}\left[\int_0^T \big|\dot{\mathbf{Y}}(t)\big|\, dt\right] < \infty \quad \textit{for each } T \in (0,\infty) \textit{ and } \mathbf{x} \in \mathbb{R}^N.$$

Then, for each $F \in C^{2,1}_{\mathrm{b}}\big(\mathbb{R}^N \times \mathbb{R}^M;\mathbb{C}\big)$ *and* $\mathbf{x} \in \mathbb{R}^N$,

$$F\big(\boldsymbol{\psi}(t),\mathbf{Y}(t,\boldsymbol{\psi})\big) - \int_0^t \Big[\tfrac{1}{2}\boldsymbol{\Delta}_{\mathbf{x}}F + \big(\dot{\mathbf{Y}}(s,\boldsymbol{\psi}),\nabla_{\mathbf{y}}F\big)_{\mathbb{R}^M}\Big]\big(s,\boldsymbol{\psi}(s)\big)\, ds$$

is a $\mathcal{W}_{\mathbf{x}}^{(N)}$*-martingale relative to* $\{\mathcal{B}_t : t \in [0,\infty)\}$. *In particular, if* $f \in C^{1,2}\big([0,\infty) \times \mathbb{R}^N; \mathbb{C}\big)$ *and, for each* $T \in (0,\infty)$, *there exists a* $C(T) < \infty$ *for which*

$$\text{(7.1.21)} \quad |f(t,\mathbf{y})| + \left|\left[\tfrac{\partial f}{\partial t} + \tfrac{1}{2}\Delta f\right](t,\mathbf{y})\right| \le C(T)e^{C(T)|\mathbf{y}|}, \quad (t,\mathbf{y}) \in [0,T] \times \mathbb{R}^N,$$

then

$$\text{(7.1.22)} \quad \left(f\big(t,\psi(t)\big) - \int_0^t \left[\tfrac{\partial f}{\partial s} + \tfrac{1}{2}\Delta f\right]\big(s,\psi(s)\big)\,ds, \mathcal{B}_t, \mathcal{W}_{\mathbf{x}}^{(N)}\right) \text{ is a continuous martingale}$$

for every $\mathbf{x} \in \mathbb{R}^N$.

PROOF: After an easy approximation procedure, we may and will restrict our attention to $F \in C_c^\infty\big(\mathbb{R}^N \times \mathbb{R}^M; \mathbb{C}\big)$ while proving the first assertion, in which case

$$F(\mathbf{x},\mathbf{y}) = (2\pi)^{-N-M} \int\limits_{\mathbb{R}^N \times \mathbb{R}^M} e_{\boldsymbol{\xi},\boldsymbol{\eta}}(\mathbf{x},\mathbf{y})\,\hat{F}(-\boldsymbol{\xi},-\boldsymbol{\eta})\,d\boldsymbol{\xi}d\boldsymbol{\eta},$$

where $\hat{F}$, the Fourier transform of F, is rapidly decreasing (i.e., tends to 0 at infinity faster than any power of $\big(1+|\boldsymbol{\xi}|+|\boldsymbol{\eta}|\big)^{-1}$) and

$$e_{\boldsymbol{\xi},\boldsymbol{\eta}}(\mathbf{x},\mathbf{y}) \equiv \exp\left[\sqrt{-1}\left(\big(\boldsymbol{\xi},\mathbf{x}\big)_{\mathbb{R}^N} + \big(\boldsymbol{\eta},\mathbf{y}\big)_{\mathbb{R}^M}\right)\right].$$

In particular, if $e_{\boldsymbol{\xi},\boldsymbol{\eta}}(t,\psi) = e_{\boldsymbol{\xi},\boldsymbol{\eta}}\big(\psi(t),\mathbf{Y}(t,\psi)\big)$ and

$$M_{\boldsymbol{\xi},\boldsymbol{\eta}}(t,\psi) = e_{\boldsymbol{\xi},\boldsymbol{\eta}}(t,\psi) + \int_0^t \left(\tfrac{|\boldsymbol{\xi}|^2}{2} - \sqrt{-1}\,\big(\boldsymbol{\eta},\dot{\mathbf{Y}}(s,\psi)\big)_{\mathbb{R}^M}\right) e_{\boldsymbol{\xi},\boldsymbol{\eta}}(s,\psi)\,ds,$$

then, by elementary Fourier considerations,

$$\begin{aligned} F\big(\psi(t),\mathbf{Y}(t,\psi)\big) - \int_0^t \left[\tfrac{1}{2}\Delta_{\mathbf{x}}F + \big(\dot{\mathbf{Y}}(s,\psi),\nabla_{\mathbf{y}}F\big)_{\mathbb{R}^M}\big(s,\psi(s)\big)\,ds\right. \\ = (2\pi)^{-N-M} \int\limits_{\mathbb{R}^N \times \mathbb{R}^M} M_{\boldsymbol{\xi},\boldsymbol{\eta}}(t,\psi)\,\hat{F}(\boldsymbol{\xi},\boldsymbol{\eta})\,d\boldsymbol{\xi}d\boldsymbol{\eta}. \end{aligned}$$

Hence, if we show that $\big(M_{\boldsymbol{\xi},\boldsymbol{\eta}}(t),\mathcal{B}_t,\mathcal{W}_{\mathbf{x}}^{(N)}\big)$ is a martingale for each $(\boldsymbol{\xi},\boldsymbol{\eta})$, then the desired result will follow immediately as an application of Fubini's Theorem. But

$$M_{\boldsymbol{\xi},\boldsymbol{\eta}}(t,\psi) = Z_{\boldsymbol{\xi}}(t,\psi)\,V_{\boldsymbol{\eta}}(t,\psi) - \int_0^t Z_{\boldsymbol{\xi}}(s,\psi)\,V_{\boldsymbol{\eta}}(ds,\psi),$$

where

$$Z_{\boldsymbol{\xi}}(t,\boldsymbol{\psi}) = \exp\Big[\sqrt{-1}\,\big(\boldsymbol{\xi},\boldsymbol{\psi}(t)\big)_{\mathbb{R}^N} + \tfrac{|\boldsymbol{\xi}|^2}{2}t\Big]$$

and

$$V_{\boldsymbol{\eta}}(t,\boldsymbol{\psi}) = \exp\Big[\sqrt{-1}\,\big(\boldsymbol{\eta},\mathbf{Y}(t,\boldsymbol{\psi})\big)_{\mathbb{R}^M} - \tfrac{|\boldsymbol{\xi}|^2}{2}t\Big].$$

Thus, by (7.1.6) and Theorem 7.1.19, we are done.

To handle the second assertion, first suppose that the derivatives of f are bounded, set

$$F(\mathbf{x},y) = f(0,\mathbf{x}) + \int_0^y \tfrac{\partial f}{\partial t}\big(|t|,\mathbf{x}\big)\,dt \quad \text{for } (\mathbf{x},y) \in \mathbb{R}^N \times \mathbb{R},$$

and apply the preceding with $Y(t) = t$. To handle f's satisfying (7.1.21), choose $\eta \in C_c^\infty\big(\mathbb{R}\times\mathbb{R}^N;[0,1]\big)$ so that $\eta \equiv 1$ on $[-1,1] \times \overline{B_{\mathbb{R}^N}(\mathbf{0},1)}$, and set

$$f_n(t,\mathbf{x}) = \eta\left(\tfrac{t}{n},\tfrac{\mathbf{x}}{n}\right)\, f(t,\mathbf{x}).$$

If $\zeta_n(\boldsymbol{\psi})$ is the smallest $t \geq 0$ such that $\big(t,\boldsymbol{\psi}(t)\big) \notin [-n,n] \times B_{\mathbb{R}^N}(\mathbf{0},n)$, then, by the case already handled (applied to f_n) and Corollary 7.1.15,

$$f\Big(t\wedge\zeta_n(\boldsymbol{\psi}),\boldsymbol{\psi}\big(t\wedge\zeta_n(\boldsymbol{\psi})\big)\Big) - \int_0^{t\wedge\zeta_n(\boldsymbol{\psi})} \Big[\tfrac{\partial f}{\partial s} + \tfrac{1}{2}\boldsymbol{\Delta} f\Big]\big(s,\boldsymbol{\psi}(s)\big)\,ds$$

is $\mathcal{W}_{\mathbf{x}}^{(N)}$-martingale relative to $\big\{\mathcal{B}_t : t \in [0,\infty)\big\}$. Hence, by (7.1.21) and the estimate in (3.3.30), it is an easy matter to see that one can let $n \to \infty$ and thereby arrive at the required conclusion. □

7.1.23 Remark: The preceding leads to an extremely appealing characterization of $\mathcal{W}_{\mathbf{x}}^{(N)}$. Namely, it says that

$$\left(f\big(\boldsymbol{\psi}(t)\big) - \int_0^t \tfrac{1}{2}\boldsymbol{\Delta} f\big(\boldsymbol{\psi}(s)\big)\,ds, \mathcal{B}_t, \mathcal{W}_{\mathbf{x}}^{(N)}\right) \tag{7.1.24}$$

is a martingale for each $f \in C_c^\infty(\mathbb{R}^N;\mathbb{R})$.

Conversely, if $P \in \mathbf{M}_1(\mathfrak{P}(\mathbb{R}^N))$ and

$$\left(f\big(\boldsymbol{\psi}(t)\big) - \int_0^t \tfrac{1}{2}\boldsymbol{\Delta} f\big(\boldsymbol{\psi}(s)\big)\,ds, \mathcal{B}_t, P\right)$$

is a martingale for each $f \in C_c^\infty(\mathbb{R}^N;\mathbb{R})$, then it is easy to see that the same is true for all $f \in C_b^\infty\big(\mathbb{R}^N;\mathbb{C}\big)$. In particular, this means that (cf. (7.1.5)) for $0 \leq s < t$ and $A \in \mathcal{B}_s$:

$$\mathbb{E}^P\Big[e_{\boldsymbol{\xi}}\big(\boldsymbol{\psi}(t)\big),A\Big] = \mathbb{E}^P\Big[e_{\boldsymbol{\xi}}\big(\boldsymbol{\psi}(s)\big),\,A\Big] - \tfrac{|\boldsymbol{\xi}|^2}{2}\int_s^t \mathbb{E}^P\Big[e_{\boldsymbol{\xi}}\big(\boldsymbol{\psi}(\tau)\big),A\Big]\,d\tau;$$

from which it is obvious that

$$\mathbb{E}^P\Big[e_{\boldsymbol{\xi}}\big(\boldsymbol{\psi}(t)\big), A\Big] = e^{-\frac{|\boldsymbol{\xi}|^2}{2}(t-s)}\mathbb{E}^P\Big[e_{\boldsymbol{\xi}}\big(\boldsymbol{\psi}(s)\big), A\Big],$$

and therefore that

$$\left(\exp\Big[\sqrt{-1}\,\big(\boldsymbol{\xi}, \boldsymbol{\psi}(t)\big)_{\mathbb{R}^N} - \tfrac{|\boldsymbol{\xi}|^2}{2}t\Big], \mathcal{B}_t, P\right) \quad \text{is a martingale.}$$

Hence, if, in addition, $P\big(\boldsymbol{\psi}(0) = \mathbf{x}\big) = 1$, then (cf. Example 7.1.4) we know that $P = \mathcal{W}_{\mathbf{x}}^{(N)}$. In other words, $\mathcal{W}_{\mathbf{x}}^{(N)}$ *is completely characterized by* (7.1.24) *and the initial condition* $\mathcal{W}_{\mathbf{x}}^{(N)}\big(\boldsymbol{\psi}(0) = \mathbf{x}\big) = 1$.

The reason why the preceding characterization is more appealing than the earlier one in Example 7.1.4 is that it lends itself to the following somewhat loose interpretation. Namely, it hints that *Wiener paths are the integral curves of* $\frac{1}{2}\boldsymbol{\Delta}$. To understand this interpretation, recall that if $\mathbf{b} = \sum_{k=1}^N b_k\mathbf{e}_k$ is a smooth vector field on $\mathbb{R}^N$, then the integral curve of $\mathbf{b}$ starting at $\mathbf{x} \in \mathbb{R}^N$ is that path $\boldsymbol{\psi}$ with the properties that $\boldsymbol{\psi}(0) = \mathbf{x}$ and

$$t \in [0,\infty) \longmapsto f\big(\boldsymbol{\psi}(t)\big) - \int_0^t \big[\mathbf{b}\cdot\nabla f\big]\big(\boldsymbol{\psi}(\tau)\big)\,d\tau$$

is constant for every $f \in C_c^\infty(\mathbb{R}^N;\mathbb{R})$. Hence, if one now accepts the proposition that *martingales are the stochastic analog of constants* (cf. Exercise 7.1.24 below), then one starts to understand the preceding interpretation of Wiener paths as the integral curves of $\frac{1}{2}\boldsymbol{\Delta}$.

Having made such a fuss about various martingale characterizations of Wiener's measure, it is only fitting that we close this section with a far deeper characterization which is due to P. Lévy. The depth of Lévy's characterization lies in the observation that one need not use all test functions $f \in C_c^\infty(\mathbb{R}^N;\mathbb{R})$ in (7.1.24) but need only use quadratic ones; an observation which is reminiscent of and closely related to the heuristic discussion with which we introduced the Central Limit Theorem in Section 2.1. Before presenting Lévy's Theorem, we need a definition. Namely, given a probability space $(\Omega, \mathcal{F}, P)$ and a map $\boldsymbol{\beta}$: $[0,\infty)\times\Omega \longrightarrow \mathbb{R}^N$ which is progressively measurable with respect to some family $\{\mathcal{F}_t : t \in [0,\infty)\}$, we say that $\big(\boldsymbol{\beta}(t), \mathcal{F}_t, P\big)$ is a **Brownian motion** if $\boldsymbol{\beta}(\,\cdot\,,\omega)$ is continuous for each $\omega \in \Omega$ and (cf. (7.1.5))

$$\mathbb{E}^P\Big[e_{\boldsymbol{\xi}}\big(\boldsymbol{\beta}(t) - \boldsymbol{\beta}(s)\big)\,\Big|\,\mathcal{F}_s\Big] = e^{-\frac{|\boldsymbol{\xi}|^2}{2}(t-s)} \quad \text{(a.s., } P)$$

$$\text{for all } 0 \le s < t \text{ and } \boldsymbol{\xi} \in \mathbb{R}^N.$$

Equivalently, by precisely the same reasoning as we used in Example 7.1.4, $\big(\boldsymbol{\beta}(t), \mathcal{F}_t, P\big)$ is a Brownian motion if an only if

$$\left(\exp\left[\sqrt{-1}\left(\boldsymbol{\xi}, \boldsymbol{\beta}(t) - \boldsymbol{\beta}(0)\right)_{\mathbb{R}^N} + \tfrac{|\boldsymbol{\xi}|^2}{2}t\right], \mathcal{F}_t, P\right) \tag{7.1.25}$$
$$\text{is a martingale for each } \boldsymbol{\xi} \in \mathbb{R}^N.$$

In particular, if $\boldsymbol{\beta} : [0,\infty) \times \Omega$ is an $\{\mathcal{F}_t : t \in [0,\infty)\}$-progessively measurable map with $\boldsymbol{\beta}(\cdot\,,\omega) \in \mathfrak{P}(\mathbb{R}^N)$ for each $\omega \in \Omega$, then $\big(\boldsymbol{\beta}(t), \mathcal{F}_t, P\big)$ is a Brownian motion if and only if the distribution of $\omega \in \Omega \longmapsto \boldsymbol{\beta}(\cdot\,,\omega) - \boldsymbol{\beta}(0,\omega) \in \mathfrak{P}(\mathbb{R}^N)$ under P is $\mathcal{W}^{(N)}$.

7.1.26 Theorem (Lévy). *Let $(\Omega, \mathcal{F}, P)$ be a probability space, $\{\mathcal{F}_t : t \in [0,\infty)\}$ a nondecreasing family of sub-σ-algebras, and $\boldsymbol{\beta} : [0,\infty) \times \Omega \longrightarrow \mathbb{R}^N$ an $\{\mathcal{F}_t : t \in [0,\infty)\}$-progressively measurable map with the property that $\boldsymbol{\beta}(\cdot\,,\omega) \in \mathfrak{P}(\mathbb{R}^N)$ for each $\omega \in \Omega$. In order that $\big(\boldsymbol{\beta}(t), \mathcal{F}_t, P\big)$ be a Brownian motion, it is necessary and sufficient that, for each $(\boldsymbol{\xi}, \boldsymbol{\eta}) \in \mathbb{R}^N \times \mathbb{R}^N$:*

$$\left(\boldsymbol{\xi}, \boldsymbol{\beta}(t) - \boldsymbol{\beta}(0)\right)_{\mathbb{R}^N} + \left(\boldsymbol{\eta}, \boldsymbol{\beta}(t) - \boldsymbol{\beta}(0)\right)_{\mathbb{R}^N}^2 - \tfrac{|\boldsymbol{\eta}|^2}{2}t$$

be a P-martingale relative to $\{\mathcal{F}_t : t \in [0,\infty)\}$.

PROOF: Because there is no loss in generality, we will assume throughout that $\boldsymbol{\beta}(0) = \mathbf{0}$.

Assume that $\big(\boldsymbol{\beta}(t), \mathcal{F}_t, P\big)$ is a Brownian motion. Because $\mathcal{W}^{(N)}$ is the distribution of $\omega \in \Omega \longmapsto \boldsymbol{\beta}(\cdot\,,\omega) - \boldsymbol{\beta}(0,\omega) \in \mathfrak{P}(\mathbb{R}^N)$, Corollary 7.1.20 says that

$$\left(f\big(t, \boldsymbol{\beta}(t)\big) - \int_0^t \left[\tfrac{\partial f}{\partial s} + \tfrac{1}{2}\Delta f\right]\big(s, \boldsymbol{\beta}(s)\big)\,ds, \mathcal{F}_t, P\right)$$

is a continuous martingale for every $f \in C^{1,2}\big([0,\infty) \times \mathbb{R}^N; \mathbb{C}\big)$ satisfying (7.1.21). In particular, this observation certainly provides a proof of the necessity. Alternatively, one can proceed more directly and note that, from

$$P\big(\boldsymbol{\beta}(t) - \boldsymbol{\beta}(s) \in \Gamma \,\big|\, \mathcal{F}_s\big) = \gamma_{t-s}^N(\Gamma),$$

it is easy to check that, for each $\boldsymbol{\xi} \in \mathbb{R}^N$,

$$\mathbb{E}^P\left[\exp\left[\left(\boldsymbol{\xi}, \boldsymbol{\beta}(t) - \boldsymbol{\beta}(s)\right)_{\mathbb{R}^N}\right] \,\Big|\, \mathcal{F}_s\right] = e^{\frac{|\boldsymbol{\xi}|^2}{2}(t-s)} \quad (\text{a.s.}, P).$$

Given the above expression for this conditional moment generating function, it is an easy matter to check that

$$\mathbb{E}^P\left[\left(\boldsymbol{\xi}, \boldsymbol{\beta}(t) - \boldsymbol{\beta}(s)\right)_{\mathbb{R}^N} \,\Big|\, \mathcal{F}_s\right] = 0 \quad (\text{a.s.}, P)$$

and

$$\mathbb{E}^P\left[\left(\boldsymbol{\xi},\boldsymbol{\beta}(t)-\boldsymbol{\beta}(s)\right)^2_{\mathbb{R}^N}\,\middle|\,\mathcal{F}_s\right]=\tfrac{|\boldsymbol{\xi}|^2}{2}(t-s)\quad(\text{a.s.},P);$$

and from these one can quickly verify the necessity of the stated condition.

We now want to prove the sufficiency. Thus, let $0\le T_1<T_2$, $A\in\mathcal{F}_{T_1}$, and $\boldsymbol{\xi}\in\mathbb{R}^N$ be given. What we have to show is that

$$\tag{7.1.27}\begin{aligned}\mathbb{E}^P&\left[\exp\left[\sqrt{-1}\left(\boldsymbol{\xi},\boldsymbol{\beta}(T_2)\right)_{\mathbb{R}^N}+\tfrac{|\boldsymbol{\xi}|^2}{2}T_2\right],A\right]\\&=\mathbb{E}^P\left[\exp\left[\sqrt{-1}\left(\boldsymbol{\xi},\boldsymbol{\beta}(T_1)\right)_{\mathbb{R}^N}+\tfrac{|\boldsymbol{\xi}|^2}{2}T_1\right],A\right]\end{aligned}$$

on the basis of the stated condition about quadratic functions. Our problem is to learn how to take full advantage of the assumed continuity (cf. Exercise 7.1.33 to understand just how essential this point is).† To this end, let $\epsilon\in(0,1]$ be given, set $\tau_0\equiv T_1$, and use induction to define

$$\tau_n=\left(\inf\left\{t\ge\tau_{n-1}:\,\left|\boldsymbol{\beta}(t)-\boldsymbol{\beta}(\tau_{n-1})\right|\ge\epsilon\right\}\right)\wedge\left(\tau_{n-1}+\epsilon\right)\wedge T_2$$

for $n\in\mathbb{Z}^+$. By the argument used in the proof of the last part of Lemma 4.3.2, we see that $\{\tau_n\}_0^\infty$ is a nondecreasing sequence of $[T_1,T_2]$-valued $\{\mathcal{F}_t:t\in[0,\infty)\}$-stopping times. Hence, by Theorem 7.1.14 and our assumption,

$$\tag{7.1.28}\begin{aligned}\mathbb{E}^P\left[\Delta_n\,\middle|\,\mathcal{F}_{\tau_{n-1}}\right]&=0\\\mathbb{E}^P\left[\Delta_n^2\,\middle|\,\mathcal{F}_{\tau_{n-1}}\right]&=\mathbb{E}^P\left[\delta_n\,\middle|\,\mathcal{F}_{\tau_{n-1}}\right]\end{aligned}\qquad(\text{a.s.},P),$$

where

$$\begin{aligned}\Delta_n(\omega)&\equiv\left(\boldsymbol{\xi},\boldsymbol{\beta}(\tau_n(\omega),\omega)-\boldsymbol{\beta}(\tau_{n-1}(\omega),\omega)\right)_{\mathbb{R}^N}\\\delta_n(\omega)&\equiv|\boldsymbol{\xi}|^2\left(\tau_n(\omega)-\tau_{n-1}(\omega)\right).\end{aligned}$$

Moreover, because $\boldsymbol{\beta}(\,\cdot\,,\omega)$ is continuous, we have that, for each $\omega\in\Omega$,

$$\tag{7.1.29}\left|\Delta_n(\omega)\right|\vee\delta_n(\omega)\le\epsilon$$

---

† Lévy's Theorem is Theorem 11.9 in Chapter VII of Doob's *Stochastic Processes*, publ. by J. Wiley (1953). Doob uses a clever but somewhat opaque Central Limit argument. The argument which we give is far simpler and is adapted from the one introduced by H. Kunita and S. Watanabe in their article "On square integrable martingales," *Nagoya Math. J.* **30** (1967).

and that $\tau_n(\omega) = T_2$ for all but a finite number of n's. In particular, we can write the difference between the left and the right sides of (7.1.17) as the sum over $n \in \mathbb{Z}^+$ of $\mathbb{E}^P\big[D_n M_n,\, A\big]$, where

$$D_n \equiv \exp\Big[\sqrt{-1}\,\Delta_n + \tfrac{\delta_n}{2}\Big] - 1$$
$$M_n \equiv \exp\Big[\sqrt{-1}\,\big(\boldsymbol{\xi}, \boldsymbol{\beta}(\tau_{n-1})\big)_{\mathbb{R}^N} + \tfrac{|\xi|^2}{2}\tau_{n-1}\Big].$$

By Taylor's Theorem,

$$\Big|D_n - \Big(\sqrt{-1}\,\Delta_n + \tfrac{\delta_n}{2}\Big) - \tfrac{1}{2}\Big(\sqrt{-1}\,\Delta_n + \tfrac{\delta_n}{2}\Big)^2\Big| \le \tfrac{1}{6}e^{\frac{|\xi|^2}{2}}\Big|\sqrt{-1}\,\Delta_n + \tfrac{\delta_n}{2}\Big|^3.$$

Hence, after rearranging terms, we see that

$$D_n = \sqrt{-1}\,\Delta_n - \tfrac{1}{2}\big(\Delta_n^2 - \delta_n\big) + E_n,$$

where (cf. (7.1.29))

$$|E_n| \le \tfrac{1}{2}|\Delta_n\delta_n| + \tfrac{\delta_n^2}{8} + \tfrac{2}{3}e^{\frac{|\xi|^2}{2}}\Big(|\Delta_n|^3 + \tfrac{\delta_n^3}{8}\Big) \le \epsilon e^{\frac{|\xi|^2}{2}}\big(\Delta_n^2 + \delta_n\big);$$

and so, after taking (7.1.28) into account, we arrive at

$$\begin{aligned}\Big|\sum_1^\infty \mathbb{E}^P\big[D_n M_m,\, A\big]\Big| &= \Big|\sum_1^\infty \mathbb{E}^P\big[E_n M_n,\, A\big]\Big| \\ &\le 2\epsilon e^{\frac{|\xi|^2}{2}}\sum_1^\infty \big|\mathbb{E}^P\big[\delta_n M_n,\, A\big]\big| \\ &\le 2\epsilon(T_2 - T_1)e^{\frac{|\xi|^2}{2}(1+T_2)}.\end{aligned}$$

In other words, we have now proved that, for every $\epsilon \in (0,1]$, the difference between the two sides of (7.1.27) is dominated by $2\epsilon(T_2 - T_1)e^{\frac{|\xi|^2}{2}(1+T_2)}$, and so the equality in (7.1.27) has been established. □

Exercises

7.1.30 Exercise: Define $\mathcal{F}_{t+}$ for $t \in [0,\infty)$ as in (7.1.3).

(i) Show that $\tau : \Omega \longmapsto [0,\infty]$ is an $\big\{\mathcal{F}_{t+} : t \in [0,\infty)\big\}$-stopping time if and only if $\{\tau < t\} \in \mathcal{F}_t$ for every $t \in [0,\infty)$.

(ii) If $\big(X(t), \mathcal{F}_t, P\big)$ is a martingale, show that $\big(X(t), \mathcal{F}_{t+}, P\big)$ is also a martingale.

(iii) Let $\big(X(t), \mathcal{F}_t, P\big)$ be a continuous martingale and, for each $(t, \omega) \in [0, \infty) \times \Omega$, use $|X|(t, \omega) \in [0, \infty]$ to denote the total variation of $X(\,\cdot\,, \omega) \restriction [0, t]$. If, for $R \in (0, \infty)$,

$$\sigma_R(\omega) \equiv \sup\Big\{t \in [0, \infty) : |X|(t, \omega) \leq R\Big\}, \qquad \omega \in \Omega,$$

and $X_R(t) \equiv X\big(t \wedge \sigma_R\big) - X(0)$, show that σ_R is an $\big\{\mathcal{F}_{t+} : t \in [0, \infty)\big\}$-stopping time, and conclude that $\big(X_R(t), \mathcal{F}_{t+}, P\big)$ is a continuous martingale with the property that $|X_R|(t) \leq R$. Next, note that, on the one hand, by Riemann integration theory,

$$X_R(t, \omega)^2 = 2 \int_0^t X_R(s, \omega)\, dX_R(s, \omega), \quad (t, \omega) \in [0, \infty) \times \Omega,$$

while, on the other hand, by Theorem 7.1.19,

$$\mathbb{E}^P\big[X_R(t)^2\big] = \mathbb{E}^P\left[\int_{(0,t]} X_R(s)\, X_R(ds)\right], \quad t \in [0, \infty).$$

Hence, since these Riemann and Lebesgue integrals coincide, conclude that $\mathbb{E}^P\big[X_R(t)^2\big] = 0$ for all $t \in [0, \infty)$. Finally, use these considerations to show that

$$P\Big(\exists t \in [0, \infty) \quad 0 < |X|(t) < \infty\Big) = 0.$$

In other words, *the paths of a nonconstant, continuous martingale must have unbounded variation.*

(iv) Just to make sure that it is understood what an essential rôle *continuity* plays in the preceding, recall the simple Poisson process $\{\mathbf{N}(t) : t \in [0, \infty)\}$ described in (3.2.23), let $\mathcal{F}_t$ denote the σ-algebra generated by $\{\mathbf{N}(s) : s \in [0, t]\}$ for each $t \in [0, \infty)$, and set $X(t) = \mathbf{N}(t) - t$. Show that $\big(X(t), \mathcal{F}_t, P\big)$ is a nonconstant martingale whose paths all have bounded variation on each finite time interval. Of course, it is also true that its paths are discontinuous!

7.1.31 Exercise: At the end of Remark 7.1.23, we asserted that the characterization of Wiener's measure $\mathcal{W}_{\mathbf{x}}^{(N)}$ in terms of the martingales

$$\left(f\big(\psi(t)\big) - \int_0^t \tfrac{1}{2}\big[\mathbf{\Delta} f\big]\big(\psi(s)\big)\, ds, \mathcal{B}_t, \mathcal{W}_{\mathbf{x}}^{(N)}\right), \quad f \in C_c^\infty(\mathbb{R}^N; \mathbb{R})$$

constituted grounds for thinking of Wiener paths as the integral curves of $\frac{1}{2}\mathbf{\Delta}$. In order to further support this contention, consider a bounded, smooth vector field $\mathbf{b} : \mathbb{R}^N \longrightarrow \mathbb{R}^N$ and suppose that $P \in \mathbf{M}_1(\mathfrak{P}(\mathbb{R}^N))$ has the properties that $P(\psi(0) = \mathbf{0}) = 1$ and that

$$\left(f(\psi(t)) - \int_0^t (\mathbf{b}, \nabla f)_{\mathbb{R}^N}(\psi(s))\, ds, \mathcal{B}_t, P \right)$$

is a continuous martingale for every $f \in C_c^\infty(\mathbb{R}^N; \mathbb{R})$. Show that $\psi(t) = \int_0^t \mathbf{b}(\psi(s))\, ds$, $t \in [0,\infty)$, for P-almost every $\psi \in \mathfrak{P}(\mathbb{R}^N)$. In other words, for a vector field, as distinguished from $\frac{1}{2}\boldsymbol{\Delta}$, *stochastic constancy* is the same as P-almost sure actual constancy.

Hint: As an application of Theorem 7.1.19, Leibniz's rule, and integration by parts, show that

$$\left(\left[f(\psi(t)) - \int_0^t (\mathbf{b}, \nabla f)_{\mathbb{R}^N}(\psi(s))\, ds \right]^2, \mathcal{B}_t, P \right)$$

is a continuous martingale for every $f \in C_c^\infty(\mathbb{R}^N; \mathbb{R})$, and conclude that

$$f(\psi(t)) - f(\mathbf{0}) = \int_0^t (\mathbf{b}, \nabla f)_{\mathbb{R}^N}(\psi(s))\, ds, \quad t \in [0,\infty), \quad (\text{a.s.}, P).$$

7.1.32 Exercise: It turns out that the most important special case of the Doob–Meyer Decomposition for submartingales is rather easy to prove. Namely, let $(\Omega, \mathcal{F}, P)$ be a probability space and $(X(t), \mathcal{F}_t, P)$ a continuous, square integrable (i.e., $X(T) \in L^2(P)$ for each $t \in [0,\infty)$) martingale, and assume that $\mathcal{F}_0$ contains every P-null $A \in \mathcal{F}$.† Then, $(X(t)^2, \mathcal{F}_t, P)$ is a continuous submartingale, and the Doob–Meyer Theorem says that there exists a P-almost surely unique continuous, $\{\mathcal{F}_t : t \in [0,\infty)\}$-progressively measurable function $A : [0,\infty) \times \Omega \longmapsto [0,\infty)$ with the properties that, for each $\omega \in \Omega$, $A(0,\omega) = 0$ and $A(\,\cdot\,,\omega)$ is nondecreasing, and $(X(t)^2 - A(t), \mathcal{F}_t, P)$ is a martingale. Here is the outline of a proof.‡

(i) Prove the uniqueness assertion as a consequence of Exercise 7.1.30.

(ii) Set $\tau_{0,\ell}(\omega) = \ell$ for $\ell \in \mathbb{N}$. Next, proceeding by induction, define $\{\tau_{n,\ell}\}_{\ell=0}^\infty$ for $n \in \mathbb{Z}^+$ so that $\tau_{n,0} \equiv 0$ and, for $\ell \in \mathbb{Z}^+$, $\tau_{n,\ell}(\omega)$ is equal to

$$\tau_{n-1,k}(\omega) \wedge \inf\left\{ t \geq \tau_{n,\ell-1}(\omega) : \left|X(t,\omega) - X(\tau_{n,\ell-1}(\omega),\omega)\right| \geq \tfrac{1}{n} \right\}$$

when $\tau_{n-1,k-1}(\omega) \leq \tau_{n,\ell-1}(\omega) < \tau_{n-1,k}(\omega)$. Check that, for each $n \in \mathbb{N}$, $\{\tau_{n,\ell}\}_{\ell=0}^\infty$ is a nondecreasing sequence of bounded, $\{\mathcal{F}_t : t \in [0,\infty)\}$-stopping times and that $\tau_{n,\ell} \nearrow \infty$ as $\ell \to \infty$. Further, note that $\{\tau_{n,\ell}\}_{\ell=0}^\infty \supseteq \{\tau_{n-1,k}\}_{k=0}^\infty$ for every $n \in \mathbb{Z}^+$.

---

† It is easy to check that this represents no real loss in generality in the sense that, if it is not already true, all P-null $A \in \mathcal{F}$ can be added to the $\mathcal{F}_t$'s without destroying the martingale property.

‡ I learned the idea for the existence proof in a lecture given by K. Itô.

(iii) Set

$$X_{n,\ell}(\omega) = X\big(\tau_{n,\ell-1}(\omega), \omega\big) \quad \text{and}$$
$$\Delta_{n,\ell}(t,\omega) = X\big(t \wedge \tau_{n,\ell}(\omega), \omega\big) - X\big(t \wedge \tau_{n,\ell-1}(\omega), \omega\big)$$

for $n \in \mathbb{N}$ and $\ell \in \mathbb{Z}^+$, and observe that

$$X(t,\omega)^2 - X(0,\omega)^2 = 2M_n(t,\omega) + A_n(t,\omega),$$

where

$$M_n(t,\omega) \equiv \sum_{\ell=1}^{\infty} X_{n,\ell}(\omega)\Delta_{n,\ell}(t,\omega) \quad \text{and} \quad A_n(t,\omega) \equiv \sum_{\ell=1}^{\infty} \Delta_{n,\ell}(t,\omega)^2.$$

Show that $\big(M_n(t), \mathcal{F}_t, P\big)$ is a continuous martingale and that $A_n : [0,\infty) \times \Omega \longrightarrow [0,\infty)$ is an $\{\mathcal{F}_t : t \in [0,\infty)\}$-progressively measurable function with the properties that, for each $\omega \in \Omega$: $A_n(0,\omega) = 0$, $A_n(\,\cdot\,,\omega)$ is a continuous, and $A_n(t,\omega) + \frac{1}{n} \geq A_n(s,\omega)$ whenever $0 \leq s < t$.

(iv) For $0 \leq m < n$ and $\ell \in \mathbb{Z}^+$, define

$$X^{(m)}_{n,\ell}(\omega) = X_{m,k}(\omega) \quad \text{when } \tau_{m,k-1}(\omega) \leq \tau_{n,\ell-1}(\omega) < \tau_{m,k}(\omega),$$

note that

$$M_n(t,\omega) - M_m(t,\omega) = \sum_{\ell=1}^{\infty} \Big(X_{n,\ell}(\omega) - X^{(m)}_{n,\ell}(\omega)\Big)\Delta_{n,\ell}(t,\omega),$$

and conclude that

$$\mathbb{E}^P\Big[\Big(M_n(t) - M_m(t)\Big)^2\Big] \leq m^{-2}\mathbb{E}^P\big[X(t)^2 - X(0)^2\big].$$

In particular, as an application of (7.1.10) with $p = 2$, show that there exists a continuous martingale $\big(M(t), \mathcal{F}_t, P\big)$ with the property that

$$\lim_{n\to\infty} \mathbb{E}^P\left[\sup_{t\in[0,T]} \big|M_n(t) - M(t)\big|^2\right] = 0 \quad \text{for each } T \in [0,\infty).$$

(v) To complete the proof, combine parts **(iii)** and **(iv)** to see that the function $A : [0,\infty) \times \Omega \longmapsto [0,\infty)$ given by

$$A(t,\omega) = 0 \vee \sup\Big\{X(s,\omega)^2 - X(0,\omega)^2 - 2M(s,\omega) : s \in [0,t]\Big\}$$

has the required properties.

7.1.33 Exercise: This exercise deals with various aspects of Lévy's Theorem, Theorem 7.1.26.

(i) As we mentioned in the course of its proof, the subtlety of Lévy's Theorem lies in its use of continuity. To see just how essential continuity is, again consider the martingale $X(t) = \mathbf{N}(t) - t$ introduced in part **(iv)** of Exercise 7.1.30, and verify that $\big(X(t)^2 - t, \mathcal{F}_t, P\big)$ is also a martingale. On the other hand, $\big(X(t), \mathcal{F}_t, P\big)$ certainly is not a Brownian motion!

(ii) An important application of Lévy's Theorem and the Doob–Meyer Theorem is the observation that *every real-valued, continuous martingale is Brownian motion run at a random rate.* Because the general statement would involve us in too many technical details, we will content ourselves with the a special case. Namely, refer to the setting in Exercise 7.1.32, and assume that, for each $\omega \in \Omega$, the function $A(\,\cdot\,, \omega)$ is a homeomorphism of $[0,\infty)$ onto itself. That is, $A(\,\cdot\,,\omega)$ is strictly increasing, and $\lim_{t\to\infty} A(t,\omega) = \infty$. Next, define $\tau : [0,\infty) \times \Omega \longrightarrow [0,\infty)$ by the equation $A\big(\tau(t,\omega),\omega\big) = t$ and check that $\big\{\tau(t) : t \in [0,\infty)\big\}$ is an increasing family of $\big\{\mathcal{F}_t : t \in [0,\infty)\big\}$-stopping times. Finally, set $\beta(t,\omega) = X\big(\tau(t,\omega),\omega\big)$, and show that $\big(\beta(t), \mathcal{F}_{\tau(t)}, P\big)$ is a Brownian motion. In other words, $X(t)$ *follows Brownian paths but does so at a rate determined by the clock* $A(t)$.

Hint: When each $\tau(t)$ is bounded, the result is an essentially trivial application of Theorems 7.1.14 and 7.1.26. To prove it when the $\tau(t)$'s are not necessarily bounded, what one has to do is show that, for each $t \in [0,\infty)$, $X\big(\tau(t) \wedge n\big) \longrightarrow \beta(t)$ in $L^2(P)$; a fact which can be proved as a consequence of $A\big(\tau(t)\wedge n\big) \nearrow t$.

(iii) Continue in the setting of part **(ii)** above, and note that any computation which does not involve path parameterization will be the same for $X(\,\cdot\,)$ as it is for $\beta(\,\cdot\,)$. Thus for example, show that if $-\infty < a < 0 < b < \infty$ and σ_a and τ_b are the first times t for which $X(t) - X(0) = a$ and $X(t) - X(0) = b$, respectively, then both σ_a and τ_b are P-almost surely finite and

$$P\big(\sigma_a < \tau_b\big) = \frac{b}{b-a}.$$

In this connection, also prove the following version of the Law of the Iterated Logarithm:

$$\varlimsup_{t\to\infty} \frac{X(t)}{\sqrt{2A(t)\log_{(2)} A(t)}} = 1 \quad (\text{a.s.}, P).$$

§7.2: Properties of Wiener Paths.

At this point we have enough general theory† to make some interesting computations. Thus, rather than continue developing abstract theory, we will devote the first part of this section to applications of the results in Section 7.1 to the analysis of Wiener's measure, and we will then point out how the same ideas can be used in the analysis of more general Markov processes.

The key to all our conclusions about Wiener's measure lies in the following variant of the last part of Corollary 7.1.20.

7.2.1 Lemma. *For $\mathfrak{G}$ an open subset of $\mathbb{R} \times \mathbb{R}^N$, set*

$$\zeta^{\mathfrak{G}}(\psi) = \inf\{t \geq 0 : (t, \psi(t)) \notin \mathfrak{G}\}. \tag{7.2.2}$$

Then, for any $f \in C^{1,2}(\mathfrak{G}) \cap C_{\mathrm{b}}(\overline{\mathfrak{G}})$ and any measurable function $g : \mathfrak{G} \longrightarrow \mathbb{R}$ which is bounded below and satisfies $g \leq \frac{\partial f}{\partial t} + \frac{1}{2}\Delta f$,

$$f\big(t \wedge \zeta^{\mathfrak{G}}(\psi), \psi(t \wedge \zeta^{\mathfrak{G}})\big) - \int_0^{t \wedge \zeta^{\mathfrak{G}}(\psi)} g\big(s, \psi(s)\big)\, ds$$

is a continuous $\mathcal{W}_{\mathbf{x}}^{(N)}$-submartingale relative to $\{\mathcal{B}_t : t \in [0, \infty)\}$ for each $\mathbf{x} \in \mathbb{R}^N$ such that $(0, \mathbf{x}) \in \mathfrak{G}$. In particular, if $\frac{\partial f}{\partial t} + \frac{1}{2}\Delta f \geq 0$, then

$$\Big(f\big(t \wedge \zeta^{\mathfrak{G}}(\psi), \psi(t \wedge \zeta^{\mathfrak{G}})\big), \mathcal{B}_t, \mathcal{W}_{\mathbf{x}}^{(N)}\big)\Big)$$

is a submartingale for each $\mathbf{x} \in \mathbb{R}^N$ with $(0, \mathbf{x}) \in \mathfrak{G}$; and if $\frac{\partial f}{\partial t} + \frac{1}{2}\Delta f$ is bounded, then

$$\begin{aligned} M_f^{\mathfrak{G}}(t, \psi) \equiv & f\big(t \wedge \zeta^{\mathfrak{G}}(\psi), \psi(t \wedge \zeta^{\mathfrak{G}})\big) \\ & - \int_0^{t \wedge \zeta^{\mathfrak{G}}(\psi)} \left[\tfrac{\partial f}{\partial s} + \tfrac{1}{2}\Delta f\right]\big(s, \psi(s)\big)\, ds \end{aligned} \tag{7.2.3}$$

is a continuous $\mathcal{W}_{\mathbf{x}}^{(N)}$-martingale for each $\mathbf{x} \in \mathbb{R}^N$ with $(0, \mathbf{x}) \in \mathfrak{G}$.

PROOF: Throughout, we assume that $(0, \mathbf{x}) \in \mathfrak{G}$.

Now, choose a sequence of open sets $\mathfrak{G}_n \ni (\mathbf{0}, \mathbf{x})$ in $\mathbb{R} \times \mathbb{R}^N$ in such a way that $\overline{\mathfrak{G}_n} \subset\subset \mathfrak{G}_{n+1}$ for each $n \in \mathbb{Z}^+$ and $\mathfrak{G}_n \nearrow \mathfrak{G}$ as $n \to \infty$. It is then an easy matter to check that, as $n \to \infty$, either $\zeta^{\mathfrak{G}_n}(\psi)$ tends to ∞ or $\big(\zeta^{\mathfrak{G}_n}(\psi), \psi(\zeta^{\mathfrak{G}_n})\big)$ converges to the boundary of $\mathfrak{G}$; and from this it is a short step to $\zeta^{\mathfrak{G}_n} \nearrow \zeta^{\mathfrak{G}}$ as

† French readers of a certain generation will probably not have recognized the content of §7.1 as *general theory*.

$n \to \infty$. Hence, it suffices for us to check that $\left(M_f^{\mathfrak{G}_n}, \mathcal{B}_t, \mathcal{W}_{\mathbf{x}}^{(N)}\right)$ is a martingale for each $n \in \mathbb{Z}^+$. Indeed, it will then be clear that

$$f\big(t \wedge \zeta^{\mathfrak{G}_n}(\psi), \psi(t \wedge \zeta^{\mathfrak{G}_n})\big) - \int_0^{t \wedge \zeta^{\mathfrak{G}_n}(\psi)} g\big(s, \psi(s)\big)\, ds$$

is a continuous $\mathcal{W}_{\mathbf{x}}^{(N)}$-submartingale relative to $\{\mathcal{B}_t : t \in [0,\infty)\}$, and so the desired conclusion will follow immediately from Fatou's Lemma after one lets $n \to \infty$.

Let $n \in \mathbb{Z}^+$ be given. To prove that $\left(M_f^{\mathfrak{G}_n}, \mathcal{B}_t, \mathcal{W}_{\mathbf{x}}^{(N)}\right)$ is a martingale, choose $\eta_n \in C_c^\infty\big(\mathfrak{G}_{n+1};[0,1]\big)$ so that $\eta_n \equiv 1$ on $\mathfrak{G}_n$, and define $f_n \in C_b^{1,2}\big(\mathbb{R} \times \mathbb{R}^N;\mathbb{R}\big)$ so that $f_n \equiv 0$ off $\mathfrak{G}$ and $f_n = \eta_n f$ on $\mathfrak{G}$. Then, by Corollary 7.1.20, $\left(M_{f_n}^{\mathbb{R}\times\mathbb{R}^N}(t), \mathcal{F}_t, \mathcal{W}_{\mathbf{x}}^{(N)}\right)$ is a martingale; and therefore the desired result follows from Doob's Stopping Time Theorem (cf. Corollary 7.1.15) and the identity

$$M_f^{\mathfrak{G}_n}(t) = M_{f_n}^{\mathbb{R}\times\mathbb{R}^N}(t) \quad \text{for } t \in \big[0, \zeta^{\mathfrak{G}_n}\big). \quad \square$$

As our first application of Lemma 7.2.1, we give the following computation.

7.2.4 Theorem. *There is an $\alpha \in (0,\infty)$ with the property that, for each $N \in \mathbb{Z}^+$ and every $\epsilon > 0$:*

$$\text{(7.2.5)} \qquad \mathcal{W}^{(N)}\left(\sup_{0 \le t \le T} |\psi(t)| < \epsilon\right) \ge \exp\left[-\frac{\alpha N^2 T}{\epsilon^2}\right], \quad T \in (0,\infty).$$

(See part **(i)** *of Exercise 7.2.31.)*

PROOF: First observe that

$$\mathcal{W}^{(N)}\left(\sup_{0 \le t \le T} |\psi(t)| < \epsilon\right) \ge \mathcal{W}^{(N)}\left(\sup_{0 \le t \le T} |\psi_j(t)| < \frac{\epsilon}{\sqrt{N}} \text{ for } 1 \le j \le N\right)$$

$$= \mathcal{W}\left(\sup_{0 \le t \le T} |\psi(t)| < \frac{\epsilon}{\sqrt{N}}\right)^N = \mathcal{W}\left(\sup_{0 \le t \le NT\epsilon^{-2}} |\psi(t)| < 1\right)^N,$$

where we have used the independence of the coordinates and Wiener scaling. Thus, it suffices for us to treat the case when $N = 1$ and $\epsilon = 1$. To handle this case, set

$$f(t,x) = e^{\frac{\pi^2 t}{8}} \sin\big(\tfrac{\pi}{2}(x+1)\big), \quad (t,x) \in \mathbb{R} \times \mathbb{R},$$

note that

$$\frac{\partial f}{\partial t} + \frac{1}{2}\frac{\partial^2 f}{\partial x^2} = 0,$$

and apply Lemma 7.2.1 to see that

$$e^{\frac{\pi^2 T}{8}} \mathcal{W}(\zeta > T) \ge \mathbb{E}^{\mathcal{W}}\Big[f\big(T \wedge \zeta, \psi(T \wedge \zeta)\big),\, \zeta > T\Big] = f(0,0) = 1,$$

where $\zeta(\psi) \equiv \inf\{t : |\psi(t)| \ge 1\}$. $\square$

Recall that the **support** of a Borel measure on a topological space is the smallest closed set whose complement has measure 0. As a consequence of the preceding, we can now characterize the support of $\mathcal{W}_{\mathbf{x}}^{(N)}$.

7.2.6 Corollary. *Let $\mathbf{H}(\mathbb{R}^N)$ be the Cameron–Martin subspace described in the introduction to Section 4.2. Then, for each $\mathbf{h} \in \mathbf{H}(\mathbb{R}^N)$ and $\epsilon > 0$,*

$$\begin{aligned}\mathcal{W}^{(N)}\left(\sup_{0\le t\le T}\left|\boldsymbol{\psi}(t)-\mathbf{h}(t)\right|<\epsilon\right)&\\ \ge \exp\left[-\frac{2\alpha N^2 T}{\epsilon^2}-\|\mathbf{h}\|^2_{\mathbf{H}(\mathbb{R}^N)}\right], &\quad T\in(0,\infty).\end{aligned}\tag{7.2.7}$$

In particular, for each $\mathbf{x} \in \mathbb{R}^N$, the support of $\mathcal{W}_{\mathbf{x}}^{(N)}$ as a Borel measure on $\mathfrak{P}(\mathbb{R}^N)$ coincides with the set of all $\boldsymbol{\psi} \in \mathfrak{P}(\mathbb{R}^N)$ with $\boldsymbol{\psi}(0) = \mathbf{x}$.

PROOF: Clearly, the last assertion reduces immediately to the case when $\mathbf{x} = \mathbf{0}$. Moreover, the first part guarantees that the support of $\mathcal{W}^{(N)}$ contains $\mathbf{H}(\mathbb{R}^N)$. Hence, since $\mathbf{H}(\mathbb{R}^N)$ is a dense subspace of $\{\boldsymbol{\psi} \in \mathfrak{P}(\mathbb{R}^N) : \boldsymbol{\psi}(0) = \mathbf{0}\}$, we will be done once we prove (7.2.7). To this end, let $\mathbf{h} \in \mathbf{H}(\mathbb{R}^N)$ be given, think of $\mathcal{W}^{(N)}$ as a measure on $\Theta(\mathbb{R}^N)$ (cf. the discussion preceding Lemma 4.2.14) and define $R_{\mathbf{h}} : \Theta(\mathbb{R}^N) \longrightarrow (0,\infty)$ as in (4.2.16). Then, by Lemma 4.2.15,

$$\begin{aligned}\mathcal{W}^{(N)}\left(\sup_{0\le t\le T}|\boldsymbol{\theta}(t)|<\epsilon\right) &= \int_{\left\{\sup_{0\le t\le T}|\boldsymbol{\theta}(t)-\mathbf{h}(t)|<\epsilon\right\}} R_{\mathbf{h}}(\boldsymbol{\theta})\,\mathcal{W}_{\mathbf{x}}^{(N)}(d\boldsymbol{\theta})\\ &\le \left\|R_{\mathbf{h}}\right\|_{L^2(\mathcal{W}^{(N)})}\mathcal{W}^{(N)}\left(\sup_{0\le t\le T}|\boldsymbol{\theta}(t)-\mathbf{h}(t)|<\epsilon\right)^{\frac{1}{2}},\end{aligned}$$

which, when combined with the estimate in Remark 4.2.17 and (7.2.5), yields (7.2.7). □

The preceding result says that Wiener *paths will mimic (at least in the sense of the uniform topology) any given continuous path arbitrarily well over any finite time interval.* From the analytic standpoint, the single most interesting consequence of this fact is the following version of **The Strong Maximum Principle.**

7.2.8 Theorem. *Given an open $\mathfrak{G}$ in $\mathbb{R}\times\mathbb{R}^N$ containing $(0,\mathbf{x})$, let $f : \overline{\mathfrak{G}} \longrightarrow \mathbb{R}$ be a measurable function which is bounded above, upper semicontinuous on $\left[(0,\infty)\times\mathbb{R}^N\right]\cap\mathfrak{G}$, and satisfies $f(0,\mathbf{x}) \ge f(t,\mathbf{y})$ for all $(t,\mathbf{y}) \in \mathfrak{G}$ with $t > 0$. If (cf. (7.2.2))*

$$f(0,\mathbf{x}) \le \int_{\mathfrak{P}(\mathbb{R}^N)} f\left(t\wedge\zeta^{\mathfrak{G}}(\boldsymbol{\psi}),\boldsymbol{\psi}(t\wedge\zeta^{\mathfrak{G}})\right)\mathcal{W}_{\mathbf{x}}^{(N)}(d\boldsymbol{\psi}) \quad \text{for } t\in[0,\infty),\tag{7.2.9}$$

then

$$\text{(7.2.10)} \qquad \begin{aligned} f\big(t,\psi(t)\big) = f(0,\mathbf{x}) \quad &\textit{for all } (t,\psi) \in [0,\infty) \times \mathfrak{P}(\mathbb{R}^N) \\ &\textit{such that } \psi(0) = \mathbf{x} \textit{ and } t \in \big[0,\zeta^{\mathfrak{G}}(\psi)\big). \end{aligned}$$

Hence, if $f \in C^{1,2}(\mathfrak{G};\mathbb{R})$ *satisfies* $f(0,\mathbf{x}) \geq f(t,\mathbf{y})$ *and* $\left[\frac{\partial f}{\partial t} + \frac{1}{2}\boldsymbol{\Delta} f\right](t,\mathbf{y}) \geq 0$ *for all* $(t,y) \in \mathfrak{G}$ *with* $t > 0$*, then* (7.2.10) *holds. In particular, if* G *is a connected, open set in* $\mathbb{R}^N$ *and* $f \in C^2(G;\mathbb{R})$ *is bounded above and satisfies* $\boldsymbol{\Delta} f \geq 0$*, then* f *achieves its maximum value if and only if* f *is constant.*

Proof: Given $\psi_0 \in \mathfrak{P}(\mathbb{R}^N)$ with $\psi_0(0) = \mathbf{x}$, suppose that

$$f\big(t_0,\psi_0(t_0)\big) < f(0,\mathbf{x}) \quad \text{for some } t_0 \in \big(0,\zeta^{\mathfrak{G}}(\psi_0)\big).$$

Then we can find a $\delta > 0$ with the property that

$$\begin{aligned} &\zeta^{\mathfrak{G}}(\psi) > t_0 \text{ and } f\big(t_0,\psi(t_0)\big) < f(0,\mathbf{x}) - \delta \text{ whenever} \\ &\qquad \psi \in A \equiv \left\{\psi \in \mathfrak{P}(\mathbb{R}^N) : \sup_{0\leq t\leq t_0+1} \big|\psi(t) - \psi_0(t)\big| < \delta\right\}. \end{aligned}$$

Hence, because Corollary 7.2.6 implies that $\mathcal{W}_{\mathbf{x}}^{(N)}(A) > 0$, (7.2.9) with $t = t_0$ leads to the contradiction

$$f(0,\mathbf{x}) \leq \big(f(0,\mathbf{x}) - \delta\big)\mathcal{W}_{\mathbf{x}}^{(N)}(A) + f(0,\mathbf{x})\big(1 - \mathcal{W}_{\mathbf{x}}^{(N)}(A)\big) < f(0,\mathbf{x}).$$

Next, let $f \in C^{1,2}(\mathfrak{G};\mathbb{R})$ be as in the second part of the theorem. Given $\psi \in \mathfrak{P}(\mathbb{R}^N)$ with $\psi(0) = \mathbf{x}$ and $T \in \big[0,\zeta^{\mathfrak{G}}\big)$, choose a bounded open set $\mathfrak{G}'$ with the properties that $\overline{\mathfrak{G}'} \subset\subset \mathfrak{G}$ and $\big(t,\psi(t)\big) \in \mathfrak{G}'$ for $t \in [0,T]$. Now apply Lemma 7.2.1, with $\mathfrak{G}$ replaced by $\mathfrak{G}'$, to $f \restriction \mathfrak{G}'$ in order to conclude first that (7.2.9) holds with $\mathfrak{G}$ replaced by $\mathfrak{G}'$ and then that $f(0,\mathbf{x}) = f\big(t,\psi(t)\big)$ for all $t \in [0,T]$.

Finally, in order to prove the last assertion, suppose that f achieves its maximum value at $\mathbf{x} \in G$ and apply the preceding to see that $\big\{\mathbf{y} \in G : f(\mathbf{y}) = f(\mathbf{x})\big\}$ is open. Since, by continuity, $\big\{\mathbf{y} \in G : f(\mathbf{y}) = f(\mathbf{x})\big\}$ is closed in G and G is connected, it follows that $G = \big\{\mathbf{y} \in G : f(\mathbf{y}) = f(\mathbf{x})\big\}$. □

The preceding results deal with the behavior of Wiener paths over finite time intervals. Our next results deal with the *long time* behavior of Wiener paths, and here we will see dimension start to play a critical rôle.

7.2.11 Theorem. *For* $r \in [0,\infty)$*, define*

$$\zeta_r(\psi) = \inf\big\{t \in [0,\infty) : |\psi(t)| = r\big\}, \quad \psi \in \mathfrak{P}(\mathbb{R}^N).$$

Then

$$\text{(7.2.12)}\qquad \begin{aligned} \mathbb{E}^{\mathcal{W}_{\mathbf{x}}^{(N)}}\big[\zeta_r\big] &= \frac{r^2-|\mathbf{x}|^2}{N} \\ \mathbb{E}^{\mathcal{W}_{\mathbf{x}}^{(N)}}\big[\zeta_r^2\big] &= \frac{(N+4)r^2-N|\mathbf{x}|^2}{N^2(N+2)}\big(r^2-|\mathbf{x}|^2\big) \end{aligned} \qquad \text{for } |\mathbf{x}|<r.$$

In addition, if $0<r<|\mathbf{x}|<R<\infty$, *then*

$$\text{(7.2.13)}\qquad \mathcal{W}_{\mathbf{x}}^{(N)}\big(\zeta_r<\zeta_R\big) = \begin{cases} \frac{R-|x|}{R-r} & \text{if } N=1 \\ \frac{\log R-\log|\mathbf{x}|}{\log R-\log r} & \text{if } N=2 \\ \left(\frac{r}{|\mathbf{x}|}\right)^{N-2}\frac{R^{N-2}-|\mathbf{x}|^{N-2}}{R^{N-2}-r^{N-2}} & \text{if } N\ge 3. \end{cases}$$

In particular,

$$\mathcal{W}_x^{(1)}\big(\zeta_0<\infty\big)=1 \quad \text{for all } x\in\mathbb{R},$$
$$\mathcal{W}_{\mathbf{x}}^{(2)}\big(\zeta_0<\infty\big)=0,\ \mathbf{x}\neq\mathbf{0}, \quad \text{but} \quad \mathcal{W}_{\mathbf{x}}^{(2)}\big(\zeta_r<\infty\big)=1,\ \mathbf{x}\in\mathbb{R}^2 \text{ and } r>0,$$

and

$$\mathcal{W}_{\mathbf{x}}^{(N)}\big(\zeta_r<\infty\big)=\left(\frac{r}{|\mathbf{x}|}\right)^{N-2},\ 0<r<|\mathbf{x}|, \quad \text{when } N\ge 3.$$

PROOF: To prove (7.2.12), set $f(t,\mathbf{x})=|\mathbf{x}|^2-Nt$, use Lemma 7.2.1 to show that

$$\Big(f\big(t\wedge\zeta_r,\boldsymbol{\psi}(t\wedge\zeta_r)\big),\mathcal{F}_t,\mathcal{W}_{\mathbf{x}}^{(N)}\Big)$$

and

$$\left(f\big(t\wedge\zeta_r,\boldsymbol{\psi}(t\wedge\zeta_r)\big)^2-4\int_0^{t\wedge\zeta_r}|\boldsymbol{\psi}(s)|^2\,ds,\mathcal{B}_t,\mathcal{W}_{\mathbf{x}}^{(N)}\right)$$

are continuous martingales, and conclude that

$$N\mathbb{E}^{\mathcal{W}_{\mathbf{x}}^{(N)}}\big[t\wedge\zeta_r\big]=\mathbb{E}^{\mathcal{W}_{\mathbf{x}}^{(N)}}\Big[\big|\boldsymbol{\psi}(t\wedge\zeta_r)\big|^2\Big]-|\mathbf{x}|^2,\quad t\in[0,\infty),$$

and

$$N^2\mathbb{E}^{\mathcal{W}_{\mathbf{x}}^{(N)}}\left[(t\wedge\zeta_r)^2\right]$$
$$= |\mathbf{x}|^4 + 4\mathbb{E}^{\mathcal{W}_{\mathbf{x}}^{(N)}}\left[\int_0^{t\wedge\zeta_r}|\psi(s)|^2\,ds\right]$$
$$+ 2N\mathbb{E}^{\mathcal{W}_{\mathbf{x}}^{(N)}}\left[(t\wedge\zeta_r)\left|\psi(t\wedge\zeta_r)\right|^2\right] - \mathbb{E}^{\mathcal{W}_{\mathbf{x}}^{(N)}}\left[\left|\psi(t\wedge\zeta_r)\right|^4\right]$$

for all $t \in [0,\infty)$. Next, pass to the limit as $t \to \infty$ in the first of these to arrive at the first equality in (7.2.12). Finally, to get the second equality in (7.2.12) from the second of these, use Lemma 7.2.1 to show that

$$\left(\left|\psi(t\wedge\zeta_r)\right|^4 - (4+2N)\int_0^{t\wedge\zeta_r}\left|\psi(s)\right|^2 ds, \mathcal{F}_t, \mathcal{W}_{\mathbf{x}}^{(N)}\right)$$

is a continuous martingale and therefore that

$$(4+2N)\mathbb{E}^{\mathcal{W}_{\mathbf{x}}^{(N)}}\left[\int_0^{t\wedge\zeta_r}\left|\psi(s)\right|^2 ds\right] = \mathbb{E}^{\mathcal{W}_{\mathbf{x}}^{(N)}}\left[\left|\psi(t\wedge\zeta_r)\right|^4\right] - |\mathbf{x}|^4,$$

plug this into the above, and pass to the limit as $t \nearrow \infty$.

To prove (7.2.13), for each $N \in \mathbb{Z}^+$ choose $f_N \in C_{\mathrm{b}}^\infty(\mathbb{R}^N;\mathbb{R})$ so that f_N is equal to the corresponding expression on the right of (7.2.13) in an open neighborhood U of the annulus $\{\mathbf{x} : r \le |\mathbf{x}| \le R\}$, note that $\mathbf{\Delta} f_N = 0$ on U, and conclude (via Lemma 7.2.1) that

$$\left(f_N(t\wedge\zeta_r\wedge\zeta_R), \mathcal{B}_t, \mathcal{W}_{\mathbf{x}}^{(N)}\right)$$

is a bounded, continuous martingale. In particular, after one lets $t \to \infty$, this leads to

$$\mathcal{W}_{\mathbf{x}}^{(N)}(\zeta_r < \zeta_R) = \mathbb{E}^{\mathcal{W}_{\mathbf{x}}^{(N)}}\left[f_N(\zeta_r\wedge\zeta_R)\right] = f_N(\mathbf{x}), \quad 0 < r < |\mathbf{x}| < R,$$

as required. Given (7.2.13), the rest of the theorem follows easily by letting $R \nearrow \infty$ and, in the case when $N = \{1,2\}$, $r \searrow 0$. □

7.2.14 Corollary. *If $G \ne \emptyset$ is open, then*

$$\text{(7.2.15)} \quad \mathcal{W}_{\mathbf{x}}^{(N)}\left(\int_0^\infty \mathbf{1}_G(\psi(t))\,dt = \infty\right) = 1, \quad \mathbf{x}\in\mathbb{R}^N, \quad \textit{when } N \in \{1,2\}.$$

(Cf. Exercise 7.2.32 below for further implications.) On the other hand,

$$\text{(7.2.16)} \qquad \mathcal{W}_{\mathbf{x}}^{(N)}\left(\lim_{t\to\infty}|\psi(t)| = \infty\right) = 1, \quad \mathbf{x}\in\mathbb{R}^N \quad \textit{when } N \ge 3.$$

PROOF: Clearly it suffices to prove (7.2.15) in the case when $G = B_{\mathbb{R}^N}(\mathbf{0}, R)$ for some $R \in (0,\infty)$, and so we will assume that this is the case. Next, set $\tau_0 \equiv 0$, and define $\{\sigma_n\}_1^\infty$ and $\{\tau_n\}_1^\infty$ inductively with respect to $n \in \mathbb{Z}^+$ so that

$$\sigma_n(\psi) = \inf\left\{t \geq \tau_{n-1} : \left|\psi(t)\right| \leq \tfrac{R}{2}\right\}$$

and

$$\tau_n(\psi) = \inf\left\{t \geq \sigma_n : \left|\psi(t)\right| \geq R\right\}.$$

It is easy to check (cf. the last part of Lemma 4.3.2) that each σ_n and τ_n is an $\{\mathcal{B}_t : t \in [0,\infty)\}$-stopping time, and that $\tau_{n-1} < \infty$ and $\sigma_n < \infty$ imply, respectively, that

$$\sigma_n = \tau_{n-1} + \sigma_1 \circ \mathbf{\Sigma}_{\tau_{n-1}} \quad \text{and} \quad \tau_n = \sigma_n + \tau_1 \circ \mathbf{\Sigma}_{\sigma_n}.$$

Hence, by the strong Markov property (cf. Theorem 4.3.3), one knows that, for every $\Gamma \in \mathcal{B}_{\mathbb{R}}$ and $\mathcal{W}_{\mathbf{x}}^{(N)}$-almost every $\psi \in \mathfrak{P}(\mathbb{R}^N)$,

$$\mathcal{W}_{\mathbf{x}}^{(N)}\left(\sigma_n - \tau_{n-1} \in \Gamma \middle| \mathcal{B}_{\tau_{n-1}}\right)(\psi) = \mathcal{W}_{\psi(\tau_{n-1})}^{(N)}\left(\sigma_1 \in \Gamma\right) \quad \text{if } \tau_{n-1}(\psi) < \infty$$

and

$$\mathcal{W}_{\mathbf{x}}^{(N)}\left(\tau_n - \sigma_n \in \Gamma \middle| \mathcal{B}_{\sigma_n}\right)(\psi) = \mathcal{W}_{\psi(\sigma_n)}^{(N)}\left(\tau_1 \in \Gamma\right) \qquad \text{if } \sigma_n(\psi) < \infty.$$

In addition, by (7.2.12) and (because $N \in \{1,2\}$) (7.2.13), we know that

$$\mathbb{E}^{\mathcal{W}_{\mathbf{x}}^{(N)}}\left[\tau_1\right] < \infty \quad \text{and} \quad \mathcal{W}_{\mathbf{x}}^{(N)}\left(\sigma_1 < \infty\right) = 1$$

for all $\mathbf{x} \in \mathbb{R}^N$; and so, by induction on $n \in \mathbb{Z}^+$, we now see that

$$\mathcal{W}_{\mathbf{x}}^{(N)}\left(\tau_n < \infty\right) = \mathcal{W}_{\mathbf{x}}^{(N)}\left(\sigma_n < \infty\right) = 1 \quad \text{for all } n \in \mathbb{Z}^+.$$

In fact, because of the rotation invariance discussed in Exercise 3.3.28, the distribution of τ_1 under $\mathcal{W}_{\psi(\sigma_n)}^{(N)}$ is the same for all $n \in \mathbb{Z}^+$ and $\psi \in \{\sigma_n < \infty\}$, and therefore we have now proved that if

$$X_n(\psi) = \begin{cases} \tau_n(\psi) - \sigma_n(\psi) & \text{when } \sigma_n(\psi) < \infty \\ 0 & \text{otherwise,} \end{cases}$$

then $\{X_n\}_1^\infty$ is a sequence of mutually independent, identically distributed, $\mathcal{W}_{\mathbf{x}}^{(N)}$-integrable random variables under $\mathcal{W}_{\mathbf{x}}^{(N)}$. Finally, since

$$\int_0^{\tau_n} \mathbf{1}_G\big(\psi(t)\big)\,dt \geq \sum_1^n X_m(\psi)$$

for $\mathcal{W}_{\mathbf{x}}^{(N)}$-almost every $\psi \in \mathfrak{P}(\mathbb{R}^N)$ and $X_1 > 0$ (a.s., $\mathcal{W}_{\mathbf{x}}^{(N)}$), an application of the Strong Law of Large Numbers completes the proof of (7.2.15).

Now assume that $N \geq 3$. Given $r > 0$, apply Lemma 7.2.1 to see that (cf. the notation in Theorem 7.2.11)

$$\left(\left|\psi(t \wedge \zeta_r)\right|^{-N+2}, \mathcal{B}_t, \mathcal{W}_{\mathbf{x}}^{(N)}\right)$$

is a bounded, nonnegative martingale for every $|\mathbf{x}| > r > 0$. Hence, for any $0 \leq s \leq t < \infty$ and $A \in \mathcal{B}_s$,

$$\begin{aligned}|\mathbf{x}|^{-N+2} &\geq \mathbb{E}^{\mathcal{W}_{\mathbf{x}}^{(N)}}\left[\left|\psi(s)\right|^{-N+2}, A \cap \{\zeta_r(\psi) > s\}\right] \\ &= \mathbb{E}^{\mathcal{W}_{\mathbf{x}}^{(N)}}\left[\left|\psi(t \wedge \zeta_r)\right|^{-N+2}, A \cap \{\zeta_r(\psi) > s\}\right];\end{aligned}$$

and, because $N \geq 3$ and therefore $\zeta_r \nearrow \infty$ $\left(\text{a.s., } \mathcal{W}_{\mathbf{x}}^{(N)}\right)$, an application of the Monotone Convergence Theorem and Fatou's Lemma leads to

$$|\mathbf{x}|^{-N+2} \geq \mathbb{E}^{\mathcal{W}_{\mathbf{x}}^{(N)}}\left[\left|\psi(s)\right|^{-N+2}, A\right] \geq \mathbb{E}^{\mathcal{W}_{\mathbf{x}}^{(N)}}\left[\left|\psi(t)\right|^{-N+2}, A\right]$$

for all $0 \leq s \leq t < \infty$, $A \in \mathcal{B}_s$, and $\mathbf{x} \neq \mathbf{0}$. In particular, this proves that

$$\left(-\left|\psi(t)\right|^{-N+2}, \mathcal{B}_t, \mathcal{W}_{\mathbf{x}}^{(N)}\right)$$

is a nonpositive submartingale for every $\mathbf{x} \neq \mathbf{0}$ and therefore, by Theorem 7.1.16, that $\lim_{t\to\infty} \left|\psi(t)\right|$ exists in $[0, \infty]$ for $\mathcal{W}_{\mathbf{x}}^{(N)}$-almost every $\psi \in \mathfrak{P}(\mathbb{R}^N)$. On the other hand,

$$\mathcal{W}_{\mathbf{x}}^{(N)}\left(\left|\psi(t)\right| \leq R\right) = \gamma_t{}^N\left(\{\mathbf{y} : |\mathbf{y} - \mathbf{x}| \leq R\}\right) \longrightarrow 0$$

as $t \to \infty$ for every $R \in (0, \infty)$ and $\mathbf{x} \in \mathbb{R}^N$; and so we now know that, at least when $\mathbf{x} \neq \mathbf{0}$, $|\psi(t)| \longrightarrow \infty$ for $\mathcal{W}_{\mathbf{x}}^{(N)}$-almost every $\psi \in \mathfrak{P}(\mathbb{R}^N)$. Finally, since

$$\mathcal{W}_{\mathbf{0}}^{(N)}\left(\inf_{t \geq T+1} \left|\psi(t)\right| \leq R\right) = \int_{\mathbb{R}^N \setminus \{\mathbf{0}\}} \mathcal{W}_{\mathbf{x}}^{(N)}\left(\inf_{t \geq T} \left|\psi(t)\right| \leq R\right) \gamma_1{}^N(d\mathbf{x}),$$

the same result also holds when $\mathbf{x} = 0$. □

The conclusions drawn in Theorem 7.2.11 and Corollary 7.2.14 are very famous. In particular, they say that only when $N = 1$ does a typical Wiener path ever hit a point at which it did not start, that when $N \in \{1, 2\}$ a typical Wiener path spends infinite time in every (cf. Exercise 7.2.32) nonempty open set, and that when $N \geq 3$ a typical Wiener path spends a finite amount of time in every bounded set. In the jargon of Markov process theory, these properties are summarized by the statement that *the Wiener process is* **recurrent** *in one and two dimensions but is* **transient** *in three and higher dimensions.* When one thinks about what underlies these facts, the following picture emerges. In one dimension, the typical Wiener path sweeps back and forth in such a way that (cf. (4.1.8) and use symmetry)

$$\varliminf_{t\to\infty} \psi(t) = -\infty \quad \text{and} \quad \varlimsup_{t\to\infty} \psi(t) = +\infty,$$

and therefore, by continuity, must visit every point infinitely often. Moreover, because a Wiener path is continuous, each time a Wiener path visits a point it must spend a positive amount of time in a neighborhood of that point, which, after an infinite number of visits, leads to its spending an infinite amount of time there. When $N = 2$, the two coordinates of the Wiener path are independent, and so they are very unlikely to cooperate well enough for them both to take prescribed values at the same time. On the other hand, even though they never actually simultaneously take on prescribed values, each spends enough time close to a specified value that the path still ends up spending infinite time in every neighborhood of every point in the plane. However, when $N \geq 3$, cooperation between the coordinates is so bad that, after a long time, at least one coordinate is so large that the path ends up tending to infinity.

In the rest of this section we will show that the ideas just introduced to study Wiener paths apply equally well to general Markov processes. Thus, without further comment, we will be assuming that $\mathbf{x} \in \mathbb{R}^N \longmapsto P_{\mathbf{x}} \in \mathbf{M}_1(\mathfrak{P}(\mathbb{R}^N))$ is a continuous mapping and that $\{P_{\mathbf{x}} : \mathbf{x} \in \mathbb{R}^N\}$ is a strong Markov family (cf. Exercises 4.3.55 and 5.1.30). Further, we will use $(t, \mathbf{x}) \in (0,\infty) \times \mathbb{R}^N \longmapsto P(t, \mathbf{x}, \cdot\,) \in \mathbf{M}_1(\mathbb{R}^N)$ to denote the transition probability function for $\{P_{\mathbf{x}} : \mathbf{x} \in \mathbb{R}^N\}$, and we will assume that

$$\int_0^\infty P(t, \mathbf{x}, G)\, dt > 0 \quad \text{for all } \mathbf{x} \in \mathbb{R}^N \text{ and open } G \neq \emptyset. \tag{7.2.17}$$

7.2.18 Lemma. *For any nonempty, open G and $K \subset\subset \mathbb{R}^N$, there exists a $T \in (0,\infty)$ and $\epsilon > 0$ such that*

$$\int_0^T P(t, \mathbf{x}, G)\, dt \geq \epsilon \quad \text{for all } \mathbf{x} \in K.$$

In particular, if G is a bounded open set and $\mathfrak{e}^G$ is the first exit time from G, then there is an $\alpha > 0$ for which

$$\sup_{\mathbf{x}\in G} \mathbb{E}^P\left[e^{\alpha \mathfrak{e}^G}\right] < \infty; \tag{7.2.19}$$

and, therefore, $P_{\mathbf{x}}(\mathfrak{e}^G < \infty) = 1$ for all $\mathbf{x} \in G$.

PROOF: Notice (cf. (7.2.17)) that

$$\mathbf{x} \in \mathbb{R}^N \longmapsto \int_0^\infty e^{-t} P(t, \mathbf{x}, G)\, dt \in (0, 1]$$

is lower semicontinuous. Hence, if $K \subset\subset \mathbb{R}^N$, then

$$\inf_{\mathbf{x} \in K} \frac{1}{2} \int_0^\infty e^{-t} P(t, \mathbf{x}, G)\, dt \geq \epsilon > 0.$$

Thus, by choosing $T \in (0, \infty)$ so that $e^{-T} \leq \epsilon$, we arrive at the first conclusion.

Now assume that G is bounded. By the preceding, there exist $T \in (0, \infty)$ and $\epsilon \in (0, 1)$ such that

$$\int_0^T P\big(t, \mathbf{x}, \mathbb{R}^N \setminus \overline{G}\big)\, dt \geq \epsilon > 0, \quad \mathbf{x} \in \overline{G}.$$

In particular,

$$T P_{\mathbf{x}}\big(\mathfrak{e}^G \leq T\big) \geq \mathbb{E}^{P_{\mathbf{x}}}\left[\int_0^T \mathbf{1}_{\mathbb{R}^N \setminus \overline{G}}\big(\psi(t)\big)\, dt\right] \geq \epsilon, \quad \mathbf{x} \in \overline{G};$$

and so

$$P_{\mathbf{x}}\big(\mathfrak{e}^G > T\big) \leq \theta \equiv 1 - \frac{\epsilon}{T} < 1, \quad \mathbf{x} \in \overline{G}.$$

But, because $\mathfrak{e}^G > s \implies \mathfrak{e}^G = s + \mathfrak{e}^G \circ \boldsymbol{\Sigma}_s$, the Markov property leads to

$$P_{\mathbf{x}}\big(\mathfrak{e}^G > (n+1)T\big) = \mathbb{E}^{P_{\mathbf{x}}}\Big[P_{\psi(nT)}\big(\mathfrak{e}^G > T\big),\, \mathfrak{e}^G > nT\Big] \leq \theta P_{\mathbf{x}}\big(\mathfrak{e}^G > nT\big)$$

for all $\mathbf{x} \in \overline{G}$. Thus, by induction,

$$P_{\mathbf{x}}\big(\mathfrak{e}^G > nT\big) \leq \theta^n, \quad n \in \mathbb{N} \text{ and } \mathbf{x} \in \overline{G},$$

and (7.2.19) follows easily from here. □

We can now prove the following general criterion for transience.

7.2.20 Theorem. *Assume that* (7.2.17) *holds. If, for each* $r \in (0, \infty)$,

$$(7.2.21) \qquad 1 > \begin{cases} \inf_{|\mathbf{x}| > r} P_{\mathbf{x}}\big(\exists t \in [0, \infty)\ |\psi(t)| \leq r\big) & \text{when } N = 1 \\ \overline{\lim}_{|\mathbf{x}| \to \infty} P_{\mathbf{x}}\big(\exists t \in [0, \infty)\ |\psi(t)| \leq r\big) & \text{when } N \geq 2, \end{cases}$$

then, the process $\{P_{\mathbf{x}} : \mathbf{x} \in \mathbb{R}^N\}$ *is transient in the sense that, for every* $\mathbf{x} \in \mathbb{R}^N$,

$$(7.2.22) \qquad P_{\mathbf{x}}\Big(\lim_{t \to \infty} |\psi(t)| = \infty\Big) = 1.$$

PROOF: We must show that, for each $r \in (0,\infty)$ and $\mathbf{x} \in \mathbb{R}^N$:

$$P_{\mathbf{x}}\left(\varliminf_{t\to\infty} |\psi(t)| \geq r\right) = 1.$$

To this end, let $r \in (0,\infty)$ be given, and suppose that there is an $R \in (r,\infty)$ for which

$$\theta \equiv \sup_{|\mathbf{x}|\geq R} P_{\mathbf{x}}\Big(|\psi(t)| \leq r \text{ for some } t \in [0,\infty)\Big) < 1.$$

Next, set $\sigma_0(\psi) = \inf\{t : |\psi(t)| \leq r\}$, and use induction to define

$$\tau_n(\psi) = \inf\{t \geq \sigma_n(\psi) : |\psi(t)| \geq R\}$$

and

$$\sigma_{n+1}(\psi) = \inf\{t \geq \tau_n(\psi) : |\psi(t)| \leq r\}.$$

By Lemma 4.3.2, all the σ_n's and τ_n's are $\{\mathcal{B}_t : t \in [0,\infty)\}$-stopping times. Moreover, because (cf. the notation in Theorem 7.2.11) $\sigma_n < \infty \implies \tau_n = \sigma_n + \zeta_R \circ \mathbf{\Sigma}_{\sigma_n}$, the strong Markov property and Lemma 7.2.18 lead to

$$\begin{aligned}
&P_{\mathbf{x}}\big(\sigma_n < \infty \text{ and } \tau_n = \infty\big) \\
&\quad = \int_{\{\sigma_n(\psi)<\infty\}} P_{\psi(\sigma_n)}\big(\zeta_R = \infty\big)\, P_{\mathbf{x}}(d\psi) = 0
\end{aligned}$$

for each $n \in \mathbb{N}$. Hence,

$$\begin{aligned}
P_{\mathbf{x}}(\sigma_n = \infty) &= P_{\mathbf{x}}(\sigma_0 = \infty) + \sum_{m=1}^{n} P_{\mathbf{x}}\big(\sigma_{m-1} < \infty \,\&\, \sigma_m = \infty\big) \\
&= P_{\mathbf{x}}(\sigma_0 = \infty) + \sum_{m=1}^{n} P_{\mathbf{x}}\big(\tau_{m-1} < \infty \,\&\, \sigma_m = \infty\big) \\
&\leq P_{\mathbf{x}}\left(\varliminf_{t\to\infty} |\psi(t)| \geq r\right).
\end{aligned}$$

Thus, $\varliminf_{t\to\infty} |\psi(t)| \geq r$ (a.s., P_x) will follow as soon as we check that $P_{\mathbf{x}}(\sigma_n = \infty) \longrightarrow 1$ as $n \to \infty$. But, $\tau_n < \infty \implies \sigma_{n+1} = \tau_n + \sigma_0 \circ \mathbf{\Sigma}_{\tau_n}$, and so, again by the strong Markov property,

$$\begin{aligned}
P_{\mathbf{x}}\big(\sigma_{n+1} < \infty\big) &= \int_{\{\tau_n(\psi)<\infty\}} P_{\psi(\tau_n)}(\sigma_0 < \infty)\, P_{\mathbf{x}}(d\psi) \\
&\leq \theta P_{\mathbf{x}}\big(\tau_n < \infty\big) \leq \theta P_{\mathbf{x}}(\sigma_n < \infty),
\end{aligned}$$

which, after a trivial induction argument, means that we are done.

The preceding argument takes care of the case when $N \geq 2$. However, when $N = 1$, there is more to be done. Namely, because we know only that

$$P_a\big(\exists t \in [0,\infty)\ |\psi(t)| \leq r\big) < 1 \text{ for some } a \notin [-r,r],$$

there are two possibilities: either

$$P_{-a}\big(\exists t \in [0,\infty)\ |\psi(t)| \leq r\big) < 1 \quad \text{or} \quad P_{-a}\big(\exists t \in [0,\infty)\ |\psi(t)| \leq r\big) = 1.$$

In the first case, we take $R = |a|$, $\sigma(\psi) = \inf\{t \geq 0 : |\psi(t)| \leq |a|\}$, and apply the strong Markov property to see that, for $|x| \geq R$:

$$\begin{aligned}P_x\big(\exists t \in [0,\infty)\ |\psi(t)| \leq r\big) &= \int_{\{\sigma(\varphi)<\infty\}} P_{\varphi(\sigma)}\big(\exists t \in [0,\infty)\ |\psi(s)| \leq r\big)\, P_x(d\varphi)\\ &\leq \theta \equiv P_{-a}\big(\exists t \in [0,\infty)\ |\psi(t)| \leq r\big) \vee P_a\big(\exists t \in [0,\infty)\ |\psi(t)| \leq r\big) < 1.\end{aligned}$$

Hence, this case is covered by the preceding. To handle the other case, we assume, without loss in generality, that for some $R > r$:

$$\theta \equiv P_R\big(\exists t \in [0,\infty)\ |\psi(t)| \leq r\big) < 1 \text{ but } P_{-R}\big(\exists t \in [0,\infty)\ |\psi(t)| \leq r\big) = 1.$$

With $\{\sigma_n\}_0^\infty$ and $\{\tau_n\}_0^\infty$ as above, it is clear that the place where the preceding argument fails is at the point where we have to show that $P_x\big(\sigma_n < \infty\big) \searrow 0$. Indeed, all that we can say now is that

$$\begin{aligned}P_x\big(\sigma_{n+1} < \infty\big) &= \int_{\{\tau_n(\psi)<\infty\}} P_{\psi(\tau_n)}\big(\sigma_0 < \infty\big)\, P_x(d\psi)\\ &= \theta P_x\big(\tau_n(\psi) < \infty\ \&\ \psi(\tau_n) = R\big) + P_x\big(\tau_n(\psi) < \infty\ \&\ \psi(\tau_n) = -R\big)\\ &= P_x\big(\sigma_n < \infty\big) - (1-\theta) P_x\big(\tau_n(\psi) < \infty\ \&\ \psi(\tau_n) = R\big).\end{aligned}$$

Thus, we must still show that, for some $\epsilon > 0$,

$$P_x\big(\tau_n(\psi) < \infty\ \&\ \psi(\tau_n) = R\big) \geq \epsilon P_x\big(\sigma_n < \infty\big),$$

and, by the strong Markov property, this comes down to checking that

$$\inf_{x\in[-r,r]} P_x\big(\eta_R < \eta_{-R}\big) > 0,$$

where $\eta_y(\psi) \equiv \inf\{t \geq 0 : \psi(t) = y\}$ for each $y \in \mathbb{R}$. To this end, note that, for any $y \in [-r,r]$:

$$\begin{aligned}P_{-r}\big(\eta_R < \eta_{-R}\big) &= P_{-r}\big(\eta_y < \eta_R < \eta_{-R}\big)\\ &= \mathbb{E}^{P_{-r}}\Big[P_x\big(\eta_R < \eta_{-R}\big),\ \eta_y < \eta_{-R}\Big] \leq P_x\big(\eta_R < \eta_{-R}\big);\end{aligned}$$

and so we need only show that

$$P_{-r}\big(\eta_R < \eta_{-R}\big) > 0.$$

But suppose that

$$P_{-r}\big(\eta_{-R} < \eta_R\big) = 1.$$

Then, by the strong Markov property:

$$P_{-r}\big(\tau_{n+1} < \eta_R \,\big|\, \tau_n < \eta_R\big) = P_{-R}(\tau_0 < \eta_R) = P_{-r}(\eta_{-R} < \eta_R) = 1 \quad \text{(a.s., } P_{-r}\text{)};$$

which, because $\tau_n \nearrow \infty$ (a.s., P_{-r}) means that $P_{-r}\big(\eta_R = \infty\big) = 1$. However, this would imply that

$$\int_0^\infty P\big(t, -r, (R,\infty)\big)\, dt = \mathbb{E}^{P_{-r}}\left[\int_{(\eta_R,\infty)} \mathbf{1}_{(R,\infty)}\big(\psi(t)\big)\, dt\right] = 0,$$

which obviously contradicts (7.2.17). □

We turn now to the problem of determining when $\{P_{\mathbf{x}} : \mathbf{x} \in \mathbb{R}^N\}$ is recurrent.

7.2.23 Theorem. *Again assume* (7.2.17), *but this time suppose that there is an* $r \in (0,\infty)$ *with the property that*

$$P_{\mathbf{x}}\Big(|\psi(t)| \le r \text{ for some } t \in [0,\infty)\Big) = 1 \quad \textit{for all } \mathbf{x} \in \mathbb{R}^N. \tag{7.2.24}$$

Then $\{P_{\mathbf{x}} : \mathbf{x} \in \mathbb{R}^N\}$ *is recurrent in the sense that for every* $\mathbf{x} \in \mathbb{R}^N$

$$P_{\mathbf{x}}\left(\int_0^\infty \mathbf{1}_G\big(\psi(t)\big)\, dt = \infty\right) = 1 \quad \textit{for all open } G \ne \emptyset. \tag{7.2.25}$$

(Cf. Exercise 7.2.32 for further information.)

PROOF: Let G be a nonempty open set and, using Lemma 7.2.18, choose $T \in (0,\infty)$ and $\epsilon > 0$ so that

$$\int_0^T P(t, \mathbf{x}, G)\, dt \ge 2\epsilon \quad \text{for all } |\mathbf{x}| \le r.$$

Next, define

$$\begin{aligned} \tau_0(\psi) &= \inf\big\{t \ge 0 : \ |\psi(t)| \le r\big\}, \\ \tau_{n+1}(\psi) &= \inf\big\{t \ge \tau_n(\psi) + T : \ |\psi(t)| \le r\big\}, \\ T_n(\psi) &= \mathbf{1}_{[0,\infty)}\big(\tau_n(\psi)\big) \int_{\tau_n(\psi)}^{\tau_n(\psi)+T} \mathbf{1}_G\big(\psi(t)\big)\, dt \end{aligned}$$

for $n \in \mathbb{N}$. By Lemma 4.3.2, τ_n and $\tau_n + T$ are $\{\mathcal{B}_t : t \in [0,\infty)\}$-stopping times for every $n \in \mathbb{N}$. In addition, by (7.2.24) and the strong Markov property,

$$P_{\mathbf{x}}(\tau_{n+1} < \infty) = \int_{\{\tau_n<\infty\}} P_{\psi(\tau_n+T)}(\tau_0 < \infty)\, P_{\mathbf{x}}(d\psi) = P_{\mathbf{x}}(\tau_n < \infty)$$

for all $n \in \mathbb{N}$ and $\mathbf{x} \in \mathbb{R}^N$. Hence, since (7.2.24) guarantees that $P_{\mathbf{x}}(\tau_0 < \infty) = 1$ for all $\mathbf{x} \in \mathbb{R}^N$, it follows that $P_{\mathbf{x}}(\tau_n < \infty) = 1$ for all $n \in \mathbb{N}$ and $\mathbf{x} \in \mathbb{R}^N$.

Since, for $|\mathbf{y}| \le r$,

$$2\epsilon \le \mathbb{E}^{P_{\mathbf{y}}}\left[\int_0^T \mathbf{1}_G(\psi(t))\, dt\right] = \mathbb{E}^{P_{\mathbf{y}}}[T_1] \le \epsilon + T P_{\mathbf{y}}(T_1 \ge \epsilon)$$

and $\tau_n < \infty \implies T_n = T_0 \circ \boldsymbol{\Sigma}_{\tau_n}$,

$$P_{\mathbf{y}}\left(T_n \ge \epsilon \,\middle|\, \mathcal{B}_{\tau_{n-1}}\right) \ge \theta \equiv \frac{\epsilon}{T} > 0 \quad \text{(a.s., } P_{\mathbf{x}}\text{) for all } \mathbf{x} \in \mathbb{R}^N.$$

Finally, set $A_n = \{T_n \ge \epsilon\}$, and apply the form of the Borel–Cantelli Lemma proved in part **(iii)** of Exercise 5.2.34 to see that:

$$\int_0^\infty \mathbf{1}_G(\psi(t))\, dt \ge \sum_0^\infty T_n \ge \epsilon \sum_0^\infty \mathbf{1}_{A_n} = \infty \quad \text{(a.s., } P_{\mathbf{x}}\text{)}$$

for all $\mathbf{x} \in \mathbb{R}^N$. □

Obviously, except when $N = 1$, there is a gap between the criteria given in Theorems 7.2.20 and 7.2.32. However, as the next result shows, there are conditions under which one can show that this apparent gap disappears.

7.2.26 Theorem. *Suppose that, for each $\mathbf{x} \in \mathbb{R}^N$, the support of $P_{\mathbf{x}}$ on $\mathfrak{P}(\mathbb{R}^N)$ coincides with $\{\psi \in \mathfrak{P}(\mathbb{R}^N) : \psi(0) = \mathbf{x}\}$. Further, assume that either $N = 1$ or $N \ge 2$ and, for each $r \in (0,\infty)$, the function*

$$\mathbf{x} \in \mathbb{R}^N \longmapsto P_{\mathbf{x}}\Big(|\psi(t)| \le r \text{ for some } t \in [0,\infty)\Big) \in [0,1]$$

is upper semicontinuous. Then, either $\{P_{\mathbf{x}} : \mathbf{x} \in \mathbb{R}^N\}$ is transient, in the sense that (7.2.22) holds for every $\mathbf{x} \in \mathbb{R}^N$, or it is recurrent, in the sense that (7.2.25) holds for all $\mathbf{x} \in \mathbb{R}^N$.

PROOF: Obviously, the support assumption guarantees that (7.2.17) holds; and so, by Theorem 7.2.20, there is nothing more to do when $N = 1$. Thus, assume that $N \ge 2$ and make the stated upper semicontinuity hypothesis. Then, for

$r \in (0,\infty)$, the function $\mathbf{x} \in \mathbb{R}^N \longmapsto f(\mathbf{x}) = P_{\mathbf{x}}(\sigma_r < \infty) \in [0,1]$ is upper semicontinuous when

$$\sigma_r(\psi) \equiv \inf\{t : |\psi(t)| \le r\}. \tag{7.2.27}$$

In addition, by the Markov property, for any $(t,\mathbf{x}) \in [0,\infty) \times \mathbb{R}^N$,

$$\begin{aligned} f(\mathbf{x}) &= P_{\mathbf{x}}(\sigma_r \le t) + P_{\mathbf{x}}\big(t < \sigma_r < \infty\big) \\ &= \int_{\{\sigma_r(\psi)\le t\}} f\big(\psi(\sigma_r)\big)\, P_{\mathbf{x}}(d\psi) + \int_{\{\sigma_r(\psi)>t\}} f\big(\psi(t)\big)\, P_{\mathbf{x}}(d\psi) \\ &= \int_{\mathfrak{P}(\mathbb{R}^N)} f\big(\psi(t\wedge\sigma_r)\big)\, P_{\mathbf{x}}(d\psi), \end{aligned}$$

where, in the passage to the second line, we have used the facts that $f \equiv 1$ on $\overline{B_{\mathbb{R}^N}(\mathbf{0},r)}$ and that

$$\sigma_r(\psi) = t + \sigma_r \circ \boldsymbol{\Sigma}_t(\psi) \quad \text{if } \sigma_r(\psi) > t.$$

Hence, by proceeding in precisely the same way as we did in the proof of The Strong Maximum Principle (cf. Corollary 7.2.8), we see that if $f(\mathbf{x}) = 1$ for some $|\mathbf{x}| > r$, then

$$f\big(\psi(t)\big) = 1 \quad \text{for all } \psi \in \mathfrak{P}(\mathbb{R}^N) \text{ with } \psi(0) = \mathbf{x} \text{ and } t \in \big[0,\sigma_r(\psi)\big).$$

But, because $N \ge 2$ and therefore $\mathbb{R}^N \setminus \overline{B_{\mathbb{R}^N}(\mathbf{0},r)}$ is connected, this means that $f \equiv 1$.

In view of the preceding, we have that, for any $r \in (0,\infty)$,

$$P_{\mathbf{x}}(\sigma_r < \infty) = 1 \quad \text{for all } \mathbf{x} \in \mathbb{R}^N \quad \text{or} \quad P_{\mathbf{x}}(\sigma_r < \infty) < 1 \quad \text{for all } |\mathbf{x}| > r.$$

Moreover, by upper semicontinuity, the latter case means that

$$\epsilon(r) \equiv \sup_{|\mathbf{x}|=2r} P_{\mathbf{x}}(\sigma_r < \infty) < 1,$$

and therefore, by the strong Markov property, that

$$P_{\mathbf{x}}(\sigma_r < \infty) = \int_{\{\sigma_{2r}(\psi)<\infty\}} P_{\psi(\sigma_{2r})}(\sigma_r < \infty)\, P_{\mathbf{x}}(d\psi) \le \epsilon(r)$$

for all $|\mathbf{x}| \ge 2r$. In other words, we now know that, for each $r \in (0,\infty)$,

$$P_{\mathbf{x}}(\sigma_r < \infty) = 1 \quad \text{for all } \mathbf{x} \in \mathbb{R}^N \quad \text{or} \quad \sup_{|\mathbf{x}|\ge 2r} P_{\mathbf{x}}(\sigma_r < \infty) < 1.$$

But, by Theorems 7.2.20 and 7.2.23, this dichotomy proves that $\{P_{\mathbf{x}} : \mathbf{x} \in \mathbb{R}^N\}$ is either recurrent or transient in the required sense. □

The condition of upper semicontinuity in Theorem 7.2.26 can often be checked as a consequence of the following.

7.2.28 Corollary. *Assume that, for each* $t \in (0,\infty)$ *and* $\Gamma \in \mathcal{B}_{\mathbb{R}^N}$,

$$\mathbf{x} \in \mathbb{R}^N \longmapsto P(t,\mathbf{x},\Gamma) \quad \textit{is continuous.} \tag{7.2.29}$$

Then, for every $r \in (0,\infty)$,

$$\mathbf{x} \in \mathbb{R}^N \setminus \overline{B_{\mathbb{R}^N}(\mathbf{0},r)} \longmapsto P_{\mathbf{x}}\Big(|\psi(t)| \le r \text{ for some } t \in [0,\infty)\Big) \in [0,1]$$

is continuous. In particular, if the support of $P_{\mathbf{x}}$ *is* $\{\psi \in \mathfrak{P}(\mathbb{R}^N) : \psi(0) = \mathbf{x}\}$ *for every* $\mathbf{x} \in \mathbb{R}^N$, *then* $\{P_{\mathbf{x}} : \mathbf{x} \in \mathbb{R}^N\}$ *is either transient or recurrent in the senses described in* (7.2.22) *or* (7.2.25).

PROOF: Clearly, we need only verify the first assertion. To this end, let $r \in (0,\infty)$ be given and, for $s \in [0,\infty)$, set

$$f_s(\mathbf{x}) = P_{\mathbf{x}}\Big(|\psi(t)| \le r \text{ for some } t \in [s,\infty)\Big).$$

Our goal is to prove that f_0 is continuous on $\mathbb{R}^N \setminus \overline{B_{\mathbb{R}^N}(\mathbf{0},r)}$. By the Markov property, we know that

$$f_s(\mathbf{x}) = \int_{\mathbb{R}^N} f_0(\mathbf{y})\, P(s,\mathbf{x},d\mathbf{y}), \quad (s,\mathbf{y}) \in (0,\infty) \times \mathbb{R}^N,$$

and therefore, by our hypothesis, f_s is continuous for each $s \in (0,\infty)$. At the same time, if $\delta > 0$ and $|\mathbf{x}| \ge r+\delta$, then, because $\mathbf{x} \in \mathbb{R}^N \longmapsto P_{\mathbf{x}} \in \mathbf{M}_1(\mathfrak{P}(\mathbb{R}^N))$ is continuous,

$$\begin{aligned} f_0(\mathbf{x}) - f_s(\mathbf{x}) &= P_{\mathbf{x}}\Big(|\psi(t)| \le r \text{ for some } t \in [0,s]\Big) \\ &\le P_{\mathbf{x}}\left(\sup_{t\in[0,s]} |\psi(t) - \psi(0)| \ge \delta\right) \longrightarrow 0 \quad \text{as } s \searrow 0 \end{aligned}$$

uniformly on compact subsets of $\mathbb{R}^N \setminus B_{\mathbb{R}^N}(\mathbf{0},r+\delta)$; and, obviously, the required continuity of f_0 follows from this. □

We close with criteria for transience and recurrence which are slightly more down to earth. In fact, the ones which we about to give are, more or less, the basis for our proof of Corollary 7.2.14.

7.2.30 Theorem. *Again assume (7.2.17), and, for $r \in [0,\infty)$, define σ_r as in (7.2.27). If, for some $r_0 \in [0,\infty)$, there is a strictly positive, continuous function f on $\mathbb{R}^N \setminus B_{\mathbb{R}^N}(\mathbf{0}, r_0)$ with the property that*

$$\Big(-f(\psi(t\wedge\sigma_{r_0})), \mathcal{B}_t, P_{\mathbf{x}}\Big) \quad \text{is a submartingale for every } |\mathbf{x}| > r_0,$$

then $\lim_{|\mathbf{x}|\to\infty} f(\mathbf{x}) = 0$ implies that $\{P_{\mathbf{x}} : \mathbf{x} \in \mathbb{R}^N\}$ is transient (in the sense that (7.2.22) holds for all $\mathbf{x} \in \mathbb{R}^N$) and $\lim_{|\mathbf{x}|\to\infty} f(\mathbf{x}) = \infty$ implies that it is recurrent (in the sense that (7.2.25) holds for all $\mathbf{x} \in \mathbb{R}^N$).

PROOF: Assume that $\lim_{|\mathbf{x}|\to\infty} f(\mathbf{x}) = 0$. By Theorem 7.2.20, all that we have to do is check that (7.2.21) holds for each $r \in (0,\infty)$, and, obviously, this will be done once we check it for $r \geq r_0$. But if $r \geq r_0$, then, by Doob's Stopping Time Theorem,

$$\begin{aligned} m(r)P_{\mathbf{x}}(\sigma_r < \infty) &= m(r) \lim_{t\to\infty} P_{\mathbf{x}}(\sigma_r \leq t) \\ &\leq \lim_{t\to\infty} \int_{\mathfrak{P}(\mathbb{R}^N)} f(\psi(t\wedge\sigma_r))\, P_{\mathbf{x}}(d\psi) \leq f(\mathbf{x}), \end{aligned}$$

where $m(r) \equiv \inf\{f(\mathbf{y}) : |\mathbf{y}| = r\} > 0$.

Next assume that $\lim_{|\mathbf{x}|\to\infty} f(\mathbf{x}) = \infty$. By Theorem 7.2.23, we need only check that $P_{\mathbf{x}}(\sigma_{r_0} < \infty) = 1$ for all $\mathbf{x} \in \mathbb{R}^N$. To this end, define ζ_R as in Theorem 7.2.11, and note that, again by Doob's Stopping Time Theorem and the fact that $f > 0$,

$$\begin{aligned} \mathbb{E}^{P_{\mathbf{x}}}\Big[f(\psi(\zeta_R)),\, \zeta_R < \sigma_{r_0}\Big] &= \lim_{T\to\infty} \mathbb{E}^{P_{\mathbf{x}}}\Big[f(\psi(\zeta_R)),\, \zeta_R < \sigma_{r_0}\wedge T\Big] \\ &\leq \varliminf_{T\to\infty} \mathbb{E}^{P_{\mathbf{x}}}\Big[f(\psi(\sigma_{r_0}\wedge\zeta_R\wedge T))\Big] \leq f(\mathbf{x}) \end{aligned}$$

for $r_0 < |\mathbf{x}| < R$. Now let $|\mathbf{x}| > r_0$ be fixed, let $R \nearrow \infty$, and use $\lim_{|\mathbf{x}|\to\infty} f(\mathbf{x}) = \infty$ to conclude that $\lim_{R\to\infty} P_{\mathbf{x}}(\zeta_R < \sigma_{r_0}) = 0$. Hence, since (cf. Lemma 7.2.18) $P_{\mathbf{x}}(\zeta_R < \infty) = 1$ for all $R \in (0,\infty)$, this proves that

$$P_{\mathbf{x}}(\sigma_{r_0} < \infty) \geq \lim_{R\to\infty} P_{\mathbf{x}}(\sigma_{r_0} \leq \zeta_R) = 1. \quad \square$$

Exercises

7.2.31 Exercise: The results in Theorem 7.2.4 and Corollary 7.2.6 can be sharpened in two obvious directions.

(i) In our derivation of (7.2.5), we actually found the rate at which a Wiener path escapes not from the ball $B_{\mathbb{R}^N}(\mathbf{0},\epsilon)$ but instead from the cube $\left(-\frac{\epsilon}{\sqrt{N}},\frac{\epsilon}{\sqrt{N}}\right)^N$. A sharper estimate would have been obtained had we used the fact, coming from elementary properties of the Laplacian, that there is a unique smooth function $f_N : B_{\mathbb{R}^N}(\mathbf{0},1) \longrightarrow (0,\infty)$ and $\alpha_N \in (0,\infty)$ with the properties that

$$\tfrac{1}{2}\boldsymbol{\Delta} f_N = -\alpha_N f, \quad f \le f(\mathbf{0}) = 1, \quad \text{and} \lim_{|\mathbf{x}|\nearrow 1} f_N(\mathbf{x}) = 0.$$

(Indeed, the constant α in (7.2.5) is, in the present notation, α_1.) Assuming this fact, show that (7.2.5) holds when αN^2 is replaced by this choice of α_N. With a little more work, one can show that this α_N is optimal in the sense that

$$\lim_{T\to\infty} e^{\alpha_N T}\mathcal{W}^{(N)}\left(\sup_{0\le t\le T}|\boldsymbol{\psi}(t)| < 1\right) > 0.$$

(ii) Think of $\mathcal{W}^{(N)}$ as a Borel measure on $\Theta(\mathbb{R}^N)$ (cf. Section 4.2) and show that, as such, its support is the whole of $\Theta(\mathbb{R}^N)$.

Hint: Observe that, for each $T \in [1,\infty)$,

$$\begin{aligned}&\mathcal{W}^{(N)}\big(\|\boldsymbol{\theta}\|_{\Theta(\mathbb{R}^N)} < \epsilon\big)\\ &\ge \mathcal{W}^{(N)}\left(\sup_{0\le t\le T}|\boldsymbol{\theta}(t)| < \frac{\epsilon}{2} \text{ and } \sup_{t\ge T}\frac{|\boldsymbol{\theta}(t)-\boldsymbol{\theta}(T)|}{1+t} < \frac{\epsilon}{2}\right),\end{aligned}$$

and use this together with (7.2.5) to see that $\mathcal{W}^{(N)}\big(\|\boldsymbol{\theta}\|_{\Theta(\mathbb{R}^N)} < \epsilon\big) > 0$ for all $\epsilon > 0$. Now, proceed as in the proof of Corollary 7.2.6.

7.2.32 Exercise: Let $\{P_{\mathbf{x}} : \mathbf{x} \in \mathbb{R}^N\}$ be a Markov family which is recurrent in the sense that (7.2.25) holds for every $\mathbf{x} \in \mathbb{R}^N$. Observe that, for each $\mathbf{x} \in \mathbb{R}^N$ and $P_{\mathbf{x}}$-almost every $\boldsymbol{\psi} \in \mathfrak{P}(\mathbb{R}^N)$,

$$\int_0^\infty \mathbf{1}_G\big(\boldsymbol{\psi}(t)\big)\,dt = \infty \quad \text{for every open } G \ne \emptyset.$$

In particular, $P_{\mathbf{x}}$-almost every path *spends an infinite amount of time in every nonempty open set.* Using this observation together with the Martingale Convergence Theorem, show that the only functions $f \in C\big(\mathbb{R}^N;[0,\infty)\big)$ with the property that

$$f(\mathbf{x}) \ge \int_{\mathfrak{P}(\mathbb{R}^N)} f\big(\boldsymbol{\psi}(t)\big)\,P_{\mathbf{x}}(d\boldsymbol{\psi}), \quad (t,\mathbf{x}) \in (0,\infty)\times\mathbb{R}^N,$$

are constant.

Hint: Show that $\left(-f(\psi(t)), \mathcal{B}_t, P_{\mathbf{x}}\right)$ is a submartingale, and conclude that

$$\lim_{t\to\infty} f(\psi(t)) \text{ exists for } P_{\mathbf{x}}\text{-almost every } \psi \in \mathfrak{P}(\mathbb{R}^N).$$

Now use recurrence to see that this is possible only if f is constant.

7.2.33 Exercise: Let $\mathbf{x} \in \mathbb{R}^N \longmapsto P_{\mathbf{x}} \in \mathbf{M}_1\left(\mathfrak{P}(\mathbb{R}^N)\right)$ be a continuous map, and assume that $\{P_{\mathbf{x}} : \mathbf{x} \in \mathbb{R}^N\}$ is a strong Markov process with transition probability function $P(t, \mathbf{x}, \cdot\,)$.

(i) Under the conditions in Theorem 7.2.20, show that for each bounded open G in $\mathbb{R}^N$ there is an $\alpha_G \in (0, \infty)$ such that

$$\sup_{\mathbf{x}\in\mathbb{R}^N} \int_{\mathfrak{P}(\mathbb{R}^N)} \exp\left[\alpha_G \int_0^\infty \mathbf{1}_G\big(\psi(t)\big)\, dt\right] P_{\mathbf{x}}(d\psi) < \infty.$$

(ii) Under the conditions in Theorem 7.2.26, show that $\{P_{\mathbf{x}} : \mathbf{x} \in \mathbb{R}^N\}$ is transient if and only if there exists an open $G \neq \emptyset$ and an $\mathbf{x} \in \mathbb{R}^N$ for which

$$\int_0^\infty P(t, \mathbf{x}, G)\, dt < \infty.$$

§7.3: Perturbations of Wiener Paths.

In this section we will introduce a class of diffusions which are obtained from Wiener paths by the introduction of a *force field.* Thus, let $\mathbf{b} : \mathbb{R}^N \longrightarrow \mathbb{R}^N$ be a "nice" vector field, and consider the map $X^{\mathbf{b}} : [0, \infty) \times \mathfrak{P}(\mathbb{R}^N) \longrightarrow \mathbb{R}^N$ which is determined by the integral equation

$$X^{\mathbf{b}}(t, \psi) = \psi(t) + \int_0^t \mathbf{b}\big(X^{\mathbf{b}}(s, \psi)\big)\, ds. \tag{7.3.1}$$

The reason for our looking at this class of processes is that it brings out an interesting *conflict* between the original Wiener paths and the perturbing force field. In particular, because the difference between the perturbed and the original Wiener paths are paths of locally bounded variation, it is reasonably clear that their local behavior will be governed by that of the highly irregular Wiener paths. On the other hand, because the Wiener paths tend to jiggle so much that they end up not traveling very far, the long time behavior of the perturbed paths ought to be determined by the force field.

In order to get started, suppose that we take seriously the interpretation (given in Remark 7.1.23) of Wiener paths as *the integral curves of* $\frac{1}{2}\boldsymbol{\Delta}$. Then it is reasonable to guess that, in the same sense, the paths $X^{\mathbf{b}}(\,\cdot\,,\psi)$ ought to be *the integral curves of*

$$\mathbf{L}^{\mathbf{b}} = \tfrac{1}{2}\boldsymbol{\Delta} + \mathbf{b}\cdot\nabla. \tag{7.3.2}$$

That is, we are predicting that, for each $f \in C_c^\infty(\mathbb{R}^N;\mathbb{R})$ and $\mathbf{x} \in \mathbb{R}^N$,

$$f(X^{\mathbf{b}}(t,\psi)) - \int_0^t [\mathbf{L}^{\mathbf{b}} f]\big(X^{\mathbf{b}}(s,\psi)\big)\,ds$$

is a $\mathcal{W}_{\mathbf{x}}^{(N)}$-martingale relative to $\{\mathcal{B}_t : t \in [0,\infty)\}$. Equivalently, what we are saying is that if $Q_{\mathbf{x}}^{\mathbf{b}} \in \mathbf{M}_1\big(\mathfrak{P}(\mathbb{R}^N)\big)$ is the distribution of $\psi \in \mathfrak{P}(\mathbb{R}^N) \longmapsto X^{\mathbf{b}}(\,\cdot\,,\psi) \in \mathfrak{P}(\mathbb{R}^N)$ under $\mathcal{W}_{\mathbf{x}}^{(N)}$ and

$$M_f^{\mathbf{b}}(t,\psi) \equiv f\big(\psi(t)\big) - \int_0^t [\mathbf{L}^{\mathbf{b}} f]\big(\psi(s)\big)\,ds,$$

then $\big(M_f^{\mathbf{b}}(t), \mathcal{B}_t, Q_{\mathbf{x}}^{\mathbf{b}}\big)$ ought to be a continuous martingale. In fact, we can even hope that, together with the initial condition $Q_{\mathbf{x}}^{\mathbf{b}}(\psi(0) = \mathbf{x}) = 1$, this property will characterize $Q_{\mathbf{x}}^{\mathbf{b}}$.

In order to test the preceding conjecture, we must first make its statement a little more precise. Thus, from now on, we will be assuming that $\mathbf{b} : \mathbb{R}^N \longrightarrow \mathbb{R}^N$ is a continuously differentiable vector field and we will be using $X^{\mathbf{b}}(\,\cdot\,,\psi)$ to denote the solution to (7.3.1) *up to the first time of explosion.* That is, for each $\ell \in \mathbb{Z}^+$, choose a $\mathbf{b}_\ell \in C_b^1\big(\mathbb{R}^N;\mathbb{R}^N\big)$ so that

$$\mathbf{b}_\ell \restriction B_{\mathbb{R}^N}(\mathbf{0},\ell+1) \equiv \mathbf{b} \restriction B_{\mathbb{R}^N}(\mathbf{0},\ell+1)$$

and determine $X_\ell^{\mathbf{b}} : [0,\infty) \times \mathfrak{P}(\mathbb{R}^N) \longrightarrow \mathfrak{P}(\mathbb{R}^N)$ by (7.3.1) with $\mathbf{b}_\ell$ replacing $\mathbf{b}$. As is well-known (via the Picard method for solving ordinary differential equations), $X_\ell^{\mathbf{b}}(\,\cdot\,,\psi)$ is uniquely determined for each $\psi \in \mathfrak{P}(\mathbb{R}^N)$ and, in fact, $\psi \in \mathfrak{P}(\mathbb{R}^N) \longmapsto X_\ell^{\mathbf{b}}(\,\cdot\,,\psi) \in \mathfrak{P}(\mathbb{R}^N)$ is a continuous map for which $X_\ell^{\mathbf{b}} : [0,\infty)\times \mathfrak{P}(\mathbb{R}^N) \longrightarrow \mathbb{R}^N$ is $\{\mathcal{B}_t : t \in [0,\infty)\}$-progressively measurable. Next, set

$$\mathfrak{e}_\ell^{\mathbf{b}}(\psi) = \inf\big\{t : \big|X_\ell^{\mathbf{b}}(t,\psi)\big| \ge \ell\big\}. \tag{7.3.3}$$

Then, for each $\ell \in \mathbb{Z}^+$ and $\psi \in \mathfrak{P}(\mathbb{R}^N)$, $X_{\ell+1}^{\mathbf{b}}(t,\psi) = X_\ell^{\mathbf{b}}(t,\psi)$ for $t \in \big[0,\mathfrak{e}_\ell^{\mathbf{b}}(\psi)\big)$ and so $\mathfrak{e}_\ell^{\mathbf{b}}(\psi) \le \mathfrak{e}_{\ell+1}^{\mathbf{b}}(\psi)$. Finally, define the **first explosion time** $\mathfrak{e}^{\mathbf{b}} : \mathfrak{P}(\mathbb{R}^N) \longrightarrow [0,\infty]$ by

$$\mathfrak{e}^{\mathbf{b}}(\psi) = \lim_{\ell\to\infty} \mathfrak{e}_\ell^{\mathbf{b}}(\psi), \quad \psi \in \mathfrak{P}(\mathbb{R}^N). \tag{7.3.4}$$

It is then an easy matter to check that, for each $\psi \in \mathfrak{P}(\mathbb{R}^N)$, the path $t \in [0, \mathfrak{e}^{\mathbf{b}}(\psi)) \longmapsto X^{\mathbf{b}}(t, \psi) \in \mathbb{R}^N$ given by

$$X^{\mathbf{b}}(t, \psi) = X^{\mathbf{b}}_\ell(t, \psi) \quad \text{for } \ell \in \mathbb{Z}^+ \text{ and } t \in [0, \mathfrak{e}^{\mathbf{b}}_\ell(\psi)) \tag{7.3.5}$$

is well-defined and is uniquely determined by the fact that it satisfies (7.3.1) for all $t \in [0, \mathfrak{e}^{\mathbf{b}}(\psi))$.

7.3.6 Lemma. *The function $\mathfrak{e}^{\mathbf{b}} : \mathfrak{P}(\mathbb{R}^N) \longrightarrow [0, \infty]$ is a lower semicontinuous, $\{\mathcal{B}_t : t \in [0, \infty)\}$-stopping time, and therefore*

$$\Omega(\mathbf{b}) \equiv \{\psi \in \mathfrak{P}(\mathbb{R}^N) : \mathfrak{e}^{\mathbf{b}}(\psi) = \infty\} \tag{7.3.7}$$

is a Borel measurable subset of $\mathfrak{P}(\mathbb{R}^N)$. Furthermore, if $\boldsymbol{\delta}_s : \mathfrak{P}(\mathbb{R}^N) \longrightarrow \mathfrak{P}(\mathbb{R}^N)$ is the time increment mapping in Lemma 4.1.4 and $\mathbf{T}_{\mathbf{x}} : \mathfrak{P}(\mathbb{R}^N) \longrightarrow \mathfrak{P}(\mathbb{R}^N)$ is the translation map described just before Lemma 4.3.7, then

$$\begin{aligned} \mathfrak{e}^{\mathbf{b}}(\psi) > s &\implies s + \mathfrak{e}^{\mathbf{b}}\big(\mathbf{T}_{X^{\mathbf{b}}(s,\psi)} \circ \boldsymbol{\delta}_s \psi\big) = \mathfrak{e}^{\mathbf{b}}(\psi) \quad \text{and} \\ \mathfrak{e}^{\mathbf{b}}(\psi) > s + t &\implies X^{\mathbf{b}}(s + t, \psi) = X^{\mathbf{b}}\big(t, \mathbf{T}_{X^{\mathbf{b}}(s,\psi)} \circ \boldsymbol{\delta}_s \psi\big). \end{aligned} \tag{7.3.8}$$

Finally, for each $T \in (0, \infty)$,

$$\psi \in \Omega_T(\mathbf{b}) \equiv \{\mathfrak{e}^{\mathbf{b}} > T\} \longmapsto X^{\mathbf{b}}(\cdot \wedge T, \psi) \in \mathfrak{P}(\mathbb{R}^N)$$

is a continuous, $\mathcal{B}_T\big[\{\mathfrak{e}^{\mathbf{b}} > T\}\big]$-measurable map with the property that

$$\mathcal{W}^{(N)}_{\varphi(0)}\left(\left\{\psi : \mathfrak{e}^{\mathbf{b}}(\psi) > T \text{ and } \sup_{0 \le t \le T} \big|X^{\mathbf{b}}(t, \psi) - \varphi(t)\big| < \epsilon\right\}\right) > 0 \tag{7.3.9}$$

for all $\varphi \in \mathfrak{P}(\mathbb{R}^N)$ and $\epsilon > 0$.

PROOF: Because $\mathfrak{e}^{\mathbf{b}}_\ell \nearrow \mathfrak{e}^{\mathbf{b}}$, the first assertion will be proved once we show that, for any $\psi \in \mathfrak{P}(\mathbb{R}^N)$, $\mathfrak{e}^{\mathbf{b}}_\ell(\psi) > t$ and $\psi_n \longrightarrow \psi$ imply that $\big|X^{\mathbf{b}}_\ell(s, \psi_n)\big| < \ell$, $s \in [0, t]$, for all sufficiently large $n \in \mathbb{Z}^+$. But, since $X^{\mathbf{b}}_\ell(\cdot, \psi_n) \longrightarrow X^{\mathbf{b}}_\ell(\cdot, \psi)$ uniformly on $[0, t]$ and $\{X^{\mathbf{b}}_\ell(s, \psi) : s \in [0, t]\}$ is a compact subset of $B_{\mathbb{R}^N}(\mathbf{0}, \ell)$, there is nothing more to do. In particular, this proves that, as the decreasing limit of the open sets $\{\mathfrak{e}^{\mathbf{b}} > n\}$, the set $\Omega(\mathbf{b})$ in (7.3.7) is a Borel set. Next, let $s \in [0, \infty)$ and $\psi \in \Omega_s(\mathbf{b})$ be given. Then $t \in [0, \mathfrak{e}^{\mathbf{b}}(\psi) - s) \longmapsto X^{\mathbf{b}}(s + t, \psi) \in \mathbb{R}^N$ satisfies (7.3.1) with ψ replaced by $\mathbf{T}_{X^{\mathbf{b}}(s,\psi)} \circ \boldsymbol{\delta}_s \psi$, and, by uniqueness, both parts of (7.3.8) follow. Finally, if $\psi \cup \{\psi_n\}_1^\infty \subseteq \Omega_T(\mathbf{b})$ and $\{\psi_n\}_1^\infty \longrightarrow \psi$, then there is an $\ell \in \mathbb{Z}^+$ such that $\mathfrak{e}^{\mathbf{b}}_\ell(\psi_n) > T$ for all $n \in \mathbb{Z}^+$. Hence,

$$\begin{aligned} &\sup_{t \in [0,T]} \big|X^{\mathbf{b}}(t, \psi_n) - X^{\mathbf{b}}(t, \psi)\big| \\ &\qquad = \sup_{t \in [0,T]} \big|X^{\mathbf{b}}_\ell(t, \psi_n) - X^{\mathbf{b}}_\ell(t, \psi)\big| \longrightarrow 0 \quad \text{as } n \to \infty; \end{aligned}$$

and this proves the required continuity result. Thus, all that remains is to check (7.3.9). But given $\varphi \in \mathfrak{P}(\mathbb{R}^N)$, set

$$\psi_0(t) = \varphi(t) - \int_0^t \mathbf{b}\big(\varphi(s)\big)\,ds, \quad t \in [0,\infty),$$

and note that, by uniqueness, $\mathfrak{e}^{\mathbf{b}}(\psi_0) = \infty$ and $\varphi(t) = X^{\mathbf{b}}(t,\psi_0)$ for all $t \in [0,\infty)$. In particular, given $\epsilon > 0$, we can find $\delta > 0$ so that

$$\mathfrak{e}^{\mathbf{b}}(\psi) > T \text{ and } \sup_{0\le t\le T} \big|X^{\mathbf{b}}(t,\psi) - \varphi(t)\big| < \epsilon$$

whenever $\sup_{0\le t\le T} |\psi(t) - \psi_0(t)| < \delta$. Hence, (7.3.9) follows as a consequence of Corollary 7.2.6. □

Now that we have a precise formulation of what we mean by a solution to (7.3.1), we turn to the problem of connecting the paths $X^{\mathbf{b}}(\,\cdot\,,\psi)$ with the operator $\mathbf{L}^{\mathbf{b}}$ in (7.3.2).

7.3.10 Theorem. *For each $\mathbf{x} \in \mathbb{R}^N$, there is at most one $P \in \mathbf{M}_1(\mathfrak{P}(\mathbb{R}^N))$ with the properties that $P\big(\psi(0) = \mathbf{x}\big) = 1$ and*

(7.3.11)
$$\left(f\big(\psi(t)\big) - \int_0^t \big[\mathbf{L}^{\mathbf{b}} f\big]\big(\psi(s)\big)\,ds, \mathcal{B}_t, P\right)$$
is a martingale for every $f \in C_c^\infty(\mathbb{R}^N;\mathbb{R})$.

Moreover, such a P exists if and only if

(7.3.12)
$$\mathcal{W}_{\mathbf{x}}^{(N)}\big(\Omega(\mathbf{b})\big) = 1,$$

in which case P is the distribution $Q_{\mathbf{x}}^{\mathbf{b}}$ of

$$\psi \in \Omega(\mathbf{b}) \longmapsto X^{\mathbf{b}}(\,\cdot\,,\psi) \in \mathfrak{P}(\mathbb{R}^N)$$

under $\mathcal{W}_{\mathbf{x}}^{(N)}$. In particular, when (7.3.12) holds, $\{\psi \in \mathfrak{P}(\mathbb{R}^N) : \psi(0) = \mathbf{x}\}$ is the support of $Q_{\mathbf{x}}^{\mathbf{b}}$ on $\mathfrak{P}(\mathbb{R}^N)$; and, for every $f \in C^{1,2}\big([0,\infty)\times\mathbb{R}^N;\mathbb{C}\big)$ with the property that both f and $\frac{\partial f}{\partial t} + \mathbf{L}^{\mathbf{b}} f$ are bounded on each strip $[0,T]\times\mathbb{R}^N$, $T \in (0,\infty)$:

(7.3.13)
$$M_f^{\mathbf{b}}(t,\psi) \equiv f\big(t,\psi(t)\big) - \int_0^t \left[\tfrac{\partial f}{\partial s} + \mathbf{L}^{\mathbf{b}} f\right]\big(s,\psi(s)\big)\,ds$$

is a $Q_{\mathbf{x}}^{\mathbf{b}}$-martingale relative to $\big\{\mathcal{B}_t : t \in [0,\infty)\big\}$. Finally, if (7.3.12) holds for each $\mathbf{x} \in \mathbb{R}^N$, then $\mathbf{x} \in \mathbb{R}^N \longmapsto Q_{\mathbf{x}}^{\mathbf{b}} \in \mathbf{M}_1(\mathfrak{P}(\mathbb{R}^N))$ is continuous and the family $\big\{Q_{\mathbf{x}}^{\mathbf{b}} : \mathbf{x} \in \mathbb{R}^N\big\}$ is strong Markov (cf. Exercises 4.3.55 and 5.1.30).

PROOF: First suppose that (7.3.12) holds, let $f \in C_c^{1,2}([0,\infty)\times\mathbb{R}^N;\mathbb{C})$ be given, set

$$F(\mathbf{x}',\mathbf{x}'',t) = f(0,\mathbf{x}+\mathbf{x}')+\int_0^t \frac{\partial f}{\partial s}(|s|,\mathbf{x}'+\mathbf{x}'')\,ds \quad \text{for } (\mathbf{x}',\mathbf{x}'',t) \in \mathbb{R}^N\times\mathbb{R}^N\times\mathbb{R},$$

and define

$$B_\ell(t,\psi) = \int_0^t \mathbf{b}_\ell\big(\mathbf{X}_\ell^{\mathbf{b}}(s,\psi)\big)\,ds \quad \text{for } \ell \in \mathbb{Z}^+ \text{ and } (t,\psi) \in [0,\infty)\times\mathfrak{P}(\mathbb{R}^N).$$

By Corollary 7.1.20, with $M = N+1$ and $\mathbf{Y}(t,\psi) = \big(B_\ell(t,\psi),t\big)$, applied to F, we see that

$$f\big(t,\mathbf{X}_\ell^{\mathbf{b}}(t,\psi)\big) - \int_0^t \left[\tfrac{\partial f}{\partial s} + \left[\mathbf{L}^{\mathbf{b}_\ell} f\right](s,\mathbf{X}_\ell^{\mathbf{b}}(s,\psi)\right)ds$$

is a $\mathcal{W}_{\mathbf{x}}^{(N)}$-martingale relative to $\{\mathcal{B}_t : t \in [0,\infty)\}$. Hence, by Corollary 7.1.15, we now know that

$$f\big(t\wedge\mathfrak{e}_\ell^{\mathbf{b}},\mathbf{X}^{\mathbf{b}}(t)\big) - \int_0^{t\wedge\mathfrak{e}_\ell^{\mathbf{b}}} \Big(\tfrac{\partial f}{\partial s} + \left[\mathbf{L}^{\mathbf{b}} f\right]\Big)\big(s,\mathbf{X}^{\mathbf{b}}(s)\big)\,ds$$

is a $\mathcal{W}_{\mathbf{x}}^{(N)}$-martingale relative to $\{\mathcal{B}_t : t \in [0,\infty)\}$. But, by (7.3.12), $\mathfrak{e}_\ell^{\mathbf{b}} \nearrow \infty$ $\mathcal{W}_{\mathbf{x}}^{(N)}$-almost surely, and so we have proved that (cf. (7.3.13)) $\big(M_f^{\mathbf{b}}(t),\mathcal{B}_t,Q_{\mathbf{x}}^{\mathbf{b}}\big)$ is a martingale for all $f \in C_{\mathrm{b}}^{1,2}\big([0,\infty)\times\mathbb{R}^N;\mathbb{R}^N\big)$. Moreover, if $f \in C^{1,2}\big([0,\infty)\times\mathbb{R}^N;\mathbb{C}\big)$ satisfies the given boundedness condition on strips, then we can apply a standard cut-off argument and Corollary 7.1.15 to see that $\big(M_f^{\mathbf{b}}(t),\mathcal{B}_t,Q_{\mathbf{x}}^{\mathbf{b}}\big)$ is a martingale in this case also.

We next suppose that P is a probability measure on $\mathfrak{P}(\mathbb{R}^N)$ for which $P\big(\psi(0) = \mathbf{x}\big) = 1$ and (7.3.11) holds. Set

$$\beta(t,\psi) = \psi(t) - \int_0^t \mathbf{b}\big(\psi(s)\big)\,ds, \quad (t,\psi) \in [0,\infty)\times\mathbb{R}^N.$$

In order to prove that $\mathcal{W}_{\mathbf{x}}^{(N)}\big(\Omega(\mathbf{b})\big) = 1$ and that $P = Q_{\mathbf{x}}^{\mathbf{b}}$, it suffices for us to show that $\big(\beta(t),\mathcal{B}_t,P\big)$ is a Brownian motion (cf. the discussion preceding Theorem 7.1.26). Indeed, because (by uniqueness) $\psi = X^{\mathbf{b}}\big(\,\cdot\,,\beta(\,\cdot\,,\psi)\big)$ for every $\psi \in \mathfrak{P}(\mathbb{R}^N)$, we will then know that the distribution of $\psi \in \mathfrak{P}(\mathbb{R}^N) \longmapsto \beta(\,\cdot\,,\psi) \in \mathfrak{P}(\mathbb{R}^N)$ under P is $\mathcal{W}_{\mathbf{x}}^{(N)}$ and therefore both that

$$\mathcal{W}_{\mathbf{x}}^{(N)}\big(\Omega(\mathbf{b})\big) = P\big(\beta(\,\cdot\,) \in \Omega(\mathbf{b})\big) = 1$$

and that $P = Q^{\mathbf{b}}_{\mathbf{x}}$. In order to prove that $\big(\boldsymbol{\beta}(t), \mathcal{B}_t, P\big)$ is a Brownian motion, we will use Theorem 7.1.26. Thus, let $\boldsymbol{\xi} \in \mathbb{R}^N$ be given and set

$$\psi_{\boldsymbol{\xi}}(t) = \big(\boldsymbol{\xi}, \boldsymbol{\psi}(t)\big)_{\mathbb{R}^N}, \quad \beta_{\boldsymbol{\xi}}(t,\boldsymbol{\psi}) = \big(\boldsymbol{\xi}, \boldsymbol{\beta}(t)\big)_{\mathbb{R}^N},$$
$$b_{\boldsymbol{\xi}}(t,\boldsymbol{\psi}) = \Big(\boldsymbol{\xi}, \mathbf{b}\big(\boldsymbol{\psi}(t)\big)\Big)_{\mathbb{R}^N}, \quad \text{and } B_{\boldsymbol{\xi}}(t,\boldsymbol{\psi}) = \int_0^t b_{\boldsymbol{\xi}}(s,\boldsymbol{\psi})\,ds.$$

What we must show is that both

$$\big(\beta_{\boldsymbol{\xi}}(t), \mathcal{B}_t, P\big) \quad \text{and} \quad \big(\beta_{\boldsymbol{\xi}}(t)^2 - |\boldsymbol{\xi}|^2 t, \mathcal{B}_t, P\big)$$

are martingales. To this end, for each $n \in \mathbb{Z}^+$, choose $f_n \in C^\infty_c(\mathbb{R}^N;\mathbb{R})$ so that $f_n(\mathbf{y}) = \big(\boldsymbol{\xi}, \mathbf{y}\big)_{\mathbb{R}^N}$ for $\mathbf{y} \in B_{\mathbb{R}^N}(\mathbf{x}, n)$, take $\zeta_n(\boldsymbol{\psi})$ to be the first time that $\boldsymbol{\psi} \in \mathfrak{P}(\mathbb{R}^N)$ exits from $B_{\mathbb{R}^N}(\mathbf{0}, n)$, and apply (7.3.11) followed by Corollary 7.1.15 to see that

$$\beta_{\boldsymbol{\xi}}(t \wedge \zeta_n) \quad \text{and} \quad \psi_{\boldsymbol{\xi}}(t\wedge\zeta_n)^2 - |\boldsymbol{\xi}|^2 t \wedge \zeta_n - 2\int_0^{t\wedge\zeta_n} \psi_{\boldsymbol{\xi}}(s)\,B_{\boldsymbol{\xi}}(ds)$$

are P-martingales relative to $\big\{\mathcal{B}_t : t \in [0,\infty)\big\}$. Next, apply Theorem 7.1.19, with $X(t) = \beta_{\boldsymbol{\xi}}(t\wedge\zeta_n)$ and $V(t) = B_{\boldsymbol{\xi}}(t\wedge\zeta_n)$, to see that

$$\beta_{\boldsymbol{\xi}}(t\wedge\zeta_n)B_{\boldsymbol{\xi}}(t\wedge\zeta_n) - \int_0^{t\wedge\zeta_n} \beta_{\boldsymbol{\xi}}(s)\,B_{\boldsymbol{\xi}}(ds)$$

is also a P-martingale relative to $\big\{\mathcal{B}_t : t \in [0,\infty)\big\}$. Hence, since $B_{\boldsymbol{\xi}}(t)^2 = 2\int_0^t b_{\boldsymbol{\xi}}(s)\,B_{\boldsymbol{\xi}}(ds)$ and therefore

$$\begin{aligned}\beta_{\boldsymbol{\xi}}(t)^2 - |\boldsymbol{\xi}|^2 t &= \psi_{\boldsymbol{\xi}}(t)^2 - |\boldsymbol{\xi}|^2 t - 2\psi_{\boldsymbol{\xi}}(t)B_{\boldsymbol{\xi}}(t) + B_{\boldsymbol{\xi}}(t)^2 \\ &= \psi_{\boldsymbol{\xi}}(t)^2 - |\boldsymbol{\xi}|^2 t - 2\beta_{\boldsymbol{\xi}}(t)B_{\boldsymbol{\xi}}(t) - B_{\boldsymbol{\xi}}(t)^2 \\ &= \psi_{\boldsymbol{\xi}}(t)^2 - |\boldsymbol{\xi}|^2 t - 2\int_0^t \psi_{\boldsymbol{\xi}}(s)\,B_{\boldsymbol{\xi}}(ds) \\ &\qquad - 2\left(\beta_{\boldsymbol{\xi}}(t)B_{\boldsymbol{\xi}}(t) - \int_0^t \beta_{\boldsymbol{\xi}}(s)\,B_{\boldsymbol{\xi}}(ds)\right),\end{aligned}$$

we conclude that both

$$\big(\beta_{\boldsymbol{\xi}}(t\wedge\zeta_n), \mathcal{B}_t, P\big) \quad \text{and} \quad \big(\beta_{\boldsymbol{\xi}}(t\wedge\zeta_n)^2 - |\boldsymbol{\xi}|^2 t\wedge\zeta_n, \mathcal{B}_t, P\big)$$

are martingales for each $n \in \mathbb{Z}^+$. Thus, we will be done as soon as we check that, for each $t \in [0,\infty)$, $\beta_{\boldsymbol{\xi}}(t\wedge\zeta_n) \longrightarrow \beta_{\boldsymbol{\xi}}(t)$ in $L^2(P)$. But, by what we have already proved:

$$\begin{aligned}\varlimsup_{m\to\infty} \mathbb{E}^P\Big[\big(\beta_{\boldsymbol{\xi}}(t)-\beta_{\boldsymbol{\xi}}(t\wedge\zeta_m)\big)^2\Big] &\le \lim_{m\to\infty}\sup_{n\ge m}\mathbb{E}^P\Big[\big(\beta_{\boldsymbol{\xi}}(t\wedge\zeta_n)-\beta_{\boldsymbol{\xi}}(t\wedge\zeta_m)\big)^2\Big]\\ &= \lim_{m\to\infty}\sup_{n\ge m}\Big(\mathbb{E}^P\big[\beta_{\boldsymbol{\xi}}(t\wedge\zeta_n)^2\big]-\mathbb{E}^P\big[\beta_{\boldsymbol{\xi}}(t\wedge\zeta_m)^2\big]\Big)\\ &= \lim_{m\to\infty}\sup_{n\ge m}|\boldsymbol{\xi}|^2\Big(\mathbb{E}^P\big[t\wedge\zeta_n\big]-\mathbb{E}^P\big[t\wedge\zeta_m\big]\Big)=0.\end{aligned}$$

Finally, assume that (7.3.12) holds for every $\mathbf{x}\in\mathbb{R}^N$. In order to show that $\mathbf{x}\in\mathbb{R}^N\longmapsto Q^{\mathbf{b}}_{\mathbf{x}}\in\mathbf{M}_1(\mathfrak{P}(\mathbb{R}^N))$ is continuous, it is sufficient for us to check that $\mathbf{x}\in\mathbb{R}^N\longmapsto\mathbb{E}^{Q^{\mathbf{b}}_{\mathbf{x}}}[F]\in\mathbb{R}$ is continuous for each bounded, continuous $F:\mathfrak{P}(\mathbb{R}^N)\longrightarrow\mathbb{R}$ which is $\mathcal{B}_T$-measurable for some $T\in[0,\infty)$. To this end, let such an F be given, and set

$$F^{\mathbf{b}}(\psi)=\mathbf{1}_{\Omega_T(\mathbf{b})}(\psi)\,F\big(X^{\mathbf{b}}(\,\cdot\wedge T,\psi)\big),\qquad \psi\in\mathfrak{P}(\mathbb{R}^N).$$

Then, by the last part of Lemma 7.3.6, we know that $F^{\mathbf{b}}$ is continuous at each $\psi\in\Omega_T(\mathbf{b})$; and, therefore, (7.3.12) guarantees that, for each $\mathbf{x}\in\mathbb{R}^N$, $F^{\mathbf{b}}$ is continuous $\mathcal{W}^{(N)}_{\mathbf{x}}$-almost surely and that $\mathbb{E}^{Q^{\mathbf{b}}_{\mathbf{x}}}[F]=\mathbb{E}^{\mathcal{W}^{(N)}_{\mathbf{x}}}[F^{\mathbf{b}}]$. Hence, by part (**vii**) of Theorem 3.1.4, the required continuity is now clear. As for the asserted Markov property, because (cf. (**iii**) in Exercise 4.3.55) $\mathbf{x}\in\mathbb{R}^N\longmapsto Q^{\mathbf{b}}_{\mathbf{x}}\in\mathbf{M}_1(\mathfrak{P}(\mathbb{R}^N))$ is continuous, the strong Markov property holds as soon as the Markov property does. Thus, let $s\in(0,\infty)$ and a bounded $\mathcal{B}_s\times\mathcal{B}_{\mathfrak{P}(\mathbb{R}^N)}$-measurable $F:\mathfrak{P}(\mathbb{R}^N)^2\longrightarrow\mathbb{R}$ be given. Then, by Lemma 7.3.6 and especially (7.3.8),

$$\int_{\mathfrak{P}(\mathbb{R}^N)} F(\psi,\boldsymbol{\Sigma}_s\psi)\,Q^{\mathbf{b}}_{\mathbf{x}}(d\psi)=\int_{\mathfrak{P}(\mathbb{R}^N)} F^{\mathbf{b}}(\mathbf{T}_{\mathbf{x}}\psi,\boldsymbol{\delta}_s\psi)\,\mathcal{W}^{(N)}(d\psi),$$

where $F^{\mathbf{b}}:\mathfrak{P}(\mathbb{R}^N)^2\longrightarrow\mathbb{R}$ is defined so that

$$F^{\mathbf{b}}(\varphi,\psi)=F\Big(X^{\mathbf{b}}(\,\cdot\,,\varphi),X^{\mathbf{b}}\big(\,\cdot\,,\mathbf{T}_{X^{\mathbf{b}}(s,\varphi)}\psi\big)\Big)$$

when $\big(\varphi,\mathbf{T}_{X^{\mathbf{b}}(s,\varphi)}\psi\big)\in\Omega_s(\mathbf{b})\times\Omega(\mathbf{b})$ and vanishes otherwise. Hence, by Theorem (4.3.3) (with $\sigma \equiv s$), we see that

$$\begin{aligned}
&\int_{\mathfrak{P}(\mathbb{R}^N)} F(\psi, \Sigma_s\psi)\, Q^{\mathbf{b}}_{\mathbf{x}}(d\psi) \\
&= \int_{\mathfrak{P}(\mathbb{R}^N)} \left(\int_{\mathfrak{P}(\mathbb{R}^N)} F^{\mathbf{b}}(\mathbf{T}_{\mathbf{x}}\varphi, \psi)\, \mathcal{W}^{(N)}(d\psi) \right) \mathcal{W}^{(N)}(d\varphi) \\
&= \int\limits_{\Omega_s(\mathbf{b})} \left(\int\limits_{\Omega(\mathbf{b})} F\Big(X^{\mathbf{b}}(\,\cdot\,, \varphi), X^{\mathbf{b}}\big(\,\cdot\,, \mathbf{T}_{X^{\mathbf{b}}(s,\varphi)}\psi\big)\Big)\, \mathcal{W}^{(N)}(d\psi) \right) \mathcal{W}^{(N)}_{\mathbf{x}}(d\varphi) \\
&= \int_{\mathfrak{P}(\mathbb{R}^N)} \left(\int_{\mathfrak{P}(\mathbb{R}^N)} F(\varphi, \psi)\, Q^{\mathbf{b}}_{\varphi(s)}(d\psi) \right) Q^{\mathbf{b}}_{\mathbf{x}}(d\varphi). \quad \square
\end{aligned}$$

The following provides a somewhat crude, but nonetheless useful, criterion for deciding when (7.3.12) holds.

7.3.14 Corollary. *If*

$$(\mathbf{x}, \mathbf{b}(\mathbf{x}))_{\mathbb{R}^N} \le A + B|\mathbf{x}|^2, \quad \mathbf{x} \in \mathbb{R}^N, \tag{7.3.15}$$

for some $A \in [0, \infty)$ and $B \in \mathbb{R}$, then, for each $\mathbf{x} \in \mathbb{R}^N$, (7.3.12) is satisfied and so $Q^{\mathbf{b}}_{\mathbf{x}}$ is the unique $P \in \mathbf{M}_1(\mathfrak{P}(\mathbb{R}^N))$ for which $P\big(\psi(0) = \mathbf{x}\big) = 1$ and (7.3.11) holds. In fact, if $p \in [2, \infty)$, $K_p \equiv p + N + 2A - 2$, and $\alpha_p \equiv \frac{(p-2)}{2} K_p + pB$, then

$$\mathbb{E}^{Q^{\mathbf{b}}_{\mathbf{x}}}\Big[|\psi(t)|^p\Big] \le e^{\alpha_p t}|\mathbf{x}|^p + K_p \frac{e^{\alpha_p t} - 1}{\alpha_p}, \quad (t, \mathbf{x}) \in [0, \infty) \times \mathbb{R}^N, \tag{7.3.16}$$

where $\frac{e^{\alpha_p t} - 1}{\alpha_p} \equiv t$ when $\alpha_p = 0$. In particular, (7.1.15) implies that the map $\mathbf{x} \in \mathbb{R}^N \longmapsto Q^{\mathbf{b}}_{\mathbf{x}} \in \mathbf{M}_1(\mathfrak{P}(\mathbb{R}^N))$ is continuous, the family $\{Q^{\mathbf{b}}_{\mathbf{x}} : \mathbf{x} \in \mathbb{R}^N\}$ is strong Markov, and (cf. (7.3.13)) $\big(M^{\mathbf{b}}_f(t), \mathcal{B}_t, Q^{\mathbf{b}}_{\mathbf{x}}\big)$ is a martingale for every $f \in C^{1,2}\big([0, \infty) \times \mathbb{R}^N; \mathbb{R}\big)$ with the properties that, for each $T \in (0, \infty)$, both

$$\sup_{t \in [0,T]} \big|f(t, \mathbf{x})\big| \quad \text{and} \quad \sup_{t \in [0,T]} \left|\left(\tfrac{\partial f}{\partial t} + \mathbf{L}^{\mathbf{b}} f\right)(t, \mathbf{x})\right|$$

have at most polynomial growth as $|\mathbf{x}| \longrightarrow \infty$.

PROOF: Clearly it suffices for us to prove that (7.3.16) holds for every $\mathbf{x} \in \mathbb{R}^N$. To this end, let $\mathbf{x} \in \mathbb{R}^N$ and $p \in [2, \infty)$ be given, and, for each $\ell \in \mathbb{Z}^+$, choose $f_\ell \in C^2_c(\mathbb{R}^N; \mathbb{R})$ so that $f_\ell(\mathbf{y}) = |\mathbf{y}|^p$ when $|\mathbf{y}| \le \ell + 1$. Then, by Theorem 7.3.10,

$$f_\ell\big(X^{\mathbf{b}}_\ell(t, \psi)\big) - \int_0^t \big[\mathbf{L}^{\mathbf{b}} f_\ell\big]\big(X^{\mathbf{b}}_\ell(s, \psi)\big)\, ds$$

is a continuous $\mathcal{W}_{\mathbf{x}}^{(N)}$-martingale relative to $\{\mathcal{B}_t : t \in [0,\infty)\}$, and therefore, by Doob's Stopping Time Theorem,

$$\begin{aligned}&\left|X^{\mathbf{b}}(t\wedge \mathfrak{e}_\ell^{\mathbf{b}})\right|^p\\&\qquad -p\int_0^{t\wedge \mathfrak{e}_\ell^{\mathbf{b}}}\left[\tfrac{p+N-2}{2}+\left(X^{\mathbf{b}}(s),\mathbf{b}\left(X^{\mathbf{b}}(s)\right)\right)_{\mathbb{R}^N}\right]\left|X^{\mathbf{b}}(s)\right|^{p-2}\,ds\end{aligned}$$

is also continuous $\mathcal{W}^{(N)}$-martingale. Hence, because

$$a^{p-2}=1^{\frac{2}{p}}\left(a^p\right)^{1-\frac{2}{p}}\le \tfrac{2}{p}+\left(1-\tfrac{p}{2}\right)a^p\quad\text{for all } a\in[0,\infty),$$

(7.3.15) leads immediately to the estimate

$$\begin{aligned}&\mathbb{E}^{\mathcal{W}_{\mathbf{x}}^{(N)}}\left[\left|X^{\mathbf{b}}(t\wedge \mathfrak{e}_\ell^{\mathbf{b}})\right|^p\right]\\&\qquad\le |\mathbf{x}|^p+K_pt+\alpha_p\int_0^t\mathbb{E}^{\mathcal{W}_{\mathbf{x}}^{(N)}}\left[\left|X^{\mathbf{b}}(s)\right|^p,\ \mathfrak{e}_\ell^{\mathbf{b}}>s\right]ds\end{aligned}\tag{7.3.17}$$

for $t\in[0,\infty)$. In particular, if

$$u_\ell(t)=\mathbb{E}^{\mathcal{W}_{\mathbf{x}}^{(N)}}\left[\left|X^{\mathbf{b}}(t\wedge \mathfrak{e}_\ell^{\mathbf{b}})\right|^p\right],$$

then $t\in[0,\infty)\longmapsto u_\ell(t)\in[0,\infty)$ is continuous,

$$u_\ell(t)\le |\mathbf{x}|^p+K_pt+|\alpha_p|\int_0^t u_\ell(s)\,ds,\quad t\in[0,\infty),$$

and therefore, by Gronwall's inequality (cf. Exercise 7.3.29 below),

$$\begin{aligned}\mathbb{E}^{\mathcal{W}_{\mathbf{x}}^{(N)}}\left[\left|X^{\mathbf{b}}(t\wedge \mathfrak{e}_\ell^{\mathbf{b}})\right|^p\right]&\le |\mathbf{x}|^p+K_p\frac{e^{|\alpha_p|t}-1}{|\alpha_p|}\\&\qquad\text{for } p\in[2,\infty)\text{ and } t\in[0,\infty).\end{aligned}\tag{7.3.18}$$

Starting from (7.3.18), we see that, for each $t\in(0,\infty)$,

$$\mathcal{W}_{\mathbf{x}}^{(N)}\left(\mathfrak{e}_\ell^{\mathbf{b}}\le t\right)\le \ell^{-2}\mathbb{E}^{\mathcal{W}_{\mathbf{x}}^{(N)}}\left[\left|X^{\mathbf{b}}(t\wedge \mathfrak{e}_\ell^{\mathbf{b}})\right|^2\right]\longrightarrow 0$$

as $\ell\to\infty$. Hence we have now proved that (7.3.12) holds and, after letting $\ell\to\infty$ in (7.3.18), that

$$\begin{aligned}\mathbb{E}^{Q_{\mathbf{x}}^{\mathbf{b}}}\left[|\psi(t)|^p\right]&\le |\mathbf{x}|^p+K_p\frac{e^{|\alpha_p|t}-1}{|\alpha_p|}\\&\qquad\text{for all } p\in[2,\infty)\text{ and } t\in[0,\infty).\end{aligned}\tag{7.3.19}$$

Clearly (7.3.19) is the same as (7.3.16) when $\alpha_p \geq 0$. Moreover, (7.3.19) shows that, for each $p \in [2,\infty)$ and $T \in (0,\infty)$, the family $\{|\psi(t)|^p : t \in [0,T]\}$ is uniformly $Q^{\mathbf{b}}_{\mathbf{x}}$-integrable and therefore that

$$t \in [0,\infty) \longmapsto u(t) \equiv \mathbb{E}^{Q^{\mathbf{b}}_{\mathbf{x}}}\left[|\psi(t)|^p\right] \in [0,\infty)$$

is continuous. In addition, by letting $\ell \to \infty$ in (7.3.17), we know that

$$u(t) \leq |\mathbf{x}|^p + K_p t + \alpha_p \int_0^t u(s)\,ds, \quad t \in [0,\infty).$$

Hence, after another application of Gronwall's inequality, we arrive at (7.3.16) even when $\alpha_p < 0$. □

Notice that when the constant B in Corollary 7.3.14 is strictly negative, (7.3.16) with $p = 2$ says that the *centripetal force* introduced by the vector field $\mathbf{b}$ eventually dominates the dispersive tendencies of Wiener paths and results in paths which tend to hang around. This is, of course, completely consistent with what one would expect: if the force field is strongly centripetal near infinity, then the paths should be unable to escape bounded sets. Actually, this is just the first indication of the phenomenon to which we alluded in the first paragraph of this section. Namely, although the introduction of a force field has negligible effect on the local (i.e., the regularity) properties of the path, its presence manifests itself in the global properties of the path. Indeed, the original Wiener path is so irregular that the addition of a differentiable term has hardly any impact on local properties. On the other hand, the Wiener path spends so much time jiggling around that it fails to travel as far as a smooth path can over long time intervals. (Both of these comments are made somewhat more quantitative by the Laws of the Iterated Logarithm in Theorem 4.1.6 and Exercise 4.3.32.) However, before we delve into these considerations further, it will be useful to know the following mild smoothing property of the semigroup $\{Q^{\mathbf{b}}_t : t > 0\}$.

7.3.20 Lemma. *Assume that* (7.3.12) *holds for every* $\mathbf{x} \in \mathbb{R}^N$. *Then, for each* $f \in B(\mathbb{R}^N;\mathbb{R})$,

$$(t,\mathbf{x}) \in (0,\infty) \times \mathbb{R}^N \longmapsto \left[Q^{\mathbf{b}}_t f\right](\mathbf{x}) \equiv \mathbb{E}^{Q^{\mathbf{b}}_{\mathbf{x}}}\left[f\left(\psi(t)\right)\right] \in \mathbb{R}$$

is continuous. In fact, for each $T \in (0,\infty)$,

(7.3.21) $\quad \left\{Q^{\mathbf{b}}_T f \restriction B_{\mathbb{R}^N}(\mathbf{0},r) : \|f\|_{\mathrm{u}} \leq 1\right\}$ *is uniformly equicontinuous*

for every $r \in (0,\infty)$.†

† Although the regularity result proved here will suffice for most of our purposes, standard elliptic regularity theory provides much more information. For an account of the basic facts, see A. Friedman's *Partial Differential Equations of the Parabolic Type*, publ. by Prentice Hall (1964). For a more recent treatment directed toward probabilistic applications, see the author's survey article "Diffusion semigroups corresponding to uniformly elliptic divergence form operators" in *Sem. de Prob.* **XXII**, publ. by Springer–Verlag's LNMS.

PROOF: Let $\eta \in C_c^\infty(\mathbb{R}^N;\mathbb{R})$ be given, and define $[S_tf](\mathbf{x})$ for $(t,\mathbf{x}) \in (0,\infty) \times \mathbb{R}^N$ and $f \in B(\mathbb{R}^N;\mathbb{R})$ to be the quantity

$$\int_{\mathbb{R}^N} \left[\tfrac{1}{2}\boldsymbol{\Delta}\eta(\mathbf{x}) + t^{-1}\big(\nabla\eta(\mathbf{x}) + \eta(\mathbf{x})\mathbf{b}(\mathbf{x}), \boldsymbol{\xi} - \mathbf{x}\big)_{\mathbb{R}^N}\right] f(\mathbf{x}-\boldsymbol{\xi})\, {\gamma_t}^N(d\boldsymbol{\xi}).$$

Clearly, $(t,\mathbf{x}) \in (0,\infty) \times \mathbb{R}^N \longmapsto S_tf(\mathbf{x}) \in \mathbb{R}$ is smooth. Moreover, the support of S_tf is contained in that of η, and there is a $K \in (0,\infty)$ depending only on η, such that

$$\|S_tf\|_{\mathrm{u}} \le \frac{K}{\sqrt{t}}\|f\|_{\mathrm{u}}, \quad t \in (0,\infty). \tag{7.3.22}$$

Finally, by (7.3.11) and elementary calculus, we see that, for each $t \in (0,\infty)$,

$$\frac{d}{ds}\big[Q_s^{\mathbf{b}}(\eta\gamma_{t-s}^N \bigstar f)\big](\mathbf{x}) = \big[Q_s^{\mathbf{b}} \circ S_{t-s}f\big](\mathbf{x}), \quad (s,\mathbf{x}) \in (0,t) \times \mathbb{R}^N.$$

Hence, at least when $f \in C_{\mathrm{b}}(\mathbb{R}^N;\mathbb{R})$, we have that

$$\big[Q_t^{\mathbf{b}}(\eta f)\big](\mathbf{x}) = \eta(\mathbf{x})\big[\gamma_t^N \bigstar f\big](\mathbf{x}) + \int_0^t \big[Q_s^{\mathbf{b}} \circ S_{t-s}f\big](\mathbf{x})\, ds, \tag{7.3.23}$$

for $(t,\mathbf{x}) \in (0,\infty) \times \mathbb{R}^N$. But, because of (7.3.22), it is clear that the set of f's for which (7.3.23) holds is closed under bounded, pointwise convergence; and, therefore, it holds for all $f \in B(\mathbb{R}^N;\mathbb{R})$ as soon as it does for continuous ones. In addition, for each $\delta > 0$, there is a $C(\delta) < \infty$ such that

$$\sup_{t \ge \delta} \|S_tf\|_{C_{\mathrm{b}}^1(\mathbb{R}^N;\mathbb{R})} \le C(\delta)\|f\|_{\mathrm{u}};$$

and so (7.3.23), the continuity of $\mathbf{x} \in \mathbb{R}^N \longmapsto Q_{\mathbf{x}}^{\mathbf{b}} \in \mathbf{M}_1(\mathfrak{P}(\mathbb{R}^N))$ plus Lemma 3.1.10, and the estimate in (7.3.22) are enough to guarantee both that

$$(t,\mathbf{x}) \in (0,\infty) \times \mathbb{R}^N \longmapsto \big[Q_t^{\mathbf{b}}\eta f\big](\mathbf{x}) \in \mathbb{R}$$

is continuous for each $f \in B(\mathbb{R}^N;\mathbb{R})$ and that

$$\big\{Q_T^{\mathbf{b}}(\eta f) \restriction B_{\mathbb{R}^N}(\mathbf{0},r) : \|f\|_{\mathrm{u}} \le 1\big\}$$

is uniformly equicontinuous for each $T \in (0,\infty)$ and $r \in (0,\infty)$.

To complete the proof, it suffices to show that there exists a sequence of $\eta_n \in C_{\mathrm{c}}(\mathbb{R}^N;\mathbb{R})$ with the property that

$$\sup_{\|f\|_{\mathrm{u}} \le 1} \Big|\big[Q_t^{\mathbf{b}}(\eta_n f)\big](\mathbf{x}) - \big[Q_t^{\mathbf{b}} f\big](\mathbf{x})\Big| \longrightarrow 0$$

uniformly for $(t, \mathbf{x})$ in compact subsets of $(0, \infty) \times \mathbb{R}^N$. To this end, choose $\eta \in C_c(\mathbb{R}^N; [0,1])$ so that $\eta = 1$ on $B_{\mathbb{R}^N}(\mathbf{0}, 1)$, set $\eta_n(\mathbf{x}) = \eta\left(\frac{\mathbf{x}}{n}\right)$ for $n \in \mathbb{Z}^+$, and note that

$$\left|\left[Q_t^{\mathbf{b}} f\right](\mathbf{x}) - \left[Q_t^{\mathbf{b}}(\eta_n f)\right](\mathbf{x})\right| \le \|f\|_{\mathrm{u}} Q_{\mathbf{x}}^{\mathbf{b}}\big(\psi(t) \notin B_{\mathbb{R}^N}(\mathbf{0}, n)\big).$$

Finally, since $\left\{Q_{\mathbf{x}}^{\mathbf{b}} : |\mathbf{x}| \le r\right\}$ is tight in $\mathbf{M}_1(\mathfrak{P}(\mathbb{R}^N))$ for each $r \in (0, \infty)$, we know that

$$\lim_{n \to \infty} Q_{\mathbf{x}}^{\mathbf{b}}\big(\psi(t) \notin B_{\mathbb{R}^N}(\mathbf{0}, n)\big) = 0$$

uniformly on bounded subsets of $[0, \infty) \times \mathbb{R}^N$. □

7.3.24 Theorem. *Assume that* (7.3.12) *holds for every* $\mathbf{x} \in \mathbb{R}^N$. *Then* $\left\{Q_{\mathbf{x}}^{\mathbf{b}} : \mathbf{x} \in \mathbb{R}^N\right\}$ *is either transient, in the sense that*

$$Q_{\mathbf{x}}^{\mathbf{b}}\left(\lim_{t \to \infty} |\psi(t)| = \infty\right) = 1, \quad \mathbf{x} \in \mathbb{R}^N, \tag{7.3.25}$$

or it is recurrent, in the sense that

$$Q_{\mathbf{x}}^{\mathbf{b}}\left(\int_0^\infty \mathbf{1}_G\big(\psi(t)\big)\, dt = \infty\right) = 1 \quad \text{for all } \mathbf{x} \in \mathbb{R}^N \text{ and open } G \ne \emptyset. \tag{7.3.26}$$

In fact, if

$$2\big(\mathbf{x}, \mathbf{b}(\mathbf{x})\big)_{\mathbb{R}^N} + N - 2 \;\begin{matrix} > 0 \text{ when } N \in \{1,2\} \\ \ge 0 \text{ when } N \ge 3 \end{matrix} \tag{7.3.27}$$

for sufficiently large $|\mathbf{x}|$, *then it is transient. On the other hand, if, for sufficiently large* $|\mathbf{x}|$,

$$2\big(\mathbf{x}, \mathbf{b}(\mathbf{x})\big)_{\mathbb{R}^N} + N - 2 \;\begin{matrix} \le 0 \text{ when } N \in \{1,2\} \\ < 0 \text{ when } N \ge 3, \end{matrix} \tag{7.3.28}$$

then $\left\{Q_{\mathbf{x}}^{\mathbf{b}} : \mathbf{x} \in \mathbb{R}^N\right\}$ *is recurrent.*

PROOF: We know (cf. Theorem 7.3.10) that, for each $\mathbf{x} \in \mathbb{R}^N$, the support of $Q_{\mathbf{x}}^{\mathbf{b}}$ coincides with the space of $\psi \in \mathfrak{P}(\mathbb{R}^N)$ satisfying $\psi(0) = \mathbf{x}$. Hence, by Theorems 7.2.20, the preceding lemma, and Corollary 7.2.28, we know that $\left\{Q_{\mathbf{x}}^{\mathbf{b}} : \mathbf{x} \in \mathbb{R}^N\right\}$ is either transient or recurrent in the asserted senses.

Now assume that (7.3.27) holds, and choose $r_0 \in (0, \infty)$ so that

$$\alpha \equiv \inf\left\{2\big(\mathbf{x}, \mathbf{b}(\mathbf{x})\big)_{\mathbb{R}^N} + N - 2 : |\mathbf{x}| \ge r_0\right\} \;\begin{matrix} > 0 \text{ if } N \in \{1,2\} \\ \ge 0 \text{ if } N \ge 3. \end{matrix}$$

Next, choose $f \in C^2_{\mathrm{b}}\big(\mathbb{R}^N;(0,\infty)\big)$ so that, when $N \in \{1,2\}$, $f(\mathbf{x}) = |\mathbf{x}|^{-\alpha}$ and, when $N \geq 3$, $f(\mathbf{x}) = |\mathbf{x}|^{2-N}$ for $|\mathbf{x}| \geq r_0$. Notice that $\mathbf{L}^{\mathbf{b}} f \leq 0$ on $\mathbb{R}^N \setminus B_{\mathbb{R}^N}(\mathbf{0}, r_0)$. Hence, after an obvious use of a cutoff function and an application of Doob's Stopping Time Theorem, we see first that (cf. the notation in Theorem 7.2.11 and (7.2.27))

$$\Big(-f\big(\psi(t \wedge \sigma_{r_0} \wedge \zeta_R)\big), \mathcal{B}_t, P_{\mathbf{x}}\Big) \quad \text{is a submartingale}$$

for each $R > r_0$ and $r_0 < |\mathbf{x}| < R$, and then, by Lebesgue's Dominated Convergence Theorem, that

$$\Big(-f\big(\psi(t \wedge \sigma_{r_0})\big), \mathcal{B}_t, P_{\mathbf{x}}\Big) \quad \text{is a submartingale for every } |\mathbf{x}| > r_0.$$

Hence, since $\lim_{|\mathbf{x}|\to\infty} f(\mathbf{x}) = 0$, the first part of Theorem 7.2.30 guarantees the required transience.

The proof of recurrence when (7.3.28) holds follows exactly the same pattern. Namely, we again use Theorem 7.2.30, only this time we take $r_0 \in (0,\infty)$ so that

$$\alpha \equiv \inf\Big\{2\big(\mathbf{x}, \mathbf{b}(\mathbf{x})\big)_{\mathbb{R}^N} + N - 2 : \, |\mathbf{x}| \geq r_0\Big\} \begin{matrix} \leq 0 \text{ if } N \in \{1,2\} \\ < 0 \text{ if } N \geq 3 \end{matrix}$$

and $f \in C^2\big(\mathbb{R}^N;(0,\infty)\big)$ so that

$$f(\mathbf{x}) = \begin{cases} |\mathbf{x}| & \text{if } N = 1 \\ \log \frac{2|\mathbf{x}|}{r_0} & \text{if } N = 2 \\ |\mathbf{x}|^{-\alpha} & \text{if } N \geq 3 \end{cases}$$

for $|\mathbf{x}| \geq r_0$. □

Exercises

7.3.29 Exercise: Gronwall's inequality has many forms, and the one with which we deal in this exercise is perhaps the most elementary. Namely, let $\alpha : [0,\infty) \longrightarrow \mathbb{R}$ be a right-continuous function of bounded variation on each compact interval, and suppose that $u : [0,\infty) \longrightarrow \mathbb{R}$ is a continuous function which satisfies

$$u(t) \leq \alpha(t) + \beta \int_0^t u(s)\, ds, \quad t \in [0,\infty), \tag{7.3.30}$$

for some $\beta \in \mathbb{R}$. Prove that

$$u(t) \le e^{\beta t}\alpha(0) + \int_0^t e^{\beta(t-s)}\, d\alpha(s), \quad t \in [0,\infty). \tag{7.3.31}$$

Hint: Set $w(t) = \int_0^t u(s)\, ds$, note that (7.3.30) implies that

$$\frac{d}{dt}\left(e^{-\beta t}\, w(t)\right) \le e^{-\beta t}\alpha(t), \quad t \in [0,\infty),$$

and conclude that

$$u(t) \le \alpha(t) + \beta \int_0^t e^{\beta(t-s)}\, \alpha(s)\, ds.$$

Finally, integrate by parts to arrive at (7.3.31).

7.3.32 Exercise: Consider $\mathbf{b}(\mathbf{x}) = -\frac{\mathbf{x}}{2}$, and show that

$$X^{\mathbf{b}}(t,\psi) = \psi(t) - \tfrac{1}{2}\int_0^t e^{-\frac{t-s}{2}}\psi(s)\, ds, \quad (t,\psi) \in [0,\infty) \times \mathfrak{P}(\mathbb{R}^N).$$

In particular, referring to Exercise 4.2.40, use this to see that $\mathcal{Q}_{\mathbf{0}}^{\mathbf{b}}\big(\Theta(\mathbb{R}^N)\big) = 1$ and that $\mathcal{Q}_{\mathbf{0}}^{\mathbf{b}} \restriction \Theta(\mathbb{R}^N)$ coincides with the Ornstein–Uhlenbeck measure $\mathcal{U}^{(N)}$. (See Exercise 7.5.36 below for more information.)

7.3.33 Exercise: Let a continuously differentiable vector field $\mathbf{b} : \mathbb{R}^N \longrightarrow \mathbb{R}$ be given, and define $\mathbf{L}^{\mathbf{b}}$ as in (7.3.2). Given an open region $\mathfrak{G} \subseteq \mathbb{R} \times \mathbb{R}^N$, $(s,\mathbf{x}) \in \mathfrak{G}$, and a function $f \in C^{1,2}(\mathfrak{G};\mathbb{R})$ satisfying

$$f(t,\mathbf{y}) \ge f(s,\mathbf{x}) \quad \text{and} \quad \left[\tfrac{\partial f}{\partial t} + \mathbf{L}^{\mathbf{b}} f\right](t,\mathbf{y}) \ge 0$$

for all $(t,\mathbf{y}) \in \big((s,\infty) \times \mathbb{R}^N\big) \cap \mathfrak{G}$, use (7.3.9) and the argument in the proof of Theorem 7.2.8 to show that (7.2.10) holds. In other words, the **strong maximum principle** holds for the operator $\mathbf{L}^{\mathbf{b}}$.

7.3.34 Exercise: For $\psi \in \mathfrak{P}(\mathbb{R}^N)$ and $\mathbf{x} \in \mathbb{R}^N$, define

$$\mathfrak{e}^{\mathbf{b}}(\mathbf{x},\psi) = \mathfrak{e}^{\mathbf{b}}\big(\mathbf{x} + \psi - \psi(0)\big) \quad \text{and} \quad \mathbf{X}^{\mathbf{b}}(t,\mathbf{x},\psi) = \mathbf{X}^{\mathbf{b}}\big(t, \mathbf{x} + \psi - \psi(0)\big)$$

for $t \in \big[0, \mathfrak{e}^{\mathbf{b}}(\mathbf{x},\psi)\big)$. The purpose of this exercise is to study $\mathbf{X}^{\mathbf{b}}$ as a function of its starting point $\mathbf{x}$.

(i) Given $T \in [0,\infty)$, define $\check{\psi}^T \in \mathfrak{P}(\mathbb{R}^N)$ for $\psi \in \mathfrak{P}(\mathbb{R}^N)$ by

$$\check{\psi}^T(t) = \psi(T - T\wedge t) - \psi(T \vee t), \quad t \in [0,\infty),$$

and note that $\mathcal{W}^{(N)}$ is invariant under $\psi \in \mathfrak{P}(\mathbb{R}^N) \longmapsto \check{\psi}^T \in \mathfrak{P}(\mathbb{R}^N)$.

(ii) Show that

$$\mathfrak{e}^{\mathbf{b}}(\mathbf{x},\psi) > T \implies \mathfrak{e}^{-\mathbf{b}}\big(\mathbf{X}^{\mathbf{b}}(T,\mathbf{x},\psi),\check{\psi}^T\big) > T \quad \text{and } \mathbf{X}^{\mathbf{b}}(T-t,\mathbf{x},\psi) = \mathbf{X}^{-\mathbf{b}}\big(t,\mathbf{X}^{\mathbf{b}}(T,\mathbf{x},\psi),\check{\psi}^T\big)$$

for $t \in [0,T]$. In particular, conclude that

$$\mathfrak{e}^{\mathbf{b}}(\,\cdot\,,\psi) \wedge \mathfrak{e}^{-\mathbf{b}}(\,\cdot\,,\check{\psi}^T) > T \implies \mathbf{X}^{\mathbf{b}}(T,\,\cdot\,,\psi) \text{ is a homeomorphism on } \mathbb{R}^N \quad \text{and } \mathbf{X}^{\mathbf{b}}(T,\,\cdot\,,\psi)^{-1} = \mathbf{X}^{-\mathbf{b}}(T,\,\cdot\,,\check{\psi}^T). \tag{7.3.35}$$

(iii) Assuming that $\mathfrak{e}^{\mathbf{b}}(\mathbf{x},\psi) > T$, note that $\mathfrak{e}^{\mathbf{b}}(\mathbf{y},\psi) > T$ for all $\mathbf{y}$ in an open ball $B_{\mathbb{R}^N}(\mathbf{0},r)$, and apply the elementary theory of ordinary differential equations to see that $\mathbf{X}^{\mathbf{b}}(T,\cdot,\psi)$ is continuously differentiable at $\mathbf{x}$, the Jacobian matrix

$$\mathbf{J}^{\mathbf{b}}(t,\mathbf{x},\psi) \equiv \frac{\partial \mathbf{X}^{\mathbf{b}}}{\partial \mathbf{x}}(t,\mathbf{x},\psi)$$

is continuous on $[0,T] \times B_{\mathbb{R}^N}(\mathbf{0},r) \times \mathfrak{P}(\mathbb{R}^N)$, and

$$\mathbf{J}^{\mathbf{b}}(T,\mathbf{x},\psi) = \mathbf{I} + \int_0^T \frac{\partial \mathbf{b}}{\partial \mathbf{x}}\big(\mathbf{X}^{\mathbf{b}}(t,\mathbf{x},\psi)\big)\mathbf{J}^{\mathbf{b}}(t,\mathbf{x},\psi)\,dt.$$

Conclude, in particular, that, when $\mathfrak{e}^{\mathbf{b}}(\mathbf{x},\psi) > T$,

$$\det\big(\mathbf{J}^{\mathbf{b}}(T,\mathbf{x},\psi)\big) = \exp\left[\int_0^T \operatorname{div}\mathbf{b}\big(\mathbf{X}^{\mathbf{b}}(t,\mathbf{x},\psi)\big)\,dt\right], \tag{7.3.36}$$

where $\operatorname{div}\mathbf{b} \equiv \sum_1^N \frac{\partial b^i}{\partial x_i}$ is the divergence of $\mathbf{b}$. When combined with (7.3.35), this leads to

$$\mathfrak{e}^{\mathbf{b}}(\mathbf{x},\psi) \wedge \mathfrak{e}^{-\mathbf{b}}(\mathbf{x},\check{\psi}^T) > T \text{ for all } \mathbf{x} \in \mathbb{R}^N \implies \mathbf{X}^{\mathbf{b}}(T,\,\cdot\,,\psi) \text{ is a diffeomorphism on } \mathbb{R}^N \quad \text{and } \mathbf{X}^{\mathbf{b}}(T,\,\cdot\,,\psi)^{-1} = \mathbf{X}^{-\mathbf{b}}(T,\,\cdot\,,\check{\psi}^T). \tag{7.3.37}$$

(iv) Assume that

$$\mathcal{W}^{(N)}\big(\mathfrak{e}^{\mathbf{b}}(\mathbf{x}) \wedge \mathfrak{e}^{-\mathbf{b}}(\mathbf{x}) = \infty \text{ for all } \mathbf{x} \in \mathbb{R}^N\big) = 1. \tag{7.3.38}$$

Given nonnegative, measurable φ and f on $\mathbb{R}^N$, use the preceding, Jacobi's change of variables formula, and Tonelli's Theorem to see that,

$$\begin{aligned}(7.3.39)\qquad &\int_{\mathbb{R}^N}\big[Q^{\mathbf{b}}_T\varphi\big](\mathbf{x})f(\mathbf{x})\,d\mathbf{x}\\ &= \int_{\mathbb{R}^N}\varphi(\mathbf{x})\mathbb{E}^{Q^{-\mathbf{b}}_{\mathbf{x}}}\left[f\big(\psi(T)\big)\exp\left(-\int_0^T \operatorname{div} f\big(\psi(t)\big)\,dt\right)\right]d\mathbf{x}.\end{aligned}$$

See Exercise 7.4.27 for a continuation of this line of reasoning.

§7.4: Elementary Ergodic Theory of Diffusions.

The results obtained in Section 7.3 provide us with fertile ground on which to exercise the results in Section 6.1 (cf. Theorem 7.4.16 below). However, before concentrating on the examples coming out of Section 7.3, it may be helpful to get some perspective by making a few general remarks about the ergodic theory of Markov processes.

Thus, suppose that $\{P_{\mathbf{x}} : \mathbf{x} \in \mathbb{R}^N\}$ is a Markov family (cf. Exercises 4.3.55 and 5.1.30) of probability measures on $\mathfrak{P}(\mathbb{R}^N)$, and introduce the associated transition probability function $(t,\mathbf{x}) \in (0,\infty)\times\mathbb{R}^N \longmapsto P(t,\mathbf{x},\cdot\,) \in \mathbf{M}_1(\mathbb{R}^N)$ and semigroup $\{P_t : t > 0\}$ of operators on $B(\mathbb{R}^N;\mathbb{R})$ determined by

$$P(t,\mathbf{x},\Gamma) = P_{\mathbf{x}}\big(\psi(t)\in\Gamma\big) \quad\text{and}\quad \big[P_t\varphi\big](\mathbf{x}) = \int \varphi(\mathbf{y})\,P(t,\mathbf{x},d\mathbf{y}).$$

Next, observe that $P(t,\mathbf{x},\cdot\,)$ also determines a *dual action* on the set of (nonnegative) Borel measures on $\mathbb{R}^N$; namely, given a Borel measure μ on $\mathbb{R}^N$, define μP_t for $t \in (0,\infty)$ by

$$(7.4.1)\qquad \big[\mu P_t\big](\Gamma) = \int_{\mathbb{R}^N} P(t,\mathbf{x},\Gamma)\,\mu(d\mathbf{x}), \quad \Gamma \in \mathcal{B}_{\mathbb{R}^N}.$$

Clearly, $\big[\mu P_t\big](\mathbb{R}^N) = \mu(\mathbb{R}^N)$, and, more generally, for any Borel measurable $\varphi : \mathbb{R}^N \longrightarrow [0,\infty)$:

$$\langle\varphi,\mu P_t\rangle = \langle P_t\varphi,\mu\rangle,$$

where we have returned to the notation (cf. Section 3.1) $\langle\varphi,\mu\rangle$ to denote $\int\varphi\,d\mu$. Finally, we say that the σ-finite Borel measure μ on $\mathbb{R}^N$ is **invariant under** $\{P_t : t > 0\}$ or, equivalently, $\{P_t : t > 0\}$**-invariant**, if $\mu P_t = \mu$ for each

$t \in (0,\infty)$. Notice that, by Jensen's inequality (cf. Exercise 2.3.35), for any $\{P_t : t > 0\}$-invariant μ and $p \in [1,\infty)$:

$$\|P_t\varphi\|_{L^p(\mu)} \le \|\varphi\|_{L^p(\mu)}, \quad \varphi \in L^p(\mu). \tag{7.4.2}$$

For reasons which are made clear in the following, we say that a $\{P_t : t > 0\}$-invariant μ is **ergodic** if the only $\varphi \in L^2(\mu)$ satisfying $\varphi = P_t\varphi$ (a.e., μ) for every $t \in (0,\infty)$ is constant.

7.4.3 Theorem. *Given a σ-finite, nonzero, Borel measure μ on $\mathbb{R}^N$, define the Borel measure P_μ on $\mathfrak{P}(\mathbb{R}^N)$ by*

$$P_\mu = \int_{\mathbb{R}^N} P_{\mathbf{x}}\,\mu(d\mathbf{x}). \tag{7.4.4}$$

Then $P_\mu \restriction \mathcal{B}_0$ is σ-finite. Moreover, (cf. Section 4.3)

$$\boldsymbol{\Sigma}_t^* P_\mu = P_{\mu P_t} \quad \text{for each } t \in [0,\infty),$$

and so the time-shift semigroup $\{\boldsymbol{\Sigma}_t : t \in [0,\infty)\}$ is P_μ-measure preserving if and only if μ is $\{P_t : t > 0\}$-invariant; in which case P_μ is $\{\boldsymbol{\Sigma}_t : t \in [0,\infty)\}$-ergodic if and only if μ is ergodic for $\{P_t : t > 0\}$.

PROOF: Because $P_\mu\big(\psi(0) \in \Gamma\big) = \mu(\Gamma)$, it is obvious that $P_\mu \restriction \mathcal{B}_0$ is σ-finite if μ is. Next, let $F \in B\big(\mathfrak{P}(\mathbb{R}^N); [0,\infty)\big)$ be given, and set $f(\mathbf{x}) = \mathbb{E}^{P_{\mathbf{x}}}[F]$ for $\mathbf{x} \in \mathbb{R}^N$. By the Markov property,

$$\mathbb{E}^{\boldsymbol{\Sigma}_t^* P_\mu}[F] = \mathbb{E}^{P_\mu}\Big[f\big(\psi(t)\big)\Big] = \langle f, \mu P_t\rangle = \mathbb{E}^{P_{\mu P_t}}[F],$$

which means that $\boldsymbol{\Sigma}_t^* P_\mu = P_{\mu P_t}$; and, as a consequence, we see that $\{\boldsymbol{\Sigma}_t : t \in [0,\infty)\}$ is P_μ-measure preserving if and only if μ is $\{P_t : t > 0\}$-invariant.

Next, assume that μ is $\{P_t : t > 0\}$-invariant. If $\{\boldsymbol{\Sigma}_t : t \in (0,\infty)\}$ is P_μ-ergodic and $f \in L^2(\mu)$ satisfies $f = P_t f$ (a.e., μ) for each $t \in (0,\infty)$, set $F(\psi) = f\big(\psi(0)\big)$ for $\psi \in \mathfrak{P}(\mathbb{R}^N)$,

$$\mathfrak{G}(f) = \left\{\psi : \int_0^T \big|f\big(\psi(t)\big)\big|\,dt < \infty \text{ for all } T \in (0,\infty)\right\},$$

and

$$F_T(\psi) = \begin{cases} \frac{1}{T}\int_0^T f\big(\psi(t)\big)\,dt & \text{if } \psi \in \mathfrak{G}(f) \\ 0 & \text{if } \psi \notin \mathfrak{G}(f). \end{cases}$$

Then (cf. the discussion preceding Theorem 6.1.26) $\mathfrak{G}(f)\complement$ is a $\{\boldsymbol{\Sigma}_t : t \in [0,\infty)\}$-invariant, P_μ-null set. In addition, by the Markov property, for each $\varphi \in L^2(\mu)$:

$$(f,\varphi)_{L^2(\mu)} = \left(\frac{1}{T}\int_0^T P_t f\,dt, \varphi\right)_{L^2(\mu)} = \mathbb{E}^{P_\mu}\Big[F_T(\boldsymbol{\psi})\varphi\big(\boldsymbol{\psi}(0)\big)\Big].$$

In particular, if $\mu(\mathbb{R}^N) = \infty$ and therefore $P_\mu(\mathfrak{P}(\mathbb{R}^N)) = \infty$, we see that

$$(f,\varphi)_{L^2(\mu)} = \lim_{T\to\infty} \mathbb{E}^{P_\mu}\Big[F_T(\boldsymbol{\psi})\varphi\big(\boldsymbol{\psi}(0)\big)\Big] = 0,$$

since, by Theorem 6.1.26, we know that $F_T \longrightarrow 0$ in $L^2(P_\mu)$. On the other hand, if $\mu(\mathbb{R}^N) < \infty$ and therefore $P_\mu(\mathfrak{P}(\mathbb{R}^N)) < \infty$, then the same theorem leads to

$$(f,\varphi)_{L^2(\mu)} = \frac{\mathbb{E}^{P_\mu}[F]\langle\varphi,\mu\rangle}{\mu(\mathbb{R}^N)}.$$

Hence, in either case, we conclude that f is μ-almost everywhere constant and, therefore, μ is $\{P_t : t > 0\}$-ergodic. Conversely, if μ is $\{P_t : t > 0\}$-ergodic and $F \in L^2(P_\mu)$ satisfies $F = F \circ \boldsymbol{\Sigma}_t$ (a.s., P_μ) for each $t > 0$, set $\mathfrak{G}(F) = \{\mathbf{x} : F \in L^2(P_\mathbf{x})\}$ and

$$f(\mathbf{x}) = \begin{cases} \mathbb{E}^{P_\mathbf{x}}[F] & \text{if } \mathbf{x} \in \mathfrak{G}(F) \\ 0 & \text{otherwise,} \end{cases}$$

note that $\mathfrak{G}(F)\complement$ is a μ-null set, and use the Markov property to check that, for each $t > 0$:

$$\big[P_t f\big]\big(\boldsymbol{\psi}(0)\big) = \mathbb{E}^{P_\mu}\big[F\circ\boldsymbol{\Sigma}_t\,\big|\,\mathcal{B}_0\big](\boldsymbol{\psi}) = \mathbb{E}^{P_\mu}\big[F\,\big|\,\mathcal{B}_0\big](\boldsymbol{\psi}) = f\big(\boldsymbol{\psi}(0)\big)$$

for P_μ-almost every $\boldsymbol{\psi}$. In other words, $f = P_t f$ (a.e., μ) for each $t \in (0,\infty)$ and therefore f is μ-almost everywhere constant. At the same time, by Theorem 5.3.3 and the Markov property,

$$\begin{aligned}\mathbb{E}^{P_\mu}\big[F^2\big] &= \lim_{n\to\infty}\mathbb{E}^{P_\mu}\Big[\mathbb{E}^{P_\mu}\big[F\,\big|\,\mathcal{B}_n\big]F\Big] \\ &= \lim_{n\to\infty}\mathbb{E}^{P_\mu}\Big[\mathbb{E}^{P_\mu}\big[F\,\big|\,\mathcal{B}_n\big]F\circ\boldsymbol{\Sigma}_n\Big] = \lim_{n\to\infty}\mathbb{E}^{P_\mu}\Big[\mathbb{E}^{P_\mu}\big[F\,\big|\,\mathcal{B}_n\big]f\big(\boldsymbol{\psi}(n)\big)\Big].\end{aligned}$$

Hence, since either $\mu(\mathbb{R}^N) = \infty$ and $f\big(\boldsymbol{\psi}(n)\big) = 0$ or $\mu(\mathbb{R}^N) < \infty$ and $f\big(\boldsymbol{\psi}(n)\big) = \mu(\mathbb{R}^N)^{-1}\mathbb{E}^{P_\mu}[F]$ P_μ-almost everywhere, we conclude that

$$\mathbb{E}^{P_\mu}\big[F^2\big] = \begin{cases} 0 & \text{when } \mu(\mathbb{R}^N) = \infty \\ \mu(\mathbb{R}^N)^{-1}\mathbb{E}^{P_\mu}[F]^2 & \text{if } \mu(\mathbb{R}^N) < \infty; \end{cases}$$

and, in either case, F is P_μ-almost everywhere constant. □

We next want to develop a few elementary facts about the structure of the set $\mathfrak{M}(\{P_t : t > 0\})$ of $\{P_t : t > 0\}$-invariant, σ-finite, Borel measures on $\mathbb{R}^N$; and a critical element in our analysis will be played by the following.

7.4.5 Lemma. *Let $\mu \in \mathfrak{M}(\{P_t : t > 0\})$, and suppose that $F \in L^2(P_\mu;\mathbb{R})$ satisfies*

$$F = F \circ \Sigma_t \text{ (a.s., } P_\mu) \quad \text{for each } t \in (0,\infty). \tag{7.4.6}$$

If $f : \mathbb{R}^N \longrightarrow \mathbb{R}$ is defined by

$$f(\mathbf{x}) = \begin{cases} \mathbb{E}^{P_{\mathbf{x}}}[F] & \text{when } F \in L^2(P_{\mathbf{x}}) \\ 0 & \text{otherwise,} \end{cases} \tag{7.4.7}$$

then $f \in L^2(\mu)$ and $f = P_t f$ (a.e., μ) for each $t > 0$. In fact, for each $t \in (0,\infty)$,

$$f\big(\psi(t)\big) = F(\psi) \quad \text{for } P_\mu\text{-almost every } \psi \in \mathfrak{P}(\mathbb{R}^N).$$

PROOF: Only the final assertion requires comment; and, to handle it, set $F_s = \mathbb{E}^{P_\mu}\big[F \,\big|\, \mathcal{B}_s\big]$ and note that, by the Markov property:

$$F_s(\psi) = \mathbb{E}^{P_\mu}\big[F \circ \Sigma_s \,\big|\, \mathcal{B}_s\big](\psi) = f\big(\psi(s)\big)$$

for P_μ-almost every $\psi \in \mathfrak{P}(\mathbb{R}^N)$. Hence, since $P_\mu = \Sigma_n^* P_\mu$ and, by Theorem 5.3.3, $F_{t+n} \longrightarrow F$ in $L^2(P_\mu)$ as $n \to \infty$:

$$\begin{aligned} \mathbb{E}^{P_\mu}\Big[\Big(F(\psi) - f\big(\psi(t)\big)\Big)^2\Big] &= \mathbb{E}^{P_\mu}\Big[\Big(F \circ \Sigma_n(\psi) - f\big(\psi(t+n)\big)\Big)^2\Big] \\ &= \mathbb{E}^{P_\mu}\big[(F - F_{t+n})^2\big] \longrightarrow 0 \quad \text{as } n \to \infty. \qquad \square \end{aligned}$$

7.4.8 Theorem. *The set $\mathfrak{M}_1\big(\{P_t : t > 0\}\big) = \mathfrak{M}\big(\{P_t : t > 0\}\big) \cap \mathbf{M}_1(\mathbb{R}^N)$ is a convex subset of $\mathbf{M}_1(\mathbb{R}^N)$. Moreover, $\mu \in \mathfrak{M}_1\big(\{P_t : t > 0\}\big)$ is extreme if and only if μ is ergodic. Finally, if μ_1 and μ_2 are extreme elements in $\mathfrak{M}_1(\mathbb{R}^N)$, then either $\mu_1 = \mu_2$ or $\mu_1 \perp \mu_2$.*

PROOF: The convexity is trivial. Next, suppose that $\mu = \theta\mu_1 + (1-\theta)\mu_2$, where $\theta \in (0,1)$ and μ_1 and μ_2 are distinct elements of $\mathfrak{M}_1\big(\{P_t : t > 0\}\big)$. Then $d\mu_1 = f\,d\mu$, where $f \in L^1\big(\mu;[0,1]\big)$ and f is not μ-almost surely constant. In addition, because both μ_1 and μ are $\{P_t : t > 0\}$-invariant, it is an easy matter to check that $f = P_t f$ (a.e., μ) for each $t > 0$. Hence, if

$$\tilde{f}(\mathbf{x}) \equiv \overline{\lim_{T\to\infty}} \frac{1}{T}\int_0^T [P_t f](\mathbf{x})\,dt, \quad \mathbf{x} \in \mathbb{R}^N,$$

then $f = \tilde{f}$ (a.e., μ) and $\tilde{f} = P_t\tilde{f}$ for each $t > 0$; and so μ is not ergodic.

Conversely, suppose that μ is not ergodic. Then, by Theorem 7.4.3, we know that $\{\Sigma_t : t \in [0,\infty)\}$ is not P_μ-ergodic. In particular, there is an $A \in \mathcal{B}_{\mathfrak{P}(\mathbb{R}^N)}$ such that $A = \Sigma_s^{-1}(A)$ for each $s \in (0,\infty)$ and $\theta \equiv P_\mu(A) \in (0,1)$. Thus, if $f(\mathbf{x}) = P_{\mathbf{x}}(A)$, $\mathbf{x} \in \mathbb{R}^N$, then, by the Markov property and Lemma 7.4.5, for each $\varphi \in B(\mathbb{R}^N;\mathbb{R})$:

$$\begin{aligned}\langle fP_t\varphi,\mu\rangle &= \mathbb{E}^{P_\mu}\Big[f(\boldsymbol{\psi}(0))\varphi(\boldsymbol{\psi}(t))\Big] = \mathbb{E}^{P_\mu}\Big[\varphi(\boldsymbol{\psi}(t)),\, A\Big] = \mathbb{E}^{\Sigma_t^* P_\mu}\Big[\varphi(\boldsymbol{\psi}(0)),\, A\Big]\\ &= \mathbb{E}^{P_\mu}\Big[\varphi(\boldsymbol{\psi}(0)),\, A\Big] = \mathbb{E}^{P_\mu}\Big[f(\boldsymbol{\psi}(0))\varphi(\boldsymbol{\psi}(0))\Big] = \langle f\varphi,\mu\rangle.\end{aligned}$$

Hence, if we define μ_1 and μ_2 by

$$d\mu_1 = \frac{f}{\theta}\,d\mu \quad \text{and} \quad d\mu_2 = \frac{1-f}{1-\theta}\,d\mu,$$

then both μ_1 and μ_2 are $\{P_t : t > 0\}$-invariant, $\mu_1 \neq \mu_2$, and $\mu = \theta\mu_1 + (1-\theta)\mu_2$. In other words, μ is not an extreme element of $\mathfrak{M}_1\big(\{P_t : t > 0\}\big)$.

Finally, suppose that $\mu_1 \neq \mu_2$ are extreme elements of $\mathfrak{M}_1\big(\{P_t : t > 0\}\big)$. Then P_{μ_1} and P_{μ_2} are distinct elements of $\mathbf{M}_1(\mathfrak{P}(\mathbb{R}^N))$ for which $\{\Sigma_t : t \in (0,\infty)\}$ is ergodic. Choose $B \in \mathcal{B}_{\mathfrak{P}(\mathbb{R}^N)}$ so that $P_{\mu_1}(B) \neq P_{\mu_2}(B)$, set

$$F = \overline{\lim_{T\to\infty}} \frac{1}{T}\int_0^T \mathbf{1}_B \circ \Sigma_t\,dt,$$

and consider $A \equiv \{\boldsymbol{\psi} : F(\boldsymbol{\psi}) = P_\mu(B)\}$. By the version of the Individual Ergodic Theorem in Theorem 6.1.26, $P_{\mu_1}(A) = 1$ and $P_{\mu_2}(A) = 0$. Thus, if $\Gamma \equiv \{\mathbf{x} : P_{\mathbf{x}}(A) > 0\}$, then $\mu_1(\Gamma) = 1$ and $\mu_2(\Gamma\complement) = 0$. □

7.4.9 Remark: It should be observed that nothing which has been said so far relies on the structure of $\mathbb{R}^N$ or the path-space $\mathfrak{P}(\mathbb{R}^N)$. Indeed, everything extends, verbatim, to the setting of a Markov family on the space of measurable paths on any measurable space.

We turn now to a few results in which topological considerations play some rôle.

7.4.10 Theorem. *Assume that every $f \in B(\mathbb{R}^N;\mathbb{R})$ satisfying $f = P_t f$ for each $t \in (0,\infty)$ is continuous and that*

$$P_{\mathbf{x}}\big(\forall s \in [0,\infty)\ \exists t \in (s,\infty)\ \boldsymbol{\psi}(t) \in G\big) = 1 \quad \text{for all } \mathbf{x} \in \mathbb{R}^N \text{ and open } G \neq \emptyset. \tag{7.4.11}$$

Then every $f \in B(\mathbb{R}^N;\mathbb{R})$ satisfying $f = P_t f$ for each $t \in (0,\infty)$ is constant. In particular, there is at most one $\mu \in \mathfrak{M}_1\big(\{P_t : t > 0\}\big)$. Moreover, if such a μ exists, then it is ergodic, and, for every $\mathbf{x} \in \mathbb{R}^N$ and $F \in B(\mathbb{R}^N;\mathbb{R})$:

$$\lim_{T\to\infty} \frac{1}{T}\int_0^T F\circ\Sigma_t\,dt = \mathbb{E}^{P_\mu}[F] \quad (\text{a.s.}, P_{\mathbf{x}}). \tag{7.4.12}$$

Thus, in fact, $\mathfrak{M}_1\big(\{P_t : t > 0\}\big) \neq \emptyset$ implies that $\mathfrak{M}_1\big(\{P_t : t > 0\}\big) = \{\mu\}$ where

$$\langle \varphi, \mu \rangle = \lim_{T\to\infty} \frac{1}{T}\int_0^T \big[P_t\varphi\big](\mathbf{x})\,dt \tag{7.4.13}$$

for all $\mathbf{x} \in \mathbb{R}^N$ and $\varphi \in B(\mathbb{R}^N;\mathbb{R})$.

PROOF: To prove the first assertion, suppose that f is a nonconstant, bounded continuous function which satisfies $f = P_t f$ for each $t \in (0,\infty)$. Then, by the Markov property,

$$\mathbb{E}^{P_{\mathbf{0}}}\big[f\big(\psi(t)\big)\,\big|\,\mathcal{B}_s\big](\varphi) = \big[P_{t-s}f\big]\big(\varphi(s)\big) = f\big(\varphi(s)\big) \quad \text{(a.s., } P_{\mathbf{0}})$$

for all $0 \le s < t < \infty$; and so $\Big(f\big(\psi(t)\big), \mathcal{B}_t, P_{\mathbf{0}}\Big)$ is a bounded, continuous martingale. In particular, for $P_{\mathbf{0}}$-almost every $\psi \in \mathfrak{P}(\mathbb{R}^N)$, $\lim_{t\to\infty} f\big(\psi(t)\big)$ exists. On the other hand, because f is continuous and nonconstant, there exist $-\infty < a < b < \infty$ such that $G = \{f < a\}$ and $H = \{f > b\}$ are both nonempty open sets. Hence, by (7.4.11),

$$\overline{\lim_{t\to\infty}}\, f\big(\psi(t)\big) \ge b \quad \text{and} \quad \underline{\lim_{t\to\infty}}\, f\big(\psi(t)\big) \le a$$

for $P_{\mathbf{0}}$-almost every $\psi \in \mathfrak{P}(\mathbb{R}^N)$; which is obviously a contradiction.

Now suppose that $\mu \in \mathfrak{M}_1\big(\{P_t : t > 0\}\big)$ and that $A \in \mathcal{B}_{\mathfrak{P}(\mathbb{R}^N)}$ is $\big\{\mathbf{\Sigma}_t : t \in [0,\infty)\big\}$-invariant. Then, by the Markov property, the function $\mathbf{x} \in \mathbb{R}^N \longmapsto f(\mathbf{x}) \equiv P_{\mathbf{x}}(A) \in [0,1]$ satisfies $f = P_t f$ for every $t \in (0,\infty)$ and is therefore, by the preceding, constant. But, by Lemma 7.4.5, this means that $\mathbf{1}_A$ is P_μ-almost surely constant, or, equivalently, that $P_\mu(A) \in \{0,1\}$. In other words, we have shown that, for every $\mu \in \mathfrak{M}_1\big(\{P_t : t > 0\}\big)$, $\big\{\mathbf{\Sigma}_t : t \in [0,\infty)\big\}$ is P_μ-ergodic, which, by Theorem 7.4.3 means that every such μ is ergodic. Since, by Theorem 7.4.8, $\mathfrak{M}_1\big(\{P_t : t > 0\}\big)$ must contain a nonergodic element if it contains more than one element, this completes the proof that $\mathfrak{M}_1\big(\{P_t : t > 0\}\big)$ is either empty or consists of precisely one element and that that element must be ergodic.

Finally, to prove (7.4.12), suppose that $\mu \in \mathfrak{M}_1\big(\{P_t : t > 0\}\big)$ and let $F \in B(\mathbb{R}^N;\mathbb{R})$ be given. Because $\big\{\mathbf{\Sigma}_t : t \in [0,\infty)\big\}$ is P_μ-ergodic, we know that (7.4.12) holds when $P_{\mathbf{x}}$ is replaced by P_μ. In particular, if

$$f(\mathbf{x}) \equiv P_{\mathbf{x}}\left(\lim_{T\to\infty}\int_0^T F\big(\psi(t)\big)\,dt = \mathbb{E}^{P_\mu}[F]\right), \quad \mathbf{x} \in \mathbb{R}^N,$$

then $f \in B(\mathbb{R}^N;\mathbb{R})$ and $f = 1$ (a.e., μ). Moreover, by the Markov property,

$$\begin{aligned}\big[P_t f\big](\mathbf{x}) &= P_{\mathbf{x}}\left(\lim_{T\to\infty}\int_t^{t+T} F\big(\psi(s)\big)\,ds = \mathbb{E}^{P_\mu}[F]\right)\\ &= P_{\mathbf{x}}\left(\lim_{T\to\infty}\int_0^T F\big(\psi(t)\big)\,dt = \mathbb{E}^{P_\mu}[F]\right) = f(\mathbf{x}).\end{aligned}$$

Hence, f is constant; and so we have now proved that $f(\mathbf{x}) = 1$ for every $\mathbf{x} \in \mathbb{R}^N$. After combining this with (7.4.12), one gets (7.4.13) by taking $F(\boldsymbol{\psi}) = \varphi(\boldsymbol{\psi}(0))$. $\square$

Thus far, all of our results are about properties of $\{P_t : t > 0\}$-invariant μ's under the assumption that any exist. The following result provides a basic existence criterion.

7.4.14 Lemma. *Assume that, for each $t \in (0,\infty)$, P_t maps $C_b(\mathbb{R}^N;\mathbb{R})$ into itself. Further, assume that there is a $\nu \in \mathbf{M}_1(\mathbb{R}^N)$ with the property that the family*

$$\nu_n \equiv \frac{1}{n}\int_0^n \nu P_s\,ds, \quad n \in \mathbb{Z}^+, \tag{7.4.15}$$

is tight. Then $\{\nu_n\}_1^\infty$ has at least one limit point and every limit point is an element of $\mathfrak{M}_1(\{P_t : t > 0\})$.

PROOF: Obviously, all that we have to do is show that if $\{\nu_{n_m}\}$ is a convergent subsequence of $\{\nu_n\}$, then its limit μ is $\{P_t : t > 0\}$-invariant. But, by the Chapman–Kolmogorov equation, for any $t \in (0,\infty)$:

$$n_m\nu_{n_m}P_t = \int_t^{t+n_m} \nu P_s\,ds = n_m\nu_{n_m} + \int_0^t \Big(\nu P_{n_m+s} - \nu P_s\Big)\,ds.$$

In particular, if $\varphi \in C_b(\mathbb{R}^N;\mathbb{R})$, then, because $P_t\varphi \in C_b(\mathbb{R}^N;\mathbb{R})$,

$$\langle\varphi,\mu\rangle - \langle P_t\varphi,\mu\rangle = \lim_{m\to\infty}\Big(\langle\varphi,\nu_{n_m}\rangle - \langle\varphi,\nu_{n_m}P_t\rangle\Big) = 0;$$

and so $\langle\varphi,\mu\rangle = \langle\varphi,\mu P_t\rangle$, first for all $\varphi \in C_b(\mathbb{R}^N;\mathbb{R})$ and thence for all $\varphi \in B(\mathbb{R}^N;\mathbb{R})$. $\square$

We will now apply these considerations to the processes produced in Sections 7.3. Thus, let $\mathbf{b} : \mathbb{R}^N \longrightarrow \mathbb{R}^N$ be a once continuously differentiable map, assume that (7.3.12) holds for every $\mathbf{x} \in \mathbb{R}^N$, and consider the associated strong Markov family $\{Q^{\mathbf{b}}_{\mathbf{x}} : \mathbf{x} \in \mathbb{R}^N\}$ described in Theorem 7.3.10. Next, let $(t,\mathbf{x}) \in (0,\infty)\times\mathbb{R}^N \longmapsto Q^{\mathbf{b}}(t,\mathbf{x},\cdot) \in \mathbf{M}_1(\mathbb{R}^N)$ and $\{Q^{\mathbf{b}}_t : t > 0\}$ be, respectively, the corresponding transition probability function and semigroup.

7.4.16 Theorem. *Either $\mathfrak{M}_1(\{Q^{\mathbf{b}}_t : t > 0\})$ is empty or it contains precisely one element μ. Moreover, in the latter case, μ is ergodic, its support is the whole of $\mathbb{R}^N$, and*

$$\lim_{T\to\infty}\frac{1}{T}\int_0^T F\circ\boldsymbol{\Sigma}_t\,dt = \mathbb{E}^{Q^{\mathbf{b}}_\mu}[F] \quad (\text{a.s., } Q^{\mathbf{b}}_{\mathbf{x}}) \tag{7.4.17}$$

for every $\mathbf{x} \in \mathbb{R}^N$ *and* $F \in B(\mathbb{R}^N;\mathbb{R})$. *In particular, if* $\mathfrak{M}_1(\{Q_t^{\mathbf{b}} : t > 0\}) = \{\mu\}$, *then, for every* $\varphi \in B(\mathbb{R}^N;\mathbb{R})$,

$$\lim_{T\to\infty} \frac{1}{T}\int_0^T [Q_t^{\mathbf{b}}\varphi](\mathbf{x})\, dt = \langle \varphi, \mu\rangle, \quad \mathbf{x} \in \mathbb{R}^N, \tag{7.4.18}$$

and $\{Q_{\mathbf{x}}^{\mathbf{b}} : \mathbf{x} \in \mathbb{R}^N\}$ *is recurrent in the sense that* (7.3.26) *holds. In fact, for every* $\Gamma \in \mathcal{B}_{\mathbb{R}^N}$ *with* $\mu(\Gamma) > 0$,

$$Q_{\mathbf{x}}^{\mathbf{b}}\left(\int_0^T \mathbf{1}_\Gamma(\psi(t))\, dt = \infty\right) = 1 \quad \textit{for all } \mathbf{x} \in \mathbb{R}^N. \tag{7.4.19}$$

PROOF: By Theorem 7.3.24, we know that $\{Q_{\mathbf{x}}^{\mathbf{b}} : \mathbf{x} \in \mathbb{R}^N\}$ is either recurrent or transient. In particular, either (7.4.11) holds and Theorem 7.4.10 applies, or (7.3.25) holds. But, in the latter case, for any $\mu \in \mathbf{M}_1(\mathbb{R}^N)$ and $r \in (0,\infty)$:

$$\lim_{T\to\infty} \mathbb{E}^{Q_\mu^{\mathbf{b}}}\left[\frac{1}{T}\int_0^T \mathbf{1}_{B_{\mathbb{R}^N}(\mathbf{0},r)}(\psi(t))\, dt\right] = 0, \tag{7.4.20}$$

which means that $\mathfrak{M}_1(\{Q_t^{\mathbf{b}} : t > 0\})$ must be empty. In other words, we have now proved that either $\mathfrak{M}_1(\{Q_t^{\mathbf{b}} : t > 0\})$ is empty or it contains precisely one element and Theorem 7.4.10 applies. Hence, all that remains is to remark that, from (7.4.18) and the fact (cf. (7.3.9)) that the support of $Q^{\mathbf{b}}(t,\mathbf{x},\cdot)$ is $\mathbb{R}^N$ for every $(t,\mathbf{x}) \in (0,\infty)\times\mathbb{R}^N$, it is obvious that the support of μ must also be $\mathbb{R}^N$. □

Although it is a rather weak statement by comparison to the ones which we will derive in the next section, we will close the present discussion with the following application of the criterion given in Lemma 7.4.14.

7.4.21 Theorem. *Assume that the inequality in* (7.3.15) *holds for some* $A \in [0,\infty)$ *and* $B \in (-\infty, 0)$. *Then, not only does* (7.3.12) *hold for every* $\mathbf{x} \in \mathbb{R}^N$, *but also* $\mathfrak{M}_1(\{Q_t^{\mathbf{b}} : t > 0\}) = \{\mu\}$, *where* μ *has all the properties derived in Theorem 7.4.16.*

PROOF: Let ν be the point-mass $\delta_{\mathbf{0}}$, and define ν_n accordingly, as in (7.4.15). All that we have to do is show that

$$\lim_{r\to\infty} \sup_{n\in\mathbb{Z}^+} \nu_n\big(B_{\mathbb{R}^N}(\mathbf{0},r)\complement\big) = 0;$$

and clearly this will follow if we know that

$$\sup_{n\in\mathbb{Z}^+} \int_{\mathbb{R}^N} |\mathbf{x}|^2\, \nu_n(d\mathbf{x}) < \infty.$$

But such an estimate follows immediately from (7.3.16) with $\mathbf{x} = \mathbf{0}$ and $p = 2$. □

Exercises

7.4.22 Exercise: Returning to the general setting at the beginning of this section, assume that, for each $\Gamma \in \mathcal{B}_{\mathbb{R}^N}$,

$$\begin{aligned}&\exists \mathbf{x} \in \mathbb{R}^N \quad \int_0^\infty e^{-t} P(t, \mathbf{x}, \Gamma)\, dt > 0 \\ &\qquad\qquad \Longrightarrow \forall \mathbf{x} \in \mathbb{R}^N \quad \int_0^\infty e^{-t} P(t, \mathbf{x}, \Gamma)\, dt > 0;\end{aligned} \tag{7.4.23}$$

and show that there is at most one ergodic element of $\mathfrak{M}_1\big(\{P_t : t > 0\}\big)$.

Hint: Define $\mathbf{x} \in \mathbb{R}^N \longmapsto R(\mathbf{x}, \cdot\,) \in \mathbf{M}_1(\mathbb{R}^N)$ by

$$R(\mathbf{x}, \Gamma) = \int_0^\infty e^{-t} P(t, \mathbf{x}, \Gamma)\, dt, \quad \Gamma \in \mathcal{B}_{\mathbb{R}^N};$$

note that, for each $\mu \in \mathfrak{M}_1\big(\{P_t : t > 0\}\big)$, $\mu = \int R(\mathbf{x}, \cdot\,)\, \mu(d\mathbf{x})$; and use this together with (7.4.23) to conclude that no two elements of $\mathfrak{M}\big(\{P_t : t > 0\}\big)$ can be singular. Finally, apply the final part of Theorem 7.1.8.

7.4.24 Exercise: Continuing in the general setting, assume this time that $\mathfrak{M}_1\big(\{P_t : t > 0\}\big)$ is a compact subset of $\mathbf{M}_1(\mathbb{R}^N)$. As a consequence of the famous Krein–Millman Theorem,† show that $\mathfrak{M}_1\big(\{P_t : t > 0\}\big)$ is the closed convex hull of its ergodic elements. In particular, if $\mathfrak{M}_1\big(\{P_t : t > 0\}\big)$ is compact and contains more than one element, then it contains elements which are mutually singular; and so, by Exercise 7.4.22, when, in addition, (7.4.23) holds, we know that $\mathfrak{M}_1\big(\{P_t : t > 0\}\big)$ can have at most one element.

7.4.25 Exercise: Again work in the general setting. If, for some $T \in (0, \infty)$ and σ-finite, Borel measure μ, one knows that $\mu = \mu P_T$ and that the only square μ-integrable f's satisfying $f = P_T f$ (a.e., μ) are μ-almost every constant, show that, for every $p \in [1, \infty)$ and $F \in L^p(P_\mu; \mathbb{R})$:

$$\lim_{n\to\infty} \frac{1}{n} \sum_{m=0}^{n-1} F \circ \mathbf{\Sigma}_{nT} = \frac{\mathbb{E}^{P_\mu}[F]}{\mu(\mathbb{R}^N)} \quad \text{(a.e., } P_\mu\text{) and in } L^p(P_\mu; \mathbb{R}),$$

† See § 8.3 of Chapter V in N. Dunford and J.T. Schwartz's *Linear Operators, Part I*, publ. by J. Wiley (1958).

where the ratio is taken to be 0 when the denominator is infinite. In particular, if one knows that every $f \in B(\mathbb{R}^N;\mathbb{R})$ satisfying $f = P_T f$ is constant, show that $\{\boldsymbol{\Sigma}_{nT} : n \in \mathbb{Z}^+\}$ is P_μ-ergodic for every $\mu \in \mathbf{M}_1(\mathbb{R}^N)$ which satisfies $\mu = \mu P_T$, conclude that there is at most one such μ, and that, if it exists,

$$\lim_{n\to\infty} \frac{1}{n} \sum_{m=0}^{n-1} F \circ \boldsymbol{\Sigma}_{mT} = \mathbb{E}^{P_\mu}[F] \quad (\text{a.s.}, P_{\mathbf{x}})$$

for every $F \in B\big(\mathfrak{P}(\mathbb{R}^N);\mathbb{R}\big)$ and $\mathbf{x} \in \mathbb{R}^N$.

7.4.26 Exercise: Assume that there is an $f \in C^2\big(\mathbb{R}^N;[0,\infty)\big)$ with the properties that

$$\mathbf{L}^{\mathbf{b}} f \le 0 \quad \text{and} \quad \lim_{|\mathbf{x}|\to\infty} f(\mathbf{x}) = \infty.$$

Show that

$$\big\{Q^{\mathbf{b}}(t,\mathbf{x},\,\cdot\,) : (t,\mathbf{x}) \in (0,\infty) \times K\big\}$$

is tight for each $K \subset\subset \mathbb{R}^N$. In particular, conclude that there is a unique $\mu \in \mathfrak{M}_1\big(\{P_t : t > 0\}\big)$.

7.4.27 Exercise: Here is another criterion for testing when $\mathfrak{M}_1\big(\{Q^{\mathbf{b}}_t : t > 0\}\big) \ne \emptyset$.

(i) Let ν be a locally finite, Borel measure on $\mathbb{R}^N$, and assume that

$$\nu Q^{\mathbf{b}}_t \le \nu \quad \text{for all } t \in (0,\infty). \tag{7.4.28}$$

(We write $\mu \le \nu$ if $\langle \varphi, \mu\rangle \le \langle \varphi, \nu\rangle$ for every nonnegative, Borel function φ.) If $\nu(\mathbb{R}^N) \in (0,\infty)$, show that $\mu \equiv \frac{\nu}{\nu(\mathbb{R}^N)} \in \mathfrak{M}_1\big(\{Q^{\mathbf{b}}_t : t > 0\}\big)$. Even if $\nu(\mathbb{R}^N) = \infty$, show that there is a unique Borel measure μ with the property that

$$\langle \varphi, \nu Q^{\mathbf{b}}_t\rangle \searrow \langle \varphi, \mu\rangle, \quad \varphi \in L^1\big(\nu;[0,\infty)\big);$$

and conclude that μ is $\{Q^{\mathbf{b}}_t : t > 0\}$-invariant. Unfortunately, in general, μ will be identically 0 and therefore trivial.

(ii) Assume that (7.3.38) holds, and let $f \in C^2\big(\mathbb{R}^N;[0,\infty)\big)$ satisfy

$$\mathbf{L}^{-\mathbf{b}} f - \mathrm{div}\mathbf{b} f \le 0. \tag{7.4.29}$$

Using Theorems 7.3.10 and 7.1.19, show that

$$\mathbb{E}^{Q^{-\mathbf{b}}_{\mathbf{x}}}\left[f\big(\psi(T)\big) \exp\left(- \int_0^T \mathrm{div}\mathbf{b}\big(\psi(t)\big)\,dt \right)\right] \le f(\mathbf{x}) \tag{7.4.30}$$

for all $(T,\mathbf{x}) \in (0,\infty)) \times \mathbb{R}^N$. Next, define ν on $\mathcal{B}_{\mathbb{R}^N}$ by $\nu(d\mathbf{x}) = f(\mathbf{x})\,d\mathbf{x}$, and use (7.3.39) to see that (7.4.28) holds. In particular, the conclusions drawn in (i) apply to this ν.

(iii) Continuing in the setting of **(ii)**, assume this time that div$\mathbf{b}$ is bounded below, $f \in C^2(\mathbb{R}^N;[0,\infty))$ is bounded, and that equality holds in (7.4.29). Proceeding as in **(ii)**, show that equality holds in (7.4.30) and therefore that the associated ν is $\{Q_t^{\mathbf{b}} : t > 0\}$-invariant.

§7.5: Perturbations by Conservative Vector Fields.

Although, as we saw in the preceding section, it is possible to say something about the ergodic properties of diffusions even when one has no explicit expression for their invariant measures, a great deal more can be said when one knows, ahead of time, what the invariant measure must be. To see how one might arrange things so that a particular measure will be the obvious candidate to be the invariant measure for a diffusion, we use the ideas developed in Exercise 7.4.27, only now we work backwards. Namely, suppose that μ is invariant for the semigroup $\{Q_t^{\mathbf{b}} : t > 0\}$ associated with the Markov family $\{Q_{\mathbf{x}}^{\mathbf{b}} : \mathbf{x} \in \mathbb{R}^N\}$ described in Theorem 7.3.10. Then, for any $\varphi \in C_c^\infty(\mathbb{R}^N;\mathbb{R})$, we have that

$$0 = \frac{d}{dt}\langle Q_t^{\mathbf{b}}\varphi, \mu\rangle\Big|_{t=0} = \langle \mathbf{L}^{\mathbf{b}}\varphi, \mu\rangle.$$

Hence, μ is a *generalized solution* (i.e., pretend that μ has a smooth density f with respect to Lebesgue's measure and integrate by parts) to the equation

$$\nabla \cdot \left(\tfrac{1}{2}\nabla f - f\mathbf{b}\right) = 0. \tag{7.5.1}$$

Thus, given an $f \in C^2(\mathbb{R}^N;(0,\infty))$, we can hope that the measure μ given by $\mu(d\mathbf{x}) = f(\mathbf{x})\,d\mathbf{x}$ will be $\{Q_t^{\mathbf{b}} : t > 0\}$-invariant only if f is a (classical) solution to (7.5.1). In particular, we can force the issue by taking $\mathbf{b} = 2\frac{\nabla f}{f} = 2\nabla \log f$; which, when we write $f = e^{-2U}$, is equivalent to taking

$$\mathbf{b} = -\nabla U \quad \text{where } U \in C^2(\mathbb{R}^N;\mathbb{R}). \tag{7.5.2}$$

In other words, we are hoping that the measure

$$\mu^U(\Gamma) \equiv \int_\Gamma e^{-2U(\mathbf{x})}\,d\mathbf{x}, \quad \Gamma \in \mathcal{B}_{\mathbb{R}^N}, \tag{7.5.3}$$

will be invariant for $\{Q_t^{\mathbf{b}} : t > 0\}$ when $\mathbf{b}$ is given by (7.5.2); and in this section we will see that, among other things, this hope is well-founded.

We will call the vector field $\mathbf{b}$ given by (7.5.2) and the measure μ^U given by (7.5.3) the **conservative vector field** and the **Gibbs state** determined by the **potential** U. Obviously, $\mathbf{b}$ is unchanged by the addition of constants to the potential U, and so we will always assume that U has been chosen so that

$$\mu^U(\mathbb{R}^N) = 1 \quad \text{whenever} \int_{\mathbb{R}^N} e^{-2U(\mathbf{x})}\, d\mathbf{x} < \infty. \tag{7.5.4}$$

Further, in order to emphasize that we are dealing with conservative $\mathbf{b}$'s, when $\mathbf{b}$ is given by (7.5.2), we will use the notation (cf. (7.3.4), (7.3.5), (7.3.2), and Theorem 7.3.10)

$$\begin{aligned} \mathfrak{e}^U(\psi) &= \mathfrak{e}^{\mathbf{b}}(\psi), \quad \psi \in \mathfrak{P}(\mathbb{R}^N), \\ X^U(t,\psi) &= X^{\mathbf{b}}(t,\psi), \quad \psi \in \mathfrak{P}(\mathbb{R}^N) \text{ and } t \in \left[0, \mathfrak{e}^U(\psi)\right) \\ \mathbf{L}^U &= \mathbf{L}^{\mathbf{b}}, \\ \text{and } Q^U_{\mathbf{x}} &= Q^{\mathbf{b}}_{\mathbf{x}} \quad \text{when } \mathcal{W}^{(N)}_{\mathbf{x}}\left(\mathfrak{e}^U = \infty\right) = 1. \end{aligned} \tag{7.5.5}$$

Finally, given $U \in C^2(\mathbb{R}^N;\mathbb{R})$, we set

$$\begin{gathered} R^U(t,\psi) = \exp\left[U\big(\psi(0)\big) - U\big(\psi(t)\big) + \int_0^t V^U\big(\psi(s)\big)\, ds\right] \\ \text{where } 2V^U \equiv -|\nabla U|^2 + \mathbf{\Delta} U. \end{gathered} \tag{7.5.6}$$

The careful reader will have already noticed that he has seen this notation before (specifically, at the end of Section 4.3) and will be relieved to learn in the following that its use here is consistent with its use there.

7.5.7 Theorem. *Let $U \in C^2\left(\mathbb{R}^N;\mathbb{R}\right)$ be given, and refer to the notation in (7.5.5) and (7.5.6). Then for each $(t,\mathbf{x}) \in [0,\infty) \times \mathbb{R}^N$ and $A \in \mathcal{B}_t$,*

$$\mathcal{W}^{(N)}_{\mathbf{x}}\left(\left\{\psi : X^U(\,\cdot\,,\psi) \in A \;\&\; \mathfrak{e}^U(\psi) > t\right\}\right) = \mathbb{E}^{\mathcal{W}^{(N)}_{\mathbf{x}}}\left[R^U(t),\, A\right]. \tag{7.5.8}$$

In particular,

$$\mathbb{E}^{\mathcal{W}^{(N)}_{\mathbf{x}}}\left[R^U(t)\right] = \mathcal{W}^{(N)}_{\mathbf{x}}\left(\mathfrak{e}^U > t\right) \le 1, \quad (t,\mathbf{x}) \in [0,\infty) \times \mathbb{R}^N. \tag{7.5.9}$$

Thus, $\mathcal{W}^{(N)}_{\mathbf{x}}\left(\mathfrak{e}^U = \infty\right) = 1$ if and only if $\left(R^U(t), \mathcal{B}_t, \mathcal{W}^{(N)}_{\mathbf{x}}\right)$ is a nonnegative, continuous martingale; in which case

$$Q^U_{\mathbf{x}}(A) = \mathbb{E}^{\mathcal{W}^{(N)}_{\mathbf{x}}}\left[R^U(t),\, A\right] \quad \textit{for all } t \in [0,\infty) \textit{ and } A \in \mathcal{B}_t. \tag{7.5.10}$$

PROOF: We begin by assuming that $U \in C^2_{\mathrm{b}}\left(\mathbb{R}^N;\mathbb{R}\right)$, in which case we know that $\mathfrak{e}^U(\psi) = \infty$ for all $\psi \in \mathfrak{P}(\mathbb{R}^N)$ and, by Theorem 7.3.10, that $Q^U_{\mathbf{x}}$ is the unique

$P \in \mathbf{M}_1(\mathfrak{P}(\mathbb{R}^N))$ satisfying $P(\psi(0) = \mathbf{x}) = 1$ and (7.3.11) with $\mathbf{b} = -\nabla U$. At the same time, if $f \in C_b^\infty(\mathbb{R}^N;\mathbb{R})$, $g \equiv e^{-U}f$, and we apply Theorem 7.1.19 with $P = \mathcal{W}_\mathbf{x}^{(N)}$,

$$X(t,\psi) = g(\psi(t)) - \int_0^t \left[\mathbf{L}^U g\right](\psi(s))\,ds,$$

and

$$V(t,\psi) = \exp\left[U(\psi(0)) + \int_0^t V^U(\psi(s))\,ds\right],$$

then we find that

$$\left(R^U(t)f(t,\psi(t)) - \int_0^t R^U(s)\left[\mathbf{L}^U f\right](\psi(s))\,ds, \mathcal{B}_t, \mathcal{W}_\mathbf{x}^{(N)}\right) \tag{7.5.11}$$

is a continuous martingale.

In particular, when $f = \mathbf{1}$, this says that $\left(R^U(t), \mathcal{B}_t, \mathcal{W}_\mathbf{x}^{(N)}\right)$ is a positive, continuous martingale with mean-value 1. Next, for each $n \in \mathbb{Z}^+$, define $P_n \in \mathbf{M}_1(\mathfrak{P}(\mathbb{R}^N))$ by

$$P_n(A) = \mathbb{E}^{\mathcal{W}_\mathbf{x}^{(N)}}\left[R^U(n),\, A\right], \quad A \in \mathcal{B}_{\mathfrak{P}(\mathbb{R}^N)}.$$

It is then clear (from the martingale property for $R^U(\,\cdot\,)$) that $P_{n+1} \restriction \mathcal{B}_n = P_n \restriction \mathcal{B}_n$ and therefore (cf. Exercise 3.3.26) that $\{P_n\}_1^\infty$ is tight in $\mathbf{M}_1(\mathfrak{P}(\mathbb{R}^N))$. Moreover, if P is any limit of this sequence, then (again by the martingale property)

$$\mathbb{E}^P[F] = \mathbb{E}^{\mathcal{W}_\mathbf{x}^{(N)}}\left[R^U(t)\,F\right],$$

first for bounded, $\mathcal{B}_t$-measurable F which are continuous and then for general ones. Hence, by (7.5.11), for every $f \in C_c^\infty(\mathbb{R}^N;\mathbb{R})$, $0 \le s < t$, and $A \in \mathcal{B}_s$,

$$\begin{aligned}
&\mathbb{E}^P\left[f(\psi(t)) - f(\psi(s)),\, A\right] \\
&= \mathbb{E}^{\mathcal{W}_\mathbf{x}^{(N)}}\left[R^U(t)f(\psi(t)) - R^U(s)f(\psi(s)),\, A\right] \\
&= \int_s^t \mathbb{E}^{\mathcal{W}_\mathbf{x}^{(N)}}\left[R^U(\xi)\left[\mathbf{L}^U f\right](\psi(\xi)),\, A\right] d\xi \\
&= \int_s^t \mathbb{E}^{\mathcal{W}_\mathbf{x}^{(N)}}\left[R^U(t)\left[\mathbf{L}^U f\right](\psi(\xi)),\, A\right] d\xi \\
&= \mathbb{E}^P\left[\int_s^t \left[\mathbf{L}^U f\right](\psi(\xi))\,d\xi,\, A\right].
\end{aligned}$$

That is, P satisfies (7.3.11) with $\mathbf{b} = -\nabla U$; and therefore, since it is obvious that $P(\psi(0) = \mathbf{x}) = 1$, we have now proved that $P = Q^U_{\mathbf{x}}$. In other words, the proof is complete for $U \in C^2_{\mathrm{b}}(\mathbb{R}^N;\mathbb{R})$.

Turning to general $U \in C^2(\mathbb{R}^N;\mathbb{R})$, choose $\{U_n\}_1^\infty \subseteq C^2_{\mathrm{b}}(\mathbb{R}^N;\mathbb{R})$ so that $U_n \restriction B_{\mathbb{R}^N}(\mathbf{0}, n+1) = U \restriction B_{\mathbb{R}^N}(\mathbf{0}, n+1)$ for each $n \in \mathbb{Z}^+$. Then, by the preceding applied to U_n, we see that, for $n \in \mathbb{Z}^+$, $t \in [0,\infty)$, and $A \in \mathcal{B}_t$,

$$\mathcal{W}^{(N)}_{\mathbf{x}}\Big(\{\psi : X^U(\,\cdot\,,\psi) \in A \,\&\, \mathfrak{e}^U_n(\psi) > t\}\Big) = \mathbb{E}^{\mathcal{W}^{(N)}_{\mathbf{x}}}\Big[R^U(t),\, A \cap \{\zeta_n > t\}\Big],$$

where

$$\zeta_n(\psi) = \inf\big\{t : |\psi(t)| \geq n\big\} \quad \text{and} \quad \mathfrak{e}^U_n(\psi) = \inf\big\{t : \big|X^{U_n}(t,\psi)\big| \geq n\big\}.$$

Hence (7.5.8) follows after we let $n \to \infty$, clearly (7.5.9) is just a special case of (7.5.8), and, obviously, (7.5.9) implies that $\mathcal{W}^{(N)}_{\mathbf{x}}\big(\mathfrak{e}^U = \infty\big) = 1$ if $\big(R^U(t), \mathcal{B}_t, \mathcal{W}^{(N)}_{\mathbf{x}}\big)$ is a martingale.

Finally, suppose that $\mathcal{W}^{(N)}_{\mathbf{x}}\big(\mathfrak{e}^U = \infty\big) = 1$. Then, for each $t \in [0,\infty)$, $\mathbb{E}^{\mathcal{W}^{(N)}_{\mathbf{x}}}\big[R^U(t)\big] = 1$, and so $R^{U_n}(t,\psi) \longrightarrow R^U(t,\psi)$ not only pointwise but also in $L^1\big(\mathcal{W}^{(N)}_{\mathbf{x}}\big)$. But this means that the martingale property for $R^U(\,\cdot\,)$ follows from that for the $R^{U_n}(\,\cdot\,)$'s; and obviously (7.5.10) becomes just a restatement of (7.5.8). □

Warning: From now on, we will be assuming that

$$\mathcal{W}^{(N)}_{\mathbf{x}}\big(\mathfrak{e}^U = \infty\big) = 1 \quad \text{for all } \mathbf{x} \in \mathbb{R}^N. \tag{7.5.12}$$

Hence, as a consequence of Theorem 7.3.10, we know that $\mathbf{x} \in \mathbb{R}^N \longmapsto Q^U_{\mathbf{x}} \in \mathbf{M}_1\big(\mathfrak{P}(\mathbb{R}^N)\big)$ is continuous and that $\big\{Q^U_{\mathbf{x}} : \mathbf{x} \in \mathbb{R}^N\big\}$ is a strong Markov family with the property that, for each $\mathbf{x} \in \mathbb{R}^N$, $Q^U_{\mathbf{x}}$ has support $\{\psi \in \mathfrak{P}(\mathbb{R}^N) : \psi(0) = \mathbf{x}\}$.

As we are about to see, it is (7.5.10) which will make it possible to carry out our program.

7.5.13 Theorem. *Define the transition probability function*

$$(t,\mathbf{x}) \in (0,\infty) \times \mathbb{R}^N \longmapsto Q^U(t,\mathbf{x},\,\cdot\,) \in \mathbf{M}_1\big(\mathfrak{P}(\mathbb{R}^N)\big)$$

by

$$Q^U(t,\mathbf{x},\Gamma) = Q^U_{\mathbf{x}}\big(\psi(t) \in \Gamma\big), \quad \Gamma \in \mathcal{B}_{\mathbb{R}^N};$$

and, for $(t, \mathbf{x}, \mathbf{y}) \in (0,\infty) \times \mathbb{R}^N \times \mathbb{R}^N$, *set (cf.* (4.3.29)*)*

$$q^U(t,\mathbf{x},\mathbf{y}) = e^{U(\mathbf{x})+U(\mathbf{y})}\,\gamma_t^N(\mathbf{y}-\mathbf{x})\,r^{V^U}(t,\mathbf{x},\mathbf{y}), \tag{7.5.14}$$

where (cf. (4.2.19)*)* $r^{V^U}(t,\mathbf{x},\mathbf{y})$ *is given by*

$$\int_{\mathfrak{P}(\mathbb{R}^N)} \exp\left[\int_0^t V^U\Big(\big(1-\tfrac{s}{t}\big)\mathbf{x} + \tfrac{s}{t}\mathbf{y} + \tilde{\psi}_t(s)\Big)\,ds\right] \mathcal{W}^{(N)}(d\psi).$$

Then $(t,\mathbf{x},\mathbf{y}) \in (0,\infty)\times\mathbb{R}^N\times\mathbb{R}^N \longmapsto q^U(t,\mathbf{x},\mathbf{y}) \in (0,\infty]$ *is lower semicontinuous,* $q^U(t,\mathbf{x},\mathbf{y}) = q^U(t,\mathbf{y},\mathbf{x})$, *and*

$$\begin{gathered} Q^U(t,\mathbf{x},d\mathbf{y}) = q^U(t,\mathbf{x},\mathbf{y})\,\mu^U(d\mathbf{y}), \quad (t,\mathbf{x}) \in (0,\infty)\times\mathbb{R}^N, \\ \text{where } \mu^U(d\mathbf{y}) \equiv e^{-2U(\mathbf{y})}\,d\mathbf{y}. \end{gathered} \tag{7.5.15}$$

In particular, if

$$V^U(\mathbf{x}) \le C\big(1+|\mathbf{x}|\big), \quad \mathbf{x}\in\mathbb{R}^N \tag{7.5.16}$$

for some $C \in [0,\infty)$, *then* $q^U \in C\big((0,\infty)\times\mathbb{R}^N\times\mathbb{R}^N;(0,\infty)\big)$.†

PROOF: Obviously, $q^U(t,\mathbf{x},\mathbf{y}) > 0$. Moreover, when (7.5.16) holds, the continuity of q^U follows immediately from the uniform integrability afforded by the estimate in (3.3.30). More generally, even when (7.5.16) fails, the lower semicontinuity of q^U becomes clear as soon as one replaces V^U on the right-hand side of (7.5.14) by $V^U \wedge n$ and then lets $n \to \infty$. Furthermore, the symmetry of $q^U(t,\,\cdot\,,\,\bullet\,)$ comes directly from the reversibility of $\tilde{\psi}_t$ proved in Theorem 4.2.18. Finally, to prove (7.5.15), let $f \in B\big(\mathbb{R}^N;[0,\infty)\big)$ be given, and use (7.5.10) together with (4.2.20) to see that

$$\int_{\mathbb{R}^N} f(\mathbf{y})\,Q^U(t,\mathbf{x},d\mathbf{y}) = \int_{\mathbb{R}^N} f(\mathbf{y})\,e^{U(\mathbf{x})-U(\mathbf{y})}\,r^{V^U}(t,\mathbf{x},\mathbf{y})\,\gamma_t^N(\mathbf{y}-\mathbf{x})\,d\mathbf{y},$$

from which (7.5.15) is an easy step. □

As the first step in our analysis of the ergodic properties of these processes, we give the following application of the results in Theorems 7.3.24 and 7.2.30.

† Although it is not evident from the present line of reasoning, elliptic regularity theory (cf. the references in the footnote to Lemma 7.3.23) says that q^U will *always* be continuous.

7.5.17 Corollary. *For every $U \in C^2(\mathbb{R}^N;\mathbb{R})$ satisfying (7.5.12), $\{Q^U_{\mathbf{x}} : \mathbf{x} \in \mathbb{R}^N\}$ is either transient or recurrent (cf. (7.3.25) or (7.3.26) with $\mathbf{b} = -\nabla U$). Moreover, it will be recurrent if*

$$\lim_{|\mathbf{x}|\to\infty} U(\mathbf{x}) = \infty \quad \text{and}$$
$$-2|\nabla U|^2 + \boldsymbol{\Delta} U \le 0 \text{ off of } B_{\mathbb{R}^N}(\mathbf{0}, r_0) \text{ for some } r_0 \in (0,\infty).$$

On the other hand, it will be transient if there exist a nonempty open G, an $\alpha \in (0,\infty)$, and $\mathbf{x}_0 \notin \overline{G}$ such that

$$-(2-\alpha)|\nabla U|^2 + \boldsymbol{\Delta} U \le 0 \quad \text{off } G \quad \text{and} \quad U(\mathbf{x}_0) < \inf_{\mathbf{x}\in G} U(\mathbf{x}).$$

PROOF: The initial statement is already covered by Theorem 7.3.24. As for the second assertion, note that U must be bounded below, set $f = U - m + 1$, where $m = \inf_{\mathbf{x}\in\mathbb{R}^N} U(\mathbf{x})$, and check that the criterion in Theorem 7.2.30 for recurrence is met. Finally, to prove the final criterion for transience, all that we have to do is show that, under the stated hypotheses, the process cannot be recurrent; and we will do this by checking that $Q^U_{\mathbf{x}_0}(\sigma_{\overline{G}} < \infty) < 1$, where $\sigma_{\overline{G}}(\psi) \equiv \inf\{t : \psi(t) \in \overline{G}\}$. To this end, take $f = e^{\alpha U}$, and, after repeating the same kind of reasoning as we used in last part of the proof of Theorem 7.2.30, check that

$$\begin{aligned}\min_{\mathbf{x}\in\overline{G}} f(\mathbf{x}) Q^U_{\mathbf{x}_0}(\sigma_{\overline{G}} < \infty) \\ \le \int_{\{\sigma_{\overline{G}}(\psi)<\infty\}} f\big(\psi(\sigma_{\overline{G}})\big)\, Q^U_{\mathbf{x}_0}(d\psi) \le f(\mathbf{x}_0).\end{aligned}$$

Hence, the required inequality follows from $U(\mathbf{x}_0) < \inf_{\mathbf{x}\in G} U(\mathbf{x})$. □

We next take up some analytic properties which derive from (7.5.15).

7.5.18 Theorem. *Define the semigroup $\{Q^U_t : t > 0\}$ of operators on $B(\mathbb{R}^N;\mathbb{R})$ by*

$$\big[Q^U_t f\big](\mathbf{x}) = \int_{\mathbb{R}^N} f(\mathbf{y}) Q^U(t,\mathbf{x},d\mathbf{y}), \quad \mathbf{x} \in \mathbb{R}^N.$$

Then $Q^U_t : B(\mathbb{R}^N;\mathbb{R}) \longrightarrow C_b(\mathbb{R}^N;\mathbb{R})$ for each $t \in (0,\infty)$, and, for each $f \in C_b(\mathbb{R}^N;\mathbb{R})$, $Q^U_t f \longrightarrow f$ uniformly on compacts as $t \searrow 0$. Finally, if μ^U is the measure defined in (7.5.5), then, for each $t \in (0,\infty)$,

$$\begin{aligned}&\int_{\mathbb{R}^N} f(\mathbf{x}) \big[Q^U_t g\big](\mathbf{x})\, \mu^U(d\mathbf{x}) \\ &\quad = \int_{\mathbb{R}^N} g(\mathbf{x}) \big[Q^U_t f\big](\mathbf{x})\, \mu^U(d\mathbf{x}), \quad f, g \in B_c(\mathbb{R}^N;\mathbb{R}),\end{aligned} \tag{7.5.19}$$

(7.5.20) $$\mu^U = \mu^U Q_t^U,$$

and so, for every $p \in [1,\infty)$, $Q_t^U \restriction C_c(\mathbb{R}^N;\mathbb{R})$ determines a unique extension to $L^p(\mu^U)$ as a contraction operator. In particular, if $\overline{Q_t^U}$ is the contraction on $L^2(\mu^U)$ determined by $Q_t^U \restriction C_c(\mathbb{R}^N;\mathbb{R})$, then $\{\overline{Q_t^U} : t \in (0,\infty)\}$ is a strongly continuous semigroup of self-adjoint contractions, and therefore there is a resolution of the identity $\{E_\lambda^U : \lambda \in (0,\infty)\}$ by orthogonal projections with the property that

(7.5.21) $$\overline{Q_t^U} = \int_{[0,\infty)} e^{-\lambda t}\, dE_\lambda^U, \quad t \in (0,\infty).$$

PROOF: Clearly, everything through (7.5.19) is covered either by Lemma 7.3.20 or by one part or another of Theorem 7.5.13. Moreover, given (7.5.19), we can prove (7.5.20) as follows. Let $f \in B_c(\mathbb{R}^N;[0,\infty))$ be given and choose a sequence $\{g_\ell\}_1^\infty$ from $B_c(\mathbb{R}^N;[0,\infty))$ so that $g_\ell \nearrow \mathbf{1}$. Then, by the Monotone Convergence Theorem, we know from (7.5.19) that, for any $f \in B_c(\mathbb{R}^N;[0,\infty))$,

$$\int f\, d\mu^U = \lim_{\ell\to\infty} \int f\, Q_t^U g_\ell\, d\mu^U = \lim_{\ell\to\infty} \int g_\ell\, Q_t^U f\, d^U = \int Q_t^U f\, d\mu^U;$$

and as soon as one has $\langle f, \mu^U\rangle = \langle Q_t^U f, \mu^U\rangle$ for $f \in B_c(\mathbb{R}^N;[0,\infty))$, it follows for all $f \in B(\mathbb{R}^N;[0,\infty))$ by another application of the Monotone Convergence Theorem.

Starting from (7.5.20), we have, by Jensen's inequality,

$$\|Q_t^U f\|_{L^p(\mu^U)} \le \left(\int Q_t^U(|f|^p)\, d\mu^U\right)^{\frac{1}{p}} = \left(\int |f|^p\, d\mu^U\right)^{\frac{1}{p}}$$

for each $p \in [1,\infty)$ and $f \in B(\mathbb{R}^N;\mathbb{R})$. Hence, the required extension of Q_t^U as a contraction on $L^p(\mu^U)$ is easy. Moreover, the semigroup property extends automatically; and, when $p = 2$, the self-adjointness of $\overline{Q_t^U}$ follows from the symmetry in (7.5.19) of Q_t^U. In addition, to see that $\{\overline{Q_t^U} : t \in (0,\infty)\}$ is strongly continuous on $L^2(\mu^U)$, it suffices to check that $\overline{Q_t^U} f \longrightarrow f$ in $L^2(\mu^U)$ as $t \searrow 0$ for any dense set of $f \in L^2(\mu^U)$. But if $f \in C_c(\mathbb{R}^N;\mathbb{R})$, then $Q_t^U f \longrightarrow f$ boundedly and pointwise, and therefore

$$\begin{aligned}\|\overline{Q_t^U} f - f\|^2_{L^2(\mu^U)} &= (Q_t^U f, Q_t^U f)_{L^2(\mu^U)} - 2(Q_t^U f, f)_{L^2(\mu^U)} + \|f\|^2_{L^2(\mu^U)} \\ &= (Q_{2t}^U f, f)_{L^2(\mu^U)} - 2(Q_t^U f, f)_{L^2(\mu^U)} + \|f\|^2_{L^2(\mu^U)}\end{aligned}$$

tends to 0 by Lebesgue's Dominated Convergence Theorem.

Finally, now that we know that $\{\overline{Q_t^U} : t \in (0,\infty)\}$ is a strongly continuous semigroup of self-adjoint contractions on $L^2(\mu^U)$, the spectral representation in (7.5.21) is a standard application of spectral theory.† □

† See, for example, the main theorem in § 141 of F. Riesz and B. Sz.-Nagy's *Functional Analysis*, publ. by Ungar (1955).

When $\mu^U(\mathbb{R}^N) = 1$, Theorem 7.4.16 applies and tells us not only that μ^U is the unique element of $\mathfrak{M}_1\big(\{Q_t^U : t > 0\}\big)$ but also that, for each $\mathbf{x} \in \mathbb{R}^N$:

$$\lim_{T\to\infty} \frac{1}{T}\int_0^T F\circ \boldsymbol{\Sigma}_t \, dt = \mathbb{E}^{Q^U_{\mu^U}}[F] \quad (\text{a.s., } Q^U_{\mathbf{x}}) \tag{7.5.22}$$

for every $F \in B(\mathbb{R}^N;\mathbb{R})$ and that

$$\lim_{T\to\infty} \frac{1}{T}\int_0^T \big[Q_t^U\varphi\big](\mathbf{x})\, dt = \langle \varphi, \mu^U\rangle \tag{7.5.23}$$

for every $\varphi \in B(\mathbb{R}^N;\mathbb{R})$. In particular, this proves that $\{Q^U_{\mathbf{x}} : \mathbf{x} \in \mathbb{R}^N\}$ is recurrent whenever $e^{-2U} \in L^1(\mathbb{R}^N)$. In addition, it gives precision to the intuition coming from the equation

$$X^U(t,\boldsymbol{\psi}) = \boldsymbol{\psi}(t) - \int_0^t \nabla U\big(X^U(s,\boldsymbol{\psi})\big)\, ds.$$

Namely, since U can only decrease along integral curves of the gradient field $-\nabla U$, solutions to $\dot{X}(t) = -\nabla U\big(X(t)\big)$ will be drawn toward the region $M(U) \subseteq \mathbb{R}^N$ in which U is minimal. Hence, at least when U grows fast enough at infinity so that $M(U)$ is bounded and stable, one suspects that the paths $X^U(\,\cdot\,,\boldsymbol{\psi})$ should, in the long run, spend most of their time near $M(U)$, which is exactly what (7.5.22) and (7.5.23) confirm.

Because (7.5.22) and (7.5.23) depend directly on (7.5.20) and only indirectly on (7.5.19), (7.5.22) and (7.5.23) can be sharpened when one takes full advantage of (7.5.19).

7.5.24 Corollary. *For each $T \in (0,\infty)$, the only $f \in L^2(\mu^U;\mathbb{R})$ satisfying $f = Q_T^U f$ (a.e., μ^U) is μ^U-almost everywhere constant. Hence, for each $\varphi \in L^2(\mu^U)$:*

$$\begin{aligned} \mu^U(\mathbb{R}^N) = 1 &\implies \lim_{t\to\infty} \big\|Q_t^{\mathbf{b}}\varphi - \langle\varphi,\mu^U\rangle\big\|_{L^2(\mu^U)} = 0 \\ \mu^U(\mathbb{R}^N) = \infty &\implies \lim_{t\to\infty} \big\|Q_t^{\mathbf{b}}\varphi\big\|_{L^2(\mu^U)} = 0. \end{aligned} \tag{7.5.25}$$

In fact, for each $r \in (0,\infty)$:

$$\begin{aligned} \mu^U(\mathbb{R}^N) = 1 &\implies \lim_{t\to\infty} \sup_{|\mathbf{x}|\le r} \big\|Q^{\mathbf{b}}(t,\mathbf{x},\,\cdot\,) - \mu^U\big\|_{\text{var}} = 0 \\ \mu^U(\mathbb{R}^N) = \infty &\implies \lim_{t\to\infty} \sup_{|\mathbf{x}|\le r} \big|Q^{\mathbf{b}}(t,\mathbf{x},B_{\mathbb{R}^N}(\mathbf{0},r))\big| = 0. \end{aligned} \tag{7.5.26}$$

Finally, if (cf. (7.5.14)) q^U is continuous (e.g., if (7.5.16) holds[†]*) on $(0,\infty)\times\mathbb{R}^N\times\mathbb{R}^N$, then, for all $(t,\mathbf{x},\mathbf{y}) \in (0,\infty)\times\mathbb{R}^N\times\mathbb{R}^N$:*

$$\|q^U(t,\,\cdot\,,\mathbf{x})\|_{L^2(\mu^U)} = \|q^U(t,\mathbf{x},\,\cdot\,)\|_{L^2(\mu^U)} = q^U(2t,\mathbf{x},\mathbf{x}), \tag{7.5.27}$$

[†] See the comment in the footnote to Theorem 7.5.13.

and, depending on whether $\mu^U(\mathbb{R}^N) = 1$ or $\mu^U(\mathbb{R}^N) = \infty$,

$$\text{(7.5.28)} \qquad \begin{aligned} \big|q^U(t+2,\mathbf{x},\mathbf{y}) - 1\big| &\le q^U(2,\mathbf{x},\mathbf{x})^{\frac{1}{2}} \big\|\overline{Q_t^U}\big(q^U(1,\,\cdot\,,\mathbf{y})\big) - 1\big\|_{L^2(\mu^U)} \\ \big|q^U(t+2,\mathbf{x},\mathbf{y})\big| &\le q^U(2,\mathbf{x},\mathbf{x})^{\frac{1}{2}} \big\|\overline{Q_t^U}\big(q^U(1,\,\cdot\,,\mathbf{y})\big)\big\|_{L^2(\mu^U)}. \end{aligned}$$

In particular, when q^U is continuous,

$$\text{(7.5.29)} \qquad \begin{aligned} \mu^U(\mathbb{R}^N) = 1 &\implies \lim_{t\to\infty} \sup_{|\mathbf{x}|\vee|\mathbf{y}|\le r} \big|q^U(t,\mathbf{x},\mathbf{y}) - 1\big| = 0 \\ \mu^U(\mathbb{R}^N) = \infty &\implies \lim_{t\to\infty} \sup_{|\mathbf{x}|\vee|\mathbf{y}|\le r} \big|q^U(t,\mathbf{x},\mathbf{y})\big| = 0. \end{aligned}$$

PROOF: We begin by noting that, because of the spectral representation in (7.5.21), the first assertion is equivalent to

$$\left.\begin{aligned} \mu^U(\mathbb{R}^N) = \infty &\implies E_0^U f = \mathbf{0} \\ \mu^U(\mathbb{R}^N) = 1 &\implies E_0^U f = \int f\,d\mu^U \end{aligned}\right. \qquad \text{for } f \in L^2(\mu;\mathbb{R}).$$

Alternatively, we must show that

$$\text{(7.5.30)} \qquad L^2(\mu^U;\mathbb{R}) \ni f = E_0^U f \implies f \text{ is } \mu^U\text{-a.e. constant.}$$

Thus, suppose that $f = E_0^U f$ for some $f \in L^2(\mu^U;\mathbb{R})$ with $\|f\|_{L^2(\mu^U)} = 1$. By (7.5.21), we know that

$$\text{(7.5.31)} \qquad f = \overline{Q_t^U} f \quad \mu^U\text{-a.e. for each } t \in (0,\infty).$$

We next note that

$$\mu^U(\{f > 0\}) = 0 \quad \text{or} \quad \mu^U(\{f < 0\}) = 0.$$

Indeed, because $q^U(1,\mathbf{x},\mathbf{y}) > 0$ for all $\mathbf{x}$ and $\mathbf{y}$, (7.5.31) would otherwise lead to

$$\begin{aligned} 1 = \|f\|^2_{L^2(\mu^U)} &= \big(\overline{Q_1^U} f, f\big)_{L^2(\mu^U)} \\ &= \iint f(\mathbf{x})\, q^U(1,\mathbf{x},\mathbf{y})\, f(\mathbf{y})\, \mu^U(d\mathbf{x})\mu^U(d\mathbf{y}) \\ &< \iint |f(\mathbf{x})|\, q^U(1,\mathbf{x},\mathbf{y})\, |f(\mathbf{y})|\, \mu^U(d\mathbf{x})\mu^U(d\mathbf{y}) \\ &= \Big(\overline{Q_1^U}(|f|), |f|\Big)_{L^2(\mu^U)} \le \|f\|^2_{L^2(\mu^U)} = 1. \end{aligned}$$

Hence, without loss in generality, we will assume that $f \geq 0$, which, because $q^U(1,\mathbf{x},\mathbf{y}) > 0$, means that

$$f(\mathbf{x}) = \int f(\mathbf{y})\, q^U(1,\mathbf{x},\mathbf{y})\, \mu^U(d\mathbf{y}) > 0 \quad \text{for } \mu^U\text{-a.e. } \mathbf{x} \in \mathbb{R}^N.$$

In particular, we now know that Range(E_0^U) cannot contain nonzero orthogonal elements and is therefore at most one-dimensional. Hence, as $t \to \infty$,

$$\overline{Q_t^U}\varphi \longrightarrow E_0^U\varphi = \big(\varphi, f\big)_{L^2(\mu^U)}\, f, \quad \varphi \in L^2(\mu^U). \tag{7.5.32}$$

But, from (7.5.32), we see that

$$\big(\varphi, f\big)_{L^2(\mu^U)}\, f(\mathbf{x}) \leq \|\varphi\|_{L^\infty(\mu^U)}, \quad \text{for } \mu\text{-a.e. } \mathbf{x} \text{ and every } \varphi \in B_c(\mathbb{R}^N;\mathbb{R}),$$

from which it follows that

$$\|f\|_{L^1(\mu^U)}\, \|f\|_{L^\infty(\mu^U)} \leq 1 = \|f\|^2_{L^2(\mu^U)}$$

and therefore that $f = \|f\|_{L^\infty(\mu^U)}$ μ^U–almost everywhere.

Given the preceding, (7.5.25) is an easy application of (7.5.21). As for (7.5.26), note that, by Lemma 7.3.20 and the fact that $Q_{t+1}^U = Q_1^U \circ Q_t^U$:

$$\big\{Q_t^U\varphi \restriction B_{\mathbb{R}^N}(\mathbf{0},r) : t \in (1,\infty) \text{ and } \|\varphi\|_{\mathrm{u}} \leq 1\big\}$$

is uniformly equicontinuous for each $r \in (0,\infty)$. Hence, if $\mu^U(\mathbb{R}^N) = \infty$ and therefore $Q_t^U \mathbf{1}_{B_{\mathbb{R}^N}(\mathbf{0},r)} \longrightarrow 0$ in $L^2(\mu^U)$, it is clear that the second line in (7.5.26) holds. On the other hand, if $\mu^U(\mathbb{R}^N) = 1$, then the same argument shows that, for each $\varphi \in B(\mathbb{R}^N;\mathbb{R})$, $Q_t^U\varphi \longrightarrow \langle\varphi,\mu^U\rangle$ uniformly on compacts. In order to improve this to the convergence in variation statement in (7.5.26), first note that, from the preceding, for each $r \in (0,\infty)$:

$$\varlimsup_{t\to\infty} \sup_{|\mathbf{x}|\leq r} Q^U\big(t,\mathbf{x}, B_{\mathbb{R}^N}(\mathbf{0},R)\complement\big) \leq \mu^U\big(B_{\mathbb{R}^N}(\mathbf{0},R)\complement\big) \longrightarrow 0 \quad \text{as } R \to \infty;$$

and so, if $\{\varphi_n\}_1^\infty \subseteq C_b(\mathbb{R}^N;\mathbb{R})$ is bounded and $\varphi_n \longrightarrow \varphi$ uniformly on compacts, then

$$\lim_{n\to\infty} \lim_{T\to\infty} \sup_{(t,\mathbf{x})\in[T,\infty)\times B_{\mathbb{R}^N}(\mathbf{0},r)} \big|\big[Q_t^U\varphi_n\big](\mathbf{x}) - \big[Q_t^U\varphi\big](\mathbf{x})\big| = 0.$$

Now suppose that the first part of (7.5.26) were not true. Then there would exist an increasing sequence $\{t_n\}_1^\infty \subseteq (1,\infty)$ tending to infinity, a sequence $\{\mathbf{x}_n\}_1^\infty \subseteq B_{\mathbb{R}^N}(\mathbf{0},r)$, and a sequence $\{\varphi_n\}_1^\infty \subseteq B(\mathbb{R}^N;\mathbb{R})$ satisfying

$$\|\varphi_n\|_{\mathrm{u}} \leq 1 \quad \text{and} \quad \big|\big[Q_{t_n}^U\varphi_n\big](\mathbf{x}_n) - \langle\varphi_n,\mu^U\rangle\big| \geq \epsilon$$

for all $n \in \mathbb{Z}^+$ and some $\epsilon > 0$. After passing to a subsequence if necessary, we could and would assume that $\mathbf{x}_n \longrightarrow \mathbf{x}$ for some $\mathbf{x} \in \overline{B_{\mathbb{R}^N}(\mathbf{0}, r)}$ and that there exists (cf. the remark with which this discussion began) a $\varphi \in C_b(\mathbb{R}^N; \mathbb{R})$ to which $\{Q_1^U \varphi_n\}_1^\infty$ converges uniformly on compacts. But, in view of the preceding, this would mean that

$$\begin{aligned}\varlimsup_{n\to\infty} \left|[Q_{t_n}^U \varphi](\mathbf{x}) - \langle \varphi, \mu^U\rangle\right| &\geq \varlimsup_{n\to\infty} \left|[Q_{t_n-1}^U(Q_1^U\varphi_n)](\mathbf{x}_n) - \langle Q_1^U\varphi_n, \mu^U\rangle\right| \\ &= \varlimsup_{n\to\infty} \left|[Q_{t_n}^U\varphi_n](\mathbf{x}_n) - \langle \varphi_n, \mu^U\rangle\right| \geq \epsilon,\end{aligned}$$

which obviously contradicts the fact that $Q_t^U \varphi \longrightarrow \langle \varphi, \mu^U\rangle$ uniformly on compacts. In other words, we have now proved that, for each $r \in (0, \infty)$,

$$\lim_{t\to\infty} \sup_{|\mathbf{x}|\leq r} \sup_{\|\varphi\|_u \leq 1} \left|[Q^U\varphi](\mathbf{x}) - \langle \varphi, \mu^U\rangle\right| = 0;$$

and this is equivalent to the first line in (7.5.26).

Finally, we turn to the case when q^U is continuous. The key to (7.5.27) and (7.5.28) is the observation that, in terms of q^U, the Chapman–Kolmogorov equation becomes

$$q^U(s+t, \mathbf{x}, \mathbf{y}) = \int q^U(s, \mathbf{x}, \boldsymbol{\xi})\, q^U(t, \boldsymbol{\xi}, \mathbf{y})\, \mu^U(d\boldsymbol{\xi});$$

and so, by the symmetry of $q^U(t, \mathbf{x}, \mathbf{y})$, (7.5.27) follows when the preceding is evaluated at the diagonal. In particular, if $f_{\mathbf{y}} \equiv q^U(1, \cdot, \mathbf{y})$, then, $\mathbf{y} \in \mathbb{R}^N \longmapsto f_{\mathbf{y}} \in L^2(\mu^U)$ is continuous; and so, for each $r \in (0, \infty)$, $\{f_{\mathbf{y}} : |\mathbf{y}| \leq R\}$ is compact in $L^2(\mu^U)$. Hence, by (7.5.25),

$$\lim_{t\to\infty} \sup_{|\mathbf{y}|\leq r} \left\|\overline{Q_t^U} f_{\mathbf{y}} - \kappa\right\|_{L^2(\mu^U)} = 0,$$

where $\kappa = 0$ or 1 depending on whether μ^U is infinite or finite. Thus, all that remains is to check (7.5.28). But, by the Chapman–Kolmogorov equation,

$$q^U(t+2, \mathbf{x}, \mathbf{y}) - \kappa = \int q^U(1, \mathbf{x}, \boldsymbol{\xi}) \left([\overline{Q_t^U} f_{\mathbf{y}}](\boldsymbol{\xi}) - \kappa\right) \mu^U(d\boldsymbol{\xi}),$$

from which (7.5.28) follows by Shwartz's inequality and (7.5.27). □

There are circumstances in which the qualitative conclusions drawn in (7.5.25) and (7.5.29) can be replaced by quantitative ones which provide a rate at which the convergence is taking place. For example, under quite general conditions, one can show that the operators $\overline{Q_t^U}$ must be compact and therefore that their spectrum must be completely discrete. Thus, if, in addition, one knows that $\mu^U(\mathbb{R}^N)=\infty$, then, because $E_0^U=0$, one can say that

$$e^{-\lambda_0}\equiv\sup\Big\{\big(\overline{Q_1^U}\varphi,\varphi\big)_{L^2(\mu^U)}:\varphi\in C_c(\mathbb{R}^N;\mathbb{R})\ \text{with}\ \|\varphi\|_{L^2(\mu^U)}=1\Big\}<1 \tag{7.5.33}$$

or, equivalently, that

$$\overline{Q_t^U}\varphi=\int_{[\lambda_0,\infty)}e^{-\lambda t}\,dE_\lambda^U\varphi\quad\text{for } t\in(0,\infty)\text{ and }\varphi\in L^2(\mu^U;\mathbb{R});$$

and therefore that

$$\big\|\overline{Q_t^U}\varphi\big\|_{L^2(\mu^U)}\le e^{-\lambda_0 t}\|\varphi\|_{L^2(\mu^U)},\quad t\in(0,\infty), \tag{7.5.34}$$

for all $\varphi\in L^2(\mu^U;\mathbb{R})$. On the other hand, when the $\overline{Q_t^U}$'s are compact and $\mu^U(\mathbb{R}^N)=1$, then, because $E_0^U\varphi=\langle\varphi,\mu^U\rangle$, we know that

$$e^{-\lambda_1}\equiv\sup\Big\{\big(\overline{Q_1^U}\varphi,\varphi\big)_{L^2(\mu^U)}:\ \varphi\in C_c(\mathbb{R}^N;\mathbb{R})\ \text{with}\ \|\varphi\|_{L^2(\mu^U)}=1\text{ and }\langle\varphi,\mu^U\rangle=0\Big\}<1 \tag{7.5.35}$$

and therefore that

$$\Big\|\overline{Q_t^U}\varphi-\langle\varphi,\mu^U\rangle\Big\|_{L^2(\mu^U)}\le e^{-\lambda_1 t}\|\varphi\|_{L^2(\mu^U)},\quad t\in(0,\infty), \tag{7.5.36}$$

for all $\varphi\in L^2(\mu^U;\mathbb{R})$. Finally, when q^U is continuous, notice that (7.5.34), (7.5.36), and (7.5.28) lead to the following quantitative version of (7.5.29):

$$\begin{aligned}\big|q^U(t+2,\mathbf{x},\mathbf{y})-1\big|&\le e^{-\lambda_1 t}\sqrt{q^U(2,\mathbf{x},\mathbf{x})q^U(2,\mathbf{y},\mathbf{y})}\\ \big|q^U(t+2,\mathbf{x},\mathbf{y})\big|&\le e^{-\lambda_0 t}\sqrt{q^U(2,\mathbf{x},\mathbf{x})q^U(2,\mathbf{y},\mathbf{y})},\end{aligned} \tag{7.5.37}$$

depending on whether $\mu^U(\mathbb{R}^N)=1$ or $\mu^U(\mathbb{R}^N)=\infty$.

As the preceding discussion makes clear, it is of considerable interest to know when the operators $\overline{Q_t^U}$'s are compact; and for this reason we give the following simple criterion.

7.5.38 Corollary. *For each $T\in(0,\infty)$:*

$$\big\|\overline{Q_T^U}\big\|_{\text{H.S.}(L^2(\mu^U))}\le(4\pi T)^{-\frac{N}{4}}\Big\|e^{TV^U}\Big\|_{L^2(\mathbb{R}^N)}, \tag{7.5.39}$$

where $\|A\|_{\text{H.S.}(E)}$ denotes the Hilbert–Schmidt norm of an operator A on a separable Hilbert E. In particular, if $e^{\rho V^U}\in L^1(\mathbb{R}^N;\mathbb{R})$ for some $\rho\in(0,\infty)$, then, depending on whether $\mu^U(\mathbb{R}^N)=\infty$ or $\mu^U(\mathbb{R}^N)=1$, either the λ_0 in (7.5.33) *is positive and* (7.5.34) *holds or the λ_1 in* (7.5.35) *is positive and* (7.5.36) *holds. Moreover, if, in addition, q^U is continuous, then* (7.5.37) *holds.*

PROOF: We need only prove (7.5.39). To this end, we use the Chapman–Kolmogorov equation and the symmetry of $q^U(t,\mathbf{x},\mathbf{y})$ to obtain

$$\begin{aligned}
\|Q_T^U\|^2_{\mathrm{H.S.}(L^2(\mu^U))} &= \iint_{\mathbb{R}^N\times\mathbb{R}^N} q^U(T,\mathbf{x},\mathbf{y})^2\,\mu^U(d\mathbf{x})\mu^U(d\mathbf{y})\\
&= \int_{\mathbb{R}^N} q^U(2T,\mathbf{x},\mathbf{x})\,\mu^U(d\mathbf{x}) = (4\pi T)^{-\frac{N}{2}}\int_{\mathbb{R}^N} r^{V^U}(2T,\mathbf{x},\mathbf{x})\,\mu^U(d\mathbf{x})\\
&= (4\pi T)^{-\frac{N}{2}}\int_{\mathbb{R}^N} \mathbb{E}^{\mathcal{W}^{(N)}}\left[\exp\left(\int_0^{2T} V^U\big(\mathbf{x}+\tilde{\psi}_T(s)\big)\right)\right]d\mathbf{x}\\
&\le (4\pi T)^{-\frac{N}{2}}\frac{1}{2T}\int_0^{2T}\mathbb{E}^{\mathcal{W}^{(N)}}\left[\int_{\mathbb{R}^N}\exp\Big(2TV^U\big(\mathbf{x}+\tilde{\psi}_T(s)\big)\Big)\,d\mathbf{x}\right]\\
&= (4\pi T)^{-\frac{N}{2}}\int_{\mathbb{R}^N} e^{2TV^U(\mathbf{x})}\,d\mathbf{x},
\end{aligned}$$

where $\tilde{\psi}_T(s) = \psi(s) - \frac{s\wedge T}{T}\psi(T)$ and, in the passage to the fourth line we have used Jensen's Inequality and Tonelli's Theorem. □

Exercises

7.5.40 Exercise: As some readers may have already guessed, the archetypical example of the situation discussed in Theorems 7.5.7, 7.5.13, and their corollaries is the one when U has the form

$$U_\alpha(\boldsymbol{\xi}) = \frac{\alpha|\boldsymbol{\xi}|^2}{4} + \tfrac{N}{4}\log\frac{2\pi}{|\alpha|}\quad\text{for some } \alpha\in\mathbb{R}\setminus\{0\}.$$

Indeed, notice that, by the criterion given at the beginning of Corollary 7.3.14, there is no question that (7.5.12) is satisfied for every choice of α. In addition, for each $\alpha\in\mathbb{R}\setminus\{0\}$, (7.5.16) is satisfied and $e^{V^{U_\alpha}}$ is integrable. Hence, by Theorem 7.5.13, we know that $q^{U_\alpha}\in C\big((0,\infty)\times\mathbb{R}^N\times\mathbb{R}^N;(0,\infty)\big)$ and, by Corollary 7.5.38, that, depending on whether $\alpha>0$ or $\alpha<0$, there exist strictly positive $\lambda_1(\alpha)$ or $\lambda_0(\alpha)$ such that

$$\big|q^{U_\alpha}(t+2,\mathbf{x},\mathbf{y})-1\big| \le e^{-\lambda_1(\alpha)t}\sqrt{q^{U_\alpha}(2,\mathbf{x},\mathbf{x})\,q^{U_\alpha}(2,\mathbf{y},\mathbf{y})} \tag{7.5.41}$$

or

$$q^{U_\alpha}(t+2,\mathbf{x},\mathbf{y}) \le e^{-\lambda_0(\alpha)t}\sqrt{q^{U_\alpha}(2,\mathbf{x},\mathbf{x})\,q^{U_\alpha}(2,\mathbf{y},\mathbf{y})} \tag{7.5.42}$$

for all $t\in(0,\infty)$.

Actually, when $\alpha > 0$, these examples were examined in Exercise 4.3.52, where it was shown that $q^{U_\alpha}(t,\mathbf{x},\mathbf{y})$ is given by the expression in (4.3.53). However, at that time we did not know yet about the connection between the measures $Q^{U_\alpha}_{\mathbf{x}}$ and the integral equation

$$X^{U_\alpha}(t,\psi) = \psi(t) - \frac{\alpha}{2}\int_0^t X^{U_\alpha}(s,\psi)\,ds, \quad (t,\psi) \in [0,\infty) \times \mathfrak{P}(\mathbb{R}^N),$$

and so we were not able to handle $\alpha \in (-\infty, 0)$. But, now that we know this connection, we can not only treat negative α's but can also give an entirely different, and probabilistically more appealing, derivation of the expression for q^{U_α}.

(i) Notice that when ψ is smooth, then the integral equation for $X^{U_\alpha}(\,\cdot\,,\psi)$ can be converted into the simple linear, ordinary differential equation whose solution is

$$X^{U_\alpha}(t,\psi) = e^{-\frac{\alpha t}{2}}\psi(0) + e^{-\frac{\alpha t}{2}}\int_0^t e^{\frac{\alpha s}{2}}\,\dot{\psi}(s)\,ds,$$

which, after integration by parts, becomes

$$X^{U_\alpha}(t,\psi) = \psi(t) - \frac{\alpha}{2}\int_0^t e^{\frac{\alpha(s-t)}{2}}\,\psi(s)\,ds. \tag{7.5.43}$$

Next, verify that, even when ψ is not smooth, (7.5.43) holds nonetheless.

(ii) Starting from (7.5.43) and remembering that $\left\{\left(\boldsymbol{\xi}, \psi(t)\right)_{\mathbb{R}^N} : (t,\boldsymbol{\xi}) \in [0,\infty) \times \mathbb{R}^N\right\}$ is a centered Gaussian family (cf. Exercise 4.2.39) under $\mathcal{W}^{(N)}$, conclude that, for each $(t,\mathbf{x}) \in (0,\infty) \times \mathbb{R}^N$, the distribution of $\psi \in \mathfrak{P}(\mathbb{R}^N) \longmapsto X^{U_\alpha}(t,\psi) \in \mathbb{R}^N$ under $\mathcal{W}^{(N)}_{\mathbf{x}}$ is Gaussian with mean $e^{-\frac{\alpha t}{2}}\mathbf{x}$ and covariance $\frac{1-e^{-\alpha t}}{\alpha}\mathbf{I}_{\mathbb{R}^N}$. In particular, when $\alpha \in (0,\infty)$, use this to verify that q^{U_α} is given by the formula in (4.3.53); and when $\alpha \in (-\infty,0)$, show that $q^{U_\alpha}(t,\mathbf{x},\mathbf{y})$ is equal to

$$\left(e^{|\alpha| t} - 1\right)^{-\frac{N}{2}} \exp\left[\frac{\alpha}{2}\,\frac{e^{|\alpha| t}|\mathbf{x}|^2 - 2e^{\frac{|\alpha| t}{2}}\left(\mathbf{x},\mathbf{y}\right)_{\mathbb{R}^N} + e^{|\alpha| t}|\mathbf{y}|^2}{e^{|\alpha| t} - 1}\right].$$

(iii) Again assume that $\alpha \neq 0$. As we already observed, Corollary 7.5.38 guarantees that $\overline{Q^{U_\alpha}_t}$ is compact for all $t \in (0,\infty)$. In fact, when $\alpha \in (0,\infty)$, note that $\mu^{U_\alpha} = \gamma_\alpha{}^N$ and that

$$q^{U_\alpha}(t,\mathbf{x},\mathbf{y}) = \prod_{j=1}^N M\left(\alpha^{\frac{1}{2}}x_j, \alpha^{\frac{1}{2}}y_j, e^{-\alpha t}\right),$$

where M is the Mehler kernel which appears in (2.3.6), and use this to conclude that

$$\overline{Q_t^{U_\alpha}}\varphi = \sum_{\mathbf{n}\in\mathbb{N}^N} e^{-\alpha\|\mathbf{n}\|t}\left(\overline{\mathbf{H}_{\mathbf{n}}^{(\alpha)}},\varphi\right)_{L^2(\gamma_\alpha{}^N)}\overline{\mathbf{H}_{\mathbf{n}}^{(\alpha)}},$$

where $\|\mathbf{n}\| = \sum_1^N n_j$,

$$\overline{\mathbf{H}_{\mathbf{n}}^{(\alpha)}}(\mathbf{x}) = \prod_{j=1}^N \frac{1}{\sqrt{n_j!}} H_{n_j}\left(\sqrt{\alpha}\, x_j\right),$$

and H_n is the Hermite polynomial given by (2.3.1). Hence, when $\alpha > 0$, $\lambda_1(\alpha) = \alpha$ and (7.5.41) becomes

$$\left|q^{U_\alpha}(t+2,\mathbf{x},\mathbf{y}) - 1\right| \le \left(1-e^{-2\alpha}\right)^{-\frac{N}{2}} \exp\left[\frac{\alpha\left(|\mathbf{x}|^2+|\mathbf{y}|^2\right)}{2\left(e^\alpha+1\right)}\right] e^{-\alpha t}.$$

(iv) Now suppose that $\alpha < 0$. Using the result obtained in **(ii)**, note that

$$q^{U_\alpha}(t,\mathbf{x},\mathbf{y}) = e^{\frac{N\alpha t}{2}} e^{\frac{\alpha|\mathbf{x}|^2}{2}} q^{U_{-\alpha}}(t,\mathbf{x},\mathbf{y})\, e^{\frac{\alpha|\mathbf{y}|^2}{2}},$$

and conclude that

$$\overline{Q_t^{U_\alpha}}\varphi = \sum_{\mathbf{n}\in\mathbb{N}^N} e^{\frac{N+2\|\mathbf{n}\|}{2}\alpha t}\left(\overline{\mathbf{H}_{\mathbf{n}}^{(\alpha)}},\varphi\right)_{L^2(\mu^U)}\overline{\mathbf{H}_{\mathbf{n}}^{(\alpha)}},$$

where

$$\overline{\mathbf{H}_{\mathbf{n}}^{(\alpha)}}(\mathbf{x}) = e^{\frac{\alpha|\mathbf{x}|^2}{2}}\,\overline{\mathbf{H}_{\mathbf{n}}^{(|\alpha|)}}(\mathbf{x}).$$

In particular, use this to see that $\lambda_0(\alpha) = \frac{|\alpha|N}{2}$ and (7.5.42) becomes

$$\left|q^{U_\alpha}(t+2,\mathbf{x},\mathbf{y})\right| \le \left(e^{2|\alpha|}-1\right)^{-\frac{N}{2}} \exp\left[\frac{\alpha\left(|\mathbf{x}|^2+|\mathbf{y}|^2\right)}{2\left(e^\alpha+1\right)}\right] e^{-\frac{|\alpha|Nt}{2}}.$$

7.5.44 Exercise: Return to the general setting described at the beginning of Section 7.4. Just as the $\{P_t : t > 0\}$-invariance relation $\mu = \mu P_t$, $t \in (0,\infty)$ has the consequence that $\{\Sigma_t : t \in [0,\infty)\}$ is the P_μ-invariant, so the $\{P_t : t > 0\}$-symmetry relation

$$\int \psi P_t\varphi\, d\mu = \int \varphi P_t\psi\, d\mu, \quad \varphi,\ \psi \in B\left(\mathbb{R}^N;[0,\infty)\right) \tag{7.5.45}$$

has implications about P_μ. Namely, show that (7.5.45) implies that the process is **reversible** in the sense that, for each $T \in (0,\infty)$,

$$\psi \in \mathfrak{P}(\mathbb{R}^N) \longmapsto \psi \restriction [0,T] \in C\left([0,T];\mathbb{R}^N\right)$$

and

$$\psi \in \mathfrak{P}(\mathbb{R}^N) \longmapsto \check{\psi}_T \equiv \psi(T-\cdot) \in C\left([0,T];\mathbb{R}^N\right)$$

have the same distribution under P_μ.

Hint: First observe that it is sufficient to show that, for each $n \in \mathbb{Z}^+$, $0 = t_0 < \cdots < t_n = T$, and $f_0, \dots, f_n \in B_c(\mathbb{R}^N; \mathbb{R})$,

$$
\begin{aligned}
&\int_{\mathfrak{P}(\mathbb{R}^N)} \prod_{m=0}^{n} f_m\big(\psi(t_m)\big)\, P_\mu(d\psi) \\
(7.5.46)\qquad &\qquad = \int_{\mathfrak{P}(\mathbb{R}^N)} \prod_{m=0}^{n} f_m\big(\psi(T - t_m)\big)\, P_\mu(d\psi).
\end{aligned}
$$

Next, note that when $n = 1$, (7.5.46) is simply a restatement of (7.5.45). Finally, use induction on n and the Markov property to prove (7.5.46) for general $n \in \mathbb{Z}^+$.

7.5.47 Exercise: Show that, for any $(T, \mathbf{x}) \in (0, \infty) \times \mathbb{R}^N$ and $F \in B\big(\mathfrak{P}(\mathbb{R}^N); \mathbb{R}\big) \cap L^1\big(Q^U_{\mu^U}; \mathbb{R}\big)$:

$$
\lim_{n \to \infty} \frac{1}{n} \sum_{m=1}^{n-1} F \circ \boldsymbol{\Sigma}_{mT} = \begin{cases} \mathbb{E}^{Q^U_{\mu^U}}[F] & \text{if } \mu^U(\mathbb{R}^N) = 1 \\ 0 & \text{otherwise} \end{cases} \qquad \big(\text{a.s., } Q^{\mathbf{b}}_{\mathbf{x}}\big).
$$

Hint: See Corollary 7.5.24 and Exercise 7.4.25.

Chapter VIII:
A Little Classical Potential Theory

§8.1: The Dirichlet Heat Kernel.

In this chapter we will be exploiting a few of the deep and beautiful relations which exist between Wiener's measure and the potential theory of the Euclidean Laplacian $\mathbf{\Delta}$.† We have already seen the origins of this relationship: namely, the fact stated in (7.1.22). Actually, (7.1.22) is itself simply the probabilistic counterpart of the purely analytic fact that the Gauss kernel γ_t^N is the *fundamental solution to the heat equation*:

$$(8.1.1) \qquad \frac{\partial \gamma_t^N}{\partial t} = \frac{1}{2}\mathbf{\Delta}\gamma_t^N \quad \text{and} \quad {\gamma_t}^N \Longrightarrow \delta_{\mathbf{0}} \text{ as } t \searrow 0;$$

and, as we will see, (7.1.22) is only the first remarkable conclusion which can be drawn by performing a clever dance with one foot in analysis and the other in probability theory. In particular, given an open set $\mathfrak{G}$ in $\mathbb{R}^N$, we will use $\mathfrak{e}^{\mathfrak{G}}(\psi)$ to denote the first exit time $\inf\{t \geq 0 : \psi(t) \notin \mathfrak{G}\}$ of $\psi \in \mathfrak{P}(\mathbb{R}^N)$ from $\mathfrak{G}$, and what we will do is use the distribution of

$$\psi \in \{\mathfrak{e}^{\mathfrak{G}} < \infty\} \longmapsto \left(\mathfrak{e}^{\mathfrak{G}}(\psi), \psi(\mathfrak{e}^{\mathfrak{G}})\right) \in [0,\infty] \times \partial\mathfrak{G}$$

to represent the solution to various boundary value problems for the operator $\frac{1}{2}\mathbf{\Delta}$ in $\mathfrak{G}$.

We begin our program with an important *a priori* regularity result, which, in turn, will require us to know the following simple fact from interpolation theory.

8.1.2 Lemma. *If $n \in \mathbb{Z}^+$ and $F \in C_{\mathrm{b}}^{n+1}\big(B_{\mathbb{R}^N}(\mathbf{0},1);\mathbb{R}\big)$, then*

$$\max_{|\alpha|=n}\left|\frac{\partial^{|\alpha|}F}{\partial \mathbf{x}^\alpha}(\mathbf{x})\right| \leq 3\|F\|_{C_{\mathrm{b}}(B_{\mathbb{R}^N}(\mathbf{0},1);\mathbb{R})}^{2^{-n}}\,\|F\|_{C_{\mathrm{b}}^{n+1}(B_{\mathbb{R}^N}(\mathbf{0},1);\mathbb{R})}^{1-2^{-n}}$$

for all $\mathbf{x} \in B_{\mathbb{R}^N}(\mathbf{0},1)$.

† For a truly complete account of the subject, the reader should consult J.L. Doob's *Classical Potential Theory and Its Probabilistic Counterpart*, publ. by Springer–Verlag in 1984. After all, Doob more or less invented the subject.

PROOF: First suppose that $n = N = 1$ and use a two place Taylor's expansion to see that

$$\left|F'(t)\right| \leq \frac{2\|F\|_{\mathrm{u}}}{r} + \frac{r\|F''\|_{\mathrm{u}}}{2}, \quad t \in (-1,1) \text{ and } r \in (0,1].$$

Thus, by taking

$$r^2 = \frac{\|F\|_{\mathrm{u}}}{\|F\|_{C^2_{\mathrm{b}}((-1,1);\mathbb{R})}},$$

we arrive at $\|F'\|_{\mathrm{u}} \leq 3\|F\|_{\mathrm{u}}\|F\|_{C^2_{\mathrm{b}}((-1,1);\mathbb{R})}$ in this special case. To get the general result, assume that the result holds for n, let $\boldsymbol{\alpha} \in \mathbb{N}^N$ be given, apply the preceding to see that

$$\left|\frac{\partial^{|\boldsymbol{\beta}|}F}{\partial \mathbf{x}^{\boldsymbol{\beta}}}(\mathbf{x})\right| \leq 3\left\|\frac{\partial^{|\boldsymbol{\alpha}|}F}{\partial \mathbf{x}^{\boldsymbol{\alpha}}}\right\|_{\mathrm{u}}^{\frac{1}{2}} \|F\|^{\frac{1}{2}}_{C^{n+1}_{\mathrm{b}}(B_{\mathbb{R}^N}(\mathbf{0},1);\mathbb{R})}$$

for any $\boldsymbol{\beta} \in \mathbb{N}^N$ obtained from $\boldsymbol{\alpha}$ by adding 1 to precisely one coordinate, and use the induction hypothesis to complete the proof. □

8.1.3 Theorem. *Let $f \in B\big([0,\infty) \times \partial\mathfrak{G};\mathbb{R}\big)$ be given, and set*

$$u^{\mathfrak{G}}_f(t,\mathbf{x}) = \mathbb{E}^{\mathcal{W}^{(N)}_{\mathbf{x}}}\Big[f\big(t - \mathfrak{e}^{\mathfrak{G}}, \psi(\mathfrak{e}^{\mathfrak{G}})\big),\, \mathfrak{e}^{\mathfrak{G}} \leq t\Big] \tag{8.1.4}$$

for all $(t,\mathbf{x}) \in [0,\infty) \times \mathfrak{G}$. Then $u \in C^\infty\big([0,\infty) \times \mathfrak{G};\mathbb{R}\big)$, for each $n \in \mathbb{Z}^+$, there exists an $M_n \in [1,\infty)$ for which

$$\max_{|\boldsymbol{\alpha}|=n}\left|\frac{\partial^{|\boldsymbol{\alpha}|}u^{\mathfrak{G}}_f}{\partial \mathbf{x}^{\boldsymbol{\alpha}}}(t,\mathbf{x})\right| \leq \frac{M_n\,\|f\|_{\mathrm{u}}}{\operatorname{dist}(\mathbf{x},\mathfrak{G})^n}\exp\left[-\frac{\operatorname{dist}(\mathbf{x},\partial\mathfrak{G})^2}{M_n t}\right], \tag{8.1.5}$$

and

$$\frac{\partial^n u^{\mathfrak{G}}_f}{\partial t^n} = \left(\tfrac{1}{2}\boldsymbol{\Delta}\right)^n u^{\mathfrak{G}}_f. \tag{8.1.6}$$

PROOF: We first want to reduce to the case when $\mathbf{x} = \mathbf{0}$ and $\mathfrak{G} = B \equiv B_{\mathbb{R}^N}(\mathbf{0},2)$. To this end, let $\mathbf{x} \in \mathfrak{G}$ be given and set $r = \frac{1}{2}\operatorname{dist}(\mathbf{x},\partial\mathfrak{G})$. Then, by the strong Markov property,

$$u^{\mathfrak{G}}_f(t,\mathbf{y}) = \mathbb{E}^{\mathcal{W}^{(N)}_{\mathbf{y}}}\Big[u^{\mathfrak{G}}_f\big(t - \mathfrak{e}^{B_{\mathbb{R}^N}(\mathbf{x},2r)}, \psi(\mathfrak{e}^{B_{\mathbb{R}^N}(\mathbf{x},2r)})\big),\, \mathfrak{e}^{B_{\mathbb{R}^N}(\mathbf{x},2r)} \leq t\Big].$$

Next, apply Wiener scaling and translation invariance and use the preceding to arrive at

$$u^{\mathfrak{G}}_f(r^2 t, \mathbf{x} + r\mathbf{y}) = u^{B}_f(t,\mathbf{y}), \quad (t,\mathbf{y}) \in [0,\infty) \times B,$$

where $\tilde{f}(t,\boldsymbol{\xi}) = u^{\mathfrak{G}}(r^2t, \mathbf{x} + r\boldsymbol{\xi})$ for $(t,\boldsymbol{\xi}) \in [0,\infty) \times \partial B$. Hence, without loss in generality, we may and will assume from now on that $\mathbf{x} = \mathbf{0}$ and that $\mathfrak{G} = B$.

Extend f to the whole of $\mathbb{R} \times \mathbb{R}^N$ by setting it equal to 0 off of $[0,\infty) \times \partial B$; and observe that then $u_f^B = f$ on $[0,\infty) \times \partial B$ and $u_f^B = 0$ off of $[0,\infty) \times \mathbb{R}^N$. Also, if

$$\mathfrak{e}_k(\boldsymbol{\psi}) = \inf\left\{t \ge \tfrac{1}{k} : |\boldsymbol{\psi}(t)| \ge 2\right\}$$

and

$$u_k(t,\mathbf{y}) = \mathbb{E}^{\mathcal{W}_{\mathbf{y}}^{(N)}}\left[f\big(t - \mathfrak{e}_k, \boldsymbol{\psi}(\mathfrak{e}_k)\big),\ \mathfrak{e}_k \le t\right], \quad (t,\mathbf{y}) \in \mathbb{R} \times \mathbb{R}^N,$$

then $u_k = 0$ off of $\left[\frac{1}{k},\infty\right) \times \mathbb{R}$ and

$$\left|u_k(t,\mathbf{y}) - u_f^B(t,\mathbf{y})\right| \vee \left|u_{k+1}(t,\mathbf{y}) - u_k(t,\mathbf{y})\right| \le 2\|f\|_{\mathrm{u}}\, \mathcal{W}_{\mathbf{y}}^{(N)}\left(\mathfrak{e}^B \le \tfrac{1}{k}\right).$$

In particular, this means that $u_k \longrightarrow u \equiv u_f^B$ uniformly on compact subsets of $[0,\infty) \times B$ and that (cf. (3.3.30))

$$\left|u_{k+1}(t,\mathbf{y}) - u_k(t,\mathbf{y})\right| \le 4N\|f\|_{\mathrm{u}} \exp\left[-\tfrac{k}{2N}\right], \quad (t,\mathbf{y}) \in [0,\infty) \times B_{\mathbb{R}^N}(\mathbf{0},1).$$

At the same time, because $\mathfrak{e}_k = \frac{1}{k} + \mathfrak{e}^B \circ \boldsymbol{\Sigma}_{\frac{1}{k}}$, the Markov property gives

$$u_k(t,\mathbf{y}) = \left[\gamma_{\frac{1}{k}}^N \bigstar u_f^B\left(t - \tfrac{1}{k}, \cdot\right)\right](\mathbf{y});$$

and therefore, for each $n \in \mathbb{Z}^+$, there is a $C_n \in (0,\infty)$ such that

$$\max_{|\alpha|=n} \left|\frac{\partial^{|\alpha|} u_k}{\partial \mathbf{y}^{\alpha}}(t,\mathbf{y})\right| \le \max_{|\alpha|=n} \left\|\frac{\partial^{|\alpha|} \gamma_{\frac{1}{k}}^N}{\partial \mathbf{y}^{\alpha}}\right\|_{L^1(\mathbb{R}^N)} \|f\|_{\mathrm{u}} \le C_n k^{\frac{n}{2}} \|f\|_{\mathrm{u}}.$$

Hence, by writing

$$u_f^B(t,\mathbf{y}) = \sum_{k \ge \frac{1}{t}} \big(u_{k+1}(t,\mathbf{y}) - u_k(t,\mathbf{y})\big)$$

when $t \in (0,1]$ and

$$u_f^B(t,\mathbf{y}) = u_1(t,\mathbf{y}) + \sum_{k=1}^{\infty} \big(u_{k+1}(t,\mathbf{y}) - u_k(t,\mathbf{y})\big)$$

when $t \in (1,\infty)$, we get the required estimate in (8.1.5) (with $\mathbf{x} = \mathbf{0}$ and $\mathfrak{G} = B$) as an application of the estimates already proved combined with Lemma 8.1.2.

To show that u_f^B is smooth as a function of t and, at the same time, derive (8.1.6), note that, for $h \in (0,1)$,

$$\begin{aligned}
u_f^B(t+h,\mathbf{x}) &= \mathbb{E}^{\mathcal{W}_{\mathbf{x}}^{(N)}}\Big[f\big(t+h-\mathfrak{e}^B,\psi(\mathfrak{e}^B)\big),\,\mathfrak{e}^B \le t+h\Big] \\
&= \mathbb{E}^{\mathcal{W}_{\mathbf{x}}^{(N)}}\Big[u_f^B\big(t,\psi(h)\big),\,\mathfrak{e}^B > h\Big] \\
&\quad + \mathbb{E}^{\mathcal{W}_{\mathbf{x}}^{(N)}}\Big[f\big(t+h-\mathfrak{e}^B,\psi(\mathfrak{e}^B)\big),\,\mathfrak{e}^B \le h\Big] \\
&= \big[\gamma_h^N \bigstar u_f^B(t,\,\cdot\,)\big](\mathbf{x}) \\
&\quad + \mathbb{E}^{\mathcal{W}_{\mathbf{x}}^{(N)}}\Big[f\big(t+h-\mathfrak{e}^B,\psi(\mathfrak{e}^B)\big) - u_f^B\big(t,\psi(h)\big),\,\mathfrak{e}^B \le h\Big];
\end{aligned}$$

and therefore (again by (3.3.30))

$$\Big|u_f^B(t+h,\mathbf{x}) - \big[\gamma_h^N \bigstar u_f^B(t,\,\cdot\,)\big](\mathbf{x})\Big| \le 4N\|f\|_{\mathrm{u}} \exp\left[-\frac{(2-|\mathbf{x}|)^2}{2Nh}\right]$$

for $\mathbf{x} \in B$. At the same time, from (8.1.1), we know that, for each $n \in \mathbb{N}$,

$$\frac{\partial^n \gamma_t}{\partial t^n} = \big(\tfrac{1}{2}\boldsymbol{\Delta}\big)^n \gamma_t;$$

and so, by Taylor's Theorem and elementary estimation, we see that, uniformly for $(t,\mathbf{x}) \in [0,\infty) \times B_{\mathbb{R}^N}(\mathbf{0},1)$,

$$u(t+h,\mathbf{x}) = \sum_{m=0}^{n} \frac{1}{m!}\Big[\big(\tfrac{h\boldsymbol{\Delta}}{2}\big)^m u(t,\,\cdot\,)\Big](\mathbf{x}) + o\big(h^{n+1}\big) \quad \text{as } h \searrow 0,$$

which proves not only that u_f^B is smooth as a function of t but that it satisfies (8.1.6) as well. □

As our first application of the preceding, we use it to study the Cauchy initial value problem for the heat equation in $[0,\infty) \times \mathfrak{G}$ with **Dirichlet data** (i.e., value 0) at the spatial boundary (cf. Theorem 8.1.23 below).

8.1.7 Corollary. *Define $P_t^{\mathfrak{G}} : B(\mathfrak{G};\mathbb{R}) \longrightarrow B(\mathfrak{G};\mathbb{R})$ for $t \in (0,\infty)$ by*

$$\big[P_t^{\mathfrak{G}} f\big](\mathbf{x}) = \mathbb{E}^{\mathcal{W}_{\mathbf{x}}^{(N)}}\Big[f\big(\psi(t)\big),\,\mathfrak{e}^{\mathfrak{G}} > t\Big], \quad \mathbf{x} \in \mathfrak{G}. \tag{8.1.8}$$

Then $P_{s+t}^{\mathfrak{G}} = P_s^{\mathfrak{G}} \circ P_t^{\mathfrak{G}}$ for all $s,\, t \in (0,\infty)$, and, for each $f \in B(\mathfrak{G};\mathbb{R})$, $(t,\mathbf{x}) \in (0,\infty) \times \mathfrak{G} \longmapsto \big[P_t^{\mathfrak{G}} f\big](\mathbf{x}) \in \mathbb{R}$ is a smooth function which satisfies

$$\frac{\partial P_t^{\mathfrak{G}} f}{\partial t} = \tfrac{1}{2}\boldsymbol{\Delta} P_t^{\mathfrak{G}} f \quad \text{on} \quad (0,\infty) \times \mathfrak{G}. \tag{8.1.9}$$

Next, define $p^{\mathfrak{G}} : (0,\infty) \times \mathfrak{G} \times \mathfrak{G} \longrightarrow \mathbb{R}$ *by*

$$\text{(8.1.10)} \qquad p^{\mathfrak{G}}(t,\mathbf{x},\mathbf{y}) = \gamma_t^N(\mathbf{y}-\mathbf{x}) - \mathbb{E}^{\mathcal{W}_{\mathbf{x}}^{(N)}}\left[\gamma^N_{t-\mathfrak{e}^{\mathfrak{G}}}\left(\mathbf{y}-\psi(\mathfrak{e}^{\mathfrak{G}})\right),\, \mathfrak{e}^{\mathfrak{G}} < t\right].$$

Then $p^{\mathfrak{G}} \in C^\infty\big((0,\infty)\times\mathfrak{G}\times\mathfrak{G};[0,\infty)\big)$ *and*

$$\text{(8.1.11)} \qquad \left[P_t^{\mathfrak{G}} f\right](\mathbf{x}) = \int_{\mathfrak{G}} f(\mathbf{y})\, p^{\mathfrak{G}}(t,\mathbf{x},\mathbf{y})\, d\mathbf{y}, \quad (t,\mathbf{x}) \in (0,\infty)\times\mathfrak{G},$$

for every $f \in B(\mathfrak{G};\mathbb{R})$. *Moreover, for all* $s,\ t \in (0,\infty)$ *and* $\mathbf{x},\ \mathbf{y} \in \mathfrak{G}$,

$$\text{(8.1.12)} \qquad p^{\mathfrak{G}}(s+t,\mathbf{x},\mathbf{y}) = \int_{\mathfrak{G}} p^{\mathfrak{G}}(s,\mathbf{x},\boldsymbol{\xi})\, p^{\mathfrak{G}}(t,\boldsymbol{\xi},\mathbf{y})\, d\boldsymbol{\xi}$$

$$\text{(8.1.13)} \qquad p^{\mathfrak{G}}(t,\mathbf{x},\mathbf{y}) = p^{\mathfrak{G}}(t,\mathbf{y},\mathbf{x});$$

and $p^{\mathfrak{G}}(t,\mathbf{x},\mathbf{y}) > 0$, $t \in (0,\infty)$, *so long as* $\mathbf{x}$ *and* $\mathbf{y}$ *lie in the same connected component of* $\mathfrak{G}$. *Finally,*

$$\text{(8.1.14)} \qquad \frac{\partial p^{\mathfrak{G}}}{\partial t}(t,\mathbf{x},\mathbf{y}) = \left[\tfrac{1}{2}\boldsymbol{\Delta}_{\mathbf{x}} p^{\mathfrak{G}}\right](t,\mathbf{x},\mathbf{y}) = \left[\tfrac{1}{2}\boldsymbol{\Delta}_{\mathbf{y}} p^{\mathfrak{G}}\right](t,\mathbf{x},\mathbf{y})$$

on $(0,\infty)\times\mathfrak{G}\times\mathfrak{G}$.

PROOF: The equation $P^{\mathfrak{G}}_{s+t} = P^{\mathfrak{G}}_s \circ P^{\mathfrak{G}}_t$ is an elementary expression of the Markov property. Next, given $f \in B(\mathfrak{G};\mathbb{R})$, extend f to $\mathbb{R}^N$ by setting it equal to 0 off of $\mathfrak{G}$, and define $F_t : \mathfrak{P}(\mathbb{R}^N)^2 \longrightarrow \mathbb{R}$ so that

$$F_t(\varphi,\psi) = \mathbf{1}_{[0,t]}\left(\mathfrak{e}^{\mathfrak{G}}(\varphi)\right) f\left(\psi(t-\mathfrak{e}^{\mathfrak{G}}(\varphi))\right).$$

Then, by the strong Markov property applied to $F = F_t$:

$$\begin{aligned}
&\left[P_t^{\mathfrak{G}} f\right](\mathbf{x}) \\
&\quad = \mathbb{E}^{\mathcal{W}_{\mathbf{x}}^{(N)}}\left[f(\psi(t))\right] - \mathbb{E}^{\mathcal{W}_{\mathbf{x}}^{(N)}}\left[f(\psi(t)),\, \mathfrak{e}^{\mathfrak{G}} \le t\right] \\
&\quad = \left[\gamma_t^N \star f\right](\mathbf{x}) \\
&\qquad - \int_{\{\varphi:\mathfrak{e}^{\mathfrak{G}}(\varphi)\le t\}} \left(\int_{\mathfrak{P}(\mathbb{R}^N)} f\left(\psi(t-\mathfrak{e}^{\mathfrak{G}}(\varphi)\right)\, \mathcal{W}^{(N)}_{\varphi(\mathfrak{e}^{\mathfrak{G}})}(d\psi)\right) \mathcal{W}_{\mathbf{x}}^{(N)}(d\varphi) \\
&\quad = \left[\gamma_t^N \star f\right](\mathbf{x}) - \mathbb{E}^{\mathcal{W}_{\mathbf{x}}^{(N)}}\left[\left[\gamma^N_{t-\mathfrak{e}^{\mathfrak{G}}} \star f\right]\left(\psi(\mathfrak{e}^{\mathfrak{G}})\right),\, \mathfrak{e}^{\mathfrak{G}} \le t\right].
\end{aligned}$$

In view of Theorem 8.1.3, this proves that

$$(t,\mathbf{x}) \in (0,\infty)\times\mathfrak{G} \longmapsto \left[P_t^{\mathfrak{G}} f\right](\mathbf{x}) \in \mathbb{R}$$

is a smooth function which satisfies (8.1.9). At the same time, it proves that (8.1.11) holds when $p^{\mathfrak{G}}$ is given by (8.1.10).

Turning to the other properties of $p^{\mathfrak{G}}$, note that

$$\frac{\partial^{|\beta|} p^{\mathfrak{G}}}{\partial \mathbf{y}^{\beta}}(t,\mathbf{x},\mathbf{y}) = \frac{\partial^{|\beta|}\gamma_t^N}{\partial \mathbf{y}^{\beta}}(\mathbf{y}-\mathbf{x}) - \mathbb{E}^{\mathcal{W}_{\mathbf{x}}^{(N)}}\left[\frac{\partial^{|\beta|}\gamma_{t-\mathfrak{e}^{\mathfrak{G}}}^N}{\partial \mathbf{y}^{\beta}}(\mathbf{y}-\psi(\mathfrak{e}^{\mathfrak{G}})),\ \mathfrak{e}^{\mathfrak{G}} \le t\right];$$

and therefore the regularity properties as well as (8.1.14) follow immediately from (8.1.1) and Theorem 8.1.3. Also, knowing that $p^{\mathfrak{G}}$ is smooth, one sees from (8.1.11) that it must be nonnegative and from $P_{s+t}^{\mathfrak{G}} = P_s^{\mathfrak{G}} \circ P_t^{\mathfrak{G}}$ that it satisfies (8.1.12). Next, let $(T,\mathbf{y}) \in (0,\infty)\times\mathfrak{G}$ be given, set $f(t,\mathbf{x}) = p^{\mathfrak{G}}(T-t,\mathbf{x},\mathbf{y})$ for $(t,\mathbf{x}) \in (-\infty,T)\times\mathfrak{G}$, and note that $\frac{\partial f}{\partial t} + \frac{1}{2}\Delta f = 0$ in $(-\infty,T)\times\mathfrak{G}$. Now suppose that $f(0,\mathbf{x}) = 0$ for some $\mathbf{x}$ in the same connected component of $\mathfrak{G}$ as $\mathbf{y}$, and choose a smooth $\psi \in \mathfrak{P}(\mathbb{R}^N)$ so that $\psi(0) = \mathbf{x}$, $\psi(T) = \mathbf{y}$, and $\psi(t) \in \mathfrak{G}$ for all $t \in [0,T]$. Then, by the second part of Theorem 7.2.8, $f\big(t,\psi(t)\big) = 0$ for all $t \in [0,T)$; which (cf. (8.1.10)) leads to the contradiction

$$0 = \lim_{t \nearrow T} p^{\mathfrak{G}}\big(T-t,\psi(t),\mathbf{y}\big) = \lim_{s \searrow 0}\gamma_s^N\big(\mathbf{y}-\psi(T-s)\big) = \infty.$$

Thus, all that remains is to check the symmetry property in (8.1.13); and, because of smoothness, this comes down to showing that $P_t^{\mathfrak{G}}$ is symmetric in the sense that

$$\big(g, P_t^{\mathfrak{G}} f\big)_{L^2(\mathfrak{G})} = \big(f, P_t^{\mathfrak{G}} g\big)_{L^2(\mathfrak{G})}$$

for all $f,\ g \in B_c(\mathfrak{G};\mathbb{R})$. But, by Theorem 4.2.18,

$$\big(g, P_t^{\mathfrak{G}} f\big)_{L^2(\mathfrak{G})} = \iint_{\mathfrak{G}\times\mathfrak{G}} g(\mathbf{x})f(\mathbf{y})\, r^{\mathfrak{G}}(t,\mathbf{x},\mathbf{y})\,\gamma_t^N(\mathbf{x}-\mathbf{y})\,d\mathbf{x}d\mathbf{y},$$

where

$$r^{\mathfrak{G}}(t,\mathbf{x},\mathbf{y}) = \mathcal{W}^{(N)}\Big(\big(1-\tfrac{s}{t}\big)\mathbf{x} + \tfrac{s}{t}\mathbf{y} + \tilde{\psi}_t(s) \in \mathfrak{G},\ s \in [0,t]\Big)$$

with $\tilde{\psi}_t(s) \equiv \psi(s) - \frac{s\wedge t}{t}\psi(t)$; and, by the reversibility proved in Theorem 4.2.18,

$$r^{\mathfrak{G}}(t,\mathbf{x},\mathbf{y}) = r^{\mathfrak{G}}(t,\mathbf{y},\mathbf{x}) \quad \text{for all } \mathbf{x},\ \mathbf{y} \in \mathfrak{G}. \quad \square$$

In Corollary 8.1.8 we learned that $(t,\mathbf{x}) \in (0,\infty)\times\mathfrak{G} \longmapsto \big[P_t^{\mathfrak{G}} f\big](\mathbf{x})$ solves the **heat equation** (8.1.9), and our next goal is to characterize analytically which solution it is. For this purpose, it will be useful to have the following variant of the result in Lemma 7.2.1.

8.1.15 Lemma. *For any* $(T,\mathbf{x}) \in (0,\infty) \times \mathfrak{G}$ *and* $f \in C^{1,2}\big((0,T) \times \mathfrak{G};\mathbb{R}\big) \cap C_b\big([0,T] \times \overline{\mathfrak{G}};\mathbb{R}\big)$ *with the property that* $\frac{\partial f}{\partial t} + \frac{1}{2}\mathbf{\Delta} f \geq 0$,

$$\Big(f\big(t \wedge T \wedge \mathfrak{e}^{\mathfrak{G}}, \psi(t \wedge T \wedge \mathfrak{e}^{\mathfrak{G}})\big), \mathcal{F}_t, \mathcal{W}_{\mathbf{x}}^{(N)}\Big)$$

is a bounded, continuous submartingale and therefore

$$f(0,\mathbf{x}) \leq \mathbb{E}^{\mathcal{W}_{\mathbf{x}}^{(N)}}\Big[f\big(T \wedge \mathfrak{e}^{\mathfrak{G}}, \psi(T \wedge \mathfrak{e}^{\mathfrak{G}})\big)\Big], \quad \mathbf{x} \in \mathfrak{G}. \tag{8.1.16}$$

In particular, if $\frac{\partial f}{\partial t} + \frac{1}{2}\mathbf{\Delta} f = 0$, *then this submartingale is a martingale and the inequality in* (8.1.16) *becomes an equality.*

PROOF: We want to reduce everything to the situation covered by Lemma 7.2.1 applied to subregions of $(-\infty,T) \times \mathfrak{G}$. To be precise, let $0 < s < T$ be given, set $\hat{\mathfrak{G}}_s = (-\infty, T-s) \times \mathfrak{G}$, and note that (cf. (7.2.2))

$$\zeta^{\hat{\mathfrak{G}}_s}(\psi) = (T-s) \wedge \mathfrak{e}^{\mathfrak{G}}(\psi).$$

Hence, by Lemma 7.2.1, we know that

$$f\Big(s + t \wedge (T-s) \wedge \mathfrak{e}^{\mathfrak{G}}, \psi\big(t \wedge (T-s) \wedge \mathfrak{e}^{\mathfrak{G}}\big)\Big)$$

is a bounded, continuous $\mathcal{W}_{\mathbf{x}}^{(N)}$-submartingale relative to $\big\{\mathcal{B}_t : t \in [0,\infty)\big\}$. Finally, let $s \searrow 0$. □

As an application of (8.1.16) we obtain the following uniqueness result.

8.1.17 Lemma. *If* u *is a bounded element of* $C^{1,2}\big((0,\infty) \times \mathfrak{G};\mathbb{R}\big)$ *which solves the Cauchy initial value problem*

$$\frac{\partial u}{\partial t} = \frac{1}{2}\mathbf{\Delta} u \quad \text{in } (0,\infty) \times \mathfrak{G}, \quad \lim_{t \searrow 0} u(t,\mathbf{x}) = f(\mathbf{x}), \quad \mathbf{x} \in \mathfrak{G},$$
$$\text{and} \quad \lim_{\mathbf{x} \to \mathbf{a}} u(t,\mathbf{x}) = 0, \quad (t,\mathbf{a}) \in (0,\infty) \times \partial G,$$

for some $f \in B(\mathfrak{G};\mathbb{R})$, *then* $u(t,\,\cdot\,) = P_t^{\mathfrak{G}} f$ *for all* $t \in (0,\infty)$.

PROOF: Given $T \in (0,\infty)$ and $\epsilon > 0$, set

$$f_{T,\epsilon}(t,\mathbf{x}) = u(T - t + \epsilon, \mathbf{x}), \quad \text{for } (t,\mathbf{x}) \in [0,T) \times \mathfrak{G},$$

note that equality holds in (8.1.16) with $f = f_{T,\epsilon}$, and conclude that

$$u(T+\epsilon,\mathbf{x}) = \big[P_T^{\mathfrak{G}} u(\epsilon,\,\cdot\,)\big](\mathbf{x}) \quad \text{for each } \epsilon \in (0,\infty).$$

Finally, let $\epsilon \searrow 0$ and use Lebesgue's Dominated Convergence Theorem to arrive at $u(T,\mathbf{x}) = \big[P_T^{\mathfrak{G}} f\big](\mathbf{x})$. □

What we want to do now is prove an existence statement which complements the preceding uniqueness statement. That is, we want to find out to what extent $(t,\mathbf{x}) \in (0,\infty) \times \mathfrak{G} \longmapsto [P_t^{\mathfrak{G}} f](\mathbf{x})$ is always a solution to Cauchy initial value problem in Lemma 8.1.17. To this end, notice that, at least when f is continuous, we already know that both the heat equation as well as the initial conditions in the above Cauchy problem are satisfied by $(t,\mathbf{x}) \in (0,\infty) \times \mathfrak{G} \longmapsto [P_t^{\mathfrak{G}} f](\mathbf{x})$. Hence, the only question is whether the correct boundary value is taken on at the lateral boundary $(0,\infty) \times \partial G$; and this comes down to the problem of finding out for which $\mathbf{a} \in \partial G$

$$\lim_{\substack{\mathbf{x}\to\mathbf{a}\\ \mathbf{x}\in\mathfrak{G}}} \mathcal{W}_{\mathbf{x}}^{(N)}\left(\mathfrak{e}^{\mathfrak{G}} \geq \delta\right) = 0, \quad \delta \in (0,\infty). \tag{8.1.18}$$

Indeed, given $\mathbf{a} \in \partial\mathfrak{G}$, it is reasonably clear that

$$\lim_{\substack{\mathbf{x}\to\mathbf{a}\\ \mathbf{x}\in\mathfrak{G}}} \left[P_t^{\mathfrak{G}} f\right](\mathbf{x}) = 0 \quad \text{for every } t \in (0,\infty) \text{ and } f \in C_{\mathrm{b}}(\mathfrak{G};\mathbb{R})$$

if (8.1.18) holds; and for this reason we will say that $\mathbf{a} \in \partial G$ is a **regular point for** $\mathfrak{G}$ and will write $\mathbf{a} \in \partial_{\mathrm{reg}}\mathfrak{G}$ when (8.1.18) is satisfied.

In order to get a handle on the problem of checking when $\mathbf{a} \in \partial\mathfrak{G}$ is regular, it will be convenient to introduce the $\{\mathcal{B}_t : t \in [0,\infty)\}$-stopping times

$$\mathfrak{e}_s^{\mathfrak{G}}(\psi) \equiv \inf\left\{t \geq s : \psi(t) \notin \mathfrak{G}\right\} = s + \mathfrak{e}^{\mathfrak{G}}\left(\Sigma_s \psi\right)$$

for $s \in (0,\infty)$ and $\psi \in \mathfrak{P}(\mathbb{R}^N)$. Clearly $s \in (0,\infty) \longmapsto \mathfrak{e}_s^{\mathfrak{G}}(\psi) \in [0,\infty]$ is nondecreasing and

$$\lim_{s\searrow 0} \mathfrak{e}_s^{\mathfrak{G}}(\psi) = \mathfrak{e}_{0+}^{\mathfrak{G}}(\psi) \equiv \inf\left\{t > 0 : \psi(t) \notin \mathfrak{G}\right\}$$

is the **first positive exit time from** $\mathfrak{G}$.

8.1.19 Lemma. *Regularity is a local property in the sense that, for each $r \in (0,\infty)$, $\mathbf{a} \in \partial_{\mathrm{reg}}\mathfrak{G}$ if and only if $\mathbf{a} \in \partial_{\mathrm{reg}}\left(\mathfrak{G} \cap B_{\mathbb{R}^N}(\mathbf{a},r)\right)$. Furthermore, given $\mathbf{a} \in \partial\mathfrak{G}$,*

$$\mathbf{a} \in \partial_{\mathrm{reg}}\mathfrak{G} \iff \mathcal{W}_{\mathbf{a}}^{(N)}\left(\mathfrak{e}_{0+}^{\mathfrak{G}} > 0\right) = 0; \tag{8.1.20}$$

and if $\mathbf{a} \in \partial_{\mathrm{reg}}\mathfrak{G}$, then, for each $\delta > 0$,

$$\lim_{\substack{\mathbf{x}\to\mathbf{a}\\ \mathbf{x}\in\mathfrak{G}}} \mathcal{W}_{\mathbf{x}}^{(N)}\Big(\left(\mathfrak{e}^{\mathfrak{G}}, \psi(\mathfrak{e}^{\mathfrak{G}})\right) \in (0,\delta) \times B_{\mathbb{R}^N}(\mathbf{a},\delta)\Big) = 1. \tag{8.1.21}$$

In particular, $\partial_{\mathrm{reg}}\mathfrak{G} \in \mathcal{B}_{\partial\mathfrak{G}}$; and for every $\mathbf{x} \in \mathfrak{G}$,

$$\mathcal{W}_{\mathbf{x}}^{(N)}\Big(\psi(\mathfrak{e}^{\mathfrak{G}}) \notin \partial_{\mathrm{reg}}\mathfrak{G} \ \&\ \mathfrak{e}^{\mathfrak{G}} < \infty\Big) = 0. \tag{8.1.22}$$

PROOF: Set $\mathfrak{G}(\mathbf{a},r) = \mathfrak{G} \cap B_{\mathbb{R}^N}(\mathbf{a},r)$. Since it is obvious that $\mathfrak{e}^{\mathfrak{G}(\mathbf{a},r)}$ is dominated by $\mathfrak{e}^{\mathfrak{G}}$, there is no question that $\mathbf{a} \in \partial_{\text{reg}}\mathfrak{G} \implies \mathbf{a} \in \partial_{\text{reg}}\mathfrak{G}(\mathbf{a},r)$. On the other hand, if $\mathbf{a} \in \partial_{\text{reg}}\mathfrak{G}(\mathbf{a},r)$ and $\epsilon > 0$, then, for all $0 < \delta < \epsilon$,

$$\begin{aligned}
\overline{\lim_{\substack{\mathbf{x}\to\mathbf{a}\\ \mathbf{x}\in\mathfrak{G}}}} \mathcal{W}_{\mathbf{x}}^{(N)}(\mathfrak{e}^{\mathfrak{G}} \geq \epsilon) &\leq \overline{\lim_{\substack{\mathbf{x}\to\mathbf{a}\\ \mathbf{x}\in\mathfrak{G}}}} \mathcal{W}_{\mathbf{x}}^{(N)}(\mathfrak{e}^{\mathfrak{G}} \geq \delta) \\
&\leq \overline{\lim_{\substack{\mathbf{x}\to\mathbf{a}\\ \mathbf{x}\in\mathfrak{G}(\mathbf{a},r)}}} \mathcal{W}_{\mathbf{x}}^{(N)}\left(\mathfrak{e}^{\mathfrak{G}(\mathbf{a},r)} \geq \delta\right) + \overline{\lim_{\substack{\mathbf{x}\to\mathbf{a}\\ \mathbf{x}\in\mathfrak{G}}}} \mathcal{W}_{\mathbf{x}}^{(N)}\left(\mathfrak{e}^{B_{\mathbb{R}^N}(\mathbf{a},r)} \leq \delta\right) \\
&\leq \mathcal{W}^{(N)}\left(\sup_{t\in[0,\delta]} |\psi(t)| \geq \frac{r}{2}\right) \longrightarrow 0 \quad \text{as } \delta \searrow 0.
\end{aligned}$$

Hence, we have now also proved that $\mathbf{a} \in \partial_{\text{reg}}\mathfrak{G}(\mathbf{a},r) \implies \mathbf{a} \in \partial_{\text{reg}}\mathfrak{G}$.

Next, let $\mathbf{a} \in \partial\mathfrak{G}$. To check the equivalence in (8.1.20), first use $\mathfrak{e}_s^{\mathfrak{G}} = s + \mathfrak{e}^{\mathfrak{G}} \circ \Sigma_s$ and the Markov property to see that

$$\mathbf{x} \in \mathbb{R}^N \longmapsto \mathcal{W}_{\mathbf{x}}^{(N)}\left(\mathfrak{e}_s^{\mathfrak{G}} \geq \delta\right) = \int_{\mathbb{R}^N} \mathcal{W}_{\mathbf{y}}^{(N)}\left(\mathfrak{e}^{\mathfrak{G}} \geq \delta - s\right) \gamma_s^N(\mathbf{y}-\mathbf{x})\, d\mathbf{y} \in [0,1]$$

is a continuous function for every $s \in (0,\infty)$, and therefore that

$$\mathbf{x} \in \mathbb{R}^N \longmapsto \mathcal{W}_{\mathbf{x}}^{(N)}\left(\mathfrak{e}_{0+}^{\mathfrak{G}} \geq \delta\right) = \lim_{s\searrow 0} \mathcal{W}_{\mathbf{x}}^{(N)}\left(\mathfrak{e}_s^{\mathfrak{G}} \geq \delta\right)$$

is upper semicontinuous for all $\delta \geq 0$. In particular, if $\mathcal{W}_{\mathbf{a}}^{(N)}\left(\mathfrak{e}_{0+}^{\mathfrak{G}} > 0\right) = 0$, it follows that

$$\overline{\lim_{\substack{\mathbf{x}\to\mathbf{a}\\ \mathbf{x}\in\mathfrak{G}}}} \mathcal{W}_{\mathbf{x}}^{(N)}\left(\mathfrak{e}^{\mathfrak{G}} \geq \delta\right) = \overline{\lim_{\substack{\mathbf{x}\to\mathbf{a}\\ \mathbf{x}\in\mathfrak{G}}}} \mathcal{W}_{\mathbf{x}}^{(N)}\left(\mathfrak{e}_{0+}^{\mathfrak{G}} \geq \delta\right) = 0$$

for every $\delta > 0$. To prove the converse, suppose that $\mathbf{a} \in \partial_{\text{reg}}\mathfrak{G}$, let positive ϵ and δ be given, and choose $r > 0$ so that

$$\mathcal{W}_{\mathbf{x}}^{(N)}\left(\mathfrak{e}^{\mathfrak{G}} \geq \delta\right) \leq \epsilon \quad \text{for } \mathbf{x} \in \mathfrak{G} \cap B_{\mathbb{R}^N}(\mathbf{a},r).$$

Then, by the Markov property and (3.3.30), for each $s \in (0,\delta)$ one has

$$\begin{aligned}
\mathcal{W}_{\mathbf{a}}^{(N)}\left(\mathfrak{e}_{0+}^{\mathfrak{G}} \geq 2\delta\right) &\leq \mathbb{E}^{\mathcal{W}_{\mathbf{a}}^{(N)}}\left[\mathcal{W}_{\psi(s)}^{(N)}\left(\mathfrak{e}^{\mathfrak{G}} \geq \delta\right),\, \psi(s) \in \mathfrak{G}\right] \\
&\leq \epsilon + \mathcal{W}_{\mathbf{a}}^{(N)}\left(\psi(s) \notin B_{\mathbb{R}^N}(\mathbf{a},r)\right) \leq \epsilon + 2Ne^{-\frac{r^2}{2Ns}};
\end{aligned}$$

from which it is evident that $\mathcal{W}_{\mathbf{a}}^{(N)}\left(\mathfrak{e}_{0+}^{\mathfrak{G}} > 0\right) = 0$.

Now, assume that $\mathbf{a} \in \partial_{\text{reg}}\mathfrak{G}$, and observe that, for each $0 < \epsilon < \delta$,

$$\begin{aligned}
&\mathcal{W}_{\mathbf{x}}^{(N)}\left(\psi\left(\mathfrak{e}^{\mathfrak{G}}\right) \notin B_{\mathbb{R}^N}(\mathbf{a},\delta) \text{ or } \mathfrak{e}^{\mathfrak{G}} \geq \delta\right) \\
&\qquad \leq \mathcal{W}_{\mathbf{x}}^{(N)}\left(\mathfrak{e}^{\mathfrak{G}} > \epsilon\right) + \mathcal{W}_{\mathbf{x}}^{(N)}\left(\sup_{t\in[0,\epsilon]} |\psi(t) - \mathbf{a}| \geq \delta\right).
\end{aligned}$$

Hence, (8.1.18) and (3.3.30) together imply that

$$\varlimsup_{\substack{\mathbf{x}\to\mathbf{a}\\ \mathbf{x}\in\mathfrak{G}}} \mathcal{W}^{(N)}_{\mathbf{x}}\Big(\boldsymbol{\psi}(\mathfrak{e}^{\mathfrak{G}}) \notin B_{\mathbb{R}^N}(\mathbf{x},\delta) \text{ or } \mathfrak{e}^{\mathfrak{G}} \geq \delta\Big) \leq 2N \exp\left[-\frac{\delta^2}{2N\epsilon}\right];$$

from which (8.1.21) follows after one lets $\epsilon \searrow 0$.

Finally, since $\mathbf{a} \in \partial\mathfrak{G} \longmapsto \mathcal{W}^{(N)}_{\mathbf{a}}\big(\mathfrak{e}^{\mathfrak{G}}_{0+} > 0\big)$ is measurable, it is clear from (8.1.20) that $\partial_{\text{reg}}\mathfrak{G} \in \mathcal{B}_{\partial\mathfrak{G}}$. Moreover, if $\mathbf{x} \in \mathfrak{G}$, then, by the strong Markov property and the relation $\mathfrak{e}^{\mathfrak{G}}_{0+} = \mathfrak{e}^{\mathfrak{G}} + \mathfrak{e}^{\mathfrak{G}}_{0+} \circ \boldsymbol{\Sigma}_{\mathfrak{e}^{\mathfrak{G}}}$ on $\{\mathfrak{e}^{\mathfrak{G}} < \infty\}$,

$$0 = \mathcal{W}^{(N)}_{\mathbf{x}}\big(\mathfrak{e}^{\mathfrak{G}} < \mathfrak{e}^{\mathfrak{G}}_{0+}\big) = \int_{\{\boldsymbol{\psi}:\mathfrak{e}^{\mathfrak{G}}(\boldsymbol{\psi})<\infty\}} \mathcal{W}^{(N)}_{\boldsymbol{\psi}(\mathfrak{e}^{\mathfrak{G}})}\big(\mathfrak{e}^{\mathfrak{G}}_{0+} > 0\big)\, \mathcal{W}^{(N)}_{\mathbf{x}}(d\boldsymbol{\psi}),$$

and therefore, by (8.1.20), we see that $\boldsymbol{\psi}(\mathfrak{e}^{\mathfrak{G}}) \in \partial_{\text{reg}}\mathfrak{G}$ for $\mathcal{W}^{(N)}_{\mathbf{x}}$-almost every $\boldsymbol{\psi} \in \{\mathfrak{e}^{\mathfrak{G}} < \infty\}$. □

With the preceding result, we can now complete our analytic characterization of the family of operators $\big\{P^{\mathfrak{G}}_t : t \in (0,\infty)\big\}$.

8.1.23 Theorem. *Let $f \in C_{\mathrm{b}}(\mathfrak{G};\mathbb{R})$ be given. Then $(t,\mathbf{x}) \in (0,\infty) \times \mathfrak{G} \longmapsto \big[P^{\mathfrak{G}}_t f\big](\mathbf{x}) \in \mathbb{R}$ is the one and only bounded $u \in C^{1,2}\big([0,\infty) \times \mathfrak{G};\mathbb{R}\big)$ with the properties that*

$$\begin{aligned} &\tfrac{\partial u}{\partial t} = \tfrac{1}{2}\boldsymbol{\Delta} u \quad \text{in } (0,\infty)\times\mathfrak{G}, \quad \lim_{t\searrow 0} u(t,\mathbf{x}) = f(\mathbf{x}), \quad \mathbf{x}\in\mathfrak{G},\\ &\lim_{\substack{(t,\mathbf{x})\to(s,\mathbf{a})\\ (t,\mathbf{x})\in(0,\infty)\times\mathfrak{G}}} u(t,\mathbf{x}) = 0 \quad \text{for } (s,\mathbf{a}) \in (0,\infty)\times\partial_{\text{reg}}\mathfrak{G}. \end{aligned} \tag{8.1.24}$$

In addition, for each $\mathbf{a} \in \partial_{\text{reg}}\mathfrak{G}$,

$$\lim_{\substack{\mathbf{x}\to\mathbf{a}\\ \mathbf{x}\in\mathfrak{G}}} p^{\mathfrak{G}}(t,\mathbf{x},\mathbf{y}) = 0 \tag{8.1.25}$$

uniformly fast for $(t,\mathbf{y})$ in compact subsets of $(0,\infty)\times\mathfrak{G}$.

PROOF: Let $f \in C_{\mathrm{b}}(\mathfrak{G};\mathbb{R})$ be given. If $u(t,\,\cdot\,) = P^{\mathfrak{G}}_t f$, Corollary 8.1.7 says that this u satisfies the first line in (8.1.24), and, by (8.1.21), it satisfies the second line as well. Conversely, let u be a solution to (8.1.24) and $(T,\mathbf{x}) \in (0,\infty)\times\mathfrak{G}$ be given. To see that $u(T,\mathbf{x}) = \big[P^{\mathfrak{G}}_T f\big](\mathbf{x})$, choose a sequence $\{\mathfrak{G}_n\}_1^\infty$ of open sets so that $\overline{\mathfrak{G}_n} \subset\subset \mathfrak{G}$ and $\mathfrak{G}_n \nearrow \mathfrak{G}$ as $n\to\infty$. Then, by the last part of Lemma 8.1.15 applied to $(t,\mathbf{x}) \in [0,T]\times\overline{\mathfrak{G}_n} \longmapsto u(T+\delta-t,\mathbf{x}) \in \mathbb{R}$, we see that

$$u(T+\delta,\mathbf{x}) = \big[P^{\mathfrak{G}_n}_T u(\delta,\,\cdot\,)\big](\mathbf{x}) + \mathbb{E}^{\mathcal{W}^{(N)}_{\mathbf{x}}}\Big[u\big(T+\delta-\mathfrak{e}^{\mathfrak{G}_n},\boldsymbol{\psi}(\mathfrak{e}^{\mathfrak{G}_n})\big),\, \mathfrak{e}^{\mathfrak{G}_n} \leq T\Big]$$

for every $\delta > 0$ and $\mathbf{x} \in \mathfrak{G}_n$. Hence, because $\mathfrak{e}^{\mathfrak{G}_n} \nearrow \mathfrak{e}^{\mathfrak{G}}$ as $n\to\infty$, (8.1.22) together with the second line of (8.1.24) yield $u(T+\delta,\mathbf{x}) = \big[P^{\mathfrak{G}}_T u(\delta,\,\cdot\,)\big](\mathbf{x})$ for all $\delta > 0$ and $\mathbf{x} \in \mathfrak{G}$. Finally, let $\delta \searrow 0$, and conclude that $u(T,\,\cdot\,) = P^{\mathfrak{G}}_T f$.

In order to prove the assertion in (8.1.25), we simply apply (8.1.21) to the second term on the right-hand side of (8.1.10). □

With Theorem 8.1.23, we have completed the identification of $p^{\mathfrak{G}}(t,\mathbf{x},\mathbf{y})$ as the **fundamental solution to the heat equation in $\mathfrak{G}$ with Dirichlet boundary conditions**; and, in the classical literature on this subject, the representation of $p^{\mathfrak{G}}(t,\mathbf{x},\mathbf{y})$ given in (8.1.10) would be considered an extension of **Duhamel's formula**.

Although we now know that our definition of $\partial_{\text{reg}}\mathfrak{G}$ is *correct* in the sense that it provides us with a well-posed boundary value problem, we still do not have any workable criterion for recognizing when a particular $\mathbf{a} \in \partial\mathfrak{G}$ is regular for $\mathfrak{G}$. In order to develop such a criterion, it will be useful to know the following interesting fact about Wiener's measure.

8.1.26 Theorem (Blumenthal's 0–1 Law). *For any $A \in \mathcal{B}_{0+}$ (cf. (7.1.3)), $\mathcal{W}_{\mathbf{x}}^{(N)}(A) \in \{0,1\}$ for every $\mathbf{x} \in \mathbb{R}^N$.*

PROOF: Let $A \in \mathcal{B}_{0+}$ be given, and note that

$$\begin{aligned}\mathbb{E}^{\mathcal{W}_{\mathbf{x}}^{(N)}}\big[F,\,A\big] &= \lim_{s\searrow 0}\mathbb{E}^{\mathcal{W}_{\mathbf{x}}^{(N)}}\big[F\circ\Sigma_s,\,A\big]\\ &= \lim_{s\searrow 0}\int_A \mathbb{E}^{\mathcal{W}_{\psi(s)}^{(N)}}\big[F\big]\,\mathcal{W}_{\mathbf{x}}^{(N)}(d\psi) = \mathcal{W}_{\mathbf{x}}^{(N)}(A)\,\mathbb{E}^{\mathcal{W}_{\mathbf{x}}^{(N)}}\big[F\big].\end{aligned}$$

for every $F \in C_{\mathrm{b}}\big(\mathfrak{P}(\mathbb{R}^N);\mathbb{R}\big)$. Since the set of $F \in B\big(\mathfrak{P}(\mathbb{R}^N);\mathbb{R}\big)$ for which

$$\mathbb{E}^{\mathcal{W}_{\mathbf{x}}^{(N)}}\big[F,\,A\big] = \mathcal{W}_{\mathbf{x}}^{(N)}(A)\,\mathbb{E}^{\mathcal{W}_{\mathbf{x}}^{(N)}}\big[F\big]$$

is closed under bounded pointwise convergence, we have now shown that A is $\mathcal{W}_{\mathbf{x}}^{(N)}$-independent of every $B \in \mathcal{B}_{\mathfrak{P}(\mathbb{R}^N)}$. In particular, this means that

$$\mathcal{W}_{\mathbf{x}}^{(N)}(A) = \mathcal{W}_{\mathbf{x}}^{(N)}(A\cap A) = \mathcal{W}_{\mathbf{x}}^{(N)}(A)^2. \quad \square$$

The application of Blumenthal's 0–1 in which we are interested is contained in the following criterion for regularity.

8.1.27 Theorem. *For all $\mathbf{x} \in \mathbb{R}^N$, $\mathcal{W}_{\mathbf{x}}^{(N)}\big(\mathfrak{e}_{0+}^{\mathfrak{G}} > 0\big) \in \{0,1\}$, and so $\mathbf{a} \in \partial\mathfrak{G}$ is regular if and only if $\mathcal{W}_{\mathbf{a}}^{(N)}\big(\mathfrak{e}_{0+}^{\mathfrak{G}} = 0\big) > 0$. In particular, if $\mathbf{a} \in \partial\mathfrak{G}$ and*

$$\overline{\lim_{\delta\searrow 0}}\,\frac{|\mathfrak{G}\complement \cap B_{\mathbb{R}^N}(\mathbf{a},\delta)|}{\delta^N} > 0,$$

then $\mathbf{a}$ is regular for $\mathfrak{G}$. (Here $|\Gamma|$ is used to denote the Lebesgue measure of Γ.)

PROOF: Since $\{\mathfrak{e}_{0+}^{\mathfrak{G}} > 0\} \in \mathcal{B}_{0+}$, the first assertion is an immediate consequence of Blumenthal's 0–1 Law. Next, because

$$\begin{aligned}\mathcal{W}_{\mathbf{a}}^{(N)}\left(\mathfrak{e}_{0+}^{\mathfrak{G}} \leq \delta\right) &\geq \mathcal{W}_{\mathbf{a}}^{(N)}\left(\psi(\delta) \notin \mathfrak{G}\right) \\ &= \frac{1}{(2\pi\delta)^{\frac{N}{2}}} \int_{\mathfrak{G}\complement} \exp\left[-\tfrac{|\mathbf{y}-\mathbf{a}|^2}{2\delta}\right] d\mathbf{y} \\ &\geq e^{-\frac{1}{2}} \frac{\left|\mathfrak{G}\complement \cap B_{\mathbb{R}^N}\left(\mathbf{a}, \delta^{\frac{1}{2}}\right)\right|}{(2\pi\delta)^{\frac{N}{2}}},\end{aligned}$$

the second assertion follows from the fact that

$$\mathcal{W}_{\mathbf{a}}^{(N)}\left(\mathfrak{e}_{0+}^{\mathfrak{G}} = 0\right) = \lim_{\delta \searrow 0} \mathcal{W}_{\mathbf{a}}^{(N)}\left(\mathfrak{e}_{0+}^{\mathfrak{G}} \leq \delta\right). \quad \square$$

8.1.28 Remark. The criterion for regularity in Theorem 8.1.27 is a quite good, albeit rather crude, general purpose test for regularity. In particular, it shows that $\mathbf{a} \in \partial\mathfrak{G}$ is regular if ∂G is *smooth* in a neighborhood of $\mathbf{a}$. In fact, it shows that smoothness is not really essential so long as $\mathbf{a}$ is not *hidden* from $\mathfrak{G}\complement$. Thus, for example, if $\mathbf{a}$ lies at the end of a very sharp spike on $\partial\mathfrak{G}$, then $\mathbf{a}$ will be regular if the spike points into $\mathfrak{G}\complement$, but it may be irregular if the spike goes into $\mathfrak{G}$. In particular, $\mathbf{a} \in \partial_{\text{reg}}\mathfrak{G}$ if there is a $\mathbf{c} \in \mathfrak{G}\complement$ and an $r > 0$ such that $B_{\mathbb{R}^N}(\mathbf{c}, r) \subseteq \mathfrak{G}\complement$ and $\mathbf{a} \in \overline{B}_{\mathbb{R}^N}(\mathbf{c}, r)$. This latter statement is sometimes called the **Poincaré exterior ball condition**. As our criterion makes clear, the ball can be replaced by any nontrivial open cone in $\mathfrak{G}\complement$ with $\mathbf{a}$ at its vertex. Moreover, because, by Lebesgue's Density Theorem (the special case of Theorem 5.3.20 applied to indicator functions),

$$\lim_{r \searrow 0} \frac{\left|\mathfrak{G}\complement \cap B_{\mathbb{R}^N}(\mathbf{a}, r)\right|}{\left|B_{\mathbb{R}^N}(\mathbf{a}, r)\right|} = 1 \quad \text{for almost every } \mathbf{a} \in \mathfrak{G}\complement,$$

we know now that $\partial\mathfrak{G} \setminus \partial_{\text{reg}}\mathfrak{G}$ always has Lebesgue measure 0. However, if one wants to get a criterion for regularity which is necessary as well as sufficient, one has to replace Lebesgue's measure by *capacity* and, in other ways, be more careful (cf. Wiener's test in (8.4.40)).

As our final goal in this section, we apply Duhamel's formula (cf. (8.1.10)) to give a simple derivation of a famous result proved originally by H. Weyl. In order to describe Weyl's result, observe that the operators $P_t^{\mathfrak{G}}$ admit unique extensions as nonnegative definite, self-adjoint contractions operators $\overline{P_t^{\mathfrak{G}}}$ on $L^2(\mathfrak{G})$. Indeed, from (8.1.11) and (8.1.13), it is clear that $P_t^{\mathfrak{G}}$ is symmetric on $L^2(\mathfrak{G})$, and so one can proceed in exactly the same way as we did in Theorem 7.5.18 to produce the desired extensions. Moreover, if $\mathfrak{G}$ has finite volume $|\mathfrak{G}|$, then, because $p^{\mathfrak{G}}(t,\mathbf{x},\mathbf{y}) \le \gamma_t^N(\mathbf{y}-\mathbf{x})$, the Chapman–Kolmogorov equation (cf. (8.1.12)) for $p^{\mathfrak{G}}(t,\mathbf{x},\mathbf{y})$ plus symmetry lead to

$$\iint_{\mathfrak{G}\times\mathfrak{G}} p^{\mathfrak{G}}(t,\mathbf{x},\mathbf{y})^2\,d\mathbf{x}d\mathbf{y} = \int_{\mathfrak{G}} p^{\mathfrak{G}}(2t,\mathbf{x},\mathbf{x})\,d\mathbf{x} \le (4\pi t)^{-\frac{N}{2}}\,|\mathfrak{G}| < \infty,$$

which means that all the $\overline{P_t^{\mathfrak{G}}}$'s are Hilbert–Schmidt operators. Hence, when $|\mathfrak{G}| < \infty$, we know that there is a nondecreasing sequence $\{\lambda_n\}_0^\infty \subseteq [0,\infty)$ which tends to ∞ and an orthonormal basis $\{\varphi_n\}_0^\infty$ in $L^2(\mathfrak{G})$ such that, for each $n \in \mathbb{N}$,

$$\overline{P_{2t}^{\mathfrak{G}}}\varphi_n = e^{-\lambda_n t}\varphi_n, \quad t \in (0,\infty). \tag{8.1.29}$$

In fact, from (8.1.29) and the regularity properties of $p^{\mathfrak{G}}(t,\mathbf{x},\mathbf{y})$, it is clear that we may take the φ_n's to be bounded, smooth functions in $\mathfrak{G}$ with the properties that

$$\boldsymbol{\Delta}\varphi_n = -\lambda_n\varphi_n \quad\text{and}\quad \lim_{\substack{\mathbf{x}\to\mathbf{a}\\ \mathbf{x}\in\mathfrak{G}}}\varphi_n(\mathbf{x}) = 0 \text{ for } \mathbf{a}\in\partial_{\text{reg}}\mathfrak{G}. \tag{8.1.30}$$

In other words, the numbers $\{\lambda_n\}_1^\infty$ are *the eigenvalues of self-adjoint extension of* $-\boldsymbol{\Delta}$ *corresponding to Dirichlet boundary conditions.*

Working under the assumption that $|\mathfrak{G}| < \infty$, what Weyl did is study the distribution function

$$\mathbf{N}^{\mathfrak{G}}(\lambda) \equiv \operatorname{card}\{n \in \mathbb{N} : \lambda_n \le \lambda\}. \tag{8.1.31}$$

In particular, his goal was to show that, as $\lambda \nearrow \infty$, the growth of $\mathbf{N}^{\mathfrak{G}}(\lambda)$ depends only on N and the volume of $\mathfrak{G}$.† To get his result from (8.1.10), first note that, on the one hand,

$$\begin{aligned}\sum_0^\infty \big(P_{2t}^{\mathfrak{G}}\varphi_n,\varphi_n\big)_{L^2(\mathfrak{G})} &= \sum_0^\infty \big(P_t^{\mathfrak{G}}\varphi_n, P_t^{\mathfrak{G}}\varphi_n\big)_{L^2(\mathfrak{G})}\\ &= \sum_0^\infty \int_{\mathfrak{G}} \big(p^{\mathfrak{G}}(t,\mathbf{x},\,\cdot\,),\varphi_n\big)^2_{L^2(\mathfrak{G})}\,d\mathbf{x}\\ &= \int_{\mathfrak{G}} p^{\mathfrak{G}}(t,\mathbf{x},\,\cdot\,)^2\,d\mathbf{x} = \int_{\mathfrak{G}} p^{\mathfrak{G}}(2t,\mathbf{x},\mathbf{x})\,d\mathbf{x},\end{aligned}$$

† As M. Kac explains in his wonderful article "Can one hear the shape of drum?," *Am. Math. Monthly* **73**, the origin of Weyl's result is a problem posed by Lorentz. What Lorentz noticed is that, if one takes Planck's theory of black body radiation seriously, then the distribution of high frequencies emitted should depend only on the volume of the radiator. Thus, the original interest in the result was that the asymptotic distribution of eigenvalues is so insensitive to the shape of the radiator. When Kac took up the problem, he turned it around. Namely, he asked what geometric information, besides the volume, is encoded in the eigenvalues. When he explained his program to L. Bers, Bers rephrased the problem in the terms which Kac adopted for his title. Audiophiles will be disappointed to learn that, according to C. Gordon, D. Webb, and S. Wolpert's 1992 announcement in *B.A.M.S.*, new series **27** (**2**), "One cannot hear the shape of a drum," even a two-dimensional one.

while, on the other hand,

$$\sum_0^\infty \left(P_{2t}^{\mathfrak{G}}\varphi_n, \varphi_n\right)_{L^2(\mathfrak{G})} = \sum_0^\infty e^{-t\lambda_n} = \int_{[0,\infty)} e^{-\lambda t}\, d\mathbf{N}^{\mathfrak{G}}(\lambda),$$

where the last integral is taken in the sense of Riemann–Stieljes with respect to the nondecreasing function $\lambda \in [0,\infty) \longmapsto \mathbf{N}^{\mathfrak{G}}(\lambda) \in \mathbb{N}$. Thus, we have

$$\int_{[0,\infty)} e^{-\lambda t}\, d\mathbf{N}^{\mathfrak{G}}(\lambda) = \int_{\mathfrak{G}} p^{\mathfrak{G}}(2t, \mathbf{x}, \mathbf{x})\, d\mathbf{x}. \tag{8.1.32}$$

At this point, we want to invoke what M. Kac used to call *the principle of not feeling the boundary.* That is, for each $\mathbf{x} \in \mathfrak{G}$, the probability that a Wiener path goes from $\mathbf{x}$ to $\mathbf{x}$ during a short time-interval $[0,t]$ *without hitting the boundary* $\partial\mathfrak{G}$ should be essentially the same as that of its going from $\mathbf{x}$ to $\mathbf{x}$ during $[0,t]$ without worrying about the boundary. To make this intuition quantitative, note that, by (8.1.10),

$$1 \geq (4\pi t)^{\frac{N}{2}}\, p^{\mathfrak{G}}(2t, \mathbf{x}, \mathbf{x}) \geq 1 - \rho(t, \mathbf{x}),$$

where

$$\rho(t,\mathbf{x}) = (4\pi t)^{\frac{N}{2}} \mathbb{E}^{\mathcal{W}_{\mathbf{x}}^{(N)}}\left[\gamma^N_{2t-\mathfrak{e}^{\mathfrak{G}}}\left(\mathbf{x} - \psi(\mathfrak{e}^{\mathfrak{G}})\right),\, \mathfrak{e}^{\mathfrak{G}} < 2t\right] \leq \frac{A_N t^{\frac{N}{2}}}{d(\mathbf{x})^N},$$

$d(\mathbf{x}) = \operatorname{dist}(\mathbf{x}, \mathfrak{G}\complement)$, and

$$A_N \equiv \sup_{s\in(0,\infty)} (4s)^{\frac{N}{2}}\, e^{-s}.$$

Hence,

$$1 \geq (4\pi t)^{\frac{N}{2}}\, p^{\mathfrak{G}}(2t, \mathbf{x}, \mathbf{x}) \longrightarrow 1 \quad \text{as } t \searrow 0$$

uniformly for $\mathbf{x}$ compact subsets of $\mathfrak{G}$; which, in conjunction with (8.1.32), leads to

$$\lim_{t\searrow 0} (4\pi t)^{\frac{N}{2}} \int_{[0,\infty)} e^{-\lambda t}\, d\mathbf{N}^{\mathfrak{G}}(\lambda) = |\mathfrak{G}|. \tag{8.1.33}$$

In particular, after an application of standard Tauberian theory,† we arrive at the following theorem.

† See, for example, Theorem 1.7.6 in Bingham, Goldie, and Teugel's *Regularly Varying Functions*, publ. by Cambridge U. Press (1987).

8.1.34 Theorem (Weyl). *Assume that* $|\mathfrak{G}| < \infty$ *and define* $\mathbf{N}^{\mathfrak{G}}(\lambda)$ *as in* (8.1.31). *Then*

$$\lim_{\lambda\nearrow\infty} \frac{\mathbf{N}^{\mathfrak{G}}(\lambda)}{\lambda^{\frac{N}{2}}} = \frac{|\mathfrak{G}|}{(4\pi)^{\frac{N}{2}}\Gamma\left(\frac{N}{2}+1\right)},$$

where $\Gamma(t)$ *is the Gamma function given in* (1.3.22).

Exercises

8.1.35 Exercise: Set $\mathbb{R}^N_+ = \{\mathbf{x} \in \mathbb{R}^N : x_N > 0\}$. Given an $f \in C_c(\mathbb{R}^N_+;\mathbb{R})$, define $\check{f} \in C_c(\mathbb{R}^N;\mathbb{R})$ so that $\check{f} \restriction \mathbb{R}^N_+ = f$ and $\check{f}(\mathbf{y}) = -f(\check{\mathbf{y}})$, where $\check{\mathbf{y}} \equiv (\bar{\mathbf{y}}, -y_N)$ for $\mathbf{y} = (\bar{\mathbf{y}}, y_N) \in \mathbb{R}^{N-1} \times (-\infty, 0]$. Show that

$$\mathbb{E}^{\mathcal{W}^{(N)}_{\mathbf{x}}}\left[\left[\gamma^N_{t-\mathfrak{e}^{\mathbb{R}^N_+}} \bigstar \check{f}\right]\left(\psi(\mathfrak{e}^{\mathbb{R}^N_+})\right),\, \mathfrak{e}^{\mathbb{R}^N_+} \le t\right] = 0$$

for all $(t,\mathbf{x}) \in (0,\infty) \times \mathbb{R}^N_+$, and use this to prove first that

$$\left[P^{\mathbb{R}^N_+}_t f\right](\mathbf{x}) = \left[\gamma^N_t \bigstar \check{f}\right](\mathbf{x}), \quad (t,\mathbf{x}) \in (0,\infty) \times \mathbb{R}^N_+, \tag{8.1.36}$$

for every $f \in B(\mathbb{R}^N_+;\mathbb{R})$ and second that

$$p^{\mathbb{R}^N_+}(t,\mathbf{x},\mathbf{y}) = \gamma^N_t(\mathbf{y}-\mathbf{x}) - \gamma^N_t(\check{\mathbf{y}}-\mathbf{x}) \tag{8.1.37}$$

for all $(t,\mathbf{x},\mathbf{y}) \in (0,\infty) \times \mathbb{R}^N_+ \times \mathbb{R}^N_+$. Finally, when $N = 1$, use either (8.1.36) or (8.1.37) to give another proof of the reflection principle in (4.3.5).

8.1.38 Exercise: Let $\mathfrak{G}$ be an open region in $\mathbb{R}^N$, f a bounded continuous function on $\mathfrak{G}$, and g a bounded continuous function on $[0,\infty) \times \partial\mathfrak{G}$. Show that the function

$$(t,\mathbf{x}) \in (0,\infty) \times \mathfrak{G} \longmapsto \left[P^{\mathfrak{G}}_t f\right](\mathbf{x}) + \mathbb{E}^{\mathcal{W}^{(N)}_{\mathbf{x}}}\left[g\left(t - \mathfrak{e}^{\mathfrak{G}}, \psi(\mathfrak{e}^{\mathfrak{G}})\right),\, \mathfrak{e}^{\mathfrak{G}} \le t\right] \in \mathbb{R}$$

is the one and only bounded $u \in C^{1,2}\left((0,\infty) \times \mathfrak{G};\mathbb{R}\right)$ with the properties that

$$\frac{\partial u}{\partial t} = \frac{1}{2}\Delta u \quad \text{in } (0,\infty) \times \mathfrak{G}, \qquad \lim_{t\searrow 0} u(t,\mathbf{x}) = f(\mathbf{x}),\ \mathbf{x} \in \mathfrak{G},$$

$$\lim_{\substack{(t,\mathbf{x})\to(s,\mathbf{a})\\(t,\mathbf{x})\in(0,\infty)\times\mathfrak{G}}} u(t,\mathbf{x}) = g(t,\mathbf{a}) \quad \text{for } (t,\mathbf{a}) \in (0,\infty) \times \partial_{\text{reg}}G.$$

8.1.39 Exercise: Our discussion of regular points extends to more general Markov processes. Indeed, let $\mathbf{x} \in \mathbb{R}^N \longmapsto P_{\mathbf{x}} \in \mathbf{M}_1\big(\mathfrak{P}(\mathbb{R}^N)\big)$ be a continuous map for which the family $\{P_{\mathbf{x}} : \mathbf{x} \in \mathbb{R}^N\}$ is strong Markov, and show that Blumenthal's 0–1 Law (cf. Theorem 8.1.26) holds; that is, $P_{\mathbf{x}}(A) \in \{0,1\}$ for each $\mathbf{x} \in \mathbb{R}^N$ and $A \in \mathcal{B}_{0+}$. Next, add the assumption that the associated transition probability function $P(t,\mathbf{x},\,\cdot\,)$ is strongly continuous in the sense that $\mathbf{x} \in \mathbb{R}^N \longmapsto P(t,\mathbf{x},\Gamma) \in [0,1]$ is a continuous function for each $t \in (0,\infty)$ and $\Gamma \in \mathcal{B}_{\mathbb{R}^N}$. Given an open $\mathfrak{G}$ in $\mathbb{R}^N$, proceed as in Lemma 8.1.19 to prove that, for each $\mathbf{a} \in \partial G$,

$$\lim_{\delta \searrow 0} \lim_{\substack{\mathbf{x} \to \mathbf{a} \\ \mathbf{x} \in \mathfrak{G}}} P_{\mathbf{x}}\big(\mathfrak{e}^{\mathfrak{G}} \geq \delta\big) = 0 \iff P_{\mathbf{a}}\big(\mathfrak{e}^{\mathfrak{G}}_{0+} > 0\big) = 0 \iff P_{\mathbf{a}}\big(\mathfrak{e}^{\mathfrak{G}}_{0+} > 0\big) < 1.$$

Finally, let $\partial_{\text{reg}}\mathfrak{G}$ be the set of $\mathbf{a} \in \partial\mathfrak{G}$ for which

$$\lim_{\delta \searrow 0} \lim_{\substack{\mathbf{x} \to \mathbf{a} \\ \mathbf{x} \in \mathfrak{G}}} P_{\mathbf{x}}\big(\mathfrak{e}^{\mathfrak{G}} \geq \delta\big) = 0;$$

and check that: $\partial_{\text{reg}}\mathfrak{G} \in \mathcal{B}_{\partial\mathfrak{G}}$,

$$\lim_{\delta \searrow 0} \lim_{\substack{\mathbf{x} \to \mathbf{a} \\ \mathbf{x} \in \mathfrak{G}}} P_{\mathbf{x}}\Big(\big(\mathfrak{e}^{\mathfrak{G}}, \psi(\mathfrak{e}^{\mathfrak{G}})\big) \in (0,\delta) \times B^N_{\mathbb{R}}(\mathbf{a},\delta)\Big) = 1, \quad \mathbf{a} \in \partial_{\text{reg}}\mathfrak{G},$$

and, for each $\mathbf{x} \in \mathfrak{G}$,

$$P_{\mathbf{x}}\Big(\psi(\mathfrak{e}^{\mathfrak{G}}) \notin \partial_{\text{reg}}\mathfrak{G} \ \&\ \mathfrak{e}^{\mathfrak{G}} < \infty\Big) = 0.$$

§8.2: The Dirichlet Problem.

Having dealt with the Cauchy initial value problem, we will now turn our attention to the classical Dirichlet problem. Namely, we say that a function u on $\mathfrak{G}$ is **harmonic** if $u \in C^\infty(\mathfrak{G};\mathbb{R})$ and $\boldsymbol{\Delta} u = 0$; and, given $f \in B(\partial\mathfrak{G};\mathbb{R})$, we say that u solves the **Dirichlet problem for f in $\mathfrak{G}$** if u is a harmonic function satisfying the boundary condition

$$\lim_{\substack{\mathbf{x} \to \mathbf{a} \\ \mathbf{x} \in \mathfrak{G}}} u(\mathbf{x}) = f(\mathbf{a}) \tag{8.2.1}$$

for $\mathbf{a} \in \partial G$. As the following result shows, we already know how to solve the Dirichlet problem in great generality.

8.2.2 Theorem. *Given a measurable $f : \partial\mathfrak{G} \longrightarrow \mathbb{R}$ which is either nonnegative or bounded, set*

$$\left[H^{\mathfrak{G}} f\right](\mathbf{x}) = \mathbb{E}^{\mathcal{W}_{\mathbf{x}}^{(N)}}\left[f\left(\psi(\mathfrak{e}^{\mathfrak{G}})\right),\, \mathfrak{e}^{\mathfrak{G}} < \infty\right], \quad \mathbf{x} \in \mathfrak{G}. \tag{8.2.3}$$

If f is bounded, then $H^{\mathfrak{G}} f$ is a bounded harmonic function on $\mathfrak{G}$ and

$$\lim_{\substack{\mathbf{x}\to\mathbf{a}\\ \mathbf{x}\in\mathfrak{G}}} \left[H^{\mathfrak{G}} f\right](\mathbf{x}) = f(\mathbf{a})$$

whenever $\mathbf{a} \in \partial_{\text{reg}}\mathfrak{G}$ is a point at which f is continuous. Furthermore, if f is nonnegative and u is an element of $C^2\big(\mathfrak{G}; [0,\infty)\big)$ which satisfies

$$\boldsymbol{\Delta} u \le 0 \quad \text{in } \mathfrak{G} \quad \text{and} \quad \varliminf_{\substack{\mathbf{x}\to\mathbf{a}\\ \mathbf{x}\in\mathfrak{G}}} u(\mathbf{x}) \ge f(\mathbf{a}) \quad \text{for } \mathbf{a} \in \partial_{\text{reg}}\mathfrak{G}, \tag{8.2.4}$$

then $H^{\mathfrak{G}} f \le u$. In particular, if $f \in C_{\text{b}}\big(\partial\mathfrak{G}; \mathbb{R}\big)$, then $H^{\mathfrak{G}} f$ is the one and only harmonic function u in $\mathfrak{G}$ with the properties that

$$\big|u(\mathbf{x})\big| \le C\mathcal{W}_{\mathbf{x}}^{(N)}\big(\mathfrak{e}^{\mathfrak{G}} < \infty\big) \quad \text{for some } C \in (0,\infty) \text{ and all } \mathbf{x} \in \mathfrak{G}, \tag{8.2.5}$$

and that

$$\lim_{\substack{\mathbf{x}\to\mathbf{a}\\ \mathbf{x}\in\mathfrak{G}}} u(\mathbf{x}) = f(\mathbf{a}) \quad \text{for all } \mathbf{a} \in \partial_{\text{reg}}\mathfrak{G}. \tag{8.2.6}$$

PROOF: Let $f \in B\big(\partial\mathfrak{G}; \mathbb{R}\big)$ be given. To prove that $H^{\mathfrak{G}} f$ is an harmonic function in $\mathfrak{G}$, set $u = H^{\mathfrak{G}} f$ and apply the strong Markov property to see that

$$u(\mathbf{x}) = \left[P_t^{\mathfrak{G}} u\right](\mathbf{x}) + \mathbb{E}^{\mathcal{W}_{\mathbf{x}}^{(N)}}\left[f\left(\psi(\mathfrak{e}^{\mathfrak{G}})\right),\, \mathfrak{e}^{\mathfrak{G}} \le t\right]$$

for all $(t,\mathbf{x}) \in (0,\infty) \times \mathfrak{G}$. Hence, by Theorem 8.1.3 and Corollary 8.1.7, we know both that $u \in C^\infty(\mathfrak{G}; \mathbb{R})$ and, after applying $\frac{1}{2}\boldsymbol{\Delta} - \frac{\partial}{\partial t}$ to both sides of the preceding, that $\boldsymbol{\Delta} u = 0$. Moreover, if $\mathbf{a} \in \partial_{\text{reg}}\mathfrak{G}$ is a point at which f is continuous, then, by (8.1.21) in Lemma 8.1.19, it is clear that $u(\mathbf{x}) \longrightarrow f(\mathbf{a})$ as $\mathbf{x} \to \mathbf{a}$ through $\mathfrak{G}$.

Next, suppose that u is an element of $C^2\big(\mathfrak{G}; [0,\infty)\big)$ which satisfies (8.2.4) for some nonnegative f. To prove that $H^{\mathfrak{G}} f \le u$, choose a sequence of bounded, open subsets $\mathfrak{G}_n$ so that $\overline{\mathfrak{G}_n} \subseteq \mathfrak{G}$ and $\mathfrak{G}_n \nearrow \mathfrak{G}$. Then, by Lemma 8.1.15, applied to $f(t,\cdot\,) = -u \restriction \overline{\mathfrak{G}_n}$, and Fatou's Lemma, we know that

$$\begin{aligned} u(\mathbf{x}) &\ge \varliminf_{T\nearrow\infty}\ \varliminf_{n\to\infty}\ \mathbb{E}^{\mathcal{W}_{\mathbf{x}}^{(N)}}\left[u\big(\psi(T \wedge \mathfrak{e}^{\mathfrak{G}_n})\big)\right] \\ &\ge \varliminf_{T\nearrow\infty}\ \varliminf_{n\to\infty}\ \mathbb{E}^{\mathcal{W}_{\mathbf{x}}^{(N)}}\left[u\big(\psi(\mathfrak{e}^{\mathfrak{G}_n})\big),\, \mathfrak{e}^{\mathfrak{G}} \le T\right] \\ &\ge \mathbb{E}^{\mathcal{W}_{\mathbf{x}}^{(N)}}\left[f\big(\psi(\mathfrak{e}^{\mathfrak{G}})\big),\, \mathfrak{e}^{\mathfrak{G}} < \infty\right] = \left[H^{\mathfrak{G}} f\right](\mathbf{x}) \end{aligned}$$

for each $\mathbf{x} \in \mathfrak{G}$, where, in the passage to the last line, we have used (8.1.22).

Finally, let $f \in C_b(\partial\mathfrak{G};\mathbb{R})$ be given. We already know that $H^{\mathfrak{G}}f$ is a bounded harmonic function in $\mathfrak{G}$ which satisfies (8.2.5) and (8.2.6). Now suppose that u is a second such function, and set $M = C + \|f\|_u$. Then, by the preceding, we have both that

$$\begin{aligned} M\mathcal{W}_{\mathbf{x}}^{(N)}(\mathfrak{e}^{\mathfrak{G}} < \infty) + u(\mathbf{x}) &\ge \big[H^{\mathfrak{G}}(M\mathbf{1} + f)\big](\mathbf{x}) \\ &= M\mathcal{W}_{\mathbf{x}}^{(N)}(\mathfrak{e}^{\mathfrak{G}} < \infty) + \big[H^{\mathfrak{G}}f\big](\mathbf{x}) \end{aligned}$$

and that

$$\begin{aligned} M\mathcal{W}_{\mathbf{x}}^{(N)}(\mathfrak{e}^{\mathfrak{G}} < \infty) - u(\mathbf{x}) &\ge \big[H^{\mathfrak{G}}(M\mathbf{1} - f)\big](\mathbf{x}) \\ &= M\mathcal{W}_{\mathbf{x}}^{(N)}(\mathfrak{e}^{\mathfrak{G}} < \infty) - \big[H^{\mathfrak{G}}f\big](\mathbf{x}), \end{aligned}$$

which means, of course, that $u = H^{\mathfrak{G}}f$. □

Although the preceding shows that the boundary condition in (8.2.6) makes the Dirichlet problem *well-posed*, it leaves open the possibility that one can do better. In order to see that, in general, one cannot, we will need the following.

8.2.7 Lemma. *Let $\mathfrak{G}$ be a nonempty, connected, open set in $\mathbb{R}^N$. Then*

$$\partial_{\text{reg}}\mathfrak{G} = \emptyset \iff \mathcal{W}_{\mathbf{x}}^{(N)}(\mathfrak{e}^{\mathfrak{G}} < \infty) = 0 \quad \textit{for all } \mathbf{x} \in \mathfrak{G}.$$

On the other hand, if $\partial_{\text{reg}}\mathfrak{G} \ne \emptyset$ and $\mathbf{b} \in \partial\mathfrak{G}$, then

$$\mathbf{b} \notin \partial_{\text{reg}}\mathfrak{G} \iff \lim_{r\searrow 0} \varlimsup_{\substack{\mathbf{x}\to\mathbf{b}\\ \mathbf{x}\in\mathfrak{G}}} \mathcal{W}_{\mathbf{x}}^{(N)}\Big(\psi(\mathfrak{e}^{\mathfrak{G}}) \notin B_{\mathbb{R}^N}(\mathbf{a},r)\,\&\,\mathfrak{e}^{\mathfrak{G}} < \infty\Big) > 0.$$

PROOF: The equivalence

$$\partial_{\text{reg}}\mathfrak{G} = \emptyset \iff \mathcal{W}_{\mathbf{x}}^{(N)}(\mathfrak{e}^{\mathfrak{G}} < \infty) = 0, \quad \mathbf{x} \in \mathfrak{G}$$

follows immediately from (8.1.21) and (8.1.22).

Now assume that $\partial_{\text{reg}}\mathfrak{G} \ne \emptyset$. If $\mathbf{b} \in \partial_{\text{reg}}\mathfrak{G}$, then

$$\lim_{r\searrow 0} \lim_{\substack{\mathbf{x}\to\mathbf{b}\\ \mathbf{x}\in\mathfrak{G}}} \mathcal{W}_{\mathbf{x}}^{(N)}\Big(\psi(\mathfrak{e}^{\mathfrak{G}}) \notin B_{\mathbb{R}^N}(\mathbf{b},r)\,\&\,\mathfrak{e}^{\mathfrak{G}} < \infty\Big) = 0$$

follows from (8.1.21). Thus, suppose that $\mathbf{b} \notin \partial_{\text{reg}}\mathfrak{G}$. Choose $\mathbf{a} \in \partial_{\text{reg}}\mathfrak{G}$, and set $B = B_{\mathbb{R}^N}(\mathbf{b},r)$ where $0 < r \le \frac{1}{2}|\mathbf{a}-\mathbf{b}|$. One can then construct an $f \in C\big(\partial\mathfrak{G};[0,1]\big)$ with the properties that $f = 0$ on $B \cap \partial\mathfrak{G}$ and $f(\mathbf{a}) = 1$. In particular, if $u = H^{\mathfrak{G}}f$, then

$$0 \le u(\mathbf{x}) \le \mathcal{W}_{\mathbf{x}}^{(N)}\Big(\psi(\mathfrak{e}^{\mathfrak{G}}) \notin B\,\&\,\mathfrak{e}^{\mathfrak{G}} < \infty\Big) \le 1 \quad \text{for all } \mathbf{x} \in \mathfrak{G},$$

and so we need only check that

$$\varlimsup_{\substack{\mathbf{x}\to\mathbf{b}\\ \mathbf{x}\in\mathfrak{G}}} u(\mathbf{x}) > 0.$$

To this end, first note that, since

$$\lim_{\substack{\mathbf{x}\to\mathbf{a}\\ \mathbf{x}\in\mathfrak{G}}} u(\mathbf{x}) = f(\mathbf{b}) = 1,$$

The Strong Maximum Principle (cf. Theorem 7.2.8) applied to the harmonic function $1-u$ in $\mathfrak{G}$ says that $u>0$ everywhere in $\mathfrak{G}$. Next, because $\mathbf{b}$ is not regular, we can find a $\delta>0$ and a sequence $\{\mathbf{x}_n\}_1^\infty \subseteq \mathfrak{G}$ such that $\mathbf{x}_n \to \mathbf{b}$ and

$$\epsilon \equiv \inf_{n\in\mathbb{Z}^+} \mathcal{W}^{(N)}_{\mathbf{x}_n}\left(\mathfrak{e}^{\mathfrak{G}} > \delta\right) > 0.$$

Moreover, by the Markov property, we know that

$$u(\mathbf{x}_n) \geq \mathbb{E}^{\mathcal{W}^{(N)}_{\mathbf{x}_n}}\left[f\left(\psi(\mathfrak{e}^{\mathfrak{G}})\right),\, \delta < \mathfrak{e}^{\mathfrak{G}} < \infty\right] = \int_{\mathfrak{G}} u(\mathbf{y})\, p^{\mathfrak{G}}(\delta, \mathbf{x}_n, \mathbf{y})\, d\mathbf{y}.$$

At the same time, by (8.1.10), we know that $p^{\mathfrak{G}}(\delta,\mathbf{x}_n,\mathbf{y}) \leq \gamma_\delta^N(\mathbf{y}-\mathbf{x}_n)$, and therefore that

$$\sup_{n\in\mathbb{Z}^+} \int_{\mathfrak{G}\setminus K} p^{\mathfrak{G}}(\delta, \mathbf{x}_n, \mathbf{y})\, d\mathbf{y} \leq \frac{\epsilon}{2}$$

for some compact subset K of $\mathfrak{G}$. Hence,

$$\varlimsup_{\substack{\mathbf{x}\to\mathbf{b}\\ \mathbf{x}\in\mathfrak{G}}} u(\mathbf{x}) \geq \varlimsup_{n\to\infty} u(\mathbf{x}_n) \geq \frac{\epsilon}{2} \inf_{\mathbf{y}\in K} u(\mathbf{y}) > 0. \quad \square$$

As a consequence of Lemma 8.2.7, we have the following negative complement to the positive result obtained in Theorem 8.2.2.

8.2.8 Theorem. *Let $\mathfrak{G}$ be a connected open set in $\mathbb{R}^N$ and assume that $\partial_{\mathrm{reg}}\mathfrak{G} \neq \emptyset$. If $\mathbf{b} \in \partial\mathfrak{G}\setminus\partial_{\mathrm{reg}}\mathfrak{G}$, then there exists an $f \in C\left(\partial\mathfrak{G};[0,1]\right)$ which has the property that for every $u \in C^2\left(\mathfrak{G};[0,\infty)\right)$ satisfying (8.2.4),*

$$f(\mathbf{b}) < \varlimsup_{\substack{\mathbf{x}\to\mathbf{b}\\ \mathbf{x}\in\mathfrak{G}}} u(\mathbf{x});$$

and so there can be no harmonic function u on $\mathfrak{G}$ which satisfies (8.2.5), (8.2.6), and

$$\lim_{\substack{\mathbf{x}\to\mathbf{b}\\ \mathbf{x}\in\mathfrak{G}}} u(\mathbf{x}) = f(\mathbf{b}).$$

PROOF: Given $\mathbf{b}$, use Lemma 8.2.7 to choose an $r \in (0,\infty)$ so that

$$\varlimsup_{\substack{\mathbf{x}\to\mathbf{b}\\ \mathbf{x}\in\mathfrak{G}}} \mathcal{W}_{\mathbf{x}}^{(N)}\Big(\psi(\mathfrak{e}^{\mathfrak{G}}) \notin B_{\mathbb{R}^N}(\mathbf{b},r)\ \&\ \mathfrak{e}^{\mathfrak{G}} < \infty\Big) > 0,$$

and construct f so that $f \equiv 1$ on $\partial\mathfrak{G} \cap B_{\mathbb{R}^N}(\mathbf{b},r)\complement$ and $f(\mathbf{b}) = 0$. Then, for any $u \in C^2\big(\mathfrak{G};[0,\infty)\big)$ satisfying (8.2.4),

$$f(\mathbf{b}) < \varlimsup_{\substack{\mathbf{x}\to\mathbf{b}\\ \mathbf{x}\in\mathfrak{G}}} \big[H^{\mathfrak{G}} f\big](\mathbf{x}) \le \varlimsup_{\substack{\mathbf{x}\to\mathbf{b}\\ \mathbf{x}\in\mathfrak{G}}} u(\mathbf{x}).$$

In particular, if u were a harmonic function on $\mathfrak{G}$ which satisfies (8.2.5) and (8.2.6), then u would equal $H^{\mathfrak{G}} f$ and therefore could not tend to $f(\mathbf{b})$ as $\mathfrak{G} \ni \mathbf{x} \to \mathbf{a}$. $\square$

We next want to take a closer look at the conditions under which we can assert the uniqueness of the solution to the Dirichlet problem. To begin, observe that the situation is quite satisfactory when we know that

$$\mathcal{W}_{\mathbf{x}}^{(N)}\big(\mathfrak{e}^{\mathfrak{G}} < \infty\big) = 1 \quad \text{for all } \mathbf{x} \in \mathfrak{G}. \tag{8.2.9}$$

Indeed, when (8.2.9) holds, the condition (8.2.5) becomes equivalent to the condition that u is bounded; and therefore we can say that, for each $f \in C_{\mathrm{b}}\big(\partial\mathfrak{G};\mathbb{R}\big)$, $H^{\mathfrak{G}} f$ is the one and only bounded harmonic u satisfying (8.2.6). In fact, this line of reasoning shows that the same conclusion holds as soon as one knows that $\mathcal{W}_{\mathbf{x}}^{(N)}\big(\mathfrak{e}^{\mathfrak{G}} < \infty\big)$ is bounded below by a positive constant; and therefore, because $\mathbf{x} \in \mathfrak{G} \longmapsto \mathcal{W}_{\mathbf{x}}^{(N)}(\mathfrak{e}^{\mathfrak{G}} < \infty)$ is a bounded harmonic function which satisfies (8.2.6) with $f = \mathbf{1}$, we see that

$$\inf_{\mathbf{x}\in\mathfrak{G}} \mathcal{W}_{\mathbf{x}}^{(N)}\big(\mathfrak{e}^{\mathfrak{G}} < \infty\big) > 0 \implies \inf_{\mathbf{x}\in\mathfrak{G}} \mathcal{W}_{\mathbf{x}}^{(N)}\big(\mathfrak{e}^{\mathfrak{G}} < \infty\big) = 1. \tag{8.2.10}$$

On the basis of these simple observations, we can now prove the following.

8.2.11 Theorem. *Let $\mathfrak{G}$ be a nonempty subset of $\mathbb{R}^N$. If (8.2.9) holds, then, for each $f \in C_{\mathrm{b}}(\partial\mathfrak{G};\mathbb{R})$, $H^{\mathfrak{G}} f$ is the unique bounded harmonic function u for which (8.2.6) holds. Moreover, even if (8.2.9) fails, when $N \ge 3$ and $f \in C_{\mathrm{c}}(\partial\mathfrak{G};\mathbb{R})$, $H^{\mathfrak{G}} f$ is the one and only bounded harmonic function on $\mathfrak{G}$ which satisfies not only (8.2.6) but also the condition*

$$\lim_{\substack{|\mathbf{x}|\to\infty\\ \mathbf{x}\in\mathfrak{G}}} u(\mathbf{x}) = 0. \tag{8.2.12}$$

PROOF: The first assertion is covered by the preceding discussion. To prove the second assertion, let $f \in C_{\mathrm{c}}(\partial\mathfrak{G};\mathbb{R})$ be given. We already know that $H^{\mathfrak{G}} f$ is a bounded harmonic function satisfying (8.2.6), but we must still show that it satisfies (8.2.12). For this purpose, choose $r \in (0,\infty)$ so that f is supported in $B_{\mathbb{R}^N}(\mathbf{0},r)$. Then (cf. the last part of Theorem 7.2.11), because $N \geq 3$,

$$\left|\left[H^{\mathfrak{G}} f\right](\mathbf{x})\right| \leq \|f\|_{\mathrm{u}} \mathcal{W}_{\mathbf{x}}^{(N)}(\zeta_r < \infty) \longrightarrow 0 \quad \text{as } |\mathbf{x}| \to \infty.$$

Finally, to prove that $H^{\mathfrak{G}} f$ is the only such function u, select bounded open sets $\mathfrak{G}_n \nearrow \mathfrak{G}$ with $\overline{\mathfrak{G}_n} \subset\subset \mathfrak{G}$, and note that, for each $T \in (0,\infty)$,

$$\begin{aligned} u(\mathbf{x}) &= \lim_{n\to\infty} \mathbb{E}^{\mathcal{W}_{\mathbf{x}}^{(N)}}\left[u\big(\psi(T \wedge \mathfrak{e}^{\mathfrak{G}_n})\big)\right] \\ &= \mathbb{E}^{\mathcal{W}_{\mathbf{x}}^{(N)}}\left[f\big(\psi(\mathfrak{e}^{\mathfrak{G}})\big),\, \mathfrak{e}^{\mathfrak{G}} \leq T\right] + \mathbb{E}^{\mathcal{W}_{\mathbf{x}}^{(N)}}\left[u\big(\psi(T)\big),\, T < \mathfrak{e}^{\mathfrak{G}} < \infty\right] \\ &\qquad + \mathbb{E}^{\mathcal{W}_{\mathbf{x}}^{(N)}}\left[u\big(\psi(T)\big),\, \mathfrak{e}^{\mathfrak{G}} = \infty\right]. \end{aligned}$$

Clearly,

$$\left[H^{\mathfrak{G}} f\right](\mathbf{x}) = \lim_{T\nearrow\infty} \mathbb{E}^{\mathcal{W}_{\mathbf{x}}^{(N)}}\left[f\big(\psi(\mathfrak{e}^{\mathfrak{G}})\big),\, \mathfrak{e}^{\mathfrak{G}} \leq T\right]$$

and

$$\lim_{T\nearrow\infty} \mathbb{E}^{\mathcal{W}_{\mathbf{x}}^{(N)}}\left[u\big(\psi(T)\big),\, T < \mathfrak{e}^{\mathfrak{G}} < \infty\right] = 0.$$

Finally, because $N \geq 3$ and therefore (cf. (7.2.16)) $\big|\psi(T)\big| \longrightarrow \infty$ as $T \nearrow \infty$ for $\mathcal{W}_{\mathbf{x}}^{(N)}$-almost every $\psi \in \mathfrak{P}(\mathbb{R}^N)$, (8.2.12) guarantees that

$$\lim_{T\nearrow\infty} \mathbb{E}^{\mathcal{W}_{\mathbf{x}}^{(N)}}\left[u\big(\psi(T)\big),\, \mathfrak{e}^{\mathfrak{G}} = \infty\right] = 0;$$

which completes the proof that $u = H^{\mathfrak{G}} f$. □

When $N \geq 3$, the result in Theorem 8.2.11 gives a complete analytic characterization of the distribution of $\psi \in \{\mathfrak{e}^{\mathfrak{G}} < \infty\} \longmapsto \psi(\mathfrak{e}^{\mathfrak{G}})$ under $\mathcal{W}_{\mathbf{x}}^{(N)}$. However, there is still more to be done when $N \in \{1,2\}$; and to remove this gap we present the following.

8.2.13 Theorem. *If $N \in \{1,2\}$, then for every nonempty open set $\mathfrak{G}$ in $\mathbb{R}^N$:*

$$\mathcal{W}_{\mathbf{x}}^{(N)}\big(\mathfrak{e}^{\mathfrak{G}} < \infty\big) = 1 \text{ for all } \mathbf{x} \in \mathfrak{G} \quad \text{or} \quad \mathcal{W}_{\mathbf{x}}^{(N)}\big(\mathfrak{e}^{\mathfrak{G}} < \infty\big) = 0 \text{ for all } \mathbf{x} \in \mathfrak{G},$$

depending on whether $\partial_{\text{reg}}\mathfrak{G} \neq \emptyset$ or $\partial_{\text{reg}}\mathfrak{G} = \emptyset$. Moreover, if $\partial_{\text{reg}}\mathfrak{G} = \emptyset$, then the only functions $u \in C^2\big(\mathfrak{G};[0,\infty)\big)$ satisfying $\mathbf{\Delta} u \leq 0$ are constant. In particular: either $\partial_{\text{reg}}\mathfrak{G} = \emptyset$, and there are no nonconstant, nonnegative harmonic functions on $\mathfrak{G}$; or $\partial_{\text{reg}}\mathfrak{G} \neq \emptyset$, and, for each $f \in C_{\mathrm{b}}(\partial\mathfrak{G};\mathbb{R})$, $H^{\mathfrak{G}} f$ is the unique bounded harmonic function on $\mathfrak{G}$ satisfying (8.2.6).

PROOF: Suppose that $\mathcal{W}^{(N)}_{\mathbf{x}_0}\big(\mathfrak{e}^{\mathfrak{G}} < \infty\big) < 1$ for some $\mathbf{x}_0 \in \mathfrak{G}$, and choose open sets $\mathfrak{G}_n \nearrow \mathfrak{G}$ so that $\mathbf{x}_0 \in \mathfrak{G}_1$ and $\overline{\mathfrak{G}_n} \subset\subset \mathfrak{G}$ for all $n \in \mathbb{Z}^+$. Given $u \in C^2\big(\mathfrak{G};[0,\infty)\big)$ with $\boldsymbol{\Delta} u \leq 0$, set

$$X_n(t,\boldsymbol{\psi}) = \mathbf{1}_{(t,\infty]}\big(\mathfrak{e}^{\mathfrak{G}_n}(\boldsymbol{\psi})\big)\, u\big(\boldsymbol{\psi}(t)\big) \quad \text{for } (t,\boldsymbol{\psi}) \in [0,\infty) \times \mathfrak{P}(\mathbb{R}^N).$$

As an elementary application of Lemma 8.1.15, one can easily check that $-X_n(t)$ is a nonpositive, continuous, $\mathcal{W}^{(N)}_{\mathbf{x}}$-submartingale relative to $\big\{\mathcal{B}_t : t \in [0,\infty)\big\}$ for each $n \in \mathbb{Z}^+$. Hence, since

$$X_n(t,\boldsymbol{\psi}) \nearrow X(t,\boldsymbol{\psi}) \equiv \mathbf{1}_{(t,\infty]}(\mathfrak{e}^{\mathfrak{G}})\, u\big(\boldsymbol{\psi}(t)\big) \quad \text{pointwise as } n \to \infty,$$

an application of The Monotone Convergence Theorem allows us to conclude that $\big(-X(t),\mathcal{B}_t,\mathcal{W}^{(N)}_{\mathbf{x}_0}\big)$ is also a nonpositive, continuous, submartingale. In particular, by Theorem 7.1.16, this means that

$$\lim_{t\to\infty} u\big(\boldsymbol{\psi}(t)\big) \text{ exists for } \mathcal{W}^{(N)}_{\mathbf{x}_0}\text{-almost every } \boldsymbol{\psi} \in \{\mathfrak{e}^{\mathfrak{G}} = \infty\}.$$

At the same time (cf. Exercise 7.2.32), we know that, for $\mathcal{W}^{(N)}_{\mathbf{x}_0}$-almost every $\boldsymbol{\psi} \in \mathfrak{P}(\mathbb{R}^N)$,

$$\int_0^\infty \mathbf{1}_U\big(\boldsymbol{\psi}(t)\big)\,dt = \infty \quad \text{for all open } U \neq \emptyset.$$

Hence, since $\mathcal{W}^{(N)}_{\mathbf{x}_0}\big(\mathfrak{e}^{\mathfrak{G}} = \infty\big) > 0$, there exists a $\boldsymbol{\psi}_0 \in \mathfrak{P}(\mathbb{R}^N)$ with the properties that $\mathfrak{e}^{\mathfrak{G}}(\boldsymbol{\psi}_0) = \infty$,

$$\int_0^\infty \mathbf{1}_U\big(\boldsymbol{\psi}_0(t)\big)\,dt = \infty \quad \text{for all open } U \neq \emptyset,$$

and

$$\lim_{t\to\infty} u\big(\boldsymbol{\psi}_0(t)\big) \text{ exists;}$$

which is possible only if u is constant. In other words, we have now proved that when $\mathcal{W}^{(N)}_{\mathbf{x}_0}(\mathfrak{e}^{\mathfrak{G}} < \infty) < 1$ for some $\mathbf{x}_0 \in \mathfrak{G}$, then the only $u \in C^2\big(\mathfrak{G};[0,\infty)\big)$ with $\boldsymbol{\Delta} u \leq 0$ are constant.

Given the preceding paragraph, the rest is easy. Indeed, if $\partial_{\text{reg}}\mathfrak{G} = \emptyset$, then (8.1.22) already implies that $\mathcal{W}^{(N)}_{\mathbf{x}}(\mathfrak{e}^{\mathfrak{G}} < \infty) = 0$ for all $\mathbf{x} \in \mathfrak{G}$. On the other hand, if $\mathbf{a} \in \partial_{\text{reg}}\mathfrak{G}$ but $\mathcal{W}^{(N)}_{\mathbf{x}_0}\big(\mathfrak{e}^{\mathfrak{G}} < \infty\big) < 1$ for some $\mathbf{x}_0 \in \mathfrak{G}$, set

$$u(\mathbf{x}) \equiv \big[H^{\mathfrak{G}}\mathbf{1}\big](\mathbf{x}) = \mathcal{W}^{(N)}_{\mathbf{x}}\big(\mathfrak{e}^{\mathfrak{G}} < \infty\big).$$

Then, by the preceding paragraph, u is constant and we derive the contradiction

$$1 > u(\mathbf{x}_0) = \lim_{\substack{\mathbf{x}\to\mathbf{a}\\ \mathbf{x}\in\mathfrak{G}}} u(\mathbf{x}) = 1. \quad \square$$

With Theorems 8.2.11 and 8.2.13, we now have a complete analytic characterization of the distribution of $\psi \in \{\mathfrak{e}^{\mathfrak{G}} < \infty\} \longmapsto \psi(\mathfrak{e}^{\mathfrak{G}}) \in \partial G$. However, for the reader who is familiar with the characterization of regular points in terms of **barrier functions**, the following little addendum may be helpful. For example, it shows that, when $\mathfrak{G}$ is bounded and $f \in C(\partial G; \mathbb{R})$, then $H^{\mathfrak{G}} f$ coincides with the solution to the Dirichlet problem for f at which one arrives via the procedure of Wiener, Perron, and Brelot (cf. Exercise 8.2.37).

8.2.14 Lemma. *Given $\mathbf{a} \in \partial\mathfrak{G}$, $\mathbf{a} \in \partial_{\text{reg}}\mathfrak{G}$ if and only if there exists an $r \in (0,\infty)$ and a bounded $\eta \in C^2\big(\mathfrak{G} \cap B_{\mathbb{R}^N}(\mathbf{a}, r); (0,\infty)\big)$ with the properties that $\Delta\eta \le -\epsilon$ for some $\epsilon > 0$ and $\eta(\mathbf{x}) \longrightarrow 0$ as $\mathbf{x} \to \mathbf{a}$ through $\mathfrak{G}$. In fact, if $\mathfrak{G}$ is bounded, and*

$$m^{\mathfrak{G}}(\mathbf{x}) = \mathbb{E}^{\mathcal{W}_{\mathbf{x}}^{(N)}}\left[\mathfrak{e}^{\mathfrak{G}}\right], \quad \mathbf{x} \in \mathfrak{G},$$

then $m^{\mathfrak{G}} \in C^\infty\big(\mathfrak{G}; [0,\infty)\big) \cap C_{\text{b}}(\mathfrak{G} : \mathbb{R})$, $\frac{1}{2}\Delta m^{\mathfrak{G}} = -1$, and

$$\lim_{\substack{\mathbf{x}\to\mathbf{a}\\ \mathbf{x}\in\mathfrak{G}}} m^{\mathfrak{G}}(\mathbf{x}) = 0 \iff \mathbf{a} \in \partial_{\text{reg}}\mathfrak{G}.$$

PROOF: Since (cf. Lemma 8.1.19) regularity is a local property, we may and will assume throughout that $\mathfrak{G}$ is bounded.

Because $\mathfrak{G}$ is bounded, we can find $R \in (0,\infty)$ so that $\overline{\mathfrak{G}} \subset\subset B \equiv B_{\mathbb{R}^N}(\mathbf{0}, R)$. Hence, by (7.2.12), we have that

$$0 \le m^{\mathfrak{G}}(\mathbf{x}) \le m^B(\mathbf{x}) = \frac{R^2 - |\mathbf{x}|^2}{N}, \quad \mathbf{x} \in \mathfrak{G}; \tag{8.2.15}$$

which certainly means that $m^{\mathfrak{G}}$ is bounded. In addition, by the strong Markov property,

$$\begin{aligned} m^B(\mathbf{x}) - m^{\mathfrak{G}}(\mathbf{x}) &= \mathbb{E}^{\mathcal{W}_{\mathbf{x}}^{(N)}}\left[\mathfrak{e}^B - \mathfrak{e}^{\mathfrak{G}}\right] \\ &= \mathbb{E}^{\mathcal{W}_{\mathbf{x}}^{(N)}}\left[m^B\big(\psi(\mathfrak{e}^B)\big),\, \mathfrak{e}^B < \infty\right] = \left[H^{\mathfrak{G}} m^B\right](\mathbf{x}) \end{aligned}$$

for all $\mathbf{x} \in \mathfrak{G}$. Hence, by Theorem 8.2.2 and the last relation in (8.2.15), we see that $m^{\mathfrak{G}}$ is not only smooth but also satisfies $\frac{1}{2}\Delta m^{\mathfrak{G}} = -1$. Finally, note that

$$\mathbf{a} \in \partial_{\text{reg}}\mathfrak{G} \iff \lim_{\substack{\mathbf{x}\to\mathbf{a}\\ \mathbf{x}\in\mathfrak{G}}} m^{\mathfrak{G}}(\mathbf{x}) = 0. \tag{8.2.16}$$

Indeed, the "if" direction is completely trivial; and to go the other way, observe that because $\mathfrak{e}^{\mathfrak{G}} \le \mathfrak{e}^B$, (7.2.12) (and, even more convincingly, Lemma 7.2.18) implies that

$$C \equiv \sup_{\mathbf{x}\in\mathfrak{G}} \mathbb{E}^{\mathcal{W}_{\mathbf{x}}^{(N)}}\left[\left(\mathfrak{e}^{\mathfrak{G}}\right)^2\right]^{\frac{1}{2}} < \infty,$$

and therefore that

$$m^{\mathfrak{G}}(\mathbf{x}) \le \delta + C\mathcal{W}_{\mathbf{x}}^{(N)}\left(\mathfrak{e}^{\mathfrak{G}} \ge \delta\right)^{\frac{1}{2}} \quad \text{for every } \delta > 0.$$

In view of (8.2.16), all that remains is to show that the existence of $r \in (0,\infty)$ and η with the prescribed properties implies that $m^{\mathfrak{G}}(\mathbf{x}) \longrightarrow 0$ as $\mathfrak{G} \ni \mathbf{x} \to \mathbf{a}$; and, because of the local nature of regularity, we may and will assume that $\mathfrak{G} = \mathfrak{G} \cap B_{\mathbb{R}^N}(\mathbf{a}, r)$. Next, set

$$M = \sup_{\mathbf{x}\in\mathfrak{G}} m^{\mathfrak{G}}(\mathbf{x}) \quad \text{and} \quad u(\mathbf{x}) = 2\eta(\mathbf{x}) + \left(M - m^{\mathfrak{G}}(\mathbf{x})\right) \text{ for } \mathbf{x} \in \mathfrak{G}.$$

Then

$$u \ge 0, \quad \boldsymbol{\Delta} u \le 0 \text{ in } \mathfrak{G}, \quad \text{and} \quad \varliminf_{\substack{\mathbf{x}\to\mathbf{b}\\ \mathbf{x}\in\mathfrak{G}}} u(\mathbf{x}) \ge M \text{ for } \mathbf{b} \in \partial_{\text{reg}}\mathfrak{G};$$

and so, by the second part of Theorem 8.2.2, we know that $u \ge M$ in $\mathfrak{G}$, from which it is clear that

$$\varlimsup_{\substack{\mathbf{x}\to\mathbf{a}\\ \mathbf{x}\in\mathfrak{G}}} m^{\mathfrak{G}}(\mathbf{x}) \le 2 \varlimsup_{\substack{\mathbf{x}\to\mathbf{a}\\ \mathbf{x}\in\mathbf{G}}} \eta(\mathbf{x}) = 0. \quad \square$$

In the classical liturature, a function η of the sort described in Lemma 8.3.14 is called a **barrier** in $\mathfrak{G}$ at $\mathbf{a}$.

We now have a rather complete abstract analysis of when the Dirichlet problem can be solved. Indeed, we know that, at least when $f \in C_c(\partial G; \mathbb{R})$, one cannot do better than take one's solution to be the function $H^{\mathfrak{G}} f$ given by (8.2.3). For this reason, we call

(8.2.17) $$\Pi^{\mathfrak{G}}(\mathbf{x}, \Gamma) \equiv \mathcal{W}_{\mathbf{x}}^{(N)}\left(\psi(\mathfrak{e}^{\mathfrak{G}}) \in \Gamma,\ \mathfrak{e}^{\mathfrak{G}} < \infty\right)$$

the **harmonic measure for $\mathfrak{G}$ based at $\mathbf{x} \in \mathfrak{G}$ of the set** $\Gamma \in \mathcal{B}_{\partial G}$; which means, of course, that

(8.2.18) $$\left[H^{\mathfrak{G}} f\right](\mathbf{x}) = \int_{\partial\mathfrak{G}} f(\boldsymbol{\eta})\, \Pi^{\mathfrak{G}}(\mathbf{x}, d\boldsymbol{\eta}).$$

This connection between harmonic measure and Wiener's measure is due to Doob,† and it is the starting point for what, in the hands of G. Hunt,‡ became an isomorphism between potential theory and the theory of Markov processes.

† Actually, S. Kakutani's 1944 article "Two dimensional Brownian motion and harmonic functions," *Proc. Imp. Acad. Tokyo* **20**, together with his 1949 article "Markoff process and the Dirichlet problem," *Proc. Imp. Acad. Tokyo* **21**, are generally accepted as the first place in which a definitive connection between the harmonic functions and Wiener's measure was established. However, it was not until with Doob's "Semimartingales and subharmonic functions," *T.A.M.S.* **77**, in 1954 that the connection was completed.

‡ In 1957, Hunt published a series of three articles: "Markov processes and potentials, parts I, II, & III," *Ill. J. Math.* **1** & **2**. In these articles, he literally created the modern theory of Markov processes and their relationship to potential theory. To see just how far Hunt's ideas can be elaborated, see M. Sharpe's *General Theory of Markov Processes*, Acad. Press Series in Pure & Appl. Math. **133** (1988).

Although (8.2.17) provides an intuitively appealing formula for the harmonic measure $\Pi^{\mathfrak{G}}(\mathbf{x}, \cdot)$, it can hardly be considered *explicit*. Thus, before closing this section, we will write down two important examples in which explicit formulae for the harmonic measure are readily available. The first case in which an explicit expression is known has already been discussed in Exercise 4.3.49: namely, when $\mathfrak{G}$ is a half-space. To be precise, if $N = 1$ and $\mathfrak{G} = (0, \infty)$, then, because one-dimensional Wiener paths hit points, it is clear that $\Pi^{(0,\infty)}(\mathbf{x}, \cdot)$ is nothing but the point mass δ_0 for all $\mathbf{x} \in (0, \infty)$. On the other hand, if $N \geq 2$ and $\mathfrak{G} = \mathbb{R}^N_+ \equiv \mathbb{R}^{N-1} \times (0, \infty)$, then we know from (8.2.17) and (4.3.50) that, for $y \in (0, \infty)$,

$$\Pi^{\mathbb{R}^N_+}\big((\mathbf{0}, y), d\boldsymbol{\eta}\big) = \frac{2}{\omega_{N-1}} \frac{y}{\big(y^2 + |\boldsymbol{\eta}|^2\big)^{\frac{N}{2}}} \lambda_{\mathbb{R}^{N-1}}(d\boldsymbol{\eta}), \quad y \in (0, \infty),$$

where we have identified $\partial\mathbb{R}^N_+$ with $\mathbb{R}^{N-1}$ and used $\lambda_{\mathbb{R}^{N-1}}$ to denote Lebesgue's measure on $\mathbb{R}^{N-1}$. Hence, after a trivial translation,

$$\Pi^{\mathbb{R}^N_+}\big((\mathbf{x}, y), d\boldsymbol{\eta}\big) = \frac{2}{\omega_{N-1}} \frac{y}{\big(y^2 + |\mathbf{x} - \boldsymbol{\eta}|^2\big)^{\frac{N}{2}}} \lambda_{\mathbb{R}^{N-1}}(d\boldsymbol{\eta}) \quad \text{for} \quad (\mathbf{x}, y) \in \mathbb{R}^{N-1} \times (0, \infty). \tag{8.2.19}$$

Moreover, by using further translation plus Wiener rotation invariance (cf. Exercise 3.3.28), one can pass easily from the preceding to an explicit expression of the harmonic measure for an arbitrary half-space.

In the preceding, we were able to derive an expression giving the harmonic measure for half-spaces directly from probabilistic considerations. Unfortunately, half-spaces are essentially the only regions for which probabilistic reasoning yields such explicit expressions. Indeed, embarrassing as it is to admit, it should still be recognized that, when it comes to explicit expressions, the time-honored techniques of clever changes of variables followed by separation of variables is more powerful than anything which comes out of (8.2.17). To wit, the author is unable to give a truly *probabilistic derivation* of the classical formula given in the following.

8.2.20 Theorem (Poisson's Formula). *Use $\lambda_{\mathbf{S}^{N-1}}$ to denote the surface measure on the unit sphere $\mathbf{S}^{N-1}$ in $\mathbb{R}^N$, and define*

$$\pi^{(N)}(\mathbf{x}, \boldsymbol{\eta}) = \frac{1}{\omega_{N-1}} \frac{1 - |\mathbf{x}|^2}{|\mathbf{x} - \boldsymbol{\eta}|^N} \quad \text{for } (\mathbf{x}, \boldsymbol{\eta}) \in B_{\mathbb{R}^N}(\mathbf{0}, 1) \times \mathbf{S}^{N-1}. \tag{8.2.21}$$

Then:

$$\Pi^{B_{\mathbb{R}^N}(\mathbf{0},1)}(\mathbf{x}, d\boldsymbol{\eta}) = \pi^{(N)}(\mathbf{x}, \boldsymbol{\eta})\, \lambda_{\mathbf{S}^{N-1}}(d\boldsymbol{\eta}), \quad \text{for } \mathbf{x} \in B_{\mathbb{R}^N}(\mathbf{0}, 1). \tag{8.2.22}$$

(Cf. Exercise 8.3.33 below.) More generally, if $\mathbf{c} \in \mathbb{R}^N$, $r \in (0, \infty)$, *and* $\mathbf{x} \in B_{\mathbb{R}^N}(\mathbf{c}, r)$, *then*

$$\Pi^{B_{\mathbb{R}^N}(\mathbf{c},r)}(\mathbf{x}, d\boldsymbol{\eta}) = \frac{1}{\omega_{N-1} r} \frac{r^2 - |\mathbf{x}-\mathbf{c}|^2}{|\mathbf{x}-\boldsymbol{\eta}|^N} \lambda_{\mathbf{S}^{N-1}(\mathbf{c},r)}(d\boldsymbol{\eta}), \tag{8.2.23}$$

where $\lambda_{\mathbf{S}^{N-1}(\mathbf{c},r)}$ *denotes the surface measure on the sphere*

$$\mathbf{S}^{N-1}(\mathbf{c}, r) \equiv \partial B_{\mathbb{R}^N}(\mathbf{c}, r).$$

Equivalently, for each open $\mathfrak{G}$ *in* $\mathbb{R}^N$, *harmonic function* u *on* $\mathfrak{G}$, $\overline{B_{\mathbb{R}^N}(\mathbf{c}, r)} \subset\subset \mathfrak{G}$, *and* $\mathbf{x} \in B_{\mathbb{R}^N}(\mathbf{c}, r)$:

$$u(\mathbf{x}) = \int_{\mathbf{S}^{N-1}} u(\mathbf{c} + r\boldsymbol{\xi})\, \pi^{(N)}\left(\tfrac{\mathbf{x}-\mathbf{c}}{r}, \boldsymbol{\xi}\right) \lambda_{\mathbf{S}^{N-1}}(d\boldsymbol{\xi}). \tag{8.2.24}$$

In particular, if $\{u_n\}_1^\infty$ *is a sequence of harmonic function on the open set* $\mathfrak{G}$ *and if* $u_n \longrightarrow u$ *boundedly and pointwise on compact subsets of* $\mathfrak{G}$, *then* u *is harmonic on* $\mathfrak{G}$ *and* $u_n \longrightarrow u$ *uniformly on compact subsets.*

PROOF: Set $B = B_{\mathbb{R}^N}(\mathbf{0}, 1)$. As an easy application of Wiener translation and scaling invariance, we see that, for any $\mathbf{x} \in B_{\mathbb{R}^N}(\mathbf{c}, r)$ and bounded measurable f on $\mathbf{S}^{N-1}(\mathbf{c}, r)$:

$$\left[H^{B_{\mathbb{R}^N}(\mathbf{c},r)} f\right](\mathbf{x}) = \left[H^B f_{\mathbf{c},r}\right]\left(\tfrac{\mathbf{x}-\mathbf{c}}{r}\right) \quad \text{where } f_{\mathbf{c},r}(\boldsymbol{\eta}) \equiv f(\mathbf{c} + r\boldsymbol{\eta}).$$

Hence, the formulae in (8.2.23) and (8.2.24) follow immediately from the one in (8.2.22).

To prove (8.2.22), first check, by direct calculation, that $\pi^{(N)}(\,\cdot\,, \boldsymbol{\eta})$ is harmonic in B for each $\boldsymbol{\eta} \in \mathbf{S}^{N-1}$. Hence, in order to complete the proof, all that we have to do is check that

$$\lim_{\substack{\mathbf{x}\to\mathbf{a}\\ \mathbf{x}\in B}} \int_{\mathbf{S}^{N-1}} f(\boldsymbol{\eta})\, \pi^{(N)}(\mathbf{x}, \boldsymbol{\eta})\, \lambda_{\mathbf{S}^{N-1}}(d\boldsymbol{\eta}) = f(\mathbf{a})$$

for every $f \in C(\mathbf{S}^{N-1}; \mathbb{R})$ and $\mathbf{a} \in \mathbf{S}^{N-1}$. Since, for each $\delta > 0$, it is clear that

$$\lim_{\substack{\mathbf{x}\to\mathbf{a}\\ \mathbf{x}\in B}} \int_{\mathbf{S}^{N-1} \cap B_{\mathbb{R}^N}(\mathbf{a},\delta)\complement} \pi^{(N)}(\mathbf{x}, \boldsymbol{\eta})\, \lambda_{\mathbf{S}^{N-1}}(d\boldsymbol{\eta}) = 0,$$

we will be done as soon as we show that

$$\int_{\mathbf{S}^{N-1}} \pi^{(N)}(\mathbf{x}, \boldsymbol{\eta})\, \lambda_{\mathbf{S}^{N-1}}(d\boldsymbol{\eta}) = 1 \quad \text{for all } \mathbf{x} \in B.$$

But, because $\pi^{(N)}(\,\cdot\,,\boldsymbol{\xi})$ is harmonic in B and, by **(i)** in Exercise 4.3.51,

$$\Pi^{B_{\mathbb{R}^N}(\mathbf{0},r)}(\mathbf{0},\,\cdot\,)=\frac{\lambda_{\mathbf{S}^{N-1}(\mathbf{0},r)}}{\omega_{N-1}r^{N-1}}\quad\text{for each } r\in(0,\infty),$$

we have that, for $r\in[0,1)$ and $\boldsymbol{\xi}\in\mathbf{S}^{N-1}$,

$$\begin{aligned}1&=\omega_{N-1}\pi^{(N)}(\mathbf{0},\boldsymbol{\xi})=\omega_{N-1}\big[H^{B_{\mathbb{R}^N}(\mathbf{0},r)}\pi^{(N)}(\,\cdot\,,\boldsymbol{\xi})\big](\mathbf{0})\\&=\int_{\mathbf{S}^{N-1}}\pi^{(N)}(r\boldsymbol{\eta},\boldsymbol{\xi})\,\lambda_{\mathbf{S}^{N-1}}(d\boldsymbol{\eta})=\int_{\mathbf{S}^{N-1}}\pi^{(N)}(r\boldsymbol{\xi},\boldsymbol{\eta})\,\lambda_{\mathbf{S}^{N-1}}(d\boldsymbol{\eta}),\end{aligned}$$

where, in the final step, we have used the easily verified identity

$$\pi^{(N)}(r\boldsymbol{\eta},\boldsymbol{\xi})=\pi^{(N)}(r\boldsymbol{\xi},\boldsymbol{\eta})\quad\text{for all } r\in[0,1)\text{ and }(\boldsymbol{\xi},\boldsymbol{\eta})\in\big(\mathbf{S}^{N-1}\big)^2.$$

Thus, by writing $\mathbf{x}=r\boldsymbol{\xi}$, we obtain the desired identity. □

When $N=2$, one gets the following dividend from Theorem 8.2.20.

8.2.25 Corollary. *Set $\mathbf{D}(r)=B_{\mathbb{R}^2}(\mathbf{0},r)$ for $r\in(0,\infty)$. Then*

$$\Pi^{\mathbb{R}^2\setminus\overline{\mathbf{D}(r)}}(\mathbf{x},d\boldsymbol{\eta})=\frac{r|\mathbf{x}|^2}{2\pi}\frac{|\mathbf{x}|^2-r^2}{\big||\mathbf{x}|^2\boldsymbol{\eta}-r^2\mathbf{x}\big|^2}\lambda_{\mathbf{S}^1(\mathbf{0},r)}(d\boldsymbol{\eta})\tag{8.2.26}$$

for each $\mathbf{x}\notin\overline{\mathbf{D}(r)}$. In particular, if $u\in C_{\mathrm{b}}\big(\mathbb{R}^2\backslash\mathbf{D}(r);\mathbb{R}\big)$ is harmonic on $\mathbb{R}^2\backslash\overline{\mathbf{D}(r)}$, then

$$u(\mathbf{x})=\frac{|\mathbf{x}|^2}{2\pi}\int_{\mathbf{S}^1}\frac{|\mathbf{x}|^2-r^2}{\big||\mathbf{x}|^2\boldsymbol{\eta}-r\mathbf{x}\big|^2}u(r\boldsymbol{\eta})\lambda_{\mathbf{S}^1}(d\boldsymbol{\eta});\tag{8.2.27}$$

and so

$$\lim_{|\mathbf{x}|\to\infty}u(\mathbf{x})=\frac{1}{2\pi}\int_{\mathbf{S}^1}u(r\boldsymbol{\eta})\,\lambda_{\mathbf{S}^1}(d\boldsymbol{\eta}).\tag{8.2.28}$$

PROOF: By an easy scaling argument, we may and will assume that $r=1$. Thus, set $\mathbf{D}=\mathbf{D}(1)$, and assume that $u\in C_{\mathrm{b}}\big(\mathbb{R}^2\setminus\mathbf{D};\mathbb{R}\big)$ is harmonic in $\mathbb{R}^2\setminus\overline{\mathbf{D}}$. Next, set

$$v(\mathbf{x})=u\left(\frac{\mathbf{x}}{|\mathbf{x}|^2}\right)\quad\text{for }\mathbf{x}\in\mathbf{D}\setminus\{\mathbf{0}\}.$$

Obviously, v is bounded and continuous. In addition, by using polar coordinates, one easily checks that v is harmonic in $\mathbf{D}\setminus\{\mathbf{0}\}$. In particular, if $\rho\in(0,1)$ and $\mathfrak{G}(\rho)\equiv B\setminus\overline{B_{\mathbb{R}^2}(\mathbf{0},\rho)}$, then (cf. the notation in Theorem 7.2.11)

$$v(\mathbf{x})=\mathbb{E}^{\mathcal{W}^{(N)}_{\mathbf{x}}}\Big[v(\boldsymbol{\psi}(\zeta_1)),\,\zeta_1<\zeta_\rho\Big]+\mathbb{E}^{\mathcal{W}^{(N)}_{\mathbf{x}}}\Big[v(\boldsymbol{\psi}(\zeta_\rho)),\,\zeta_\rho<\zeta_1\Big]$$

for all $\mathbf{x}\in\mathfrak{G}(\rho)$. Hence, because (cf. Theorem 7.2.11) $\zeta_\rho\nearrow\infty$ (a.s., $\mathcal{W}^{(N)}_{\mathbf{x}}$) as $\rho\searrow 0$, this leads to

$$v(\mathbf{x})=\mathbb{E}^{\mathcal{W}^{(N)}_{\mathbf{x}}}\Big[v(\boldsymbol{\psi}(\zeta_1)),\,\zeta_1<\infty\Big]=\frac{1}{2\pi}\int_{\mathbf{S}^1}\frac{1-|\mathbf{x}|^2}{|\boldsymbol{\eta}-\mathbf{x}|^2}u(\boldsymbol{\eta})\,\lambda_{\mathbf{S}^1}(d\boldsymbol{\eta})$$

for all $\mathbf{x}\in B\setminus\{\mathbf{0}\}$. Finally, given the preceding, the rest comes down to a simple matter of bookkeeping. □

As a second application of Poisson's formula, we give the following famous observation, which can be viewed as a quantitative version of Strong Maximum Principle (cf. Theorem 7.2.8) for harmonic functions.

8.2.29 Corollary (Harnack's Principle). *For any* $\mathbf{c} \in \mathbb{R}^N$ *and* $r \in (0, \infty)$*:*

$$\begin{aligned}&\frac{r^{N-2}\big(r - |\mathbf{x}-\mathbf{c}|\big)}{\big(r + |\mathbf{x}-\mathbf{c}|\big)^{N-1}}\,\Pi^{B_{\mathbb{R}^N}(\mathbf{c},r)}(\mathbf{c},\,\cdot\,)\\ &\qquad \le \Pi^{B_{\mathbb{R}^N}(\mathbf{c},r)}(\mathbf{x},\,\cdot\,) \le \frac{r^{N-2}\big(r + |\mathbf{x}-\mathbf{c}|\big)}{\big(r - |\mathbf{x}-\mathbf{c}|\big)^{N-1}}\,\Pi^{B_{\mathbb{R}^N}(\mathbf{c},r)}(\mathbf{c},\,\cdot\,).\end{aligned} \tag{8.2.30}$$

for all $\mathbf{x} \in B_{\mathbb{R}^N}(\mathbf{c},r)$. *Hence, if* u *is a nonnegative, harmonic function on* $B_{\mathbb{R}^N}(\mathbf{c},r)$*, then*

$$\frac{r^{N-2}\big(r - |\mathbf{x}-\mathbf{c}|\big)}{\big(r + |\mathbf{x}-\mathbf{c}|\big)^{N-1}}\,u(\mathbf{c}) \le u(\mathbf{x}) \le \frac{r^{N-2}\big(r + |\mathbf{x}-\mathbf{c}|\big)}{\big(r - |\mathbf{x}-\mathbf{c}|\big)^{N-1}}\,u(\mathbf{c}). \tag{8.2.31}$$

In particular, if $\mathfrak{G}$ *is a connected region in* $\mathbb{R}^N$ *and* $\{u_n\}_1^\infty$ *is a nondecreasing sequence of harmonic functions on* $\mathfrak{G}$*, then either* $\lim_{n\to\infty} u(\mathbf{x}) = \infty$ *for every* $\mathbf{x} \in \mathfrak{G}$ *or there is a harmonic function* u *on* $\mathfrak{G}$ *such that* $u_n \longrightarrow u$ *uniformly on compact subsets of* $\mathfrak{G}$.

PROOF: The inequalities in (8.2.30) are immediate consequences of Poisson's formula and the triangle inequality; and, given (8.2.30), the inequalities in (8.2.31) comes from integrating the inequalities in (8.2.30). Finally, let a connected $\mathfrak{G}$ and a nondecreasing sequence $\{u_n\}_1^\infty$ of harmonic functions be given. By replacing u_n with $u_n - u_0$ if necessary, we may and will assume that all the u_n's are nonnegative. Next, for each $\mathbf{x} \in \mathfrak{G}$, set $u(\mathbf{x}) = \lim_{n\to\infty} u_n(\mathbf{x}) \in [0,\infty]$. Because (8.2.31) holds for each of the u_n's and $\overline{B_{\mathbb{R}^N}(\mathbf{c},r)} \subset\subset \mathfrak{G}$, the Monotone Convergence Theorem allows us to conclude that it also holds for u itself. Hence, we know that both

$$\{\mathbf{x} \in \mathfrak{G} : u(\mathbf{x}) = \infty\} \quad \text{and} \quad \{\mathbf{x} \in \mathfrak{G} : u(\mathbf{x}) < \infty\}$$

are open subsets of $\mathfrak{G}$, and so one of them must be empty. Finally, assume that $u < \infty$ everywhere on $\mathfrak{G}$, and suppose that $\overline{B_{\mathbb{R}^N}(\mathbf{c},2r)} \subset\subset \mathfrak{G}$. Then, by the right-hand side of (8.2.31), the u_n's are uniformly bounded on $\overline{B_{\mathbb{R}^N}\left(\mathbf{c},\frac{3r}{2}\right)}$; and so, by the last part of Theorem 8.2.20, we know that u is harmonic and that $u_n \longrightarrow u$ uniformly on $B_{\mathbb{R}^N}(\mathbf{c},r)$. □

Notice that, by taking $\mathbf{c} = \mathbf{0}$ and letting $r \nearrow \infty$ in (8.2.31), one gets an easy derivation of the following general statement, of which we already know a sharper version when $N \in \{1,2\}$ (cf. Theorem 8.2.13).

8.2.32 Corollary (Liouville's Theorem). *The only nonnegative harmonic functions on $\mathbb{R}^N$ are constant.*

Exercises

8.2.33 Exercise: As a consequence of (8.2.28), note that if u is a bounded harmonic function in the exterior of a compact subset of $\mathbb{R}^2$, then u has a limit as $|\mathbf{x}| \to \infty$. Show (by counterexample) that the analogous result is false in dimensions greater than two.

8.2.34 Exercise: Once we reduced the problem to that of studying v on $\bar{B}\setminus\{\mathbf{0}\}$, the rest of the argument which we used in the proof of (8.2.28) was based on a general principle. Namely, given an open $\mathfrak{G}$, a $K \subset\subset \mathfrak{G}$, and a harmonic function on $\mathfrak{G} \setminus K$, one says that K is a **removable singularity** for u in $\mathfrak{G}$ if u admits a unique harmonic extension to the whole of $\mathfrak{G}$.

(ii) Let $K \subset\subset \mathbb{R}^N$ and use $\sigma_K(\psi)$ to denote the first entrance time of $\psi \in \mathfrak{P}(\mathbb{R}^N)$ into K. Given an open $\mathfrak{G} \supset\supset K$, show that

$$\mathcal{W}^{(N)}_{\mathbf{x}}\big(\sigma_K < \mathfrak{e}^{\mathfrak{G}}\big) = 0 \quad \text{for all } \mathbf{x} \in \mathfrak{G} \setminus K \tag{8.2.35}$$

if and only if $K \cap \partial_{\text{reg}}(\mathfrak{G} \setminus K) = \emptyset$, and use the locality proved in Lemma 8.1.19 to conclude that (8.2.35) for some $\mathfrak{G} \supset\supset K$ is equivalent to $K \cap \partial_{\text{reg}}(\mathfrak{G} \setminus K) = \emptyset$ for all $\mathfrak{G} \supset\supset K$. In particular, conclude that (8.2.35) holds for some $\mathfrak{G} \supset\supset K$ if and only if

$$\mathcal{W}^{(N)}_{\mathbf{x}}\Big(\exists t \in [0,\infty) \;\; \psi(t) \in K\Big) = 0 \quad \text{for all } \mathbf{x} \notin K. \tag{8.2.36}$$

(ii) Let $K \subset\subset \mathbb{R}^N$ be given, and assume that (8.2.36) holds. Given $\mathfrak{G} \supset\supset K$ and a harmonic function u on $\mathfrak{G} \setminus K$ which is bounded in a neighborhood of ∂K, show that K is a removable singularity for u in $\mathfrak{G}$.

Hint: Begin by choosing a bounded open set $\mathfrak{H} \supset\supset K$ so that $\overline{\mathfrak{H}} \subset\subset \mathfrak{G}$. Next, set

$$\sigma_n(\psi) = \inf\Big\{t \geq 0 : \operatorname{dist}\big(\psi(t), K\big) \leq \tfrac{1}{2n}\operatorname{dist}\big(K, \mathfrak{H}\mathbf{C}\big)\Big\},$$

and define u_n on $\mathfrak{H}$ by

$$u_n(\mathbf{x}) = \mathbb{E}^{\mathcal{W}^{(N)}_{\mathbf{x}}}\Big[u\big(\psi(\mathfrak{e}^{\mathfrak{H}})\big),\, \mathfrak{e}^{\mathfrak{H}} < \sigma_n\Big]$$

Show that, on the one hand, $u_n \longrightarrow u$ on $\mathfrak{H} \setminus K$, while, on the other hand,

$$\lim_{n\to\infty} u_n(\mathbf{x}) = \mathbb{E}^{\mathcal{W}^{(N)}_{\mathbf{x}}}\Big[u\big(\psi(\mathfrak{e}^{\mathfrak{H}})\big) : \mathfrak{e}^{\mathfrak{H}} < \infty\Big]$$

for all $\mathbf{x} \in \mathfrak{H}$.

(iii) Finally, let K be a compact subset of $\mathbb{R}^N$ and a connected $\mathfrak{G} \supset\supset K$ be given. Assuming either that $N \geq 3$ or that $\partial_{\text{reg}}\mathfrak{G} \neq \emptyset$, show that (8.2.36) holds if K is a removable singularity in $\mathfrak{G}$ for every bounded, harmonic function on $\mathfrak{G} \setminus K$.

Hint: Consider the function $\mathbf{x} \in \mathfrak{G} \setminus K \longmapsto \mathcal{W}_{\mathbf{x}}^{(N)}\big(\sigma_K < \mathfrak{e}^{\mathfrak{G}}\big) \in [0,1]$, and use the Strong Maximum Principle.

8.2.37 Exercise: Although probability theory is not much help when it comes to writing down explicit expressions for $H^{\mathfrak{G}} f$, it is a rich source of schemes with which to approximate $H^{\mathfrak{G}} f$. For example, let $\mathfrak{G}$ be a region in $\mathbb{R}^N$ for which (8.2.9) holds. Given an $\mathbf{x} \in \mathfrak{G}$, use induction to construct, on an appropriate probability space $(\Omega, \mathcal{F}, P)$, random variables $\mathbf{Z}_n$, $n \in \mathbb{N}$, so that $\mathbf{Z}_0 \equiv \mathbf{x}$ and, for $n \in \mathbb{Z}^+$, $\mathbf{Z}_n(\omega)$ is uniformly distributed on the sphere $\mathbf{S}^{N-1}\big(R_n(\omega)\big)$, where $R_n(\omega) \equiv \big|\mathbf{Z}_{n-1}(\omega) - \mathfrak{G}\mathbf{C}\big|$. Proceeding as in Exercise 4.3.51, show that, for any $F \in C_{\mathrm{b}}\big(\mathbb{R}^N;\mathbb{R}\big)$,

$$\big[H^{\mathfrak{G}} f\big](\mathbf{x}) = \lim_{n\to\infty} \mathbb{E}\Big[F\big(\mathbf{Z}_n\big)\Big]$$

when $f = F \restriction \partial\mathfrak{G}$.

8.2.38 Exercise: As a second example of the way in which probability theory provides approximation schemes for solving the Dirichlet problem, consider the following prescription based on the Invariance Principle. Namely, let $\{\mathbf{X}_n\}_1^\infty$ be a sequence of independent, identically distributed $\mathbb{R}^N$-valued random variables on some probability space $(\Omega, \mathcal{F}, P)$. Further, assume that $\mathbf{X}_1$ is square P-integrable, has mean-value $\mathbf{0}$, and covariance $\mathbf{I}_{\mathbb{R}^N}$; and define $\omega \in \Omega \longmapsto \mathbf{S}_n(\,\cdot\,,\omega) \in \mathfrak{P}(\mathbb{R}^N)$ accordingly, as in (3.3.9). By Theorem 3.3.20, we know that $\mathbf{S}_n^* P \Longrightarrow \mathcal{W}_{\mathbf{x}}^{(N)}$ as $n \to \infty$.

(i) Referring to the preceding paragraph, let $\mathfrak{G}$ be an open region in $\mathbb{R}^N$, $\mathbf{x}$ an element of $\mathfrak{G}$, and set

$$\tau_n^{\mathfrak{G}}(\mathbf{x},\omega) = \inf\big\{t \geq 0 : \mathbf{x} + \mathbf{S}_n(t,\omega) \notin \mathfrak{G}\big\}.$$

Under the condition that

$$\lim_{\varphi\to\psi} \mathfrak{e}^{\mathfrak{G}}(\varphi) = \mathfrak{e}^{\mathfrak{G}}(\psi) < \infty \quad \text{for } \mathcal{W}_{\mathbf{x}}^{(N)}\text{-almost every } \psi \in \{\mathfrak{e}^{\mathfrak{G}} < \infty\}, \tag{8.2.39}$$

show that, for any $F \in C_{\mathrm{b}}(\mathbb{R}^N;\mathbb{R})$,

$$\big[H^{\mathfrak{G}} f\big](\mathbf{x}) = \lim_{n\to\infty} \mathbb{E}^P\Big[f\Big(\mathbf{x} + \mathbf{S}_n\big(\tau_n^{\mathfrak{G}}(\mathbf{x})\big)\Big),\ \tau_n^{\mathfrak{G}}(\mathbf{x}) < \infty\Big] \tag{8.2.40}$$

where $f = F \restriction \partial\mathfrak{G}$.

(ii) Part **(i)** raises the question of determining when (8.2.39) holds. That is, we need to develop a criterion with which to determine when $\mathfrak{e}^{\mathfrak{G}}$ is $\mathcal{W}_{\mathbf{x}}^{(N)}$-almost surely continuous on $\{\mathfrak{e}^{\mathfrak{G}} < \infty\}$. To this end, first observe that $\mathfrak{e}^{\mathfrak{G}}$ is always lower semicontinuous. Next, introduce

$$\mathfrak{e}^{\overline{\mathfrak{G}}}(\psi) \equiv \inf\{t \geq 0 : \psi(t) \notin \overline{\mathfrak{G}}\},$$

note that $\mathfrak{e}^{\overline{\mathfrak{G}}}$ is upper semicontinuous, and conclude that $\mathfrak{e}^{\mathfrak{G}}$ is continuous on the $\{\mathfrak{e}^{\mathfrak{G}} = \mathfrak{e}^{\overline{\mathfrak{G}}} < \infty\}$. Hence, (8.2.39) will hold if

$$\mathcal{W}_{\mathbf{x}}^{(N)}\left(\mathfrak{e}^{\mathfrak{G}} = \mathfrak{e}^{\overline{\mathfrak{G}}} < \infty\right) = 1. \tag{8.2.41}$$

Finally, under condition (8.2.9), show that (8.2.41) holds for every $\mathbf{x} \in \mathfrak{G}$ if

$$\mathcal{W}_{\mathbf{a}}^{(N)}\left(\mathfrak{e}^{\overline{\mathfrak{G}}} = 0\right) = 1 \tag{8.2.42}$$

for all $\mathbf{a} \in \partial_{\text{reg}}\mathfrak{G}$.

(iii) Using Blumenthal's 0–1 Law (cf. Theorem 8.1.26), first show that, for any $\mathbf{a} \in \mathbb{R}^N$, $\mathcal{W}_{\mathbf{a}}^{(N)}\left(\mathfrak{e}^{\overline{\mathfrak{G}}} = 0\right) \in \{0, 1\}$, and then proceed as in the proof of Theorem 8.1.27 to prove that (8.2.42) will hold for any $\mathbf{a}$ satisfying

$$\varlimsup_{\delta \searrow 0} \frac{\left|\mathfrak{G}\complement \cap B_{\mathbb{R}^N}(\mathbf{a}, \delta)\right|}{\delta^N} > 0.$$

After collecting these results together, we see that if $\mathfrak{G}$ is a bounded region whose boundary is minimally smooth, then (8.2.40) will hold for all $\mathbf{x} \in \mathfrak{G}$ and $f \in C_{\text{b}}(\partial\mathfrak{G}; \mathbb{R})$.†

§8.3: Poisson's Problem and Green's Functions.

Let $\mathfrak{G}$ be an open subset of $\mathbb{R}^N$ and f a smooth function on $\mathfrak{G}$. The basic problem which motivates the contents of this section is that of analyzing solutions u to **Poisson's problem**

$$\tfrac{1}{2}\mathbf{\Delta} u = -f \text{ in } \mathfrak{G} \quad \text{and} \quad \lim_{\mathbf{x}\to\mathbf{a}} u(\mathbf{x}) = 0 \text{ for } \mathbf{a} \in \partial_{\text{reg}}\mathfrak{G}. \tag{8.3.1}$$

† This type of approximation was carried out originally by H. Phillips and N. Wiener in "Nets and Dirichlet problem," *J. Math. Phys.* **2** in 1923. Ironically, the authors do not appear to have made the connection between this procedure and probability theory. In 1928, a more complete analysis was carried out in the famous article "Über die partiellen Differenzengleichungen der Phsik," *Ann. Math.* **5 (2)**, of R. Courant, K. Friedrichs, and H. Lewy. Interestingly, these authors do allude to a possible probabilistic interpretation, although their method (based on energy considerations) makes no direct use of probability theory.

Notice that, at least when $\mathfrak{G}$ is bounded, or, more generally, whenever (8.2.9) holds, there is at most one bounded $u \in C^2(\mathfrak{G};\mathbb{R})$ which satisfies (8.3.1). Indeed, if there were two, then their difference would be a bounded harmonic function on $\mathfrak{G}$ satisfying boundary condition 0 at $\partial_{\text{reg}}\mathfrak{G}$, which, because of (8.2.9), means that this difference vanishes. Moreover, when $N \geq 3$, even if (8.2.9) fails, one can (cf. Theorem 8.2.11) recover uniqueness by adding to (8.3.1) the condition that

$$\lim_{\substack{|\mathbf{x}|\to\infty \\ \mathbf{x}\in\mathfrak{G}}} u(\mathbf{x}) = 0. \tag{8.3.2}$$

In view of the preceding discussion, the *problem* in Poisson's problem is that of proving that solutions exist. In order to get a feeling for what is involved, recall the operators $P_t^{\mathfrak{G}}$ defined in (8.1.8). Given $f \in C_b(\mathfrak{G};\mathbb{R})$, define

$$u_T(\mathbf{x}) = \int_{[T^{-1},T]} \big[P_t^{\mathfrak{G}} f\big](\mathbf{x})\,dt \quad \text{for } T \in (1,\infty) \text{ and } \mathbf{x} \in \mathfrak{G}. \tag{8.3.3}$$

It is then an immediate consequence of Corollary 8.1.7 (especially (8.1.9)) and Theorem 8.1.23 that u_T is a smooth function on $\mathfrak{G}$ which satisfies

$$\tfrac{1}{2}\boldsymbol{\Delta} u_T = P_T^{\mathfrak{G}} f - P_{\frac{1}{T}}^{\mathfrak{G}} f \quad \text{and} \quad \lim_{\substack{\mathbf{x}\to\mathbf{a} \\ \mathbf{x}\in\mathfrak{G}}} u_T(\mathbf{x}) = 0 \text{ for } \mathbf{a} \in \partial_{\text{reg}}\mathfrak{G}. \tag{8.3.4}$$

Hence, at least when (8.2.9) holds and therefore $P_T^{\mathfrak{G}} f \longrightarrow 0$ as $T \nearrow \infty$, it is reasonable to hope that $u = \lim_{T\to\infty} u_T$ exists and will be the desired solution to (8.3.1). On the other hand, it is neither obvious that the limit will exist nor, even if it does exist, in what sense the smoothness properties and (8.3.2) will persist in the limit.

Motivated by the preceding considerations, we now define the **Green's operator** $G^{\mathfrak{G}}$ by

$$G^{\mathfrak{G}} f(\mathbf{x}) = \int_{(0,\infty)} \big[P_t^{\mathfrak{G}} f\big](\mathbf{x})\,dt, \quad \mathbf{x} \in \mathfrak{G}, \tag{8.3.5}$$

for nonnegative, measurable f on $\mathfrak{G}$; and our first order of business will be to prove that there are reasonable conditions under which we can show that $G^{\mathfrak{G}} f$ is finite.

8.3.6 Lemma. *If either* (8.2.9) *holds or* $N \geq 3$, *then* $G^{\mathfrak{G}}$ *has a unique extension as a nonnegative, linear map from* $B_c(\mathfrak{G};\mathbb{R})$ *into* $B(\mathfrak{G};\mathbb{R})$. *In fact, in both cases, for each* $K \subset\subset \mathfrak{G}$ *and* $p \in [1,\infty)$,

$$\sup_{\mathbf{x}\in\mathfrak{G}} \mathbb{E}^{\mathcal{W}_{\mathbf{x}}^{(N)}}\left[\left(\int_0^{\mathfrak{e}^{\mathfrak{G}}} \mathbf{1}_K\big(\psi(t)\big)\,dt\right)^p\right] < \infty. \tag{8.3.7}$$

PROOF: Clearly, it suffices to check (8.3.7); and when $N \geq 3$, we may restrict our attention to the case when $\mathfrak{G} = \mathbb{R}^N$. Thus, let $N \geq 3$, and set

$$f_n(\mathbf{x}) = \mathbb{E}_{\mathbf{x}}\left[\left(\int_0^\infty \mathbf{1}_K(\psi(t))\,dt\right)^n\right].$$

Because $N \geq 3$,

$$f_1(\mathbf{x}) = \int_0^\infty \left(\int_K \gamma_t^N(\mathbf{y}-\mathbf{x})\,d\mathbf{y}\right)\,dt \leq 1 + (2\pi)^{-\frac{N}{2}}|K|\int_1^\infty t^{-\frac{N}{2}}\,dt \leq 1 + \frac{2|K|}{N-2};$$

and, by the Markov property, $f_{n+1}(\mathbf{x})$ is equal $(n+1)!$ times

$$\mathbb{E}^{\mathcal{W}_{\mathbf{x}}^{(N)}}\left[\underset{0<t_1<\cdots<t_{n+1}<\infty}{\int\cdots\int} \mathbf{1}_K(\psi(t_1))\cdots\mathbf{1}_K(\psi(t_{n+1}))\,dt_1\cdots dt_{n+1}\right]$$

$$= \mathbb{E}^{\mathcal{W}_{\mathbf{x}}^{(N)}}\left[\underset{0<t_1<\cdots<t_n<\infty}{\int\cdots\int} \mathbf{1}_K(\psi(t_1))\cdots\mathbf{1}_K(\psi(t_n))f_1(\psi(t_n))\,dt_1\cdots dt_n\right].$$

Hence, by induction, (8.3.7) is proved when $N \geq 3$.

Now assume that $N \in \{1,2\}$ and that (8.2.9) holds. In order to prove the desired result, it suffices to show that for any $\mathbf{c} \in \mathfrak{G}$ and $r \in (0,\infty)$ with $\overline{B_{\mathbb{R}^N}(\mathbf{c},2r)} \subset\subset \mathfrak{G}$, we have

$$\sup_{\mathbf{x}\in\mathfrak{G}} \mathbb{E}^{\mathcal{W}_{\mathbf{x}}^{(N)}}\left[\left(\int_0^{\mathfrak{e}^{\mathfrak{G}}} \mathbf{1}_{B_{\mathbb{R}^N}(\mathbf{c},r)}(\psi(t))\,dt\right)^p\right] < \infty;$$

and during the proof of this, we may and will assume that $\mathfrak{G}$ is connected and will use the notation $B = B_{\mathbb{R}^N}(\mathbf{c},r)$ and $2B = B_{\mathbb{R}^N}(\mathbf{c},2r)$. Assuming that $N=2$, set $\tau_0 \equiv 0$, and define $\{(\sigma_n,\tau_n)\}_1^\infty$ inductively by

$$\sigma_n(\psi) = \inf\left\{t \geq \tau_{n-1}(\psi) : \psi(t) \in \overline{B}\right\}$$

and

$$\tau_n(\psi) = \inf\left\{t \geq \sigma_n(\psi) : \psi(t) \notin 2B\right\}$$

for $n \in \mathbb{Z}^+$. Notice that if $u(\mathbf{x}) \equiv \mathcal{W}_{\mathbf{x}}^{(2)}\left(\sigma_1 < \mathfrak{e}^{\mathfrak{G}}\right)$ for $\mathbf{x} \in \mathfrak{G}$, then u is a nonnegative, harmonic function on $\mathfrak{G} \setminus \overline{B}$ and

$$\lim_{\substack{\mathbf{x}\to\mathbf{a}\\ \mathbf{x}\in\mathfrak{G}\setminus B}} u(\mathbf{x}) = \begin{cases} 0 & \text{if } \mathbf{a} \in \partial_{\text{reg}}\mathfrak{G} \\ 1 & \text{if } \mathbf{a} \in \partial B. \end{cases}$$

Thus, because $N = 2$ and therefore $\mathfrak{G} \setminus \overline{B}$ is connected, an application of The Strong Maximum Principle (Theorem 7.2.8) shows that $u(\mathbf{x}) \in (0,1)$ for every $\mathbf{x} \in \mathfrak{G} \setminus \overline{B}$. In particular, this means that

$$\alpha \equiv \max_{|\mathbf{x}-\mathbf{c}|=2r} u(\mathbf{x}) \in (0,1).$$

At the same time, by the strong Markov property,

$$\mathcal{W}_{\mathbf{x}}^{(2)}\left(\sigma_{n+1} < \mathfrak{e}^{\mathfrak{G}}\right) = \mathbb{E}^{\mathcal{W}_{\mathbf{x}}^{(2)}}\left[u\left(\psi(\tau_n)\right),\, \tau_n < \mathfrak{e}^{\mathfrak{G}}\right] \le \alpha \mathcal{W}_{\mathbf{x}}^{(2)}\left(\sigma_n < \mathfrak{e}^{\mathfrak{G}}\right)$$

for all $n \in \mathbb{Z}^+$ and $\mathbf{x} \in \mathfrak{G}$, and therefore

$$\mathcal{W}_{\mathbf{x}}^{(2)}\left(\sigma_n < \mathfrak{e}^{\mathfrak{G}}\right) \le \alpha^{n-1}, \quad \text{for all } n \in \mathbb{Z}^+ \text{ and } \mathbf{x} \in \mathfrak{G}.$$

Finally, observe that

$$\begin{aligned}
&\mathbb{E}^{\mathcal{W}_{\mathbf{x}}^{(2)}}\left[\left(\int_0^{\mathfrak{e}^{\mathfrak{G}}} \mathbf{1}_B\left(\psi(t)\right) dt\right)^p\right]^{\frac{1}{p}} \\
&\qquad \le \sum_{n=1}^{\infty} \mathbb{E}^{\mathcal{W}_{\mathbf{x}}^{(2)}}\left[\left(\int_{\sigma_n}^{\tau_n} \mathbf{1}_B\left(\psi(t)\right) dt\right)^p,\, \sigma_n < \infty\right]^{\frac{1}{p}} \\
&\qquad \le \sum_{n=1}^{\infty} \mathbb{E}^{\mathcal{W}_{\mathbf{x}}^{(2)}}\left[f_p\left(\psi(\sigma_n)\right),\, \sigma_n < \infty\right]^{\frac{1}{p}},
\end{aligned}$$

where

$$f_p(\mathbf{y}) \equiv \mathbb{E}^{\mathcal{W}_{\mathbf{y}}^{(2)}}\left[\left(\mathfrak{e}^{2B}\right)^p\right] \quad \text{for } \mathbf{y} \in 2B.$$

But, by Lemma 7.2.18, we know that f_p is uniformly bounded by some constant $C(p,r) \in (0,\infty)$, and so, after combining this with the preceding, we arrive at the estimate

$$\mathbb{E}^{\mathcal{W}_{\mathbf{x}}^{(2)}}\left[\left(\int_0^{\mathfrak{e}^{\mathfrak{G}}} \mathbf{1}_B\left(\psi(t)\right) dt\right)^p\right]^{\frac{1}{p}} \le \frac{C(p,r)^{\frac{1}{p}}}{1-\alpha^{\frac{1}{p}}} < \infty.$$

Turning to the case when $N = 1$, we assume, without loss in generality, that $\mathfrak{G} = (0,\infty)$ and that $B = (c-r, c+r)$, where $c > 2r$. We then define $\left\{(\sigma_n, \tau_n)\right\}_1^{\infty}$ as in the preceding, only this time we take

$$\tau_n(\psi) = \inf\left\{t \ge \sigma_{n-1}(\psi) : \psi(t) = c - 2r\right\} \quad \text{for } n \in \mathbb{Z}^+.$$

Because (cf. (8.2.23) with $N = 1$)

$$\begin{aligned}
\mathcal{W}_x^{(1)}\left(\sigma_{n+1} < \mathfrak{e}^{\mathfrak{G}}\right) &= \mathbb{E}^{\mathcal{W}_x^{(1)}}\left[\mathcal{W}_{\psi(\tau_n)}^{(1)}\left(\sigma_1 < \mathfrak{e}^{\mathfrak{G}}\right),\, \tau_n < \mathfrak{e}^{\mathfrak{G}}\right] \\
&\le \frac{c-2r}{c-r} \mathcal{W}_x^{(1)}\left(\sigma_n < \mathfrak{e}^{\mathfrak{G}}\right),
\end{aligned}$$

the rest of the argument goes through without change. □

With the preceding result in mind, we will assume from now on that either $N \geq 3$ or (8.2.9) holds. Our next goal is to represent the Green's operator $G^{\mathfrak{G}}$ in terms of a kernel. But clearly,

$$\text{(8.3.8)} \qquad \left[G^{\mathfrak{G}} f\right](\mathbf{x}) = \int_{\mathfrak{G}} g^{\mathfrak{G}}(\mathbf{x},\mathbf{y})\, f(\mathbf{y})\, dy, \quad \mathbf{x} \in \mathfrak{G} \text{ and } f \in B\big(\mathfrak{G};[0,\infty)\big),$$

where (cf. (8.1.11)) the **Green's function**† $g^{\mathfrak{G}}$ is given by

$$\text{(8.3.9)} \qquad g^{\mathfrak{G}}(\mathbf{x},\mathbf{y}) \equiv \int_{(0,\infty)} p^{\mathfrak{G}}(t,\mathbf{x},\mathbf{y})\, dt.$$

In particular, if

$$\text{(8.3.10)} \qquad \Gamma_N(\mathbf{x}) \equiv \int_{(0,\infty)} \gamma_t^N(\mathbf{x})\, dt = \frac{2|\mathbf{x}|^{2-N}}{(N-2)\omega_{N-1}} \quad \text{for } N \geq 3 \text{ and } \mathbf{x} \in \mathbb{R}^N \setminus \{\mathbf{0}\},$$

then, for all $N \geq 3$, open $\mathfrak{G} \subseteq \mathbb{R}^N$, and distinct $\mathbf{x}$ and $\mathbf{y}$ from $\mathfrak{G}$, we have **Duhamel's formula**

$$\text{(8.3.11)} \qquad g^{\mathfrak{G}}(\mathbf{x},\mathbf{y}) = \Gamma_N(\mathbf{x}-\mathbf{y}) - \mathbb{E}^{\mathcal{W}_{\mathbf{x}}^{(N)}}\left[\Gamma_N\big(\psi(\mathfrak{e}^{\mathfrak{G}}) - \mathbf{y}\big),\ \mathfrak{e}^{\mathfrak{G}} < \infty\right]$$

as a consequence of (8.1.10). Notice that, given (8.3.11), or equivalently, for each $f \in B_{\mathrm{c}}\big(\mathfrak{G};\mathbb{R}^N\big)$,

$$\text{(8.3.12)} \qquad \left[G^{\mathfrak{G}} f\right](\mathbf{x}) = \left[\Gamma_N \bigstar f\right](\mathbf{x}) - \left[H^{\mathfrak{G}}\big(\Gamma_N \bigstar f\big)\right](\mathbf{x}), \quad \mathbf{x} \in \mathfrak{G},$$

most of our problems about finding a solution to (8.3.1) disappear. Indeed, assume that $N \geq 3$, and suppose that $f \in C^2_{\mathrm{c}}(\mathbb{R}^N;\mathbb{R})$ is given. Because differentiation commutes with convolution, it is clear that $\Gamma_N \bigstar f \in C^2_{\mathrm{b}}(\mathbb{R}^N;\mathbb{R})$. Thus,

$$\tfrac{1}{2}\left[\Delta\big(\Gamma_N \bigstar f\big)\right](\mathbf{x}) = \lim_{t \searrow 0} \frac{\left[\gamma_t^N \bigstar \Gamma_N \bigstar f\right](\mathbf{x}) - \Gamma_N \bigstar f(\mathbf{x})}{t} = -f(\mathbf{x}),$$

where, in the first equality we have used (8.1.1), and, in the second, we have used the first expression for Γ_N in (8.3.10). Hence, by (8.3.12), we see that $G^{\mathfrak{G}} f$ is a solution to (8.3.1). Moreover, since $\Gamma_N \bigstar |f| \longrightarrow 0$ as $|\mathbf{x}| \to \infty$ and, by (8.3.12),

$$\left|G^{\mathfrak{G}} f\right| \leq G^{\mathfrak{G}} |f| \leq \Gamma_N \bigstar |f|,$$

we also know that $G^{\mathfrak{G}} f$ satisfies (8.3.2). Thus, at least for $f \in C^2_{\mathrm{c}}(\mathbb{R}^N;\mathbb{R})$, we have now shown that in three or more dimensions the unique solution to (8.3.1) satisfying (8.3.2) is given by $G^{\mathfrak{G}} f$.

† The author has chosen this terminology in spite of the stern admonition by Doob in the historical comments at his end of his book referred to at the beginning of this chapter. At least (or, perhaps, worse), the present author has used the possessive in references to Lebesgue's and Wiener's measures.

Notice that Duhamel's formula (8.3.11) could have been guessed. To be precise, Γ_N is a *fundamental solution* for $-\frac{1}{2}\boldsymbol{\Delta}$ in $\mathbb{R}^N$ in the sense that

$$\tfrac{1}{2}\boldsymbol{\Delta}(\Gamma_N \star \varphi) = -\varphi \quad \text{for all test functions } \varphi \text{ on } \mathbb{R}^N;$$

and $g^{\mathfrak{G}}$ is to be a fundamental solution for $-\frac{1}{2}\boldsymbol{\Delta}$ in $\mathfrak{G}$ with 0 boundary data in the sense that it is the kernel for the solution operator which solves the Poisson problem in (8.3.1). With this in mind, one should guess that a reasonable approach to the construction of $g^{\mathfrak{G}}$ would be to *correct* $\Gamma_N(\,\cdot - \mathbf{y})$ for each $\mathbf{y} \in \mathfrak{G}$ by subtracting off a harmonic function which has $\Gamma_N(\,\cdot - \mathbf{y})$ as boundary data, and this is, of course, precisely what is being done in (8.3.11).

Obviously, what accounts for the simplicity in the cases $N \geq 3$ is the fact that we are dealing there with a transient process. Hence, when $N \in \{1, 2\}$ and process is recurrent, we should have to work a little harder. In particular, we cannot simply apply here the reasoning which led us from (8.1.10) to (8.3.11) because, when $N \in \{1, 2\}$, the integral in (8.3.10) diverges at every point. On the other hand, starting from (8.1.37) with $N = 1$, we see that,

$$\int_{(0,\infty)} p^{(0,\infty)}(t, x, y)\, dt = \Gamma_1(x - y) - \Gamma_1(x + y)$$

where

$$\Gamma_1(x) \equiv (2\pi)^{-\frac{1}{2}} \int_0^\infty t^{-\frac{1}{2}} \left(e^{-\frac{x^2}{2t}} - 1 \right) dt = -|x| \quad \text{for } x \in \mathbb{R}; \tag{8.3.13}$$

and therefore we know that

$$g^{(0,\infty)}(x, y) = \Gamma_1(x - y) - \Gamma_1(x + y) = 2x \wedge y \quad \text{for } x,\ y \in (0, \infty). \tag{8.3.14}$$

Similarly, from (8.1.37) with $N = 2$, we see that

$$\begin{aligned} 2\pi & \int_{(0,\infty)} p^{\mathbb{R}^2_+}(t, \mathbf{x}, \mathbf{y})\, dt \\ &= \lim_{T \nearrow \infty} \int_0^T \frac{1}{t} \left(\exp\left[-\frac{|\mathbf{y} - \mathbf{x}|^2}{2t} \right] - \exp\left[-\frac{|\check{\mathbf{y}} - \mathbf{x}|^2}{2t} \right] \right) dt \\ &= \lim_{T \nearrow \infty} \int_{|\check{\mathbf{y}} - \mathbf{x}|^{-2}}^{|\mathbf{y} - \mathbf{x}|^{-2}} \frac{1}{t} e^{-\frac{1}{2tT}}\, dt, \end{aligned}$$

and therefore that

$$g^{\mathbb{R}^2_+}(\mathbf{x}, \mathbf{y}) = \Gamma_2(\mathbf{x} - \mathbf{y}) - \Gamma_2(\mathbf{x} - \check{\mathbf{y}}) = \tfrac{1}{\pi} \log \frac{|\mathbf{x} - \mathbf{y}|}{|\mathbf{x} - \check{\mathbf{y}}|}, \tag{8.3.15}$$

where

$$(8.3.16) \qquad \Gamma_2(\mathbf{x}) \equiv -\frac{1}{\pi}\log\bigl[|\mathbf{x}|\bigr] \quad \text{for } \mathbf{x} \in \mathbb{R}^2 \setminus \{0\}.$$

Next, suppose that $\mathfrak{G} \subseteq \mathbb{R}^N_+$. Then, just as in the derivation of (8.1.10), we know that

$$p^{\mathfrak{G}}(t,\mathbf{x},\mathbf{y}) = p^{\mathbb{R}^N_+}(t,\mathbf{x},\mathbf{y}) - \mathbb{E}^{\mathcal{W}^{(N)}_{\mathbf{x}}}\Bigl[p^{\mathbb{R}^N_+}\bigl(t-\mathfrak{e}^{\mathfrak{G}}, \psi(\mathfrak{e}^{\mathfrak{G}}),\mathbf{y}\bigr),\ \mathfrak{e}^{\mathfrak{G}} < \infty\Bigr],$$

which, in turn, leads to

$$g^{\mathfrak{G}}(\mathbf{x},\mathbf{y}) = g^{\mathbb{R}^N_+}(\mathbf{x},\mathbf{y}) - \mathbb{E}^{\mathcal{W}^{(N)}_{\mathbf{x}}}\Bigl[g^{\mathbb{R}^N_+}\bigl(\psi(\mathfrak{e}^{\mathfrak{G}}),\mathbf{y}\bigr),\ \mathfrak{e}^{\mathfrak{G}} < \infty\Bigr]$$

for distinct $\mathbf{x}$, $\mathbf{y} \in \mathfrak{G}$. In particular, if $N \in \{1,2\}$ and we use (8.3.14) and (8.3.15), then we arrive at

$$\begin{aligned} g^{\mathfrak{G}}(\mathbf{x},\mathbf{y}) = \Gamma_N(\mathbf{x}-\mathbf{y}) &- \bigl[H^{\mathfrak{G}}\Gamma_N(\,\cdot\,-\mathbf{y})\bigr](\mathbf{x}) \\ &+ \bigl[H^{\mathfrak{G}}\Gamma_N(\,\cdot\,-\check{\mathbf{y}})\bigr](\mathbf{x}) - \Gamma_N(\mathbf{x}-\check{\mathbf{y}}). \end{aligned}$$

But $\Gamma_N(\,\cdot\,-\check{\mathbf{y}})$ is itself a bounded harmonic in $\mathfrak{G}$, and therefore the last two terms in the preceding cancel. Hence, we have now proved that, even when $N \in \{1,2\}$, (8.3.11) (cf. (8.3.13) and (8.3.16)) and therefore also (8.3.12) continue to hold first whenever $\mathfrak{G}$ is a subset of $\mathbb{R}^N_+$ and then, after an obvious translation and rotation, whenever $\mathfrak{G}$ is bounded. In particular, this means that, by exactly the same sort of argument which we used before in the case when $N \geq 3$: *for every bounded* $\mathfrak{G}$ *and every* $f \in C^2_c(\mathbb{R}^N;\mathbb{R})$, $G^{\mathfrak{G}}f$ *is the unique solution to* (8.3.1). Thus, all that remains is the situation covered in the following theorem.

8.3.17 Theorem. *Let $\mathfrak{G}$ be a nonempty, open subset of $\mathbb{R}^2$ for which $\partial_{\text{reg}}\mathfrak{G} \neq \emptyset$. Then, (8.1.9) holds,*

$$(8.3.18) \qquad \sup_{\mathbf{x},\mathbf{y}\in K} \mathbb{E}^{\mathcal{W}^{(2)}_{\mathbf{x}}}\Bigl[\bigl|\log|\psi(\mathfrak{e}^{\mathfrak{G}})-\mathbf{y}|\bigr|,\ \mathfrak{e}^{\mathfrak{G}} < \infty\Bigr] < \infty \quad \textit{for } K \subset\subset \mathfrak{G},$$

and

$$(\mathbf{x},\mathbf{y}) \in \mathfrak{G}^2 \longmapsto \mathbb{E}^{\mathcal{W}^{(2)}_{\mathbf{x}}}\Bigl[\log|\psi(\mathfrak{e}^{\mathfrak{G}})-\mathbf{y}|,\ \mathfrak{e}^{\mathfrak{G}} < \infty\Bigr] \in \mathbb{R}$$

is a smooth function which is harmonic with respect to $\mathbf{x} \in \mathfrak{G}$ *for fixed* $\mathbf{y}$ *and with respect to* $\mathbf{y} \in \mathfrak{G}$ *for fixed* $\mathbf{x}$. *In addition, for each* $\mathbf{c} \in \mathfrak{G}$, *the limit*

$$(8.3.19) \qquad h^{\mathfrak{G}}(\mathbf{x}) \equiv \lim_{r\to\infty} \frac{\log r}{\pi}\mathcal{W}^{(2)}_{\mathbf{x}}\Bigl(\mathfrak{e}^{B_{\mathbb{R}^2}(\mathbf{c},r)} \leq \mathfrak{e}^{\mathfrak{G}}\Bigr), \quad \mathbf{x} \in \mathfrak{G},$$

exists, is uniform with respect to $\mathbf{x}$ *in compact subsets of* $\mathfrak{G}$ *and independent of* $\mathbf{c} \in \mathfrak{G}$, *and determines a harmonic function of* $\mathbf{x} \in \mathfrak{G}$. *Finally,*

$$g^{\mathfrak{G}}(\mathbf{x},\mathbf{y}) = \Gamma_2(\mathbf{x}-\mathbf{y}) - \mathbb{E}^{\mathcal{W}_{\mathbf{x}}^{(2)}}\Big[\Gamma_2\big(\psi(\mathfrak{e}^{\mathfrak{G}})-\mathbf{y}\big),\ \mathfrak{e}^{\mathfrak{G}} < \infty\Big] + h^{\mathfrak{G}}(\mathbf{x}) \tag{8.3.20}$$

for all distinct $\mathbf{x}$ *and* $\mathbf{y}$ *from* $\mathfrak{G}$; *and so*

$$g^{\mathfrak{G}}(\,\cdot\,,\mathbf{y}) \longrightarrow h^{\mathfrak{G}} \quad \text{uniformly on compacts as } |\mathbf{y}| \to \infty. \tag{8.3.21}$$

PROOF: Note that, because $N = 2$, Theorem 8.2.13 guarantees that (8.2.9) follows from $\partial_{\text{reg}}\mathfrak{G} \neq \emptyset$.

In proving the rest of the theorem, we may and will assume that $\mathfrak{G}$ is connected. For $\mathbf{c} \in \mathfrak{G}$ and $r \in (0,\infty)$, set

$$B_r(\mathbf{c}) = B_{\mathbb{R}^2}(\mathbf{c},r), \quad \mathfrak{G}_r(\mathbf{c}) = \mathfrak{G} \cap B_r(\mathbf{c}), \quad \text{and } g_{\mathbf{c},r}(\,\cdot\,,\mathbf{y}) = g^{\mathfrak{G}_r(\mathbf{c})}(\,\cdot\,,\mathbf{y}).$$

Because $\mathfrak{G}_r(\mathbf{c})$ is bounded (and therefore (8.3.11) holds), we know that

$$g_{\mathbf{c},r}(\mathbf{x},\mathbf{y}) = \Gamma_2(\mathbf{x}-\mathbf{y}) - \mathbb{E}^{\mathcal{W}_{\mathbf{x}}^{(2)}}\Big[\Gamma_2\big(\psi(\mathfrak{e}^{\mathfrak{G}_r(\mathbf{c})})-\mathbf{y}\big),\ \mathfrak{e}^{\mathfrak{G}_r(\mathbf{c})} < \infty\Big]. \tag{8.3.22}$$

In particular, this shows that $g_{\mathbf{c},r}(\,\cdot\,,\mathbf{y})$ is a harmonic in $\mathfrak{G}_r(\mathbf{c}) \setminus \{\mathbf{y}\}$ for each $r > 0$ and $\mathbf{y} \in \mathfrak{G}(\mathbf{c},r)$. At the same time, because $p^{\mathfrak{G}_r(\mathbf{c})}(t,\,\cdot\,,\mathbf{y})$ is nondecreasing in r for each $(t,\mathbf{y}) \in (0,\infty)\times\mathfrak{G}$, we know that $g_{\mathbf{c},r}(\,\cdot\,,\mathbf{y})$ is nondecreasing in r for each $\mathbf{y} \in \mathfrak{G}$. Hence, by Harnack's Principle (cf. Corollary 8.2.29), either $\lim_{r\nearrow\infty} g_{\mathbf{c},r}(\mathbf{x},\mathbf{y}) = \infty$ for every $\mathbf{x} \in \mathfrak{G}\setminus\{\mathbf{y}\}$ or $g_{\mathbf{c},r}(\,\cdot\,,\mathbf{y})$ converges uniformly on compact subsets of $\mathfrak{G}\setminus\{\mathbf{y}\}$ to a harmonic function. But, because

$$\begin{aligned} g^{\mathfrak{G}}(\mathbf{x},\mathbf{y}) &= \int_{(0,\infty)} p^{\mathfrak{G}}(t,\mathbf{x},\mathbf{y})\,dt \\ &= \lim_{r\nearrow\infty} \int_{(0,\infty)} p^{\mathfrak{G}_r(\mathbf{c})}(t,\mathbf{x},\mathbf{y})\,dt = \lim_{r\nearrow\infty} g_{\mathbf{c},r}(\mathbf{x},\mathbf{y}), \end{aligned}$$

we conclude from Lemma 8.3.6 that only the second alternative is possible. Thus, we now know that, for each $\mathbf{y} \in \mathfrak{G}$, $g^{\mathfrak{G}}(\,\cdot\,,\mathbf{y})$ is a nonnegative (in fact positive) harmonic function on $\mathfrak{G}\setminus\{\mathbf{y}\}$. But, by symmetry, this means that, for each $\mathbf{x} \in \mathfrak{G}$, $g^{\mathfrak{G}}(\mathbf{x},\,\cdot\,)$ is a nonnegative harmonic function on $\mathfrak{G}\setminus\{\mathbf{x}\}$, which leads, in turn, to the conclusions that $g^{\mathfrak{G}}$ is a smooth function on $\widehat{\mathfrak{G}} \equiv \big\{(\mathbf{x},\mathbf{y}) \in \mathfrak{G}^2 : \mathbf{x} \neq \mathbf{y}\big\}$ and so, by Dini's Lemma, that

$$g_{\mathbf{c},r}(\mathbf{x},\mathbf{y}) \nearrow g^{\mathfrak{G}}(\mathbf{x},\mathbf{y}) \quad \text{uniformly on compact subsets of } \widehat{\mathfrak{G}}. \tag{8.3.23}$$

To go further, first notice that (8.3.22) can be rewritten as

$$\begin{aligned} g_{\mathbf{c},r}(\mathbf{x},\mathbf{y}) =& \Gamma_2(\mathbf{x}-\mathbf{y}) - \mathbb{E}^{\mathcal{W}_{\mathbf{x}}^{(2)}}\Big[\Gamma_2\big(\psi(\mathfrak{e}^{\mathfrak{G}})-\mathbf{y}\big),\ \mathfrak{e}^{\mathfrak{G}} < \mathfrak{e}^{B_r(\mathbf{c})}\Big] \\ &- \mathbb{E}^{\mathcal{W}_{\mathbf{x}}^{(2)}}\Big[\Gamma_2\big(\psi(\mathfrak{e}^{B_r(\mathbf{c})})-\mathbf{y}\big),\ \mathfrak{e}^{B_r(\mathbf{c})} \le \mathfrak{e}^{\mathfrak{G}} < \infty\Big]. \end{aligned} \tag{8.3.24}$$

Next, set

$$u_r(\mathbf{x},\mathbf{y}) = -\mathbb{E}^{\mathcal{W}_{\mathbf{x}}^{(2)}}\left[\Gamma_2\big(\psi(\mathfrak{e}^{\mathfrak{G}})-\mathbf{y}\big),\, \mathfrak{e}^{\mathfrak{G}} < \mathfrak{e}^{B_r(\mathbf{c})}\right] \quad \text{for } (\mathbf{x},\mathbf{y}) \in \mathfrak{G}_r(\mathbf{c})^2.$$

We want to prove that, as $r \nearrow \infty$, the u_r's tend uniformly on compacts to a function which is harmonic in $\mathbf{x}$ and $\mathbf{y}$ separately. To this end, let a connected, open subset A of $\mathfrak{G}$ with $\overline{A} \subset\subset \mathfrak{G}$ be given, set $D(A) = \operatorname{dist}\big(A, \mathfrak{G}\complement\big)$, and observe that, for each $(\mathbf{x},\mathbf{y}) \in A^2$,

$$\begin{aligned} r \in (0,\infty) \longmapsto & u_r(\mathbf{x},\mathbf{y}) - \frac{\log\big(D(A)\big)}{\pi}\mathcal{W}_{\mathbf{x}}^{(2)}\left(\mathfrak{e}^{\mathfrak{G}} < \mathfrak{e}^{B_r(\mathbf{c})}\right) \\ &= \frac{1}{\pi}\mathbb{E}^{\mathcal{W}_{\mathbf{x}}^{(2)}}\left[\log\left(\frac{|\psi(\mathfrak{e}^{\mathfrak{G}})-\mathbf{y}|}{D(A)}\right),\, \mathfrak{e}^{\mathfrak{G}} < \mathfrak{e}^{B_r(\mathbf{c})}\right] \in [0,\infty) \end{aligned}$$

is nondecreasing. Hence, since

$$(\mathbf{x},\mathbf{y}) \in \mathfrak{G}_r(\mathbf{c})^2 \longmapsto u_r(\mathbf{x},\mathbf{y}) - \frac{\log\big(D(A)\big)}{\pi}\mathcal{W}_{\mathbf{x}}^{(2)}\left(\mathfrak{e}^{\mathfrak{G}} < \mathfrak{e}^{B_r(\mathbf{c})}\right)$$

is harmonic in $\mathbf{x}$ and $\mathbf{y}$ separately while

$$\mathcal{W}_{\mathbf{x}}^{(2)}\left(\mathfrak{e}^{\mathfrak{G}} < \mathfrak{e}^{B_r(\mathbf{c})}\right) \nearrow \mathcal{W}_{\mathbf{x}}^{(2)}\left(\mathfrak{e}^{\mathfrak{G}} < \infty\right) = \left[H^{\mathfrak{G}}\mathbf{1}\right](\mathbf{x}) = \mathbf{1},$$

Harnack's Principle says that either u_r tends to ∞ everywhere on A^2 or it tends uniformly on compacts to a function which is also harmonic in $\mathbf{x}$ and $\mathbf{y}$ separately. But, by (8.3.24),

$$u_r(\mathbf{x},\mathbf{y}) \le g^{\mathfrak{G}}(\mathbf{x},\mathbf{y}) - \Gamma_2(\mathbf{x}-\mathbf{y}) \quad \text{if } r > |\mathbf{y}|+1,$$

and therefore we have now proved that

$$(\mathbf{x},\mathbf{y}) \in \mathfrak{G}^2 \longmapsto \mathbb{E}^{\mathcal{W}_{\mathbf{x}}^{(N)}}\left[\Gamma_2\big(\psi(\mathfrak{e}^{\mathfrak{G}})-\mathbf{y}\big),\, \mathfrak{e}^{\mathfrak{G}} < \infty\right]$$

is a smooth function which is harmonic in $\mathbf{x}$ and $\mathbf{y}$ separately. In particular, this completes the proof of the estimate in (8.3.18). At the same time, in conjunction with (8.3.24), we also know that, as $r \nearrow \infty$,

$$\begin{aligned} &-\mathbb{E}^{\mathcal{W}_{\mathbf{x}}^{(N)}}\left[\Gamma_2\big(\psi(\mathfrak{e}^{B_r(\mathbf{c})})-\mathbf{y}\big),\, \mathfrak{e}^{B_r(\mathbf{c})} < \mathfrak{e}^{\mathfrak{G}}\right] \\ &\qquad \longrightarrow g^{\mathfrak{G}}(\mathbf{x},\mathbf{y}) - \Gamma_2(\mathbf{x}-\mathbf{y}) + \mathbb{E}^{\mathcal{W}_{\mathbf{x}}^{(N)}}\left[\Gamma_2\big(\psi(\mathfrak{e}^{\mathfrak{G}})-\mathbf{y}\big),\, \mathfrak{e}^{\mathfrak{G}} < \infty\right] \end{aligned}$$

uniformly on compact subsets of $\mathfrak{G}^2$. Thus, both the uniform existence on compacts of the limit in (8.3.19) as well as the equality in (8.3.20) come down to the trivial observation that, as $r \nearrow \infty$,

$$-\mathbb{E}^{\mathcal{W}_{\mathbf{x}}^{(2)}}\left[\Gamma_2\big(\psi(\mathfrak{e}^{B_r(\mathbf{c})}) - \mathbf{y}\big),\, \mathfrak{e}^{B_r(\mathbf{c})} < \mathfrak{e}^{\mathfrak{G}}\right] - \frac{\log r}{\pi}\mathcal{W}_{\mathbf{x}}^{(N)}\left(\mathfrak{e}^{B_r(\mathbf{c})} < \mathfrak{e}^{\mathfrak{G}}\right)$$
$$= \frac{1}{\pi}\mathbb{E}^{\mathcal{W}_{\mathbf{x}}^{(N)}}\left[\log\left(\frac{|\psi(\mathfrak{e}^{B_r(\mathbf{c})}) - \mathbf{y}|}{r}\right),\, \mathfrak{e}^{B_r(\mathbf{c})} < \mathfrak{e}^{\mathfrak{G}}\right] \longrightarrow 0$$

uniformly for $(\mathbf{x}, \mathbf{y})$ in compact subsets of $\mathfrak{G}^2$. Moreover, by the last part of Theorem 8.2.20, we know that $h^{\mathfrak{G}}$ (cf. (8.3.19)) is harmonic; and, given (8.3.20), it is clear that the limit in (8.3.19) does not depend on the choice of $\mathbf{c} \in \mathfrak{G}$. Finally, to prove (8.3.21), use (8.3.20) to write

$$h^{\mathfrak{G}}(\mathbf{x}) = g^{\mathfrak{G}}(\mathbf{x}, \mathbf{y}) + \tfrac{1}{\pi}\mathbb{E}^{\mathcal{W}_{\mathbf{x}}^{(2)}}\left[\log\left(\frac{|\psi(\mathfrak{e}^{\mathfrak{G}}) - \mathbf{y})|}{|\mathbf{x} - \mathbf{y}|}\right),\, \mathfrak{e}^{\mathfrak{G}} < \infty\right],$$

and apply Lebesgue's Dominated Convergence Theorem together with the estimate in (8.3.18) to see that, as $|\mathbf{y}| \to \infty$, the second term tends to 0 boundedly and pointwise for each $\mathbf{x} \in \mathfrak{G}$. Hence, after another application of the last part of Theorem 8.2.20, we see that the pointwise convergence is, in fact, uniform on compacts. □

8.3.25 Corollary. *Let everything be as in Theorem 8.3.17. Then, for each* $K \subset\subset \mathfrak{G}$ *and* $r > 0$,

$$\sup\left\{g^{\mathfrak{G}}(\mathbf{x}, \mathbf{y}) : |\mathbf{x} - \mathbf{y}| \ge r \text{ and } \mathbf{y} \in K\right\} < \infty, \tag{8.3.26}$$

and

$$\lim_{\substack{\mathbf{x}\to\mathbf{a}\\ \mathbf{x}\in\mathfrak{G}}} \sup_{\mathbf{y}\in K} g^{\mathfrak{G}}(\mathbf{x}, \mathbf{y}) = 0 \quad \text{for each } \mathbf{a} \in \partial_{\text{reg}}\mathfrak{G}. \tag{8.3.27}$$

Moreover, for each $f \in C^2_c(\mathfrak{G}; \mathbb{R})$, $G^{\mathfrak{G}} f$ *is the unique bounded solution to* (8.3.1).

PROOF: To prove (8.3.26) and (8.3.27), let $\mathbf{c} \in \mathfrak{G}$ and $r > 0$ satisfying $\overline{B_{\mathbb{R}^2}(\mathbf{c}, 2r)} \subset\subset \mathfrak{G}$ be given, set $B = B_{\mathbb{R}^2}(\mathbf{c}, r)$, and define the first entrance time $\sigma(\psi)$ of ψ into $\overline{B}$ by

$$\sigma(\psi) = \inf\left\{t \ge 0 : \psi(t) \in \overline{B}\right\}.$$

By the strong Markov property, we see that, for any $f \in B_c\big(B; [0, \infty)\big)$:

$$\int_{\mathfrak{G}} g^{\mathfrak{G}}(\mathbf{x}, \mathbf{y}) f(\mathbf{y})\, d\mathbf{y}$$
$$= \mathbb{E}^{\mathcal{W}_{\mathbf{x}}^{(2)}}\left[\int_{\sigma}^{\mathfrak{e}^{\mathfrak{G}}} f\big(\psi(t)\big)\, dt,\ \sigma < \mathfrak{e}^{\mathfrak{G}}\right]$$
$$= \mathbb{E}^{\mathcal{W}_{\mathbf{x}}^{(2)}}\left[\int_{\mathfrak{G}} g^{\mathfrak{G}}\big(\psi(\sigma), \mathbf{y}\big) f(\mathbf{y})\, d\mathbf{y},\ \sigma < \mathfrak{e}^{\mathfrak{G}}\right].$$

Hence, if $\mathbf{x} \notin 2B \equiv B_{\mathbb{R}^2}(\mathbf{c}, 2r)$ and therefore $g^{\mathfrak{G}}(\mathbf{x}, \cdot\,) \upharpoonright B$ is continuous, we find that

$$g^{\mathfrak{G}}(\mathbf{x}, \mathbf{y}) = \mathbb{E}^{\mathcal{W}_{\mathbf{x}}^{(2)}}\Big[g^{\mathfrak{G}}\big(\psi(\sigma), \mathbf{y}\big),\ \sigma < \mathfrak{e}^{\mathfrak{G}}\Big] \quad \text{for all } y \in B.$$

But, because $g^{\mathfrak{G}} \upharpoonright \big(\partial(2B)\big) \times B$ is bounded, we now see that

$$\sup_{\mathbf{y}\in B} g^{\mathfrak{G}}(\mathbf{x}, \mathbf{y}) \le C\mathcal{W}_{\mathbf{x}}^{(2)}\big(\sigma < \mathfrak{e}^{\mathfrak{G}}\big), \quad \mathbf{x} \notin 2B,$$

for some $C \in (0, \infty)$. In particular, this, combined with the obvious Heine–Borel argument, proves (8.2.26). In addition, if $\mathbf{a} \in \partial_{\text{reg}}\mathfrak{G}$, then, for each $\delta > 0$,

$$\begin{aligned}\varlimsup_{\substack{\mathbf{x}\to\mathbf{a}\\ \mathbf{x}\in\mathfrak{G}}} \mathcal{W}_{\mathbf{x}}^{(2)}\big(\sigma < \mathfrak{e}^{\mathfrak{G}}\big) &\le \varlimsup_{\substack{\mathbf{x}\to\mathbf{a}\\ \mathbf{x}\in\mathfrak{G}}} \mathcal{W}_{\mathbf{x}}^{(2)}\big(\sigma \le \delta\big) + \varlimsup_{\substack{\mathbf{x}\to\mathbf{a}\\ \mathbf{x}\in\mathfrak{G}}} \mathcal{W}_{\mathbf{x}}^{(2)}\big(\mathfrak{e}^{\mathfrak{G}} > \delta\big)\\ &= \varlimsup_{\substack{\mathbf{x}\to\mathbf{a}\\ \mathbf{x}\in\mathfrak{G}}} \mathcal{W}_{\mathbf{x}}^{(2)}\big(\sigma \le \delta\big).\end{aligned}$$

Hence, since the last expression obviously tends to 0 as $\delta \searrow 0$, we have now proved that

$$\varlimsup_{\substack{\mathbf{x}\to\mathbf{a}\\ \mathbf{x}\in\mathfrak{G}}} \sup_{\mathbf{y}\in B} g^{\mathfrak{G}}(\mathbf{x}, \mathbf{y}) = 0;$$

which (again after the obvious Heine–Borel argument) means that we have also proved (8.3.27).

Turning to the second part of the statement, let $f \in C_c^2(\mathfrak{G}, \mathbb{R})$ be given. Using (8.3.9) and the first equality in (8.3.15), check that $\frac{1}{2}\Delta\Gamma_2\bigstar f = -f$ for $f \in C_c^2(\mathbb{R}^2; \mathbb{R})$. Because of (8.2.9) and the observation made in the discussion following (8.3.1), we know that there is at most one bounded solution to (8.3.1). Secondly, as an application of Lemma 8.3.6, we know that $G^{\mathfrak{G}} f$ is bounded; and, from the representation of $g^{\mathfrak{G}}$ in (8.3.20) and the properties (proved in Theorem 8.3.17) of the quantities on the right-hand side of (8.3.20), it is now an easy matter to check that $G^{\mathfrak{G}} f \in C^2(\mathfrak{G}; \mathbb{R})$ and that $\frac{1}{2}\Delta G^{\mathfrak{G}} f = -f$. Finally, because of (8.3.27), it is clear that $G^{\mathfrak{G}} f$ goes to 0 at $\partial_{\text{reg}}\mathfrak{G}$. □

The appearance of the extra term $h^{\mathfrak{G}}$ in (8.3.20) is, of course, a reflection of the fact that, for unbounded regions in $\mathbb{R}^2$, we do not know *a priori* which harmonic function (cf. the paragraph following the derivation of (8.3.11)) should be used to correct $\Gamma_2(\cdot - \mathbf{y})$; when $N \ge 3$, the obvious choice was the one which behaved the same way at ∞ as Γ_N itself (i.e., the one which tends to 0 at ∞). Actually, as (8.3.21) makes explicit, the same principle applies to the case $N = 2$, although now (8.3.11) will not, in general, do the job. To get a feeling for when the extra correction term $h^{\mathfrak{G}}$ actually appears, consider the case when $\mathfrak{G}$ is the half-space $\mathbb{R}^2_+$. By combining (8.3.15) with (8.3.21), we see that $h^{\mathbb{R}^2_+} \equiv 0$, and, obviously, the same will be true of any half-space and therefore of any $\mathfrak{G}$ which is contained in a half-space. On the other hand, if $\mathbf{D}(R)$ is the open disk $\{\mathbf{x} : |\mathbf{x}| \le R\}$ and $\mathfrak{G} = \mathbb{R}^2 \setminus \overline{\mathbf{D}(R)}$, then it is an easy matter to check that, for $R < |\mathbf{x}| < r$,

$$\mathcal{W}_{\mathbf{x}}^{(2)}\left(\mathfrak{e}^{\mathbf{D}(r)} < \mathfrak{e}^{\mathfrak{G}}\right) = \left[H^{\mathbf{D}(r)\setminus\overline{\mathbf{D}(R)}}\mathbf{1}_{\mathbf{S}^1(\mathbf{0},r)}\right](\mathbf{x}) = \frac{\log\frac{|\mathbf{x}|}{R}}{\log\frac{r}{R}}.$$

Hence, by (8.3.19), we see that

$$h^{\mathbb{R}^2\setminus\overline{\mathbf{D}(R)}}(\mathbf{x}) = \frac{1}{\pi}\log\frac{|\mathbf{x}|}{R}, \quad \mathbf{x} \notin \overline{\mathbf{D}(R)}. \tag{8.3.28}$$

In fact, as the following result shows, for $\mathfrak{G}$'s whose complements are compact, the form of the expression in (8.3.28) is typical, at least as $|\mathbf{x}| \to \infty$.

8.3.29 Theorem. *Let $\mathfrak{G}$ be an open region in $\mathbb{R}^2$ for which $\partial_{\text{reg}}\mathfrak{G} \neq \emptyset$ and $K \equiv \mathbb{R}^2 \setminus \mathfrak{G}$ is compact. Then, for each $R \in (0,\infty)$ with the property that $K \subseteq \mathbf{D}(R)$, one has that*

$$\begin{aligned} h^{\mathfrak{G}}(\mathbf{x}) - \frac{1}{\pi}\log\frac{|\mathbf{x}|}{R} &= \frac{|\mathbf{x}|^2}{2\pi}\int_{\mathbf{S}^1}\frac{|\mathbf{x}|^2 - R^2}{\big||\mathbf{x}|^2\boldsymbol{\eta} - R\mathbf{x}\big|^2}h^{\mathfrak{G}}(R\boldsymbol{\eta})\,\lambda_{\mathbf{S}^1}(d\boldsymbol{\eta}) \\ &\longrightarrow \frac{1}{2\pi}\int_{\mathbf{S}^1}h^{\mathfrak{G}}(R\boldsymbol{\eta})\,\lambda_{\mathbf{S}^1}(d\boldsymbol{\eta}) \end{aligned} \tag{8.3.30}$$

as $|\mathbf{x}| \to \infty$.

PROOF: In view of Corollary 8.2.25, all that we have to do is check that

$$\mathbf{x} \in \mathbb{R}^2 \setminus \overline{\mathbf{D}(R)} \longmapsto h^{\mathfrak{G}}(\mathbf{x}) - \frac{1}{\pi}\log\frac{|\mathbf{x}|}{R} \in \mathbb{R}$$

is bounded. To this end, define $\sigma : \mathfrak{P}(\mathbb{R}^N) \longrightarrow [0,\infty]$ to be the first entrance time into $\overline{\mathbf{D}(R)}$, and note (cf. the preceding discussion) that, for each $r > R$ and $R < |\mathbf{x}| < r$:

$$\begin{aligned} &\mathcal{W}_{\mathbf{x}}^{(2)}\left(\mathfrak{e}^{\mathbf{D}(r)} < \mathfrak{e}^{\mathfrak{G}}\right) \\ &= \mathcal{W}_{\mathbf{x}}^{(2)}\left(\mathfrak{e}^{\mathbf{D}(r)} < \sigma\right) \\ &\qquad + \mathbb{E}^{\mathcal{W}_{\mathbf{x}}^{(2)}}\left[\mathcal{W}_{\psi(\sigma)}^{(2)}\left(\mathfrak{e}^{\mathbf{D}(r)} < \mathfrak{e}^{\mathfrak{G}}\right),\ \sigma < \mathfrak{e}^{\mathbf{D}(r)}\right] \\ &= \frac{\log\frac{|\mathbf{x}|}{R}}{\log\frac{r}{R}} + \mathbb{E}^{\mathcal{W}_{\mathbf{x}}^{(2)}}\left[\mathcal{W}_{\psi(\sigma)}^{(2)}\left(\mathfrak{e}^{\mathbf{D}(r)} < \mathfrak{e}^{\mathfrak{G}}\right),\ \sigma < \mathfrak{e}^{\mathbf{D}(r)}\right]. \end{aligned}$$

Hence, after multiplying the preceding through by $\frac{\log r}{\pi}$, using (8.3.19), and letting $r \to \infty$, we arrive at

$$h^{\mathfrak{G}}(\mathbf{x}) = \frac{1}{\pi}\log\frac{|\mathbf{x}|}{R} + \frac{1}{\pi}\mathbb{E}^{\mathcal{W}_{\mathbf{x}}^{(2)}}\left[h^{\mathfrak{G}}(\psi(\sigma)),\ \sigma < \infty\right], \quad \mathbf{x} \in \mathbb{R}^2 \setminus \overline{\mathbf{D}(R)},$$

which certainly implies the required boundedness. □

The number on the right-hand side of (8.3.30) plays an important rôle in classical two-dimensional potential theory, where it is known as the **Robin's constant** for $\mathfrak{G}$.

Exercises

8.3.31 Exercise: In this exercise, we give an explicit expression for the Green's function $g^{B_{\mathbb{R}^N}(\mathbf{c},R)}$. To this end, first use translation and scaling to see that

$$g^{B_{\mathbb{R}^N}(\mathbf{c},R)}(\mathbf{x},\mathbf{y}) = R^{2-N} g^{B_{\mathbb{R}^N}(\mathbf{0},1)}\left(\frac{\mathbf{x}-\mathbf{c}}{R}, \frac{\mathbf{y}-\mathbf{c}}{R}\right)$$

for distinct $\mathbf{x}$, $\mathbf{y}$ from $B_{\mathbb{R}^N}(\mathbf{c},R)$. Second, observe that

$$|\mathbf{x}-\mathbf{y}| = \left| |\mathbf{y}|\mathbf{x} - \tfrac{\mathbf{y}}{|\mathbf{y}|}\right| \quad \text{for } \mathbf{x} \in \mathbf{S}^{N-1} \text{ and } \mathbf{y} \in B_{\mathbb{R}^N}(\mathbf{0},1) \setminus \{\mathbf{0}\},$$

set $\mathbf{e} = (1,0,\dots,0) \in \mathbf{S}^{N-1}$, and use the preceding observation together with (8.3.11) to conclude that

$$g^{B_{\mathbb{R}^N}(\mathbf{0},1)}(\mathbf{x},\mathbf{y}) = \Gamma_N(\mathbf{x}-\mathbf{y}) - \begin{cases} \Gamma_N\left(|\mathbf{y}|\mathbf{x} - \frac{\mathbf{y}}{|\mathbf{y}|}\right) & \text{for } \mathbf{y} \neq \mathbf{0} \\ \Gamma_N(\mathbf{e}) & \text{for } \mathbf{y} = \mathbf{0} \end{cases}. \tag{8.3.32}$$

8.3.33 Exercise: The derivation which we gave of Poisson's formula (cf. Theorem 8.2.20) required us to already know the answer and simply verify that it is correct. Here we give another approach, which is the basis for a quite general procedure. To begin with, recall the classical **Green's Identity**

$$\int_{\mathfrak{G}} \left(u\Delta v - v\Delta u\right) d\mathbf{x} = \int_{\partial\mathfrak{G}} \left(u\tfrac{\partial v}{\partial \mathbf{n}} - v\tfrac{\partial u}{\partial \mathbf{n}}\right) d\lambda_{\partial\mathfrak{G}}$$

for bounded, smooth regions $\mathfrak{G}$ in $\mathbb{R}^N$ and functions u and v which are smooth in a neighborhood of $\overline{\mathfrak{G}}$. (In the preceding, $\frac{\partial w}{\partial \mathbf{n}}(\boldsymbol{\xi})$ is used to denote the normal derivative $\left(\nabla w(\boldsymbol{\xi}), \mathbf{n}(\boldsymbol{\xi})\right)_{\mathbb{R}^N}$, where $\mathbf{n}(\boldsymbol{\xi})$ is the outer unit normal at $\boldsymbol{\xi} \in \partial\mathfrak{G}$, and $\lambda_{\partial\mathfrak{G}}$ is the standard surface measure for $\mathfrak{G}$.) Next, let $\mathbf{c}$ be an element of $B_{\mathbb{R}^N}(\mathbf{0},1)$, suppose $r > 0$ satisfies $\overline{B_{\mathbb{R}^N}(\mathbf{c},r)} \subset\subset B_{\mathbb{R}^N}(\mathbf{0},1)$, and let u be a function which is harmonic in a neighborhood of $\overline{B_{\mathbb{R}^N}(\mathbf{0},1)}$. By applying Green's Identity with $\mathfrak{G} = B_{\mathbb{R}^N}(\mathbf{0},1) \setminus \overline{B_{\mathbb{R}^N}(\mathbf{c},r)}$ and $v = g^{B_{\mathbb{R}^N}(\mathbf{0},1)}(\mathbf{c},\,\cdot\,)$, use (8.3.32) to verify

$$\begin{aligned} u(\mathbf{c}) &= \lim_{r\searrow 0} r^{N-1} \int_{\mathbf{S}^{N-1}} \big(\boldsymbol{\xi}, \nabla v(\mathbf{c}+r\boldsymbol{\xi})\big)_{\mathbb{R}^N} u(\mathbf{c}+r\boldsymbol{\xi})\, \lambda_{\mathbf{S}^{N-1}}(d\boldsymbol{\xi}) \\ &= \int_{\mathbf{S}^{N-1}} \big(\boldsymbol{\xi}, \nabla v(\boldsymbol{\xi})\big)_{\mathbb{R}^N} u(\boldsymbol{\xi})\, \lambda_{\mathbf{S}^{N-1}}(d\boldsymbol{\xi}) \\ &= \int_{\mathbf{S}^{N-1}} u(\boldsymbol{\xi}) \pi^{(N)}(\mathbf{c},\boldsymbol{\xi})\, \lambda_{\mathbf{S}^{N-1}}(d\boldsymbol{\xi}), \end{aligned}$$

where $\pi^{(N)}$ is the Poisson kernel given in (8.3.21). Finally, given $f \in C(\partial\mathfrak{G};\mathbb{R})$, extend f to $B_{\mathbb{R}^N}(\mathbf{0},1)\complement$ so that it is constant on rays, take

$$u_R = H^{B_{\mathbb{R}^N}(\mathbf{0},R)} f \quad \text{for } R > 1,$$

check that $u_R \longrightarrow H^{B_{\mathbb{R}^N}(\mathbf{0},1)} f$ uniformly on $\overline{B_{\mathbb{R}^N}(\mathbf{0},1)}$, and use the preceding to conclude that

$$\big[H^{B_{\mathbb{R}^N}(\mathbf{0},1)} f\big](\mathbf{c}) = \int_{\mathbf{S}^{N-1}} f(\boldsymbol{\xi})\, \pi^{(N)}(\mathbf{c},\boldsymbol{\xi})\, \lambda_{\mathbf{S}^{N-1}}(d\boldsymbol{\xi});$$

which is, of course, Poisson's Formula.

§8.4: Green's Potentials, Riesz Decompositions, and Capacity.

The origin of Green's functions lies in the theory of electricity and magnetism. Namely, if $\mathfrak{G}$ is a region in $\mathbb{R}^N$ whose boundary is *grounded* and $\mathbf{y} \in \mathfrak{G}$, then $g^{\mathfrak{G}}(\,\cdot\,,\mathbf{y})$ should be the *electrical potential* in $\mathfrak{G}$ which results from placing a unit point charge at $\mathbf{y}$. More generally, if μ is any *distribution of charge in* $\mathfrak{G}$ (i.e., a nonnegative Radon measure on $\mathfrak{G}$), then one can consider the **potential** $G^{\mathfrak{G}}\mu$ given by

$$\text{(8.4.1)} \qquad \big[G^{\mathfrak{G}}\mu\big](\mathbf{x}) = \int_{\mathfrak{G}} g^{\mathfrak{G}}(\mathbf{x},\mathbf{y})\, \mu(dy), \quad \mathbf{x} \in \mathfrak{G},$$

where we have implicitly assumed about $\mathfrak{G}$ either that $N \geq 3$ or that (8.2.9) holds (cf. Lemma 8.3.6 and (8.3.9)). In this concluding section, we will prove a general theorem (cf. Corollary 8.4.16 below) which characterizes functions which arise in this way (i.e., are potentials) and will close with a probabilistic interpretation of a special but particularly important class of potentials.

Throughout this section, $\mathfrak{G}$ will be a nonempty, connected, open region in $\mathbb{R}^N$, and we will be assuming either that $N \geq 3$ or that (8.2.9) holds. Thus, by the results obtained in Section 8.3, the Green's function (cf. (8.3.9)) $g^{\mathfrak{G}}$ satisfies (depending on whether $N \in \mathbb{Z}^+ \setminus \{2\}$ or $N = 2$) either (8.3.11) or (8.3.20) (cf. (8.3.10), (8.3.16), and (8.3.13)); and, in order to have $g^{\mathfrak{G}}$ defined everywhere on $\mathfrak{G}^2$, we will take $g^{\mathfrak{G}}(\mathbf{x},\mathbf{x}) = \infty$, $\mathbf{x} \in \mathfrak{G}$, when $N \geq 2$. Next, we will say that u is **excessive on** $\mathfrak{G}$ and will write $u \in \mathcal{E}(\mathfrak{G})$ if u is a lower semicontinuous, $[0,\infty]$-valued function with the *super mean-value property* that

$$u(\mathbf{x}) \geq \frac{1}{\omega_{N-1}} \int_{\mathbf{S}^{N-1}} u(\mathbf{x}+r\boldsymbol{\xi})\,\lambda_{\mathbf{S}^{N-1}}(d\boldsymbol{\xi}) \quad \text{whenever } \overline{B_{\mathbb{R}^N}(\mathbf{x},r)} \subseteq \mathfrak{G}. \tag{8.4.2}$$

As the next lemma shows, there are lots of excessive functions.

8.4.3 Lemma. *$\mathcal{E}(\mathfrak{G})$ is closed under nonnegative linear combinations and nondecreasing limits. Moreover, if u and v are excessive, so is $u \wedge v$. Finally, for each nonnegative Radon measure μ on $\mathfrak{G}$ and each harmonic function $h : \mathfrak{G} \longrightarrow [0,\infty)$, $G^{\mathfrak{G}}\mu + h$ is an excessive function.*

PROOF: The first two assertions are easy; and, clearly, the third assertion comes down to showing that $G^{\mathfrak{G}}\mu$ is excessive. Moreover, by Fatou's Lemma and Tonelli's Theorem, we will know that $G^{\mathfrak{G}}\mu$ is excessive as soon as we show that, for each $\mathbf{y} \in \mathfrak{G}$, $g^{\mathfrak{G}}(\,\cdot\,,\mathbf{y})$ is excessive. To this end, set $f_n = p^{\mathfrak{G}}\left(\frac{1}{n},\,\cdot\,,\mathbf{y}\right)$ and (cf. (8.3.5)) $u_n = G^{\mathfrak{G}} f_n$. Obviously, u_n is lower semicontinuous. In addition, by the strong Markov property and rotation invariance: $\overline{B_{\mathbb{R}^N}(\mathbf{x},r)} \subset\subset \mathfrak{G}$ implies

$$\begin{aligned} u_n(\mathbf{x}) &\geq \mathbb{E}^{\mathcal{W}_{\mathbf{x}}^{(N)}}\left[\int_{\mathfrak{e}_r}^{\mathfrak{e}^{\mathfrak{G}}} f_n(\psi(t))\,dt\right] = \mathbb{E}^{\mathcal{W}_{\mathbf{x}}^{(N)}}\left[u_n(\psi(\mathfrak{e}_r)),\, \mathfrak{e}_r < \infty\right] \\ &= \frac{1}{\omega_{N-1}} \int_{\mathbf{S}^{N-1}} u_n(\mathbf{x}+r\boldsymbol{\xi})\,\lambda_{\mathbf{S}^{N-1}}(d\boldsymbol{\xi}), \end{aligned}$$

where we have introduced the notation

$$\mathfrak{e}_r(\psi) = \inf\left\{t : \left|\psi(t) - \psi(0)\right| \geq r\right\}. \tag{8.4.4}$$

Hence, each u_n is excessive, and therefore, since

$$u_n(\mathbf{x}) = \int_{\frac{1}{n}}^{\infty} p^{\mathfrak{G}}(t,\mathbf{x},\mathbf{y})\,dt \nearrow g^{\mathfrak{G}}(\mathbf{x},\mathbf{y}),$$

we are done. □

Our next goal is to prove that, apart from the trivial case when $u \equiv \infty$, every excessive function on $\mathfrak{G}$ admits a unique representation in the form $G^{\mathfrak{G}}\mu + h$ for an appropriate choice of μ and h.

8.4.5 Lemma. *If $u \in \mathcal{E}(\mathfrak{G})$, then either $u \equiv \infty$ or u is integrable on each compact subset of $\mathfrak{G}$. Next, choose some reference point $\mathbf{c} \in \mathfrak{G}$, set $R = \frac{1}{2}|\mathbf{c} - \mathfrak{G}\complement|$, and, for each $n \in \mathbb{Z}^+$, take*

$$\mathfrak{G}_n = \left\{\mathbf{x} \in \mathfrak{G} \cap B_{\mathbb{R}^N}(\mathbf{c}, n) : \left|\mathbf{x} - \mathfrak{G}\complement\right| > \tfrac{R}{n}\right\}. \tag{8.4.6}$$

Given $u \in \mathcal{E}(\mathfrak{G})$ which is not identically infinite, define

$$u_n(\mathbf{x}) = \int_{\mathfrak{G}_{4n}} \rho_n(\mathbf{x} - \mathbf{y})\, u(\mathbf{y})\, d\mathbf{y}, \tag{8.4.7}$$

where $\rho_n(\mathbf{x}) \equiv n^N \rho(n\mathbf{x})$ for $n \in \mathbb{Z}^+$ and $\mathbf{x} \in \mathbb{R}^N$ and ρ is a rotationally invariant element of $C_c^\infty\left(B_{\mathbb{R}^N}(\mathbf{0}, \frac{R}{4}); [0,\infty)\right)$ which has total integral 1. Then, $u_n \in C_c^\infty(\mathfrak{G}; [0,\infty))$, $u_n \le u$ and $\boldsymbol{\Delta} u_n \le 0$ on $\mathfrak{G}_n$, and $u_n(\mathbf{x}) \longrightarrow u(\mathbf{x})$ as $n \to \infty$ for each $\mathbf{x} \in \mathfrak{G}$. In particular, there exists a unique nonnegative Radon measure μ on $\mathfrak{G}$ with the property that

$$\tfrac{1}{2}\int_{\mathfrak{G}} \boldsymbol{\Delta}\varphi\, u\, d\mathbf{x} = -\int_{\mathfrak{G}} \varphi\, d\mu \quad \text{for all } \varphi \in C_c^2(\mathfrak{G}; \mathbb{R}); \tag{8.4.8}$$

and, in fact, for all $\varphi \in C_c(\mathfrak{G}; \mathbb{R})$,

$$\int_{\mathfrak{G}} \varphi\, d\mu = \lim_{n\to\infty} \int_{\mathfrak{G}} \varphi\, d\mu_n \quad \text{where } \mu_n(d\mathbf{y}) = -\tfrac{1}{2}\left[\mathbf{1}_{\mathfrak{G}_n} \boldsymbol{\Delta} u_n\right](\mathbf{y})\, d\mathbf{y}. \tag{8.4.9}$$

PROOF: To prove the first assertion, let $\mathcal{U}$ denote the set of all $\mathbf{x} \in \mathfrak{G}$ with the property that

$$\int_{B_{\mathbb{R}^N}(\mathbf{x},r)} u(\mathbf{y})\, d\mathbf{y} < \infty \quad \text{for some } r > 0 \text{ with } \overline{B_{\mathbb{R}^N}(\mathbf{x}, r)} \subset\subset \mathfrak{G}.$$

Obviously, $\mathcal{U}$ is an open subset of $\mathfrak{G}$. At the same time, if $\mathbf{x} \in \mathfrak{G} \setminus \mathcal{U}$ and $r > 0$ is chosen so that $\overline{B_{\mathbb{R}^N}(\mathbf{x}, 2r)} \subset\subset \mathfrak{G}$, then, for each $\mathbf{y} \in B_{\mathbb{R}^N}(\mathbf{x}, r)$ and $s \in (0, r)$:

$$u(\mathbf{y}) \ge \frac{1}{\omega_{N-1}} \int_{\mathbf{S}^{N-1}} u(\mathbf{y} + s\boldsymbol{\xi})\, \lambda_{\mathbf{S}^{N-1}}(d\boldsymbol{\xi})$$

and so, after integrating this with respect to $Ns^{N-1}\, ds$ over $(0, r)$, we get

$$u(\mathbf{y}) \ge \frac{1}{\boldsymbol{\Omega}_{N-1} r^N} \int_{B_{\mathbb{R}^N}(\mathbf{y},r)} u(\boldsymbol{\xi})\, d\boldsymbol{\xi} \ge \frac{1}{\boldsymbol{\Omega}_{N-1} r^N} \int_{B_{\mathbb{R}^N}(\mathbf{x},\delta)} u(\boldsymbol{\xi})\, d\boldsymbol{\xi} = \infty,$$

where $\delta \equiv r - |\mathbf{y} - \mathbf{x}|$. Hence, we now see that $\mathfrak{G} \setminus \mathcal{U}$ is also open, and therefore that either $\mathcal{U} = \mathfrak{G}$ or $\mathcal{U} = \emptyset$ and $u \equiv \infty$.

Now assume that $u \in \mathcal{E}(\mathfrak{G})$ is not identically infinite. If u is smooth at some $\mathbf{x} \in \mathfrak{G}$, then an easy computation shows that

$$\boldsymbol{\Delta} u(\mathbf{x}) = \lim_{r \searrow 0} \frac{N}{\omega_{N-1} r^2} \int_{\mathbf{S}^{N-1}} \Big(u(\mathbf{x} + r\boldsymbol{\xi}) - u(\mathbf{x}) \Big) \lambda_{\mathbf{S}^{N-1}}(d\boldsymbol{\xi}) \le 0.$$

Next, define $\mathfrak{G}_n$ and u_n as in (8.4.6) and (8.4.7), respectively. Obviously, $u_n \in C_c^\infty\big(\mathfrak{G}; [0, \infty)\big)$. In addition, if $\mathbf{x} \in \mathfrak{G}_n$, then, by taking advantage of the rotation invariance of ρ, we see that

$$\begin{aligned} u_n(\mathbf{x}) &= \int_{(0, \frac{R}{4})} t^{N-1} \tilde{\rho}(t) \left(\int_{\mathbf{S}^{N-1}} u\left(\mathbf{x} + \tfrac{t}{n}\boldsymbol{\xi}\right) \lambda_{\mathbf{S}^{N-1}}(d\boldsymbol{\xi}) \right) dt \\ &\le u(\mathbf{x})\, \omega_{N-1} \int_{(0, \frac{R}{4})} t^{N-1} \tilde{\rho}(t)\, dt = u(\mathbf{x}), \end{aligned}$$

where $\tilde{\rho} : \mathbb{R} \longrightarrow [0, \infty)$ is taken so that $\rho(\boldsymbol{\xi}) = \tilde{\rho}\big(|\boldsymbol{\xi}|\big)$. Similarly, if $\overline{B_{\mathbb{R}^N}(\mathbf{x}, r)} \subset\subset \mathfrak{G}_n$, then

$$\begin{aligned} &\int_{\mathbf{S}^{N-1}} u_n(\mathbf{x} + r\boldsymbol{\eta})\, \lambda_{\mathbf{S}^{N-1}}(d\boldsymbol{\eta}) \\ &\quad = \int_{B_{\mathbb{R}^N}(\mathbf{0}, \frac{R}{4})} \rho(\boldsymbol{\xi}) \left(\int_{\mathbf{S}^{N-1}} u\left(\mathbf{x} + \tfrac{1}{n}\boldsymbol{\xi} + r\boldsymbol{\eta}\right) \lambda_{\mathbf{S}^{N-1}}(d\boldsymbol{\eta}) \right) d\boldsymbol{\xi} \\ &\quad \le \omega_{N-1} \int_{B_{\mathbb{R}^N}(\mathbf{0}, \frac{R}{4})} \rho(\boldsymbol{\xi}) u\left(\mathbf{x} + \tfrac{1}{n}\boldsymbol{\xi}\right) d\boldsymbol{\xi} = \omega_{N-1} u_n(\mathbf{x}). \end{aligned}$$

Hence, $u_n \restriction \mathfrak{G}_n$ is a smooth element of $\mathcal{E}(\mathfrak{G}_n)$, and therefore we know that $\boldsymbol{\Delta} u_n \le 0$ on $\mathfrak{G}_n$. To see that $u_n \longrightarrow u$ pointwise, observe that we already know that $u(\mathbf{x}) \ge \overline{\lim}_{n\to\infty} u_n(\mathbf{x})$. On the other hand, because u is lower semicontinuous, an application of Fatou's Lemma yields

$$u(\mathbf{x}) \le \varliminf_{n\to\infty} \int_{\mathfrak{G}} \rho(\boldsymbol{\xi})\, u\left(\mathbf{x} + \tfrac{1}{n}\boldsymbol{\xi}\right) d\boldsymbol{\xi} = \varliminf_{n\to\infty} u_n(\mathbf{x}).$$

To complete the proof, let μ_n be the measure defined in (8.4.9) and note that

$$\begin{aligned} u_n(\mathbf{x}) &= \mathbb{E}^{\mathcal{W}_{\mathbf{x}}^{(N)}}\Big[u_n\big(\boldsymbol{\psi}(t \wedge \mathfrak{e}^{\mathfrak{G}_n})\big) \Big] - \mathbb{E}^{\mathcal{W}_{\mathbf{x}}^{(N)}}\left[\int_0^{t \wedge \mathfrak{e}^{\mathfrak{G}_n}} \tfrac{1}{2} \boldsymbol{\Delta} u_n\big(\boldsymbol{\psi}(s)\big)\, ds \right] \\ &\ge -\mathbb{E}^{\mathcal{W}_{\mathbf{x}}^{(N)}}\left[\int_0^{t \wedge \mathfrak{e}^{\mathfrak{G}_n}} \tfrac{1}{2} \boldsymbol{\Delta} u_n\big(\boldsymbol{\psi}(s)\big)\, ds \right] \\ &= \int_0^t \left(\int_{\mathfrak{G}_n} p^{\mathfrak{G}_n}(s, \mathbf{x}, \mathbf{y})\, \mu_n(d\mathbf{y}) \right) ds \end{aligned}$$

for all $n \in \mathbb{Z}^+$ and $(t, \mathbf{x}) \in (0,\infty) \times \mathfrak{G}_n$. Hence, after letting $t \nearrow \infty$, we see that

$$u(\mathbf{x}) \geq u_n(\mathbf{x}) \geq \int_{\mathfrak{G}_n} g^{\mathfrak{G}_n}(\mathbf{x},\mathbf{y})\,\mu_n(d\mathbf{y}), \quad n \in \mathbb{Z}^+ \text{ and } \mathbf{x} \in \mathfrak{G}_n.$$

In particular, because $u(\mathbf{x}) < \infty$ for Lebesgue almost every $\mathbf{x} \in \mathfrak{G}$, this proves that, for each $K \subset\subset \mathfrak{G}$, $\sup_{n\in\mathbb{Z}^+} \mu_n(K) < \infty$; and therefore (cf. part (**iv**) of Exercise 3.1.17 and use a Cantor diagonalization procedure) $\{\mu_n\}_1^\infty$ is relatively compact in the sense that every subsequence $\{\mu_{n_m}\}$ admits a subsequence $\{\mu_{n_{m_k}}\}$ and a Radon measure μ on $\mathfrak{G}$ with the property that

$$\lim_{k\to\infty} \int_{\mathfrak{G}} \varphi\, d\mu_{n_{m_k}} = \int_{\mathfrak{G}} \varphi\, d\mu \quad \text{for all } \varphi \in C_c(\mathfrak{G};\mathbb{R}).$$

At the same time, using integration by parts followed and Lebesgue's Dominated Theorem, we see that

$$\lim_{n\to\infty} \int_{\mathfrak{G}} \varphi\, d\mu_n = -\tfrac{1}{2} \lim_{n\to\infty} \int_{\mathfrak{G}} \mathbf{\Delta}\varphi\, u_n\, d\mathbf{x} = -\tfrac{1}{2} \int_{\mathfrak{G}} \mathbf{\Delta}\varphi\, u\, d\mathbf{x}, \quad \varphi \in C_c^2(\mathfrak{G};\mathbb{R}),$$

and therefore any limit μ of $\{\mu_n\}$ must satisfy (8.4.8), which proves not only that there is such a μ but also that (8.4.9) is satisfied. □

8.4.10 Lemma. *For any lower semicontinuous $u : \mathfrak{G} \longrightarrow [0,\infty]$, $u \in \mathcal{E}(\mathfrak{G})$ if and only if*

$$\text{(8.4.11)} \qquad \mathbb{E}^{\mathcal{W}_{\mathbf{x}}^{(N)}}\Big[u\big(\psi(\tau)\big),\ \tau < \mathfrak{e}^{\mathfrak{G}}\Big] \leq \mathbb{E}^{\mathcal{W}_{\mathbf{x}}^{(N)}}\Big[u\big(\psi(\sigma)\big),\ \sigma < \mathfrak{e}^{\mathfrak{G}}\Big]$$

for every pair σ and τ of $\{\mathcal{B}_t : t \in [0,\infty)\}$-stopping times with $\sigma \leq \tau$. In particular, if $u \in \mathcal{E}(\mathfrak{G})$ and $\overline{B_{\mathbb{R}^N}(\mathbf{x},r)} \subset\subset \mathfrak{G}$, then for any rotationally symmetric $\rho \in C_c\big(B_{\mathbb{R}^N}(\mathbf{0},r);[0,\infty)\big)$ with total integral 1,

$$t \in (0,1) \longmapsto \int_{B_{\mathbb{R}^N}(\mathbf{0},r)} \rho(\mathbf{y})\, u(\mathbf{x}+t\mathbf{y})\, d\mathbf{y} \in [0,\infty]$$

is a nonincreasing function.

PROOF: Let $u \in \mathcal{E}(\mathfrak{G})$ be given. Clearly (8.4.11) is trivial in the case when $u \equiv \infty$. Thus, assume that $u \not\equiv \infty$, and define $\mathfrak{G}_n$ and u_n for $n \in \mathbb{Z}^+$ as in (8.4.6) and (8.4.7). Because $\mathbf{\Delta}u_n \restriction \mathfrak{G}_n \leq 0$, we know (cf. Lemma 8.1.15 and apply Theorem 7.1.15) that

$$\mathbb{E}^{\mathcal{W}_{\mathbf{x}}^{(N)}}\Big[u_n\big(\psi(\tau\wedge\mathfrak{e}^{\mathfrak{G}_m}\wedge T)\big),\ \sigma\wedge T < \mathfrak{e}^{\mathfrak{G}_m}\Big] \leq \mathbb{E}^{\mathcal{W}_{\mathbf{x}}^{(N)}}\Big[u_n\big(\psi(\sigma\wedge T)\big),\ \sigma\wedge T < \mathfrak{e}^{\mathfrak{G}_m}\Big]$$

for all $1 \le m \le n$, $\mathbf{x} \in \mathfrak{G}_m$, and $T \in [0,\infty)$. Next, after noting that $\mathfrak{e}^{\mathfrak{G}_m} < \infty$ $\mathcal{W}_{\mathbf{x}}^{(N)}$-almost surely, let $T \nearrow \infty$ in the preceding and arrive at

$$\mathbb{E}^{\mathcal{W}_{\mathbf{x}}^{(N)}}\Big[u_n\big(\psi(\tau \wedge \mathfrak{e}^{\mathfrak{G}_m})\big),\, \sigma < \mathfrak{e}^{\mathfrak{G}_m}\Big] \le \mathbb{E}^{\mathcal{W}_{\mathbf{x}}^{(N)}}\Big[u_n\big(\psi(\sigma)\big),\, \sigma < \mathfrak{e}^{\mathfrak{G}_m}\Big].$$

But, because $\sigma \le \tau$ and $u \ge u_n \ge 0$, this means that

$$\mathbb{E}^{\mathcal{W}_{\mathbf{x}}^{(N)}}\Big[u_n\big(\psi(\tau)\big),\, \tau < \mathfrak{e}^{\mathfrak{G}_m}\Big] \le \mathbb{E}^{\mathcal{W}_{\mathbf{x}}^{(N)}}\Big[u\big(\psi(\sigma)\big),\, \sigma < \mathfrak{e}^{\mathfrak{G}_m}\Big],$$

which, because $0 \le u_n \longrightarrow u$ pointwise, leads, via Fatou's Lemma, first to

$$\mathbb{E}^{\mathcal{W}_{\mathbf{x}}^{(N)}}\Big[u\big(\psi(\tau)\big),\, \tau < \mathfrak{e}^{\mathfrak{G}_m}\Big] \le \mathbb{E}^{\mathcal{W}_{\mathbf{x}}^{(N)}}\Big[u\big(\psi(\sigma)\big),\, \sigma < \mathfrak{e}^{\mathfrak{G}_m}\Big],$$

and thence, by the Monotone Convergence Theorem, to (8.4.11).

From here, the rest is easy. Given a lower semicontinuous $u : \mathfrak{G} \longrightarrow [0,\infty]$ and $\overline{B_{\mathbb{R}^N}(\mathbf{x},r)} \subset\subset \mathfrak{G}$ we have (cf. (8.4.4))

$$\frac{1}{\omega_{N-1}} \int_{\mathbf{S}^{N-1}} u(\mathbf{x} + r\boldsymbol{\xi})\, \lambda_{\mathbf{S}^{N-1}}(d\boldsymbol{\xi}) = \mathbb{E}^{\mathcal{W}_{\mathbf{x}}^{(N)}}\Big[u\big(\psi(\mathfrak{e}_r)\big),\, \mathfrak{e}_r < \mathfrak{e}^{\mathfrak{G}}\Big].$$

Thus, if, in addition, (8.4.11) holds, then

$$t \in [0,1] \longmapsto \frac{1}{\omega_{N-1}} \int_{\mathbf{S}^{N-1}} u(\mathbf{x} + tr\boldsymbol{\xi})\, \lambda_{\mathbf{S}^{N-1}}(d\boldsymbol{\xi}) \in [0,\infty]$$

is nonincreasing; and, therefore, not only is u excessive but also (after passing to polar coordinates) one gets the monotonicity described in the final assertion is true. □

8.4.12 Theorem (Riesz Decomposition). *Let $\mathfrak{G}$ be a nonempty, connected open subset of $\mathbb{R}^N$, and assume either that $N \ge 3$ or that (8.2.9) holds. If $u \in \mathcal{E}(\mathfrak{G})$ is not identically infinite, then there exists a unique nonnegative Radon measure μ on $\mathfrak{G}$ and a unique nonnegative harmonic function h with the property that*

$$u(\mathbf{x}) = \big[G^{\mathfrak{G}}\mu\big](\mathbf{x}) + h(\mathbf{x}) \quad \textit{for all } \mathbf{x} \in \mathfrak{G}. \tag{8.4.13}$$

In fact, μ is uniquely determined by (8.4.8), and h is the unique harmonic function on $\mathfrak{G}$ with the property that $h \ge w$ for every nonnegative harmonic w which is dominated by u. (Cf. Exercise 8.4.53 as well.)

PROOF: Define $\mathfrak{G}_n$, u_n, and μ_n as in (8.4.6), (8.4.7), and (8.4.9). As an application of Lemma 7.2.1, we see that, for each $1 \le m \le n$ and $\mathbf{x} \in \mathfrak{G}_m$:

$$u_n(\mathbf{x}) = \int_{\mathfrak{G}_m} g^{\mathfrak{G}_m}(\mathbf{x}, \mathbf{y})\, \mu_n(d\mathbf{y}) + \big[H^{\mathfrak{G}_m} u_n\big](\mathbf{x}).$$

By the last part of Lemma 8.4.10, we know that $u_m \le u_n \le u$ on $\mathfrak{G}_m$ for each $1 \le m \le n$. Hence, after combining this with the pointwise convergence result in Lemma 8.4.5, we see first that $u_{k \vee m} \nearrow u$ pointwise on $\mathfrak{G}_m$ as $k \to \infty$, and then, by the Monotone Convergence Theorem, that the preceding leads to

$$\begin{aligned} \int_{\mathfrak{G}_m} u(\mathbf{x})\, \nu(d\mathbf{x}) = \lim_{k\to\infty} \iint_{\mathfrak{G}_m^2} g^{\mathfrak{G}_m}(\mathbf{x},\mathbf{y})\, \nu(d\mathbf{x}) d\mu_k(\mathbf{y}) \\ + \int_{\mathfrak{G}_m} \big[H^{\mathfrak{G}_m} u\big](\mathbf{x})\, \nu(\mathbf{x}) \end{aligned} \tag{8.4.14}$$

for every nonnegative Radon measure ν on $\mathfrak{G}_m$.

Notice (cf. Harnack's Principle) that, as the nondecreasing limit of nonnegative harmonic functions, $H^{\mathfrak{G}_m} u$ is either identically infinite or itself a nonnegative harmonic function on $\mathfrak{G}$; and so, since $u(\mathbf{x}) < \infty$ Lebesgue almost everywhere, (8.4.14) shows that the latter must be the case. Now let $\mathbf{a}$ be a fixed element of $\mathfrak{G}_m$, let ρ_n be the function described in the statement of Lemma 8.4.5, and define

$$\varphi_n(\mathbf{y}) = \begin{cases} \int_{\mathfrak{G}_m} \rho_n(\mathbf{x} - \mathbf{a}) g^{\mathfrak{G}_m}(\mathbf{x}, \mathbf{y})\, d\mathbf{x} & \text{if} \quad \mathbf{y} \in \mathfrak{G}_m \\ 0 & \text{otherwise.} \end{cases}$$

By taking $\nu(d\mathbf{x}) = \rho_n(\mathbf{x} - \mathbf{a})\, d\mathbf{x}$ in (8.4.14), we see that, for $n \ge m$,

$$\begin{aligned} \int_{\mathfrak{G}_m} \rho_n(\mathbf{x} - \mathbf{a})\, u(\mathbf{x})\, d\mathbf{x} = \lim_{k\to\infty} \int_{\mathfrak{G}} \varphi_n(\mathbf{y})\, \mu_k(d\mathbf{y}) \\ + \int_{\mathfrak{G}_m} \rho_n(\mathbf{x} - \mathbf{a}) \big[H^{\mathfrak{G}_m} u\big](\mathbf{x})\, d\mathbf{x}. \end{aligned}$$

But, since every element of $\partial\mathfrak{G}_m$ satisfies Poincaré's exterior ball condition (cf. Remark 8.1.28) and is therefore regular, the fact that $\rho_n(\,\cdot - \mathbf{a}) \in C_c^\infty(\mathfrak{G}_m; [0,\infty))$ for all n's larger than some $n(\mathbf{a})$ implies that φ_n is continuous for all large $n \ge n(\mathbf{a})$. In particular, by (8.4.9), we can now say that

$$\int_{\mathfrak{G}_m} \rho_n(\mathbf{x} - \mathbf{a})\, u(\mathbf{x})\, d\mathbf{x} = \int_{\mathfrak{G}} \varphi_n(\mathbf{x})\, \mu(d\mathbf{x}) + \int_{\mathfrak{G}_m} \rho_n(\mathbf{x} - \mathbf{a}) \big[H^{\mathfrak{G}_m} u\big](\mathbf{x})\, d\mathbf{x}$$

for all $n \ge n(\mathbf{a})$. In addition (by the results in Lemmas 8.4.5 and 8.4.10), as $n \to \infty$, the term on the left tends to $u(\mathbf{a})$, the second term on the right goes to

$\left[H^{\mathfrak{G}_m}u\right](\mathbf{a})$, and $\{\varphi_n(\mathbf{y}) : n \geq n(\mathbf{a})\}$ tends nondecreasingly to $g^{\mathfrak{G}_m}(\mathbf{a},\mathbf{y})$. Thus, we have now proved that

$$u = G^{\mathfrak{G}_m}\mu + H^{\mathfrak{G}_m}u \quad \text{on } \mathfrak{G}_m \text{ for every } m \in \mathbb{Z}^+. \tag{8.4.15}$$

Starting from (8.4.15), the rest of the proof is quite simple. Namely, fix $\mathbf{x} \in \mathfrak{G}$, choose m so that $\mathbf{x} \in \mathfrak{G}_m$, note that, $g^{\mathfrak{G}_n}(\mathbf{x},\cdot)$ is nondecreasing as $n \geq m$ increases, and conclude that $\left[G^{\mathfrak{G}_{n\vee m}}\mu\right](\mathbf{x}) \nearrow \left[G^{\mathfrak{G}}\mu\right](\mathbf{x})$. Hence, by (8.4.15) (alternatively, by (8.4.11)), we know that $\left[H^{\mathfrak{G}_{m\vee n}}u\right](\mathbf{x})$ tends nonincreasingly to a limit $h(\mathbf{x})$, which Harnack's Principle guarantees to be harmonic as a function of $\mathbf{x} \in \mathfrak{G}$. Thus, after passing to the limit as $m \to \infty$ in (8.4.15), we conclude that (8.4.13) holds with the μ satisfying (8.4.8) and $h = \lim_{m\to\infty} H^{\mathfrak{G}_m}u$. To prove that these quantities are unique, note that if ν is any nonnegative Radon measure on $\mathfrak{G}$ for which $u - G^{\mathfrak{G}}\nu$ is a nonnegative harmonic function, then, for every $\varphi \in C_c^\infty(\mathfrak{G};\mathbb{R})$, simple integration parts plus the symmetry of $g^{\mathfrak{G}}$ shows that

$$-\tfrac{1}{2}\int_{\mathfrak{G}} \Delta\varphi u\,d\mathbf{x} = -\tfrac{1}{2}\int_{\mathfrak{G}} \Delta G^{\mathfrak{G}}\varphi\,d\nu = \int_{\mathfrak{G}} \varphi\,d\nu.$$

That is, ν must satisfy (8.4.8); and so we have now derived the required uniqueness result. Finally, to check the asserted characterization of h, suppose that w is a nonnegative harmonic function which is dominated by u on $\mathfrak{G}$. We then have

$$w(\mathbf{x}) = \left[H^{\mathfrak{G}_m}w\right](\mathbf{x}) \leq \left[H^{\mathfrak{G}_m}u\right](\mathbf{x}) \quad \text{for all } m \in \mathbb{Z}^+ \text{ and } \mathbf{x} \in \mathfrak{G}_m;$$

and therefore the desired conclusion follows from the fact that $H^{\mathfrak{G}_m}u$ tends to h. □

By combining Lemma 8.4.3 with Theorem 8.4.12, we arrive at the following characterization of potentials.

8.4.16 Corollary. *Let everything be as in Theorem 8.4.12, and suppose that $u : \mathfrak{G} \longrightarrow [0,\infty]$ is not identically infinite. Then a necessary and sufficient condition for u to be the potential $G^{\mathfrak{G}}\mu$ of some nonnegative Radon measure μ on $\mathfrak{G}$ is that u be excessive on $\mathfrak{G}$ and have the property that the constant function 0 is the only nonnegative harmonic function which is dominated by u on $\mathfrak{G}$.*

Let u be an excessive function on $\mathfrak{G}$ which is not identically infinite. In keeping with the electrostatic metaphor, we will call the Radon measure μ entering the Riesz decomposition (8.4.13) of u the **charge determined by** u. Thinking more mathematically, we can give the charge another interpretation. Namely, by Lemma 8.4.5, we know that u is locally integrable on $\mathfrak{G}$, and, as such, can be viewed as a distribution (in the sense of L. Schwartz) on $\mathfrak{G}$. Hence, in the language of distribution theory, (8.4.8) is the statement that the charge μ of u is $-\frac{1}{2}\boldsymbol{\Delta}u$. On the other hand, the general theory of distributional derivatives is a little too crude to be very useful here, and it is better to approach the problem of computing μ with an approximation scheme which is intimately connected to the problem at hand. To this end, note that, for any $\varphi \in C^2_{\mathrm{c}}(\mathfrak{G};\mathbb{R})$:

$$\text{(8.4.17)} \qquad \frac{\varphi - P^{\mathfrak{G}}_\epsilon \varphi}{\epsilon} \longrightarrow -\tfrac{1}{2}\boldsymbol{\Delta}\varphi \quad \text{uniformly as } \epsilon \searrow 0.$$

Hence, because it is obviously closely related to the analysis which we have been doing, it is reasonable to suppose that (8.4.17) might provide a good way to construct μ from u. Thus, after noting that, by (8.4.11) with $\sigma \equiv 0$ and $\tau \equiv \epsilon$, $u \geq P^{\mathfrak{G}}_\epsilon u$, define

$$\text{(8.4.18)} \qquad f_\epsilon(\mathbf{x}) = \frac{1}{\epsilon}\begin{cases} u(\mathbf{x}) - \big[P^{\mathfrak{G}}_\epsilon u\big](\mathbf{x}) & \text{if} \quad u(\mathbf{x}) < \infty \\ \infty & \text{if} \quad u(\mathbf{x}) = \infty. \end{cases}$$

Clearly, each f_ϵ is a nonnegative, locally integrable function on $\mathfrak{G}$. In fact, if $u(\mathbf{x}) < \infty$, then

$$\begin{aligned} &\int_{\mathfrak{G}} g^{\mathfrak{G}}(\mathbf{x},\mathbf{y})\, f_\epsilon(\mathbf{y})\, d\mathbf{y} \\ &\quad = \lim_{T \nearrow \infty} \epsilon^{-1}\left(\int_0^\epsilon \big[P^{\mathfrak{G}}_t u\big](\mathbf{x})\, dt - \int_T^{T+\epsilon} \big[P^{\mathfrak{G}}_t u\big](\mathbf{x})\, dt\right) \\ &\quad \leq \epsilon^{-1}\int_0^\epsilon \big[P^{\mathfrak{G}}_t u\big](\mathbf{x})\, dt \leq u(\mathbf{x}). \end{aligned}$$

Hence, since $u \not\equiv \infty$, we see that, for each $K \subset\subset \mathfrak{G}$,

$$\text{(8.4.19)} \qquad \sup_{\epsilon > 0} \mu_\epsilon(K) < \infty \quad \text{where } \mu_\epsilon(d\mathbf{x}) \equiv f_\epsilon(\mathbf{x})\, d\mathbf{x}.$$

In particular, this means that the family of Radon measures $\{\mu_\epsilon : \epsilon > 0\}$ is relatively compact in the sense that (cf. the discussion near the end of the proof of Lemma 8.4.5) every subsequence admits a subsequence which converges when tested against functions $\varphi \in C_{\mathrm{c}}(\mathfrak{G};\mathbb{R})$. At the same time, by the symmetry of $p^{\mathfrak{G}}(t,\mathbf{x},\mathbf{y})$ and (8.4.17), for any $\varphi \in C^2_{\mathrm{c}}(\mathfrak{G};\mathbb{R})$:

$$\int_{\mathfrak{G}} \varphi\, d\mu_\epsilon = \int_{\mathfrak{G}} \frac{\varphi - P^{\mathfrak{G}}_\epsilon \varphi}{\epsilon}\, u\, d\mathbf{x} \longrightarrow -\tfrac{1}{2}\int_{\mathfrak{G}} \boldsymbol{\Delta}\varphi\, u\, d\mathbf{x}$$

as $\epsilon \searrow 0$. In other words, every limit point of $\{\mu_\epsilon : \epsilon > 0\}$ as $\epsilon \searrow 0$ satisfies (8.4.8), and therefore we have now proved the following result.

8.4.20 Theorem. *Let $\mathfrak{G}$ be as in Theorem 8.4.12 and u an element of $\mathcal{E}(\mathfrak{G})$ which is not identically infinite. Then the Radon measure μ entering (8.4.13) is determined by*

$$\int_{\mathfrak{G}} \varphi\, d\mu = \lim_{\epsilon \searrow 0} \int_{\mathfrak{G}} \varphi\, d\mu_\epsilon, \quad \varphi \in C_c(\mathfrak{G};\mathbb{R}), \tag{8.4.21}$$

where the measures μ_ϵ are the ones described in (8.4.19) (cf. (8.4.18) as well).

We will now apply these considerations to a special class of potentials. Namely, given $\mathfrak{G}$ and $K \subset\subset \mathfrak{G}$, set

$$p_K^{\mathfrak{G}}(\mathbf{x}) = \sup\left\{ \left[G^{\mathfrak{G}}\mu\right](\mathbf{x}) : \mu(\mathfrak{G}\setminus K) = 0 \text{ and } G^{\mathfrak{G}}\mu \le 1 \right\}, \quad \mathbf{x} \in \mathfrak{G}. \tag{8.4.22}$$

For reasons justified in the following, the function $p_K^{\mathfrak{G}}$ is called the **capacitory potential of K in $\mathfrak{G}$.**

8.4.23 Theorem. *Let $\mathfrak{G}$ be as in Theorem 8.4.12, $K \subset\subset \mathfrak{G}$, and $p_K^{\mathfrak{G}}$ the function given by* (8.4.22). *Then*

$$p_K^{\mathfrak{G}}(\mathbf{x}) = \mathcal{W}_{\mathbf{x}}^{(N)}\left(\exists t \in \left(0, \mathfrak{e}^{\mathfrak{G}}\right)\ \psi(t) \in K\right) \quad \text{for all } \mathbf{x} \in \mathfrak{G}. \tag{8.4.24}$$

In particular, $p_K^{\mathfrak{G}}$ is a potential whose charge $\mu_K^{\mathfrak{G}}$ is supported on K.†

PROOF: Let $u(\mathbf{x})$ be the right-hand side of (8.4.24). We begin by checking that u is excessive. Obviously, u takes its values in $[0,1]$. Furthermore, by the Markov property, for each $\epsilon > 0$:

$$u_\epsilon(\mathbf{x}) \equiv \left[P_\epsilon^{\mathfrak{G}} u\right](\mathbf{x}) = \mathcal{W}_{\mathbf{x}}^{(N)}\left(\exists t \in \left(\epsilon, \mathfrak{e}^{\mathfrak{G}}\right)\ \psi(t) \in K\right). \tag{8.4.25}$$

Hence, $u_\epsilon \nearrow u$; and therefore, because each u_ϵ is continuous, we now know that u is lower semicontinuous. At the same time, by the strong Markov property, if $\overline{B_{\mathbb{R}^N}(\mathbf{x}, r)} \subset\subset \mathfrak{G}$, then (cf. (8.4.4))

$$\begin{aligned}\frac{1}{\omega_{N-1}} \int_{\mathbf{S}^{N-1}} u(\mathbf{x} + r\boldsymbol{\xi})\, \lambda_{\mathbf{S}^{N-1}}(d\boldsymbol{\xi}) &= \mathbb{E}^{\mathcal{W}_{\mathbf{x}}^{(N)}}\left[u(\psi(\mathfrak{e}_r)),\ \mathfrak{e}_r < \mathfrak{e}^{\mathfrak{G}}\right] \\ &= \mathcal{W}_{\mathbf{x}}^{(N)}\left(\exists t \in \left(\mathfrak{e}_r, \mathfrak{e}^{\mathfrak{G}}\right)\ \psi(t) \in K\right) \le u(\mathbf{x}).\end{aligned}$$

Hence, we have now proved that u is excessive.

† It is interesting to note that, although Wiener's 1924 article "Certain notions in potential theory," *J. Math. Phys. M.I.T.* **4**, contains the first proof that an arbitrary compact set is *capacitable*, it contains no reference to Wiener's measure.

The next step is to prove that u is a potential. For this purpose, choose the sets $\mathfrak{G}_n$ as in (8.4.6), and let w be any nonnegative harmonic function which is dominated by u. Then, for each $\mathbf{x} \in \mathfrak{G}$,

$$\begin{aligned} w(\mathbf{x}) = \left[H^{\mathfrak{G}_n} w\right](\mathbf{x}) \leq \left[H^{\mathfrak{G}_n} u\right](\mathbf{x}) &= \mathcal{W}_{\mathbf{x}}^{(N)}\left(\exists t \in \left(\mathfrak{e}^{\mathfrak{G}_n}, \mathfrak{e}^{\mathfrak{G}}\right)\ \psi(t) \in K\right) \\ &= \mathcal{W}_{\mathbf{x}}^{(N)}\left(\exists t \in \left(\mathfrak{e}^{\mathfrak{G}_n}, \mathfrak{e}^{\mathfrak{G}}\right)\ \psi(t) \in K \ \&\ \mathfrak{e}^{\mathfrak{G}} < \infty\right) \\ &\quad + \mathcal{W}_{\mathbf{x}}^{(N)}\left(\exists t \in \left(\mathfrak{e}^{\mathfrak{G}_n}, \infty\right)\ \psi(t) \in K \ \&\ \mathfrak{e}^{\mathfrak{G}} = \infty\right). \end{aligned}$$

Obviously,

$$\lim_{n\to\infty} \mathcal{W}_{\mathbf{x}}^{(N)}\left(\exists t \in \left(\mathfrak{e}^{\mathfrak{G}_n}, \mathfrak{e}^{\mathfrak{G}}\right)\ \psi(t) \in K \ \&\ \mathfrak{e}^{\mathfrak{G}} < \infty\right) = 0;$$

and when (8.2.9) holds,

$$\mathcal{W}_{\mathbf{x}}^{(N)}\left(\exists t \in \left(\mathfrak{e}^{\mathfrak{G}_n}, \mathfrak{e}^{\mathfrak{G}}\right)\ \psi(t) \in K \ \&\ \mathfrak{e}^{\mathfrak{G}} = \infty\right) \leq \mathcal{W}_{\mathbf{x}}^{(N)}\left(\mathfrak{e}^{\mathfrak{G}} = \infty\right) = 0$$

for all $n \in \mathbb{Z}^+$. Thus, when (8.2.9) holds, we have shown that $w(\mathbf{x}) = 0$. To handle the case when $N \geq 3$ but (8.2.9) fails, observe that

$$\begin{aligned} \lim_{n\to\infty} \mathcal{W}_{\mathbf{x}}^{(N)}&\left(\exists t \in \left(\mathfrak{e}^{\mathfrak{G}_n}, \infty\right)\ \psi(t) \in K \ \&\ \mathfrak{e}^{\mathfrak{G}} = \infty\right) \\ &\leq \mathcal{W}_{\mathbf{x}}^{(N)}\left(\forall s \in (0,\infty)\ \exists t \in (s,\infty)\ \psi(t) \in K\right) = 0, \end{aligned}$$

since $\mathfrak{e}^{\mathfrak{G}_n}(\psi) \nearrow \mathfrak{e}^{\mathfrak{G}}(\psi)$ and $\lim_{t\to\infty} |\psi(t)| = \infty$ for $\mathcal{W}_{\mathbf{x}}^{(N)}$-almost every $\psi \in \mathfrak{P}(\mathbb{R}^N)$. Hence, in this case also, $w(\mathbf{x}) = 0$.

We now know that $u = G^{\mathfrak{G}}\mu$ for the Radon measure μ on $\mathfrak{G}$ satisfying (8.4.8). To see that $\mu(\mathfrak{G} \setminus K) = 0$, note that $u \upharpoonright \mathfrak{G} \setminus K = H^{\mathfrak{G}\setminus K}\mathbf{1}_K$ and is therefore harmonic. Thus, if $\varphi \in C^2_c(\mathfrak{G} \setminus K; \mathbb{R})$, then, after elementary integration by parts, one has (cf. (8.4.8))

$$\int_{\mathfrak{G}} \varphi\, d\mu = -\tfrac{1}{2}\int_{\mathfrak{G}} \Delta\varphi\, u\, d\mathbf{x} = -\tfrac{1}{2}\int_{\mathfrak{G}\setminus K} \varphi\, \Delta u\, d\mathbf{x} = 0;$$

from which it is clear that $\mu(\mathfrak{G} \setminus K) = 0$.

So far, we have proved that $u = G^{\mathfrak{G}}\mu$ with $\mu(\mathfrak{G} \setminus K) = 0$. In particular, since $u \leq 1$, we know that $u \leq p_K^{\mathfrak{G}}$. To prove the opposite inequality, let $w = G^{\mathfrak{G}}\nu$, where $\nu(\mathfrak{G} \setminus K) = 0$, and assume that $w \leq 1$. Clearly w is harmonic on $\mathfrak{G} \setminus K$ and $w(\mathbf{x}) \longrightarrow 0$ as $\mathbf{x} \in \mathfrak{G} \to \mathbf{a} \in \partial_{\text{reg}}\mathfrak{G}$. Hence, for any $\delta > 0$, we know that

$$w(\mathbf{x}) = \mathbb{E}^{\mathcal{W}_{\mathbf{x}}^{(N)}}\left[w\left(\psi(\sigma^{(\delta)})\right),\ \sigma^{(\delta)} < \mathfrak{e}^{\mathfrak{G}}\right] \quad \text{for } \mathbf{x} \in \mathfrak{G},$$

where

$$\sigma^{(\delta)}(\psi) \equiv \inf\{t \geq 0 : |\psi(t) - K| \leq \delta\}.$$

At the same time, because $w \leq 1$,

$$\mathbb{E}^{\mathcal{W}_{\mathbf{x}}^{(N)}}\left[w\big(\psi(\sigma^{(\delta)})\big),\ \sigma^{(\delta)} < \mathfrak{e}^{\mathfrak{G}}\right] \leq \mathcal{W}_{\mathbf{x}}^{(N)}\left(\sigma^{(\delta)} < \mathfrak{e}^{\mathfrak{G}}\right),$$

and so

$$w(\mathbf{x}) \leq \mathcal{W}_{\mathbf{x}}^{(N)}\left(\sigma^{(\delta)} < \mathfrak{e}^{\mathfrak{G}}\right) \quad \text{for } \delta > 0 \text{ and } \mathbf{x} \in \mathfrak{G}.$$

But, for $\mathbf{x} \in \mathfrak{G} \setminus K$,

$$\begin{aligned}&\lim_{\delta \searrow 0} \mathcal{W}_{\mathbf{x}}^{(N)}\left(\sigma^{(\delta)} < \mathfrak{e}^{\mathfrak{G}}\right)\\ &\qquad = u(\mathbf{x}) + \mathcal{W}_{\mathbf{x}}^{(N)}\left(\forall \delta > 0\ \sigma^{(\delta)} < \infty \text{ but } \lim_{\delta \searrow 0} \sigma^{(\delta)} = \infty = \mathfrak{e}^{\mathfrak{G}}\right),\end{aligned}$$

and (by the same sort of reasoning as we used earlier), the second term on the right vanishes when either (8.2.9) holds or $N \geq 3$. Hence, we now know that $w \restriction \mathfrak{G} \setminus K \leq u \restriction \mathfrak{G} \setminus K$. To handle $\mathbf{x} \in K$, first note that $u \restriction K = 1 \geq w$ when $N = 1$; and so we may and will assume that $N \geq 2$. Because, when $N \geq 2$, $g^{\mathfrak{G}}(\mathbf{x}, \mathbf{x}) = \infty$ and therefore (since $w \leq 1$) $\nu(\{\mathbf{x}\}) = 0$, it suffices for us to show that

$$w_\delta(\mathbf{x}) \equiv \int_{\mathfrak{G} \setminus B_{\mathbb{R}^N}(\mathbf{x}, \delta)} g^{\mathfrak{G}}(\mathbf{x}, \mathbf{y})\, \nu(d\mathbf{y}) \leq u(\mathbf{x}) \quad \text{for all } \delta > 0.$$

However, by the argument just given, we know that

$$w_\delta(\mathbf{x}) \leq \mathcal{W}_{\mathbf{x}}^{(N)}\left(\exists t \in \left(0, \mathfrak{e}^{\mathfrak{G}}\right)\ \psi(t) \in K \setminus B_{\mathbb{R}^N}(\mathbf{x}, \delta)\right) \leq u(\mathbf{x}),$$

which completes the proof. □

The charge $\mu_K^{\mathfrak{G}}$ is called the **capacitory distribution for K in $\mathfrak{G}$** and its total mass

(8.4.26) $$\operatorname{Cap}(K; \mathfrak{G}) \equiv \mu_K^{\mathfrak{G}}(K)$$

is called the **capacity of K in $\mathfrak{G}$**. As a dividend from Theorem 8.4.23, we get the following important connection between properties of Wiener paths and classical potential theory.

8.4.27 Corollary. *Let everything be as in the statement of Theorem 8.4.23. Then the following are equivalent:*

(i) *For every* $\mathbf{x} \in \mathfrak{G}$,

$$\mathcal{W}^{(N)}_{\mathbf{x}}\Big(\exists t \in \big(0, \mathfrak{e}^{\mathfrak{G}}\big)\ \boldsymbol{\psi}(t) \in K\Big) > 0;$$

(ii) *There is an* $\mathbf{x} \in \mathfrak{G}$ *for which*

$$\mathcal{W}^{(N)}_{\mathbf{x}}\Big(\exists t \in \big(0, \mathfrak{e}^{\mathfrak{G}}\big)\ \boldsymbol{\psi}(t) \in K\Big) > 0;$$

(iii) *There exists a nonzero, bounded potential on* $\mathfrak{G}$ *whose charge is supported in* K*;*

(iv) $\mathrm{Cap}(K;\mathfrak{G}) > 0$.

(Cf. Exercise 8.4.55 for additional information.)†

Proof: The only implications which are not completely trivial are the ones (iii) $\Longrightarrow$ (iv) and (iv) $\Longrightarrow$ (i). But, by (8.4.23), (iii) implies that $p^{\mathfrak{G}}_K \not\equiv 0$ and therefore that $\mu^{\mathfrak{G}}_K \neq 0$. Similarly, (iv) implies that $\mu^{\mathfrak{G}}_K \neq 0$, and therefore, since $g^{\mathfrak{G}} > 0$ throughout $\mathfrak{G}^2$, that $p^{\mathfrak{G}}_K > 0$ throughout $\mathfrak{G}$. Hence, by (8.4.24), (iv) $\Longrightarrow$ (i). □

As our final goal, we will present a couple of calculations which, if nothing else, should leave no doubt about the intimacy of the connection between potential theory and Wiener paths. The first of these calculations gives a probabilistic interpretation of the capacitory distribution $\mu^{\mathfrak{G}}_K$; and for this purpose it will be convenient to have introduced the function $\ell^{\mathfrak{G}}_K : \mathfrak{P}(\mathbb{R}^N) \longrightarrow [0,\infty]$ given by

$$\begin{aligned}\ell^{\mathfrak{G}}_K(\boldsymbol{\psi}) &= \sup\Big\{t \in \big(0, \mathfrak{e}^{\mathfrak{G}}(\boldsymbol{\psi})\big) : \boldsymbol{\psi}(t) \in K\Big\}\\ &\qquad \Big(\equiv 0 \text{ if } \Big\{t \in \big(0, \mathfrak{e}^{\mathfrak{G}}(\boldsymbol{\psi})\big) : \boldsymbol{\psi}(t) \in K\Big\} = \emptyset\Big).\end{aligned} \tag{8.4.28}$$

Notice that the **quitting time** $\ell^{\mathfrak{G}}_K$ *is not a stopping time.* On the other hand, it transforms nicely under the time-shift maps $\boldsymbol{\Sigma}_t$ (cf. the discussion preceding Lemma 4.3.7). Namely,

$$\ell^{\mathfrak{G}}_K \circ \boldsymbol{\Sigma}_t = \big(\ell^{\mathfrak{G}}_K - t\big)^+. \tag{8.4.29}$$

8.4.30 Theorem (Chung).‡ *If everything is as in Theorem 8.4.23, then, for all* $\varphi \in B(\mathfrak{G};\mathbb{R})$ *and every* $\mathbf{c} \in \mathfrak{G}$,

$$\int_{\mathfrak{G}} \varphi\, d\mu^{\mathfrak{G}}_K = \mathbb{E}^{\mathcal{W}^{(N)}_{\mathbf{c}}}\left[\frac{\varphi\big(\boldsymbol{\psi}(\ell^{\mathfrak{G}}_K)\big)}{g^{\mathfrak{G}}\big(\mathbf{c}, \boldsymbol{\psi}(\ell^{\mathfrak{G}}_K)\big)},\ \ell^{\mathfrak{G}}_K \in (0,\infty)\right]. \tag{8.4.31}$$

† This result is anticipated in the article by S. Kakutani's 1944 article referred to after (8.2.18).

‡ This result appeared originally in K.L. Chung's "Probabilistic approach in potential theory to the equilibrium problem," *Ann. Inst. Fourier Gren.* **23** (**3**) (1973). It gives the first direct probabilistic interpretation of the capacitory measure.

PROOF: Take $u = p_K^{\mathfrak{G}}$, and define f_ϵ for $\epsilon > 0$ as in (8.4.18). From (8.4.24) and (8.4.25), it is clear that

$$\epsilon f_\epsilon(\mathbf{x}) = \mathcal{W}_{\mathbf{x}}^{(N)}\Big(0 < \ell_K^{\mathfrak{G}} \le \epsilon\Big);$$

and so, when μ_ϵ is defined as in (8.4.19), we find that, for any $\varphi \in C_{\mathrm{b}}(\mathfrak{G};\mathbb{R})$:

$$\begin{aligned}
\int_{\mathfrak{G}} g^{\mathfrak{G}}(\mathbf{c},\mathbf{y})\varphi(\mathbf{y})\,\mu_\epsilon(d\mathbf{y}) &= \mathbb{E}^{\mathcal{W}_{\mathbf{c}}^{(N)}}\left[\int_0^{\mathfrak{e}^{\mathfrak{G}}} \varphi\big(\psi(t)\big)\,f_\epsilon\big(\psi(t)\big)\,dt\right]\\
&= \frac{1}{\epsilon}\int_0^\infty \mathbb{E}^{\mathcal{W}_{\mathbf{c}}^{(N)}}\left[\varphi\big(\psi(t)\big)\mathcal{W}_{\psi(t)}^{(N)}\big(0 < \ell_K^{\mathfrak{G}} \le \epsilon\big),\ \mathfrak{e}^{\mathfrak{G}} > t\right]dt\\
&= \frac{1}{\epsilon}\int_0^\infty \mathbb{E}^{\mathcal{W}_{\mathbf{c}}^{(N)}}\left[\varphi\big(\psi(t)\big),\ t < \ell_K^{\mathfrak{G}} \le t+\epsilon\right]dt\\
&= \mathbb{E}^{\mathcal{W}_{\mathbf{c}}^{(N)}}\left[\frac{1}{\epsilon}\int_{(\ell_K^{\mathfrak{G}}-\epsilon)^+}^{\ell_K^{\mathfrak{G}}} \varphi\big(\psi(t)\big)\,dt,\ \ell_K^{\mathfrak{G}} \in (0,\infty)\right]\\
&\longrightarrow \mathbb{E}^{\mathcal{W}_{\mathbf{c}}^{(N)}}\left[\varphi\big(\psi(\ell_K^{\mathfrak{G}})\big),\ \ell_K^{\mathfrak{G}} \in (0,\infty)\right] \quad \text{as } \epsilon \searrow 0,
\end{aligned}$$

where, in the passage to the second line we have applied the Markov property and in the passage to the third we have used (8.4.29). Next, let $\eta \in C_{\mathrm{c}}(\mathfrak{G};\mathbb{R})$ be given, note that $\varphi = \frac{\eta}{g^{\mathfrak{G}}(\mathbf{c},\cdot)}$ is again an element of $C_{\mathrm{c}}(\mathfrak{G};\mathbb{R})$, and conclude from (8.4.21) and the preceding that (8.4.31) holds first for φ's in $C_{\mathrm{c}}(\mathfrak{G};\mathbb{R})$ and then for all $\varphi \in B(\mathfrak{G};\mathbb{R})$. □

Aside from its intrinsic beauty, (8.4.31) has the virtue that it simplifies the proofs of various important facts about capacity. To illustrate what we have in mind, we must first define the **energy form** $\mathcal{E}^{\mathfrak{G}}(\mu,\nu)$, for nonnegative Radon measures μ and ν on $\mathfrak{G}$, given by

$$\mathcal{E}^{\mathfrak{G}}(\mu,\nu) = \iint_{\mathfrak{G}^2} g^{\mathfrak{G}}(\mathbf{x},\mathbf{y})\,\mu(d\mathbf{x})\nu(d\mathbf{y}) \tag{8.4.32}$$

and prove the following Schwarz inequality.

8.4.33 Lemma. *For any pair of nonnegative Radon measures μ and ν,*

$$\mathcal{E}^{\mathfrak{G}}(\mu,\nu) \le \sqrt{\mathcal{E}^{\mathfrak{G}}(\mu,\mu)}\,\sqrt{\mathcal{E}^{\mathfrak{G}}(\nu,\nu)};$$

and, when the factors on the right are both finite, equality holds if and only if $a\mu - b\nu = 0$ for some pair $(a,b) \in [0,\infty)^2 \setminus (0,0)$.

PROOF: For each $(t,\mathbf{x}) \in (0,\infty) \times \mathfrak{G}$, set

$$f(t,\mathbf{x}) = \int_{\mathfrak{G}} p^{\mathfrak{G}}(t,\mathbf{x},\mathbf{y})\,\mu(dy) \quad \text{and} \quad g(t,\mathbf{x}) = \int_{\mathfrak{G}} g^{\mathfrak{G}}(t,\mathbf{x},\mathbf{y})\,\nu(d\mathbf{y}),$$

and note that, by (8.3.9), the Chapman–Kolmogorov equation (8.1.12), Tonelli's Theorem, and Schwarz's inequality:

$$\begin{aligned}
\mathcal{E}^{\mathfrak{G}}(\mu,\nu) &= \int_{(0,\infty)} \left(\iint_{\mathfrak{G}^2} p^{\mathfrak{G}}(t,\mathbf{x},\mathbf{y})\,\mu(d\mathbf{x})\nu(d\mathbf{y}) \right) dt \\
&= \iint_{(0,\infty)\times\mathfrak{G}} f\left(\tfrac{t}{2},\boldsymbol{\xi}\right) g\left(\tfrac{t}{2},\boldsymbol{\xi}\right) dtd\boldsymbol{\xi} \\
&\leq \left(\iint_{(0,\infty)\times\mathfrak{G}} f\left(\tfrac{t}{2},\boldsymbol{\xi}\right)^2 dtd\boldsymbol{\xi} \right)^{\frac{1}{2}} \left(\iint_{(0,\infty)\times\mathfrak{G}} g\left(\tfrac{t}{2},\boldsymbol{\xi}\right)^2 dtd\boldsymbol{\xi} \right)^{\frac{1}{2}} \\
&= \left(\iint_{(0,\infty)\times\mathfrak{G}} f(t,\mathbf{x})\,dtd\mathbf{x} \right)^{\frac{1}{2}} \left(\iint_{(0,\infty)\times\mathfrak{G}} g(t,\mathbf{x})\,dtd\mathbf{x} \right)^{\frac{1}{2}} \\
&= \sqrt{\mathcal{E}^{\mathfrak{G}}(\mu,\mu)}\sqrt{\mathcal{E}^{\mathfrak{G}}(\nu,\nu)}.
\end{aligned}$$

Furthermore, when f and g are squareintegrable, then equality holds if and only if they are linearly dependent in the sense that $af - bg = 0$ Lebesgue almost everywhere for some nontrivial choice of a, $b \in [0,\infty)$. But this means that

$$\begin{aligned}
a\int_{\mathfrak{G}} \varphi\,d\mu &= \lim_{T\searrow 0} \frac{a}{T} \int_0^T \left(\int_{\mathfrak{G}} [P_t^{\mathfrak{G}}\varphi](\boldsymbol{\xi})\,\mu(d\boldsymbol{\xi}) \right) dt \\
&= \lim_{T\searrow 0} \frac{a}{T} \iint_{(0,T]\times\mathfrak{G}} \varphi(\boldsymbol{\xi})\,f(t,\boldsymbol{\xi})\,dtd\boldsymbol{\xi} \\
&= \lim_{T\searrow 0} \frac{b}{T} \iint_{(0,T]\times\mathfrak{G}} \varphi(\boldsymbol{\xi})\,g(t,\boldsymbol{\xi})\,dtd\boldsymbol{\xi} \\
&= \lim_{T\searrow 0} \frac{b}{T} \int_0^T \left(\int_{\mathfrak{G}} [P_t^{\mathfrak{G}}\varphi](\boldsymbol{\xi})\,\nu(d\boldsymbol{\xi}) \right) dt = b\int_{\mathfrak{G}} \varphi\,d\nu
\end{aligned}$$

for every $\varphi \in C_c(\mathfrak{G};\mathbb{R})$, and so $a\mu - b\nu = 0$. □

8.4.34 Corollary. *Continuing with the setting in Theorem 8.4.23, let $\{K_n\}$ be a nonincreasing sequence of compact subsets of $\mathfrak{G}$ and set $K = \bigcap_1^\infty K_n$. Then, for every measurable $\varphi : \mathfrak{G} \longrightarrow \mathbb{R}$ which is continuous in a neighborhood of K_1,*

$$\lim_{n\to\infty} \int_{\mathfrak{G}} \varphi \, d\mu_{K_n}^{\mathfrak{G}} = \int_{\mathfrak{G}} \varphi \, d\mu_K^{\mathfrak{G}}; \tag{8.4.35}$$

and so,

$$\operatorname{Cap}(K;\mathfrak{G}) = \lim_{n\to\infty} \operatorname{Cap}(K_n;\mathfrak{G}). \tag{8.4.36}$$

Finally, if μ is any nonnegative Radon measure on $\mathfrak{G}$ satisfying $\mu(\mathfrak{G} \setminus K) = 0$ and $G^{\mathfrak{G}}\mu \le 1$, then

$$\mathcal{E}^{\mathfrak{G}}(\mu,\mu) \le \operatorname{Cap}(K;\mathfrak{G}) \quad \text{and equality holds} \iff \mu = \mu_K^{\mathfrak{G}}. \tag{8.4.37}$$

PROOF: Let $\mathbf{c} \in \mathfrak{G} \setminus K_1$ be given. In view of (8.4.31), checking (8.4.35) comes down to showing that, for $\mathcal{W}_{\mathbf{c}}^{(N)}$-almost every $\psi \in \mathfrak{P}(\mathbb{R}^N)$:

$$\begin{gathered} \ell_{K_n}^{\mathfrak{G}}(\psi) \longrightarrow \ell_K^{\mathfrak{G}}(\psi) \in \big(0, \mathfrak{e}^{\mathfrak{G}}(\psi)\big) \quad \text{if either} \\ \{\ell_{K_n}^{\mathfrak{G}}(\psi)\}_1^\infty \subseteq \big(0, \mathfrak{e}^{\mathfrak{G}}(\psi)\big) \quad \text{or} \quad \ell_K^{\mathfrak{G}}(\psi) \in \big(0, \mathfrak{e}^{\mathfrak{G}}(\psi)\big). \end{gathered}$$

To this end, let $\psi \in \mathfrak{P}(\mathbb{R}^N)$ with $\psi(0) = \mathbf{c}$ be given. If $\{\ell_{K_n}^{\mathfrak{G}}(\psi)\} \subseteq \big(0, \mathfrak{e}^{\mathfrak{G}}(\psi)\big)$, then it is clear that

$$\ell_{K_n}^{\mathfrak{G}}(\psi) \searrow T \ge \ell_K^{\mathfrak{G}}(\psi) \quad \text{where } T \in \big(0, \mathfrak{e}^{\mathfrak{G}}(\psi)\big).$$

In addition, by continuity, $\psi(T) \in K$; which means first that $T \le \ell_K^{\mathfrak{G}}(\psi)$ and then that $\ell_{K_n}^{\mathfrak{G}}(\psi) \longrightarrow \ell_K^{\mathfrak{G}}(\psi) \in \big(0, \mathfrak{e}^{\mathfrak{G}}(\psi)\big)$. Next, observe that

$$0 < \ell_K^{\mathfrak{G}}(\psi) < \mathfrak{e}^{\mathfrak{G}}(\psi) < \infty \implies \ell_{K_n}^{\mathfrak{G}}(\psi) \in \big[\ell_K^{\mathfrak{G}}(\psi), \mathfrak{e}^{\mathfrak{G}}(\psi)\big) \quad \text{for all } n \in \mathbb{Z}^+.$$

Hence, we are done if (8.2.9) holds. On the other hand, if $N \ge 3$, then, because $\lim_{t\to\infty} |\psi(t)| = \infty$ for $\mathcal{W}_{\mathbf{c}}^{(N)}$-almost all $\psi \in \mathfrak{P}(\mathbb{R}^N)$, we know that, for $\mathcal{W}_{\mathbf{c}}^{(N)}$-almost every $\psi \in \mathfrak{P}(\mathbb{R}^N)$:

$$\mathfrak{e}^{\mathfrak{G}}(\psi) = \infty \text{ and } \ell_K^{\mathfrak{G}}(\psi) \in (0,\infty) \implies \{\ell_{K_n}^{\mathfrak{G}}(\psi)\}_1^\infty \subseteq (0,\infty);$$

and so we have now completed the proof of (8.4.35).

Obviously, (8.4.36) is just a special case of (8.4.35). Finally, to prove (8.4.37), first choose compact K_n's in $\mathfrak{G}$ so that $K \subset\subset K_n^\circ$ for each $n \in \mathbb{Z}^+$ and $K_n \searrow K$ as $n \to \infty$. Because (cf. (8.4.24)) $p^{\mathfrak{G}}_{K_n} \upharpoonright K \equiv 1$ and $p^{\mathfrak{G}}_{K_n} \leq 1$, we then have that

$$\begin{aligned}\operatorname{Cap}(K;\mathfrak{G}) &= \int_{\mathfrak{G}} p^{\mathfrak{G}}_{K_n}(\mathbf{x})\,\mu^{\mathfrak{G}}_K(d\mathbf{x}) = \mathcal{E}^{\mathfrak{G}}\left(\mu^{\mathfrak{G}}_K,\mu^{\mathfrak{G}}_{K_n}\right)\\ &\leq \mathcal{E}^{\mathfrak{G}}\left(\mu^{\mathfrak{G}}_K,\mu^{\mathfrak{G}}_K\right)^{\frac{1}{2}}\mathcal{E}^{\mathfrak{G}}\left(\mu^{\mathfrak{G}}_{K_n},\mu^{\mathfrak{G}}_{K_n}\right)^{\frac{1}{2}}\\ &= \mathcal{E}^{\mathfrak{G}}\left(\mu^{\mathfrak{G}}_K,\mu^{\mathfrak{G}}_K\right)^{\frac{1}{2}}\left(\int_{\mathfrak{G}} p^{\mathfrak{G}}_{K_n}(\mathbf{x})\,\mu^{\mathfrak{G}}_{K_n}(d\mathbf{x})\right)^{\frac{1}{2}}\\ &\leq \mathcal{E}^{\mathfrak{G}}\left(\mu^{\mathfrak{G}}_K,\mu^{\mathfrak{G}}_K\right)^{\frac{1}{2}}\operatorname{Cap}(K_n;\mathfrak{G})^{\frac{1}{2}} \longrightarrow \mathcal{E}^{\mathfrak{G}}\left(\mu^{\mathfrak{G}}_K,\mu^{\mathfrak{G}}_K\right)^{\frac{1}{2}}\operatorname{Cap}(K;\mathfrak{G})^{\frac{1}{2}}\end{aligned}$$

as $n \to \infty$. Hence, $\operatorname{Cap}(K;\mathfrak{G}) \leq \mathcal{E}^{\mathfrak{G}}\left(\mu^{\mathfrak{G}}_K,\mu^{\mathfrak{G}}_K\right)$. On the other hand, if $\mu(\mathfrak{G}\backslash K) = 0$ and $G^{\mathfrak{G}}\mu \leq 1$, then, since (cf. (8.4.22)) $G^{\mathfrak{G}}\mu \leq p^{\mathfrak{G}}_K \leq 1$:

$$\begin{aligned}\mathcal{E}^{\mathfrak{G}}(\mu,\mu) &= \int_{\mathfrak{G}} G^{\mathfrak{G}}\mu\,d\mu \leq \int_{\mathfrak{G}} p^{\mathfrak{G}}_K\,d\mu = \mathcal{E}^{\mathfrak{G}}\left(\mu^{\mathfrak{G}}_K,\mu\right)\\ &\leq \mathcal{E}^{\mathfrak{G}}\left(\mu^{\mathfrak{G}}_K,\mu^{\mathfrak{G}}_K\right)^{\frac{1}{2}}\mathcal{E}^{\mathfrak{G}}(\mu,\mu)^{\frac{1}{2}}\\ &= \left(\int_{\mathfrak{G}} p^{\mathfrak{G}}_K\,d\mu^{\mathfrak{G}}_K\right)^{\frac{1}{2}}\mathcal{E}^{\mathfrak{G}}(\mu,\mu)^{\frac{1}{2}} \leq \sqrt{\operatorname{Cap}(K;\mathfrak{G})}\sqrt{\mathcal{E}^{\mathfrak{G}}(\mu,\mu)},\end{aligned}$$

and equality can hold only if $a\mu^{\mathfrak{G}}_K - b\mu = 0$ for some nontrivial pair $(a,b) \in [0,\infty)^2$. When one takes $\mu = \mu^{\mathfrak{G}}_K$, this, in conjunction with the preceding, proves that $\operatorname{Cap}(K;\mathfrak{G}) = \mathcal{E}^{\mathfrak{G}}\left(\mu^{\mathfrak{G}}_K,\mu^{\mathfrak{G}}_K\right)$. In addition, for any μ with $\mu(\mathfrak{G}\setminus K) = 0$ and $G^{\mathfrak{G}}\mu \leq 1$, it shows that $\mathcal{E}^{\mathfrak{G}}(\mu,\mu) \leq \operatorname{Cap}(K;\mathfrak{G})$ and that equality can hold only if μ and $\mu^{\mathfrak{G}}_K$ are related by a nontrivial linear equation, in which case $\mu = \mu^{\mathfrak{G}}_K$ follows immediately from the equality $\mathcal{E}^{\mathfrak{G}}\left(\mu^{\mathfrak{G}}_K,\mu^{\mathfrak{G}}_K\right) = \mathcal{E}^{\mathfrak{G}}(\mu,\mu)$. □

As a second example of the way in which Wiener paths can be used to provide insight into purely potential theoretic questions, we will now give a probabilistically motivated proof of Wiener's famous test for regularity (cf. (8.1.18)). Thus, let $N \geq 2$, $\mathfrak{G}$ an open subset of $\mathbb{R}^N$, and an $\mathbf{a} \in \partial\mathfrak{G}$ be given, set

(8.4.38) $$K_n = \left\{\mathbf{y} \notin \mathfrak{G} : 2^{-n-1} \leq |\mathbf{y}-\mathbf{a}| \leq 2^{-n}\right\},$$

and define

(8.4.39) $$W_n(\mathbf{a},\mathfrak{G}) = \operatorname{Cap}\big(K_n; B_{\mathbb{R}^N}(\mathbf{a},1)\big)\tilde{\Gamma}_N\left(2^{-n}\right) \quad \text{for } n \in \mathbb{Z}^+,$$

where (cf. (8.3.10) and (8.3.16))

$$\tilde{\Gamma}_N(t) \equiv \Gamma_N(t\mathbf{e}) \quad \text{with } \mathbf{e} = (1,0,\dots,0) \in \mathbf{S}^{N-1}.$$

Then **Wiener's test** says that

$$\mathbf{a} \in \partial_{\text{reg}}\mathfrak{G} \iff \sum_{n=1}^{\infty} W_n(\mathbf{a}, \mathfrak{G}) = \infty. \tag{8.4.40}$$

Notice that, at least qualitatively, (8.4.40) is what one would expect (cf. Remark 8.1.28) in the sense that the divergence of the series is some sort of statement to the effect that $\mathfrak{G}\complement$ *is robust at* **a**.

The key to our proof of Wiener's test is the trivial observation that because

$$\begin{aligned} p_n(\mathbf{x}) &\equiv \mathcal{W}_{\mathbf{x}}^{(N)}\left(\exists t \in \left(0, \mathfrak{e}^{B_{\mathbb{R}^N}(\mathbf{a},1)}\right) \psi(t) \in K_n\right) \\ &= \int_{K_n} g^{B_{\mathbb{R}^N}(\mathbf{a},1)}(\mathbf{x}, \mathbf{y})\, \mu_{K_n}^{B_{\mathbb{R}^N}(\mathbf{a},1)}(d\mathbf{y}), \end{aligned} \tag{8.4.41}$$

and (cf. Exercise 8.3.31) there exists an $\alpha_N \in (0,1)$ such that

$$\alpha_N \Gamma_N(\mathbf{x} - \mathbf{y}) \le g^{B_{\mathbb{R}^N}(\mathbf{a},1)}(\mathbf{x}, \mathbf{y}) \le \Gamma_N(\mathbf{x} - \mathbf{y}) \quad \text{for } \mathbf{x}, \mathbf{y} \in B_{\mathbb{R}^N}\left(\mathbf{a}, \tfrac{1}{2}\right), \tag{8.4.42}$$

we know that

$$\alpha_N W_n(\mathbf{a}, \mathfrak{G}) \le p_n(\mathbf{a}) \le W_n(\mathbf{a}, \mathfrak{G}), \quad n \in \mathbb{Z}^+.$$

Hence, in probabilistic terms, Wiener's test comes down to the assertion that (cf. Theorem 8.1.27)

$$\mathcal{W}_{\mathbf{a}}^{(N)}\left(\mathfrak{e}_{0+}^{\mathfrak{G}} = 0\right) = 1 \iff \sum_{1}^{\infty} \mathcal{W}_{\mathbf{a}}^{(N)}(A_n) = \infty, \tag{8.4.43}$$

where A_n is the set of $\psi \in \mathfrak{P}(\mathbb{R}^N)$ which visit K_n before leaving $B_{\mathbb{R}^N}(\mathbf{a}, 1)$. Actually, although (8.4.43) may not be immediately obvious, the closely related statement

$$\mathcal{W}_{\mathbf{a}}^{(N)}\left(\mathfrak{e}_{0+}^{\mathfrak{G}} = 0\right) = 1 \iff \mathcal{W}_{\mathbf{a}}^{(N)}\left(\varlimsup_{n\to\infty} A_n\right) > 0 \tag{8.4.44}$$

is essentially immediate. Indeed, if $\psi(0) = \mathbf{a}$ and $\mathfrak{e}_{0+}^{\mathfrak{G}}(\psi) = 0$, then there exists a sequence of times $t_m \searrow 0$ with the property that $\psi(t_m) \in B_{\mathbb{R}^N}(\mathbf{a}, 1) \cap \mathfrak{G}\complement$ for all m; from which it is clear that ψ visits infinitely many K_n's before leaving $B_{\mathbb{R}^N}(\mathbf{a}, 1)$. Hence, the "$\implies$" in (8.4.44) is trivial. As for the opposite implication, suppose that $\psi \in \mathfrak{P}(\mathbb{R}^N)$ has the properties that $\psi(0) = \mathbf{a}$, ψ leaves $B_{\mathbb{R}^N}(\mathbf{a}, 1)$ in a finite time, $\psi(t) \ne \mathbf{a}$ for any $t > 0$, but ψ visits infinitely many K_n's before leaving $B_{\mathbb{R}^N}(\mathbf{a}, 1)$. We can then find a subsequence $\{n_m\}_1^\infty$ and a convergent sequence of times $t_m > 0$ such that $\psi(t_m) \in K_{n_m}$ for each m.

But clearly, if $T = \lim_{m\to\infty} t_m$, then $\psi(T) = \mathbf{a}$, and therefore we would know that $\lim_{m\to\infty} t_m = 0$ which, in turn, would mean that $\mathfrak{e}_{0+}^{\mathfrak{G}}(\psi) = 0$. Hence, since (remember that $N \neq 1$ and recall Theorem 7.2.11)

$$\mathcal{W}_{\mathbf{a}}^{(N)}\Big(\Big\{\psi : \mathfrak{e}^{B_{\mathbb{R}^N}(\mathbf{a},1)}(\psi) < \infty \text{ and } \{t : \psi(t) = \mathbf{a}\} = \{0\}\Big\}\Big) = 1,$$

we now see that

$$\mathcal{W}_{\mathbf{a}}^{(N)}\Big(\mathfrak{e}_{0+}^{\mathfrak{G}} = 0\Big) \geq \mathcal{W}_{\mathbf{a}}^{(N)}\Big(\varlimsup_{n\to\infty} A_n\Big);$$

and therefore, because $\mathcal{W}_{\mathbf{a}}^{(N)}\big(\mathfrak{e}_{0+}^{\mathfrak{G}} = 0\big) \in \{0,1\}$, we have now proved the equivalence in (8.4.44).

In view of the preceding paragraph, the proof of Wiener's test reduces to the problem of showing that

$$\text{(8.4.45)} \qquad \mathcal{W}_{\mathbf{a}}^{(N)}\Big(\varlimsup_{n\to\infty} A_n\Big) > 0 \iff \sum_{1}^{\infty} \mathcal{W}_{\mathbf{a}}^{(N)}(A_n) = \infty.$$

By the trivial part of the Borel–Cantelli Lemma, the " $\Longrightarrow$ " implication in (8.4.45) is easy. On the other hand, because the events $\{A_n\}_1^\infty$ are not mutually independent, the nontrivial part of that lemma does not apply and therefore cannot be used to go in the opposite direction. Nonetheless, as we will see, the following interesting variation on the theme of the Borel–Cantelli Lemma does apply and gives us the "$\Longleftarrow$" implication in (8.4.45).

8.4.46 Lemma. *Let $(\Omega, \mathcal{F}, P)$ be a probability space and $\{A_n\}_1^\infty$ a sequence of $\mathcal{F}$-measurable sets with the property that*

$$P(A_m \cap A_n) \leq C P(A_m) P(A_n), \quad m \in \mathbb{Z}^+ \text{ and } n \geq m + d,$$

for some $C \in [1,\infty)$ and $d \in \mathbb{Z}^+$. Then

$$\sum_{1}^{\infty} P(A_n) = \infty \implies P\Big(\varlimsup_{n\to\infty} A_n\Big) \geq \frac{1}{4C}.$$

PROOF: Because

$$\sum_{n=1}^{\infty} P(A_n) = \infty \implies \sum_{n=1}^{\infty} P(A_{nd+k}) = \infty \quad \text{for some } 0 \leq k < d,$$

whereas

$$P\Big(\varlimsup_{n\to\infty} A_n\Big) \geq P\Big(\varlimsup_{n\to\infty} A_{nd+k}\Big) \quad \text{for each } 0 \leq k < d,$$

we may and will assume that $d = 1$. Further, since

$$P\left(\overline{\lim_{n\to\infty}} A_n\right) \geq \overline{\lim_{n\to\infty}} P(A_n),$$

we will assume that $P(A_n) \leq \frac{1}{4C}$ for all $n \in \mathbb{Z}^+$. In particular, these assumptions mean that, for each $m \in \mathbb{Z}^+$, we can find an $n_m > m$ such that

$$s_m \equiv \sum_{\ell=m}^{n_m} P(A_\ell) \in \left[\tfrac{3}{4C}, \tfrac{1}{C}\right].$$

Indeed, simply take n_m to be the largest $n > m$ for which

$$\sum_{\ell=m}^{n} P(A_\ell) \leq \frac{1}{C}.$$

At the same time, by an easy induction on $n > m$, one has that, for all $n > m \geq 1$:

$$P\left(\bigcup_{\ell=m}^{n} A_\ell\right) \geq \sum_{\ell=m}^{n} P(A_\ell) - \frac{1}{2} \sum_{m\leq k\neq\ell\leq n} P(A_k)P(A_\ell);$$

and therefore

$$P\left(\bigcup_{\ell=m}^{\infty} A_\ell\right) \geq P\left(\bigcup_{\ell=m}^{n_m} A_\ell\right) \geq s_m - \frac{Cs_m^2}{2} \geq \frac{1}{4C}$$

for all $m \in \mathbb{Z}^+$. □

Proof of Wiener's Criterion: All that remains is to check that the sets A_n appearing in (8.4.45) satisfy the hypothesis in Lemma 8.4.46 when $P = \mathcal{W}_{\mathbf{a}}^{(N)}$. To this end, set

$$\tau_n(\psi) = \inf\left\{t \in (0,\infty) : \psi(t) \in K_n\right\}.$$

Clearly, $A_n = \left\{\tau_n < \mathfrak{e}\right\}$, where $\mathfrak{e} \equiv \mathfrak{e}^{B_{\mathbb{R}^N}(\mathbf{a},1)}$; and so

$$\mathcal{W}_{\mathbf{a}}^{(N)}(A_m \cap A_n) \leq \mathcal{W}_{\mathbf{a}}^{(N)}\left(\tau_m < \tau_n < \mathfrak{e}\right) + \mathcal{W}_{\mathbf{a}}^{(N)}\left(\tau_n < \tau_m < \mathfrak{e}\right)$$

for all $m, n \in \mathbb{Z}^+$ satisfying $|m - n| \geq 2$. But, by the strong Markov property, (cf. (8.4.41))

$$\mathcal{W}_{\mathbf{a}}^{(N)}\left(\tau_m < \tau_n < \mathfrak{e}\right) \leq \mathbb{E}^{\mathcal{W}_{\mathbf{a}}^{(N)}}\left[p_n(\psi(\tau_m)),\ \tau_m < \mathfrak{e}\right] \leq \beta(m,n)p_m(\mathbf{a}),$$

where we have introduced the notation $\beta(m,n) \equiv \max_{\mathbf{x}\in K_m} p_n(\mathbf{x})$. Finally, by (8.4.41) and (8.4.42), it is easy to check that there exists a $C_N \in (0,\infty)$ such that

$$\beta(m,n) \leq C_N p_n(\mathbf{a}) \quad \text{for all } |m-n| \geq 2.$$

Hence, since $p_n(\mathbf{a}) = \mathcal{W}_{\mathbf{a}}^{(N)}(A_n)$, we have now shown that

$$\mathcal{W}_{\mathbf{a}}^{(N)}(A_m \cap A_n) \leq 2C_N \mathcal{W}_{\mathbf{a}}^{(N)}(A_m)\mathcal{W}_{\mathbf{a}}^{(N)}(A_n) \quad \text{for all } |m-n| \geq 2,$$

which means that Lemma 8.4.46 applies with $C = 2C_N$ and $d = 2$. □

We have just seen that Wiener paths can be used to gain insight into the purely potential theoretic notion of capacity. Thus, it is only fitting that we turn the tables and see whether capacity cannot be used to make some interesting computations which are of probabilistic significance. As our first example of such a computation, we give the following calculation, which was made originally by A. Joffe.†

8.4.47 Theorem. *Assume that* $N \geq 3$, *let* $K \subset\subset \mathbb{R}^N$, *and define the first positive entrance time* $\tau_K : \mathfrak{P}(\mathbb{R}^N) \longmapsto [0, \infty]$ *by*

$$\tau_K(\boldsymbol{\psi}) = \inf\{t > 0 : \boldsymbol{\psi}(t) \in K\}.$$

Then, as $t \nearrow \infty$:

$$t^{\frac{N}{2}-1}\mathcal{W}_{\mathbf{x}}^{(N)}\Big(\tau_K \in (t,\infty)\Big) \longrightarrow \frac{2\mathrm{Cap}(K;\mathbb{R}^N)\big(1 - p_K^{\mathbb{R}^N}(\mathbf{x})\big)}{(2\pi)^{\frac{N}{2}}(N-2)}$$

uniformly as $\mathbf{x}$ *runs over compacts.*

PROOF: Without loss in generality (cf. Corollary 8.4.27), we will assume that $\mathrm{Cap}(K;\mathbb{R}^N) > 0$. Next, set

$$p_K(t,\mathbf{x}) = \mathcal{W}_{\mathbf{x}}^{(N)}\big(\tau_K \in (t,\infty)\big), \quad p_K(\mathbf{x}) = p_K^{\mathbb{R}^N}(\mathbf{x}),$$
$$\text{and } p_K(t,\mathbf{x},\mathbf{y}) = p^{K\complement}(t,\mathbf{x},\mathbf{y});$$

and note that, by the Markov property:

$$p_K(t,\mathbf{x}) = \int_{K\complement} p_K(\mathbf{y})\, p_K(t,\mathbf{x},\mathbf{y})\, d\mathbf{y}.$$

Thus, since $p_K(t,\mathbf{x},\mathbf{y}) \leq (2\pi t)^{-\frac{N}{2}}$, we know that

$$\varlimsup_{t\to\infty} \sup_{\mathbf{x}\in\mathbb{R}^N} t^{\frac{N}{2}-1}\left|p_K(t,\mathbf{x}) - \int_{|\mathbf{y}|\geq R} p_K(\mathbf{y})\, p_K(t,\mathbf{x},\mathbf{y})\, d\mathbf{y}\right| = 0$$

for every $R > 0$ with $K \subset\subset B_{\mathbb{R}^N}(\mathbf{0},R)$. At the same time, from (cf. (8.3.10))

$$p_K(\mathbf{y}) = \int_K \Gamma_N(\mathbf{y}-\boldsymbol{\xi})\, \mu_K^{\mathbb{R}^N}(d\boldsymbol{\xi}),$$

it is clear that

$$\lim_{|\mathbf{y}|\to\infty} |\mathbf{y}|^{N-2} p_K(\mathbf{y}) = \frac{2\mathrm{Cap}(K;\mathbb{R}^N)}{(N-2)\omega_{N-1}}.$$

† Although Joffe was the first to verify this conjecture made by Kac, we will follow F. Spitzer (cf. the article cited in the footnote to Exercise 8.4.56).

Hence, we have now shown that

$$\varlimsup_{t\to\infty}\sup_{\mathbf{x}\in\mathbb{R}^N} t^{\frac{N}{2}-1}\left|p_K(t,\mathbf{x})-\frac{2\mathrm{Cap}(K;\mathbb{R}^N)}{(N-2)\omega_{N-1}}\int_{|\mathbf{y}|\ge R}\frac{p_K(t,\mathbf{x},\mathbf{y})}{|\mathbf{y}|^{N-2}}\,d\mathbf{y}\right|=0$$

for each $R\in(0,\infty)$ with $K\subset\subset B_{\mathbb{R}^N}(\mathbf{0},R)$.

Now choose and fix $R>0$ so that $K\subset\subset B_{\mathbb{R}^N}(\mathbf{0},R)$, and set

$$q(t,\boldsymbol{\xi})\equiv\int_{|\mathbf{y}|\ge R}\frac{\gamma_t^N(\boldsymbol{\xi}-\mathbf{y})}{|\mathbf{y}|^{N-2}}\,d\mathbf{y}\quad\text{for } (t,\boldsymbol{\xi})\in(0,\infty)\times\mathbb{R}^N.$$

After using polar coordinates and a change of variables, one can easily check that, for each $r\in(0,\infty)$ and $T\in[0,\infty)$:

$$\lim_{t\to\infty}\sup_{\substack{0<s\le T\\|\boldsymbol{\xi}|\le r}}\left|t^{\frac{N}{2}-1}q(t-s,\boldsymbol{\xi})-\frac{\omega_{N-1}}{(2\pi)^{\frac{N}{2}}}\right|=0.$$

Moreover, by (8.1.10):

$$\int_{|\mathbf{y}|\ge R}\frac{p_K(t,\mathbf{x},\mathbf{y})}{|\mathbf{y}|^{N-2}}\,d\mathbf{y}=q(t,\mathbf{x})-\mathbb{E}^{\mathcal{W}_{\mathbf{x}}^{(N)}}\Big[q\big(t-\tau_K,\boldsymbol{\psi}(\tau_K)\big),\ \tau_K<t\Big].$$

Thus, for any $r\in(0,\infty)$ and $T\in[0,\infty)$:

$$\begin{aligned}&\varlimsup_{t\to\infty}\sup_{|\mathbf{x}|\le r}\left|t^{\frac{N}{2}-1}\int_{|\mathbf{y}|\ge R}\frac{p_K(t,\mathbf{x},\mathbf{y})}{|\mathbf{y}|^{N-1}}\,d\mathbf{y}-\frac{\omega_{N-1}}{(2\pi)^{\frac{N}{2}}}\mathcal{W}_{\mathbf{x}}^{(N)}\Big(\tau_K\in[T,\infty]\Big)\right|\\&\qquad\le\varlimsup_{t\to\infty}\sup_{|\mathbf{x}|\le r}\mathbb{E}^{\mathcal{W}_{\mathbf{x}}^{(N)}}\Big[t^{\frac{N}{2}-1}q\big(t-\tau_K,\boldsymbol{\psi}(\tau_K)\big),\ \tau_K\in[T,t)\Big];\end{aligned}$$

and so we will be done as soon as we show that

$$\begin{aligned}&\lim_{T\to\infty}\sup_{\mathbf{x}\in\mathbb{R}^N}\mathcal{W}_{\mathbf{x}}^{(N)}\Big(\tau_K\in(T,\infty)\Big)=0\quad\text{and}\\&\lim_{T\to\infty}\sup_{\substack{t>T\\\mathbf{x}\in\mathbb{R}^N}}t^{\frac{N}{2}-1}\mathbb{E}^{\mathcal{W}_{\mathbf{x}}^{(N)}}\Big[q\big(t-\tau_K,\boldsymbol{\psi}(\tau_K)\big),\ \tau_K\in[T,t]\Big]=0.\end{aligned}\tag{8.4.48}$$

To this end, first note that, by the Markov property, (8.1.10), and the estimate in (3.3.30),

$$\begin{aligned}\mathcal{W}_{\mathbf{x}}^{(N)}\Big(\tau_K\in(T,T+2]\Big)&=\int_{K\complement}p_K(T,\mathbf{x},\mathbf{y})\mathcal{W}_{\mathbf{y}}^{(N)}\Big(\tau_K\le 2\Big)\,d\mathbf{y}\\&\le(2\pi T)^{-\frac{N}{2}}\int_{\mathbb{R}^N}\mathcal{W}_{\mathbf{y}}^{(N)}\Big(\tau_K\le 2\Big)\,d\mathbf{y}\le CT^{-\frac{N}{2}},\end{aligned}$$

where $C = C(N, R) \in (0, \infty)$. Hence, after writing

$$\mathcal{W}_{\mathbf{x}}^{(N)}\Big(\tau_K \in (T, \infty)\Big) = \sum_{n=0}^{\infty} \mathcal{W}_{\mathbf{x}}^{(N)}\Big(\tau_K \in (T+n, T+n+1]\Big),$$

we see that, as $T \to \infty$, $\mathcal{W}_{\mathbf{x}}^{(N)}\big(\tau_K \in (T, \infty)\big) \longrightarrow 0$ uniformly with respect to $\mathbf{x} \in \mathbb{R}^N$. To handle the second expression in (8.4.48), note that there is a constant $A \in (0, \infty)$ for which

$$q(t, \mathbf{y}) \le A\,(t \vee 1)^{1-\frac{N}{2}}, \quad (t, \mathbf{y}) \in (0, \infty) \times K,$$

and therefore

$$\begin{aligned} t^{\frac{N}{2}-1}\mathbb{E}^{\mathcal{W}_{\mathbf{x}}^{(N)}}&\Big[q\big(t-\tau_K, \boldsymbol{\psi}(\tau_K)\big),\ \tau_K \in (T, t)\Big] \\ &\le A t^{\frac{N}{2}-1}\Bigg(\mathcal{W}_{\mathbf{x}}^{(N)}\Big(\tau_K \in \big([t]-1, t\big)\Big) \\ &\qquad + \sum_{\ell=[T]}^{[t]-1}(t-\ell)^{1-\frac{N}{2}}\mathcal{W}_{\mathbf{x}}^{(N)}\big(\tau_K \in (\ell-1, \ell)\big)\Bigg) \\ &\le ACt^{\frac{N}{2}-1}([t]-1)^{-\frac{N}{2}} + ACt^{\frac{N}{2}-1}\sum_{\ell=[T]}^{[t]-1}(t-\ell)^{1-\frac{N}{2}}\ell^{-\frac{N}{2}}. \end{aligned}$$

Thus, the second part of (8.4.48) will follow once we show that

$$\lim_{m\to\infty}\ \sup_{n>m} n^{\frac{N}{2}-1}\sum_{\ell=m}^{n-1}(n-\ell)^{1-\frac{N}{2}}\ell^{\frac{N}{2}} = 0.$$

But by taking $\epsilon_m = m^{\frac{2}{N}-1}$ and considering

$$\sum_{m\le\ell\le(1-\epsilon_m)n}(n-\ell)^{1-\frac{N}{2}}\ell^{-\frac{N}{2}} \quad \text{and} \quad \sum_{(1-\epsilon_m)n\le\ell\le n}(n-\ell)^{1-\frac{N}{2}}\ell^{-\frac{N}{2}}$$

separately, one finds that there is a $B \in (0, \infty)$ such that

$$n^{\frac{N}{2}-1}\sum_{\ell=m}^{n-1}(n-\ell)^{1-\frac{N}{2}}\ell^{-\frac{N}{2}} \le B\epsilon_m. \quad \square$$

As one might guess, on the basis of (8.3.19), the analogous situation in $\mathbb{R}^2$ is considerably more delicate in that it involves logarithms.

8.4.49 Theorem (Hunt).[†] *Let K be a compact subset of $\mathbb{R}^2$, define τ_K as in Theorem 8.4.47, assume that $\mathcal{W}_{\mathbf{x}}^{(2)}(\tau_K < \infty) = 1$ for all $\mathbf{x} \in \mathbb{R}^2$, and use h_K to denote the function $h^{\mathfrak{G}}$ given in (8.3.19) when $\mathfrak{G} = \mathbb{R}^2 \setminus K$. Then*

$$\lim_{t\to\infty} \frac{\log t}{2\pi} \mathcal{W}_{\mathbf{x}}^{(2)}\left(\tau_K > t\right) = h_K(\mathbf{x}) \quad \text{for each } \mathbf{x} \in \mathbb{R}^2 \setminus K. \tag{8.4.50}$$

PROOF: Set $\mathfrak{G} = \mathbb{R}^2 \setminus K$. By assumption, $\mathfrak{G}$ satisfies the hypotheses of Theorem 8.3.17. Now let $\mathbf{x} \in \mathfrak{G}$ be given, and choose $\mathbf{y} \in \mathfrak{G}\setminus\{\mathbf{x}\}$ from the same connected component of $\mathfrak{G}$ as $\mathbf{x}$. Then $p^{\mathfrak{G}}(t,\mathbf{x},\mathbf{y}) > 0$ for all $t \in (0,\infty)$. In addition, by (8.1.10), for each $\alpha \in (0,\infty)$,

$$\begin{aligned}&\int_0^\infty e^{-\alpha t} p^{\mathfrak{G}}(t,\mathbf{x},\mathbf{y})\,dt \\ &= \int_0^\infty e^{-\alpha t}\gamma_t^2(\mathbf{x}-\mathbf{y})\,dt - \mathbb{E}^{\mathcal{W}_{\mathbf{x}}^{(N)}}\left[e^{-\alpha\tau_K}\int_0^\infty e^{-\alpha t}\gamma_t^2\left(\psi(\tau_K)-\mathbf{y}\right)dt\right].\end{aligned}$$

Next observe that

$$\int_0^\infty e^{-\alpha t}\gamma_t^2(\mathbf{z})\,dt = f\left(\frac{\alpha|\mathbf{z}|^2}{2}\right)$$
$$\text{where } f(\beta) \equiv \frac{1}{2\pi}\int_0^\infty t^{-1}\exp\left[-\beta t - t^{-1}\right]dt \text{ for } \beta > 0.$$

Writing

$$\begin{aligned}2\pi f(\beta) = \int_0^1 t^{-1}\exp\left[-\beta t - t^{-1}\right]dt + \int_1^\infty t^{-1}e^{-\beta t}\left(\exp\left[-t^{-1}\right]-1\right)dt \\ + \int_\beta^\infty t^{-1}e^{-t}\,dt\end{aligned}$$

integrating by parts, and performing elementary manipulations, we find that

$$f(\beta) = \frac{\log\frac{1}{\beta}}{2\pi} + \kappa + o(1) \quad \text{as } \beta \searrow 0,$$

where

$$\kappa = \frac{1}{\pi}\int_0^\infty e^{-t}\log t\,dt.$$

[†] This theorem is taken from G. Hunt's article "Some theorems concerning Brownian motion," *T.A.M.S.* **81**. With breathtaking rapidity, it was followed by the articles referred to after (8.2.18).

At the same time, we have that

$$\int_0^\infty e^{-\alpha t} p^{\mathfrak{G}}(t, \mathbf{x}, \mathbf{y})\, dt \longrightarrow g^{\mathfrak{G}}(\mathbf{x}, \mathbf{y}) \quad \text{as } \alpha \searrow 0.$$

Hence, when we plug these into the preceding, we find (cf. (8.3.16)) that

$$\begin{aligned} g^{\mathfrak{G}}(\mathbf{x}, \mathbf{y}) = \Gamma_2(\mathbf{x} - \mathbf{y}) &- \mathbb{E}^{\mathcal{W}^{(2)}_{\mathbf{x}}}\Big[\Gamma_2\big(\psi(\tau_K) - \mathbf{y}\big),\ \tau_K < \infty\Big] \\ &+ \frac{\log \frac{1}{\alpha}}{2\pi}\Big(1 - \mathbb{E}^{\mathcal{W}^{(2)}_{\mathbf{x}}}\big[e^{-\alpha \tau_K}\big]\Big) + o(1) \end{aligned}$$

as $\alpha \searrow 0$. But, by (8.3.20), this means that

$$\lim_{\alpha \searrow 0} \frac{\log \frac{1}{\alpha}}{2\pi}\Big(1 - \mathbb{E}^{\mathcal{W}^{(N)}_{\mathbf{x}}}\big[e^{-\alpha \tau_K}\big]\Big) = h_K(\mathbf{x}),$$

or, equivalently, that

$$\lim_{\alpha \searrow 0}\left(\frac{\alpha \log \frac{1}{\alpha}}{2\pi}\right) \int_0^\infty e^{-\alpha t} \mathcal{W}^{(N)}_{\mathbf{x}}\big(\tau_K > t\big)\, dt = h_K(\mathbf{x}).$$

Thus, by Karamata's Tauberian Theorem† we have first that

$$\lim_{T \to \infty} \frac{\log T}{2\pi T} \int_0^T \mathcal{W}^{(N)}_{\mathbf{x}}\big(\tau_k > t\big)\, dt = h_K(\mathbf{x})$$

and then, because $t \in (0, \infty) \longmapsto \mathcal{W}^{(N)}_{\mathbf{x}}\big(\tau_K > t\big) \in [0, 1]$ is monotone, (8.4.44). □

8.4.51 Remark. Let $K \subset\subset \mathbb{R}^N$ be as in Theorem 8.4.49 and $\mathbf{c} \in K\complement$ be given. By comparing (8.4.50) with (8.3.19), we see that

$$\text{(8.4.52)} \qquad \lim_{t \to \infty} \frac{\mathcal{W}^{(2)}_{\mathbf{x}}\big(\tau_K > t\big)}{\mathcal{W}^{(2)}_{\mathbf{x}}\big(\tau_K > \mathfrak{e}^{B_{\mathbb{R}^2}(\mathbf{c}, t)}\big)} = 2 \quad \text{for each } \mathbf{x} \in K\complement.$$

It would be interesting to know if there is a more direct route to this conclusion; in particular, one which avoids a Tauberian argument. Indeed, given (8.3.19), (8.4.50) and (8.4.52) become equivalent; and so a direct proof of (8.4.52) might point the way to the next term in the asymptotic expansion of $\mathcal{W}^{(2)}\big(\tau_K > t\big)$.

Exercises

8.4.53 Exercise: Let $\mathfrak{G}$ be a connected open set in $\mathbb{R}^N$, and suppose that $N \in \{1, 2\}$. If (8.2.9) fails, show that every excessive function on $\mathfrak{G}$ is constant. Hence, the only cases not already covered by Riesz's Decomposition Theorem are trivial anyhow.

† See the reference given in the footnote to (8.1.33).

Hint: Using the reasoning employed to prove the first part of Lemma 8.4.5, reduce to the case when u is smooth and satisfies $\mathbf{\Delta} u \le 0$, and in this case apply the result in Exercise 7.2.32.

8.4.54 Exercise: Referring to Corollary 8.4.34, show that for every μ with $\mu(\mathfrak{G} \setminus K) = 0$ and $G^{\mathfrak{G}}\mu \le 1$,

$$\mu(K) \le \mathrm{Cap}(K;\mathfrak{G}) \quad \text{and equality holds} \iff \mu = \mu_K^{\mathfrak{G}}; \tag{8.4.55}$$

and use this to conclude both that

$$\mathrm{Cap}(K;\mathfrak{G}) \le \mathrm{Cap}(K;\mathfrak{G}') \quad \text{whenever } K \subset\subset \mathfrak{G}' \subseteq \mathfrak{G}$$

and that

$$p_K^{\mathfrak{G}}(\mathbf{x}) = 1 \quad \text{for } \mu_K^{\mathfrak{G}} \text{ almost every } \mathbf{x} \in K.$$

8.4.56 Exercise: Given a $K \subset\subset \mathbb{R}^N$, one says that K has positive capacity if $N \ge 3$ and $\mathrm{Cap}(K;\mathbb{R}^N) > 0$ or $N \in \{1,2\}$ and $\mathrm{Cap}(K;\mathfrak{G}) > 0$ for some open $\mathfrak{G} \supset\supset K$ satisfying (8.2.9). More generally, given any subset A of $\mathbb{R}^N$, one says that A **has positive capacity** and writes $\mathrm{Cap}(A) > 0$ if A contains a compact set of positive capacity.

(i) Show that $\mathrm{Cap}(A) > 0$ if and only if there is a nonzero, compactly supported Radon measure μ such that $\mathrm{supp}(\mu) \subseteq A$ and (cf. (8.3.10), (8.3.16), and (8.3.13))

$$\mathbf{x} \in \mathbb{R}^N \longmapsto \int_{\mathbb{R}^N} \Gamma_N(\mathbf{x}-\mathbf{y})\,\mu(d\mathbf{y}) \in \mathbb{R}$$

is bounded above. When $N = 1$, conclude that $\mathrm{Cap}(A) > 0$ if and only if $A \ne \emptyset$ and, when $N \ge 2$, $\mathrm{Cap}(A) > 0$ whenever A contains a set of positive Lebesgue measure. (Actually, for those who know what it means, it will be clear that $\mathrm{Cap}(A) > 0$ as soon as A has positive inner $(N-1)$-dimensional Hausdorff measure.)

(ii) Given $K \subset\subset \mathbb{R}^N$, show that

$$\mathrm{Cap}(K) > 0 \iff \partial_{\mathrm{reg}}(\mathbb{R}^N \setminus K) \ne \emptyset.$$

Also, given a connected $\mathfrak{G} \supset\supset K$ with $\partial_{\mathrm{reg}}\mathfrak{G} \ne \emptyset$, show that $\mathrm{Cap}(K) = 0$ if and K is a removable singularity in $\mathfrak{G}$ for every bounded, harmonic u on $\mathfrak{G} \setminus K$. (Cf. Exercise 8.2.34.)

(iii) Given a nonempty open set $\mathfrak{G}$ in $\mathbb{R}^N$, show that $\partial\mathfrak{G} \setminus \partial_{\text{reg}}\mathfrak{G}$ has capacity 0 (i.e., it does not have positive capacity). In view of part **(ii)** above, this represents a considerable sharpening of the comment at the end of Remark 8.1.28.

8.4.57 Exercise: Let $N \geq 3$, $K \subset\subset \mathbb{R}^N$, and τ_K (cf. Theorem 8.4.47) the first positive entrance time into K. Following Spitzer,† we consider the quantity

$$E_K(t) \equiv \int_{K\complement} \mathcal{W}_{\mathbf{x}}^{(N)}(\tau_K \leq t)\, d\mathbf{x},$$

which can be thought of as the *amount of heat which flows from* $K\complement$ *into* K *during* $[0,t]$.

(i) Check that $t \in [0,\infty) \longmapsto E_K(t)$ is a continuous, nonnegative, nondecreasing function. Further, given $0 < h < t$, show that

$$E_K(t) - E_K(t-h) = \int_{\mathbb{R}^N} \mathcal{W}_{\mathbf{x}}^{(N)}(t-h < \tau_K \leq t)\, d\mathbf{x}.$$

Hint: The first part is a more or less immediate consequence of Theorem 8.1.3 and the estimate in (3.3.30). As for the second part, the only real problem is to show that the integration over $K\complement$ can be replaced by integration over the whole of $\mathbb{R}^N$, and this comes down to showing that $\mathcal{W}_{\mathbf{x}}^{(N)}\big(\tau_K > 0\big) = 0$ for Lebesgue almost every $\mathbf{x} \in K$. Obviously, there is no question about $\mathbf{x} \in K^\circ$. Thus, the problem is only whether almost every $\mathbf{x} \in \partial K$ is a regular point for $\mathbb{R}^N \setminus K$, and this is handled by the comment at the end of Remark 8.1.28 or by parts **(ii)** and **(iii)** of the preceding exercise.

(ii) Building on part **(i)**, show that, for $t > h$,

$$E_K(t) - E_K(t-h) = \int_{\mathbb{R}^N} \mathcal{W}_{\mathbf{y}}^{(N)}\Big(\tau_K \leq h \text{ and } \tau_K^h > t\Big)\, d\mathbf{y}, \tag{8.4.58}$$

where

$$\tau_K^h(\psi) \equiv \inf\Big\{s \in (h,\infty) : \psi(s) \in K\Big\}$$

is the first entrance time into K after h.

Hint: Perhaps the most intuitively appealing way to see (8.4.58) is to set (cf. (4.2.19))

$$\psi_t^{(\mathbf{x},\mathbf{y})}(s) = \frac{t-s}{t}\mathbf{x} + \tilde{\psi}_t(s) + \frac{s}{t}\mathbf{y}, \quad s \in [0,t],$$

† This exercise is adapted from F. Spitzer's beautiful "Electrostatic capacity, heat flow, and Brownian motion," *Z. Wahrsh. Gebiete.* **3**.

use the reversibility proved in Theorem 4.2.18 to write

$$\begin{aligned}
\mathcal{W}_{\mathbf{x}}^{(N)}&(t-h<\tau_K\le t)\\
&=\int_{\mathbb{R}^N}\mathcal{W}^{(N)}\Big(t-h<\tau_K\big(\psi_t^{(\mathbf{x},\mathbf{y})}\big)\le t\Big)\,\gamma_t^N(\mathbf{y}-\mathbf{x})\,d\mathbf{y}\\
&=\int_{\mathbb{R}^N}\mathcal{W}^{(N)}\Big(\tau_K\big(\psi_t^{(\mathbf{y},\mathbf{x})}\big)\le h \text{ and } \tau_K^h\big(\psi_t^{(\mathbf{y},\mathbf{x})}\big)>t\Big)\,\gamma_t^N(\mathbf{y}-\mathbf{x})\,d\mathbf{y},
\end{aligned}$$

and then integrate with respect to $\mathbf{x}$ to arrive at (8.4.58) after an application of Tonelli's Theorem.

(iii) Starting from (8.4.58), show that, for each $h\in[0,\infty)$

$$\begin{aligned}
\Delta_K(h)&\equiv\lim_{t\to\infty}\big(E_K(t+h)-E_K(t)\big)\\
&=\int_{\mathbb{R}^N}\mathcal{W}_{\mathbf{y}}^{(N)}\Big(\tau_K\le h \text{ and } \tau_K^h=\infty\Big)\,d\mathbf{y}
\end{aligned}\tag{8.4.59}$$

and that the convergence is uniform for h in compacts. In particular, conclude that Δ_K is a nonnegative, continuous function which is additive (i.e., $\Delta_K(h_1+h_2)=\Delta_K(h_1)+\Delta_K(h_2)$.) Hence, by standard results about additive functions, we can now say that $\Delta_K(h)=h\Delta_K(1)$.

(iv) In order to evaluate $\Delta_K(1)$, first observe that (cf. (3.3.30))

$$\mathcal{W}_{\mathbf{y}}^{(N)}\Big(\tau_K\le h \text{ and } \tau_K^h=\infty\Big)\le 2N\exp\left[-\frac{|\mathbf{y}-K|^2}{2Nh}\right],$$

and therefore that

$$\lim_{R\nearrow\infty}\sup_{h\in(0,1]}\int_{\{|\mathbf{y}|\ge R\}}\mathcal{W}_{\mathbf{y}}^{(N)}\Big(\tau_K\le h \text{ and } \tau_K^h=\infty\Big)\,d\mathbf{y}=0.$$

Second, note that

$$\begin{aligned}
\mathcal{W}_{\mathbf{y}}^{(N)}\Big(\tau_K\le h \text{ and } \tau_K^h=\infty\Big)&=\mathcal{W}_{\mathbf{y}}^{(N)}\Big(\tau_K^h=\infty\Big)-\mathcal{W}_{\mathbf{y}}^{(N)}\Big(\tau_K=\infty\Big)\\
&=p_K^{\mathbb{R}^N}(\mathbf{y})-\big[\gamma_h^N\bigstar p_K^{\mathbb{R}^N}\big](\mathbf{y}).
\end{aligned}$$

Finally, combine these with Theorem 8.4.20 and the preceding to see that

$$\Delta_K(1)=\operatorname{Cap}\big(K;\mathbb{R}^N\big),\quad h\in(0,\infty).\tag{8.4.60}$$

(v) By writing

$$E_K(t)=E_K\big(]t[\big)+\sum_{n=1}^{[t]}\Big(E_K\big(]t[+n\big)-E_K\big(]t[+n-1\big)\Big),\quad \text{where }]t[\equiv t-[t],$$

use the preceding to show that the *asymptotic rate of heat transfer* is equal to capacity: that is,

$$\lim_{t\to\infty}\frac{E_K(t)}{t}=\operatorname{Cap}\big(K;\mathbb{R}^N\big).$$

Notation

Notation	Description	See†
$A^{\complement}$	The complement of the set A	
$A^{(\delta)}$	The δ-hull around the set A	§3.1
$\mathbf{1}_A$	The indicator function of the set A.	§1.1
$a \wedge b \,\&\, a \vee b$	The minimum and the maximum of a, $b \in \mathbb{R}$.	
$a^+ \,\&\, a^-$	The positive part $a \vee 0$ and negative part $(-a) \vee 0$ of $a \in \mathbb{R}$.	
$K \subset\subset E$	To be read: K *is a compact subset of* E.	
$\bigvee_{i \in \mathcal{I}} \mathcal{F}_i$	The σ-algebra generated by $\bigcup_{i \in \mathcal{I}} \mathcal{F}_i$	§1.1
$\mu \ll \nu$	The measure μ is absolutely continuous with respect to the measure ν.	
$\mu \perp \nu$	The measures μ and ν are singular.	**L.** 3.1.1
$\prod_{i \in \mathfrak{I}} \mu_i$	The product measure with factors μ_i, $i \in \mathfrak{I}$.	**E.** 1.1.14
$\mathbb{Q}$	The set of rational numbers.	
$\langle \varphi, \mu \rangle$	Alternative notation for $\int \varphi \, d\mu$	§3.1
$\mu_n \Longrightarrow \mu$	$\{\mu_n\}$ converges weakly to μ	§3.1
$[t]$	Integer part of $t \in \mathbb{R}$	
$B(E; \mathbb{R})$	Space of bounded, Borel measurable functions from E into $\mathbb{R}$.	§3.1
$B_{\mathrm{c}}(G; \mathbb{R})$	Space of bounded, Borel measurable functions with compact support in the open set G	

† This column points to the place in the text where the notation is used first. **L**=Lemma, **E**=Example, **T**= Theorem, and numbers enclosed in parentheses refer to equations.

$\mathcal{B}_s$	The σ-algebra over $\mathfrak{P}(\mathbb{R}^N)$ generated by $\psi \in \mathfrak{P}(\mathbb{R}^N) \mapsto \psi(t) \in \mathbb{R}^N$ for $t \in [0, s]$.	(4.1.3)
$\mathcal{B}_E$	The Borel field over the topological space E.	
$\|\cdot\|_u$	The uniform or "sup" norm on functions	§3.1
$C_b(E;\mathbb{R})$	Space of bounded continuous functions from E into $\mathbb{R}$.	
$C_c(G;\mathbb{R})$	Space of continuous, $\mathbb{R}$-valued functions having compact support in the open set G.	
$C^{1,2}(\mathbb{R}^M \times \mathbb{R}^N;\mathbb{R})$	Space of functions $(\mathbf{x},\mathbf{y}) \in \mathbb{R}^M \times \mathbb{R}^N \longrightarrow \mathbb{R}$ which are continuously differentiable once in $\mathbf{x}$ and twice in $\mathbf{y}$.	
δ_s	The time increment map on $\mathfrak{P}(\mathbb{R}^N)$.	§4.1
$\mathbb{E}^\mu[X,\, A]$	To be read *the expectation value of* X *with respect to* μ *on* A. Equivalent to $\int_A X\,d\mu$. When A is unspecified, it is assumed to be the whole space.	
$\mathbb{E}^\mu[X\,\|\,\mathcal{F}]$	To be read *the conditional expectation value of* X *given the* σ*-algebra* $\mathcal{F}$.	§5.1
$f \star g$	The convolution product of functions f and g.	
$\Phi^*\mu$	The pushforward (image) of the measure μ under Φ.	(1.1.4)
$\Gamma(t)$	Euler's Gamma function.	(3.2.23)
$\Gamma_N(\mathbf{x})$	The fundamental solution in $\mathbb{R}^N$: $N \geq 3$, $N = 1$, and $N = 2$.	(8.3.10), (8.3.13), & (8.3.16)
γ	The Gauss kernel.	§1.3
$g^{\mathfrak{G}}(\mathbf{x},\mathbf{y})$	Green's function in the region $\mathfrak{G}$.	(8.3.9)
$G^{\mathfrak{G}}$	Green's operator in the region $\mathfrak{G}$.	(8.3.8)
$\gamma_t{}^N(d\mathbf{x})$ & $\gamma_t^N(\mathbf{x})$	The $\mathfrak{N}(\mathbf{0}, t\mathbf{I})$-measure on $\mathbb{R}^N$ and its density.	§2.2 & §4.3
$\partial_{\text{reg}}\mathfrak{G}$	The regular points of $\partial\mathfrak{G}$.	§8.1
$\mathbf{H}(\mathbb{R}^N)$	The Cameron–Martin subspace for Wiener's measure on $\Theta(\mathbb{R}^N)$.	§4.2
$H^{\mathfrak{G}}f$	The harmonic extention of $f \in C_b(\partial\mathfrak{G};\mathbb{R})$ to $\mathfrak{G}$.	(8.2.3)
$\mathcal{I}(h)$	The Paley–Wiener integral of h.	**T.** 4.2.4

$L^p(\mu; E)$	The Lebesgue space of E-valued functions f for which $\|f\|_E^p$ is μ-integrable. When E is omitted, it is understood to be $\mathbb{R}$.	§5.1
$\mathbf{M}_1(E)$	Space of Borel probability measures on E.	§3.1
$\mathrm{med}(Y)$	The set of medians of the random variable Y.	§1.4
$\hat{\mu}$	The characteristic function (Fourier transform) of μ.	§2.2 & §4.2
$\mu\bigstar\nu$	The convolution product for $\mathbf{M}_1(\mathbb{R}^N)$.	§3.1
$\mathbf{N}(t)$	The simple Poisson process.	(3.2.23)
$\mathbb{N}$	The non-negative integers: $\mathbb{N} = \{\not{\!0}\} \cup \mathbb{Z}^+$.	
$\mathfrak{N}(\mathbf{m}, \mathbf{C})$	The Gaussian (normal) distribution with mean-value $\mathbf{m}$ and covariance $\mathbf{C}$.	§2.2
ω_{N-1}	The surface area of the unit sphere $\mathbf{S}^{N-1}$.	**E.** 2.1.40
$\mathfrak{P}(E)$	The space $C([0,\infty); E)$ of continuous E-valued paths.	§3.2
p'	The Hölder conjugate $\frac{p}{p-1}$ of $p \in [1, \infty]$.	
$p_y^{(N-1)}(\mathbf{x})$	The Poisson kernel for $\mathbb{R}_+^N$.	**E.** 4.3.50
$p^{\mathfrak{G}}(t, \mathbf{x}, \mathbf{y})$	The Dirichlet heat kernel for the region $\mathfrak{G}$.	**C.** 8.1.6
$\{P_t^{\mathfrak{G}} : t > 0\}$	The Dirichlet heat flow semigroup in the region $\mathfrak{G}$.	(8.1.7)
$\{Q_t^{\mathbf{b}} : t > 0\}$	Heat flow semigroup with force field $\mathbf{b}$	§4.3
$\{Q_t^U : t > 0\}$	Heat flow semigroup with conservative force field $-\nabla U$	§4.3 & §7.5
$\{P_t^V : t > 0\}$	Feynman–Kac semigroup	§4.3
$\tilde{\psi}_T$	The pinned Wiener path.	**T.** 4.2.4
$\mathbf{S}^{N-1}$	The unit sphere in $\mathbb{R}^N$.	
$\mathbf{S}_\alpha$	The Wiener scaling transformation.	§4.3
Σ_s	The time shift transformation on $\mathfrak{P}(\mathbb{R}^N)$.	§4.3
S_n	The sum $\sum_1^n X_k$ of n random variables.	§1.2
$\overline{S}_n$	The average $\frac{1}{n} S_n$ of n random variables.	§1.2

$\tilde{S}_n$	Sum scaled for Law of the Iterated Logarithm.	§1.5
$\hat{S}_n$	Central limit scaling of n random variables.	§2.1
$\mathbf{S}_n(t,\omega)$	Piecewise linearization of central limit scaling.	§3.3
$\Theta(\mathbb{R}^N)$	The space of $\mathbb{R}^N$-valued Wiener paths.	§4.2
$U^{\rho}_{\mathrm{b}}(E;\mathbb{R})$	Space of bounded, ρ-uniformly continuous $\varphi : E \longrightarrow \mathbb{R}$	§3.1
$\mathcal{W}$	Wiener's measure on $\mathfrak{P}(\mathbb{R})$ or $\Theta(\mathbb{R})$.	§3.3
$\mathcal{W}^{(N)}$	Wiener's measure on $\mathfrak{P}(\mathbb{R}^N)$ or $\Theta(\mathbb{R}^N)$.	§3.3
$\mathcal{W}^{(N)}_{\mathbf{x}}$	The distribution of $\psi \in \mathfrak{P}(\mathbb{R}^N) \mapsto \mathbf{x} + \psi \in \mathfrak{P}(\mathbb{R}^N)$ under $\mathcal{W}^{(N)}$.	§4.3
$\mathbb{Z}^+$	The strictly positive integers.	

Index

G

H

I

J

K

L

M

T

U

V

W